Marine Biology

Marine Biology

Comparative Ecology of Planet Ocean

Roberto Danovaro
Polytechnic University of Marche

Paul Snelgrove
Memorial University of Newfoundland

This edition first published 2024
© 2024 John Wiley & Sons Ltd

All rights reserved. No part of this publication may be reproduced, stored in a retrieval system, or transmitted, in any form or by any means, electronic, mechanical, photocopying, recording or otherwise, except as permitted by law. Advice on how to obtain permission to reuse material from this title is available at http://www.wiley.com/go/permissions.

The right of Roberto Danovaro and Paul Snelgrove to be identified as the authors of this work has been asserted in accordance with law.

Registered Offices
John Wiley & Sons, Inc., 111 River Street, Hoboken, NJ 07030, USA
John Wiley & Sons Ltd, The Atrium, Southern Gate, Chichester, West Sussex, PO19 8SQ, UK

For details of our global editorial offices, customer services, and more information about Wiley products visit us at www.wiley.com.

Wiley also publishes its books in a variety of electronic formats and by print-on-demand. Some content that appears in standard print versions of this book may not be available in other formats.

Trademarks: Wiley and the Wiley logo are trademarks or registered trademarks of John Wiley & Sons, Inc. and/or its affiliates in the United States and other countries and may not be used without written permission. All other trademarks are the property of their respective owners. John Wiley & Sons, Inc. is not associated with any product or vendor mentioned in this book.

Limit of Liability/Disclaimer of Warranty
While the publisher and authors have used their best efforts in preparing this work, they make no representations or warranties with respect to the accuracy or completeness of the contents of this work and specifically disclaim all warranties, including without limitation any implied warranties of merchantability or fitness for a particular purpose. No warranty may be created or extended by sales representatives, written sales materials or promotional statements for this work. This work is sold with the understanding that the publisher is not engaged in rendering professional services. The advice and strategies contained herein may not be suitable for your situation. You should consult with a specialist where appropriate. The fact that an organization, website, or product is referred to in this work as a citation and/or potential source of further information does not mean that the publisher and authors endorse the information or services the organization, website, or product may provide or recommendations it may make. Further, readers should be aware that websites listed in this work may have changed or disappeared between when this work was written and when it is read. Neither the publisher nor authors shall be liable for any loss of profit or any other commercial damages, including but not limited to special, incidental, consequential, or other damages.

A catalogue record for this book is available from the Library of Congress

Paperback ISBN: 9781394200078; ePub ISBN: 9781394200108; ePDF ISBN: 9781394200092

Cover Image: Courtesy of Gabriella Luongo
Cover Design: Wiley

Set in 9.5/12.5pt STIXTwoText by Integra Software Services Pvt. Ltd, Pondicherry, India
Printed and bound by CPI Group (UK) Ltd, Croydon, CR0 4YY

C9781394200078_230324

For Cinzia Corinaldesi and Michele DuRand – friends, muses, partners

Contents

Acknowledgments *xvii*
Preface *xix*
About the Companion Website *xxi*

Part I The Ocean Domain: Introduction to Planet Ocean *1*

1   The Life Aquatic *3*
1.1   Introduction *3*
1.2   Comparison Between Sea and Land *3*
1.3   Fractal Complexity of Marine and Terrestrial Ecosystems *6*

2   The Seabed *13*
2.1   Ocean Basins *13*
2.2   Ocean Bottom: A (Mostly) Static Habitat of Ocean Life *15*
2.3   Characteristics of Sediments *20*
2.4   Boundary Layers and Their Characteristics *20*
2.5   Sediment Movement *23*
2.6   Characteristics of Hard Substrata *26*
2.7   Characteristics of Soft Sediments *26*

3   The Water Column *29*
3.1   Properties and Characteristics of Seawater *29*
3.1.1   Hydrogen Bonds *31*
3.1.2   Salinity *31*
3.1.3   Temperature *34*
3.1.4   Density *35*
3.1.5   Viscosity *38*
3.1.6   Pressure *38*
3.1.7   Sound *39*
3.1.8   Light *40*
3.1.9   Inorganic Nutrients and Trace Elements *42*
3.1.10 Oxygen *43*
3.1.11 Dissolved Gases *43*
3.2   An Ocean In Motion *44*
3.2.1   Ocean Currents *44*

Part II Life in Seas and Oceans: Fundamentals of Marine Biology 53

4 General Adaptations in Marine Organisms I: From the Ocean Surface to the Seabed 55
4.1 Adaptations to Temperature 55
4.2 Adaptations to Low Oxygen Concentrations 60
4.3 Adaptation to Salinity 64
4.4 Adaptation to Pressure 68
4.5 Adaptations to Light 69
4.5.1 Photosynthesis 69
4.5.2 Vision and Bioluminescence 69
4.6 Adaptations to Nutrients 73
4.7 Electrical Conductivity Adaptations 74
4.8 Ectocrine Adaptations 74
4.9 Adaptations to Produce Sound and Communicate in Water 74

5 Adaptations in Marine Organisms II: Life in a Fluid Habitat 79
5.1 Adaptions to Life in the Water Column 79
5.1.1 Density, Shape, and Buoyancy 79
5.1.2 Swimming and Dispersal 86

6 Adaptations in Marine Organisms III: Benthic Biota between a Rock and a Soft Place 91
6.1 Adaptations to Life on the Seafloor 91
6.2 Support and Protection Structures 91
6.2.1 Bioconstructors 93
6.3 Adaptation to Waves and Energy 94
6.4 Feeding and Nutrition 95
6.5 Adaptation to Aerial Exposure 98
6.6 Adaptation to Extreme Temperatures and Potentially Toxic Chemicals 98

Part IIB Life in Seas and Oceans: Fundamentals of Marine Biology 101

7 Marine Biodiversity 103
7.1 Introduction 103
7.2 Origin and Evolution of Marine Life 103
7.2.1 Theories on the Origin of Life 103
7.2.2 Evolution of Marine Biodiversity 106
7.3 Mechanisms of Marine Speciation 107
7.4 Quantifying Marine Organism Biodiversity 113
7.4.1 Definition of Biodiversity 113
7.4.2 Defining Different Ways of Measuring Biodiversity: Species Richness 113
7.4.3 Comparison Between Marine and Terrestrial Biodiversity 115
7.4.4 Measures of Biodiversity 119
7.4.5 Composite Measures of Species Diversity 121

8 Biodiversity Patterns 125
8.1 Broad-Scale Biodiversity Patterns 125
8.2 Processes Controlling the Distribution of Marine Biodiversity 125
8.2.1 Biodiversity Hotspots 125
8.2.2 Latitudinal Gradient of Biodiversity 126
8.2.3 Longitudinal Gradients in Tropical Biodiversity 128
8.2.4 Bathymetric Patterns in Marine Biodiversity 128

8.3	Marine Biogeography	*128*
8.3.1	Terrestrial and Marine Biogeography	*130*
8.3.2	Biogeographical Regions	*130*
8.3.3	Species Distributions Within Biogeographic Regions	*131*
8.3.4	Biogeography of the Mediterranean	*132*
8.4	Theories on Evolution and Maintenance of Biodiversity: The Deep-Sea as Examplar	*136*
9	**Biodiversity of the Benthos** *141*	
9.1	Introduction: Benthos and Plankton	*141*
9.2	Benthic Biota	*141*
9.3	Classification of Benthos Based on Size	*144*
9.3.1	Femtobenthos	*144*
9.3.2	Picobenthos	*147*
9.3.3	Nanobenthos	*148*
9.3.4	Microbenthos	*149*
9.3.5	Meiobenthos (Meiofauna)	*151*
9.3.6	Macrobenthos	*152*
9.3.7	Megabenthos	*153*

Part IIC Life in Seas and Oceans: Fundamentals of Marine Biology *157*

10	**Ecology of Benthos** *159*	
10.1	Ecology of the Benthos: From Microbes to Megafauna	*159*
10.2	Trophic Groups: Classification of Benthos Based on Diet	*164*
10.3	Comparison Between Hard and Soft Bottom Benthos	*165*
10.4	Ecology of Benthos Inhabiting Soft Bottoms	*167*
10.5	Changes in Benthos in Space and Time	*170*
10.6	Organization of Benthic Assemblages	*171*
10.7	Zonation of Benthic Organisms	*172*
11	**Biodiversity of the Plankton** *179*	
11.1	Introduction to the Plankton	*179*
11.2	Planktonic Organisms	*180*
11.3	Planktonic Classification Based on Water Column Distributions	*181*
11.4	Plankton Classification Based on Life Cycles	*182*
11.5	Plankton Size Classes	*184*
11.5.1	Virioplankton	*186*
11.5.2	Picoplankton	*189*
11.5.3	Nanoplankton	*192*
11.5.4	Microplankton	*193*
11.5.5	Mesozooplankton	*194*
11.5.6	Macro and Mega-zooplankton	*195*
11.6	Abundance Comparisons Among Different Planktonic Components	*197*
12	**Ecology of the Plankton** *201*	
12.1	Plankton Distribution	*201*
12.1.1	Physical Control of Macro-scale Plankton Distribution: The Case of El Niño	*201*
12.1.2	Small-scale Plankton Distribution: The Vertical Migration Example	*202*
12.2	Ecology of Plankton	*208*
12.3	How Many Phytoplankton Species Coexist in a Volume of Water? "Homage to Santa Rosalia"	*211*
12.4	Zooplankton Nutritional Mode	*211*

Part IID Life in Seas and Oceans: Fundamentals of Marine Biology 215

13   Biodiversity of the Nekton 217
13.1   Species Contributing to the Nekton 217
13.2   Main Organisms and Characteristics of Nekton 218
13.3   Fishes 218
13.3.1   Osteichthyes – The Bony Fishes 220
13.3.2   Chondricthyes – The Cartilaginous Fishes 222
13.3.3   Agnatha – The Jawless Fishes 224
13.4   Marine Mammals 224
13.5   Cephalopods 227
13.6   Reptiles – The "Land-based" Marine Species 228
13.7   Seabirds 228
13.7   Patterns of Biodiversity in Nekton 229

14   Ecology of the Nekton 233
14.1   Introduction 233
14.2   Fishes and Formation of Fish Shoals 233
14.3   Ecology of Chondrichthyes 235
14.4   Sharks at Risk of Extinction from Indiscriminate Hunting 236
14.4.1   Measuring Abundance of Sharks 236
14.4.2   Shark Attacks – How Often Do Sharks Attack Humans? 236
14.5   Ecology of Cephalopods 236
14.6   Marine Reptile Ecology 239
14.7   Ecology of Seabirds 241
14.8   Ecology of Marine Mammals 242
14.9   Great Migrations of Nekton 247
14.10   Role of Top Predators in Food Webs 255

15   Life Cycles and Larval Ecology 261
15.1   Life Cycles and Reproduction 261
15.2   Larval Ecology 264
15.3   Life History Strategies 267
15.4   Supply Side Ecology 270
15.5   Forms of Resistance and Benthic-pelagic Coupling 270

Part III Comparative Marine Ecology: Habitat Types, Their Biodiversity, and Their Functioning 275

16   Ecosystem Functioning I: Primary and Secondary Production 277
16.1   Introduction 277
16.2   Primary Production 277
16.3   Chemosynthetic Primary Production in the Ocean 278
16.4   Photosynthetic Primary Production 280
16.4.1   Primary Producers 286
16.5   Secondary Production 287
16.5.1   Methods of Measurement of Secondary Production in the Sea 288
16.6   Respiration 290

17	**Ecosystem Functioning II: Organic Matter Recycling** *293*
17.1	Introduction – Extra-Specific Processes *293*
17.2	Organic Matter and Detritus in the Ocean *293*
17.3	Dissolved Organic Matter in the Ocean (DOM) *299*
17.4	Pelagic-Benthic Coupling *300*
17.4.1	Organic Aggregates in the Ocean *302*
17.5	Consequences of Organic Matter Export to the Seabed *307*

18	**Interspecific Interactions and Trophic Cascades** *313*
18.1	Biodiversity and Ecosystem Functioning *313*
18.2	Facilitation and Cooperation – Positive Interactions *315*
18.3	Symbiosis *319*
18.4	Complex Biotic Interactions: Trophic Networks and Cascades *323*
18.4.1	Trophic Networks *323*
18.4.2	Detrital Trophic Network *326*
18.4.3	Trophic Networks Based on Dissolved Organic Matter *327*
18.4.4	Microbial Loop *327*
18.4.5	Viral Shunt *328*
18.4.6	Bottom-Up Control of Trophic Food Webs *329*
18.4.7	Top-Down Control on Trophic Food Webs *331*
18.4.8	Mixed Wasp-Waist Control *332*
18.5	Keystone Species *333*
18.6	Trophic Cascades *333*

Part IIIB Comparative Marine Ecology: Habitat Types, Their Biodiversity, and Their Functioning *341*

19	**Interspecific Interactions II: Negative Interactions** *343*
19.1	Predation *343*
19.2	Methods to Escape Predation *348*
19.3	Competition *349*
19.4	Parasitism *351*
19.5	Diseases of Marine Organisms *354*
19.5.1	Coral Diseases *355*

20	**Intertidal Ecosystems and Lagoons** *359*
20.1	Rocky Intertidal Habitats *359*
20.1.1	Survival Strategies for Rocky Intertidal Environments *361*
20.1.2	Rocky Intertidal Zonation *361*
20.1.3	Rocky Intertidal Primary Producers *363*
20.1.4	Rocky Intertidal Consumers *364*
20.1.5	Competition in Rocky Intertidal Environments *366*
20.1.6	Predation in Intertidal Habitats and the Intermediate Disturbance Hypothesis *366*
20.1.7	Keystone Species *369*
20.1.8	Rocky Intertidal Trophic Food Webs *370*
20.1.9	Comparison Between Soft and Hard Bottom Intertidal Environments *370*
20.1.10	Niche Displacement to Reduce Competition *371*
20.2	Transitional Environments Between Land and Ocean *372*
20.2.1	Lagoons *372*

20.2.2 Ecology of Coastal Lagoons *375*
20.2.3 Lagoon Functioning *376*
20.2.4 Models of Functional Zonation of Coastal Lagoons *377*
20.2.5 Lagoon Biodiversity *378*
20.3 Mangroves *380*
20.3.1 Biodiversity Associated with Mangals *382*
20.3.2 Mangal Ecosystem Functioning *384*
20.4 Salt Marshes *385*
20.4.1 Biodiversity Associated with Salt Marshes *387*
20.4.2 Salt Marsh Ecosystem Functioning *387*
20.5 Summary *388*

21 Subtidal Hard Substrata Ecosystems *391*
21.1 Introduction *391*
21.2 Subtidal Distributions *391*
21.2.1 Effect of Physical Variables and Disturbance on Benthic Communities *391*
21.2.2 Biotic Factors *393*
21.3 Kelp Forests *394*
21.3.1 Biodiversity Associated with Kelp *396*
21.3.2 Trophic Networks *397*
21.3.3 Macroalgal Forests in the Mediterranean *398*
21.4 Coral Reefs *399*
21.4.1 Zonation within Coral Reefs *400*
21.4.2 Types of Coral Reef *401*
21.4.3 Theory of Coral Reef Formation *401*
21.4.4 Characteristics of Reef Building (Bioconstructor) Corals *402*
21.4.5 Coral Reproduction *402*
21.4.6 Coral Feeding and Symbiosis with Zooxanthellae *404*
21.4.7 Primary Factors Limiting Coral Growth *407*
21.4.8 Coral Reef Biodiversity *407*
21.4.9 Coral Reefs Functioning and Trophic Food Webs *412*
21.4.10 Primary Consumers *412*
21.4.11 Deposit Feeders / Scavengers *413*
21.4.12 Secondary Consumers *413*
21.4.13 Tertiary Consumers *413*
21.4.14 Competition for Space in Coral Reefs *414*
21.4.15 Interactions Between Coral Reefs and Adjacent Ecosystems *417*
21.5 Coralligenous Habitats *418*
21.6 Rhodolith Beds (Maërl) *422*
21.7 Underwater Caves *423*
21.7.1 Cave Biodiversity *425*
21.7.2 Adaptations in Marine Invertebrates to Life in Caves *426*
21.7.3 Food Webs and Functioning in Caves *426*
21.8 Summary *429*

Part IIIC Comparative Marine Ecology: Habitat Types, Their Biodiversity, and Their Functioning *431*

22 Estuarine, Seagrass, and Sedimentary Habitats *433*
22.1 Estuaries *433*
22.1.1 The Complexity of Estuarine Environments *435*

22.1.2	Survival Strategies for Living in Estuaries	*435*
22.1.3	Estuarine Food Webs	*437*
22.1.4	Why are Estuaries Important?	*438*
22.1.5	Pressures on Estuaries	*438*
22.2	Seagrass Beds	*438*
22.2.1	Biodiversity Associated With Seagrasses	*442*
22.2.2	Seagrass Functioning	*446*
22.3	Sedimentary Habitats	*446*
22.3.1	Food Sources for Sedimentary Fauna	*449*
22.3.2	Sedimentary Environments and Ecosystem Functioning	*451*

23 Polar Ecosystems *455*

23.1	Biogeography and Characteristics	*457*
23.2	Biodiversity	*460*
23.3	Biodiversity Within Sea Ice	*460*
23.4	Pelagic Biodiversity	*462*
23.5	Fishes	*464*
23.6	Marine Mammals	*465*
23.7	Benthic Biodiversity	*465*
23.8	Food Webs and Functioning	*466*
23.9	Antarctica	*467*
23.9.1	Zonation, Extent, and Distribution	*467*
23.9.2	Antarctic Habitats	*472*
23.9.3	Biodiversity	*474*
23.9.4	Birds and Mammals	*478*
23.9.5	Trophic Webs and Functioning	*478*
23.10	Summary	*483*

24 Neritic Aquatic Ecosystems *485*

24.1	Introduction	*485*
24.2	Zonation, Extent, and Distribution	*485*
24.3	Biogeography and Characteristics	*485*
24.4	Biological Characteristics	*487*
24.4.1	Primary Producers	*488*
24.4.2	Zooplankton	*488*
24.4.3	Nekton	*489*
24.5	Ecosystem Functioning in the Neritic Zone	*490*
24.6	Fisheries Production	*492*
24.7	Factors Influencing Functioning of Neritic Systems	*493*
24.7.1	Intertwining of Vertical and Horizontal Currents	*493*
24.7.2	Physical and Chemical Factors	*493*
24.7.3	Large-Scale Currents	*493*
24.7.4	Terrestrial Inputs	*494*
24.7.5	Coastal Upwelling	*494*
24.7.6	Large-Scale Changes in Water Masses	*494*
24.7.7	River Plumes	*495*
24.7.8	Fronts	*495*
24.7.9	Neritic Food Webs	*495*
24.8	Summary	*497*

Part IIID Comparative Marine Ecology: Habitat Types, Their Biodiversity, and Their Functioning *499*

25 Deep-Sea Ecosystems along Continental Margins *501*
25.1 Introduction to the Deep Sea *501*
25.1.1 Is the Deep Sea on a Diet? *503*
25.1.2 Extreme and Harsh Conditions? *503*
25.1.3 Are Deep-Sea Ecosystems Depauperate? *505*
25.1.4 Metabolism and Functioning of Deep Ecosystems *506*
25.2 Deep-Sea Biodiversity *508*
25.2.1 Oases or Biological Deserts? *508*
25.2.2 Deep Faunal Origins *510*
25.2.3 Mechanisms of Generation and Maintenance of Deep-Sea Biodiversity *511*
25.3 Deep-Sea Habitats *511*
25.4 Submarine Canyons *512*
25.4.1 Canyon Biodiversity *513*
25.4.2 Canyon Functioning *514*
25.5 Deep-Water Corals *515*
25.6 Cold Seep (Hydrocarbon-Based) Ecosystems *517*
25.7 Cold Seep Biodiversity and Symbiotic Organisms *518*
25.7.1 Functioning of Cold Seep Ecosystems *521*
25.8 Hypoxic and Anoxic Systems (Dead Zones) *522*
25.9 Oxygen Minimum Zones, OMZs *522*
25.10 Summary *526*

26 Deep Ocean Basins *531*
26.1 Introduction *531*
26.2 Abyssal Plains *531*
26.3 Abyssal Biodiversity and Adaptations *531*
26.4 Abyssal Gigantism and Dwarfism *533*
26.5 Functioning of Abyssal Systems *535*
26.5.1 Seamounts *535*
26.5.2 Seamount Biodiversity *538*
26.6 Deep-Sea Hydrothermal Vents *541*
26.6.1 Biodiversity Associated with Deep–Sea Hydrothermal Vents *543*
26.6.2 Ecosystem Functioning at Hydrothermal Vents *546*
26.7 Whale Carcasses *551*
26.7.1 Whale Carcass Biodiversity *552*
26.7.2 Functioning of Whale Carcass Systems *552*
26.8 Affinities Between Vent and Seep Communities *554*
26.9 Anoxic Basins *554*
26.9.1 Hypersaline Anoxic Systems *555*
26.10 Ocean Trenches *556*
26.10.1 Hadal Biodiversity *558*
26.11 Summary *562*

27 Oceanic Ecosystems *567*
27.1 Introduction *567*
27.2 Factors Influencing the Life and Distribution of Pelagic Organisms *567*
27.2.1 Light, Darkness, and Nutrients *567*
27.2.2 Pressure *568*
27.2.3 Shallow-Deep Connectivity *568*

27.2.4	Vertical Migrations 568
27.2.5	Feeding and Recruitment in the Deep 569
27.2.6	Body Size 569
27.2.7	Biodiversity 570
27.3	Classification of Pelagic Regions 571
27.4	Functional Classification of Pelagic Systems 573
27.5	Vertical Zonation in Pelagic Ecosystems 575
27.6	Biodiversity of Pelagic Systems 577
27.6.1	Epipelagic Biodiversity 577
27.6.2	Mesopelagic Biodiversity 578
27.6.3	Bathypelagic Biodiversity 579
27.6.4	Abyssopelagic Biodiversity 579
27.6.5	Hadopelagic Biodiversity 579

Part IV Human Impacts and Solutions for Planet Ocean: Applied Marine Biology 581

28	**Human Impacts on Marine Ecosystems** 583
28.1	Historical Data 583
28.1.1	Marine Animal Populations in Human History 584
28.2	Biodiversity Loss 587
28.3	The Main Threats to Marine Life and Ecosystems 590
28.3.1	Contamination 590
28.3.2	Habitat Degradation, Fragmentation, and Destruction 590
28.3.3	Overfishing 590
28.3.4	Extraction of Abiotic Resources 592
28.3.5	Non-indigenous or Alien Species 592
28.3.6	Global Climate Change 593
28.4	Synergistic Impacts on Marine Ecosystems 597

29	**Marine Biodiversity Conservation** 603
29.1	Introduction 603
29.2	Conservation Objectives 603
29.3	The Third Dimension of Marine Conservation 606
29.4	Conservation Strategies 606
29.4.1	Access to Fisheries: Who and How 607
29.4.2	When to Fish: Time-Based Approaches 608
29.4.3	Where to Fish: Area-Based Tools 608
29.5	Marine Protected Areas 608
29.5.1	Criteria for Prioritizing Marine Areas to Protect 609
29.5.2	What Have We Learned from Existing Marine Protected Areas? 611
29.6	Cumulative Impacts and Biodiversity Conservation 613
29.7	Conservation Frameworks 614
29.8	Legal Instruments 615
29.9	Science Challenges and Solutions – Moving Science to Policy? 616
29.10	How Science Can Contribute 616

30	**Restoring Marine Habitats** 619
30.1	A Decade For Ecosystem Restoration 619
30.2	Defining Ecological Restoration 620
30.3	A Global Plan for Marine Ecosystem Restoration 623
30.4	Restoring Fragile Marine Habitats 623

30.5	Restoration of Coral Reefs	*625*
30.6	Restoration of Seagrass Meadows	*626*
30.7	Restoration of Macroalgal Forests	*627*
30.7.1	Restoration of Kelp Forests	*627*
30.7.2	Restoration of *Cystoseira* spp. Forests	*627*
30.8	Restoring Ecosystem Engineers: The Case of Coralligenous Outcrops	*629*
30.9	Restoration of Deep-Sea Habitats	*630*
30.10	Perspectives of Marine Ecosystem Restoration	*631*
31	**How Far We Have Come: Past, Present, and Future Research on the Marine Biology of Planet Ocean**	*633*
31.1	Introduction	*633*
31.2	The Birth of Marine Biology	*634*
31.3	The History of Ocean Exploration	*636*
31.4	Present and Future of Marine Biology	*637*
31.4.1	Sampling Platforms	*637*
31.4.2	Implementation of Technologies Enabling Biological Observations At Sea	*641*
31.4.3	Sensors	*643*
31.5	Application of Marine Technologies	*644*
31.6	Marine Biology Research in the Next Decade	*645*

Glossary *649*

Index *675*

Acknowledgments

We thank Philippe Archambault, Lisandro Benedetti-Cecchi, Judy Grassle, Nadia Papadopoulou, Pierre Pepin, Murray Roberts, Martin Solan, Chris Smith, and Derek Tittensor for helpful comments on various chapters of this book. Giovanni Lanteri provided support for image acquisition and Michael Tangherlini is thanked for the huge contribution in the drawing of the images

Preface

The concept of modern marine biology as a scientific discipline owes its beginnings to Charles Darwin's expedition on the *Beagle* (1831–1836), which enabled Darwin's understanding of the role of different environments in how marine organisms adapt to different conditions and what factors drive natural selection and speciation. A few years later, the *Challenger* expedition (1872–1876), which explored the ocean well beyond the continental shelf, undertook the first systematic attempt to document the presence of life in the deep sea and the factors controlling their distribution. The ensuing scientific publications documented the rich biodiversity of marine life and boosted ecological approaches to the study of marine ecosystems. After World War II, taking advantage of new technologies, explorer Jacques Cousteau showed a curious public the multifaceted beauty and mystery of underwater seascapes and life in the global ocean. Cousteau's work and other spectacular documentaries in the 1960s and 1970s attracted many dreamers, enthusiasts, and scholars to marine biology.

Fascination with the immensity and mystery of the ocean has long attracted human societies in many different ways, leaving an indelible mark in science, art, and culture. This enthusiasm spans from scientists to artists to directors and writers, and, of course, children. Some 50 years after the expedition of the Trieste bathyscaphe, the film director James Cameron dove in his new-generation bathyscaphe to explore the bottom of the Mariana Trench, at almost 11,000 meters depth. Marine organisms have frequently inspired fantastic or alien creatures in countless films, including tube worms that become huge retractable plants and bioluminescent jellyfish that become flying creatures. Many "monsters" are thinly (or un-) disguised small sea creatures, such as sea urchin larvae or turbellarians. Filmmakers have also celebrated the idea of the "marine biologist" as an adventurous (and even courageous!) figure. Countless films have celebrated the sea and its life forms, from a spate of horror films in the 1950s (such as *Monsters from the Ocean Floor* and *The Phantom from 10,000 Leagues*) to the 1987 blockbuster *Jaws*, with countless sequels and variations on the theme (such as *Sharknado*, *Leviathan*, and *The Deep*), marked a turning point in the relationship between humans and the ocean, portraying a marine biologist able to interpret the predator's behavior and save the day. Subsequently, *Sphere* and *Abyss* presented marine life and marine biologists as central protagonists, and *Free Willy* promoted ocean conservation ideas. Even the television series *Baywatch* (1991) featured a marine biologist (Megan), and many readers will remember George Costanza's marine biology ambitions on *Seinfeld*. Marine biology has featured heavily in animation film production, from the *Little Mermaid* (Disney) to *Finding Nemo*, from *Oceans* to *Flipper*, and the turtle *Sammy* (and sequels). Many now recognize marine biology as a potential profession: the "marine biologist," who plays a key role in the protection of marine life (marine protected areas, aquariums), helps to solve various environmental problems (cleanup, management of marine resources), and collaborates in the production of resources (aquaculture and mariculture plants, marine biotechnologies).

Why This Marine Biology Book Focused on the *Comparative Ecology of Planet Ocean*?

This book illustrates the different components of Planet Ocean, from marine habitats to their biodiversity, prioritizing scientific issues and communicating ecological relevance, particularly highlighting new knowledge and discoveries within the last few years. The book opens with an introduction to the ocean (Chapters 1–3) and then illustrates the fundamentals of marine biology (Chapters 4–19). The remainder of the book focuses on marine ecosystems (Chapters 20–31), which provides readers with a comprehensive summary of current knowledge on marine organisms, on the factors controlling their patterns and life cycles, and on the biodiversity, structure, and functioning of different marine ecosystems and habitats. It also addresses human interactions with the ocean. Through a comparative approach, the book aims to impart an understanding of different marine ecosystems. In a sense, we consider this text a sort of manual of "comparative marine ecology," which contrasts different ecosystems and their biota to understand better their characteristics, peculiarities, and threats, and how human society can develop potential solutions.

Detailed Summary of the Book Structure

Of the major parts in which we organized "*Marine Biology: the Comparative Ecology of Planet Ocean*," Part I (*The Ocean Domain*, Chapters 1 to 3) introduces the reader to the global ocean environment. Chapter 1 describes the characteristics of the marine environment, from a perspective in which we compare physical factors and environmental characteristics with the terrestrial realm, presented and explored from the point of view of marine organisms. Chapter 2 illustrates ocean basins and the complex three-dimensionality of the seabed, from the continental shelves to the abyssal plains and hadal depths. It also describes how scientists map the seafloor. Chapter 3 describes the special properties of salt water and how environmental constraints, such as temperature, density, pressure, and dissolved gasses change across water bodies and depths. This chapter also explores how marine organisms respond to dynamic water masses. Boxes within the chapter illustrate the methodologies used to measure key ocean variables.

Part II (*Life in Seas and Oceans*, Chapters 4–15) addresses the fundamentals of marine biology. It begins by placing marine organisms in the ocean and assessing how they have adapted to their environment, and the spectacular variety of conditions that characterize ocean ecosystems. These ecosystems vary from the extraordinary pressures at great depths to anoxic pools and hypoxic seabed environments, from the extreme cold temperatures of Antarctica to the superheated waters of hydrothermal vents. These chapters also describe how marine organisms swim, "breathe", eat, disperse, and migrate, and illustrate predatory and defensive strategies displayed by marine biota. Chapters 7 and 8 describe marine biodiversity, starting from the origin and evolution of ocean life, and then moving to marine biogeography and the mechanisms that regulate its patterns in nature. Chapters 9–14 describe the components of biodiversity, including the patterns and ecology of benthos, plankton, and nekton and the characteristics of all living components, from microscopic viruses to the great whales. Chapters 15–19 focus specifically on the study of ecosystem functions and species interactions, including intraspecific and extra-specific processes, with particular reference to life cycles and reproductive strategies.

Part III (*Comparative Marine Ecology*, Chapters 16–27) focuses on ocean function and the rich array of marine ecosystems. Chapters 20–27 take the reader through a journey beginning on the shoreline and ending at the deepest depths of the ocean, passing through the most remote and unexplored marine habitats. We structure all chapters similarly: (i) an introduction to the characteristics of the environment or ecosystem, (ii) a section on its biodiversity and (iii) a section describing how the ecosystem functions. Most chapters describe different types of habitats or ecosystems. The chapters conclude with a synopsis for each habitat comparing their extent, geographical location, depth, type of substrate, biodiversity, endemism, functioning, trophic networks, and unique features. Chapter 20 illustrates the major coastal environments, from intertidal habitats (mesolittoral or mid-littoral) to lagoons, soft and hard bottom subtidal environments, and ends with a description of mangrove forests and salt marshes. Chapter 21 describes the most common hard substrate coastal habitats on the planet in mid- and low latitudes, such as coral reefs, macroalgae forests, coralligenous habitats, rhodolith beds, and caves. Chapter 22 illustrates the complexity of estuarine environments, seagrass beds, and shelf sedimentary habitats. Chapter 23 describes polar pelagic and benthic habitats, including sympagic (ice) biodiversity, whereas Chapter 24 considers the main features of the neritic (continental shelf waters) zone, its biodiversity, its importance in fisheries production, and the complex factors that influence its functioning such as terrestrial inputs, river plumes, coastal upwelling, and fronts. Chapter 25 addresses the study of deep-sea ecosystems, reviewing biodiversity analytical tools and the characteristics of canyons, deep-water corals, cold seeps, dead zones, and oxygen minimum zones. Chapter 26 examines "extreme" marine environments, from abyssal plains to seamounts, from deep-sea hydrothermal vents to coastal vents, whale carcasses, hypoxic and anoxic zones, deep hypersaline basins, and ocean trenches. Chapter 27 finishes off the major habitats by examining open ocean waters and their vertical zonation.

Part IV (*Human Impacts and Solutions for Planet Ocean*, Chapters 28–31) introduces applied marine biology. Chapter 28 investigates human impacts on marine organisms and ecosystems, beginning with early human exploitation of marine resources and subsequent progressive biodiversity alteration and habitat destruction. This chapter also addresses basic concepts regarding the ocean under global climate change and the synergistic impacts of multiple stressors on marine ecosystems. Chapter 29 illustrates the importance of marine biodiversity conservation and current strategies to protect marine habitats. Chapter 30 presents innovative approaches for restoration of degraded marine habitats. Chapter 31 reports on the birth of marine biology and its development through to today, including sampling and monitoring tools and future challenges in marine biological research. The chapter also identifies the study and research topics we predict future marine biologists will be called upon to address.

Roberto Danovaro
Paul Snelgrove

About the Companion Website

This book is accompanied by a companion website which includes a number of resources created by the authors for students and instructors that you may find helpful.

http://www.wiley.com/go/danovaro/marinebiology

The Student website includes the following resources for each chapter:

- PowerPoint Slides

Part I

The Ocean Domain: Introduction to Planet Ocean

The title of this book reflects the dominant feature of our planet, the 70% covered by ocean. Most humans interact with a very small component of the ocean, whether walking or swimming along the shoreline, but that limited relationship biases our view of ocean life and ocean processes. It also undervalues the importance of the ocean to all life on our planet, including humans, and the diversity of ocean life and habitats that dominate and contribute significantly to our planet's life support system.

When we consider that 70% of our planet, most people lack any sense of the immense diversity of habitats it encompasses, and the environmental variables that help to drive the spectacular diversity of species, habitats, and processes. Similarly, the 1.35 billion cubic kilometers of seawater (including ~5×10^{16} tons of sodium chloride, or salt) that fills the ocean basins is far from uniform, again supporting different types of "habitat" defined largely by multiple characteristics of the seawater and how it moves. In contrast, seabed characteristics such as geology (among others) define seafloor habitats. Collectively the ocean seabed and water column habitats above it encompass >95% of the **biosphere**, the portion of our planet that supports life. Little wonder we've only explored a few percent of the dominant feature on this planet.

The three chapters that comprise *Introduction to Planet Ocean* introduce and describe the nature of ocean basins (*The Life Aquatic*), the bottom of the ocean (*The Seabed*), and the water that fills the basins (*The Water Column*). We identify the features and key variables that define the seabed and water column, and the myriad environments they form. In essence we use this section to build the theater in which we present our "play" on marine biology and marine ecology. So let's dive in and start the "show"!

1

The Life Aquatic

1.1 Introduction

The global ocean covers 70% of the Earth's surface, from the **land–sea interface**, where land masses meet the ocean, to depths greater than 11,000 m. The science fiction writer Arthur C. Clarke had it right when he said that we should call our home **Planet Ocean,** an alternative name for Planet Earth, (Figure 1.1). Encompassing more than 95% of the **biosphere**, the total habitable portion of Earth, ocean **ecosystems**, or biological communities of living organisms that live in and interact with each other in a specific environment, dominate our planet by far. All life on Earth depends on ocean ecosystems because of the major roles they play in climate regulation, gas, and nutrient cycling, and from a human-centric perspective, as a vital source of protein for our ~7.7 billion inhabitants. But the ocean is far from homogenous, and scientists therefore divide the **seabed**, referring to the bottom of the ocean, and overlying water column into different depth or **bathymetric** zones based on their unique physical and geochemical characteristics, which profoundly influence the composition of the resident biota. These organisms must adapt to the unique properties of the seawater or seabed in which they live, whose characteristics vary among and even within bathymetric zones, including temperature and **salinity** (or salt content) and thus **density** (or mass per volume), dissolved oxygen, dissolved nutrients and food, water clarity, **viscosity** or "thickness" of water, and the nature of the seafloor **substratum**, the material that makes up the seabed.

The fluid nature of much of the marine environment contributes to its dynamic nature. Marine currents determined both by winds and by temperature differences (and, therefore, seawater density), affect many physical, geochemical, and biological processes, including the transport of organisms (and nutrients and energy) among different ocean environments. Although marine biology links to the ecology of Earth, many general ecological principles developed from terrestrial studies rarely apply to the study of marine habitats. The challenge of transferring concepts extends beyond ecological study. As a land-based species, we see Earth from a terrestrial perspective that has little meaning in the viscous, mostly dark, cold, and vast three-dimensional ocean world. Many people never see the ocean, and many fewer see below the ocean surface by diving with scuba, and practically no one, except some lucky scientists, can explore using submersibles.

Below, we describe how understanding marine biology and ecology requires resetting our frame of reference. In subsequent chapters, we assess how the chemical, geological, and physical peculiarities of the marine environment set the stage for the biology of marine organisms and marine ecological processes that form the basis for the remainder of the book; understanding how marine organisms live and relate to their environment requires appreciating the properties of the seawater and ocean basin that define their "world."

1.2 Comparison Between Sea and Land

Some general ecological principles apply both to marine and terrestrial ecosystems, but the two domains differ in many ways in how they function. Many ecologists believe that, given the physicochemical properties of water and particularly seawater, the ocean gave rise to the first forms of life on Earth. The specific conditions that occur in hydrothermal vents, including high temperature and pressure, water, abundant methane, ammonia, and hydrogen, replicate those that Stanley Miller and Harold Urey used in their famous laboratory experiments in 1952 that produced amino acids, the basic building

Marine Biology: Comparative Ecology of Planet Ocean, First Edition. Roberto Danovaro and Paul Snelgrove.
© 2024 John Wiley & Sons Ltd. Published 2024 by John Wiley & Sons Ltd.
Companion Website: www.wiley.com/go/danovaro/marinebiology

Figure 1.1 Image of the eastern and western hemispheres of Earth from space illustrating why Earth is truly an ocean planet. *Source:* NASA / Public Domain.

blocks of proteins (and also life on Earth). Organisms in marine and terrestrial communities differ in the wide range of adaptations they have evolved to live on land or in the ocean. In Table 1.1 we summarize some of these major differences.

Primary producers: The unicellular algae, or phytoplankton, are the dominant photosynthetic organisms in most ocean environments, which contrasts with the higher plants (trees, shrubs, grasses) that dominate the primary producers in terrestrial environments. Higher plants, such as seagrasses, **macroalgae** (referring to large, multicellular algae, such as kelps and seaweeds, visible to the naked eye), mangal (mangrove forests), and salt marsh vegetation, play an important role in some land–sea interfaces. Much like rooted plants that dominate terrestrial ecosystems, these plants and algae that occur at the land–sea interface create habitat for a wide range of biota. In the ocean, the unicellular algae that drift through the upper ~200 m of the ocean provide no equivalent habitat structure.

Connectivity: Unlike the terrestrial environment, where mountains and rivers create visible physical barriers between habitats, the circulation of water masses that flow as giant conveyor belts potentially connect all ocean habitats. In the ocean, differences in environmental characteristics, such as temperature, salinity and/or depth represent the main obstacles to global mobility of marine organisms, but they are minimal and subtle. The currents that distribute suspended particles in the ocean also negate the need for organisms analogous to pollinating insects, which are essential in terrestrial environments.

Three-dimensionality: the massive third dimension of the marine environment conferred by depths averaging ~3800 m and up to 11,000 m adds a range of habitats not matched by the tallest trees (~100 m) or even the greatest heights at which birds fly (1000s of m). Most life on land lives within 10's of m of the ground or within it, in contrast to the plethora of life in the ocean from the surface through thousands of meters of water to the greatest depths, and deep into seafloor

Table 1.1 Comparison between land and ocean.

Environmental Factor	Terrestrial Environment	Marine Environment
Temperature	Varies greatly between night and day, winter and summer, and between habitats at different latitudes. The low heat capacity of air results in little protection against air-cooling and heating.	Changes minimally between day and night, summer and winter therefore are limited. Most biota do not expend energy to maintain body temperature, they instead track the temperature of the surrounding environment; thus, most are cold-blooded animals.
Salinity	Only applies in a few environments such as salt lakes and salt mines.	Marine invertebrates use physiological adaptations (e.g., filtration, re-sorption and secretion mechanisms) in order to produce urine that differs in osmolarity compared to body fluids. Many organisms cannot tolerate variation in salinity.
Winds vs currents	Winds can transport seeds and small organisms, though over modest distances Winds also spray saltwater onto coastal vegetation, favoring tolerant species in those locations.	Marine currents transport food, nutrients, plankton (organisms living in the water column that cannot make significant headway against currents), eggs and larvae of seafloor species, and even sediments over distances up to 1000s of km.

Table 1.1 (Continued)

Environmental Factor	Terrestrial Environment	Marine Environment
Wind and wave action	Winds can shape landscape vegetation composition and growth forms to cope with conditions. Some organisms (insects, birds) have evolved structures that take advantage of the wind or to limit its possible negative effects.	Noting that water is more than 800 times denser than air, waves can cause major damage along the land–sea interface, creating drag and moving sediments, rocks, and debris. Wave activity requires another set of adaptations in addition to those to cope with desiccation and temperature change
Light	Although light intensity and seasonality vary with latitude, they vary little otherwise, except from shading by clouds and other vegetation.	Quantity and quality of light vary greatly with depth, concentrations of plankton and particles, as well as wave intensity. Light filtering limits photosynthetic production to the upper 70–130 m depth in the clearest water.
Substratum for primary producers	Plants require humid soil and nutrients; they transport water and nutrients from the roots to the leaves, where photosynthesis takes place.	Kelps and seaweeds take up nutrients and water from the surrounding seawater and use root-like structures to adhere to the substratum. Phytoplankton, the predominant photosynthetic organisms in the ocean, drift and use water as habitat.
Substratum for non-photosynthetic organisms	Soils provide habitat for many organisms, particularly microbes and invertebrates. Most plants that create bio-habitat for other species depend on soils for substratum. Mobile fauna move across solid bedrock, but relatively few organisms use it as habitat.	Sediments, like their soil counterparts, provide habitat for a rich diversity of microbes and invertebrates. Hard substrata support a rich diversity of fauna specifically adapted to that environment, including habitat-forming species such as corals and kelp forests.
Habitat height/depth	Less than 100 m (tallest trees) to 1000s of m (migrating birds), largely bi-dimensional (2D)	Up to more than 11,000 m (deepest point of the global ocean), strongly 3 dimensional (3D)
Openness	Lower (many dispersal barriers, such as mountains, rivers, and deserts)	Higher (few visible dispersal barriers)
Electricity	Air is a poor conductor of electricity, and few terrestrial organisms use electricity as sense organs or as weapons.	Seawater conducts electricity well and diverse marine species from sharks to **mollusks** (a phylum of (mostly) shelled invertebrate animals that includes snails, slugs, mussels, and octopuses) can perceive electricity produced by moving organisms or geomagnetic fields.
Body weight	**Determinate growth,** where growth stops once a genetically pre-determined structure has completely formed or growth stage has been reached, characterizes most terrestrial organisms. Gravity and the low density of air place limits on size. Bones cannot sustain massive weights such as in whales, and many organisms exhibit different growth patterns, where organisms cannot grow beyond a certain size.	Because of the similarity in density between the water that comprises the bulk of living tissue and the surrounding seawater, gravitational effects limit growth much less in marine than in terrestrial systems. Massive whales and meters-long gelatinous zooplankton can exist in ocean habitats.
Body shape	Other than birds, most organisms do not require elaborate aerodynamic design because air resistance has little effect on speed of movement.	Water viscosity resists the movement of organisms much more than air (try walking across the bottom of a swimming pool!) Fast-moving fishes therefore require hydrodynamically efficient shapes (e.g., torpedo) to move effectively, as well as efficient fins for thrust, in contrast to paws for running or climbing on land.
Base of food web	Large habitat-forming trees, shrubs, grasses	**Unicellular** (organisms consisting of a single cell), floating **protists**, referring to **eukaryotic organisms** (those whose cells contain a cell nucleus) that are not animals, plants, or fungi
Population turnover in primary producers	Slow	Rapid
Food webs	Pyramidal	Inverse pyramidal often

(Continued)

Table 1.1 (Continued)

Environmental Factor	Terrestrial Environment	Marine Environment
Diet	Many animals are born relatively large, and feed on relatively few different types of food as they grow. Adults may be 10–50 times larger than juveniles.	Many animals produce and release large quantities of eggs, which change dramatically in size as they develop. For example, a fish can grow from 0.1 mm to 5 m (a 50,000x increase!) and to several hundreds or thousands of kilograms in weight. Dietary sources change many times during this transition.
Animal diversity	Low phyletic diversity (few phyla, hyper-diverse insects)	High phyletic and species diversity except in plants. Many animal phyla
Respiration	Terrestrial animals use lungs or lung-like structures to exploit oxygen from air; air passes through smaller and smaller structures eventually to exchange oxygen between air and internal fluids, oxygenating the organism.	Most marine vertebrates and invertebrates use gills or gill-like structures. Water passes **unidirectionally** (moving in just one direction) through gills to conserve energy. As with lungs, blood flows through gills in tiny vessels that facilitate exchange of gases, in this case in dissolved form from fluid to fluid.
Olfaction	Odors must be strong (highly concentrated) in order to be sensed through olfaction.	High concentrations of odors can occur in the ocean and persist longer; some organisms with specialized sensory organs can sense odor plumes from great distances.
Hearing	For organisms on land, to perceive sounds, vibrations must transfer from a highly compressible medium (air) to a rigid medium (the auditory system).	For organisms in the ocean to perceive sounds, vibrations must transfer from a non-compressible medium (water) to a rigid receptor. Sound travels four times faster in seawater than in air, and acoustic waves can easily pass through hundreds to even thousands of kilometers. Many strong vibrations and noises beyond the sensitivity of the human ear occur in the ocean.
Domestication	Humans have domesticated a wide variety of plants and a modest number of animals, so that food production depends primarily on domesticated species.	Historically, wild fisheries for fishes and invertebrates provided most of the nutrition that humans extracted from the ocean. Marine aquaculture production has increased greatly in recent decades though still lags behind wild fisheries.

sediments. Much like the terrain and trees in terrestrial habitats, underwater landscape features add to the complexity and heterogeneity of the **substratum** on which many organisms live, and thus accommodate the **biodiversity**, or the large variety of ecosystems, species or genes, that characterize a place. The total volume of marine environments provides 300 times more space for life than land and fresh water combined.

1.3 Fractal Complexity of Marine and Terrestrial Ecosystems

How marine organisms perceive space. A fractal refers to the fractional dimension of a geometric object; the greater the complexity in the number of branches (nodes) and its three-dimensionality, the greater its "ecological volume." Ecological volume denotes the theoretical volume ideally obtained from the connection of all apexes and the space provided for the establishment and protection of other organisms. High fractal complexity characterizes the woods and forests of terrestrial environments, in contrast to the low complexity in bare bedrock or in deserts. Likewise, the biological components in marine environments may confer greater complexity to the habitat; kelp forests (large macroalgae), seagrass beds, and bioconstructions (coral reefs, **coralligenous assemblages**, referring to calcareous formations produced by the accumulation of encrusting algae growing in dim light) all create high heterogeneity. Moreover, different levels of fractal

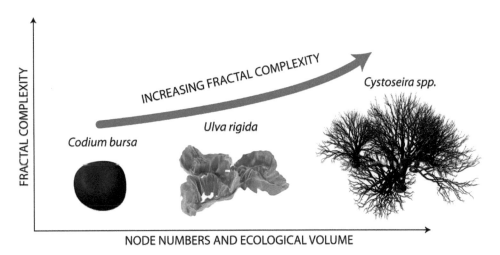

Figure 1.2 Diagram showing how different algal morphologies can increase fractal complexity within their habitats.

complexity characterize algae, from the flat frond of sea lettuce (*Ulva lactuca*) that occupies largely two-dimensional space to the complex structures with many nodes and internodes of *Cystoseira*, with its wider total surface area (Figure 1.2), and much higher "ecological volume." Fractal dimension also varies with scale. What large organisms may perceive as a flat surface may appear enormously complex for the microscopic organisms that comprise the majority of marine life. Nematodes, animals smaller than one millimeter in size and the most abundant **metazoans** (multicellular animals) in the ocean, can use the smallest leafy structures as a refuge from predation.

Evolutionary history and antiquity: Stromatolites, the laminar fossil structures produced by marine cyanobacterial activity and discovered in Shark Bay, Australia and elsewhere, represent the oldest known fossils and first evidence of life, dating to ~3,500 million years ago (Figure 1.3). The first traces of simple marine animals date back to 800 million years ago, but the "true" animal fossils date to ~640 million years ago. The oldest terrestrial fossils, putatively spores of bryophytes, date back to the Ordovician (~450 million years ago). The first animals appeared in the terrestrial fossil record around 400 million years ago. Therefore, marine organisms have had considerably more time to diversify than their terrestrial counterparts (about twice as long for animals), but science has described far fewer animal species in the ocean than on land. Scientists often attribute this paradox to the enormous dispersal potential of marine animal propagules, and eggs and larvae in particular, reducing speciation by increasing gene flow in an ocean environment that lacks obvious dispersal barriers such as rivers and mountains on land. However, as we discuss below, this difference may simply reflect accessibility of specimens from terrestrial versus marine habitats.

Size range: Marine organisms vary immensely in size; unicellular primary producers (from <1 to several microns) at the base of the food web contrast the largest consumers, such as blue whales (up to 35 m length and 200 metric tons). By comparison, African elephants grow to lengths up to 7 m and weights of 5 metric tons. Nonetheless, some macroalgae grow up to 30 m in length but with considerably lower biomass than large trees on land.

Biochemical composition of organisms: Aquatic habitats support structurally simple and fragile forms of life, such as comb jellies and feather dusters, because comparatively dense water provides greater support, buoyancy, and transport, than air, reducing the need for strong structural support. Marine kelps and seaweeds also benefit from this support in that they do not need to invest substantial energy in cellulose support structures comparable to the trunks of trees on land, and can use simple gas bladders to keep their stems erect in the water where they can optimize light use. Marine macroalgae can reduce investment in structural carbohydrate and invest more in protein content. This difference explains the greater nutritional value of macroalgae, and marine algae in general, compared to terrestrial higher plants, with implications for nutrition and food webs of their respective ecosystems. These large terrestrial plants, with their major structural support, grow slowly and live long compared to rapidly growing and relatively "protein-rich" marine primary producers. In general, along the coasts, changes in environmental characteristics such as temperature and salinity are more pronounced in time and space than those in the open ocean.

Trophic levels and renewal of biomass: The small size and rapid reproduction of marine primary producers result in much more rapid replacement of the base of the food web than occurs on land. Thus, the turnover of biomass, the large

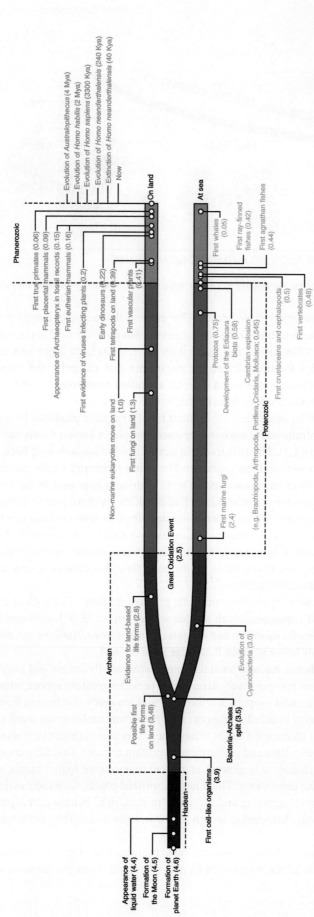

Figure 1.3 Timeline of history of life on Earth, highlighting major events and eons. Events that occurred on land shown on the upper timeline, those that occurred in the ocean on the lower one. Time points are expressed as billion years, unless specifically noted (Mya: million years; Kya: thousand years).

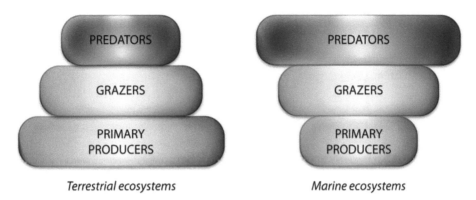

Figure 1.4 Left: typical biomass pyramid of terrestrial systems; right, typical biomass pyramid of many marine ecosystems.

ratio of primary producer production to biomass (P/B) in the ocean greatly exceeds that on land. Phytoplankton community composition can change in days, whereas development of terrestrial plant communities spans decades to centuries. Although the time scales of turnover in macrophytes, **benthic** (organisms living on seafloor) invertebrates, and fish are similar to those of equivalent terrestrial biota, food web turnover rates differ. In terrestrial ecosystems, primary producers (e.g., trees) typically develop over much longer time scales than the herbivores that graze on them (e.g., insects and mammals), whereas primary producers (e.g., algae) develop over shorter or comparable time scales to their consumers (e.g., grazers). This difference explains "classic" pyramids of biomass in terrestrial systems (Figure 1.4), in which the biomass of the plants exceeds that of the herbivores that graze on them, and the biomass of grazers exceeds that of predators. Marine environments also support inverted biomass pyramids, in which herbivore biomass per unit volume exceeds that of primary producers. The constant renewal of marine primary producers through rapid reproduction supports a larger biomass of herbivores at a given time in what would otherwise seem energetically impossible.

Number of trophic levels: generally, the number of **trophic levels**, the position occupied in a food web by different organisms in marine environments exceeds that in terrestrial habitats because the small size of primary producers can support multiple trophic levels of tiny grazers, (such as protozoa, nanoflagellates) and predators on those grazers (e.g., protozoan ciliates) only one tenth of a millimeter in size, eventually leading up the food web to large pelagic organisms (Figure 1.5) such as tuna or sharks (top predators) that may grow to several meters in size (Figure 1.6). Many marine environments support 6 or 7 trophic levels in contrast to the 3 or 4 trophic levels that characterize many terrestrial ecosystems, where large herbivores graze large plants and become prey for large predators.

Environmental fluctuations: The properties of water (see Chapter 2) help to "buffer" aquatic environments with respect to environmental variation compared to their terrestrial counterparts. Generally, chemical and physical parameters vary less in space and time in marine environments compared to on land. For example, constant and stable environmental conditions characterize the deep waters that occupy more than 90% of the biosphere. Marine environments, though diverse over broad spatial scales, appear relatively homogeneous on small spatial and temporal scales compared to patterns on land. Differences in oxygen concentrations and availability of various key components (such as nutrients) promoted the evolution of very different primary producers than those of terrestrial systems. We discuss the most important adaptations of marine organisms in more detail in subsequent chapters in order to understand better the magnitude of differences in adaptations on land and in the ocean.

Marine and terrestrial biodiversity: Marine and terrestrial scientists debate the numbers of species that occur in the respective domains, but why does it matter? Whether land or sea supports more diversity, the real issue is appreciating what we know and do not know, and specifically what biodiversity we may lose as human impacts on ocean environments escalate. The question is harder than it sounds because scientists have yet to describe most species on land and in the ocean and must therefore extrapolate this information from indirect calculations based on mathematical models, adding controversy because of a high level of uncertainty.

The most recent credible estimate, from 2011, examines how we classify organisms for taxonomic purposes: Taxonomists organize species into higher taxa (from genus to phylum) according to a consistent pattern that allows prediction of the total number of species in each taxonomic group. Based on this model, up to 8.7 million species occur on Earth (give or take 1.3 million), including 750,000–2.2 million from the ocean (Table 1.2). This model also estimates that 86% of terrestrial species and 91% of marine species remain undiscovered. Scientists have already described about 1.5 million species of

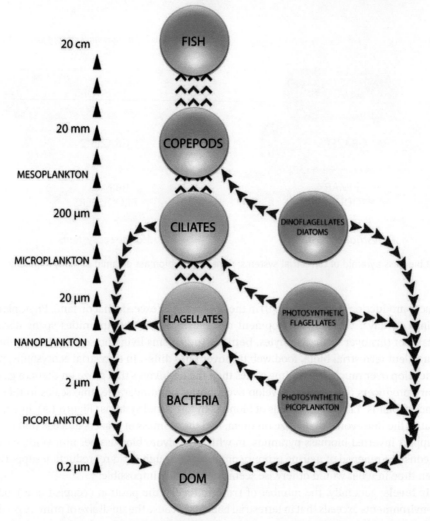

Figure 1.5 Example of a marine food chain starting from dissolved organic matter (DOM) and moving upward through increasingly larger microbial components such as cyanobacteria, bacteria, protozoa, nanoflagellates (microbes with flagella within the size range 2 to 20 μm), and ciliates (including photosynthetic algae), passing through zooplankton such as copepods, and up the food web to fish and sometimes on to marine mammals such as seals. All of these components create an intricate web of interactions that increase the complexity of marine food webs compared with terrestrial food webs. Arrows indicate flow of energy and nutrients.

Figure 1.6 Example of large predators in the sea. Tuna reach 3 m in length and several hundred kg, but they produce eggs only a few microns in length (millionths of a mm). OAR/National Undersea Research Program (NURP) / Wikipedia Commons / Public Domain.

Table 1.2 Number of species currently cataloged and projected on Earth and in the ocean.

	Earth		Oceans	
	Described	Expected	Described	Expected
EUKARYOTES				
Animalia	953,434	7,770,000	171,082	**2,150,000**
Chromista	13,033	27,500	4,859	**7,400**
Fungi	43,271	611,000	1,097	**5,320**
Plantae	215,644	298,000	8,600	**16,600**
Protozoa	8,118	36,400	8,118	**36,400**
Total	**1,233,500**	**8,740,000**	**193,756**	**2,210,000**

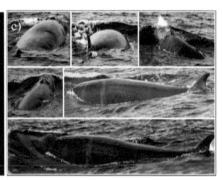

Figure 1.7 Examples of some large, recently discovered taxa: (a) *Aureophycus aleuticus*, (b) *Kiwa hirsuta*, (c) *Balaenoptera omurai*. *Sources:* Golden V kelp (*Aureophycus aleuticus*) near Kagamil Island, taken 7/19/2007. Credit: Max Hoberg, Fifis Alexis/Ifremer.

terrestrial plants and animals, in contrast to ~250,000–270,000 for the ocean. In the terrestrial realm, insects (phylum Arthropoda) dominate the animals in comparison to 8 phyla (Porifera, Cnidaria, Nematoda, Annelida, Arthropoda, Mollusca, Bryozoa, and Chordata) that comprise ~90% of marine animals. Of the 30 animal phyla described to date, 15 (including Echinodermata, Urochordata and Ctenophora) occur only in the ocean whereas only a single phylum, the Onychophora (velvet worms), occurs only in terrestrial environments (although with marine fossil ancestors). Scientists continue to discover new marine phyla, indicating that major biodiversity discoveries are possible. The most recent new phylum, the Cycliophora (discovered in 2000), live **commensally** (benefitting from the host but not harming them) in the mouthparts of some lobster species. In 1983, scientists discovered rotifer-like animals of the new phylum Loricifera living in unconsolidated sediments across a wide depth range. Even in relatively accessible surface waters, discovery of new species occurs frequently, particularly for the smallest animals. For example, our knowledge of the **meiofauna** (animals <44 μm in size; Chapter 4) remains fragmented, even in European countries with a long tradition of natural history research. Discoveries of marine species also include some relatively large in size, such as the 3-m long golden V kelp (*Aureophycus aleuticus*), discovered in Alaska in 2008; Figure 1.7(a), the "yeti crab" (*Kiwa hirsute*), discovered at a hydrothermal vent in 2005, Figure 1.7(b), and a new species of whale (*Balaenoptera omurai*), discovered in 2003, Figure 1.7(c). Researchers recently projected that another ~50 large marine species (>2 m) remain to be discovered. Scientists only recently first observed one of the 10 known species of giant squid (>20 m) in situ, despite their widespread distribution.

Questions

1) What are the main differences between marine and terrestrial environments?
2) What are the main differences between marine and terrestrial organisms?

Suggested Reading

Acha, E.M., Piola, A., Iribarne, O., and Mianzan, H. (2015) *Ecological Processes at Marine Fronts*. Springer, Cham.

Baird, D., McGlade, J.M., and Ulanowicz, R.E. (1991) The comparative ecology of six marine ecosystems. *Philosophical Transactions of the Royal Society B: Biological Sciences*, 333, pp. 15–29.

Barnes, R.S.K. and Mann, K.H. (1991) *Fundamental of Aquatic Ecology*. Blackwell Publishing.

Bar-On, Y.M. and Milo, R. (2019) The biomass composition of the oceans: A blueprint of our blue planet. *Cell*, 179, pp. 1451–1454.

Cochran, J.K., Bokuniewicz, H.J., and Yager, P.L. (2019) *Encyclopedia of Ocean Science*. 3rd Edition. Academic Press.

Cornell, C.E., Black, R.A., Xue, M. et al. (2019) Prebiotic amino acids bind to and stabilize prebiotic fatty acid membranes. *Proceedings of the National Academy of Sciences of the United States of America*, 116, pp. 17239–17244.

Costello, M.J. and Chaudhary, C. (2017) Marine biodiversity, biogeography, deep-sea gradients, and conservation. *Current Biology*, 27, pp. R511–R527.

Eddy, T.D., Bernhardt, J.R., Blanchard, J.L. et al. (2021) Energy flow through marine ecosystems: Confronting transfer efficiency. *Trends in Ecology and Evolution*, 36, pp. 76–86.

Lalli, C.M. and Parsons, T.R. (1993) *Biological Oceanography: An Introduction*. Elsevier.

McIntyre, A. ed. (2010) *Life in the World's Oceans: Diversity, Distribution, and Abundance*. Wiley-Blackwell.

Miller, C.B. and Wheeler, P.A. (2012) *Biological Oceanography*. 2nd Edition. Wiley-Blackwell.

Miller, S.L. (1953) A production of amino acids under possible primitive Earth conditions. *Science*, 117, pp. 528–529.

Mora, C., Tittensor, D.P., Adl, S. et al. (2011) How many species are there on Earth and in the ocean? *PLOS Biology*, 9(8), p. e1001127.

Parker, E.T., Zhou, M., Burton, A.S. et al. (2014) A plausible simultaneous synthesis of amino acids and simple peptides on the primordial Earth. *Angewandte Chemie International Edition*, 28, pp. 8132–8136.

2

The Seabed

2.1 Ocean Basins

The ocean covers most of the surface of our planet, but the percent coverage differs between the southern and northern hemispheres (Figure 2.1). Despite the interconnectedness of the global ocean, scientists recognize five relatively distinct oceans: Pacific, Atlantic, Indian, Arctic, and Antarctic (also known as the Southern Ocean). Smaller marine basins such as the Mediterranean Sea, the Caribbean Sea, the Baltic Sea, the China Sea, the North Sea, and others border the ocean, but their distinct biogeochemical characteristics merit separate consideration. New calculations, based on satellite measurements that detect seafloor geological formations and depth more accurately than in the past, indicate a total ocean volume of ~1.33 billion km^3, slightly less than previous estimates. No definitive data currently exist on the volumes and average and maximum depths of the five main oceans. Recalculation of average ocean depth based on this approach yields a value of 3,682 m (slightly less than previous estimates of 3,734 m). This estimate dwarfs the average altitude of terrestrial environments of ~840 m (Figure 2.2).

The largest and deepest ocean, the Pacific Ocean, borders all of the continents except Europe and Africa (Table 2.1).

The deepest point of the ocean globally occurs in the northwestern Pacific Ocean between the Mariana Trench (~11,033 m) and the Philippines Trench (also called the Mindanao Trench, ~10,540 m); the uncertainty in depth measures in these remote areas adds ambiguity to any bathymetric differences between the Mariana and Philippines Trenches (Figure 2.2). The deepest point in the South Pacific Ocean occurs near the Chilean coast, in the Atacama Trench (8,065 m). The Atlantic Ocean averages ~3,930 m in depth, with a large area of **continental shelf**, referring to the region of the seafloor that gently slopes (1°) from the subtidal (or sublittoral) zone to the edge of the shelf at 150–200 m depth. The Atlantic reaches its maximum depth in the Puerto Rico Trench (8,648 m). An underwater "mountain chain" known as a **mid-ocean ridge**, effectively divides both the Atlantic and Pacific Ocean floors longitudinally into two sub-basins.

The mid-ocean ridges provide an important contribution to the 3-D shape of the seabed and increase the heterogeneity of seafloors globally (Figure 2.3). The Mid-Atlantic Ridge, for example, runs the entire length of the Atlantic Ocean, typically hundreds of meters below sea level, merging above the sea as the volcanic islands of Azores and Iceland. With an average depth similar to that of the Atlantic Ocean, the Indian Ocean reaches its deepest point in the Java Trench, at 7,725 m. The Southern (Antarctic) Ocean, with a maximum depth of 7,235 m, differs from the other oceans in that its northern border merges with other oceans rather than abutting a continent (Figure 2.4).

The smallest and shallowest ocean, the Arctic Ocean, averages ~1,205 m in depth, with its deepest point at 5,450 m in the Eurasian Basin. For comparison, the Mediterranean Sea, which is one fifth of the Arctic Ocean in area, averages ~1,450 m in depth with a maximum of 5,092 m in the Matapan Trench between Italy and Greece.

These data reflect a fundamental role for the oceans in global processes, because all ocean environments from the land-sea interface to the deepest trenches teem with microorganisms and other biota, both above and below the seabed. These summary data also suggest that the study of ocean life must consider both the seabed and the water column. Broadly speaking, the ocean encompasses a **pelagic** domain (the totality of the water volume) and a **benthic** domain (the totality of the seabed). Scientists divide each domain into different zones. The pelagic domain consists of a **neritic province**, the

Marine Biology: Comparative Ecology of Planet Ocean, First Edition. Roberto Danovaro and Paul Snelgrove.
© 2024 John Wiley & Sons Ltd. Published 2024 by John Wiley & Sons Ltd.
Companion Website: www.wiley.com/go/danovaro/marinebiology

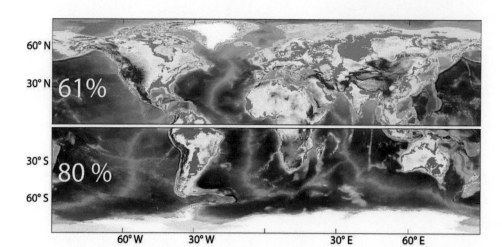

Figure 2.1 Distribution of the total area covered by the global ocean. Percentages indicate volume of each hemisphere occupied by ocean.

Table 2.1 Average depth (m) and total area (km²) of five oceans.

Average Depth	Ocean	Surface km²	Percentage of Earth's surface
4,028	Pacific	155,557,000	30.5
3,926	Atlantic	76,762,000	15.1
3,963	Indian	68,556,00	13.4
4,000–5,000	Southern	20,327,000	4
1,205	Arctic	14,056,000	2.8

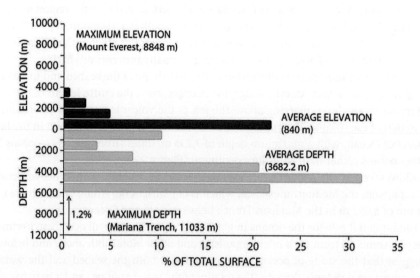

Figure 2.2 Oceanic depths and land elevations expressed as percentage of total marine and terrestrial areas.

relatively shallow portion of the ocean that extends from the land-sea interface to the edge of the continental shelf at 200 m, and where light often penetrates all the way to the seafloor. The **oceanic** (or **pelagic**) **province**, refers to the water masses beyond the continental shelf (Figure 2.5). Below we describe these domains in detail. We begin by describing the basins themselves and then discuss the fluid that fills them.

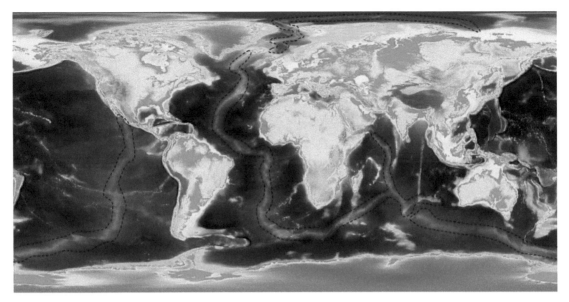

Figure 2.3 Mid-ocean ridges (highlighted by black dashed lines) that increase seafloor elevation, contribute to bottom heterogeneity and offer important opportunities for creation of new habitats and biodiversity hot spots.

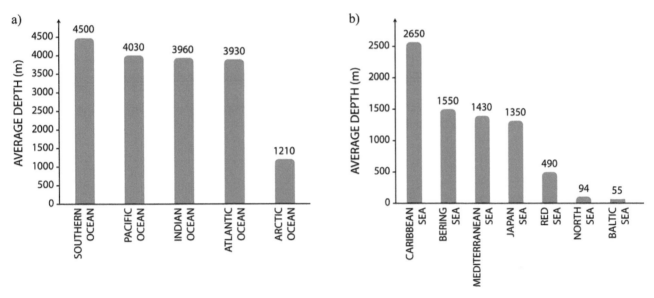

Figure 2.4 Average depths (m) of: (a) oceans; (b) major seas.

2.2 Ocean Bottom: A (Mostly) Static Habitat of Ocean Life

Seabeds are covered either by unconsolidated sediments (soft bottom) that originated from the accumulation of organic and inorganic particles on the bottom, or by consolidated sediments (hard bottom), consisting of rocky outcrops, often originating from underwater volcanic eruptions. Soft bottoms cover >60% of the global seabed, and hard bottoms the remaining ~40%. Seabed topography often varies greatly, including habitats of interspersed rock and sediment. This intermingling of substrates has important implications for adaptations of the associated biota (see Chapter 6), because completely different forms of life colonize the two types of substrates. Moreover, the spaces between the particles of unconsolidated sediments create a complex, three-dimensional environment. For this reason, biota occur not only as **epifauna**, referring to organisms that live atop sediment (or hard substratum) surfaces, but also as **infauna**, referring to organisms that live

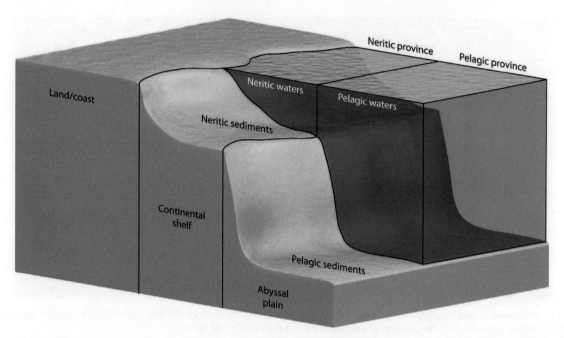

Figure 2.5 Distribution of neritic province (portion of ocean over the continental shelves, including water column and seabed) and pelagic province (portion of ocean overlying the deep sea, beyond the continental shelves, including the water column and seabed).

below the sediment surface. Sediment thickness varies among ocean regions from thin veneers of just a few millimeters over bedrock to 150 m to more than 10-km thick layers (Figure 2.6). The thickest sediments occur in the 11-km thick Indian Ocean seafloor crust. This variation is important because although multicellular organisms colonize only the first meter of the sediment or less, microorganisms such as viruses, bacteria, and archaea may occur up to several kilometers deep in the seabed. At these depths the proximity of heat emitted from the Earth's core may drive temperatures close to 100 °C. We describe these extreme environments in detail elsewhere (Chapter 26).

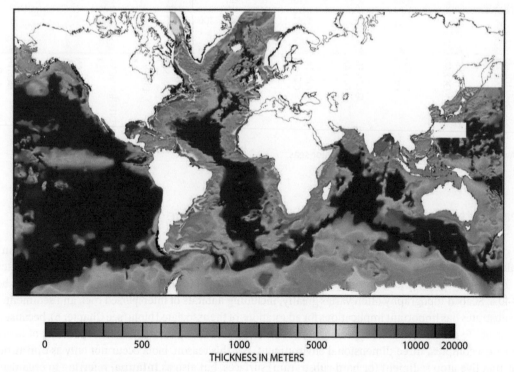

Figure 2.6 Thickness of sediments along oceanic margins.

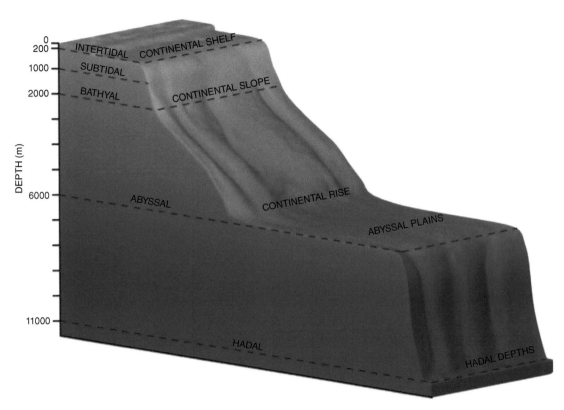

Figure 2.7 Subdivision of oceans into four main zones: continental shelf, continental slope, abyssal plains, and oceanic trenches. The combined continental shelf and slope form continental margins.

Ocean floor topography varies greatly, but encompasses four broad regions, spanning from the coast to the open ocean: (1) continental shelf; (2) continental slope or **bathyal** region; (3) abyssal plains; 4) hadal seafloor from 6,000 to 10,000 m known as **ocean trenches** (Figure 2.7).

Previous estimates of ≈ 361,254,000 km² for the area of the Earth's surface covered by the ocean assumed a flat bottom, but by also including the topographic complexities of the seafloor, the estimated area available to benthic organisms increases by almost 20%, producing a total "bioavailable" surface of ~434,386,264 km². However, this correction has little effect on the proportional contributions of continental shelf, slope, abyssal plains, and trenches (Figure 2.8). Box 2.1 explains how we map the seafloor.

Box 2.1   How we map ocean depth

Historically, seafloor mapping was achieved from ships, initially by lowering a weighted lead line to the seafloor at specific geographic locations. Building on earlier exploratory ocean cruises, such as James Cook's expeditions around the world that were motivated by safe exploration of new lands and emerging trade routes, the first truly interdisciplinary oceanographic expedition, the *Challenger* (1872–1876), achieved 492 "measurements" of ocean depths. More recently, satellite altimetry, which measures small differences in sea surface height, has provided a fuzzy, but global-scale seafloor map. Modern seafloor mapping uses acoustics, such as echo sounders built into ship hulls that generate sound waves that bounce off the bottom and use a receiver to measure the time delay and nature of the return signal. The echo sounder transmits sound pulses at 10–30 kHz and analyzes the return time from the bottom. The time from transmission to reception, when multiplied by the speed of sound, equals twice the depth of the seafloor. Alternatively, scientists may also tow "**side scan sonar**" (Figure 2.13) behind a ship, "scanning" the seafloor with two divergent sonar beams that acoustically image two strips of seabed that vary in width from a few meters to 30 km, depending on bottom depth. The difference in density between the water and ocean seabed returns a reflected signal that elucidates the morphology, and therefore depth, of the seabed. More recently, researchers have placed multibeam acoustics on autonomous underwater vehicles (AUVs) such as REMUS to provide more rapid and comprehensive seabed maps.

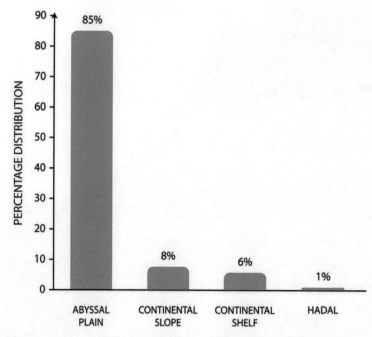

Figure 2.8 Percentage distribution of different topographic features of global ocean seabed.

Seafloor ecologists divide the benthic domain into the land-sea interface (including the intertidal zone) and the **continental shelf**, which covers ~8% of the seabed globally and gently slopes (1°) from the subtidal (or **sublittoral**) zone to the edge of the shelf at 150–200 m depth (the **shelf break**). Beyond the shelf break, the **continental slope** extends from the edge of the continental shelf at ~200 m to the continental rise at ~3000-m depth, covering ~6% of the global seafloor. The **upper bathyal zone** of this division slopes downward at an ~4° grade (and sometimes up to 20°, comparable to the steepest hill a car can climb) to ~3,000 m depth. It then transitions into the **continental rise** at 3,000-to 4,000 m and then slowly grades into the **abyssal zone**, which covers 85% of the global seafloor and extends 4,000–6,000 m depth. Finally, the **hadal zone**, the portion of the seabed below 6,000 m depth, includes **oceanic trenches**, which extend to more than 11,000-m depth (Figure 2.8).

These ocean zones also occupy vastly different proportions of ocean volume.

1) The continental shelf, particularly shelf regions near river mouths, often experiences intense sedimentation of organic and inorganic material, mostly from **terrigenous** (referring to terrestrial) sources. In contrast, the continental shelves and coastal platforms in isolated regions lack significant terrestrial input. For example, the island of Crete, and portions of the Azores and southern Cuba almost completely lack land-based input. Inorganic and inert material generally dominate marine sedimentation, but **biogenic material**, referring to material resulting from biological activity can provide important sources of sediments in some ocean regions. In this case, biogenic sediments may represent a fraction >30% of the seabed, such as the coral sands of Fiji or the calcareous sediments of the Bahamas and Persian Gulf. Wave action and currents largely control the distribution of **surficial** (surface) sediments on continental shelves and determine the transport of fine (silts and clays) to coarse (sand) sediments from source (supply) areas such as rivers to areas of accumulation (sinks). Transport not only influences abiotic (e.g., bulk sediments) and biogenic, non-living components, but it also affects the distribution of larvae, juveniles, and adults (see Chapter 15). The mechanical action of waves, currents, and tides can erode entire portions of the coastal seascape, whether altering beaches and barrier islands, or eroding limestone or other rocky coasts. Sea level change further alters coastline development. From a morpho-dynamic point of view, the high energy (currents and waves) that characterizes the land-sea interface favors the evolution of life forms specifically adapted to cope with high energy (see Chapter 6 on intertidal ecosystems). Environmentally, the land-sea interface experiences some of the most rapid and extreme fluctuations in temperature and salinity on Earth, producing major impacts on the biology and ecology of the resident biota and determining which species can thrive within these "extreme" marine environments. Abundant light supports dense cover of plants (seagrasses, mangroves, salt marsh plants) and algae (e.g., kelps, seaweeds, calcareous algae) along many coastlines that attach to the bottom and, in the case of some calcareous algae, can survive to depths up to 150–170 m.

2) The continental slope consists mostly of mud and sand originating from erosion of the land and transport across the shelf onto the continental slope. Deep canyons furrow the slope, formed by riverine erosion during glaciation or when unstable, unconsolidated sediments slump downslope, opening a discontinuity along the continental margin. This erosion, driven by the interaction of currents, waves, and instability, creates a characteristic V-shaped canyon incision. Relatively narrow, steep-sided, and deep furrows characterize most canyons (Figure 2.9), which often transport large amounts of sediment and organic material from the shelf via dense, cold currents that originate at the surface and sink rapidly in the canyons at velocities of up to 1 m s^{-1}, excavating the seabed down slope (e.g., Figure 2.9). Generally, the bathyal platform develops at 2,000–4,000 m depth, where the slope of the seabed decreases drastically.

3) The **abyssal plains**, the gently rolling deep seabed that extends over 65–70% of the ocean floor and covers >40% of the Earth's surface, represent the most extensive ecosystem in the world in total area (Figure 2.10). The mid-ocean ridges that bisect the abyssal plains collectively constitute the most pronounced structure on the Earth's surface, extending for 60,000 km over an area equivalent to 23% of the Earth's surface. This long strip of recently formed oceanic crust arches upward, demarcating the edges of Earth's tectonic plates where new crust forms. Volcanic islands, including some volcanically active sites such as Hawaii, sometimes emerge from these ridges. At tropical latitudes, coral reefs may surround these islands, as seen in the coral atolls in the Pacific and Indian Ocean. In some cases, individual formations or

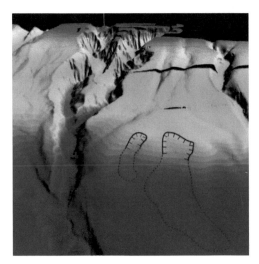 

Figure 2.9 Three-dimensional reconstruction of seabed, showing submarine canyons (left) and seabed (furrows (right)), created by erosion by bottom currents. *Source:* Miquel Canals.

Figure 2.10 Map of global ocean seafloor. Regions at >3000 m depth in dark blue.

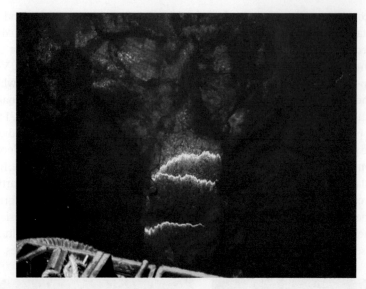

Figure 2.11 Lava flows from an underwater eruption. About 75% of all vulcanoes globally occur underwater and their eruptions cause major changes in the seascape. *Source:* NOAA/ Public Domain.

portions of oceanic ridges never extend above the surface, and remain submerged where they form **seamounts**, referring to geological features that rise 1,000 m or more above the seafloor. Over time, eruption and/or wave erosion can flatten the summit of the seamount to create table-like structures referred to as **guyots**. Because these fully submerged structures typically occur at substantial depth, shallow corals that require light (see Chapter 21) cannot colonize their flanks, although deep-water corals that do not require light sometimes do.

4) Oceanic trenches, the long narrow incisions of the abyssal plains that often extend to depths > 6,000 m (and up to 11,000 m), occur along the margins of tectonic plates where one plate subducts another plate downward into Earth's **mantle**, referring to the intermediate layer between the crust and core.

Underwater geological features strongly influence ocean circulation. "**Sills**," referring to oceanic ridges that partly separate basins, may limit the exchange of water between basins. The eruptions of underwater volcanoes cause major changes in seascape morphology and composition, and greatly influence marine life (Figure 2.11). Canyons may provide conduits for water masses both upward and downward, and bathymetry, such as that along the coast, strongly influences the movement paths of coastal currents and tides. The tens of thousands of seamounts scattered in the ocean basins, whether isolated or clustered, alter the trajectory of ocean currents, increasing turbulence and enhancing mixing of oceanic water.

2.3 Characteristics of Sediments

Size of sediment grains represents the single most used geological descriptor of sediment type (Figure 2.12). Sediments can vary in size from boulders of more than 26 cm in diameter to fine clays and colloids < 0.2 microns. Scientists group sediments using the Wentworth grain size scale (Table 2.2), which considers average diameter of sediment grains, often expressed as a $\log_2$ scale or **phi** (ϕ) scale. Sediments also vary in **sorting** (Figure 2.12), which refers to the degree to which a given sediment sample contains a mixture of different grain size categories (**poorly sorted**) or just one size category (**well sorted**). Sediment diameter relates strongly to how quickly particles settle and how easily near-bottom flow can **resuspend** particles or move them off the bottom so they are suspended in the water. Understanding how particles move and settle helps in understanding how sediments cover the seafloor and the role sediments plays in creating habitat.

2.4 Boundary Layers and Their Characteristics

Understanding sediment transport requires understanding benthic boundary layers. When any fluid (including water or air) moves across a fixed surface, friction slows down that movement, resulting in changes in fluid velocity with increasing distance away from the surface, a change referred to as **shear**. When wind blows over land, when streams flow over

Table 2.2 Wentworth grain-size scale.

Sediment	Type	Diameter (mm)
Gravel	Boulder	>256
	Cobble	64.0–256.0
	Pebble	4.0–64.0
	Granule	2.0–4.0
Sand	Very coarse	1.0–2.0
	Coarse	0.50–1.0
	Medium	0.25–0.50
	Fine	0.125–0.25
	Very fine	0.0625–0.125
Mud	Silt	0.0039–0.0625
	Clay	0.002–0.0039
Colloid		<0.0002

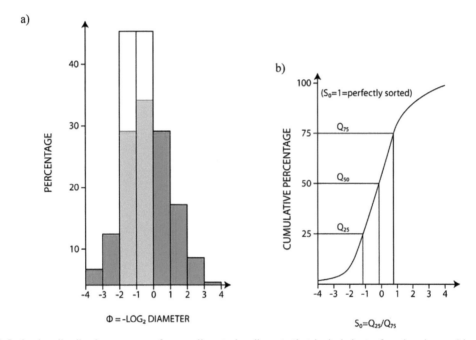

Figure 2.12 (a) Grain size distributions can vary from well-sorted sediments that include just a few size classes (shown in lighter shading) to poorly sorted sediments that encompass multiple size classes (shown in brown). Researchers typically express these distributions as relative percentages of the total sample. (b) Sorting S_o is calculated numerically as the cumulative proportion of sediments in the first 25% quartile of size classes divided by the first 75% quartile.

streambeds, or when currents or waves move across the ocean floor, such velocity change occurs, resulting in a **boundary layer** (Figure 2.14), a region where velocity increases with increasing distance from a surface. Bottom currents, tides, and waves can all form boundary layers, which may vary in thickness from millimeters, such as those caused by waves, to 10s of meters thick, such as those formed by tides. These boundary layers are important because they result in substantially weaker flows close to the bottom compared to flows only centimeters to meters away; organisms or sediments on the seafloor experience much weaker flows as a result. This weakening effect means bottom flows are less likely to erode the bottom than in the absence of the boundary later.

The stationary bottom slows the movement of water flowing across its surface by creating **drag**, or frictional resistance, thus called **frictional drag** or viscous drag. This drag creates **shear stress**, which is a force per unit area τ. Shear stress τ describes the flux of momentum to the bottom, whereby the boundary slows the momentum through frictional drag, which ultimately dissipates as heat. Boundary layers occur over any solid surface, but boundary layers vary in their thickness, the

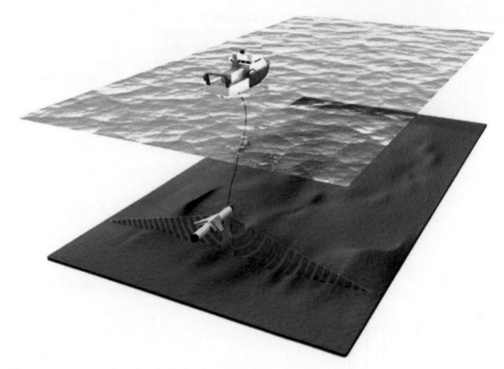

Figure 2.13 Schematic representation of seabed mapping using side scan sonar.

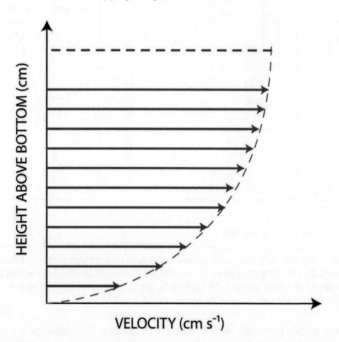

Figure 2.14 Schematic representation of a benthic boundary layer showing how velocity changes with height above bottom as well as how different layers of the boundary display different characteristics. The figure shows units to provide a sense of scale, but boundary layers may vary from mm to 10s of m in thickness.

specific shape of the velocity profile, and in their internal flow characteristics such as turbulence; these characteristics depend on the fluid, the flow itself, and the surface over which it flows.

Thus, all boundary layers share the characteristics of velocity slowing to zero at some point near the seafloor and increasing with distance from the bottom to some point above the bottom where bottom friction no longer has any slowing effect on the flow; this height above bottom corresponds to the boundary layer thickness. Velocity changes with height above bottom within the boundary but not in a uniform way. Immediately above the sediment-water interface over smooth bottoms, a thin **viscous sublayer** occurs. Within this layer, which is mm in thickness, velocity (u) increases linearly with height above bottom (z). Immediately above this layer, the **log layer** makes up about 15% of the total boundary layer thickness but velocity varies with the logarithm of height above bottom. Finally, within the uppermost layer, the **log-deficit layer**, the difference between the **free-stream velocity** (the velocity of the fluid above the boundary layer that experiences no frictional drag from the bottom) minus the velocity at a given height above bottom in the boundary layer varies with the logarithm of height above bottom. The log layer provides an opportunity to calculate a number called **shear velocity** known as **u*** because plotting velocity and the logarithm of height above bottom produces a linear relationship in which the inverse of the slope of the line equals u*, a constant number within the log layer.

So why care about u*? This one parameter provides a measure of momentum transfer within the boundary layer with units of velocity and thus a means to quantify or reproduce u* or friction velocity or shear velocity, an index of flow turbulence or momentum exchange in a fluid. The larger the u* the greater the turbulent mixing, meaning the more well-mixed the boundary layer and the more likely it will resuspend bottom material, whether that material is sediment grains, organic matter, recently settled larvae, or even adult benthic fauna. This characteristic of the boundary layer plays an important role in sediment transport, which defines the environment for many benthic species. Knowing u* also allows calculation of boundary layer thickness based on the formula:

$$\delta = \kappa \, u^*/\sigma$$

where κ = von Karmann's constant = 0.4, δ is boundary layer thickness, and

$$\sigma = 2\pi \, P$$

where P = the period of the forcing function responsible for the boundary layer, such as tides or waves

This calculation has several implications. First, a larger period or larger u* both contribute to thicker boundary layers. Second, noting wave periods on the order of seconds and tidal periods on the order of ~12 h, this means that waves produce very thin boundary layers (centimeters thick) that concentrate turbulence compared to the much thicker boundary layers produced by tides (10s of m thick). This observation leads to the conclusion that waves can easily resuspend sediments, but tides move them much more effectively.

2.5 Sediment Movement

The resuspension of sediments begins when sediments attain their **critical erosion velocity**, known as u^*_{crit}, which refers to the lowest velocity at which sediment grains of a given size (diameter) will move (Figure 2.15). Note that u^*_{crit} characterizes the specific sediment in question rather than the flow, but it relates to u*, which characterizes the flow itself. Sediment transport takes place in three steps:

1) Initiation of particle motion (u^*_{crit}) called "critical erosion velocity" or "threshold velocity," where particles begin to move when flow (characterized by u*) exceeds u^*_{crit}.
2) Particles move as **bedload transport** (rolling along the bottom) or **suspended load transport** (particles in suspension), which depends on the ratio of particle fall velocity to critical erosion velocity. Fine particles tend to move directly into suspended load transport once flow exceeds u^*_{crit} but coarser particles move first as bedload along the bottom and then as flow increases they move into suspension as suspended load.
3) Net transport by mean flow depends on the velocity profile and the sediment particle concentration profile.

The **Rouse Parameter** (ω) refers to the ratio of the tendency of particles to sink to the seafloor to the tendency for flow turbulence to keep them in suspension. Turbulence mixes areas of high concentrations with areas of low concentrations, leading to more vertically uniform concentration profiles. Thus, the relevant formula is:

$$\omega = \kappa \, u^*$$

Whereas boundary layer flow resuspends particles, gravity tends to make them sink, but heavy particles (sand) settle much more quickly than lighter particles (silts and clays); collectively the interplay between bottom flow and particle sinking velocities lead to very complex patterns of sediment distribution.

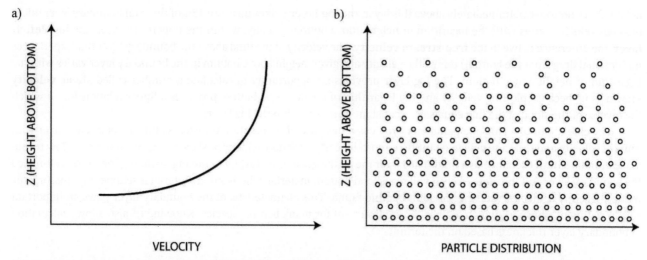

Figure 2.15 (a) Typical velocity profile above the bottom interacts with (b) the distribution of suspended particles, which varies with height above bottom. Collectively, this interaction determines sediment transport.

Grain size (or sediment texture) alone does not define sedimentary environments. Accurate description of the environmental characteristics (Figure 2.16 and Table 2.3) that can alter organism interaction with the seabed also requires consideration of the following sediment parameters: (a) **porosity**, which refers to the amount of pore space available between sediment grains; (b) **water content**, links closely to the porosity of sediment, and may vary from <30% in sand to >95% in some muds; (c) **permeability** also relates to porosity, and represents the rate at which water percolates through sediment. Collectively, these attributes define the available habitat within sediments; **angularity** of sediment grains (which can influence microbial colonizers), **redox potential (Eh)**, a measure of how easily a metal (or other ion) will give up electrons or retain electrons, and **oxygen concentration** depend primarily on the rates at which water can move

Figure 2.16 Sampling sediments seems simple but their geochemical complexity and mobility complicate efforts to obtain samples. Tools for sampling sedimentary environments include. (a) Push cores deployed either by diver or submersible, or instruments lowered over the side of a ship that include a (b) Van Veen grab, (c) box corer, (d) multicorer. In situ approaches include (e) remotely operated vehicles such as Canada's Remotely Operated Platform for Ocean Science (ROPOS) or (f) submersibles such as ALVIN. *Source:* NOAA / Wikipedia Commons / Public Domain.

Table 2.3 Sampling tools for marine sediments and their pros and cons.

Gear	Deployment	Advantages	Disadvantages
Grabs	Small boat	Cheap, any platform	semi-quantitative (bow wave, shape), shallow depths, sampling blindly
Hand-held corers	Scuba or wading depth	Quantitative, location precise	time, labor intense, shallow depths, small surface area
Box corer	Larger boat	Quantitative, any depth	Need big ship, sampling blindly
Alvin corers	Submersible	Quantitative, location precise	Submersibles are very expensive
Towed Video/Film	Small boat, Submersible, Diver	Quantitative, broad coverage	Expensive, only good for epifauna and **megafauna** (referring to organisms visible in bottom photographs)

within sediments, as well as rates of respiration in sediments. Oxygen plays a major role in limiting the distribution of biota: metazoans require oxygen to survive whereas **obligate anaerobic microbes** (which survive only in the absence of oxygen) and **facultative anaerobic microbes** (which survive with or without oxygen) differ fundamentally from **obligate aerobes** that survive only in the presence of oxygen. Oxygen typically penetrates just a few cm below the sediment water interface, but if that penetration depended on diffusion alone, oxygen would disappear at sediment depths of just a few mm. Invertebrates living in sediments play a critical role in moving pore water and sediment grains around, and thereby oxygenating the sediment; (d) **organic content** also closely relates to the grain size of sediments, and refers to the amount of organic matter present, typically expressed as carbon to nitrogen ratios, total organic carbon, or biochemical components of organic matter. Organic content generally increases with decreases in average grain size of sediments. Smaller grain size results in a larger surface per unit of volume and mass, characterized by a proportionally higher organic matter content. This higher organic content, in turn, results in increased oxygen consumption through oxidation of organic molecules and reduced oxygen penetration into deeper sediment layers. As a result, oxygen penetrates only the top few mm or cm of muddy sediments, whereas in sandy sediments oxygen can penetrate to a depth of 20 cm or more. These differences influence the volume of sediments available for colonization. Organic matter represents a crucial food source for seafloor biota, a topic that future chapters will explore in greater detail.

These variations in sediment characteristics beg the question of why such patterns exist and what variables determine sediment composition. In the simplest sense, the sediments at a given location reflect three drivers: history, sediment supply, and **hydrodynamics**, referring to the study of fluid motion.

From a historical perspective, some sediment deposits reflect historical events such as glaciation. **Relict sediment** refers to shelf sediment deposited during the recent geologic past that is not in hydrodynamic equilibrium with the present-day environment – a leftover from past events. Glacial marine sediments formed during glaciation and glacier retreat resulted in significant cobble and gravel substrate over much of the continental shelves at high latitudes.

From a sediment supply perspective, consider that sediments arise from five primary sources. **Terrigenous sediment** forms from weathering and erosion on land, and occurs in highest abundance near the mouths of rivers. These sediments dominate many continental shelf regions. **Biogenous sediments** form from the hard parts of living organisms, such as silica diatom frustules, sands derived from coral calcium carbonate skeletons, and calcium carbonate tests from foraminiferans, Although, such sediments occur on continental shelves, they are much more dominant at greater depths with reduced terrigenous inputs. **Hydrogenous sediments** result from chemical reactions in seawater and represent a relatively minor sediment source. Another minor source, **cosmogenous sediments**, form from tiny sediment grains from outer space delivered as space debris breaks up entering the atmosphere. The final source, **volcanogenous sediments**, arise from material ejected from volcanoes; such sediments can dominate in specific locations.

Hydrodynamics plays the single greatest role in determining patterns of sediments on continental shelves because the sediments in most environments achieve an equilibrium with local hydrodynamics. Fine sediments can only persist in weak flows, so that coarser sediments dominate stronger flows.

2.6 Characteristics of Hard Substrata

The spatial distribution and structure of marine benthic hard substrate communities also depend on multiple abiotic factors; the presence of organisms, in turn, also often influences the substratum, in a mutual exchange. Among abiotic factors, researchers increasingly recognize the importance of mineralogy of hard substrata, introducing the term **bio-mineralogy** to explain the interrelationships between biological systems at different hierarchical levels (cells, organisms, species, communities) and minerals. Bio-mineralogy can influence hard-bottom assemblages and explain some "anomalies" in the structure of communities growing on different types of rocks. The species that colonize rocks with a high percentage of quartz differ completely from those that colonize other rock types; different species may be slightly more attracted to a particular substrate (e.g., basalt versus carbonate rock) which could, in turn, influence the succession of organisms colonizing the substrate and subsequent interactions between biotic and abiotic components. Studies of artificial substrates reported a similar effect, with differences in both species composition and abundance. Studies also show that roughness, angularity, and inclination of rock substrate can favor or hamper colonization, through selective larval settlement, available surface, availability of refugia from predation, or grazing and food availability (e.g., turbulence and local currents that influence availability of drifting suspended particulate material or plankton). A relatively large number of studies reported on the influence of substrate mineralogy on **boring** or **endolytic** organisms that drill into bedrock and other hard substrata. For example, rocks with a high percentage of quartz or on rocks covered by **pelitic** (fine-grained sediment, diameter <63 μm) layers prevent the colonization of endolytic species. Bio-mineralogy likely plays a major role in defining benthic hard substrate communities, selecting biogenic species with long lasting effects on associated biological assemblages. A large portion of endolytic species (e.g., clionid sponges and bivalves) prefer carbonate rocks, which they can bore into by producing carbonic acid that penetrates the limestone. Some bivalves, such as date mussels (e.g., *Lithophaga lithophaga*), use sulfuric acid and/or neutral mucoproteins to bore into rocks, including limestone. The interaction between organisms and rocks, in turn, has important consequences for rock texture and roughness.

2.7 Characteristics of Soft Sediments

The characteristics of soft sediments differ completely from hard bottoms. First, soft sediments (also called mobile sediments) create a highly three-dimensional habitat whose characteristics (geochemistry, grain size) often change rapidly with distance below the sediment-water interface. Although most biota, which includes diverse invertebrates, algae, and microbes, live in the upper few cm, biota sometimes live to depths of 1 m or more. Conversely, other than the borers that penetrate into rock and the presence of cracks and crevices on some surfaces, hard bottoms offer largely two-dimensional habitat. Soft sediments, much like desert dunes, move; waves, strong currents, or benthic bottom storms can resuspend, move, and reshape sedimentary habitats. The size and distribution of grain sizes within sediments largely reflect the hydrodynamics and mechanical energy or their environment. Coarse sand or gravel characterize areas exposed to high energy (e.g., splash zones), whereas smaller and uniform (few size classes) distributions (i.e., well-sorted sediments) characterize sheltered areas. Benthic **psammon** (fauna able to move within interstitial spaces in sand) can displace individual grains of sediment, illustrating the phenomenon of **bioturbation**, where living biota, and larger sedimentary organisms in particular, move sediment grains. As we shall see in future chapters, this bioturbation activity has profound impacts on ocean processes.

Questions

1) How do scientists divide up the seafloor based on bathymetric ranges?
2) What are the relative proportion of land and ocean in the two hemispheres?
3) Is the ocean seafloor homogeneous?
4) What is the average depth of the global ocean?
5) What are the deepest points in the ocean and what do we call them?
6) What is the proportion of hard and soft sediments in the global seafloor?
7) What are canyons and mid-oceanic ridges?
8) How does substrate mineralogy influence benthic biota?

9) What main physical characteristics of hard substrata influence colonization?
10) What are the main properties of soft sediments and how do they influence colonization?
11) What are the main differences between hard and soft bottoms?
12) What is the benthic boundary layer?

Suggested Reading

Amblas, D., Ceramicola, S., Gerber, T.P. et al. (2018) Submarine canyons and gullies, in Micallef, A., Krastel, S., and Savini, A. (eds.), *Submarine Geomorphology*. Springer Geology. Springer, Cham. pp. 251–272.

Charette, M.A. and Smith, W.H.F. (2010) The volume of Earth's ocean. *Oceanography*, 23, pp. 112–114.

Geersen, J., Voelker, D., and Behrmann, J.H. (2018) Oceanic trenches, in Micallef, A., Krastel, S., and Savini, A. (eds.), *Submarine Geomorphology*. Springer Geology. Springer, Cham. pp. 409–424.

Harris, P.T., MacMillan-Lawler, M., Rupp, J., and Baker, E.K. (2014) Geomorphology of the oceans. *Marine Geology*, 352, pp. 4–24.

Hu, M., Li, L., Jin, T. et al. (2021) A new $1' \times 1'$ global seafloor topography model predicted from satellite altimetric vertical gravity gradient anomaly and ship soundings BAT_VGG2021. *Remote Sensing*, 13, p. 3515.

Judd, A. and Hovland, M. (2007) *Seabed Fluid Flow: The Impact on Geology, Biology and the Marine Environment*. Cambridge University Press.

Micallef, A., Krastel, S., and Savini, A. (2018) *Submarine Geomorphology*. Springer Geology. Springer, Cham.

Minliang, D., Tong, S., Chen, J. et al. (2022) Distribution and geological controls of the seabed fluid flow system, the central-western Bohai Sea: A general overview. *Basin Research*, 34. doi: 10.1111/bre.12666.

Shepard, F.P. (1936) The underlying causes of submarine canyons. *Proceedings of the National Academy of Sciences of the United States of America*, 22, pp. 496–502.

Smith, W. and Sandwell, D.T. (1997) Global sea floor topography from satellite altimetry and ship depth soundings. *Science*, 277, pp. 1956–1962.

Smoot, N. (2015) *Marine Geomorphology*. 3rd Edition. Mindstir Media.

Spietz, R.L., Butterfield, D.A., Buck, N.J. et al. (2018) Deep-sea volcanic eruptions create unique chemical and biological linkages between the subsurface lithosphere and the oceanic hydrosphere. *Oceanography*, 31, pp. 128–135.

Straudigal, H. and Clauge, D.A. (2010) The geological history of deep-sea volcanoes: Biosphere, hydrosphere, and lithosphere interactions. *Oceanography*, 23, pp. 58–71.

Sun, Q., Magee, C., Jackson, C. et al. (2020) How do deep-water volcanoes grow? *Earth and Planetary Science Letters*, 542, p. 116320.

Sverdrup, H.U., Johnson, M.W., and Flemming, R.H. (1942) *The Oceans, Their Physics, Chemistry, and General Biology*. Prentice-Hall.

Ulla, F.A., Ramirez-Llodra, E., Aguzzi, J. et al. (2017) Ecological role of submarine canyons and need for canyon conservation: A review. *Frontiers in Marine Science*, 4. doi: 10.3389/fmars.2017.00005.

Venkatesan, R., Tandon, A., D'Asaro, E., and Atmanand, M.A. (2018) *Observing the Oceans in Real Time*. Springer Oceanography.

Wang, X., Kneller, B., and Sun, Q. (2023) Sediment waves control origins of submarine canyons. *Geology*, 51, pp. 310–314.

Wölfl, A.-C., Snaith, H., Amirebrahimi, S. et al. (2019) Seafloor mapping – The challenge of a truly global ocean bathymetry. *Frontiers in Marine Science*, 6, pp. 283, 16.

Zhu, C., Li, Q., Li, Z. et al. (2023) Seabed fluid flow in the China Seas. *Frontiers in Marine Science*, 10, pp. 1158685, 13.

3

The Water Column

The water that fills the ocean basins, known as the **water column**, represents the fluid habitat for ocean life that lacks visible habitat features beyond water itself. Nonetheless, water may vary in depth (and therefore pressure, presence of light, temperature, salinity, nutrient content, clarity and, of course, its biota that respond to these variables). **Estuaries** include those water column environments in which water from rivers measurably dilutes the seawater. The **neritic zone** refers to the relatively shallow portion of the ocean that extends from the land-sea interface to the edge of the continental shelf at 200 m, and where light often penetrates all the way to the seafloor. This zone contrasts the **oceanic zone**, which refers to the open ocean beyond 200 m, including the **pelagic zone**, or the water column portion of the oceanic zone.

The pelagic zone contains four major distinct vertical layers based on the depth of light penetration (Figure 3.1):

1) the **euphotic** (or simply **photic**) **zone**, from the surface to a depth of 20–50 m depending on water clarity, refers to the highly illuminated portion of the water column where intensive photosynthesis can take place and where energy produced by photosynthetic activity exceeds energy burned through respiration.
2) the **mesophotic zone**, typically spanning from 50- to 150-/200-m depth, refers to the weakly illuminated layer of the water column with low light penetration and insufficient amounts of light to support significant photosynthetic activity.
3) the **disphotic zone**, typically spanning 200–500/600 m, refers to the minimally illuminated layer of the water column with insufficient light penetration to support photosynthetic activity that balances respiration needs.
4) the **aphotic zone**, beginning at 500/600–1,000 m depth and extending to the bottom of the trenches, describes the deep, perpetually dark portion of the water column where no light penetrates nor photosynthetic activity occurs.

The water column can be vertically divided into layers based on depth. The so-called "dark zone" occupies the largest volume of the ocean, and encompasses the **mesopelagic, bathypelagic, abyssopelagic,** and **hadopelagic zones** (Table 3.1).

3.1 Properties and Characteristics of Seawater

The chemical and physical properties of water primarily determine seawater characteristics. The water molecule (H_2O) asymmetrically binds two atoms of hydrogen (H), the most common and lightest element in the universe, to an atom of oxygen (O), the most abundant element in Earth's crust (Figure 3.2a). The simple chemical composition of water belies its complexity and special physical properties; no other substance coexists in nature in all three states of matter (gas, liquid, and solid). The two atoms of hydrogen bond **covalently** (bonds with shared electrons) to the atom of oxygen, forming an asymmetrical molecule with a 105° angle between the hydrogen atoms, resulting in a shape similar to that of the head of a cartoon mouse! This asymmetry results in a polarity akin to a magnet, so although electrically neutral, water molecules bond somewhat weakly to one another **via hydrogen bonds**, electrostatic attractions between molecules involving a hydrogen atom. These hydrogen bonds provide cohesion but also fluidity (Figure 3.2b).

Marine Biology: Comparative Ecology of Planet Ocean, First Edition. Roberto Danovaro and Paul Snelgrove.
© 2024 John Wiley & Sons Ltd. Published 2024 by John Wiley & Sons Ltd.
Companion Website: www.wiley.com/go/danovaro/marinebiology

3 The Water Column

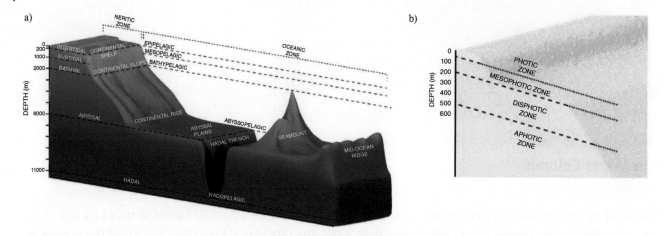

Figure 3.1 Diagram illustrating: (a) the different zones of the benthic domain and those of the pelagic domain; (b) vertical layers based on the depth of light penetration.

Table 3.1 Depths occupied by pelagic and benthic domains.

	Depth (meters)	Volume (%)	Zone
PELAGIC DOMAIN			
Epipelagic	0–200	3	Photic
Mesopelagic	200–1,000	28	Disphotic
Bathypelagic	1,000–2,000	15	Aphotic
Abyssopelagic	2,000–6,000	54	Aphotic
Hadopelagic	>6,000	<1	Aphotic
BENTHIC DOMAIN	Depth (meters)	Surface(%)	Zone
Continental Shelf	0–200	6	Photic
Upper Bathyal (continental slope)	200–2,000	8	Disphotic-Aphotic
Lower Bathyal (continental rise)	2,000–4,000	56	Aphotic
Abyssal	4,000–6,000	29	Aphotic
Hadal	>6,000	<1	Aphotic

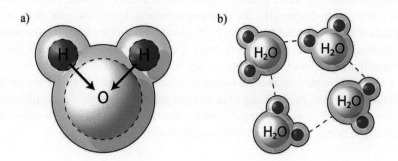

Figure 3.2 (a) The structure of a water molecule, and (b) and the hydrogen bonds that form between water molecules.

3.1.1 Hydrogen Bonds

Hydrogen bonds result in the peculiar physical properties of water that make life on Earth, as we know it, possible (Table 3.2), The high melting and boiling points of water mean that transformation between states requires substantially more energy than it would otherwise because hydrogen bonds must break before ice melts or liquid transforms into gas.

Even at the greatest ocean depth, the lowered freezing point conferred by salinity (and pressure) enables marine waters to remain fluid at cold temperatures approaching –2 °C as found in Antarctica, supporting a rich diversity of life. High **heat capacity**, which refers to the amount of heat required to raise water temperature by 1 °C, means that ocean temperature rises and drops slowly. When water temperature decreases, it releases a large amount of heat per unit of mass. Seawater requires 3.3 times more energy than air per unit mass to increase temperature by 1 °C. Indeed, the upper 10 m of the ocean alone contains more heat than the entire atmosphere, and the ocean has always mediated temperature fluctuations on land, particularly in maritime regions. The oceans have absorbed ~ 84–90% of the excess heat humans have added to the atmosphere or, stated differently, the ocean has absorbed the equivalent of a staggering 7 Hiroshima-class nuclear explosions *every second* for the last 25 years. In addition, since historical times the global ocean has absorbed 30–40% of the excess carbon dioxide produced by humans; carbon dioxide is the molecule primarily responsible for current global climate change.

Ice, the solid state of water, is less dense than water, the liquid state, because ice molecules organize into an open crystal with wider angles between the hydrogen atoms. The lower density of ice means that it floats on liquid (think of your favorite cold beverage!) and melts rapidly with solar radiation and warming. If ice were denser than seawater, it would reside on the seabed for long periods of time without dissolving, with important consequences for how marine ecosystems work.

The high surface tension of water refers to how its surface behaves as a tight, elastic film that can support organisms such as Portuguese man o' war with extensive body surface areas and low specific density.

3.1.2 Salinity

Salinity, defined by the solutes in water resulting from terrestrial rock dissolution and volcanic emissions, represents a key variable for ocean life. The polarity of water creates an excellent solvent that can dissolve the majority of the 92 natural chemical elements that occur in seawater. Water dissolves salt crystals and surrounds sodium and chlorine ions, bringing them into solution and preventing their precipitation. Sodium chloride, the dominant salt in seawater, forms by combining the positive ion (**cation**) sodium (Na+), and the negative ion (**anion**) chlorine, Cl^-: the Na^+ (Figure 3.3a) and Cl^- ions (Figure 3.3b) comprise more than 85% of all the substances dissolved in seawater, followed by sulfate (SO_4^{2-}), magnesium (Mg^{2-}), calcium (Ca^{2+}) and potassium (K^+) ions. Together, these elements constitute 99.3% of the solutes present in seawater (Figure 3.4); interestingly, their relative amounts vary little, even with changes in overall salinity values.

Table 3.2 Main characteristics of seawater and implications of interest for ocean life.

Characteristic of Water	Implication
High melting and boiling points.	Large volumes of water are liquid at typical temperatures on Earth
Higher viscosity than air	Marine organisms typically are more streamlined, and can be larger in size (e.g., whales) or require little structural support given the structural support water helps provide.
High heat capacity	Ocean temperature varies less temporally than on land, helping to dampen climate variation.
Maximum density at 4 °C	Water freezes from top down rather than bottom up, allowing life to persist beneath ice.
Salt reduces freezing temperature	Ocean less likely to freeze than freshwater and remain fluid over a greater range of conditions.
Minimally compressible	Water remains consistently fluid across a wide depth (and thus pressure) range.
Freezing and evaporation points occur in typical conditions on Earth	Water can cycle between different compartments, resulting in a continuity of availability on a global scale
Polarity of water molecules	Water is an excellent solvent for nutrients and other ions essential for life.
Increased sound transmission	Sound travels longer distances than in air, allowing marine mammals, for example, to communicate over long distances.

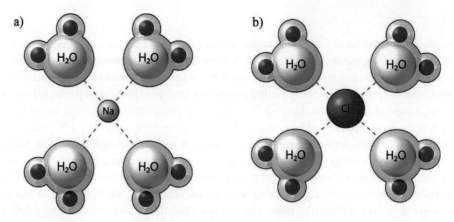

Figure 3.3 Molecules of water, a polar compound, surround ions such as: (a) sodium and (b) chlorine, orienting in order to neutralize the charges and thereby weakening the attraction between ions.

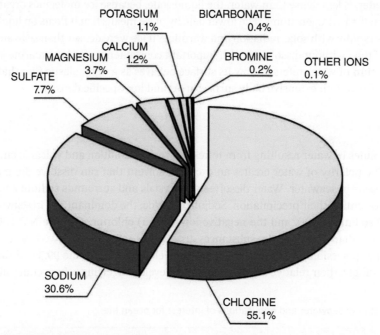

Figure 3.4 Percentages of dissolved elements in seawater.

Until a few years ago, researchers expressed salinity in parts per thousand, but now typically express it as an absolute value that requires no units, or more recently as practical salinity units, or **PSU**. The average salinity of seawater of 35 describes a volume of seawater with a weight of 1.000 grams that contains an average of 35 g of solute, producing a total mass of ~1.035 grams. Box 3.1 explains measurement of salinity.

Anyone who has ever tasted seawater appreciates just how salty it can be; indeed, seawater is 4 times saltier than the water used to cook spaghetti! In marine environments, salinity is one of the most stable parameters, but there are vast regions where salinity differs dramatically from typical open ocean values (Table 3.3). In fact, in closed or semi-enclosed seas, where evaporation exceeds fresh water input, salt concentrations and thus salinity increases. For example, the Red Sea, where the arid climate of the surrounding regions and the scarcity of inflowing rivers result in salinities of 40–42, which pales compared to the salinities of some enclosed seas such as the Dead Sea, where salinities reach 280 (Figure 3.5).

In the Mediterranean Sea, an example of an **evaporite basin** where evaporation exceeds freshwater input, salinity typically varies between 37.5 and 39 or more. However, salinity values can decrease significantly near river mouths or where major runoff from land occurs. Generally, salinity in the open ocean exceeds that of coastal regions, with higher salinity with increasing distance from river mouths or the coast. In cold locations, where large volumes of water freeze during the

Figure 3.5 The high salinity of the Dead Sea makes swimming difficult because we float so high in the water relative to other marine or freshwater bodies. Gulping water is extremely unpleasant!

Table 3.3 Examples of salinity values of different marine environments.

	Average Salinity
"Typical" open ocean	35
Red Sea	40
Mediterranean	38
Black Sea	18
Baltic Sea	8

winter, salinity increases because ice formation releases salt. At the other extreme, where significant contribution river inflow and low evaporation coincide, such as in the Black Sea, salinity averages only 18 (Table 3.3). Salinity also typically increases with depth. A **halocline** refers to changes in salinity across water masses, as often shown in salinity-depth profiles (Figure 3.6).

When seawater begins to freeze, the dissolved ions interfere with the lattice arrangement of the molecules into ordered crystals, lowering the freezing point to about –2 °C. Therefore, some species can live in environments characterized by temperatures close to 0 °C, where seawater reaches its maximum density. At full salinity (35) seawater freezes at temperatures below –1.9/–2.0 °C, expelling the salt to form ice that is largely freshwater, thereby increasing the salinity and density of surrounding water (Box 3.1). Because salt increases seawater density, the denser waters sink, creating a **thermohaline** current, a circulation feature driven by temperature and salinity. Seawater also alters light and sound transmission, as discussed below.

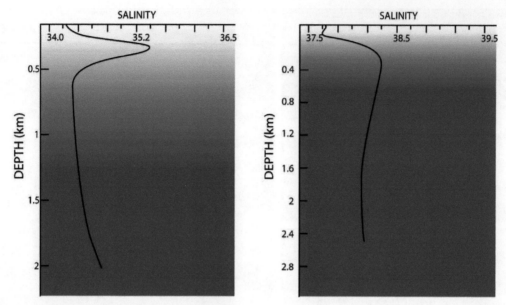

Figure 3.6 Vertical profiles of sality in different oceanographic regions.

Box 3.1 Measures of salinity in the sea

Scientists in the past measured salinity by completely drying a seawater sample and weighing the remaining salts. In some cases, scientists now use an instrument called a **hydrometer**, which determines the density of a liquid of known temperature by measuring how high the hydrometer floats. The denser the liquid, the greater the buoyant force it exerts, so the line on the hydrometer corresponding to the surface of the liquid corresponds to its density. Density easily converts to salinity using a standard equation. Today, scientists attach conductivity probes, which measure the degree to which seawater conducts electricity, known as the **conductivity** of seawater (which converts to salinity with a simple algorithm). Often conductivity probes form part of multi-parameter **CTD probes** (*Conductivity, Temperature, Depth*) that simultaneously collect high precision measurements of temperature, conductivity, and pressure. Lowering CTD probes with a cable or conductor cable into the ocean, a memory card or on-board computer in the pressure housing of the CTD records the data. Modern instruments can now detect salinity differences to the second decimal place.

3.1.3 Temperature

The action of the wind and atmospheric temperature interact to determine ocean surface temperature. Seawater absorbs solar radiation, warming surface layers that heat adjacent waters through diffusion and through mixing by currents, waves, and tides. As with atmospheric temperature, ocean surface temperature can vary with time and space, depending on latitude, changes in season, distance from continents, and even time of day. However, the slow heating and cooling processes in the ocean linked to its high heat capacity result in much less temperature variation in a given locale than occurs on land.

Temperature varies similarly with latitude in the northern and southern hemispheres, except that the larger continental masses in the northern hemisphere contribute to slightly warmer surface waters; at 60°N or S large differences in land mass cover contribute to striking temperature differences. At the poles, seawater temperature can drop to nearly –2 °C, and rarely warms to more than 20 °C at the very surface. In contrast, consistently warm temperatures between 20 and 28 °C characterize low latitude surface waters, with warmer temperatures to depths of ~500 m than at similar depths elsewhere in the ocean. At mid-latitudes, such as in the North mid-Atlantic or Mediterranean, surface water temperatures fluctuate seasonally, with typical tropical water temperatures in summer and mid-latitude water temperatures in winter (Figure 3.7).

In spring, especially along the coasts, increased input of solar radiation increases surface water temperatures. Then, a warmer surface layer develops, increasing with depth as waves mix warmer waters downward. Eventually at depths below wave mixing, the solar warming effect becomes negligible and ocean temperatures remain nearly constant throughout the year. Between the warmer water masses that typically sit above colder and denser deeper water masses, a water layer

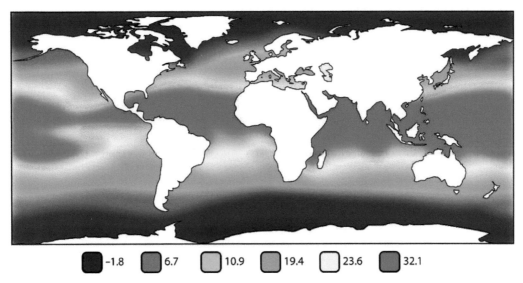

Figure 3.7 Distribution of average annual ocean surface temperature (°C). Reported are the upper temperature limits for each geographic region.

characterized by rapid temperature change with depth – the **seasonal thermocline** – occurs, effectively stratifying or layering the two water masses (Figure 3.8a). In addition to this seasonal thermocline, which may extend to depths up to ~40 m, a **permanent thermocline** may co-occur along with the seasonal thermocline but extend to depths of 1000 m in tropical and temperate latitudes, separating surface and deep ocean layers (Figure 3.8b). At high latitudes, a reverse thermocline may occur, with colder temperatures (but very salty water) on the surface and warmer temperatures (and less salty water) at depth. Alternatively, two thermoclines may develop (Figure 3.8d) when, in addition to a first surface thermocline, a wedge of denser subsurface waters with different temperature also occurs.

The presence of a strong thermocline, such as in summer at mid- and high latitudes, limits vertical exchange between water masses and therefore the exchange of nutrients needed for phytoplankton growth. Thus, as phytoplankton take up nutrients, die, and sink from the surface layer, they deplete nutrients seasonally. Because this invisible thermocline represents a strong density interface, it can also create a barrier to the movement of zooplankton and non-swimming biota. In autumn, as air temperatures begin to cool the water surface and reduce the density differences between the upper and lower layers, increased winds and storms expedite mixing and the seasonal thermocline breaks down; mixing then reinjects nutrients from deeper waters into the surface layers. In some cases, this nutrient reinjection may also cause a fall phytoplankton bloom but in most cases the largest bloom occurs in the following spring as sunlight increases, warming begins, and stratification sets up. Box 3.2 explains how temperature measurement has evolved over time.

3.1.4 Density

The combined effects of salinity, temperature, and pressure determine seawater density (mass per unit volume, $g \cdot cm^{-3}$). One liter of seawater weighs ~2–3% more per liter than pure water because of the additional salts. Generally, density decreases with increasing temperature and increases with increasing salinity (assuming constant pressure; Figure 3.10a) and depth (Figure 3.10b). Vertical gradients in temperature and salinity therefore also determine density gradients because cold and/or salty waters are typically denser than less salty ("fresher") and warmer waters, and tend to sink, resulting in less dense upper layers of the water column. The **pycnocline** describes an abrupt change in the density profile, which often coincides with a thermocline or halocline.

Water masses that different in density tend to **stratify**, or form layers with less dense water rising above denser water, which sinks. Temperature and salinity do not contribute equally to density variation in the ocean. Generally, temperature plays a greater role in variation in seawater density. For this reason, sudden cooling of surface water often causes it to sink, even for water with relatively low salt content. As a result, the coldest waters occur at depth. Therefore, for most of the ocean a surface mixed layer (in which waters are relatively warm and low in density), sits above the pycnocline, where

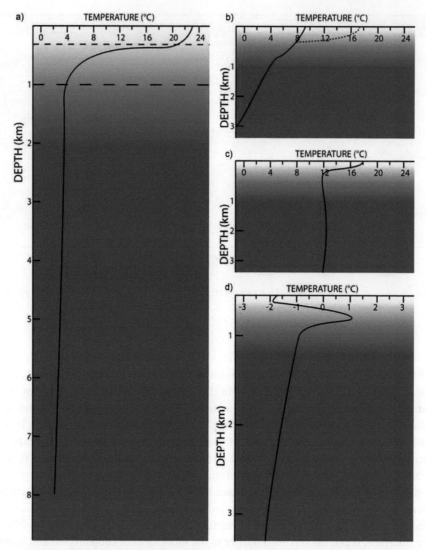

Figure 3.8 Idealized comparison of the thermocline at various latitudes, contrasting: (a) mid-latitude oceanic waters; (b) mid-high latitudes waters of cold seas; (c) mid-latitude waters of warm seas such as the Mediterranean; (d) high latitude waters with a freshwater subsurface peak (due to melting ice) and a progressive decrease with increasing water depth.

Box 3.2 Measuring the temperature of the sea

Scientists use various systems for measuring water temperature. One of the oldest systems, the "cockpit thermometer," affixes a classic mercury thermometer to a support and immerses the bulb in a small metallic glass cockpit. Generally, researchers lower the thermometer on a string into the water, where the glass fills with water. This thermometer works effectively for surface water samples because users can retrieve the water-filled cockpit quickly enough to determine the temperature with accuracy to a tenth of a degree. For subsurface waters, researchers in the 19th century used an ingenious device known as a **reversing thermometer** (Figure 3.9). In these special thermometers, a relatively large mercury reservoir connects through a thin capillary to a smaller bulb. Just above the reservoir, the capillary forms a spiral bottleneck that then feeds in a straight line to the upper bulb. With the thermometer in the upright position, temperature determines the volume occupied by the mercury. However, when the thermometer reverses the mercury column separates, locking the mercury remaining in the small bulb, which connects to a graduated mercury column. The height of the mercury column indicates the water temperature at the time the thermometer reversed. A second thermometer mounted next to the reversing thermometer measures water temperature at the surface.

Box 3.2 (Continued)

In the 1930s, scientists at MIT and Woods Hole Oceanographic Institution collaborated with the US Navy to develop a means to quickly profile the water column and identify the thermocline; the navy was interested in the thermocline because submarines could "hide" from the echosounders of enemy ships by sitting at the density interface. The bathythermograph, a torpedo-shaped instrument measures temperature by means of a spiral metal nib, which is sensitive to temperature because of thermal expansion. The nib engraves a trace on a microscope slide, coated initially with skunk oil but eventually replaced with a carbon black film. The slide links to a support connected to a barometric box that moves in response to pressure (and therefore depth), orthogonal (perpendicular) to the movement of the thermal recording stylus, to produce a depth-temperature diagram. The 1950s saw the development of the CTD probe system (Conductivity, Temperature, Depth) described above (see "How We Measure Salinity"), which users lower into the water on a cable, either storing digital data internally as it lowers or sending the data through a conductor cable where researchers can observe the data on board in real time. These increasingly sophisticated and miniaturized sensors today can detect temperature differences of 0.01 °C. The recent extreme miniaturization of CTDs to probes smaller than a ballpoint pen allows attachment of these instruments to migratory organisms or insertion into the body cavity of fishes. These living "sentinels" can carry the sensors across the ocean and to great depths. Attachment of these sensors to mammals (whales, seals), penguins, sharks, and marine turtles, has yielded novel information on dive durations, locations, depths, temperatures, and salinities encountered by these organisms, and increased sampling opportunities in remote ocean environments at a tiny fraction of the cost of a ship.

Figure 3.9 Reversing thermometer.

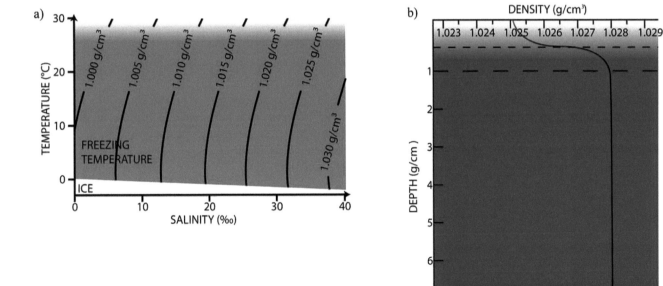

Figure 3.10 Change in seawater density as a function of (a) temperature and salinity and (b) depth.

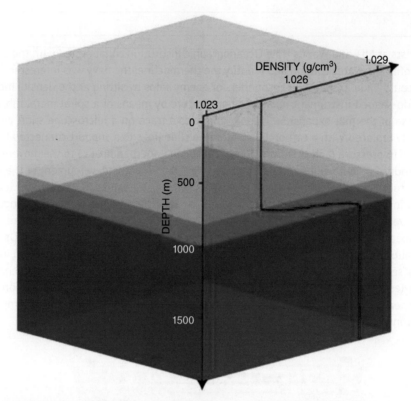

Figure 3.11 Division of the water column in three layers characterized by different densities: surface, pycnocline, and deep layers. The pycnocline exhibits the strongest gradients (variations) in density.

density increases rapidly with depth, which in turn sits above a layer of deep, dense water that constitutes ~80% of the total ocean volume (Figure 3.11).

Differences in density caused by temperature and salinity can also result in vertical currents as dense waters sink, resulting in dense-water cascading events. At the same time less dense water masses rise, upwelling and redistributing nutrients in coastal waters, and sometimes driving distributions of organisms, their migrations, and their life cycles. Box 3.3 describes how researchers determine density.

3.1.5 Viscosity

Salinity and temperature influence water **viscosity**, or "thickness" of water, with increased viscosity in cold and salty water. From a biological perspective, viscosity matters – particularly for small organisms – because viscosity lowers the velocity at which organisms sink or move in water, slowing movement but improving flotation. Viscosity plays also a role in the sinking of particles and detritus, and in the propagation of sounds and noise.

3.1.6 Pressure

Pressure strongly affects many marine organisms. Pressure increases by one atmosphere for each 9.8 m of depth, meaning that an organism living on the abyssal plains may experience pressures in excess of ~400 atm. Although water is relatively

---

Box 3.3 How we measure density

Typically, scientists calculate density mathematically from *in situ* salinity, temperature, and pressure measurements often carried out with a CTD probe. Once salinity, temperature, and pressure are measured, density σ can be derived from: $\sigma(S, T, p) = \rho(S, T, p) - 1.000 \text{ kg} \cdot \text{m}^{-3}$, in which ρ is the density of pure water measured at 4 °C and at a pressure of 1 bar, and S and T denote salinity and temperature, respectively. Less commonly, seawater density can be measured directly with **pycnometers** or **hydrometers**.

incompressible, air compresses greatly. This means that any air pockets in organisms, such as lungs in mammals, air bladders in fishes, or sinuses in skulls may encounter extreme pressures at depth and extreme changes in volume with changes in depth. Proteins also fold in different ways depending on pressure, creating physiological problems for organisms changing depth.

Scientists must also deal with pressure. Pressure normally limits technical scuba diving to approximately 100 m or less because of the risks associated with decompression, and to approximately 200 m because of the crushing pressure itself. Submersibles that dive to significant depths require pressure reinforced hulls to avoid crushing the scientists on board. Deep-sea scientists sometimes send Styrofoam cups to depth because the high pressures crush the air pockets in the Styrofoam to produce a miniaturized cup that makes a great souvenir. Sending instruments that contain electronics to great depths requires pressure housings that can keep water out. Leakage would not only ruin the instrument, and particularly its electronics, but also create potential danger once retrieved because any pressurized water within the housing could release explosively.

3.1.7 Sound

Sound propagates in seawater at speeds almost 5 times higher than those in air (ranging from 340 m · s^{-1} on land to ~1.500 m · s^{-1} in aquatic environments, with "typical" sound propagation speeds in seawater of 1.480 m · s^{-1}). Propagation speed mainly depends on temperature (increasing ~4.5 m · s^{-1} per degree Celsius), pressure (increasing ~1.7 m · s^{-1} per 100 m depth) and, to a lesser degree, salinity (increasing about 1.3 m · s^{-1} for each thousandth of change in salinity; Figure 3.12). Water absorbs ultrasonic frequencies more than audible frequencies, and water attenuates an acoustic signal 10 to 130 times less than air. In the absence of absorption, reflection, and dispersion, sound intensity decreases proportionally to the square of the distance of the source, meaning that sound can propagate large distances through seawater, even up to thousands of kilometers.

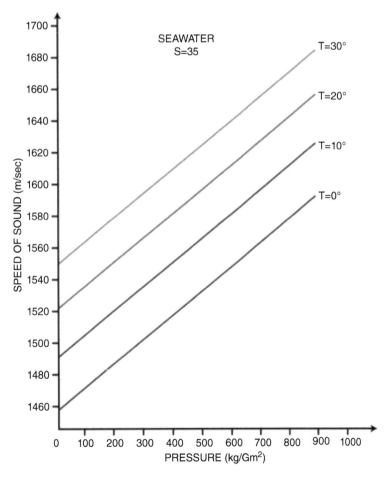

Figure 3.12 Speed of sound in seawater in relation to temperature and pressure (or depth).

3.1.8 Light

Light enters seawater at a speed of 2.2×10^8 m·s^{-1} but its intensity and duration vary greatly spatially and temporally. The amount of solar radiation arriving at the sea surface depends on various factors (Figure 3.13): (1) daily (day vs night) changes depending on the presence of sunlight, noting that such changes occur at the poles daily rather than seasonally; (2) cloud cover, together with **sea state** (degree of wave motion) and the angle of the sun on the horizon at a given time, influences the percentage of light reflected rather than absorbed; (3) season, in that sun elevation on the horizon changes seasonally, increasing in variability with increasing latitude; (4) latitude, in that solar radiation at the equator does not vary seasonally, in contrast to polar regions that receive near constant light at some times of year and none at other times of year (Figure 3.13). Moreover, the angle at which the sun hits the Earth varies with latitude.

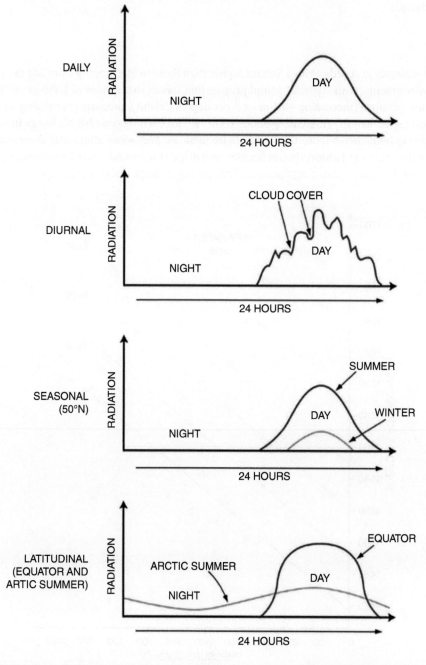

Figure 3.13 Variability in solar radiation over 24 hours under different sets of environmental conditions that include latitude, cloud cover, and seasonal changes in sun elevation.

$$\mu = \frac{sen\Theta i}{sen\Theta r}$$

Depth adds an additional factor because dissolved and particulate materials present in seawater progressively absorb light. In contrast to air, water absorbs light strongly and selectively, resulting in an exponential decrease in light intensity and selective absorption of wavelengths as light passes through the ocean's surface to depth. Light attenuation depends on two main processes: absorption and scattering. Absorption occurs with the conversion of electromagnetic energy into other forms, such as algal photosynthesis, which intercepts light photons as they pass through water. Water **scatters,** or changes light direction, as it reflects off suspended particles. As a result, water rapidly absorbs and scatters light, limiting penetration depth. The photic zone, where most photosynthesis occurs, typically extends to tens of meters in depth, but light can support net photosynthesis up to 150–200 m of depth in the most transparent water (where coralline algae have been reported). Photons can penetrate seawater to depths up to 1000 m in the clearest topical waters, but not in sufficient amounts to support net photosynthesis. The bathymetric band between the photic zone and 200 m of depth, known as the **mesophotic or twilight zone** is where light might become limiting for photosynthetic primary productivity. In addition to absorbing light, seawater refracts light. **Refraction**, which refers to the bending of light as it passes through media of different densities, changes direction as moves from air into seawater. The refractive index (μ) equals the ratio between the sine of the incident angle and the cosine of the refraction angle, yielding a refractive index for water of approximately 1.33:

$$Rf = 1/2 \left[\frac{\sin^2(\Theta i - \Theta r)}{\sin^2(\Theta i + \Theta r)} + \frac{\tan^2(\Theta i - \Theta r)}{\tan^2(\Theta i + \Theta r)} \right]$$

$$Rf = 1/2 \left[\frac{\sin^2(\Theta i - \Theta r)}{\sin^2(\Theta i + \Theta r)} + \frac{\tan^2(\Theta i - \Theta r)}{\tan^2(\Theta i + \Theta r)} \right]$$

Incident light reflectivity is calculated from Fresnel's law:

μ = refractive index
sin = sine
Θi = angle of incidence
Θr = angle of refraction

By bending light, refraction can sometimes extend the path through which light passes to reach a given depth than it would otherwise, thereby increasing attenuation.

Intensity of solar radiation passing through the water column decreases exponentially following Bird's law, described by the following equation:

$$I_z = I_0 e^{-\eta M}$$

η = coefficient of extinction
I_z = intensity of energy at z (depth in meters)
I_0 = intensity of incident energy
M = light path through the water column (a function of depth and refraction light angle). The distance at which light passes through water from the surface to z, therefore increases.

$$M = zb$$

where $b = 1 / \cos \Theta r$ and z = depth in meters

Radiation intensity at a given depth z can be calculated knowing surface radiation I_0 and the coefficient of extinction (n) for pure water (0.035):

$$dI/dz = -kI \quad I_z = I_0 e^{(-nbz)}$$

dz = infinitesimal slab of sample
dI = the intensity absorbed in the slab
b = path length
η = coefficient of extinction
I_z = intensity of energy at z (depth in meters)
I_0 = intensity of incident energy

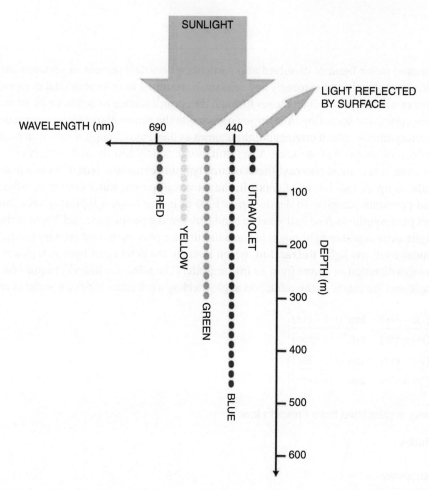

Figure 3.14 Diagram of solar radiation at the ocean's surface, which partially reflects some light and also absorbs some wavelengths that penetrate to greater depths.

The wavelengths that comprise light range from ultraviolet radiation (200–400 nm) to infrared (>750 nm). However, seawater differentially absorbs different wavelengths of light, strongly absorbing ultraviolet rays in surface layers and dramatically altering UV wavelength composition with depth. The first few meters of seawater also rapidly absorb infrared (IR) wavelengths. Clear waters preferentially transmit low wavelengths (light blue, between ~400 and 550 nm), resulting in the blue color typical of seawater, whereas waters rich in dissolved organic matter and suspended particles transmit higher wavelengths (yellow-green, between ~500 and 590 nm), resulting in a beige color. Light intensity typically decreases 1.5 orders of magnitude for every 100 m of depth (Figure 3.14. Box 3.4 describes how we measure light in the ocean).

3.1.9 Inorganic Nutrients and Trace Elements

in seawater provide essential nutrients required by photosynthetic primary producers (**autotrophs**). which **assimilate** or take up salts of nitrogen (~46 µmol · kg^{-1}), phosphorus (~3.5 µmol · kg^{-1}), and silicon (~150 µmol · kg^{-1}), in the form of **nitrate** (NO_3^-) and **phosphate** (PO_4^{3-}), whereas unicellular microalgae known as diatoms also require silicates (Si^{2-}) to build their protective shells known as **thecae**. Primary producers assimilate and release these comparatively abundant macronutrients, resulting in rapid changes in concentration over time and space. For, example, inorganic nutrients typically occur at higher concentrations in coastal waters, close to rivers, in the deep sea, or at the water-sediment interface where decomposition of organic matter releases nutrients through **remineralization**. In contrast, low nutrient concentrations characterize open ocean surface waters far from the continents. The trace elements that occur in concentrations less than 1 ppm (part per million) in seawater, include manganese, iron, cobalt, copper, zinc, and strontium. Despite their low concentrations, marine organisms critically require these trace nutrients and some species require vitamins in small amounts to support development and survival, as well as important processes such as recruitment, mobility of large vertebrates, or photosynthesis). Vertical currents transport nutrients up from seafloor sediments and deep water to the surface,

> **Box 3.4 How we measure light intensity and transmission in the ocean**
>
> Light measurements in the ocean
>
> 1) **Measuring sunlight irradiance at the surface.** Sunlight intensity is measured with a **pyranometer**, an instrument in which a sensor measures global solar radiation (W · m^{-2}), the energy collectively produced by direct and diffuse radiation reflected from the sky and clouds. Continuous measurements with pyranometers provide the measurements necessary to obtain an accurate estimate of light conditions at a site, by accounting for the high variability.
> 2) **Measuring irradiance at depth.** Irradiance, the amount of solar radiation arriving at a given location, can be measured using **photoelectric sensors**. These instruments consist of a radiation collector-diffuser, a filter that selects the range of measured wavelengths, a sensor (usually a silicon diode), an amplifier, and a direct reading or recording system. The instruments commonly used to study photosynthetic systems measure the full range of **photosynthetically active radiation** (PAR), the wavelengths used in photosynthesis. Animal studies select wavelengths corresponding to the spectral sensitivity of the organisms of interest. Finally, spectroradiometers filter a specific portion of the spectrum (for example UV) to study specific wavelengths bands.
> 3) **Instantaneous measures.** For vertical profiles in the water column, researchers can obtain direct irradiance measurements either from a diver-operated sensor or a support boat using a Teflon sphere of ~2 cm.
> 4) **Measuring transparency using a Secchi disk.** This instrument, first used by an Italian priest named Father Secchi in 1865 during an early Mediterranean oceanographic cruise, consists of a weighted, white circular disc 30 cm in diameter, lowered on a rope into the water until it disappears. This empirical measure depends on the sensitivity of human eyes, which can detect differences in intensity of 1/133. Researchers today continue to use Secchi disks in long- and medium-term monitoring of transparency. Researchers usually lower the disk from the shady side of the boat in order to minimize reflected light, and keep the rope as vertical as possible to measure accurately the depth at which it disappears from sight (Secchi depth Zsd in meters). By relating the Secchi depth (Zsd) and the coefficient (η), we obtain η = k/DS (k = constant 0.61–1.9). Oceanographers use a k = 1.7.

allowing the floating, single-celled primary producers known as **phytoplankton** to thrive. These nutrients also support macroalgae, such as kelps and seaweeds, and marine plants such as seagrasses and mangles.

3.1.10 Oxygen

Oxygen represents approximately 21% of the gases in air, in contrast to at least 20 × lower concentrations in water, i.e., average 5–10 ml · L^{-1}, or ~0.5–1.0%. Moreover, the concentration of dissolved oxygen in water (not to be confused with oxygen bonded in molecules such as H_2O) varies greatly because oxygen saturation depends on water temperature and salinity; oxygen dissolution in water decreases with increasing temperature and, less dramatically, with increasing salinity, and because organism respiration or chemical oxidation of organic molecules can rapidly consume oxygen. The dissolved oxygen content of a given water mass also reflects two processes that increase O_2 – atmospheric exchange and photosynthesis. Exchanges at the air-water interface saturate the ocean surface with oxygen, and photosynthesis sometimes results in super-saturation. Photosynthetic oxygen production occurs in the **photic** or **euphotic** zone, the surface depths with sufficient light to support **net photosynthesis**, meaning the proportion of photosynthesis that exceeds respiration in primary producers. Vertical profiles of O_2 concentration decrease with depth, often rapidly to 100–200 m. In some ocean regions with high O_2 concentrations, below this layer O_2 may further slowly decrease to reach a minimum (generally between 500 and 1,500 m, known as the **Oxygen Minimum Zone**), which occurs in regions of high oxygen consumption. At greater depths, oxygen concentrations increase again because of the intrusion of deep, oxygenated waters in which limited biological consumption occurs. Oxygen concentrations below saturation (typically 8–10 ml · L^{-1}) characterize most ocean water masses. Box 3.5 describes oxygen measurement techniques.

3.1.11 Dissolved Gases

In addition to salts and nutrients, seawater also contains dissolved gases. The composition of the gases dissolved in seawater is the same of those in the air, but their relative importance changes when dissolved in seawater. Nitrogen represents ~48% of the total volume of dissolved gas (compared to 78% in air); oxygen (~0.5–1% compared to 21% in air) and carbon dioxide (~15% compared to 0.4% in air) also differ. Photosynthesis and respiration require oxygen and CO_2 respectively, and both

> **Box 3.5 Measures of dissolved (free) oxygen in the sea**
>
> Historically, researchers used an analysis known as the **Winkler method**, which measures dissolved oxygen in a precise volume of water (generally 250 mL) in glass, stoppered bottles, avoiding any bubbling of the water during collection. Oxygen reacts with manganese sulfate, sodium hydroxide, and sodium azide that bind and precipitate all oxygen on the bottom of the bottle. A laboratory titration follows the addition of sulfuric acid with sodium thiosulfate, allowing calculation of the total dissolved oxygen present.
>
> Nowadays, dissolved oxygen measurements typically use **Polarographic Oxygen Sensors** that work on the polarographic principle, which determines ions and molecules based on the change in electrical current when applying electric potential across two electrodes immersed in an electrolyte. A thin Teflon, polyethylene, or silicone membrane separates the electrolyte solution (usually with saturated potassium chloride) from the seawater sample, selectively allowing diffusion of the molecule of interest (in this case O_2). The voltage applied between anode and cathode reduces the oxygen that diffuses into the electrolyte to OH^-, and the resulting current is directly proportional to the amount of dissolved oxygen. In contrast, **Mackeret cells,** galvanic electrodes consisting of a noble metal cathode and a lead or zinc anode, generate their own electrical potential in the presence of oxygen, with no need for an external voltage source. Within the last decade, many researchers have switched to novel **optodes**, which use the optical properties of an indicator to measure oxygen changes. This approach allows easy determination of the concentration of dissolved oxygen by using a colorimetric test kit. Despite the development of more advanced technologies, the delicate nature of dissolved oxygen measurements often requires the inter-calibration of sensor data with Winkler titrations or chemical standards. Furthermore, oxygen concentrations can change rapidly when manipulating samples, requiring great care during processing.

processes significantly change oxygen and CO_2 concentrations through the biological activities of primary producers and consumers respectively. Physicochemical conditions also alter gas solubility, which increases with decreasing water temperature. However, nitrogen and other **noble** (inert) gases (argon, neon, and helium, which together comprise an additional 1.4% of gases) are largely irrelevant to most biological processes.

3.2 An Ocean In Motion

The surface and deep currents that mix and transport water masses, in tandem with waves and tides, result in an ocean in constant motion. Forces outside the ocean ultimately drive this movement. The large masses and associated gravitational attraction of the moon and sun result in tides, although the shape of ocean basins and land masses surrounding them influence the magnitude and intensity of tides among different geographic locations. Solar radiation from the sun heats the Earth's surface differently between night and day and among latitudes, leading to the formation of the winds that drive waves and currents, and generate water masses of different densities that sink or rise when they encounter water masses of different density.

3.2.1 Ocean Currents

The term **current** refers to any persistent movement of water masses, induced by various causes, capable of mobilizing or transporting large volumes of water and sometimes sediments. Currents transport organic and inorganic matter, nutrients, and organisms among different ocean locations, and they also affect physicochemical parameters (temperature, salinity, and density); these variables all influence biology (Box 3.6). Moreover, currents help to replenish oxygen, nutrients, and food, and remove waste products **Suspension feeders**, animals that feed by filtering water and removing suspended particles or prey, have no equivalent in terrestrial ecosystems. Currents can also transport eggs, spores, larvae, and juveniles over substantial distances, and thus play an important role in reproduction and geographical distribution of species.

> **Box 3.6 Effect of currents on biology**
>
> Currents can have multiple and important effects on marine life:
>
> - Currents influence salinity and temperature distribution, creating favorable or unfavorable conditions for primary production;
> - Vertical currents in upwelling areas promote primary production, enriching surface layers;
> - Currents facilitate the movement of marine organisms, and plankton in particular;
> - Currents support benthic organisms in the deep ocean by transporting oxygen and nutrients and eliminating catabolites or waste products from metabolism;
> - Currents influence the transport and settlement of benthic organisms in suitable habitats;
> - Currents affect climate and the abiotic variables that influence where many marine organisms occur.

Ocean currents vary in origin and form, and include gradient currents, drift currents, tidal currents, and geostrophic currents. Moreover, depending on temperature differences between the specific current and the surrounding water masses, currents may be warm or cold. Classification of currents also relates to the depths at which they occur: **surface currents** move the portion of the water column spanning from the surface to 200-m depth, **internal currents** move water at depths ranging from below 200 m to the near bottom, and **bottom currents** move water immediately above the seabed.

3.2.1.1 Surface Currents

Two primary factors determine ocean surface currents, broad-scale wind patterns and Earth's rotation. **Winds**, the air masses that move from areas of high pressure to areas of low pressure, result from the rising of hot air and descent of cold air (Figure 3.15). The rotation of the Earth, from West to East, deflects air currents as they ascend and descend through the atmosphere, modulating wind direction as they drive the major wind patterns that characterize the globe. Moreover, because wind moves surface water, friction also acts on subsurface water layers much like pushing the top card on a deck of playing cards drags the cards below it. In this case, the layers of subsurface water dragged along extend to ~100 m depth, but when currents move toward or away from the poles across latitudes, the different rotational velocities at different latitudes result in the **Coriolis effect**, which deflects the layers further and further to the right with increasing depth in the northern hemisphere and to the left in the southern hemisphere. The magnitude of this deflection also increases with latitude and velocity of the water. The resulting current structure, known as an **Ekman spiral** (Figure 3.16), deflects surface

Figure 3.15 Simplified diagram of the major wind patterns on the Earth's surface.

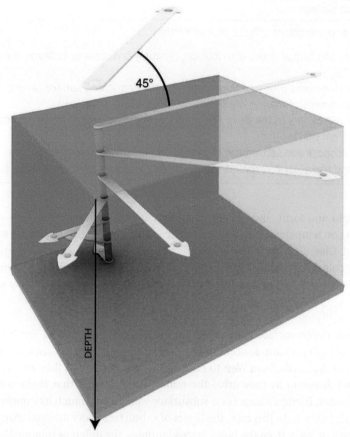

Figure 3.16 Ekman transport occurs when winds (wide arrow on top) blow the surface water and drag the layers beneath, deflecting them at a 45° angle, producing a net transport whose direction differs completely from the wind direction.

currents by 45° to the right of the wind direction in the northern hemisphere, where current deviation from wind direction increases with depth and current intensity decreases exponentially with depth.

Ekman transport, integrated across the Ekman spiral, transports the ~100 m surface layer ~90° with respect to the wind (Figure 3.16). The combined effect of the trade winds that continuously blow at ~0–30° latitude in both the northern and the southern hemispheres and Coriolis force move huge volumes of surface water, creating large circular current systems that move clockwise in the northern hemisphere and anti-clockwise in the southern hemisphere. Currents moving along the equator do not cross latitudes (and rotational velocity therefore does not change), and the absence of the Coriolis effect means that water follows the same direction as the wind. Indeed, the equatorial currents of the Atlantic and Pacific Ocean run parallel to the equator and transport water west.

Thermohaline circulation, driven by differences in temperature and salinity, and thus density, drives much of ocean circulation at depths greater than 1,000 m, and results from the sinking of cold water at high latitudes to considerable depths where it flows towards the equator (Figure 3.17).

Density determines the depths to which deep currents sink. Imagine the water column between the surface and bottom of the ocean as a series of stacked layers; those with greater density flow beneath less dense layers. The surface waters of high latitudes link to deep layers of low latitudes in that the cold and dense, high latitude (70–80°S) waters of the Southern Ocean (the Weddell Sea in Antarctica) sink and move north, flowing across the bottom as **Antarctic Bottom Water** – ABW. At latitudes near the Antarctic Circle (60°S), the less dense water flowing above the Antarctic bottom moves northward at intermediate depths between the deep, dense Antarctic water and the shallow waters. A similar pattern occurs in the northern hemisphere, where cold waters along the Norwegian and Greenland coasts and Labrador Sea form an intermediate and deep-water layer that flows southward (**North Atlantic Deep Water** – NADW). In this region, the cooling of the low salinity surface water accelerates sinking. In fact, as surface water freezes and ejects salt to the water below, the newly formed ice increases the salinity and therefore the density of surface water. The water sinks and moves south, causing saltier water from the southwest to move north to replace it, generating the Gulf Stream. Therefore, deep

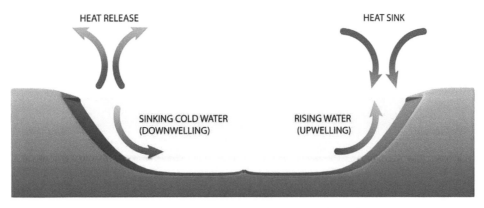

Figure 3.17 Simplified diagram heat storage and release in relation to deep-water formation.

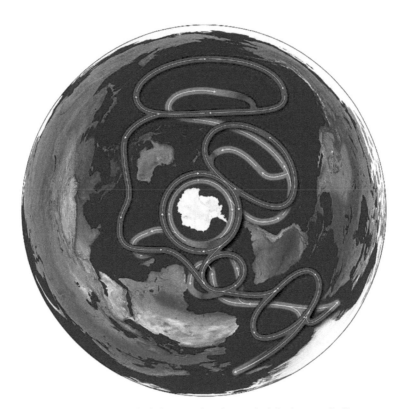

Figure 3.18 Schematic of the deep ocean conveyor belt (centered on Antarctica). Red arrows indicate warmer surface movement and blue arrows indicate colder deep movement.

water sinking in the North Atlantic forms part of a global transport mechanism. The interconnection of these processes forms a single large system of global circulation called the **deep ocean conveyor belt** (Figure 3.18) that requires ~1,000 years to complete its global tour.

Life in the deep ocean depends on this sinking water to supply oxygen to resident biological communities. Indeed, the higher solubility of oxygen in cold waters results in saturation, with little consumption of the oxygen during sinking; abyssal plains thereby generally avoid anoxia. The renewal of oxygen to the inner ocean depends on these processes and the massive transport volumes involved. For example, the Mediterranean Sea, an ocean system in miniature, requires only 80–100 years for complete renewal, or **turnover**, of water.

Vertical currents: As described above, steady winds blowing across the ocean surface displaces surface water along coasts when the Coriolis effect deflects the current produced by the wind at 90°. This process transports surface waters offshore, "piling up" water and creating a pressure gradient that pushes deep waters to the surface to replace it, a process

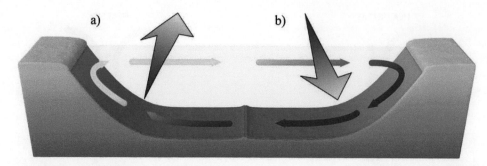

Figure 3.19 Diagram of northern hemisphere upwelling (a) and southern hemisphere downwelling (b) generated by winds along the coast.

called **upwelling** (Figure 3.19a). Conversely, consistent, strong winds blowing along the coast move offshore water masses to the coast, causing surface waters to sink through the process of **downwelling** (Figure 3.19b). The rising and sinking of ocean waters in response to wind strongly influence the climate of the Earth, as well as ocean biology and productivity. In fact, phytoplankton blooms follow the upwelling of deep, nutrient-rich waters. Moreover, the large biomass and intense primary production support high secondary production (zooplankton), and high fish production. One of the most productive coastal fisheries occurs off Chile and Peru, where the trade winds from the southeast cause constant upwelling that supports abundant production of anchovies ("Peruviana anchoveta", *Engraulis ringens*) and the largest pelagic fishery in the world. We discuss the biology associated with these processes in detail in Chapters 12 and 14. Low productivity characterizes downwelling areas because nutrient- (and phytoplankton) poor water offshore water masses replace sinking surface waters. This sinking also limits exchange of nutrients with bottom waters, severely reducing water column production.

Waves: Winds also cause wave formation (Figure 3.20). The duration, intensity, consistency of direction, and **fetch**, the distance over which the wind blows, all influence the size of waves. Ocean waves form when the wind begins to blow, pushing the crest of the wave forward. Locations away from where the wind blows may experience long waves, or **swell**, as waves moving long distances catch up with one another and combine or cancel each other out. As waves approach the coast and shoal at shallower depths, particle movement changes from a circular to an elliptical motion, wavelength decreases, and wave height increases. Eventually growing waves becomes unstable and break. The wave does not move the water particles themselves but instead transmits energy in a circular motion; waves, in fact, do not transport water masses, but transfer energy horizontally across the ocean surface.

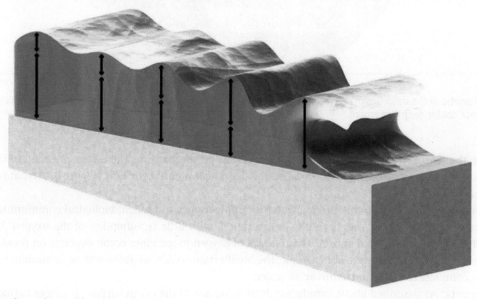

Figure 3.20 The formation and breaking of a wave.

Figure 3.21 The ebb and flood of the tide.

Tides: Tides are a periodic phenomenon of rising (**flood** or high tide) and lowering (**ebb** or low tide) sea level that result from the combined effects of the gravitational pull of the moon and sun and the centrifugal forces (akin to swinging a bucket in a circle and the water stays in the bucket) of the Earth-Sun-Moon rotational system (Figure 3.21). Although the sun is far greater in size than the moon, the greater proximity of the moon results in a relatively stronger gravitational force. Tides are most visible at the land sea-interface, but their effect occurs through the ocean, though with varying intensity depending on latitude, basin shape, and land mass configuration.

Imagine the Earth as a water-covered sphere: the gravitational pull on the ocean causes a "bulging" of some parts of this hydrosphere closest to the sun and moon and a corresponding flattening in those regions furthest away.

The arrangement of the sun and moon relative to the Earth creates three scenarios:

1) When the gravitational attraction of the sun and the moon align on the same axis during a **new moon** (Figure 3.22a, right), they pull together to create a bulge on the closest side of Earth while centripetal force produces an equal bulge on the opposite side; with this orientation, the hydrosphere reaches its highest **tidal excursion (depth range)** within a given month, referring to the difference in tidal height between low and high tide. These high tidal excursions are referred to as **spring tides**;
2) When the gravitational attraction of the moon and sun align on opposite sides of Earth along the same axes, they again produce a wide tidal range because the gravitational and centripetal forces continue to act in tandem though in opposite directions (Figure 3.22a, left).
3) When the moon and the sun orient at an angle of 90° with respect to the Earth (Figure 3.22b), the tidal and centripetal forces act in perpendicular axes, and this weaken one another. This scenario results in the weakest tides with the lowest range between low and high tides known as **neap tides**, and they occur during the first and the last quarter phases of the moon.

Tidal ranges vary from a few centimeters (such as in many parts of the Mediterranean Sea) up to more than 10 meters. In Normandy and the coast of England, tides can exceed 3 m in range, which still pales in comparison with the largest tidal excursions in the world, which occur in the Bay of Fundy and Ungava Bay in Canada where tidal ranges can exceed 11 m.

Tidal range has important effects on the biology of the organisms that colonize intertidal habitats. Indeed, low tide exposes organisms to the air and thus to **desiccation** or drying out, with consequences for respiration. Greater temperature fluctuations characterize the intertidal zone with important consequences on metabolism and other processes. We examine these intertidal adaptations in detail in Chapter 20.

Ice: Although many parts of the global ocean never experience ice, the role of ice is significant in polar and many subpolar regions. Ice can influence marine life in many ways. **Pack ice,** referring to ice formed from by freezing of seawater but not attached to land and their precursor platelets in various forms, influences light penetration and seawater mixing. Ice melting influences the stratification of the water column and its salinity. Pack ice melting releases nutrients and **sympagic algae** that live attached to the underside of the ice into the water column. Floating ice provides a surface platform for temporary activities of many bird and mammal species. Anchor ice, ice tongues, and icebergs cause abrasion along coastlines down to impressive depths, often scraping away most of the biota (see Chapter 23).

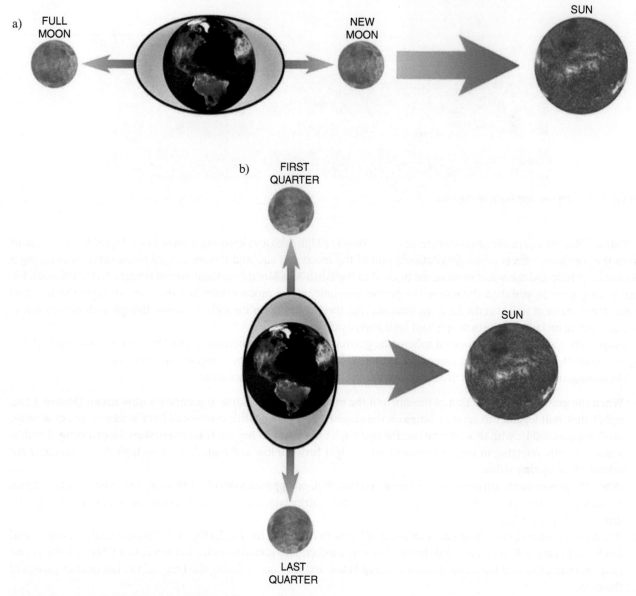

Figure 3.22 (a) Spring tides and (b) neap tides.

Questions

1) How can the water column be divided from the surface to the bottom and what environmental conditions characterize these divisions?
2) How can the water column be divided in terms of light penetration?
3) What are the physical properties of water molecules and what factors contribute to those properties?
4) How do the properties of air and water differ and what are the implications for marine organisms?
5) How do currents influence life and distribution of dissolved components?
6) Which factors influence oxygen content of seawater?
7) Which are the main chemical components of water and how do they affect the dynamics of oceanic water masses?
8) What are the main types of seawater movement?
9) How do tides influence marine life?
10) How does ice influence marine life?

Suggested Reading

Hummels, R. (2018) Introduction to physical oceanography, in Salomon, M. and Markus, T. (eds.), *Handbook on Marine Environment Protection*. Springer Cham.pp. 3–35.

Open University Course Team (2001) *Ocean Circulation*. 2nd Edition, Colling, A. (ed.). Open University, Butterworth-Heinemann.

Siedler, G., Griffies, S.M., Gould, J., and Church, J.A. eds. (2013) Ocean circulation and water masses, in *Ocean Circulation and Climate: A 21st Century Perspective*. International Geophysics, Academic Press, p. 103.

Stewart, R.H. (2006) *Introduction to Physical Oceanography*. Texas A & M University.

Part II

Life in Seas and Oceans: Fundamentals of Marine Biology

Charles Darwin's famous expedition on HMS *Beagle* (1831–1836) helped him understand adaptations to living in specific environments, and the factors that drive natural selection and speciation in species. The initial chapters in this section summarize the impressive array of adaptations that organisms evolved to live in the ocean, beginning with basic adaptations common to living in a wide range of aquatic habitats from the surface to the seafloor. These adaptations include communicating, coping with variations or extremes of temperature, oxygen, salinity, pressure, light availability, nutrients, and sound and communication. However, some adaptations apply to specific environments. The adaptations required to effectively drift or swim in the water or feed on suspended food items differ greatly from the adaptations required to burrow through or attach to the seabed, to avoid exposure to air or seafloor toxins, or to feed on sediment particles or obtain energy from chemical (non-organic) compounds. Specialized adaptations in marine organisms include those related to reproducing, feeding, moving, and protection from predators, disturbances, and toxic substances.

In short, numerous morphological, physiological, and behavioral adaptations enable marine organisms to live in the ocean, and although some of those adaptations apply both to seabed and water column organisms, life on a fixed seafloor fundamentally differs from the moving and fluid environment of the water column. Let's explore adaptation to life below sea level. As for our "play" analogy, these adaptations correspond to how our actors use the stage to tell their story!

We'll also broadly describe the "cast" of our "play" and look at global biodiversity, the patterns of distributions of diverse marine biota and where they actually live their lives and why. Marine scientists estimate that we have sampled perhaps 5% of the global ocean, and as a result cannot even agree on whether the ocean contains half a million species or perhaps 10 million or more, not including the microbes. Recognizing this great uncertainty, scientists nonetheless accept that the ocean harbors a spectacular range of biodiversity, from its surface to its greatest depths, including many unknown species. Many researchers believe that the primary diversification of life on Earth occurred in the ocean, with the notable exceptions of vascular plants and insects, which diversified on land. This ocean diversification helps explain the total absence of many major groups of organisms from land and freshwater.

Despite many unsampled or poorly sampled ocean environments, scientists have documented patterns of biodiversity and species distributions that are far from uniform. They have also shown that contrasting habitats support very different numbers and types of species over various spatial scales. Typically, seabed habitats support a much more diverse biota than the waters above, in large part because of the variety of available seabed habitats created by a mix of geology, biology, chemistry, and physics. For the most part, water characteristics such as temperature and salinity, as well as nutrients and food, define the habitat of organisms that live in the water column, thus limiting the opportunities for specialization. The chapters that address biodiversity patterns describe the different approaches we use to measure biodiversity, how we use that information to identify broad-scale biodiversity patterns, and what those patterns tells us about the evolutionary drivers of marine biodiversity.

If the previous chapters defined the "theater" for our play, then the chapters on biodiversity and ecology of different environments define the diverse "actors" that form the core elements of marine biology. These actors span from microbes to the great whales, from the seafloor to strong swimmers and to surface drifters. These chapters describe the composition and ecology of the dizzying array of biota that live on the seafloor and swim in ocean waters. The organisms that live their lives swimming and drifting in ocean water differ from those that live on the seafloor in several fundamental ways that extend well beyond basic habitat differences. First, organisms that live in the water experience a more three-dimensional and nomadic life, where the water that defines their habitat generally moves with them as ocean currents transport them around ocean basins. This lifestyle, where conditions may quickly change, often leads to short lifespans (noting major exceptions such as whales and some fishes) that may span months or less to just a few years. Organisms on the seafloor often live a much more static existence, experiencing less change; lifespans therefore vary from annual cycles to multiyear and even multi-decadal durations. One other key difference links to energy cycling, in that the most seafloor environments occur at depths far below where light can penetrate and where photosynthesis occurs. In contrast, organisms in surface waters coexist with abundant photosynthetic microbes. This difference in energy cycles means organisms that live in the upper water column can rely on relatively "local" food supplies, but those food resources can change quickly. In contrast, organisms on the seafloor must generally depend on food supplied from the water above and they therefore ingest more energy than they produce, resulting in a net negative energy deficit. These major differences result in some general similarities in the types of organisms that live in the water and on the seafloor, but also major differences in composition, abundance, and ecology that these chapters explore.

4

General Adaptations in Marine Organisms I: From the Ocean Surface to the Seabed

Biological adaptation refers to any alteration in the structure or function of an organism or any of its parts that results from natural selection and by which the organism becomes better suited to survive and multiply in its environment. Adaptive strategies, or **adaptations**, thus refer to biological characteristics, evolved over time, that allow them to colonize and expand their population in a given environment.

The range of abiotic variables described in the previous chapters constrains which organisms can survive in which environments. Indeed, they must either adapt, move, or die. Furthermore, extreme values of abiotic variables that differ greatly from "average" ocean conditions characterize some oceanic environments and create some of the most inhospitable habitats on Earth; survival in these environments requires adaptations. On the abyssal seafloor, the largest biome on Earth in terms of surface area, pressures range from approximately 400 to 600 atmospheres. These pressures would instantly crush human lungs and sinuses, which are adapted to pressures of just 1 atmosphere. Ocean temperatures range from −1.9 °C (the freezing point of seawater) near the poles to the >400 °C fluids that spill out of some **hydrothermal vents**, the cracks in the Earth's crust near spreading and subduction zones along the edges of ocean plates where superheated seawater rich in dissolved minerals and chemicals spews out. Salinity varies in time and space, from values near 0 at river inflows to 40 or more in some enclosed seas. These extremes, as well as salinity variation over time, create **osmotic** challenges, where ion concentrations inside the cells of biota may differ greatly from the surrounding seawater. On land, light is wholly absent only in caves or at the poles during the winter but, in the ocean, water quickly filters out light with increasing depth and, as light disappears, organisms that rely on vision must evolve large eyes to adapt to low light conditions. At depths beyond light penetration, some species detect predators and prey by scent or vibration or produce **bioluminescence**, referring to biologically produced light, to communicate with each other, to lure prey, or to deter predators. Utilizing the relatively high speed of sound transmission in water (about five times faster than in air), some marine cetaceans **"vocalize"** by producing sounds at high frequency wavelengths.

These few examples illustrate how adaptive strategies may be reproductive, behavioral, morphological, physiological, or biochemical. This chapter describes different strategies of adaptation of marine organisms to life in the ocean, including the difference between adaptation and **acclimatization**, the process by which an organism adjusts to changes in its environment in order to survive. Humans, for example, acclimatize to lower oxygen concentrations at high altitude by breathing faster and deeper, however, Incas in the Peruvian Andes have **adapted** by evolving larger lung capacity and hemoglobin affinity to oxygen.

4.1 Adaptations to Temperature

The majority of aquatic organisms (fishes and marine invertebrates) are **poikilotherms** (or **heterotherms**), meaning that their body temperature tracks that of their environment. In contrast, **homeotherms** such as marine mammals and seabirds (and us!) maintain a relatively constant body temperature that may often differ from their environment. The body temperatures of some other organisms neither conforms to that of the environment nor maintains a constant value; the muscles of some strong swimming fishes, for example, generate metabolic heat, thus warming the muscles and tissues.

Marine Biology: Comparative Ecology of Planet Ocean, First Edition. Roberto Danovaro and Paul Snelgrove.
© 2024 John Wiley & Sons Ltd. Published 2024 by John Wiley & Sons Ltd.
Companion Website: www.wiley.com/go/danovaro/marinebiology

These two extreme metabolic strategies have important implications, because maintaining a warm-blooded body (homeothermy) consumes substantial calories (or energy), requiring that the animal feed much more intensely and frequently than poikilotherms. Poikilotherms comprise the majority of marine organisms, and simply slow their metabolism as temperatures decrease, and can therefore invest a higher portion of their energetic reserves in biomass production (i.e., **somatic** or body growth) or in reproduction, rather than generating (and inevitably dispersing) heat to maintain constant body temperature. Variations in environmental temperature result in corresponding variation in poikilotherm metabolism, where increased temperatures result in increased metabolism and corresponding increased oxygen consumption to support that metabolism; the opposite pattern occurs when temperature decreases.

In ecology, metabolic theory identifies three factors control flows of energy and matter at the level of single organisms (i.e., their metabolic rate): body size, ambient temperature, and resource availability. Marine organisms, including species that live at temperatures as cold as −2 °C and as warm as >100 °C, cannot escape these rules. Physiologists use a general index known as Q_{10} to measure temperature effects on metabolism:

$$Q_{10} = R_2 / R_1^{\frac{10}{t_2 - t_1}}$$

R_1 and R_2 denote the metabolic rates at temperatures t_1 and t_2, respectively. Q_{10} represents the factor by which we must multiply R for every 10 °C increase in temperature. Q_{10} values of 2 to 3 characterize most marine poikilotherms, though some may be as great as 5; this means that a temperature increase of 10 °C results in a doubling, tripling, or even quintupling of oxygen consumption, depending on the metabolic characteristics of the species of interest.

Body size and temperature are primary determinants of metabolic rate for marine organisms. The **Metabolic Theory of Ecology** examines the relative roles of temperature and food availability in influencing metabolic rate, growth rate, lifespan, body size, abundance, biomass, and biodiversity of life. The relative impacts of thermal and chemical energy change across organizational scales in marine organisms, suggesting that individual metabolic rates, growth, and turnover proceed as quickly as temperature-influenced biochemical kinetics allow, but that chemical energy limits higher-order community structure and function.

Extreme temperatures, whether too high or too low, can physically and chemically alter an organism. Extreme cold physically damages cell membrane structure when ice crystals form, leading to chemical damage from high concentrations of solutes. In contrast, O_2 supply at high temperatures cannot meet the high metabolic demand, and rapid deterioration of substrates follows, including potential reduction of activity or even protein **denaturing** (alteration or breakdown), resulting in changes to metabolic pathways.

Deep-sea environments, one of the most food-poor habitats on Earth, offer a unique opportunity to assess the relative importance of temperature and food resource availability on individual energetics. Given the absence of light, and therefore photosynthetic primary production, the vast majority of the seabed depends on food sinking from surface waters. These conditions, exacerbated by low temperatures (consistently <4 °C below 1000-m depth in most of the global ocean) and high pressures (up to 1,100 atm), can complicate metabolic activities for species inhabiting deep waters. Pressure may influence metabolism, including activity levels. Some studies demonstrate generally lower metabolic rates in deep environments compared to similar species living in coastal or shallow environments (even after correcting for temperature effects), however, other analyses indicate similar metabolic rates in related species that occupy shallow and deep depths.

Biochemical adaptations also play a role. For example, marine species also invest more in lipids than their terrestrial counterparts. In addition to their role as thermal insulation for warm-blooded animals (seal and whale lipids may represent 20–30% of body weight), lipids in heterothermic organisms also decrease body density, thus increasing buoyancy, as seen in many species of amphipod crustaceans and oily prawns (Figure 4.1).

Many organisms that live in deep environments also invest in lipids as an adaptive strategy in the face of uncertainty. In fact, these organisms accumulate fats from abundant food (e.g., sinking phytoplankton blooms) to cope with long periods of fasting. Low proportions of carbohydrates in marine organisms contrast the carbohydrate-rich support structures typical of the higher plants on land, replaced by carbonate or chitinous exoskeletons in marine biota. This replacement elevates the importance of protein in marine organisms.

Organisms complete metabolic activities within temperature limits specific to each species, resulting in a **thermal optimum** for many marine organisms, defined as the temperature that supports maximum growth. Consequently, the thermocline not only influences nutrient exchange in the water column but also directly affects reproductive period, larval development and mortality, juvenile survival, and movements of marine of organisms living in its vicinity. Each species has a specific thermal optimum, but many marine organisms experience considerable temperature variation during different seasons. **Eurythermal** organisms, referring to those that tolerate wide variations in temperature contrast **stenothermal** organisms intolerant to such variation (Figure 4.2). Most stenothermal species live in the relatively constant

Figure 4.1 A thick layer of brown blubber (vascularized adipose, or fatty tissue) covers the body of most marine mammals, preventing the rapid cooling of internal organs of these homeotherms. In Antarctic seals, fats can represent more than 30% of their biomass, providing greater insulation and tolerance to cold temperatures: (a); the same strategy applies to elephant seals that have, on average, 40% of their biomass as fats (b), in other species, such as sea otters, a thick layer of hair isolates their body from the surrounding cold waters. *Source:* Jan Roletto / Wikimedia Commons / Public domain, National Marine Sanctuaries / Wikimedia Commons / Public Domain.

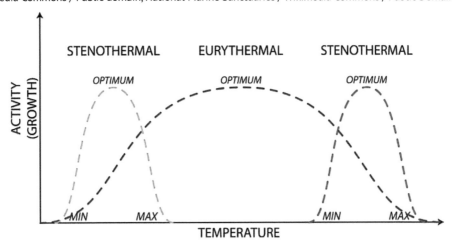

Figure 4.2 Diagram showing how temperature influences growth of different types of organisms.

thermal environments of the deep sea or tropical surface waters, whereas eurythermal species often occupy a wide range of latitudinal and bathymetric habitats, as well as highly seasonal habitats.

Global warming has affected both marine and terrestrial fauna. Recent studies have shown that marine ectotherms experience hourly body temperatures closer to their upper thermal limits than for terrestrial ectotherms across all latitudes. This thermal safety margin provides an index of the physiological stress caused by warming. On land, the smallest thermal safety margins occurred historically in species at mid-latitudes where the hottest hourly body temperatures occurred. In contrast, the marine species with the smallest thermal safety margins occurred near the equator. This difference results in local extirpations related to warming occurring nearly twice as often in the ocean as on land. The higher sensitivity to warming suggests that extirpations will be more frequent, and faster rates of colonization in the marine realm by newly arriving species may result in more rapid species turnover.

In variable temperature environments, organisms must **acclimate** or adapt their physiological processes to changes in climatic conditions. Some species are eurythermic at some latitudes but stenothermal at others. Scleractinians (stony corals of the Phylum Cnidaria), which are eurythermal and thus tolerant of a wide range of temperatures in the mid-latitude conditions of the Mediterranean Sea, contrast the stenothermal corals that occupy tropical seas with temperature conditions favorable for the development of coral reefs. However, when temperatures exceed the tolerated values for several days ("heat waves"), massive mortality of sessile species occurs (Figure 4.3).

Mussels illustrate the capacity of some organisms to acclimate their metabolism to sudden changes in temperature (Figure 4.4). The partial overlap in oxygen consumption after two weeks, despite very different temperatures, indicates that mussels acclimate and return to a standard metabolic rate under different temperature regimes.

Figure 4.3 Heat waves are responsible for the increasingly frequent bleaching events in stony corals and coral reefs. *Sources:* Brett Monroe Garner/Moment/Getty Images, Roberto Danovaro (Book Author).

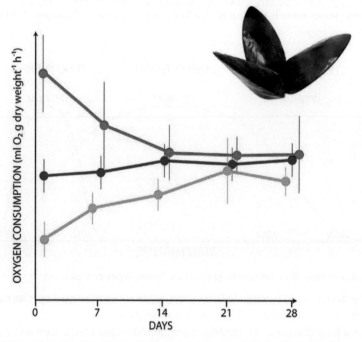

Figure 4.4 Acclimatization in *Mytilus galloprovincialis* to temperature variation. Three pools of mussels (n = 5 individuals) of uniform size were taken from one location and incubated at three different temperatures: in blue, the metabolic rates of the control pool were maintained at ambient temperature (15 °C); in red, the metabolic rates of the pool were maintained at 20 °C; in green, the data from the pool of mussels incubated at 10 °C. The vertical bars indicate the standard deviation (i.e., the average variability between repeated measurements).

In other cases, species adapt rather than acclimate to low temperatures. For example, very low temperature, the presence of ice, and seasonal primary production in Antarctica waters create inhospitable conditions that would prove lethal for most fishes because cold temperatures retard physiological processes, changing protein-protein interactions, reducing cell membrane fluidity, and increasing the viscosity of biological fluids, eventually freezing the animal. However, a group of teleost fishes, the Notothenioidei, have adapted well to these cold conditions, and dominate the cold coastal regions of the Southern (Antarctic) Ocean both in terms of species numbers (> 50%) and biomass (90–95%) of fishes. Notothenioidei evolved a reduced **hematocrit** (the percentage of blood volume occupied by the cellular component) and reduced concentrations of **hemoglobin** (Hb, the protein responsible for oxygen transport), so that temperature increases of just a few degrees would prove lethal. Moreover, the modified cardiovascular system in these **ice fishes** increases blood volume (2–4 times greater), heart size, blood vessel diameter, and **cutaneous vasculature** (blood supply to the skin) compared to similar fishes elsewhere. These changes ensure adequate oxygenation of tissues, decreased metabolic rates, and increased gas exchange through the gills and through cutaneous respiration. These specific cold-water adaptations reduce body metabolism and increase solubility of oxygen in liquids and tissues.

Hemoglobin reduction reaches extreme levels within the family Channichthyidae (also known as ice fish; Figure 4.5), which lack hemoglobin to the point that their blood is colorless or white, replaced by a series of proteins that avoid viscosity problems at low temperatures. Antifreeze proteins actually prevent body fluids from freezing in these teleosts (whose freezing point equilibrium, between -0.7 to -1 °C, exceeds that of seawater, -1.9 °C), preventing blood from freezing. Simple repetitions of a glycotripeptide comprise **Anti-Freeze Glycoproteins** (AFGLPs), but at least three other different types of antifreeze proteins exist, spanning 11 phylogenetically (the evolutionary diversification of a taxonomic group) distant families. These differences in glycoprotein structure suggest independent evolution of the ability to produce antifreezes in different Antarctic fishes. Ice fishes exemplify how temperature has selectively influenced polar evolution: the genome of these fishes lacks the genes to encode hemoglobin, and although they lack functionally active **erythrocytes** (red blood cells), hemoglobin, and carbonic anhydrase, these fish have evolved adaptations to compensate for normal metabolic functions of erythrocytes and hemoglobin for oxygen transport.

Low temperatures in polar environments drive adaptations within organisms at all levels of biological organization. These include **psychrotrophic**, or "cold loving" bacterial communities, usually restricted to permanently cold habitats; obligate psychrophiles actually require cold conditions and die at temperatures above 4 °C.

Many organisms respond to temperature stress by decreasing synthesis of normal proteins and instead producing specific proteins called **heat-shock proteins** (HSP), which form enzyme complexes that protect the enzyme from denaturation. Such HSPs also occur at the other end of the heat spectrum, where some marine organisms tolerate relatively high temperatures. Microbes at hydrothermal vents survive in temperatures greater than 113 °C; we discuss hydrothermal vent adaptations in Chapter 26.

In conclusion, although the majority of organisms perform metabolic functions optimally at temperatures near 20 °C, some organisms have evolved adaptations to much lower or much higher temperatures, and some organisms can tolerate a wide range of temperatures (Figure 4.6).

Figure 4.5 Two examples of ice fish: (a) *Macropteris maculatus* and (b) *Dissostichus mawsoni*. *Sources:* Konrad Meister, NOAA / Valerie Loeb / Public Domain, Pcziko / Wikimedia Commons / CC BY 2.5.

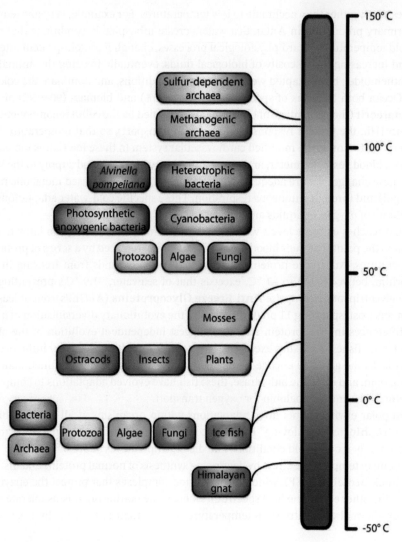

Figure 4.6 Temperature limits for the survival of organisms.

4.2 Adaptations to Low Oxygen Concentrations

Marine organisms have evolved sophisticated systems to utilize the comparatively low concentrations of oxygen dissolved in water. Most organisms are **aerobic**, meaning they require oxygen to break down or **oxidize** organic substances. However, **anaerobic** respiration, which releases energy by splitting organic molecules into simpler forms, requires no free oxygen. Anaerobic respiration occurs in some protozoa, bacteria, and archaea, as well as in some nematodes and parasitic flatworms within their host. Aerobic marine organisms take up oxygen dissolved in water through integuments, gills, and areas of the digestive tract. In primitive groups such as protozoa, sponges, and cnidarians, eggs and embryos lack specialized respiratory structures, and oxygen uptake occurs through simple diffusion across the **integument** (skin) from oxygenated water to the less oxygenated cytoplasm, the gelatinous liquid inside cells.

A similar process removes carbon dioxide in the opposite direction from the organism to their environment. However, larger organisms require structures or mechanisms to move water and thus renew oxygen. Hence, organisms require specific morphological, physiological, and behavioral adaptations to respire, such as small size or a large respiratory surface. Animals with specialized organs such as **gills** use these highly vascularized structures to exchange gas. In addition to their respiratory function, gills may collect suspended organic particles, a key food source for some animals (Figure 4.7). However, the presence of gills does not preclude respiratory exchange through the integument or in portions of the digestive system.

An organism's oxygen consumption relates proportionally to feeding and other metabolic activities. Respiratory pigments in blood plasma differ chemically from other molecules in the plasma and may be dissolved in the circulating liquid or located within specific cells.

In coastal environments, particularly in habitats periodically exposed to air, some species (including barnacles, mussels, gastropods, chitons) prevent excessive dehydration and reduce their metabolism and respiration by hermetically sealing their valves or shells, entrapping some seawater inside (Figure 4.8). Sometimes, fish species can tolerate exposure to air through specific adaptations to life in intertidal habitats. In this case, these teleosts develop specific air breathing ability that uses lung-like structures to enable gas exchange.

In this way they can live attached to a rock or laying on a sandy beach, escaping competitors and predators (Figure 4.8). In extreme low oxygen conditions, organisms need particularly efficient respiratory pigments. For example, at hydrothermal vents, high temperatures reduce oxygen solubility and result in low oxygen concentrations, so animals require respiratory pigments with high affinity for oxygen (Figure 4.9).

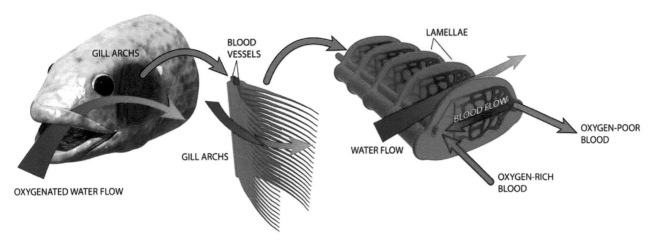

Figure 4.7 Diagram of gill circulation in fish.

Figure 4.8 Examples of organisms adapted to aerial exposure: (a) barnacles; (b) mussels (*Mytilus californianus*); (c) gastropods and other barnacles; (d) a fish, (the periophtalm, *Periophthalmus variabilis* of the Indonesian intertidal zone); (e) clingfish, *Sicyases sanguineus*, attached to rocks on the islands of Robinson Crusoe and Selkirk in Chile.

Figure 4.9 The bamboo worm *Riftia pachyptila* is an annelid of the family Siboglinidae (previously described as Vestimentifera, hypothesized as a new Phylum) that contains hemoglobin with high affinity for oxygen, enabling delivery of oxygen to its tissues, even in environments with high temperatures and therefore lower oxygen concentrations. *Source:* C. Van Dover / Wikimedia Commons / Public Domain.

Siboglinidae, a family of polychaetes, support a rich bacterial flora of chemoautotrophic symbionts within their **trophosome**, a specialized organ that may represent half the body weight of the worm. Their rich circulatory system delivers hemoglobin-filled fluid and gives the retractable trophosome its bright red color.

Several groups of respiratory pigments occur within different groups of marine organisms, some based on iron and some based on copper (Figure 4.10) These include:

- chlorocruorin (iron, Fe, green) in polychaetes (sabellid worms and serpulids);
- hemocyanin (copper, Cu, light blue) in crustaceans and cephalopods;
- hemovanadin in tunicates;
- hemerythrin in sipunculid worms, in brachiopods and in priapulid worms;
- hemoglobin (Fe, red) in fishes, in some bivalves, in polychaetes (including Siboglinidae; Figure 4.9).

Some subsurface worms living within sediments, such as *Arenicola marina*, can survive up to nine days without oxygen.

When the low oxygen concentrations of oxygen minimum zones (OMZs) intersect the seabed, they profoundly alter benthic communities, reducing diversity over broad areas that can extend 8,000 to 285,000 km². At these large spatial scales, many factors also help to structure benthic communities, but oxygen availability dominates. Organisms adapted to life in **hypoxic** (i.e., oxygen concentrations less than $2.0\ ml \cdot L^{-1}$) or **anoxic** (no oxygen) conditions thrive in these environments. Few OMZs occur in European waters; however hypoxic systems occur off the west coasts of South and North America, western Africa, Pakistan margin, Gulf of Oman, and in areas of the Adriatic and Baltic Sea.

Gobies (*Sufflogobius bibarbatus*) live in the oxygen minimum environments of **upwelling** areas of South West Africa, and migrate daily from the oxygenated water column where they feed on jellyfish, to the hypoxic/anoxic bottom where they can escape predators. When on the bottom, they immediately reduce their ventilation rate in response to reduced oxygen availability, and lactate progressively builds up in their blood (Figure 4.11). An exceptional heart pumping capacity and high resistance to hydrogen sulfide that often occurs in anoxic or hypoxic environments means that these organisms can move smoothly between oxic and anoxic systems, a capacity few other species can match.

Some species of Loricifera recently discovered in the deep hypersaline anoxic sediments of the eastern basin of the Mediterranean Sea exhibit extreme adaptation to lack of oxygen. **Deep hypersaline anoxic systems** (DHAB) have been discovered not only in the Mediterranean but also in the Red Sea and in the Gulf of Mexico. We currently know of 6 permanently anoxic basins located along the Mediterranean ridge within the Mediterranean Sea at depths between 3,200 and 3,600 m; these basins contain brine of varying thickness that prevents oxygenation of bottom water. The high

Figure 4.10 Examples of organisms containing different respiratory pigments: (a) sabellid worm with chlorocruorin; (b) cephalopod with hemocyanin; (c) tunicate with hemovanadin; (d) priapulid worm with hemerythrin; (e) polychaete with hemoglobin. *Sources:* Shunkina Ksenia / Wikimedia Commons / CC BY 3.0, Brocken Inaglory / Wikimedia Commons / CC BY SA 3.0.

concentrations (close to saturation) of salt contribute to the dense, viscous seawater, which acts in tandem with hydrostatic pressure, absence of light, complete anoxia, and a strong **chemocline** (chemical gradient) to make these basins among the most extreme habitats on Earth. Even so, bacterial diversity in these basins exceeds that in the overlying waters. Loricifera in L'Atalante basin live, eat, and breed in permanently anoxic conditions, a capacity unknown in any other metazoan. This discovery documents the first multicellular organisms that live their entire life cycle without oxygen. The hydrogenosomes that apparently replace mitochondria in these organisms produce hydrogen used by symbiotic prokaryotes that live in association with these cellular structures (Figure 4.12).

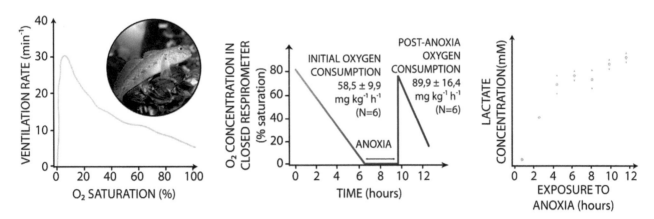

Figure 4.11 Adaptations of gobies to life in anoxic conditions. Graphs show: (a) reduced ventilation (gas exchange) rate at low oxygen concentrations; (b) rapid increase in oxygen consumption when transitioning from anoxic to oxic conditions; (c) increased lactate concentrations in blood with reduced oxygen. *Source:* Eric Engbretson / Wikimedia Commons / Public Domain

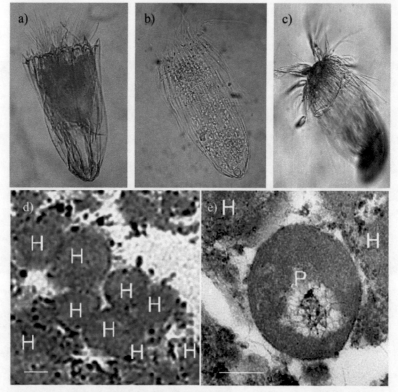

Figure 4.12 Three new species of the Phylum Loricifera discovered in the anoxic L'Atalante hypersaline basin (Eastern Mediterranean): (a) *Spinoloricus* sp.; (b) *Rugiloricus* sp.; (c) *Pliciloricus* sp. In *Spinoloricus* and *Rugiloricus* an oocyte (egg) inside the body appears as a pink thickening and suggests reproductive capacity in these organisms despite permanent anoxic conditions. Transmission electron microscope analysis of the ultrastructures of these organisms highlights the presence of (d) hydrogenosomes (H) rather than mitochondria, and (e) prokaryotes (P) associated with these hydrogenosomes.

4.3 Adaptation to Salinity

The global ocean encompasses a wide range of salinities (Figure 4.13). The ability to tolerate salinity fluctuations varies greatly among marine species. In contrast to high sensitivity to salinity changes in **stenohaline** organisms, **euryhaline** species tolerate a wide range of salinities. Euryhaline organisms have evolved specific mechanisms for osmotic regulation within certain salinity limits.

Some organisms undertake extensive migrations between sea and fresh water to reproduce. **Catadromous** species such as eels spawn in the sea (Figure 4.14a). In this species, larvae (leptocephali) that hatch in the ocean (specifically the Sargasso Sea) metamorphose into small, blind, unpigmented eels (60–90 mm) that return to freshwater to develop and mature. In contrast, **anadromous** species such as salmon live in marine waters and return to rivers to spawn (Figure 4.14b, c). **Amphidromous** species such as some species of mullets (Figure 4.14d) move between fresh water and seawater but not for reproduction.

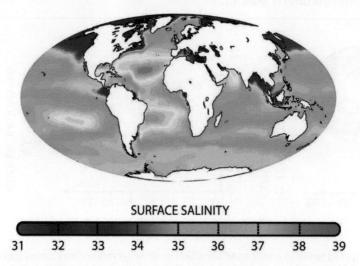

Figure 4.13 Distribution of salinity in the oceans. The dominant green color indicates salinities close to 35. The cooler blue-violet colors, particularly in the Arctic, indicate dilution with freshwater inflow from continents and/or high rates of precipitation. Salinities in excess of 38, which characterizes locations with little freshwater input and high rates of evaporation, occur in the Red Sea, Arabian Gulf, and Mediterranean Sea.

Figure 4.14 Examples of fishes that are: (a) catadromous (eel); (b) anadromous (blue back herring, a species endemic to the Hudson River estuary); (c) a salmon; and (d) amphidromous (*Chelon labrosus*). *Sources:* GerardM / Wikipedia Commons / CC BY-SA 3.0, Alessandro Duci/Etrusko25/Wikimedia Commons, Hans-Petter Fjeld / Wikimedia Commons / Public domain.

Homeosmotic organisms or **osmotic regulators** maintain a constant internal salt concentration (fish, mollusks, crustaceans) whereas the internal salt concentration of **osmotic conformers** such as flatworms and polychaetes effectively track their environment. Within the organism's body fluids, nutrients and other solutes help regulate osmotic pressure. Active transport across cell membranes results in different concentrations of some solutes internally compared to those outside the cell. Active elimination helps to maintain concentration gradients (Figure 4.15).

Higher concentrations of solutes in **hypertonic** marine organisms compared to seawater mean that in cases of osmotic stress, organisms change cell internal or **turgor** pressure. Subsequently, osmotic regulation selectively incorporates or eliminates certain ions and **osmolytes**, the small organic molecules produced by cells. *Dunaliella salina* (Figure 4.16), for example, survive in salt marshes by forming glycerol as an osmolyte.

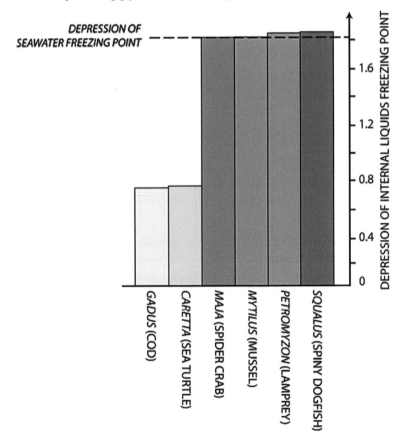

Figure 4.15 Solute concentration of internal fluids of different marine species. For most heterothermic marine animals other than reptiles and teleosts, the freezing point of bodily fluids closely matches that of seawater.

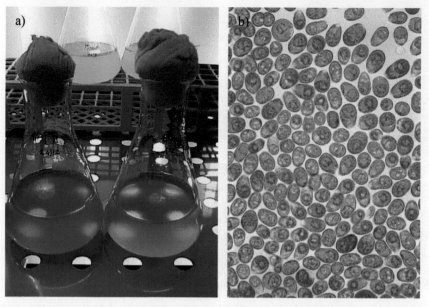

Figure 4.16 Some algae can adapt to salinity: (a) cultures of algae in flasks; (b) *Dunaliella salina* under a compound microscope.

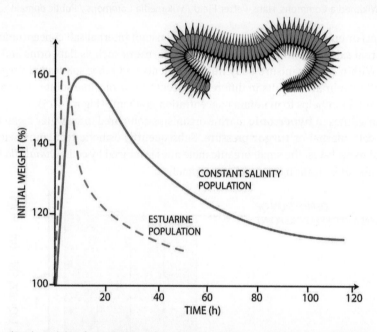

Figure 4.17 Variation over time in weight of populations of the marine polychaete, *Hediste diversicolor*. The dotted line denotes a population subject to frequent freshwater input, whereas the solid line denotes a population from an area of constant salinity (~37). Both populations increase in weight when placed in seawater with a salinity of 20, as a result of an osmotic response. The graph shows population response over time. The euryhaline population (dotted line) recovers their original weight quickly, whereas the stenohaline population recovers slowly.

Many protozoa osmoregulate with a contractile vacuole, whereas euryhaline polychaetes such as *Hediste diversicolor* (Figure 4.17) regulate through slightly permeable integuments, which give the animal time to adjust internal solute concentrations. Homeosmotic decapod crustaceans maintain hypertonic internal fluids through active absorption of salts through the gills.

Moreover, some animals, such as the copepod genus *Tigriopus*, can live in particularly salty water such as tidepools (Figure 4.18).

Most mollusks are good osmoregulators, in contrast to stenohaline echinoderms, few of which can tolerate even modest salinity variation. Within vertebrates, sharks (Order Selachii) modulate osmotic pressure by adjusting urea concentrations in their blood (in fact, the concentration of urea in blood makes it almost iso-osmotic to seawater; Figure 4.19).

Hypotonic blood in marine teleosts helps them avoid dehydration by ingesting large volumes of water through their alimentary canal and removing excess salt through the gills, producing small volumes of hypertonic urine. Marine **amniotes** (reptiles, birds, and mammals that lay eggs on land or retain them) osmoregulate by minimizing seawater ingestion, minimizing water losses, and producing hypertonic urine. The marine invertebrates on which amniotes feed provide a water source. Reptiles osmoregulate by producing a saline solution in their orbital gland, a modified **lacrimal** (tear) gland (Figure 4.20). Seabirds osmoregulate through a salt gland located in their nasal cavity.

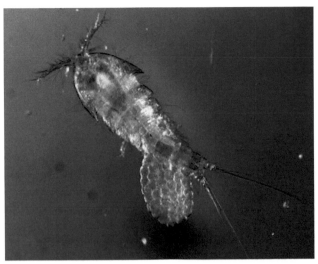

Figure 4.18 Copepods of the genus *Trigriopus* inhabit intertidal rock pools that change rapidly in salinity. *Source:* Morgan Kelly/University of California Davis.

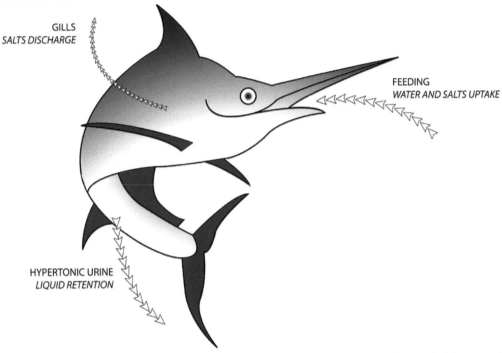

Figure 4.19 Marine fishes typically osmoregulate by producing small quantities of hypertonic urine, minimizing water loss while expelling maximum salts.

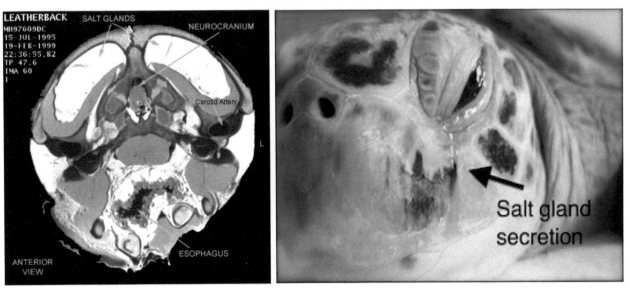

Figure 4.20 Sea turtle salt gland and salt gland secretion.

4.4 Adaptation to Pressure

In contrast to the ubiquitous and **barotolerant** (or **piezotolerant**) organisms that tolerate a wide range of pressures, many species can withstand only narrow pressure ranges. Therefore, internal fluid pressure of the organism must counterbalance external pressure. However, the compressibility and dissociation of several salts increase with depth, and only **eurybathic** species that span wide depth ranges can withstand these conditions. Specific adaptations in these species include resistant cell membranes and structurally stable enzymes that are insensitive to pressure increases. Eurybathic species include the polychaetes *Notomastus latericeus* (intertidal to 3,000 m deep) and *Amphicteis gunneri* (20–5,000 m depth), the cumaceans *Diastylis laevis*, *Eudorella truncatula* (9–2,820 m depth), and the seastar *Henricia sanguinolenta* (0–2,450 m depth). The siboglinid worm, *Siboglinum caulleryi*, spans a staggering depth range of 20 to 8,000 m. Moreover, some species that live only at shallow depths, such as the common mussel *Mytilus galloprovincialis*, can survive at considerable depth (600–800 m) for several days, but presumably they do not colonize deep environments because of limited food availability.

High pressure influences the viscosity, ionic balance, conductivity, and relative gas solubility of seawater and body fluids. Organisms adapt to high pressure physiologically and biochemically, rather than through pronounced morphological changes. Bony fishes (Class/InfraClass **Teleostei**), use their swim bladder not only as a flotation aid, but also as a means to detect pressure and automatically adjust buoyancy. Swim bladders presumably detect water vibrations, and, in some species, function as a hydrophone by connecting with the ear.

High pressure noticeably affects organism mobility. In surface-living animals, pressure above certain levels can disturb neuromuscular activities, resulting in uncoordinated movements and, with further pressure increase, complete paralysis. At abyssal depths, fishes and other organisms typically move slowly and gradually, with no rapid, pulsed movement. Fishes such as some macrourids (grenadiers) living at depths of 4,000–5,000 m lack a swim bladder and live in close contact with the seabed. The deepest shark species, the Portuguese dogfish (*Centroscymnus coelolepis*), was captured in traps at 3,750 m. This species typically does not occur at depths less than ~500 m (Figure 4.21).

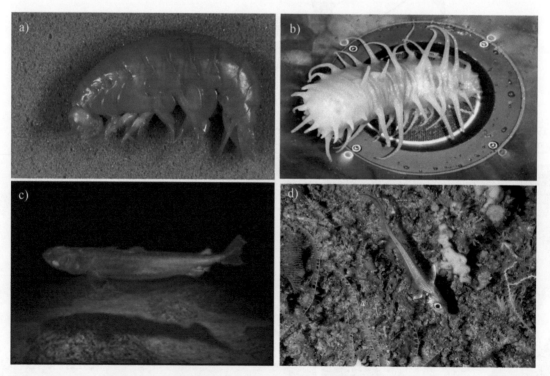

Figure 4.21 Some examples of barotolerant or piezotolerant deep-sea organisms. These species are adapted to life at depth and high pressure but can move to shallower depths or tolerate depth ranges >1000 m: (a) amphipod crustaceans (*Eurythenes gryllus*); (b) holothurians; (c) macrourid fish; (d) rat-tail fish (*Nezumia* sp.). *Source:* National Oceanic and Atmospheric Administration (NOAA) / Public Domain.

4.5 Adaptations to Light

Algal photosynthesis requires light wavelengths between 450–700 nm, a band known as "**photosynthetically active radiation**" or PAR. The existence of life on land and in seawater depends on photosynthesis. In terrestrial environments, light is ubiquitous, at least during some seasons and times of day. In contrast, >95% of ocean volume experiences complete and perpetual darkness, irrespective of time of day or season. Most primary production occurs in surface waters, particularly along the continental margins. Low offshore primary production compares with productivity levels in Arctic tundra or deserts in terrestrial ecosystems, whereas productivity in nearshore environments, estuaries, and lagoons in particular, compares favorably with tropical rainforests.

4.5.1 Photosynthesis

Photosynthesis requires light, whether in macrophytes attached to the seabed or phytoplankton drifting in surface waters. As noted in Chapter 3, although we tend to think of water as transparent, the reality is that light dissipates fairly quickly, limiting photosynthesis to depths typically less than 100 meters. This rapid decline in light explains why benthic primary producers such as kelp, benthic diatoms, and seaweeds typically decline rapidly in abundance below 10s of meters depth, leaving only benthic photosynthetic species adapted to dim light (e.g., some corals and coralline algae). Different photosynthetic pigments vary in their sensitivity and effectiveness at capturing light of different wavelengths. Chlorophyll a and b absorb light within the blue and red wavelengths, carotenoids in the blue green spectrum, phycoerythrin absorbs maximally in the blue, green, and yellow spectrum, and phycocyanin absorbs maximally in the yellow-red spectrum. Because water selectively absorbs different wavelengths of light (see Chapter 2), availability varies with depth and photosynthesizing organisms must adapt accordingly. Thus, prokaryotes such as *Prochlorococcus* spp. and *Synechococcus* spp. possess pigments that capture the relatively faint light well below the surface whereas the dominant pigments in eukaryotic groups such as diatoms and dinoflagellates fare best in near surface waters.

4.5.2 Vision and Bioluminescence

Over evolutionary time, some marine organisms have developed spectacular coloration. The wide variety of colors and morphologies of fishes on coral reefs, in tandem with behavioral analysis and physiological evidence of photoreceptors in reef fishes that sense different wavelengths, suggests color perception capacity in at least some reef fishes.

Environmental light conditions in the ocean differ greatly from those on land, and pose different challenges to visual systems. Optical mechanisms and retinal structures in marine mammal eyes show specific adaptations for vision in water. Recent studies demonstrate unusual vision systems in cetaceans and pinnipeds, with apparently complete loss of visual cones to detect short wavelengths. However, behavioral data indicate that cetaceans, pinnipeds, and manatees can discriminate colors. A keen sense of vision in fishes helps them find food, shelter, mates, and avoid predators. The recent discovery of ultraviolet components of reef fish colors, and that some of these fishes appear sensitive to ultraviolet light or have some photoreceptors sensitive to ultraviolet wavelengths, suggests not only that fishes see color but that they also use ultraviolet signals for communication (e.g., tropical angelfish, *Pomacentrus amboinensis*; Figure 4.22).

Figure 4.22 The tropical angelfish, *Pomacentrus amboinensis*.

Marine organisms see differently in deep environments compared to surface waters. In surface waters light illuminates the seabed, available light easily supports vision, and the visual field perceived by animals extends in all directions. Because light barely penetrates to depths >200 m (depending on water transparency; Figure 4.23), and considering an average ocean depth of almost 4,000 m, most marine organisms have evolved specific adaptations for life in low light or aphotic conditions, such as blind organisms lacking pigment, or organisms with huge telescopic and specialized eyes and **photophores** or light organs to detect prey, escape predators, or attract mates. In the "twilight zone" characterized by weak light, the eyes of fishes typically increase in size relative to body length and pupil width. But at greater depths, where light levels decrease exponentially, bioluminescence replaces sunlight as a light source. For example, full daylight at 700 m depth is insufficient for even the most specialized vision. At these depths, bioluminescent signals produced by other animals produce the only visible light.

About two-thirds of abyssal species possess luminous organs. These marine organisms use luciferin and luciferase that, in the presence of oxygen, react to produce light (Figure 4.24). Whereas bacteria produce continuous light in the presence

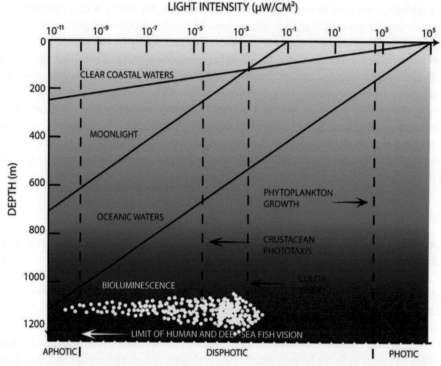

Figure 4.23 Dissipation of light radiation in the ocean, highlighting the relationship between water transparency and penetration depth of light radiation. The vertical dotted line delineates the boundaries between the photic, disphotic, and aphotic zones.

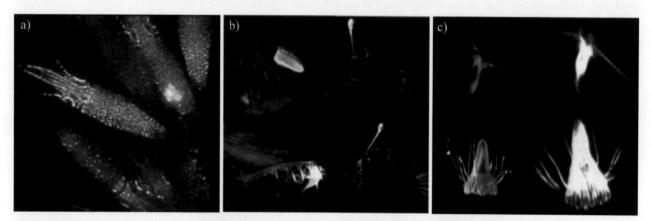

Figure 4.24 Examples of deep pelagic organisms that emit bioluminescence: (a) cephalopods; (b) fishes with bioluminescent organs in different parts of the body but always located apically to facilitate prey capture; (c) a bioluminescent copepod at the top of the image produces bioluminescence that provides a defense mechanism that temporarily blinds potential predators (bottom of the image); a bioluminescent jellyfish appears at the bottom of the image.

of oxygen, allowing them to emit bioluminescence both at night and during the day, other organisms produce light only in response to stimulus. Bioluminescent signals often occur in flashes that last from hundreds of milliseconds to several seconds. Below 1,000 m, the frequency of emitted flashes decays considerably, becoming very infrequent below 2,000 m depth. The intensity and color of these flashes varies greatly, but typically span blue wavelengths between 10^7 and 10^{13} photons in intensity, creating a highly visible stimulus in the lightless deep ocean.

At bathypelagic depths, organisms require large pupils to capture available light, much of which arises from bioluminescence. These eyes help to identify partners, predators, or prey, in tandem with numerous highly developed senses such as smell and **lateral line systems**, the tactile sense organs on the sides of fishes (and some other vertebrates) that detect movements and pressure changes in the surrounding water. Collectively, these adaptations increase the probability that bathypelagic fishes will "intercept" a partner to reproduce, or prey to consume, within a dark, dilute world. Bathypelagic fish eyes can detect bioluminescent flashes tens of meters away. This distance seems modest, but represents a considerable distance for a pelagic fish with weak muscles and a light skeleton. At abyssopelagic depths, small eyes characterize most fishes because of the limited food supply, whereby low respiration rates cannot provide the energy necessary to support large eyes. Nonetheless, large pupil sizes compared to eye size differentiates deep-sea fishes from their surface counterparts.

Luminescence in marine animals arises from three sources: (1) extracellular secretions of luminescent material produced by glands; (2) intracellular processes, and (3) symbiotic bacteria. The latter two cases often utilize complex light producing organs called **photophores**. Bioluminescence occurs in various species of cnidarians and ctenophores, polychaetes, shrimps, copepods, mollusks (bivalves, gastropods), cephalopods, and fishes (Figure 4.25). Photophores of varying complexity produce light in many species of crustaceans, cephalopods, and fishes. These organs are particularly well developed in species that live at great depths. The simplest photophores consist of a mass of photogenic cells on a layer of connective tissue that reflects light. Symbiotic luminous bacteria also cause luminescence in cephalopods and teleosts (see Chapter 7). Although most fishes depend on bioluminescent bacteria or control production of their own light by means of their nervous system, fishes of the genus *Photoblepharon* use complex "gates" to open and close their photophores. This mechanism allows fish to use light for underwater communication.

Other luminous phenomena occur at the sea surface, such as in pelagic unicellular radiolarians and dinoflagellates. For example, *Noctiluca miliaris* produces frequent luminescence at the ocean surface. In this organism, light originates from granules scattered in the periphery of the cell body. Some marine organisms use luminescence to recognize conspecifics for reproduction. Indeed, bioluminescent structures can also exhibit marked **sexual dimorphism** where the morphologies of a species differ between sexes; bioluminescence can therefore facilitate reproduction at depth by differentiating individuals by sex.

In well-lit surface and most freshwater environments, visual signals aid in sexual recognition, attraction, and mating (as seen in obvious ornamentation in many male fishes). Bioluminescence serves the same purpose in dark or dimly lit habitats by providing visual stimuli. Light producing organs, both in males and females, can differ in size and position, or may be lost in one of the sexes (typically males). Secondary sexual characteristics may include "bioluminescent" sexual dimorphism.

Bioluminescent ovaries or eggs offer the simplest example of sexual dimorphism. In the copepod *Oncaea conifer* considerably smaller males occur less frequently and in somewhat different distributions than females, although both have numerous cuticular bioluminescent glands. Female glandular bioluminescence may offer a defensive response to potential predators. Numerous ventral photophores in different groups of decapods presumably produce **counter-shading** that reduces decapod visibility to predators against surface lighting.

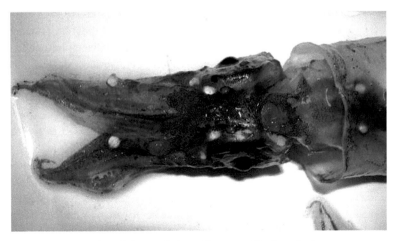

Figure 4.25 Photophore position in a cephalopod. *Source:* Roberto Danovaro (Book author).

Figure 4.26 The Humpback anglerfish, *Melanocetus johnsonii*, a small demersal fish attracts prey with a photophore that sits on a mobile spine on its dorsal fin. *Source:* Masaki et al., 2010 / Springer Nature/ CC BY 2.0.

Figure 4.27 Cloud of liquid emitted by different bioluminescent deep crustaceans such as decapods and copepods in order to confuse potential predators and facilitate escape.

In cephalopods and euphausiid crustaceans, particular photophores appear only in mature females, although male photophores outnumber those in females, especially on their tentacles. In dragon fishes, 17 of 24 genera exhibit sexual dimorphism in bioluminescent organs but only in photophores around the eyes. The most pronounced bioluminescent sexual dimorphism occurs in lantern fish through the presence and/or extent of specialized photophores at the base of their tail (Figure 4.26). Bioluminescent signals contribute to an individual's reproductive success, thus illustrating sexual selection, which can influence mate choice by females or competition among males.

Low population abundance in oceanic depths for most species reduces the frequency of competition among males, and attraction of males by females drives many bioluminescence strategies. Males and females can produce different bioluminescent sexual signals without sexual dimorphism in bioluminescent structures, as occurs in some lantern fishes. Female marine polychaetes (Family Sillidae), in Bermuda produce bioluminescent secretions as they swim in surface waters, attracting males that produce additional bioluminescent secretions. In California, the males start the party!

Predators sometimes use photophores as a lure. In some families of monkfishes, only females possess a bright lure (a complex structure of protuberances and filaments with a central core of symbiotic luminescent bacteria (Figure 4.26)). Bright lures occur in only a few species, but females of many of these species emit light. In some cases, animals emit bioluminescent fluid to confuse predators (Figure 4.27). In other cases, they use bioluminescence to direct the predator to a body region they can easily regenerate. Finally, some animals use bioluminescent structures to illuminate the environment and locate prey.

4.6 Adaptations to Nutrients

Whether attached to the seabed or drifting in surface waters, photosynthetic organisms need **macronutrients** (nitrate, phosphate, silicon) required in high concentrations as well as **micronutrients** required in low concentrations, such as iron, magnesium, and zinc. Some species such as kelps and diatoms thrive in high nutrient conditions whereas other species such as small photosynthetic flagellates thrive in nutrient replete waters, where their large surface area to volume ratio favors nutrient update at low concentrations. The distribution of nutrients is not homogeneous in seawater. We can imagine a cube of organic particles, surrounded by a plume of nutrient released by degradation of these particles and by bacteria and phytoplankton cells that exploit these molecules (Figure 4.28). The oxidation of dissolved organic nutrients leads to the production of inorganic nutrients, such as nitrates, ammonium, phosphates, and silicates. Along with these molecules, the degradation of organic material results in the release of micronutrients that are essential for the metabolism of all organisms (Chapter 16 provides a more detailed explanations of nutrient-phytoplankton interactions). To sustain their production, primary producers do not need only the major nutrients ones such as carbon, nitrogen, or phosphate, but also micronutrients such as iron (Fe), manganese, copper (Cu), cobalt (Co) and magnesium (Mg), zinc (Zn), nickel (Ni) and molybdenum (Mo). Micronutrients must be sufficiently available to supply the required elemental composition of marine phytoplankton species. Phytoplankton need micronutrients because they play an important role in their metabolism: for example, Co, Cd, and Zn in carbon dioxide acquisition; Fe and Mn in carbon fixation; Zn, Cd and Se in silica uptake; Fe and Mo in nitrogen fixation, and Fe, Cu, and Ni in organic N utilization. Fe plays a key role in primary productivity, and limits primary productivity in wide oceanic regions, such as in the central oceans. However, many other trace elements also occur at concentrations close to their potentially limiting values, such as cobalt whose scarcity can influence phytoplankton species composition in the Peru upwelling region. Fe, Zn and Co mediate phytoplankton community structure in the Southern Ocean and Mn, Cu and Zn co-limit productivity in the subarctic Pacific Ocean. Fe, Co, and Cu can sometimes limit diatom growth, contrary to the generally accepted view that phosphorus (phosphate) or nitrogen (nitrate) limit growth in most ocean regions.

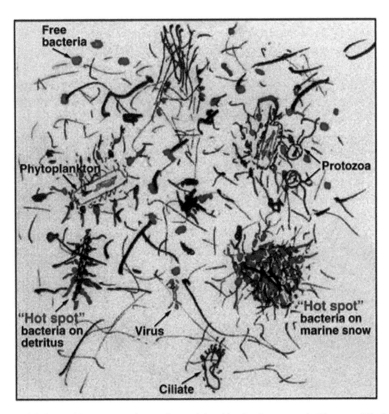

Figure 4.28 An "impressionistic" view of interspersed organic particles (detritus), surrounded by a multitude of microscopic organisms, including bacteria, viruses, protozoa and phytoplankton cells.

4.7 Electrical Conductivity Adaptations

The ions of dissolved salts in seawater give seawater its ability to conduct electricity more than air or freshwater, and some marine animals can detect the electrical activity of other distant animals, an ability terrestrial animals lack. For example, the **ampoules of Lorenzini** (Figure 4.29), specialized electro-reception organs in the heads of sharks filled with a gelatinous electrical conductor connect nerve endings to the outside through small holes, and allow them to sense electrical fields produced by other animals and potentially orient themselves relative to Earth's magnetic field.

4.8 Ectocrine Adaptations

Marine **tele-mediators** or **ectocrines** refer to a broad class of organic compounds synthesized and released into seawater by marine plants and animals that influence the behavior or biological characteristics of conspecifics. They can also sometimes affect other organisms. For example, red tides, or blooms of dinoflagellate algae such as *Gymnodinium breve* and *Gonyaulax polyhedra*, release vitamin B12 (cyanocobalamin) and thiamine, increasing cell replication once the bloom starts. However, knowledge of natural marine tele-mediators remains scant.

4.9 Adaptations to Produce Sound and Communicate in Water

Sounds fundamentally contribute to how ecosystems function, and can provide an indicator for assessing species presence/absence, abundance and, given the impact of anthropogenic noise on some life forms, to assess marine environmental health (Table 4.1). Seawater transmits sounds over impressively long distances (at 1500 m · sec^{-1}), noting that high frequency emissions can propagate for 10–16 km.

Many marine organisms (such as cetaceans) take advantage of efficient sound transmission to communicate over long distance. Whales can communicate with each other across ocean basins. Blue whales, for example, communicate over distances greater than 1,500 km. By comparison, the human voice can communicate distances up to 400–600 m over land. Besides marine mammals, many other marine species also produce sounds that propagate for short-distances (e.g., fishes, crustaceans, and other invertebrates) in relation to their behavioral and bio- ecological functions (e.g., communication, mating, defending territories). Animals depend on ambient sounds as a source of environmental cues for coping with several needs, such as foraging, navigation, predation (or detection of predators or competitors), and habitat selection.

However, the complex nature of sound production in marine organisms leaves many questions unanswered. Marine organisms communicate using a variety of vocal and non-vocal signals. Some animals produce non-vocal sounds such as the slaps of a whale tail, or snapping sounds in some crustaceans that originate from the production of air bubbles by cavitation. **Cavitation** refers to the formation of gas bubbles in liquids that move when pressure drops below that of water vapor while the surrounding temperature remains constant. Jerky movements of the claws produce cavitation, and the characteristic noises associated with snapping shrimp on coral reefs. In addition to cavitation, marine organisms produce sounds through surface vibrations or the movement and rubbing of different body parts (fins, tail) to produce sound. Sound production occurs primarily in three groups of marine animals:

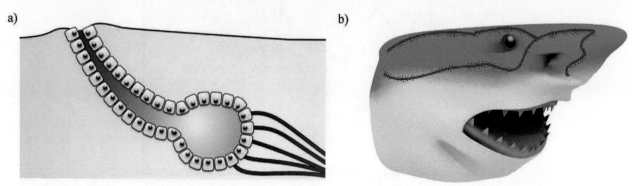

Figure 4.29 Example of sensory organs that detect mechanical waves caused by motion: (a) the structure of an ampoule of Lorenzini, (b) the distribution of these ampoules around a shark's head.

Table 4.1 Sounds emitted by some biological sources.

Animal	Frequency (Hz)	I_A (W/m² a R=1 m)
Human voice	100–10.000	10^{-10}–10^{-4}
Crangonidae (shrimp)	1.000–50.000	?
Fishes with swim bladders	350–1.500	?
Sharks	20–30	$<10^{-6}$
Tursiops truncates (Bottlenose dolphin): Whistles	5.000–10.000	10^{-5}
Tursiops truncates (Bottlenose dolphin): Clicks	25.000–60.000	10^4
Finback whale (*Balaenoptera physalus*)	20	0,1–30
Bowhead whale (*Balaena mysticetus*)	100–2.000	0,01–10

a) crustaceans, especially shrimps that produce cavitation by rubbing body parts;
b) some fishes (especially teleosts, and members of the fish Families Sciaenidae, Triglidae and Batrachoididae) that produce vibrations with their swim bladder and other groups (such as Balistidae and Carangidae, *Balistes* and *Caranx*) by rubbing body parts;
c) marine mammals (cetaceans) utilize muscles in their airways.

Highly developed sound production and reception represent an important adaptation to pelagic life. Cetacean sound production varies greatly, from 10s of Hertz for baleen whales songs, to 150 kHz in dolphin echolocation pulses. Sound production differs from species to species, depending on the purpose of the sound. Cetaceans, using the most advanced marine communicators currently known, can communicate the desire to mate, approaching risks such as predators, and the presence of food. Moreover, they can exchange information on navigation, aggression, or territoriality. The acoustic sound waves produced by different animals allow rapid recognition (including sexual discrimination) of conspecifics or of different species, and communication of dangerous situations. Many fishes and all marine mammals can recognize the direction from which acoustic waves come. Dolphins can distinguish shapes, materials, sizes, and even the thickness of targets. However, other marine animals also likely use sound as a form of sonar (seals, catfish, and possibly some baleen whales). Animals also emit sounds, using **echolocation** to judge distances and speeds of moving objects. Echolocation, or biosonar, can be thought of as an auditory "imaging system" used by various species to navigate and to estimate the location of prey, particularly in environments where visual cues are ineffective. The animal vocalizes and echolocates by detecting the reflected echo to produce three-dimensional information on the environment.

Sperm whales (*Physeter macrocephalus*), the largest of the toothed whales (Order Odontoceti), feed primarily on cephalopods (mainly squid and occasionally octopus) at depths up to 100s of meters, where light is weak or completely absent. These whales spend about three-quarters of their time foraging, diving to depths of nearly 1,000 m for average dive durations of 45 minutes, after which they emerge on the surface for about 9 minutes before the next dive. They emit a series of clicks during these dives to locate prey (mainly squid) using echolocation (Figure 4.30). As sperm whales descend to depth, they increase the frequency of clicking, giving them detailed information on prey location and movement. This long-range biological sonar, clicking with increasing frequency as the sperm whale approaches the prey, becomes a steady hum of clicks as they zero in on their target.

Acoustic sounds for echolocation vary widely. **Narrowband signals**, which span a narrow range of frequencies, last longer, and detect targets over long distances. **Broadband signals** span a wide range of frequencies, typically lasting less than 5 milliseconds, and primarily aid in localizing targets. Dolphins produce two main types of echolocation signals: broadband signals of short duration, or narrowband signals of long duration (Table 4.2). Dolphin produce echolocation signals from their nasal sacs located in the upper part of the brain. A structure called the **melon** acts as a lens, focusing acoustic waves in a narrow beam projected forward. A bone in the dolphin jaw senses the echo, and the adipose tissue behind the jaw transmits the acoustic waves to the middle ear and then to the brain. The teeth may also help in transmitting echoes to the dolphin brain.

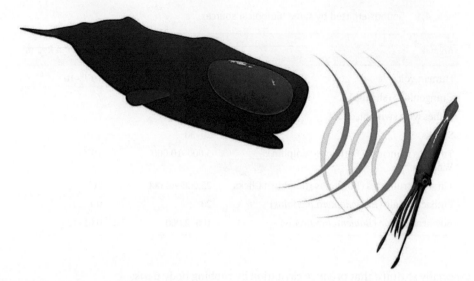

Figure 4.30 Echolocation of sperm whales to locate prey.

Table 4.2 Maximum transmission distances calculated for some marine mammals.

Source	Frequency (Hz)	Distance (Km)
Dolphins: click	25000	30
Dolphins: whistles	10000	70
Finback whale	20	10000

Many marine animals likely produce and detect different sounds, but marine soundscapes remain a topic of ongoing research. Aquatic animals can be divided into four groups based on their ability to hear sound:

a) crustaceans, especially lobster, with modest ability to produce and detect sound;
b) non-specialist fishes, including salmon and cod, which hear sounds especially by using their swim bladder;
c) specialist fishes, including carp, which also use *otoliths* or ear bones to detect sounds;
d) marine mammals, produce sounds as much as any known animals living on Earth.

Questions

1) How do marine organisms cope with low temperatures?
2) How have some animals adapted to high temperatures?
3) How do marine organisms tolerate low oxygen concentrations?
4) What is bioluminescence, how it is produced, and what is it used for?
5) How have some marine organisms adapted to aerial exposure?
6) How have some animals adapted to varying light intensity?
7) Which organisms depend on macro and micronutrients?
8) What are the mechanisms of protection/defense from high light intensity?
9) What sounds do animals produce to identify the presence of potential predators and prey?
10) Which organisms produce sounds and how do they use their vocalizations?

Suggested Reading

Davis, R.W. (2019) *Marine Mammals – Adaptations for an Aquatic Life*. Springer.

Elshahed, M.S., Najar, F.Z., Bruce, A.R., Oren, A., Dewers, T.A., and Krumholz, L.R. (2004) Survey of archaeal diversity reveals an abundance of halophilic Archaea in a low-salt, sulfide- and sulfur-rich spring. *Applied and Environmental Microbiology*, 70, pp. 2230–2239.

ESF (2007) *Investigating Life in Extreme Environments: A European Perspective*.

Gillooly, J.F., Brown, J.H., West, G.B. et al. (2001) Effects of size and temperature on metabolic rate. *Science*, 293, p. 2248.

Kato, C., Li, L., Nogi, Y., Nakasone, K., and Bartlett, D.H. (2002) Marine microbiology: Deep sea adaptations, in Taniguchi, Y., Stanley, H.E., and Ludwig, H. (eds.), *Biological Systems Under Extreme Conditions*. Biological and Medical Physics Series. Springer. pp. 205–220.

Lynn, J., Rothschild, L.J., and Mancinelli, R.L. (2001) Life in extreme environments. *Nature*, 409, pp. 1092–1101.

Miller, D.D., Ota, Y., Sumaila, R.U., Cisneros-Montemayor, A.M., and Cheung, W.W.L. (2018) Adaptation strategies to climate change in marine systems. *Global Change Biology*, 24, pp. e1–e14.

Morrissey, J. and Sumich, J. (2012) *Introduction to the Biology of Marine Life*. Jones and Bartlett.

Sayed, A.M., Hassan, M.H.A., Alhadrami, H.A., Hassan, H.M. et al. (2020) Extreme environments: Microbiology leading to specialized metabolites. *Journal of Applied Microbiology*, 128, pp. 630–657.

Somero, G.N. (2022) Solutions: How adaptive changes in cellular fluids enable marine life to cope with abiotic stressors. *Marine Life Science and Technology*, 4, pp. 389–413.

Walter, N. (2005) *Investigating Life in Extreme Environments – A European Perspective*. European Science Foundation.

5

Adaptations in Marine Organisms II: Life in a Fluid Habitat

5.1 Adaptions to Life in the Water Column

The differences between marine and terrestrial environments influence both animals and plants at a structural level (abundance, species composition) and at a functional level (size, biomass, production). In contrast to air in terrestrial environments, seawater density and viscosity (830 and 65,000 times greater than in air, respectively) help suspend even very large organisms in the water column without requiring large energy expenditure. **Archimedes principle** describes this phenomenon and refers to the fact that an object displaces its own volume in water, producing a counter effect of uplift equivalent to the mass displaced. Thus, the buoyancy of the organisms depends on their density relative to water density. The buoyancy provided by water enables reduced support structures in many aquatic organisms relative to terrestrial organisms and the evolution of relatively large marine creatures. The largest known terrestrial animal, a reptile of the Mesozoic, was less than 30 tons, and pales in comparison to the largest animal on Earth (the blue whale, *Balaenoptera musculus*), which weighs in at an impressive 190 tons and grows to lengths up to 35 m. Organisms as large as these marine cetaceans could not live in the terrestrial environment because the limbs necessary to support such massive weight could not allow mobility, and the animal would collapse and suffocate under its own weight. However, an aquatic environment does not limit movement as strongly, and the largest cetaceans move quickly and gracefully through the ocean; their hydrodynamic shape exemplifies adaptation to the aquatic environment.

The ocean therefore supports organisms varying much more in size than organisms on land, from viruses (from 20 to 100 nm) to blue whales (over 35 m) that span approximately 9 orders of magnitude (Figure 5.1). This means we would need to place one billion viruses one after the other to match the length of a single large whale. These differences vary even in mass, spanning from <1 fg (1 femtogram = 10^{-15} g) to over 100 tons (= 100,000 kg = 10^7 g), representing ~22 orders of magnitude in biovolume or biomass.

5.1.1 Density, Shape, and Buoyancy

Generally, seawater density varies within narrow limits (between 1.020 and 1.040 g · cm^{-3}), typically increasing with depth and distance from coastlines where riverine inputs and runoff from land dilute waters. Animal cytoplasm density varies between 1.03 and 1.10 g · cm^{-3}, compared to fish tissue with a density of 1.07 g · cm^{-3}. The density of most skeletal structures, including bone and carbonate skeletons such as shells, exceeds that of seawater and limits the distributions of most shelled biota to seabed habitat. However, high-density structures do appear in some microscopic organisms in the water column. The density of the theca of microalgae, such as diatoms, may be 2.60 g · cm^{-3}, slightly less dense than the calcite or aragonite plates of coccolithophores at ~2.70–2.90 g · cm^{-3}. Typically, the density (also called **specific gravity**) of many organisms, including those living in sediments, resembles that of seawater. Plankton, in particular, illustrate the specific adaptation of reduced specific body weight. Terrestrial environments lack any real counterpart to plankton, which, thanks to their low density, live suspended within the water column; density of most planktonic organisms (e.g. zooplankton) varies from 1.05 to 1.15 g · cm^{-3}.

The density of some of these planktonic species exceeds that of seawater, resulting in gradual sinking that can be problematic for both phytoplankton and zooplankton. However, the reduced size and flattened shape of planktonic organisms helps to maintain their position in the water column by reducing their sinking. When phytoplankton sink below the

Marine Biology: Comparative Ecology of Planet Ocean, First Edition. Roberto Danovaro and Paul Snelgrove.
© 2024 John Wiley & Sons Ltd. Published 2024 by John Wiley & Sons Ltd.
Companion Website: www.wiley.com/go/danovaro/marinebiology

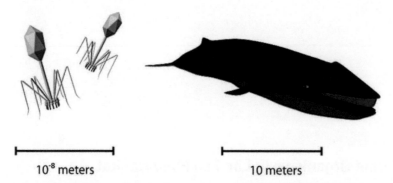

Figure 5.1 Comparison between the smallest marine organisms, viruses (left) and whales (right), representing a size range between 1/10,000 mm and tens of meters (factor 10^8). Differences in linear measurements typically increase with biomass cubed.

euphotic zone, they cannot photosynthesize, whereas zooplankton may sink to depths where food resources become scarce. Planktonic organisms have therefore adopted numerous mechanisms to reduce sinking, mainly exploiting friction or buoyancy to offset gravitational forces. In order to utilize friction, organisms have adapted by:

1) generally decreasing size, thereby increasing their surface/volume ratio which increases frictional resistance;
2) increasing body surface area or flattening of the body, sometimes by expanding various appendices (such as elongate appendages, thorns or feathery bristles in crustaceans).

In the first instance, smaller size increases the ratio of total surface area to body volume (known as the **surface/volume ratio**, or S/V). Spherical shapes illustrate this point. But body shapes characterized by long and/or extended appendages help reduce sinking. Imagine dropping a marble and a flat coin of equal weight in water; the marble sinks much more rapidly because the larger surface area of the coin resists movement through the fluid. This resistance depends on the amount of water that a body in motion must push against as it sinks, and thus the total "friction" between the water and the body surface. For small organisms, resistance depends mainly on their surface area: the larger the surface area, the greater the resistance from water and the slower the sinking rate. Compared to larger animals that propel themselves, the typically small size of planktonic organisms results in a larger surface area and greater resistance to sinking per unit of volume. Furthermore, small organisms can exchange critical substances (oxygen, metabolic waste) with their outside environment more easily, often through diffusion across their surface, whereas larger animals must develop additional mechanisms to increase surface area for such exchange, such as respiratory (e.g., gills) and excretory systems (e.g., kidneys).

In the second case, animals with similar mass or biovolume but differing in shape tend to sink at different rates. In fact, higher resistance and a lower sinking speed typically characterize a greater body surface. For this reason, planktonic organisms have evolved flattened body shapes and structures to maintain their buoyancy (think again of the coin and marble example). These structures reduce sinking speed and increase the surface area/volume ratio (Figure 5.2a,b). For example, the long antennae of planktonic copepods decrease their sinking rates when they stop swimming (Figure 5.2a). The same applies for the flattened body of the planktonic **phyllosoma** larval stages produced by lobsters and some other crustaceans (Figure 5.2b). In addition to their flattened bodies, the flattened lateral parapodia of the planktonic polychaete *Tomopteris* further reduce sinking rates (Figure 5.3).

Sea jellies ("jellyfish") further illustrate adaptations for floatation (Figure 5.4a-c). Rhythmic contraction of the body expels water from the bell and pushes the jellyfish upward. Moreover, the bell shape and tentacles create a "parachute effect" that resists sinking.

Organisms may also "lighten" their body to reduce density. By secreting gas into a cavity, animals can maintain neutral buoyancy, much as scuba divers blow air into their buoyancy compensator to avoid sinking. Because gas is less dense than water at equal volume, this addition of air increases buoyancy. Moreover, some species produce real floating structures that contain gas that they can increase or decrease. Perhaps the best known example of this strategy, the Portuguese man o' war (*Physalia physalis*; Figure 5.5a) uses an actual floating bag, much like many other species belonging to this group (such as the siphonophore *Forskalia edwardsi*; Figure 5.5b). This bag (a modified body called a "pneumatophore" 10 to 30 cm long) fills with air, adjusting in volume as needed. Although some forms of planktonic algae or **vesicles** (sacs or bladders of some macroalgae, such as *Fucus* sp.; Figure 5.5c) utilize this strategy, such buoyancy strategies rarely occur in photosynthesizing organisms relative to its common use among animal species, such as in *Nautilus* sp. (an animal that uses gas-filled chambers in its shell to float (Figure 5.5d)).

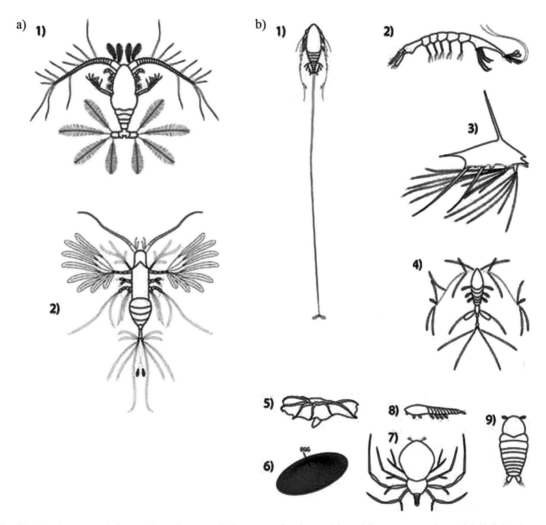

Figure 5.2 Planktonic copepods have adapted to remain in suspension by evolving elaborate appendages. (a) 1) *Calocalanus pavo*, 2) *Euaugaptilus filigerus*; (b) some examples of adaptations: (1) copepod (*Aegisthus* sp.), (2) decapod (*Lucifer* sp.), and (3) nauplius of barnacles, (4) copepod (*Oithona* sp.), (5) sea cucumber (*Pelagothuria* sp.), (6) pelagic copepod egg (*Tortanus* sp.); (7) larval lobster phyllosoma, (8, 9) copepod (*Sapphirina* sp.), dorsal and lateral views.

Figure 5.3 The polychaete *Tomopteris* sp., one of the few planktonic polychaete species, uses it parapodia for swimming. *Source:* uwe kils/Wikimedia Commons.

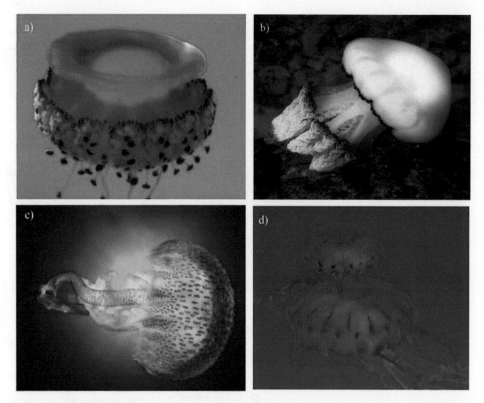

Figure 5.4 Examples of adaptations used by marine organisms to maintain themselves suspended in water. The similar density of sea jellies to seawater, combined with a morphology that slows their descent, helps them maintain their position in the water column without using much energy: (a) the Mediterranean jelly, *Cotylorhiza tuberculata*; (b) the barrel jelly, *Rhizostoma pulmo*; (c) the mauve stinger, *Pelagia noctiluca*; (d) the sea nettle, *Chrysaora* sp. *Source:* Ales Kladnik / CC BY 2.0 / Public Domain.

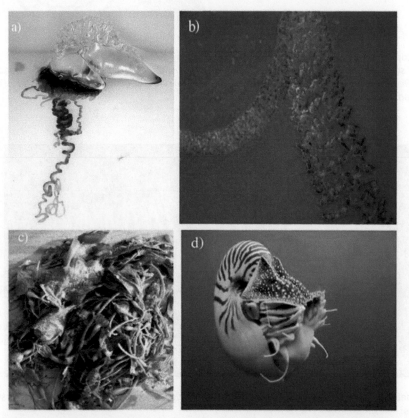

Figure 5.5 Many organisms have evolved floatation strategies that incorporate gas and/or produce substances that reduce their density: (a) *Physalia physalis*; (b) *Forskalia edwardsi*; (c) the rockweed, *Fucus* sp.; (d) *Nautilus* sp. *Source:* NOAA., Profberger / Wikipedia Commons / CC BY-SA 3.0.

 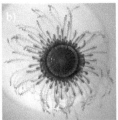  

Figure 5.6 Examples of flotation mechanisms: (a) *Velella velella*; (b) *Porpita porpita*; (c) *Janthina janthina*; (d) the blue glaucus, *Glaucus atlanticus*. *Sources:* Bruce Moravchik / NOAA / Wikimedia Commons / Public Domain, Rcz242 / Wikimedia Commons / Public Domain, Taro Taylor / Flickr / CC BY 2.0.

 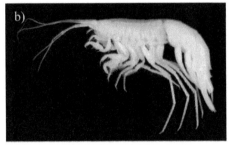 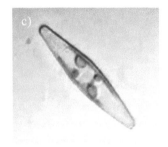

Figure 5.7 Examples of organisms that utilize high lipid content to reduce their density and facilitate neutral buoyancy in seawater: (a) A specimen of mature shrimp in the Mediterrean; (b) the amphipod, *Princaxelia* sp.; (c) a pennate diatom with clearly visible drops of intracellular oil. The relative average density of these oils (~0.91 g·cm^{-3}) may be considerably lighter in some species. Generally, deposits of lipid drops within phytoplankton cytoplasm expand and contract with temperature, altering the position of the organism within the water column over a daily cycle. Many species of zooplankton also fill vacuoles or cavities with oil. *Source:* Simonpietro Canesie; Roberto Danovaro (Book Author).

A special structure in the small floating jellyfish *Velella velella* (the "Boat of St. Peter", a colonial hydroid, Figure 5.6a) floats above the water's surface and acts as a sail directed by the wind, whereas the basal portion provides some stabilization. The siphonophore *Porpita porpita* lacks a sail, and instead stabilizes itself by hanging a crown of tentacles from its bell (Figure 5.6b). The mollusk *Janthina* sp., the violet seasnail (Figure 5.6c), secretes a series of bubbles on the surface of the water that enables the animal to float upside down below the ocean surface.

Once *Janthina* sp. detaches from its "raft," it sinks, apparently unable to stay on the surface. Another mollusk, the pelagic nudibranch *Glaucus atlanticus* (Figure 5.6d), floats hanging from an air bubble. Some diatoms use vacuoles and cavities filled with air to maintain their position in the water column, whereas others deposit oil droplets (Figure 5.7c) or carbon dioxide bubbles within their cytoplasm.

Many other organisms use lipid inclusions to reduce their density because the low density of lipids relative to water increases buoyancy; the high energy content of lipids also offers an energy reserve. Many planktonic crustacean species deposit oil droplets inside their carapace (Figure 5.7a, b), and diatoms accumulate intracellular drops of oil (Figure 5.7c).

The body fluids of marine fishes are generally more dilute and less dense than seawater. These dilute fluids are particularly advantageous to species that accumulate fat to achieve neutral buoyancy. Body densities in sharks, which are generally large pelagic predators, approach neutral buoyancy thanks to large livers (sometimes a third of the animal's biomass) rich in lipids (particularly squalene, a low density oil: 0.86 g·cm^{-3} at 20 °C), and to a mass of gelatinous tissue located in their nose. This buoyancy mechanism allows sharks to move easily and quickly, despite small pectoral fins. For example, the small fins of sharks such as *Cetorhinus maximus* (basking shark; Figure 5.8a) or *Rhincodon typus* (whale sharks; Figure 5.8b) provide modest lift, but an overall density close to that of seawater results in neutral buoyancy even when swimming slowly. Other elasmobranchs, such as giant manta ray, *Manta bisostris* (Figure 5.8c), use large fins that provide thrust and lift to compensate for their relatively higher density.

In deep-sea teleost fishes of the family Gonostomatidae, adipose (fatty) tissue in the bristlemouth, *Cyclothone* spp., represents ~15% of body volume. Spermaceti, a waxy substance rich in fat in the head of sperm whales, *Physester macrocephalus* acts as a stabilizer by adding substantial buoyancy. In fact, the spermaceti crystallizes when the sperm whale plunges into deep water, thereby reducing its buoyancy, which would otherwise interfere with diving.

Cellular level adaptations can also reduce density. During late stages of development, some fish eggs become buoyant by the mother secreting liquid less salty than seawater from follicular cells or by depositing high concentration of lipids. Similarly, the lower relative density of the external vacuolated protoplasm in radiolarians compared to seawater aids flotation; by varying numbers of vacuoles, cells can ascend or descend. Contraction and expansion of protoplasm in pseudopods

Figure 5.8 Adaptations to facilitate swimming in sharks: (a) high amounts of squalene in shark livers decreases their density, as in *Cetorhinus maximus*; (b) sharks with smaller livers control their buoyancy by swallowing air, such as in whale sharks, *Rhincodon typus*; or alternatively, (c) they develop major surfaces for thrust, such as in *Manta bisostris*. Sources: Chris Gotschalk / Wikimedia Commons / Public Domain, Zac Wolf / Wikimedia Commons / CC BY-SA 2.5, MASSIMO BOYER.

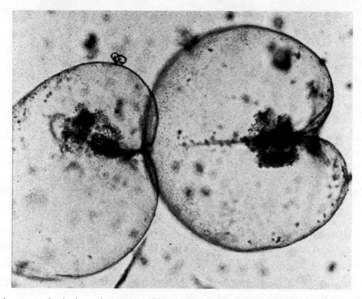

Figure 5.9 Adaptations in algae may include replacement of heavy ions with lighter ions, as in the dinoflagellate, *Noctiluca miliaris*.

alters total density, allowing radiolarians to adjust their depth within the water column. Thinner frustules in pelagic diatoms compared to benthic diatoms reduce sinking. Some organisms vary ion composition in body fluids, replacing heavier ions with lighter ions, thus maintaining osmotic balance but at a lower density than seawater. The dinoflagellate *Noctiluca* (*Noctiluca scintillans*; Figure 5.9) reduces its density by accumulating low specific weight ions such as ammonium chloride. Similarly, many species of squid accumulate ammonium chloride in specific fat and bulbous organs. The gelatinous tissues of sea jellies, including siphonophores, ctenophores, salps, doliolids, heteropods, and pteropods, support their denser tissues, and movement requires little positive thrust. Some gelatinous organisms, such as salps and ctenophores, and some planktonic mollusks such as heteropods, actively expel heavy ions, such as sulfate SO_4^{-2} and magnesium Mg^{+2}, replacing them with **homeo-osmotic ions**, or ions similar to those in the surrounding environment, such as chloride Cl^- and ammonium NH^{4+}. *Noctiluca* uses high concentration of ammonium ions and excludes relatively heavy **divalent** (ions with a valence of two) ions, such as sulfate. Unusually low levels of sulfate and an ionic composition composed almost entirely of chlorides characterize tissues of the "moon jelly," *Aurelia aurita*, and cuttlefish, *Sepia officinalis*. Several taxonomically diverse animals, such as ctenophores, cnidarians, tunicates, and mollusks, also extrude sulfates.

Organisms in cold and salty water face greater resistance to movement than those in warm, less salty water because of the greater density, and thus viscosity, of seawater. In calm waters, some small organisms have developed "ornamentation" that reduces sinking; cold-water organisms generally lack such ornamentation because comparatively denser water reduces sinking and limits the need for such structures. Hence, viscosity influences the shapes and forms of marine biota. In fact, the streamlined shapes and loss of unnecessary appendages in most fishes and other **nekton**, referring to actively swimming organisms, reduces resistance to movement through a viscous and dense seawater medium. As an adaptation to viscosity, many marine organisms that move quickly or across long distances have evolved strategies such as the smooth skin of dolphins or slime layer on salmon.

Some fishes possess a **swim bladder**, an elastic structure filled with air or other gases, which essentially works as a hydrostatic organ (akin to a buoyancy compensator used by scuba divers, as noted earlier). This structure occurs dorsally in the visceral cavity, below the spinal column. By varying swim bladder gas content, fishes can change their specific weight and attain neutral buoyancy at a given depth, maintaining their position vertically with little or no active movement and associated energy expenditure. Because they no longer need to use their lateral fins for constant swimming, fishes with swim bladders can instead modify their lateral fins for sensory, defensive, or signaling functions during courtship. Fishes lacking swim bladders must generally live on the seabed, largely moving in short bursts in an otherwise sedentary lifestyle and often evolving pelvic fins as "feet" (Figure 5.10). Some bladderless redfish, for example, push themselves off the bottom with their pectoral fins. Sharks also lack a swim bladder but, as noted above, solve the buoyancy problem with their lipid-rich liver. Enlarged gas bladders characterize many marine teleost (bony) fishes living at great depth, such as myctophids (lantern fish), sternoptychids (hatchet fish), and gonostomatids (bristlemouths). Other teleost species living below 1,000 m depth, such as some macrourids (grenadiers or rattails) (Figure 5.11), may lack a swim bladder as adults, although it may occur in larval stages.

However, highly developed swim bladders in some fishes that inhabit the hadal environment below 7,000 m allow them to swim close to the bottom at neutral buoyancy. In contrast, some cephalopods obtain neutral buoyancy by adjusting internal gases. Rigid walls limit gas compression so that gas volume does not change with variation in depth. For example, the cephalopod *Sepia* spp. (cuttlefish) partially fills its cuttlebone chambers with gas and other chambers with liquid. The overall density of cuttlefish bone varies between 0.5 and 0.7, but the mass of gas inside apparently remains constant, and they control density by adjusting the amount of liquid. By maintaining concentrations of liquids in this bone below that of blood, cuttlefish generate sufficient osmotic pressure to balance that of the surrounding water. As with *Nautilus* spp., a related group of cephalopods that use their shell in a similar way, cuttlefish maintain their orientation in the water by varying the distribution of gas inside the cuttlebone.

Figure 5.10 A frog fish on the seafloor using its pectoral fins as "legs".

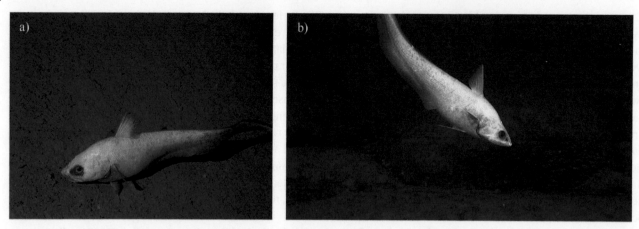

Figure 5.11 Macrourids (grenadiers) photographed below 1,000 meters depth. *Source:* NOAA / Wikimedia Commons / Public Domain.

5.1.2 Swimming and Dispersal

The active movement of the largest planktonic organisms maintains their body in suspension, which would otherwise tend to sink. A wide variety of marine organisms (small and large) swims more or less actively to regulate their "position" in the water column. Body shape and locomotion strategy closely link and depend on the lifestyle of a given species. Slender shape typically characterizes swimmers, with elongated shape in the fastest pelagic swimmers such as teleost fishes and marine mammals. These organisms require significant power to overcome resistance by water, which increases with speed (again, try walking across the bottom of a swimming pool!).

Many crustaceans, fish, tetrapods, turtles, and birds use locomotory apparatus (wings and appendices) for propulsion (Figure 5.12).

The energy expense of swimming demands simplification of body shape. The typical elongate shape of pelagic species increases their hydrodynamic efficiency. Some species ambush prey through rapid acceleration (e.g., barracuda), and thus evolved a dart-shaped body form with posterior placement of unpaired fins to increase precision when shifting direction. Researchers classify swimming movements into two broad categories according to time: 1) **sustained swimming** and 2) **temporary (transient) movement** (sometimes referred to as **burst swimming** or **cruising swimming**). Cyclical repetition of propulsive movements characterizes sustained swimming, typically used for relatively long journeys at constant speed. Large pelagic fish and migratory marine species adopt this swimming mode. In contrast, temporary movements include short darting movement to escape predators, as well as other more or less random small movements. Many planktonic organisms utilize this strategy, particularly, small zooplankton crustaceans such as copepods, as well as larger crustaceans such as euphausiids, mysids and sergestid shrimp, which swim almost incessantly but in jerky movements. These organisms tend to sink rapidly when they stop actively swimming.

Fishes are classified into 3 groups based on vertices that represent morphological adaptations (Figure 5.13), whose vertices denote morpho-functional specializations: (1) specialists in **fast acceleration**, (2) specialists in **maneuvering**, and (3) specialists in **cruising**, which refers to continuous long-term movement.

Figure 5.12 (a) Large pectoral fins of the flying gurnard, *Dactylopterus volitans*, (b) a seaturtle, and (c) the wings of a cormorant. *Sources:* Beckmannjan / Wikipedia Commons / CC BY-SA 3.0, Gabriella Luongo, Shahin Abasov / Flickr / CC BY-ND 2.0.

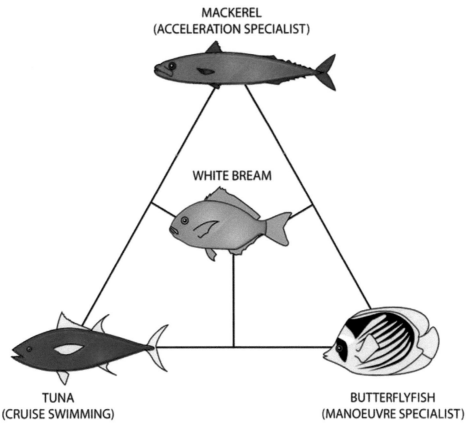

Figure 5.13 Relationship between fish body shape and their movements. Elongated fishes can attain high swim speeds, whereas laterally flattened species specialize in maneuvering and in stationary buoyancy.

Size also plays a fundamental role: rather than actively swimming, the smallest organisms drift passively with ocean currents. In fact, some small planktonic organisms exploit movements of the water column, such as vertical convection in **Langmuir cells**, the orbital surface circulation cells created by wind. Viscosity, the resistance of the fluid to flow, plays a key role in the movement of very small animals (including the larval stages of many organisms) that utilize ciliary structures or small appendages in movement, in contrast to large animals that depend on muscular movement.

Small animals generally swim slower compared to larger organisms. Some large tuna swim at burst speeds in excess of 20 m·s^{-1} (72 km·h^{-1}) in contrast to their cruising speed of 1–2 m·s^{-1} (between 3.6 and 7.2 km·h^{-1}). Small copepods, by comparison, generally move at speeds of a few cm s^{-1}. An empirical formula relates an organism's body length to its maximum speed of movement:

$$V_{max} = K \times L^2$$

where V_{max} describes the maximum speed, K denotes a constant that depends on the species, and L denotes body length. This formula suggests that a 2-cm long fish can swim four times faster than a 1 cm long fish.

Viscous forces dominate the movement of small organisms, whereas frictional forces dominate movement of large organisms. **Reynolds Number**:

$$Re = l \times v \times \rho/\mu$$

describes the ratio between inertial forces and frictional forces, taking into account the length of the animals (l), their speed (v), seawater density (ρ), and the dynamic viscosity of water (μ). We humans operate at large Reynolds numbers so when we place a baseball mitt out to catch a baseball, the ball (hopefully) lands in our mitt; when copepods reach out an appendage to capture a passing particle, the particle follows the flow streamlines and goes around the appendage rather than flowing into the appendage like the baseball sails into our mitt.

Figure 5.14 Trigger fish and box fish use both their dorsal and anal fin to maintain great control of their swimming. *Source:* Norbert Potensky / Wikimedia Commons / CC BY-SA 3.0, Jan Derk/Wikimedia commons.

Researchers also classify swimming as undulatory or oscillatory. Long-based fins (e.g. dorsal fins) typically generate undulating swimming propulsive waves, but forms may vary. Gurnards, for example, use undulatory motion of their large pectoral fins to move across the bottom, whereas wrasses move in small steps by oscillating their pectoral fins. Balistids (trigger fish) use both dorsal and anal fins to swim. In all these fishes, the movements of fins produce propulsion, and adapt the fishes to slow swimming and maneuverability in structurally complex environments, such as rocky bottoms, coral reefs, or seagrass beds (Figure 5.14). Many such species can move quickly and accelerate through muscular wave movements of their body.

Muscles strongly influence swimming. Low percentage of myoglobin, low mitochondrial volume, and poor vascularization characterize the white muscle that dominates most fishes. These fibers work mainly under anaerobic conditions and work best for short, intense swimming efforts. Some teleosts also use red muscle fibers; their high concentration of myoglobin increases oxygen transport from the blood to the muscles, aided by large numbers of mitochondria, and support the prolonged aerobic activity of good swimmers including large pelagic and highly migratory species such as tuna. Swimming adaptation therefore also dictates the relative proportion of the two types of fibers, with high percentages of red fibers in tuna (with typical "red meat") and other large migratory pelagic species. The highly vascularized network of red muscles in tuna (the ***rete mirabilis***) occurs as masses nested within white fibers in order to conserve the heat produced by red muscle movement and thus increase efficiency. Tuna muscle temperature may therefore exceed that of the surrounding water by 6–8 °C.

Low percentages of red muscle characterize most **demersal** (near-bottom living) fishes, which typically move in small, transitional steps. Some fishes (e.g. butterfly fish) largely lack red muscle. These fishes can quickly dart short distances but cannot maintain prolonged swimming. The flying fish group includes more than fifty species of the family Exocoetidae, typically in tropical and sub-tropical areas. The most common species, *Cypselurus heterurus* (Figure 5.15), lives in warm North and South Atlantic waters. As with some other marine fishes and mammalian species, these fish "fly" by leaping out and across the ocean surface, often for unknown reasons. Some fishes leap out of the water (often in shoals) to confuse predators. Other species, such as white sharks and baleen whales, may leap to ambush prey or to increase effective swim speed. In some marine mammals, the behavior may simply indicate a form of play.

Every year, gray whales (*Eschrichtius robustus*) undertake a migration spanning nearly 21,000 kilometers from feeding areas in the Arctic to breeding regions in the subtropics; this annual total migration of 40,000 km annually represents the longest known mammal migration.

Most scientists define dispersal as the movement of an individual from one location to another, often from areas of spawning or hatching to juvenile or adult habitat. This phenomenon occurs widely among almost all marine and terrestrial organisms. Plants and algae disperse largely passively; currents, wind, or other organisms transport their **seeds** and **propagules** (the unit of dispersal, whether a spore, egg, larva, or other life history stage). For **sessile** or **sedentary** (non-mobile, often attached) adults of plants or animals, dispersal occurs almost exclusively during the planktonic larval stage. Currents can disperse eggs and larvae during the hours/days/weeks they remain suspended in the water, potentially over distances of 100s and even 1000s of kilometers. Once (largely) passive dispersal transports larvae to a location with suitable environmental conditions for development, they settle and grow, maturing over time.

Figure 5.15 Examples of organisms that emerge from the water. (a) *Cypselurus heterurus* (flying fish) can sail above water for 45 seconds, sometimes landing on the decks of boats; (b) mobula rays (genus *Mobula*) can jump more than two meters out of the water, as shown in the Gulf of California; (c) a great white shark (*Carcharodon carcharias*) jumping out of the water to prey on a seal. *Sources:* NOAA / Wikipedia Commons / Public Domain, BBC.

Figure 5.16 Some examples of phoresis and rafting in marine environment: (a) shellfish on the surface of cetaceans; (b) hydroids growing on a hermit crab shell occupied by a hermit crab; (c) barnacles on plastic and glass bottles; (d) buoy colonized by tubicolous polychaetes. *Sources:* Scripps Institution of Oceanography, Dan Hershman / Flickr / CC BY 2.0.

Phoresis describes a dispersal mechanism in which transport of a (generally small) organism occurs by adhering to the body of a larger animal. Typical examples of phoresis include barnacle crustaceans that colonize the head and tail of cetaceans (Figure 5.13a), or the algae and hydroids and/or sea anemones that settle on the carapace of decapod crustaceans (Figure 5.13b). In some cases, these interactions yield **symbiotic relationships**, where both participants benefit from the association. For example, transport of the anemone, *Adamsia palliate*, on the front claws of the hermit crab, *Pagurus arrosor*, benefits the crab in that the stinging anemones help defend the crab against possible attackers. In turn, the sea anemone benefits from crab movement that exposes them to a greater range of food, potentially including prey debris from crab feeding. Generally, hermit crabs without "passengers" actively look for an anemone to place on their shell; even molting rarely interrupts this symbiosis in that the molting process provides for the "transfer" of the sea anemone onto its new shell. **Rafting**, describes a similar mechanism of passive transport whereby small organisms settle on or adhere to floating objects, both natural (floating tree trunks, porous stones) and artificial (plastic bottles, navigation buoy; Figure 5.16).

Questions

1) Describe buoyancy mechanisms used by planktonic organisms.
2) How do fish control their buoyancy?
3) What is the effect of shape on organism buoyancy?
4) Are swim bladders present in all deep-sea fishes?
5) How do fishes use their fins?
6) Which organisms can jump out of the water? Why do they do so?
7) What are the differences between sessile and sedentary organisms?
8) What is the effect of shape on swimming ability?
9) How do non-fish species use fins?
10) Which are the differences between phoresis and rafting in the marine environment?

Suggested Reading

Bailey, D.M., Wagner, H.J., Jamieson, A.J. et al. (2007) A taste of the deep-sea: The roles of gustatory and tactile searching behavior in the grenadier fish *Coryphaenoides armatus*. *Deep Sea Research Part I: Oceangraphic Research Papers*, 54, pp. 99–108.

Griebel, U. (2002) Color vision in marine mammals: A review, in Bright, M., Dworschak, P.C., and Stachowitsch, M. (eds.), *The Vienna School of Marine Biology: A Tribute to Jörg Ott*. Facultas Universitätsverlag, Wien, pp. 73–87.

Herring, P.J. (2007) Review: Sex with the lights on? A review of bioluminescent sexual dimorphism in the sea. *Journal of the Marine Biological Association of the United Kingdom*, 87, pp. 829–842.

Huang, W., Chen, W., Liu, Y., and Gao, X. (2006) The evolution of the cavitation bubble driven by different sound pressure. *Ultrasonics*, 44, pp. 407–410.

Li, F., Qiao, Z., Duan, Q., and Nevo, E. (2021) Adaptation of mammals to hypoxia. *Animal Models and Experimental Medicine*, 4, pp. 311–318.

Macdonald, A. (2021) *Life at High Pressure – In the Deep-Sea and Other Environments*. Springer.

Marshall, J., Carleton, K.L., and Cronin, T. (2015) Colour vision in marine organisms. *Current Opinions in Neurobiology*, 34, pp. 86–94.

Martinez, C.M., Friedman, S.T., Corn, K.A. et al. (2021) The deep sea is a hot spot of fish body shape evolution. *Ecology Letters*, 24, pp. 1788–1799.

Montgomery, J.C., Jeffs, A., Simpson, S.D. et al. (2006) Sound as an orientation cue for the pelagic larvae of reef fishes and decapod crustaceans. *Advances in Marine Biology*, 51, pp. 143–196.

Montgomery, J.C. and Radford, C.A. (2017) Marine bioacoustics. *Current Biology*, 27, pp. R502–R507.

Rees, J.F., de Wergifosse, B., Noiset, O. et al. (1998) The origins of marine bioluminescence: Turning oxygen defence mechanisms into deep-sea communication tools. *Journal of Experimental Biology*, 201, pp. 1211–1221.

Shimomura, O. (2006) *Bioluminescence: Chemical Principles and Methods*. World Scientific.

Simpson, S.D., Meekan, M., Montgomery, J. et al. (2005) Homeward sound. *Science*, 308, pp. 221.

Suryan, R.M. and Harvey, J.T. (1999) Variability in reactions of Pacific harbor seals, *Phoca vitulina richardsi*, to disturbance. *Fishery Bulletin*, 97, pp. 332–339.

Warrant, E. (2000) The eyes of deep-sea fishes and the changing nature of visual scenes with depth. *Philosophical Transactions of the Royal Society B: Biological Sciences*, 355, pp. 1155–1159.

Widder, E.A. (2010) Bioluminescence in the ocean: Origins of biological, chemical, and ecological diversity. *Science*, 328, pp. 704–708.

6

Adaptations in Marine Organisms III: Benthic Biota between a Rock and a Soft Place

6.1 Adaptations to Life on the Seafloor

In contrast to organisms that live in the water column, those that live on the seafloor do not require specific adaptations to cope with sinking and transport, and instead must adapt to live in or on substrata spanning from bedrock to fine muds or even attached to living organisms such as eelgrass. Organisms living in these environments require different types of adaptations in order to thrive and survive. These adaptations refer primarily to the need of protect their body from injuries or the need to maintain their position on the substrate (anchorage). Living on or in the seafloor also means exploiting different food sources or exposure to fluids and chemicals released from the sediments. Future chapters address adaptations and processes within specific seafloor environments but here we focus on general adaptations for living on the seabed.

6.2 Support and Protection Structures

Marine benthic organisms have adapted a wide variety of support and protection structures. For example, the internal skeleton of poriferans (sponges) consists of calcareous or siliceous spicules. One group of sponges, demosponges (Figure 6.1a), combines fibers of spongin, a proteinaceous substance, with siliceous spicules (Figure 6.1). Other marine organisms, such as mollusks, form a calcareous shell or exoskeleton, whereas other organisms build their skeleton within the **dermis** (skin or integument), covered by the **epidermis** (outer skin layer). Some animals develop plates in the integument (Holothuroidea), or they form articulated plates (Asteroideae), or weld them together to form a solid armor (Echinodermata). Others build an internal cartilaginous skeleton, calcified by depositing crystals of calcium phosphate on the cartilage surface as in **Selachii**, or sharks, or an internal bony skeleton (**Teleostei**, the bony fishes). Some cnidarians build a rigid exoskeleton covered by epidermis.

In octocorals, cells of the **mesoglea**, the jelly layer of tissue in jellyfish, secrete a calcareous or horny skeleton. The epidermis in stony corals secretes a calcareous exoskeleton. In nematodes, a **hydro-pneumatic skeleton**, a structure that contains fluid under pressure (not unlike a car tire), fills the whole body of the animal, providing the body with support and rigidity.

In Arthropods, which include the class Crustacea, a chitinous cuticle forms an exoskeleton, a rigid shell covering the epidermis that secretes it. Because of this exoskeleton, crustaceans must molt during their development; during the post-embryonic period, they slip from the cuticle and grow while renewing the cuticular coating. Prior to rejecting their existing chitinous cuticle, the animal changes its blood chemistry and then stops feeding just before the molt. Once the animal slips out of its old cuticle, it secretes a new, soft and delicate shell, which can expand somewhat as the animal grows and before it subsequently hardens. In contrast, many protists, such as calcareous foraminifera (Figure 6.1b) produce protective shells. In these animals, numerous pores often perforate the shell to allow material exchange (e.g., feeding). Even diatoms (Figure 6.1c) develop a protective external structure, the siliceous theca (a sort of protective case similar in appearance to a petri dish). Three layers make up the shell of bivalve mollusks: (1) **periostracum** (consisting of a substance similar to chitin), (2) **ostracum** (comprising an organic substance, conchiolin, and calcium carbonate), and (3) **hypostracum** (comprising alternating organic lamellae and calcareous layers). Specialized cells along the margin of the mantle between the

---

Marine Biology: Comparative Ecology of Planet Ocean, First Edition. Roberto Danovaro and Paul Snelgrove.
© 2024 John Wiley & Sons Ltd. Published 2024 by John Wiley & Sons Ltd.
Companion Website: www.wiley.com/go/danovaro/marinebiology

epithelium and periostracum secrete calcium carbonate. This characteristic allows direct measurement of molluscan growth rate by incubating individuals in a water bath with calcein fluorescent dye that the animal takes up as it forms a new shell, resulting in fluorescent growth rings in shell cross sections visible with fluorescent microscopy. Serpulid polychaetes (Figure 6.1f) live in calcareous tubes they secrete. These tubes vary in shape, depending on the species. The polychaetes obtain the necessary calcium carbonate from seawater absorbed in the anterior region of their digestive tract.

Some species secrete mucous to cement tubes from deposited foreign materials, such as grains of sand, mud, and shell fragments. Other polychaetes, such as sabellids (Figure 6.1e), also produce mucus as they add mineral particles, creating a non-calcareous and flexible tube. Many species of algae, especially the Rhodophyceae, create support systems impregnated with calcium carbonate that roughly resemble corals. For example, the coralline algae *Corallina mediterranea* (Figure 6.1g) has a consistency and appearance similar to corals that gives rise to its name.

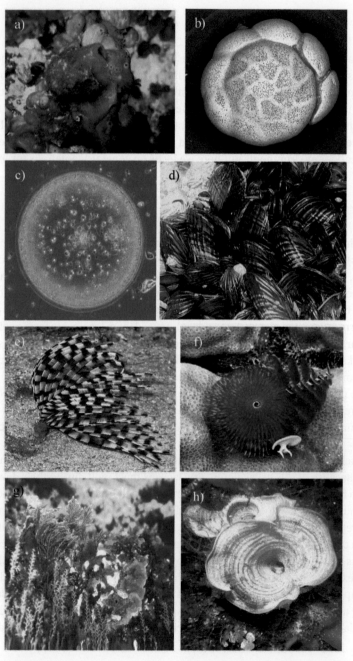

Figure 6.1 Some examples of support and protection structures: (a) demosponges; (b) foraminiferans; (c) diatoms – *Thalassiosira eccentrica*; (d) bivalves; (e) sabellid polychaetes; (f) serpulid polychaetes; (g) Mediterranean coralline alga (*Corallina mediterranea*); (h) the brown alga, (*Padina pavonica*).

6.2.1 Bioconstructors

Many plant and animal **bioconstructor** species in the ocean produce carbonate **bio-constructions**, macroscopic structures they maintain and often expand over time. Generally, these structures develop through progressive accumulation of calcification products produced through multiple generations. Bio-construction development typically follows two strategies: (1) **gregarious species**, where propagules of a given species settle in proximity to other organisms of the same species, as seen in serpulid polychaetes or worm snails; (2) **colonial species** such as corals, which reproduce clonally and form expanding "colonies." However, mechanical damage by waves, or biological breakdown by **bioeroders** such as boring snails, may counterbalance bioconstruction growth. Examples of bioconstructions include coral reefs, vermetidae gastropod reefs, coralligenous assemblages, **maerl** (rhodolith) beds, bryozoan beds, and mussel and oyster beds/reefs. These biological structures add new habitat and thereby increasing habitat spatial heterogeneity that increases biodiversity (**biological habitat provisioning**).

Given their ecological relevance, these habitat-forming species create important three-dimensional structures that host a large number of endemic species, represent nursery or spawning locations, and often create hot spots of biodiversity (Figure 6.2)

The carbonate skeletons of bryozoans can produce bio-constructions that may date back hundreds or even thousands of years; their modularity, plasticity, and longevity contribute to a wide range of bio-constructions that peak in extent at temperate latitudes, extending hundreds of meters from the shoreline (Figure 6.3).

Coralligenous habitats represent an excellent example of complex interactions between bio-constructors where both photosynthetic and non-photosynthetic organisms produce calcium carbonate structures that modify the seafloor landscape, creating different micro-habitats. These features typically develop on rocky substrates, generally beginning with the formation of calcareous algal clusters in both **infralittoral** (intertidal) and **circalittoral** (sublittoral to depths with sufficient light to support photosynthesis) zones, i.e., down to 80–100 m depth (Figure 6.3).

The word "coralligenous" (coralligène in French) was first used by Marion (1883) to describe the hard bottoms that fishermen from Marseilles called "broundo" and which occur at depths between 30 and 70 m, below seagrass meadows of *Posidonia oceanica* and above coastal muddy bottoms. Photosynthetic calcareous algae (e.g., *Lithophyllum bissoides*), which create calcium carbonate structures and largely build this habitat, and the presence of facies of red corals determined the origin of the name. Thus, light primarily limits coralligenous growth and development because of the primary role of macroalgae as macroconstructors. This habitat offers opportunity for the colonization of non-photosynthetic organisms (e.g., various coral species) that support a broad diversity of other species. We now know of hundreds of species that contribute to the biodiversity of these habitats, including bryozoans (62%), polychaetes (23%), cnidarians (4%), sponges (4%), mollusks (4%), and crustaceans (about 2%), supporting a known 315 species of algae, 61 protozoan taxa, and 1290 animal species (including 110 fishes).

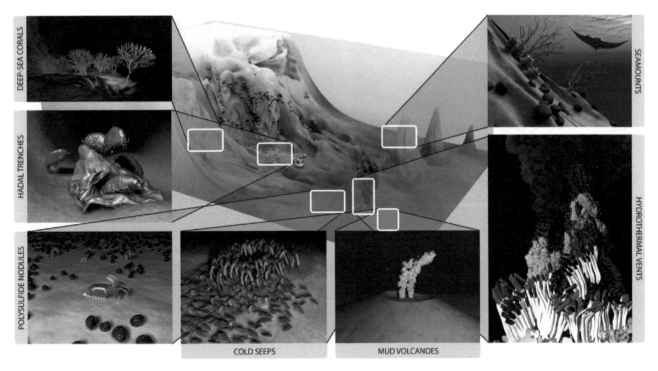

Figure 6.2 Examples of various marine habitats created by habitat-forming species, in most cases represented by bioconstructors.

Figure 6.3 Examples of bioconstructors in coastal and shallow water environments that create habitat for many other species: (a) coralligenous habitat, which is primarily formed by calcareous macro-algae and other carbonatic species; (b) tropical stony corals; (c) gorgonian and coral forests; (d) oyster reef. *Source:* Oyster Recovery Partnership.

Figure 6.4 A "trottoir" or Vermetidae reef with a detailed view of the high density of habitat-forming mollusks. *Source:* C.R.E.A. / Wikimedia Commons / CC BY-SA 3.0.

Gastropod mollusks such as *Dendropoma petraeum* build Vermetidae reefs in association with encrusting red algae such as *Neogoniolithon brassica-florida*. These bio-constructions alter the intertidal zone of rocky coasts of the southern Mediterranean, where average temperatures range from 14–15 °C in winter to >24 °C in summer. The largest platforms may extend tens to hundreds of meters in length within a few meters of depth, creating carbonate sidewalks (worm snail pavements) at the ocean surface (Figure 6.4). Even algae contribute significantly to bio-constructions with >100 habitat-forming species. In fact, the calcareous red alga *Neogoniolithon brassica-florida* produces bioconstructions by cementing together tubes of the gastropod *D. petraeum* noted above, and the red alga *Lithophyllum byssoides* fouls both ends of the platform, enhancing structural complexity. The encrusting brown alga *Padina pavonica* and encrusting green alga *Halimeda tuna* add further complexity.

6.3 Adaptation to Waves and Energy

High-energy environments, and especially intertidal habitats, experience wave action that may be particularly strong during storms. Organisms living in such environments may face abrasion from suspended sediments, logs, and even rocks moved around by the waves. The waves contain a massive amount of energy and displace huge rocks that can also damage large infrastructure during storms. These same waves also exert **drag**, or pull, on attached structures, biological or otherwise. Attached organisms must therefore adapt to survive in such conditions. Many species such as limpets develop low, flattened profiles that keep them close to the substratum, whereas others such as anemones develop simplified morphologies that reduce drag. Some organisms such as barnacles develop thick shells that protect them from debris, and they cement their shells onto rocks to avoid being swept away. Common mussels attach **byssus threads**, or a series of tethers to the rocks to keep them affixed. Mobile organisms such as crabs hide in cracks and crevices during periods of intense wave action. In contrast to the rigid structure of barnacles, kelps and seaweeds bend with the flow, thus dissipating energy. Some tropical corals develop morphotypes that never attach to the substratum, but simply settle on soft sediments (e.g., coral fungia) and eventually roll with the waves. Suspension feeders living on the seafloor typically feed most effectively within a particular range of current speeds and have evolved feeding structures appropriate to those conditions. Finally, animals in sediments may also face wave energy, particularly in sandflats or mudflats along the shoreline. In these cases, organisms typically burrow deeper in the sediment, or allow themselves to be thrown around by the waves; bivalves such as juvenile soft-shelled clams, *Mya arenaria* illustrate this strategy.

6.4 Feeding and Nutrition

In terrestrial environments, access to food sources mostly occurs close to the soil. In contrast, food in marine environments occurs in three dimensions, from the surface of the ocean through depths up to thousands of meters to seafloor sediments, though far from uniformly.

Early metazoans living at depth had only prokaryotes and protists available as food sources, and therefore had to develop techniques and structures to utilize microorganisms. The most primitive forms of Platyhelminthes (flatworms, the first important benthic predators to appear in the evolution of life) fed on bacteria and protists associated with sediments and rocks. This "ancestral" diet has remained dominant in all descendants of the Platyhelminthes, including terrestrial forms. Filtration (which evolved separately in sponges) followed as a specialization. Despite their abundance and wide distribution, bacteria and protists influenced the evolution of greater mobility of organisms to search for other types of food sources and eliminate the need to remain small in size. The larger an animal, the greater its metabolic needs; a large animal cannot survive on a diet of bacteria or protists because these sources alone cannot provide sufficient food. Increased size allows animals such as fish to produce more and more offspring, reducing their vulnerability to predators and increasing their competitive success compared to smaller organisms.

In marine environments much of the available food comes not as live organisms such as phytoplankton and zooplankton, but rather as **detritus**, or particles of debris (i.e., nonliving organic matter, in the form of feces or cells or tissues from bodies of marine organisms). A significant portion of organic matters occurs in dissolved form (for example, carbohydrates such as glucose dissolved in seawater) and many marine organisms can absorb dissolved organic matter directly through their body wall.

Detritus also occurs as particles, referred to as **particulate organic matter**, or POM, typically in concentrations ten times lower than in dissolved form. Many marine organisms, including larval stages, planktonic organisms, and corals absorb these particles through specific filtration systems, both active and passive, and use them as food. Organisms that capture suspended particles fall into two fundamental categories:

1) **microphages** include suspension feeders that feed on suspended particles, and scavengers that feed on microscopic organic particles consisting of debris, bacteria, protists, small planktonic organisms (Figure 6.3), capturing food with either pseudopodia or tentacles;
2) **macrophages** include masticators and shredders that feed on larger organisms, animals and, less frequently, on organisms Figure 6.4, Box 6.1).

Box 6.1 Carnivorous sponges
Carnivorous sponges feed on small crustaceans and other invertebrates, using small hooks to trap them on their body surfaces (Figure 6.5). Sponge cells then migrate to the prey, presumably utilizing extracellular digestion that combines sponge and bacterial action to break down the prey. **Archaeocytes** and **bacteriocytes**, specialized amoeba-like sponge cells, phagocytize and digest the prey fragments. This unusual example illustrates how a multicellular animal with no digestive cavity can digest relatively large prey, and likely typifies all carnivorous sponges (family Cladorhizidae).

(Continued)

Box 6.1 (Continued)

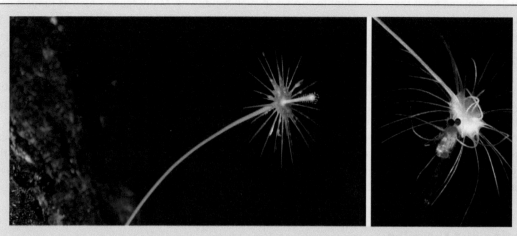

Figure 6.5 A carnivorous sponge shows that in food limited systems, such as caves or in the deep sea, filter feeders can adapt their feeding habit to become carnivorous. *Source:* NOAA / Wikimedia Commons / Public Domain.

Filter feeders (or suspension feeders) often use filtering structures to capture food particles from the water surrounding them, and they typically live where strong currents carry abundant suspended particles but not where high concentrations of suspended sediment could damage their delicate filtering structures. The filter apparatus can be external in some organisms, such as bryozoans, or internal, as in sponges. Many different types of planktonic larvae (mollusk veligers, polychaete trochophores, ribbon worm pilidium larva, echinoderm pluteus larvae) utilize this type of strategy, although their adult stages may feed very differently. Ciliates, sponges, bryozoans, mollusks, tunicates, crustaceans, cartilaginous fishes, bony fishes, and mammals all have filtering larval stages.

Suspension feeding may be passive or active. **Passive suspension feeders** depend on currents to bring particles into contact with feeding structures, whereas **active suspension feeders** create feeding currents that bring particles to them. Sessile animals can actively move water continuously across their filtering structures in order to avoid fully depleting the water of food particles, using cellular flagellae (such as in specialized cells in sponges called **choanocytes**) and modified gill lamellae (bivalve mollusks). Two filter systems work in tandem in some highly specialized suspension feeders: a mucous secretion aggregates particles, whereas other structures eliminate undesirable particles.

Deposit feeders such as sea cucumbers, nematodes, and some polychaetes feed on organic matter and microbes attached to sediment grains, expediting decomposition as they continually rework and move the sediment (a process calls **bioturbation**) while moving or feeding. They ingest significant volumes of sediment, often indiscriminately, separating food and inorganic matter in their digestive system. Through this activity, they oxygenate sediment and sometimes also stabilize the substrate by secreting mucus to construct cemented walls to line their burrows. In many cases these deposit feeders feed on living organisms associated with detritus, which may provide a vehicle for ingestion and transport of microorganisms into their intestine. An example of these feeding types is illustrated in Figure 6.6.

Scavengers (e.g., some polychaetes, mollusks, and crustaceans) feed on large particles or carcasses of dead organisms. **Limivorous** animals, or deposit feeders, ingest sediment particles and strip off associated food. In this sense they may also be considered scavengers, although microbes may be the more critical resource for many of these organisms. **Grazers** are **epistrate feeders**, meaning organisms that feed on primary producers, whether by scraping microbial films from hard surfaces (e.g., limpets) or even detaching small fragments from species such as kelp (e.g., sea urchins). These animals feed primarily upon sessile organisms (e.g., macroalgae). Specialized structures facilitate grazing, such as the hard radula of gastropods, or Aristotle's lantern in echinoids.

Predators refer to mobile animals that attack, kill, and consume individual and generally mobile prey. Predators can be grouped as: (1) **Stealth predators** that seek highly mobile prey; (2) **Searching predators** that look for less mobile prey; (3) **Ambush predators** that are less mobile and wait for prey to come close to them. Specialized structures often allow stealth and ambush predators to capture and immobilize prey. Ctenophores, major predators in the planktonic food-web, capture prey using sticky cells on their tentacles called **colloblasts**. Voracious predators such as sharks, predators *par excellence*, and cephalopods can capture prey much larger than themselves. The genus *Octopus* paralyzes decapods (crabs and lobsters) with poison before devouring them. Predation also occurs widely among fishes (both pelagic and benthic, Figure 6.7),

Figure 6.6 Examples of microphages: (a) a sponge in Indonesian coral reefs (sponges obtain nourishment from the water current that enters through the thin pores perforating the entire surface; (b) *Spirobranchus* sp. polychaetes that feed on debris and on small plankton; in this case threadlike gills covered with cilia and mucus glands in the cephalic area trap particles on which the polychaete feeds; (c) The branchial basket of *Polycarpa* sp. ascidian tunicates serves as a respiratory organ within a large cavity. Water enters into the basket, passes through the openings of the cavity, and exits through another opening, the cloacal siphon. A veil of mucus retains the particles while cilia on the walls of the cavity move the particles to the esophagus; (d) the sea cucumber, *Thelenota anax*, like many other benthic animals, ingests food and sediment particles as it move through the sediment.

Figure 6.7 Some examples of macrophages: (a) *Cassis cornuta*, the horned helmut, feeding on the banded urchin, *Echinothrix calamaris*: the **radula**, a toothed, rasping structure that means these gastropods may be voracious predators; (b) *Chromodoris gleniei*: these predatory nudibranchs, a type of shell-less gastropod mollusk, typically restrict their diet to a few prey species; (c) Moray eels, *Murena helena*, prey on fish, mollusks (mainly cephalopods) and shellfish; (d) the cephalopod *Octopus macropus* preys voraciously on animals much larger than they are. *Source:* Roberto Danovaro (Book author).

echinoderms (many from the class Asteroidea), polychaetes, cnidarians, and toothed whales (Odontoceti), with different jaw and mouth adaptations, depending on their target prey. For example, a long and narrow rostrum with numerous sharp teeth on both jaws characterize **ichthyophagous** species, referring to those taxa that feed on fish, in contrast to a shorter rostrum with fewer teeth in predators of bottom fishes (**bentho-ichthyophagous** species).

6.5 Adaptation to Aerial Exposure

Organisms living in intertidal environments experience periodic **emersion**, or exposure to air. These organisms, as well as the floating plankton or **pleuston** that live at the air-sea interface, have evolved special adaptations to resist drying during exposure to the atmosphere. The most common strategies include:

1) Some organisms trap water between their valves or inside the shell, thus resisting drying. Examples include bivalve mollusks (including common mussels that occupy frequently exposed surfaces) and gastropods such as periwinkles (e.g., *Littorina littorea*) and limpets (e.g., *Patella coerulea*). Some crustaceans, including barnacles (e.g., *Lepas lepas*), utilize a similar strategy to survive exposure to air;
2) Other organisms, such as Cnidarians (e.g., common anemone, *Actinia equina*), simply tolerate water loss for short periods, although with less resistance to drying and high temperatures than bivalves and gastropods. Some algae and some species of lichens (genus *Verrucaria*) utilize a similar strategy;
3) Mobile organisms can simply move into tidepools or damp cracks or crevices until the tides rise again (e.g., intertidal crabs such as rock crabs, *Cancer irroratus*).
4) A variety of organisms of varying mobility live primarily in tide pools (e.g., amphipods) or in damp cracks and crevices (e.g., some barnacles, the dog whelk *Nucella lapillus*), or on the damp undersides of seaweed (e.g., the polychaete *Spirorbis spirorbis*, or isopod crustaceans such as *Idotea*).
5) Though relatively uncommon among truly marine species, some species can breathe air. For example, the clingfish *Sicyases sanguineus* adheres to smooth rocky intertidal surfaces along the Pacific Ocean shore and gulps air into its opercular cavity as the tide ebbs. They then seal the air inside the operculum and exchange gases across the gill surface using their mouth and ventral fins as a sucker for attachment.

Despite the challenges associated with exposure to air, species that have evolved adaptations to cope with this environment gain numerous benefits that include reduced susceptibility to marine predators, reduced competition for substrate, abundant sunlight (for photosynthetic species), and high productivity and potential food resources.

6.6 Adaptation to Extreme Temperatures and Potentially Toxic Chemicals

Hydrothermal vents typically occur close to mid-ocean ridges and subduction zones at depths of around 3,000 m, where hot water enriched in chemical compounds and minerals percolates through cracks in the basalt. Water temperature in hydrothermal vents may vary between 5 and 250 °C, but some flows exceed 300 °C. Moreover, temperatures may vary widely and suddenly, even on daily or hourly time scales. Some highly specialized **hyper-thermophilic** (organism that thrive in extremely hot environments) prokaryotes can develop at temperatures greater than 113 °C, whereas eukaryotes normally cannot survive temperatures >50 °C, because their enzymes become unstable. Most enzymes and proteins lose their three-dimensional structure at temperatures >70 °C; temperatures >90 °C destabilize the double helix structure of DNA and a wide variety of metabolites rapidly hydrolyze.

However, special adaptions allow animals living close to hydrothermal vents to cope with high temperatures and encounter water temperatures of 100–150 °C (areas of hot plume diffusion) for short periods of time. Furthermore, these animals may experience sudden changes in temperature; currents may move vent plumes and expose animals to water masses of 2 °C. In some cases, different parts of the body simultaneously experience different temperatures. Furthermore, seafloor species may disperse through planktonic larval stages, so that even some hydrothermal vent species must tolerate low-temperature conditions (~2 °C) for extended periods of time.

The chemistry of hydrothermal vents adds further challenge. Hydrothermal fluids typically contain high concentrations of hydrogen sulfide, which is highly toxic to most organisms. Methane can be equally harmful. Hydrothermal vent organisms either avoid toxic concentrations or adapt; specialized **chemoautotrophic** microbes can utilize chemicals such as hydrogen sulfide or methane as an energy source to create organic carbon to form the base of vent food webs, and the main source of "dark" (in the absence of light) primary production.

Figure 6.8 *Alvinella pompejana*: a polychaete worm that lives at hydrothermal vents. *Source:* National Science Foundation / Public Domain.

Despite chemicals and temperatures toxic to most organisms, hydrothermal vents support an abundant and taxonomically unique fauna adapted to the conditions these habitats offer, with a biomass 500–1,000 times greater than "typical" deep-sea environments. In this sense, hydrothermal vents represent oases on an otherwise biomass poor seabed. Among hydrothermal vent animals, the small annelid, *Alvinella pompejana*, spectacularly illustrates adaptation to high temperatures (Figure 6.8). This polychaete lives close to the warmest flows (which can reach 450 °C) and appears to tolerate temperatures in excess of 100–150 °C. *A. pompejana* forms thin-layered tubes along the external walls of vent "chimneys," produced by the accumulation of precipitating metal sulfides. These worms spend most of the time within their tube, but exemplify eurythermal adaptation by tolerating wide variations in temperature; indeed this worm is the most **thermotolerant** (able to tolerate high temperatures) metazoan known to date. Although the mechanisms by which it has adapted physiologically to cope with these extreme temperatures remain unclear, *A. pompejana* has apparently evolved forced blood circulation that cools in the colder regions of the animal, and exchanges heat with the hottest regions. Within its chetae and expanded posterior parapodia, this worm supports a rich chemoautotrophic bacterial community that utilizes chemicals in the vent fluid, providing food to the polychaete.

Unusual enzymes with modified molecular structure characterize hyperthermophiles, but their biochemistry otherwise resembles **mesophiles** that live in moderate temperatures. Most hyper-thermophilic enzymes are extremely stable at high temperatures, with optimum catalytic activity at temperatures above 100 °C. The same 20 amino acids occur in hyperthermophilic and conventional enzymes, and proteins sequences are essentially identical, pointing to a need to study sequences in order to understand their stability.

Questions

1) How do marine organisms protect themselves?
2) Which are the main bioconstructors?
3) How have some animals adapted to high wave energy?
4) How do marine organisms cope with extremely high temperatures?
5) How do marine organisms modify benthic habitats?
6) How so some benthic organisms adapt to aerial exposure?
7) What are the mechanisms of feeding of benthic organisms?
8) Describe the different feeding types that occur in benthic organisms.

Suggested Reading

Cary, S.C., Shank, T., and Stein, J. (1998) Worms bask in extreme temperatures. *Nature*, 391, pp. 545–546.

Catarino, D., Knutsen, H., Veríssimo, A. et al. (2015) The Pillars of Hercules as a bathymetric barrier to gene-flow promoting isolation in a global deep-sea shark (*Centroscymnus coelolepis*). *Molecular Ecology*, 24, pp. 6061–6079.

De Busserolles, F., Fogg, L., Cortesi, F., and Marshall, J. (2020) The exceptional diversity of visual adaptations in deep-sea teleost fishes. *Seminars in Cell Development Biology*, 106, pp. 20–30.

De Busserolles, F. and Marshall, N.J. (2017) Seeing in the deep-sea: Visual adaptations in lanternfishes. *Philosophical Transactions of the Royal Society of London B Biological Sciences*, 372(1717), p. 20160070.

Falcucci, G., Amati, G., Fanelli, P. et al. (2021) Extreme flow simulations reveal skeletal adaptations of deep-sea sponges. *Nature*, 595, pp. 537–541.

Frank, T.M., Johnsen, S. and Cronin, T.W. (2012) Light and vision in the deep-sea benthos: II. Vision in deep-sea crustaceans. *Journal of Experimental Biology*, 215, pp. 3344–3353.

Giere, O. (1992) Benthic life in sulfidic zones of the sea – Ecological and structural adaptations to a toxic environment. *Verhandlungen der Deutschen Zoologischen Gesellschaft*, 85, pp. 77–93.

Jebbar, M., Franzetti, B., Girard, E., and Oger, P. (2015) Microbial diversity and adaptation to high hydrostatic pressure in deep-sea hydrothermal vents prokaryotes. *Extremophiles*, 19, pp. 721–740.

Johnsen, S. (2005) The red and the black: Bioluminescence and the color of animals in the deep sea. *Integrative and Comparative Biology*, 45, pp. 234–246.

Jönsson, K.I., Rabbow, E., Schill R.O. et al. (2008) Tardigrades survive exposure to space in low Earth orbit. *Current Biology*, 18, pp. R729–R731.

Maugeri, T.L., Gugliandolo, C., Caccamo, D. et al. (2002) A halophilic thermotolerant bacillus isolated from a marine hot spring able to produce a new exopolysaccharide. *Biotechnology Letters*, 24, pp. 515–519.

Minic, Z., Serre, V., and Hervé, G. (2006) Adaptation of organisms to extreme conditions of deep-sea hydrothermal vents. *Comptes Rendues Biologies*, 329, pp. 527–540.

Orsi, W.D. (2018) Ecology and evolution of seafloor and subseafloor microbial communities. *Nature Reviews in Microbiology*, 16, pp. 671–683.

Schiaparelli, S., Rowden, A.A., and Clark, M.R. (2016). Deep-sea fauna, in Clark, M.R., Consalvey, M. and Rowden, A.A. (eds.), *Biological Sampling in the Deep-Sea*. Wiley, pp. 16–35.

Seibel, B.A. and Drazen, J.C. (2007) The rate of metabolism in marine animals: Environmental constraints, ecological demands, and energetic opportunities. *Philosophical Transactions of the Royal Society B: Biological Sciences*, 362, pp. 2061–2078.

Siebeck, U.E., Wallis, G.M., and Litherland, L. (2008) Colour vision in coral reef fish. *Journal of Experimental Biology*, 211, pp. 354–360.

Sun, J., Zhang, Y., Xu, T. et al. (2017) Adaptation to deep-sea chemosynthetic environments as revealed by mussel genomes. *Nature Ecology and Evolution*, 1, p. 0121.

Timofeev, S.F. (2001) Bergmann's principle and deep-water gigantism in marine crustaceans. *Biological Bulletin*, 28, pp. 646–650.

Part IIB

Life in Seas and Oceans: Fundamentals of Marine Biology

7

Marine Biodiversity

7.1 Introduction

Life on Earth almost certainly began in the ocean. The presence of many exclusively marine taxa, including several Phyla (Ctenophora, Loricifera, Phoronida, Brachiopoda, Siboglinidae, Chaetognatha, and Echinodermata), supports the assumption of a marine origin for much of Earth's biodiversity. All animal kingdom classes, with the exception of myriapods and amphibians, occur in the ocean, and even for these taxa, their phyla include marine species. Only the Onycophora (velvet worms) lack living marine representatives, despite marine fossil forms.

Biodiversity changes not only with evolution over geological time (Figure 7.1), but also with specific historical events, including the appearance of humans. In 2015, Douglas McCauley demonstrated that humans have caused 15 documented extinctions in the ocean (likely an underestimate) and numerous regional extirpations and major population declines, particularly in large marine vertebrates, such as bowhead whales, *Balaena mysticetus*, grey whales, *Eschrichtius robustus*, right whales, *Eubalena* spp., as well as multiple species of sea turtles, sirenians, and seabirds. This chapter describes the origin of life, biological diversification in the ocean, the structural (alpha, beta, gamma, delta, and epsilon) and functional measures of biodiversity, as well as methods for measuring biodiversity (species richness, expected number of species, equitability indices, dominance).

7.2 Origin and Evolution of Marine Life

7.2.1 Theories on the Origin of Life

The creation of life requires two basic elements: the presence of liquid water and the presence of organic polymers such as nucleic acids and proteins, the molecules responsible for the central biological functions of replication and catalysis. Without these two conditions, life as we know it would be impossible. Approximately 4.56 billion years ago, molten rocks released gases such as hydrogen, ammonia, methane, carbon dioxide, and water vapor. Once these gases, reached the surface of the Earth, they provided the raw materials for the formation of the atmosphere and the primitive ocean. Researchers believe most ocean water originated from degassing processes. The melting of Earth's core also segregated different layers of the planet as a function of differing densities: heavy elements, such as iron, moved toward the center of the Earth forming a molten core, whereas lighter silicates moved to the surface. This stratification of the planet, in turn, gave rise to **tectonics**, the geological process that forms and destroys continents and seabed.

The estimated age of the moon of about 4.5 billion years makes it 30–50 million years younger than the Earth. Following its creation (probably as a result of a collision between an asteroid and Earth), the moon was much closer to the Earth, and its proximity led to huge tides (about 300 times more intense than those of today). Tidal friction gradually dissipated the rotational energy of the Earth, slowing its movement. These tides may have played a key role in some aspects of the formation of life, especially by periodically flooding lagoons, islands, and other coastal regions.

The early atmosphere likely contained predominantly methane, ammonia, and hydrogen, subsequently augmented with carbon dioxide, nitrogen, and water vapor. During this first period, the Earth was frequently bombarded by comets and asteroids, and the sun was less luminous than today.

Marine Biology: Comparative Ecology of Planet Ocean, First Edition. Roberto Danovaro and Paul Snelgrove.
© 2024 John Wiley & Sons Ltd. Published 2024 by John Wiley & Sons Ltd.
Companion Website: www.wiley.com/go/danovaro/marinebiology

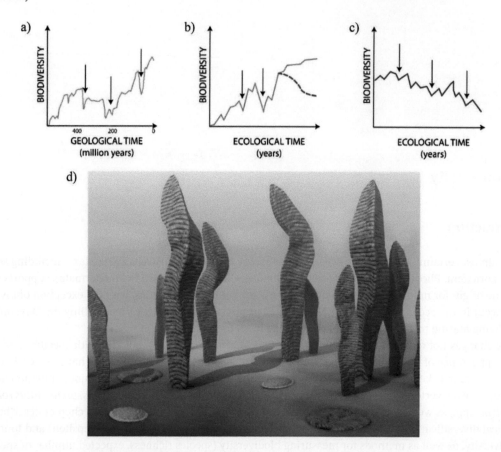

Figure 7.1 General trends of marine biodiversity over geological and evolutionary time. The arrows indicate drastic decreases in biodiversity: (a) general increases over geological time, with periods of decline corresponding to mass extinctions (b); trends in biodiversity associated with chronic human disturbance; continuous line shows general trend in marine biodiversity (species richness, equitability, functional diversity) over a geological time scale in the absence of human disturbance. The dotted line represents the decline of biodiversity during the last community successional stage as dominant competitors take over; (c) zoom in of b; (d) depiction of a possible Precambrian underwater landscape.

Recall from Chapter 2 the unique properties of water essential for life. Ocean formation played a crucial role in laying the foundations for the origin of life. The Earth, after its formations may have generated the ocean in a few hundred million years. Today, oceans contain about 10^{21} liters of water, but the original volume was probably significantly smaller. Some 4 billion years ago, ocean water was likely as salty as today's (perhaps even more so), but also highly toxic, with high concentrations of lethal compounds such as hydrogen cyanide and formaldehyde. The **"primordial soup"** (or **primordial sea**) refers to the primitive environment from which life originated, and the evolutionary process that led to various forms of life. The heterotrophic hypothesis proposes that the first living organisms appeared within the primitive ocean as masses of complex molecules that fed on organic compounds (Figure 7.2).

Polymers (molecules formed from many similar subunits bonded together), such as nucleic acids and proteins, formed the basis for life on Earth, as we know it today, allowing the essential biological functions of replication and catalysis. These molecules, which have biological functions, were synthesized by assembling simple **monomers** (molecules comprised of a single subunit). In the subsequent period after the formation of Earth, high temperatures from massive volcanic eruptions precluded the possibility of stable organic compounds. As the Earth's surface cooled and the oceans formed, organic compounds may have begun to accumulate on the seabed. Environmental conditions on the seafloor 4.2 billion years ago included numerous plumes of hot (~150 °C) and acidic (pH of ~4.5) seawater spewing from the seabed.

The theory that a "primordial soup" (Figure 7.2) of prebiotics formed from non-living material was proposed independently by Aleksandr Oparin and John Haldane in 1920. Later, in 1953, Harold Urey and his student Stanley Miller successfully tested the Oparin-Haldane theory and published a study about the natural synthesis of amino acids under conditions that simulated those of early Earth. Water, methane, ammonia, and molecular hydrogen were used as reagents and, when

Figure 7.2 A schematic view of the hypothesis of the primordial soup and the combination of monomers to form polymers and the first organic molecules present in the primordial ocean (see text below for details).

subjected to electric pulses, these molecules reacted to form hydrogen cyanide (HCN), aldehydes, and ketones. Had this process taken place on early Earth, these reagents could have produced amino acids and other compounds; however, the associated accumulation of HCN in the oceans would have quickly become highly toxic to any organism. In 1961, Juan Oro demonstrated that mildly heating a highly concentrated HCN solution produced adenine, a basic component of DNA and RNA, and an important part of **adenosine triphosphate** (ATP), an organic compound that provides energy to drive and support many processes in living cells.

By 1861, Alexander Butlerov had already demonstrated the synthesis of different sugars from formaldehyde, carbonate, and aluminum. Moreover, in the presence of various mineral catalysts, production of sugar phosphates may occur that may have been incorporated into the first nucleic acids. These reactions primarily produced an oily residue that, on early Earth, over a few million years, would have formed an oil layer 1–10 m thick that could protect molecules from the sun's ultraviolet light. Amino acids and other biochemical monomers, synthesized by reactions very similar to those in Urey and Miller's experiment, occur in meteorites.

Molecules in the prebiotic soup gradually increased in complexity, some developing the ability to catalyze reactions. At some point in time, a molecule formed capable of producing the "blueprint" for duplicating complementary molecules and the key property that characterizes and defines life: the capacity for replication. This event transitioned prebiotic chemistry to biochemistry and marked the beginning of life on Earth.

The key features that distinguish living and nonliving forms include the ability to:

- Duplicate themselves, though often inexactly,
- Create "active" molecule that help with replication,
- Persist long enough to ensure replication and thus the survival of "offspring" molecules.

One of the most widely accepted scenarios for the transition from an abiotic world to one characterized by biological chemistry is that simple monomeric compounds, present in the "prebiotic soup" gave rise in some way to increasingly longer and complex polymer chains, perhaps as a result of the presence of clays and minerals. Over time, some of these polymers acquired structures and specific properties that prolonged their survival and, eventually, the fundamental ability to replicate. This transition marked the turning point from molecular entities capable of multiplication to the origin of life and evolution as we know it today.

Experimental tests demonstrate the capacity of clays, metal cations, and other organic components to have potentially catalyzed many prebiotic reactions, including polymerization. The selective adsorption of molecules on various mineral surfaces provides a potential mechanism to promote polymerization. Given that adsorption involves weak non-covalent bonds, aquatic habitats with moderate temperatures and high rates of evaporation would favor this process, particularly in tidal and lagoon environments. Recent evidence demonstrates the particular effectiveness of this adsorption in the enzymatic synthesis of nucleotide chains, ribonucleic acid (RNA, an important biological macromolecule that converts the genetic information of DNA into proteins in most living organism), and nucleic acid peptide analogs. RNA molecules that promoted self-replication would, in turn, have evolved into the first proteins and eventually DNA (deoxyribonucleic acid, the helical molecule that carries genetic information for the development and functioning of most organisms). From 7.5 billion years ago, life presumably evolved in unicellular organisms resembling modern cyanobacteria.

7.2.2 Evolution of Marine Biodiversity

Life diversified quickly at the beginning of the Cambrian period (540 million years ago), with a variety of marine taxa that included some complex forms, to produce the more than ~250,000 currently known marine species. The fundamental events of life evolution and development are illustrated in Box 7.1 (Figure 7.3). Paleontological data show increasing numbers of taxa during the Phanerozoic, with notable drastic declines in diversity during five major extinction events (Table 7.1) when biodiversity collapsed. Several such declines (at the end of the Ordovician, Permian, and Cretaceous Periods, for example), were associated with "mass extinctions," whereas decreases in diversity in the late Devonian and at the end of the Triassic resulted from low rates of speciation in tandem with high extinction rates. However, recent studies question the paradigm of steady increase in diversity over time, interrupted only by mass extinction events, and some experts believe that taxonomic diversity remained fairly stable during some geological periods. Even ecosystems themselves change over geological time scales as result of processes that significantly alter physical properties and biogeochemical characteristics of environments. Acknowledging limited information on ecosystem and taxonomic diversity over geologic time, the number of different types of marine ecosystems has clearly increased compared to the pre-Cambrian period.

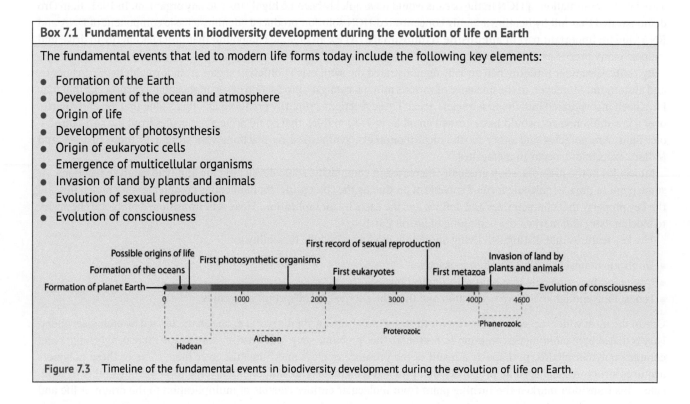

Box 7.1 Fundamental events in biodiversity development during the evolution of life on Earth

The fundamental events that led to modern life forms today include the following key elements:

- Formation of the Earth
- Development of the ocean and atmosphere
- Origin of life
- Development of photosynthesis
- Origin of eukaryotic cells
- Emergence of multicellular organisms
- Invasion of land by plants and animals
- Evolution of sexual reproduction
- Evolution of consciousness

Figure 7.3 Timeline of the fundamental events in biodiversity development during the evolution of life on Earth.

Table 7.1 Temporal chart of the fundamental events of evolution of life.

Million years ago	Eras	Periods	Ages	Main events
0.1	Cenozoic	Quaternary	Holocene	First human
2		Tertiary	Pleistocene	
5			Pliocene	
25			Miocene	
40			Oligocene	First hominids, heteropod mollusks
55			Eocene	Marine mammals, pteropods mollusks and seabirds
60				
65			Paleocene	Dinosaurs extinction
100	Mesozoic	Cretaceous		Diatoms and silico-flagellates
140				Ammonoid cephalopods extinction
				Coccolithophores and foraminifera; increasing of dinoflagellates number
200		Jurassic		Marine reptiles
				Nautiloid cephalopods extinction
240		Triassic		Dinosaurs and mammals; nautiloid cephalopods extinction begins
290		Permian		Trilobites
300				Teleost fish
360		Carboniferous		
410		Devonian		Terrestrial diversification
420				
435		Silurian		Barnacles and dinoflagellates
				Explosion of mollusks bivalve
450	Paleozoic			Elasmobranchs (sharks), cephalopods (ammonoids), and tintinnids
500		Ordovician		evolve, creating coral reefs
				Primitive fishes
550				Trilobites, ostracods, echinoderms, nautiloid cephalopods dominate
570		Cambrian		Marine algae differ considerably
600				Radiolarians
				Abundance of marine invertebrates fossils (jellyfish, sponges, mollusks)
650				The oldest marine fossils metazoans
				Green algae
700				
1,000		Precambrian		Calcareous algae
1,500				First photosynthetic organisms
2,000				First heterotrophic bacteria
3,000				
4,000				Ocean and atmosphere form

7.3 Mechanisms of Marine Speciation

The process of **speciation** leads to the emergence of new species. Exposure of existing species to environmental pressures that differ from those experienced in the past can result in modification of species characteristics and progressively "select" for specific traits. If this happens, heritable characteristics may vary from generation to generation according to population genetic models. Organisms vary in their sensitivity to environmental changes, and their capacity to adapt may reflect

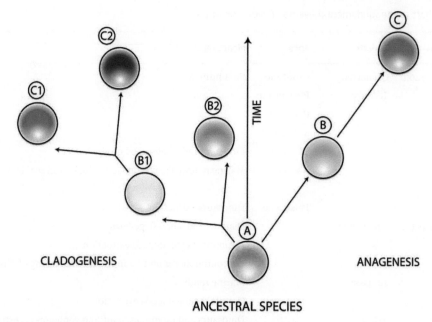

Figure 7.4 Schematic representations of cladogenesis (left) and anagenesis (right).

genetic variability or phenotypic plasticity. **Phenotypic plasticity** reflects the tolerance spectrum of an organism. However, if environmental change selects for particular genotypes of individuals living in a given habitat, speciation may occur, even in the absence of a reproductive barrier with other genotypes of that species.

A new species may emerge in two ways: if an ancestral population undergoes a gradual change over time without splitting, this is called **phyletic evolution** or **anagenesis** (Figure 7.4); if the original population splits into two or more sub-populations that represent new species, we call the process speciation or **cladogenesis** (Figure 7.4, left). Cladogenesis represents the only biological mechanism leading to evolutionary diversification and increased species number.

Speciation results in reproductive discontinuity between previously interbreeding populations and can develop gradually or rapidly. Successful speciation fundamentally requires isolation, whether geographic or behavioral, from the genes of individuals from surrounding populations in order for species differentiation to occur (i.e., sympatric speciation).

Gene flow refers to the spread of genes through migration and subsequent reproduction between individuals belonging to different populations. In fact, gene flow tends to reduce local differences and homogenizes populations. When reproductive isolation of groups of individuals interrupts or slows the flow of genes, then the probability of formation of a new species increases. Speciation probabilities depend on the dispersal capacity and population structure of single species. In marine environments, heterogeneous environments favor rapid speciation, because of greater spatial isolation possibilities and divergent selection pressures, which increase the frequency of divergent characteristics among populations rather than intermediate characteristics. Consistently heterogeneous habitats occur more frequently in benthic environments than in pelagic habitats, and even considering just seafloor environments, infaunal organisms living within sediments experience more consistent environments than the epifauna living upon the bottom. Epifaunal organisms such as gastropods therefore vary geographically more so than polychaetes living in sediments, the latter often characterized by cosmopolitan distributions and lower species diversity. But dispersal also strongly influences speciation: restricted gene flow occurs in species with **direct development**, where adults release juveniles rather than eggs or larvae that might disperse large distances. Similarly, species with pelagic **planktotrophic larvae**, referring to larvae that feed as they develop, exhibit higher gene flow than species with pelagic **lecithotrophic larvae** that depend on internal food reserves and do not feed as they develop and therefore typically disperse shorter distances. Fossil records provide the only direct evidence of long-term morphological evolution and speciation, however, noting that many fossil records of species persist for 10^6 or 10^7 generations without noticeable changes in morphology.

The paleontologists Niles Eldredge and Stephen Jay Gould proposed the controversial theory of **punctuated equilibrium** (Figure 7.5a), in 1972, which proposes periods of rapid and relevant genetic and phenotypic changes alternating with long periods of stability and genetic invariance. Such theory offered an alternative view to the pervasive idea of **phyletic gradualism**, in which evolution occurs as a slow and gradual process (Figure 7.5b). Evolutionary data on marine organisms support the theory of punctuated equilibrium, where rapid diversification occurs in fits and starts.

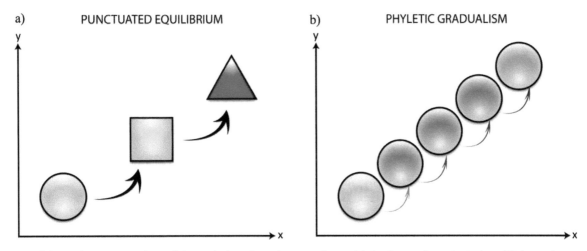

Figure 7.5 Schematic representations of the evolution of species according to (a) the theory of punctuated equilibrium, where changes occur rapidly between long periods of stability and (b) the theory of phyletic gradualism, where slow and steady change occurs. *Source:* Michael Tangherlini.

Some data confirm widespread cladogenesis, or diversification through multiple splits from ancestral forms (living in the sediments), others point to anagenesis or evolution with little splitting into multiple species (e.g., organisms living suspended in the seawater). Moreover, numerous examples in the ocean illustrate allopatric, sympatric, and parapatric speciation, concepts we will consider shortly.

The fossil record appears to support phyletic gradualism in that most species show evolutionary stasis with no net changes for long periods of time, in contrast to scenarios where new species appear suddenly over geological time scales from a common ancestor that may or may not persist. Therefore, in addition to steady change, a few major upheavals marked evolutionary history, causing sudden environmental change that allowed some species to prevail over others, sometimes randomly.

Charles Darwin invoked gaps in the fossil records to explain the absence of intermediate forms in the evolutionary story, but now the theory of punctuated equilibrium accounts for fossil gaps as evidence for rapid change. The bryozoan genera *Metrarabdotos* and *Stylopoma* (Figure 7.6a, b) offer excellent support for punctuated equilibrium: 11 of the 19 species, including abundant ones, remained morphologically unchanged for 2–16 million years, but these same 11 species appeared "suddenly" in the geological record, with no evidences of morphological intermediates. Other studies with phylogenetic data suggest gradual morphological evolution in fossil species.

The oldest species of the first evolutionary line, *Melanopsis impressa* (Figure 7.6c), persisted without clear morphological changes for about 7 million years, until the last representative of the marine fauna went extinct and two new species appeared by anagenesis within two million years. At the same time, however, six new species appeared in the second evolutionary line through rapid cladogenesis.

The gastropod *Prunum coniforme* (Figure 7.6d) has remained unchanged for 11 million years, both before and after the appearance of *P. christineladdae* through cladogenesis. Three species of the gastropod *Amalda* spp., 19 species of bivalves, and 12 species of corals also illustrate periods of stasis. Several species of coral demonstrate modest morphological changes for many millions of years, but with less net change during the entire history of the species than during recent intraspecific changes.

Plankton speciation time varies more than that of the benthos because of the enormous abundance and generally wide geographical distribution of planktonic species. Plankton species that preserve well in the fossil record, such as radiolarians and foraminifera, offer great advantages for speciation studies in spatial and temporal resolution of the samples (a little less than a thousand years), based on more than a thousand worldwide discoveries in deep oceans. The clade of the foraminifera of the genus *Globorotalia* (*Globoconella*; Figure 7.6e) from samples collected in the South Pacific show both anagenesis and punctuated speciation with cladogenesis at different times. Overall, 29 of the 31 benthic species in the Neogene, for which phylogenetic data are available, exhibit rapid changes in morphology (cladogenesis) that support punctuated equilibrium. Speciation patterns in planktonic protists vary greatly, even if problems of morphologically similar species complicate interpretation of some cases of anagenesis.

Small morphological changes associated with speciation do not necessarily imply changes in behavior, development, or life cycle. Moreover, larval development often differs among phylogenetically "close" species with near identical adult morphology. For example, the sea urchin *Heliocidaris tuberculata* produces small eggs that develop into drifting larvae (Figure 7.7) that feed on plankton and disperse for weeks before settling and metamorphosing. In contrast, the sympatric species

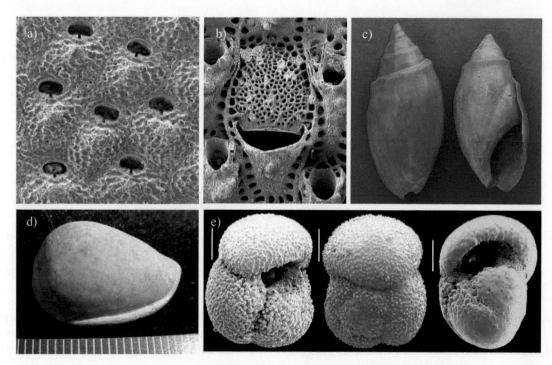

Figure 7.6 Examples of speciation in marine plankton and benthos: (a) bryozoans of the genus *Metrarabdotos* and (b) *Stylopoma* spp. offer excellent models for punctuated equilibrium. *Source:* James Klaus; (c) the gastropod *Melanopsis impressa* has persisted unchanged for 7 million years; (d) similarly, the gastropod *Prunum coniforme* remained unchanged for about 11 million years before giving rise through cladogenesis, to *P. christineladdae*; (e) the clade of the forams *Globorotalia* spp. demonstrate both cladogenesis and anagenesis. *Source:* SylvainGirardot / Wikimedia Commons / CC BY-SA 4.0.

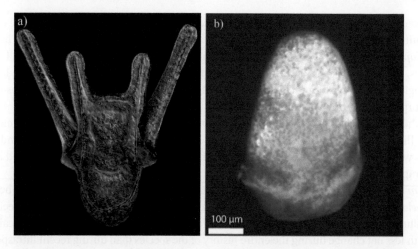

Figure 7.7 Two phylogenetically similar species differ significantly in development and life cycle: (a) planktonic pluteus larva of *Heliocidaris tuberculata* (the larva metamorphoses into an adult 6 weeks after fertilization); (b) larva of *Heliocidaris erythrogramma*, this species originates from a larva that develops directly (the larva metamorphoses 3–4 days after fertilization) and the internal characteristics closely resemble those of the adult. *Source:* Bruno Vellutini / Flickr / CC BY 2.0.

Heliocidaris erythrogramma produces eggs 100 times larger, which develop into non-feeding larvae (Figure 7.6), that disperse for only a few days before metamorphosing into small adults.

The two species also differ in how they form embryonic axes and in gene expression patterns during their development. However, the fossil record for *Heliocidaris* is not sufficient to estimate the time of divergence in development and speciation.

Scientific evidence supports several possible speciation processes using marine organisms as study models. The main mechanisms and processes can be summarized as follows:

1) **Allopatric or geographic speciation** (Figure 7.8) occurs when geographic isolation caused by physical barriers (oceans, seas, submarine mountain chains) prevents contact between two populations initially belonging to the same

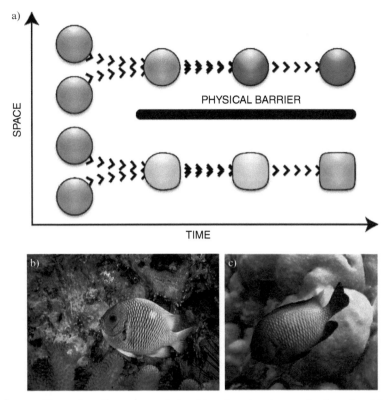

Figure 7.8 Conceptual diagram illustrating allopatric speciation (a) and a related example of coral reef fish species: (b) *Dascyllus strasburgi* (c) *Dascillus albisella*. *Source:* SylvainGirardot / Wikimedia Commons / CC BY-SA 4.0.

species. In this case, genetic variation between the two populations evolves quickly, facilitating the creation of a new species (the famous Galapagos finches that led Charles Darwin to formulate the Theory of Evolution represent the best-known example of allopatric speciation).

The coelacanth *Latimeria chalumnae* illustrates spatial segregation between a population in Madagascar/Comoros Islands and an Indonesian species (Island of Sulawesi), demonstrating allopatric speciation (Box 7.2).

Geographic speciation can follow a dicopatric model or a peripatric model. In a **dicopatric model**, a geographic barrier that individuals cannot overcome initially divides a large ancestral population with a wide distribution range into two similar sized subpopulations. Isolation then promotes the accumulation of genetic differences between the populations, which can contribute to reproductive isolation should they ever reestablish contact, for example, by extending their ranges or if the barrier disappears (e.g., Bering Strait).

In the **peripatric model**, speciation involves just a few individuals geographically isolated from the original population through fragmentation of dispersal into a new location. The genetic composition of the new population therefore differs from the original one because of the **founder effect**, where the frequency of genes in the new population differs because it was founded by a small subset of individuals (and genes) from the larger source population. In both cases, if the geographic barrier disappears, a true test of speciation becomes possible with three possible outcomes:

a) if **pre-zygotic** (before egg fertilization occurs) mechanisms of reproductive isolation evolved, the two populations do not hybridize, and complete speciation has occurred;
b) if the two populations **hybridize** or mix genetically with no reduction in fitness of the hybrids, then the two populations will merge and speciation has failed;
c) if the two populations hybridize but with reduced hybrid viability or fertility (from post-zygotic mechanisms), natural selection will favor assortative intersections between members of the same population and the development of pre-zygotic reproductive isolation (the "theory of reinforcement"). In this case, speciation has occurred.

2) **Sympatric speciation** (Figure 7.10) occurs when two groups of individuals from the same species become reproductively isolated but without geographic barrier. In contrast to allopatric speciation, the new species evolves within the same geographic area as the parental population, without spatial isolation.
3) **Semi-geographic or parapatric speciation** occurs in widely distributed species where adjacent populations become reproductively isolated in the absence of geographical barriers (Figure 7.11). In this case, reproductive isolation arises because of reduced fitness of hybrids. In other cases, chromosomal mutations may block gene flow between parapatric populations.

> **Box 7.2** *Latimeria chalumnae* (the coelacanth) a living fossil fish
>
> The primitive fish coelacanth (Figure 7.9) was thought to have gone extinct with the dinosaurs 65 million years ago. But its discovery in 1938 by a South African museum curator on a local fishing trawler fascinated the world and ignited new studies on the evolution of these animals and terrestrial counterparts. Subsequently another coelacanth species was found in the waters off Sulawesi, in Indonesia. Many scientists believe that the unique characteristics of the coelacanth represent an early step in the evolution of fish to terrestrial four-legged animals such as amphibians. Through a phylogenomic analysis, scientists concluded that the lungfish, and not the coelacanth, is the closest living relative of tetrapods. Coelacanth protein-coding genes evolve at a significantly slower rate than those of tetrapods, unlike other genomic features. Functional assays of enhancers involved in the fin-to-limb transition and in the emergence of extra-embryonic tissues show the importance of the coelacanth genome as a blueprint for understanding tetrapod evolution. Coelacanths are deep-sea animals, living at depths to 700 m. They can reach 2 m in length or more, weighing almost 100 kg. Scientists estimate they can live up to 60 years or more. Paired lobe fins represent the most striking feature of this "living fossil," extending away from its body like legs and moving in an alternating pattern. Other unique characteristics include a hinged joint in the skull that allows the fish to widen its mouth for large prey; an oil-filled tube, called a notochord, which serves as a backbone; thick scales common only in extinct fish; and an electro-sensory rostral organ in its snout likely used to detect prey. 
>
> **Figure 7.9** *Latimeria chalumnae* (the coelacanth). *Source:* Bruce A.S. Henderson / Wikimedia Commons / CC BY 4.0.

Figure 7.10 Conceptual diagram illustrating (a) sympatric speciation, and (b) related example for coral reef fish species: *Hexagrammos agrammus*, and (c) *Hexagrammos otakii*.

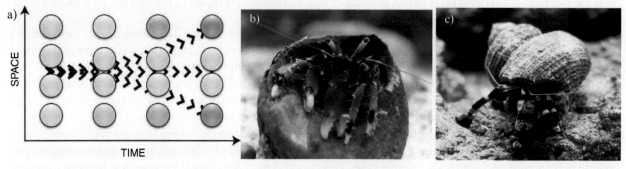

Figure 7.11 (a) Conceptual diagram illustrating the mechanism of parapatric speciation, and related example for coral reef crab species: (b) *Calcinus latens*, and (c) *Calcinus elegans*. *Source:* Fafner/Wikimedia Commons.

7.4 Quantifying Marine Organism Biodiversity

7.4.1 Definition of Biodiversity

Diversity comprises every living system at levels of organization spanning from molecules to ecosystems, complicating any single definition. For this reason, experts have defined, organized, and expressed biodiversity in multiple ways over the last thirty years, including: (1) variety and variability of living organisms and of their ecological, (2) any kind of variability between living organisms, and (3) total variety of life on Earth. The most widely accepted definition of biodiversity, which was developed by the Convention on Biology Diversity, defines biodiversity (or biological diversity) as "the variability among living organisms from all sources, including inter alia, terrestrial, marine and other aquatic ecosystems and the ecological complexes of which they are part; this includes diversity within species, between species and of ecosystems." Different types of researchers focus on different aspects of biodiversity. Geneticists and molecular biologists work at different genetic and molecular levels, thus recognizing biodiversity within different populations of the same species, and even among siblings, as well as in an ecological context (Figure 7.10). Molecular tools help to differentiate **cryptic** species, referring to two or more reproductively incompatible species that are morphologically indistinguishable from one another. Zoologists and botanists traditionally differentiate biodiversity based on morphology. Ecologists sometimes consider biodiversity as the total number of species (including those not yet described), as well as evenness, habitat diversity, and regional diversity. Experimental ecologists focus on the number of species they can manipulate for experiments. Marine ecologists represent biodiversity as the different communities interacting with the environment and affecting marine ecosystems. Biogeographers consider biodiversity as the variety of biomes and the different biogeographic regions with which native species associate (in contrast to non-native or alien species). Socio-economists consider the amount of biodiversity as the number of species from which society can extract commercial value. Thus, depending on the goals and participants in a study, perceptions of both biodiversity "quantity" and "quality" may differ (Figure 7.12).

7.4.1.1 Genetic and Morphologic Diversity

Closely related, morphologically similar but reproductively isolated species occur in all major taxonomic groups and marine habitats, such as in the genera *Gammarus* (amphipods), *Jaera* (isopods), and *Calanus* (copepods).

The widespread occurrence of sibling species reflects both the inadequate study of morphological characteristics of living organisms ("pseudo-sibling species") and the divergence in habitats, life cycles, and in recognizing chemical or behavioral change without parallel divergence in morphology. Sibling species may be impossible to distinguish based on morphological characters, despite major genetic differences (Figure 7.13).

Humans typically study biodiversity because of the direct and indirect economic benefits derived from it (through resource exploitation such as fishing, natural products, etc.). But many people recognize the intrinsic value of biodiversity in just knowing it exists, for aesthetic recreational values (e.g., tourism, sports), aesthetic values (e.g., artistic inspiration in photographs and paintings), and, finally, for scientific knowledge. Therefore, biodiversity offers intrinsic, anthropocentric, aesthetic, and recreational value, but important scientific value as well. The complexity and number of levels at which biodiversity can be classified require different definitions: (a) the diversity of the parts that form a discrete biological system (an individual or sub-individual (genetic)); (b) the diversity among individuals in a population or populations that collectively form a species; (c) the diversity of species that collectively form ecological communities; (d) the diversity of communities that collectively form ecosystems, and (e) the diversity of the habitats/ecosystems that collectively form the biosphere.

Consequently, biodiversity studies address different diversity levels: (1) genetic – representing variation in genetic composition among individuals of a given species; (2) systematic or population – representing phylogenetic variation within a species; (3) taxonomic – referring to the number of species (or families or phyla) within a local community; (4) regional – referring to the number of species (or families or phyla) within a region; (5) functional – referring to the different trophic or other functional groups to which different species belong (e.g., detritivores, herbivores, carnivores); (6) biocenotic and biogeographical – referring to the number and relative extent of different types of habitat or ecosystems in a given region.

7.4.2 Defining Different Ways of Measuring Biodiversity: Species Richness

Even at the species level, ecologists measure species richness in different ways (Figure 7.14):

1) alpha diversity (α) represents the diversity of a community or a sample (e.g., all species of organisms living within a single sample, such as a sediment core from a seamount). Ecologists often measure alpha diversity as the total number of species in a sample;

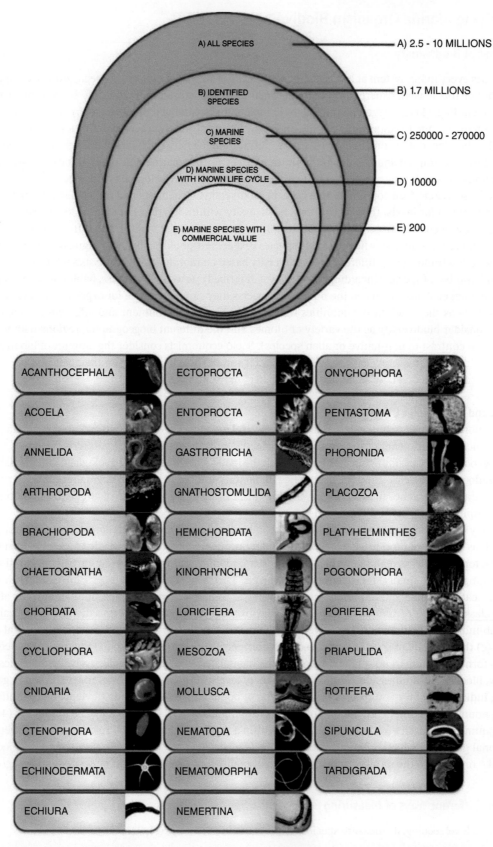

Figure 7.12 Diagram showing different levels of: (a) all species (which various estimates place between 2.5 and 10 million species); (b) all known species (1,700,000); (c) all known marine species (250,000 to 270,000); (d) known species in relation to strategies and life cycles (assumed to be ~10,000); (e) current commercially harvested marine species (~200). List of known marine phyla at the bottom.

Figure 7.13 Example of "sister species" (sibling species), the nudibranchs (a) *Chromodoris geometrica* and (b) *Chromodoris conchyliata*. *Source:* Kevin Bourdon / Flickr / CC BY 2.0.

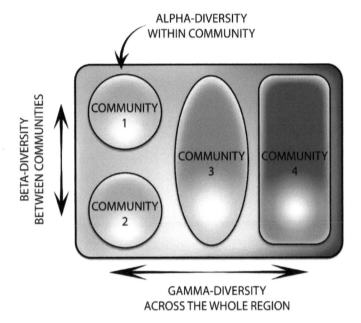

Figure 7.14 Schematic representation of different biodiversity levels.

2) beta diversity (β) represents the degree of change (or turnover) in species composition between communities within a habitat (e.g., species living on two sides of a seamount). β-diversity can be measured as dissimilarity between the species present in two communities;
3) gamma diversity (γ) represents the total number of species in a habitat or landscape (for example, all species on a seamount). In this case, γ-diversity can be measured as $\gamma = \alpha + \beta$ (when expressing β-diversity as the number of species not shared by both communities);
4) delta diversity (δ) represents the degree of dissimilarity (turnover) of species diversity between two habitats, or two γ-diversity estimates, such as the overall diversity from two seamounts;
5) epsilon diversity (ε) represents changes in broadly regional diversity, such as the change in species diversity between marine and terrestrial ecosystems (e.g., all species of tropical forests vs coral reefs species);
6) omega diversity (ω) represents changes in species diversity at a global level.

Below, we also consider different measures of biodiversity as well as evenness, which represent other aspects of biodiversity comparison beyond species richness.

7.4.3 Comparison Between Marine and Terrestrial Biodiversity

Marine and terrestrial biodiversity clearly differ. First, marine biota exhibit much higher diversity at the phylum level. Terrestrial environments support many species per genus, on average, in contrast to fewer species per genus but many more phyla in the ocean. This means greater variation in marine organisms, both at the genetic and at phylum level, than on land. Table 7.2 summarizes the current complete list of marine kingdoms and phyla.

Table 7.2 List of currently recognized kingdoms and Phyla reported by the World Register of Marine Species.

Kingdom Animalia

- Phylum Acanthocephala
- Phylum Annelida
- Phylum Arthropoda
- Phylum Brachiopoda
- Phylum Bryozoa
- Phylum Chaetognatha
- Phylum Chordata
- Phylum Cnidaria
- Phylum Ctenophora
- Phylum Cycliophora
- Phylum Dicyemida
- Phylum Echinodermata
- Phylum Entoprocta
- Phylum Gastrotricha
- Phylum Gnathifera
- Phylum Gnathostomulida
- Phylum Hemichordata
- Phylum Kinorhyncha
- Phylum Loricifera
- Phylum Mollusca
- Phylum Nematoda
- Phylum Nematomorpha
- Phylum Nemertea
- Phylum Orthonectida
- Phylum Phoronida
- Phylum Placozoa
- Phylum Platyhelminthes
- Phylum Porifera
- Phylum Priapulida
- Phylum Rotifera
- Phylum Sipuncula
- Phylum Tardigrada
- Phylum Xenacoelomorpha

Kingdom Bacteria

- Phylum Acidobacteria
- Phylum Actinobacteria
- Phylum Aquificae
- Phylum Bacteroidetes
- Phylum Caldiserica
- Phylum Chlamydiae
- Phylum Chlorobi
- Phylum Chloroflexi
- Phylum Cyanobacteria
- Phylum Deferribacteres
- Phylum Deinococcus-Thermus
- Phylum Elusimicrobia
- Phylum Fibrobacteres
- Phylum Firmicutes
- Phylum Fusobacteria
- Phylum Gemmatimonadetes
- Phylum Lentisphaerae
- Phylum Nitrospirae
- Phylum Planctomycetes
- Phylum Proteobacteria
- Phylum Spirochaetes
- Phylum Synergistetes
- Phylum Tenericutes
- Phylum Thermodesulfobacteria
- Phylum Thermotogae
- Phylum Verrucomicrobia

Kingdom Chromista

- Phylum Bigyra
- Phylum Cercozoa
- Phylum Ciliophora
- Phylum Cryptophyta
- Phylum Foraminifera
- Phylum Haptophyta
- Phylum Heliozoa
- Phylum Myzozoa
- Phylum Ochrophyta
- Phylum Oomycota
- Phylum Radiozoa

Kingdom Plantae

- Phylum (Division) Anthocerotophyta
- Phylum (Division) Bryophyta
- Phylum (Division) Charophyta
- Phylum (Division) Chlorophyta
- Phylum (Division) Glaucophyta
- Phylum (Division) Marchantiophyta
- Phylum (Division) Rhodophyta
- Phylum (Division) Tracheophyta

Kingdom Protozoa

- Phylum Acritarcha
- Phylum Amoebozoa
- Phylum Apusozoa
- Phylum Choanozoa
- Phylum Euglenozoa
- Phylum Loukozoa

Table 7.2 (Continued)

	Phylum Metamonada		Phylum (Division) Basidiomycota
	Phylum Percolozoa		
	Phylum Picozoa		Phylum (Division) Chytridiomycota
Kingdom Archaea			
	Phylum Crenarchaeota		Phylum (Division) Glomeromycota
	Phylum Euryarchaeota		Phylum (Division) Microsporidia
	Phylum Korarchaeota		
	Phylum Thaumarchaeota		Phylum (Division) Zygomycota
Kingdom Fungi			
	Phylum (Division) Ascomycota		

The current number of known species on Earth is somewhere between 1.4 and 1.7 million (excluding prokaryotes and viruses). Overall, insects represent the most diverse group (750–800,000 known species). Recent projections of terrestrial plants and animals estimate ~9 million species on Earth, once all species have been described, some 91% of which come from a single phylum (Figure 7.15), the arthropods. Discoveries of new, large species are rare, and typically occur in extremely remote areas, or where human conflicts or political regimes have hindered scientific exploration. The situation differs for marine biodiversity, with 250,000–270,000 described species where eight animal phyla comprise about 90% of known marine species, in contrast to the one dominant phylum on land, the Arthropoda.

The number of marine species described to date is relatively low compared with terrestrial environments, even with ~1,300–1,500 new species discovered each year in the ocean (Figure 7.16). The fraction of undescribed species in a marine sample typically exceeds that from a terrestrial sample. These hundreds of new species of marine organisms described each year do not include the many microbial forms from viruses to bacteria and fungi. Nonetheless, much of marine biodiversity remains undiscovered.

Note that most species in a given phylum occur primarily (some exclusively) in marine, benthic habitats, with only some occupying the entire range of habitats (Table 7.3).

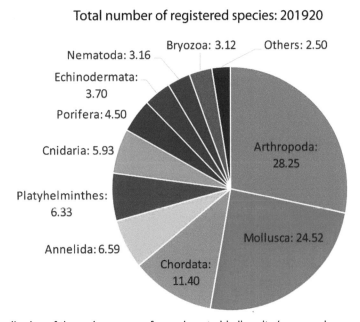

Figure 7.15 Percentage contribution of the major groups of organisms to biodiversity (expressed as number of species).

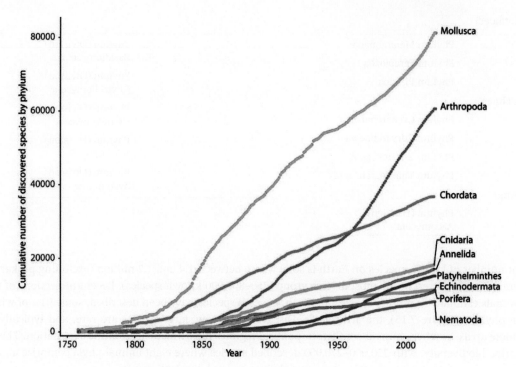

Figure 7.16 Number of animal species discovered for the major marine phyla. Note that none of these curves shows any sign of levelling off, indicating no decline in discovery rates.

Table 7.3 Phyla present in different aquatic and terrestrial habitats. Note that the majority of described animal taxa live in marine sediments.

Phyla	Marine		Freshwater		Terrestrial		Symbiont	
Subphyla	Benthic	Pelagic	Benthic	Pelagic	Wet	Dry	Ecto	Endo
Porifera	+++		+				+	
Placozoa	+							
Orthonectida								+
Dicyemida								+
Cnidaria	+++	++	+	+			+	+
Ctenophora	+	+						
Platyhelminthes	+++	+	+++		++		+	++++
Gnathostomulida	++							
Nemertea	++	+	+		+		+	
Nematoda	+++	+	+++	+	+++	+	+++	+++
Nematomorpha								++
Acanthocephala								++
Rotifera	+	+	++	++	+		+	+
Gastrotricha	++		++					
Kinorhyncha	++							
Loricifera	+							
Tardigrada	+		++		+			
Priapula	+							
Mollusca	++++	+	+++		+++	+	+	+

Table 7.3 (Continued)

Phyla	Marine	Freshwater	Terrestrial	Symbiont
Kamptozoa	+	+		+
Sipuncula	++		+	
Echiura	++			
Anellida	++++	+ ++	+++	++
Onychophora				
Arthropoda				
Crustacea	++++	+++ +++ ++	++	++ ++
Chelicerata	++	+ ++ ++	++++ +++	++ +
Uniramia	+	+ +++ ++	++++ +++	++ ++
Phoronida	+			
Brachiopoda	++			
Bryozoa	+++	+		
Echinodermata	+++	+		
Hemichordata	+			
Chordata				
Urochordata	+++	+		
Cephalochordata	+			
Vertebrata	+++	+++ ++ +++	+++ +++	+ +

The symbol "+" indicates the approximate abundance of the described living species: $+ = 1–100$; $++ = 100–1,000$; $+++ = 10^3–10^4$; $++++ = 10^4–10^5$; $+++++ =$ more than 10^5.

7.4.4 Measures of Biodiversity

Measuring biodiversity requires identifying phenotypic traits (or genes) that identify a species. Species diversity measures fall into three broad categories. First, **indices of species richness** essentially measure the number of species in a defined sampling unit. The values of diversity indices are not always comparable and depend on sampling effort and distribution of sampling effort within a location. Second, **species abundance** describes the number of individuals per species. Third, **indices of relative species abundance** (such as the Shannon index and Simpson index) describe apportioning of individuals among species. **Species richness**, the simplest measure to obtain, describes the number of species present in a sample, community, or taxonomic group. Relative species abundance also incorporates **evenness**, which describes whether similar or different numbers of individuals represent different species. Counts of the number of species as a descriptor of an ecological community oversimplify diversity, even though obtaining such a number for an ecosystem requires considerable effort, given the goal of collecting and correctly identifying sufficient organisms to represent all of the species in a community. The relative abundances of different species (the percentages of different species in the sample), however, can effectively define the diversity of an ecosystem. In fact, some measures of diversity often take into account the relative abundance of individual species, a key aspect of biodiversity comparison.

Many studies that identify organisms, a process known as **taxonomy**, often do not identify individuals to the species level because of the complexity and time-consuming analysis involved. However, even higher-level taxonomic analysis (e.g., family, even phylum) can provide significant information on biodiversity patterns. Some studies focus on the relationship between abundances of two different taxa (usually comparing a taxon tolerant of specific environmental conditions with one susceptible to changes in such conditions). For example, ecologists sometimes compare the ratio of nematodes to copepods, which requires adequate knowledge of the **auto-ecology**, or individual ecologies and life cycle of the two taxa. The presence of some taxa may indicate particular environmental conditions. Some oligochaetes, for example, dominate environments with reduced concentrations of oxygen or the presence of oil, and among the polychaetes, species belonging to the genus *Capitella* can indicate organic enrichment, such as in sewer outfalls.

Once we known the species present in an environment, further analyses become possible; combining the total number of species with the relative percentages of species yields "K-dominance curves," a method that highlights which species dominate others, thus creating an easy tool for baseline comparisons (Figure 7.17a). Methods of multivariate clustering

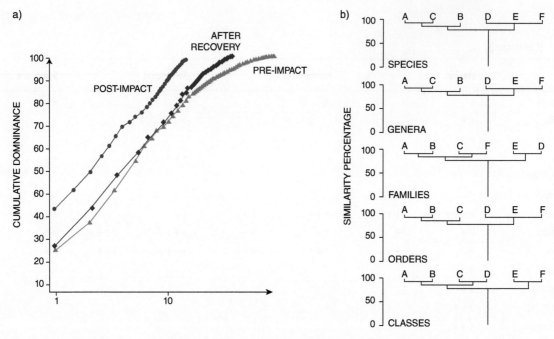

Figure 7.17 Graphic representation of biodiversity: (a) Example of a K-dominance curve, and (b) a classic representation of a dendrogram of dissimilarity between samples, based on species composition, showing closer similarity of some communities than others.

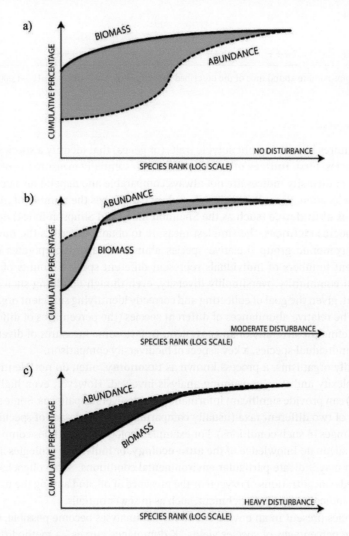

Figure 7.18 Theoretical ABC curves showing patterns of abundance and biomass within a community: (a) unperturbed; (b) moderately disturbed, and (c) highly disturbed.

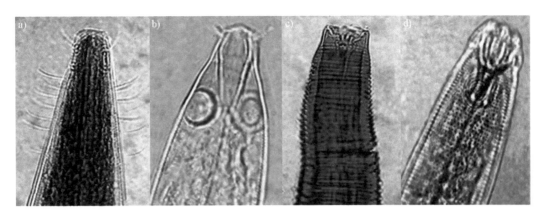

Figure 7.19 Nematode trophic groups: from left to right (a) selective deposit feeders (bacterivorous, with small mouth adapted to feeding on microbial components); (b) non-selective deposit feeder (with larger mouth that ingest portions of sediment in order to utilize the associated organic matter; (c) epibiontic microalgal grazer, with mouth equipped with sclerotic structures to scrape the surfaces on which their prey adhere, and (d) predator/omnivore with mouth equipped with large teeth, claws, or swords capable of drilling, biting, or swallowing any prey.

Dividing organisms into different functional groups (such as trophic groups) can produce a measure of functional diversity. Some studies of benthic biodiversity typically use the Index of Trophic Diversity (IDT) for this purpose. In reality, the relationship between taxonomic and functional diversity can change quickly when important functional traits are homogeneous for a particular phylum within a sample. Individuals may be divided into functional groups (Figure 7.19), distributed in the community after taxonomic sorting of organisms from an entire sample, for example based on different feeding strategies.

Functional diversity measures a variety of functions carried out by the organisms in a system by dividing species within a community into different functional types on the basis of different feeding patterns (filter feeders or predators), life cycles, or growth modes (especially in the case of plants), for example.

7.4.5 Composite Measures of Species Diversity

Composite indices of diversity combine aspects of numbers of species with apportioning of individuals among them. Below we identify a few examples of such measures, but note that there are many others, each with its strengths and weaknesses.

Shannon-Wiener Index (or Shannon-Weaver): examines the characteristics of a community by proportionately weighting all species present in a sample. It can distinguish between two areas (or samples) with the same number of species and with the same number of individuals, but in different proportions:

$$H' = -\sum_{j=1}^{s} p_j \log p_j$$

where:
s denotes the total number of species,
$p_j = n_j/N$,
n_j denotes the number of individuals of the jth species,
N denotes the number of total individuals.

The logarithm can be natural (**ln**, as indicated in the formula) or base 2 ($\log_2$).

This index varies between values approaching 0 and approaching ∞, with a minimum value when all individuals belong to the same species and a maximum when individuals are equally distributed among all of the species. This number represents the average amount of information per individual, according to a criterion that every individual of a species contains more information when that species is rare. The index therefore takes into account the number of species and the evenness of the species.

Simpson's Index expresses the probability that two individuals randomly chosen from a habitat (in a sample) belong to the same species. A high value indicates that one or a few species monopolize the resources. Like the Shannon index, this index is based on the ratio **p/j** (proportion of a given species in a community):

$$D = \frac{1}{\sum_{i=1}^{s} p_i^2}$$

where:

S = total number of total species,
p$_i$ = proportional contribution of each species to total number.

Margalef's Index: considers the cumulative rate between the number of species and the number of individuals in a community:

$$D = \frac{s-1}{\ln(N)}$$

where

S = total number of species,
N = number of individuals present in a sample. The more species present in a sample, the higher the value of the index.

Equitability Index, or Pielou's Index (Evenness, J) does not represent an index of diversity in the strict sense, but it compares the distribution of the individuals among the different species within a sample:

$$J = \frac{H}{H_{max}}$$

where;

H = Shannon-Wiener index
H$_{max}$ = theoretical maximum value of H (where all species in a sample are equally abundant).

Pielou's index measures the distribution of abundances among species, with a maximum value of 1 when individuals are perfectly equitably distributed among species, with a minimum value approaching 0 when a single species dominates all others.

Questions

1) How and where did life likely originate?
2) Which are the main steps of evolution for ocean life?
3) How many phyla occur in the ocean and how many of these are endemic to ocean environments?
4) How do scientists measure marine biodiversity?
5) What are the main mechanisms of speciation in the ocean?
6) Which are the differences between sympatric and parapatric biodiversity?
7) What factors affect marine biodiversity?
8) How do we measure and quantify biodiversity?
9) What does the Pielou index measure?

Suggested Reading

Amann, R. and Roselló-Móra, R. (2016) After all, only millions? *mBio*, 7(4), pp. e00999–16.
Amato, A., Kooistra, W.H.C.F., Levialdi Ghiron, J.H. et al. (2007) Reproductive isolation among sympatric cryptic species in marine diatoms. *Protist*, 158, pp. 193–207.

Appeltans, W., Ahyong, S.T., Anderson, G. et al. (2012) The magnitude of global marine species diversity. *Current Biology*, 22, pp. 2189–2202.

Beaugrand, G., Edwards, M., and Legendre, L. (2010) Marine biodiversity, ecosystem functioning, and carbon cycles. *Proceedings of the National Academy of the United States of America*, 107, pp. 10120–10124.

Bernard, L., Schaler, H., Joux, F. et al. (2000) Genetic diversity of total, active and culturable marine bacteria in coastal seawater. *Aquatic Microbial Ecology*, 23, pp. 1–11.

Bhat, A. and Magurran, A.E. (2006) Taxonomic distinctness in a linear system: A test using a tropical freshwater fish assemblage. *Ecography*, 29, pp. 104–110.

Bierne, N., Bonhomme, F., and David, P. (2003) Habitat preference and the marine-speciation paradox. *Proceedings of the Royal Society B: Biological Sciences*, 270, pp. 1399–1406.

Boero, F. (2010) The study of species in the era of biodiversity: A tale of stupidity. *Diversity*, 2, pp. 115–126.

Boeuf, G. (2011) Marine biodiversity characteristics. *Comptes Rendues Biologies*, 334, pp. 435–440.

Bowen, B.W., Rocha, L.A., Toonen, R.J. et al. (2013) The origins of tropical marine biodiversity. *Trends in Ecology and Evolution*, 28, pp. 359–366.

Chapin, F.S., III, Zavaleta, E.S., Eviner, V.T. et al. (2000) Consequences of changing biodiversity. *Nature*, 405, pp. 234–242.

Danovaro, R. (2012) Extending the approaches of biodiversity and ecosystem functioning to the deep ocean, in Solan, M., Aspden, R.J., and Paterson, D.M. (eds.), *Marine Biodiversity and Ecosystem Functioning: Frameworks, Methodologies and Integration*. Oxford University Press, pp. 115–126.

Danovaro, R., Company, J.B., Corinaldesi, C. et al. (2010) Deep-sea biodiversity in the Mediterranean Sea: The known, the unknown, and the unknowable. *PLOS One*, 5(8), p. e11832.

Danovaro, R., Dell'Anno, A., Pusceddu, A. et al. (2010) The first metazoa living in permanently anoxic conditions. *BMC Biology*, 8, p. 30.

Fonseca, V.G., Carvalho, G.R., Sung, W. et al. (2010) Second-generation environmental sequencing unmasks marine metazoan biodiversity. *Nature Communactions*, 1, p. 98.

Forbes, E. (1856) Map of the distribution of marine life, in Johnston, A.K. (ed.), *The Physical Atlas of Natural Phenomena*. William Blackwood and Sons, pp. 99–102 and plate 131.

Funch, P. and Kristensen, R.M. (1995) Cycliophora is a new phylum with affinities to Entoprocta and Ectoprocta. *Nature*, 378, pp. 711–714.

Gagné, T.O., Reygondeau, G., Jenkins, C.N. et al. (2020) Towards a global understanding of the drivers of marine and terrestrial biodiversity. *PLOS One*, 15(2), p. e0228065.

García Molinos, J., Halpern, B., Schoeman, D. et al. (2016) Climate velocity and the future global redistribution of marine biodiversity. *Nature Climate Change*, 6, pp. 83–88.

Gaston, K.J. (2000) Global patterns in biodiversity. *Nature*, 405, pp. 220–227.

Giovannoni, S.J. and Stingl, U. (2005) Molecular diversity and ecology of microbial plankton. *Nature*, 437, pp. 343–348.

Gotelli, N.J. and Colwell, R.K. (2001) Quantifying biodiversity: Procedures and pitfalls in the measurement and comparison of species richness. *Ecology Letters*, 4, pp. 379–391.

Gould, S.J. and Eldredge, N. (1972) Punctuated equilibria: An alternative to phyletic gradualism, in Schopf, T.J.M. (ed.), *Models in Paleobiology*. Freeman, Cooper, pp. 82–115.

Hewitt, J.E., Thrush, S.F., Halliday, J., and Duffy, C. (2005) The importance of small-scale habitat structure for maintaining beta diversity. *Ecology*, 86, pp. 1619–1626.

Hooper, D.U., Chapin, F.S., III, Ewel, J.J. et al. (2005) Effects of biodiversity on ecosystem functioning: A consensus of current knowledge. *Ecological Monographs*, 75, pp. 3–35.

Hutchinson, G.E. (1959) Homage to Santa Rosalia or why are there so many kinds of animals? *The American Naturalist*, 93, pp. 145–159.

Isaac, N.J., Turvey, S.T., Collen, B. et al. (2007) Mammals on the EDGE: Conservation priorities based on threat and phylogeny. *PLOS One*, 2(3), p. e296.

Jackson, J.B.C. and Cheetham, A.H. (1999) Tempo and mode of speciation in the sea. *Trends in Ecology and Evolution*, 14, pp. 72–77.

Knowlton, N. (1993) Sibling species in the sea. *Annual Review of Ecology and Systematics*, 24, pp. 189–216.

Longhurst, A. (1998) *Ecological Geography of the Sea*. Academic Press.

Loreau, M., Naeem, S., Inchausti, P. et al. (2001) Biodiversity and ecosystem functioning: Current knowledge and future challenges. *Science*, 294, pp. 804–808.

MacArthur, R.H. and Wilson, E.O. (1967) *The Theory of Island Biogeography*. Princeton University Press.

Magurran, A.E. (2013) *Measuring Biological Diversity*. Wiley-Blackwell.

May, R.M. (1994) Biological diversity: Differences between land and sea. *Philosophical Transactions of the Royal Society B: Biological Sciences*, 343, pp. 105–111.

Miya, M. and Nishida, M. (1997) Speciation in the open ocean. *Nature*, 389, pp. 803–804.

Naeem, S., Thompson, L.J., Lawler, S.P. et al. (1994) Declining biodiversity can alter the performance of ecosystems. *Nature*, 368, pp. 734–736.

O'Connor, N.E. and Crowe, T.P. (2005) Biodiversity loss and ecosystem functio- ning: Distinguishing between number and identity of species. *Ecology*, 86, pp. 1783–1796.

Ocean Biodiversity Information System (OBIS)[www.obis.org]*. Intergovernmental Oceanographic Commission of UNESCO.

Ormond, R.F., Gage, J.D., and Angel, M.V. eds. (1997) *Marine Biodiversity. Patterns and Processes*. Cambridge University Press.

Palumbi, S.R. (1994) Genetic divergence, reproductive isolation, and marine speciation. *Annual Review of Ecology and Systematics*, 25, pp. 547–572.

Ruxton, G.D. and Humphries, G. (2008) Can ecological and evolutionary arguments solve the riddle of the missing marine insects? *Marine Ecology*, 29, pp. 72–75.

Schulze, E.-D. and Mooney, H.A. (1993) *Biodiversity and Ecosystem Function*. Springer Verlag.

Sherman, K. and Alexander, L.M. eds. (1986) *Variability and Management of Large Marine Ecosystems*. Westview Press.

Snelgrove, P.V.R. (2010) *The Census of Marine Life: Making Ocean Life Count*. Cambridge University Press.

Sogin, M.L., Morrison, H.G., Huber, J.A. et al. (2006) Microbial diversity in the deep sea and the underexplored "rare biosphere". *Proceedings of the National Academy of Sciences of the United States of America*, 103, pp. 12115–12120.

Solan, M., Raffaelli, D.G., Paterson, D.M. et al. (2006) Marine biodiversity and ecosystem function: Empirical approaches and future research needs – Introduction. *Marine Ecology Progress Series*, 311, pp. 175–178.

Spalding, M.D., Fox, H.E., Allen, G.R. et al. (2007) Marine ecoregions of the world: A bioregionalization of coastal and shelf areas. *BioScience*, 57, pp. 573–583.

Sutton, T.T., Clark, M.R., and Dunn, D.C. (2017) A global biogeographic classification of the mesopelagic zone. *Deep-Sea Research Part I: Oceanographic Research Papers*, 126, pp. 85–102.

Terlizzi, A., Bevilacqua, S., Fraschetti, S., and Boero, F. (2003) Taxonomic sufficiency and the increasing insufficiency of taxonomic expertise. *Marine Pollution Bulletin*, 46, pp. 556–561.

Thébault, E. and Loreau, M. (2006) The relationship between biodiversity and ecosystem functioning in food webs. *Ecological Research*, 21, pp. 17–25.

Tilman, D., Reich, P.B., Knops, J. et al. (2001) Diversity and productivity in a long-term grassland experiment. *Science*, 294, pp. 843–845.

Watling, L., Guinotte, J., Clark, M.R., and Smith, C.R. (2013) A proposed biogeography of the deep ocean floor. *Progress in Oceanography*, 111, pp. 91–112.

Wilson, E.O. (1988) *Biodiversity*. National Academies Press.

Worm, B. and Tittensor, D.P. (2018) *A Theory of Global Biodiversity*. Princeton University Press.

8

Biodiversity Patterns

8.1 Broad-Scale Biodiversity Patterns

Marine biodiversity generally decreases moving from the equator to the poles, but researchers debate whether this decrease is simply monotonic or whether bimodal peaks occur near the subtropics. Different taxa adhere to this pattern to different degrees, noting for example that many polar systems show significant endemism and particularly high diversity of marine mammals, and deep-sea brittle stars peaks at 30–50° latitude. Longitudinal gradients have received less attention, although the coral triangle between the Pacific and Indian Ocean contains the highest concentration of marine species on the planet. In all oceans and seas, bathymetric biodiversity patterns show a bell-shaped trend with maximum values at intermediate depths, between 1,500 and 2,500 m. The diversity of macroinvertebrates and fishes generally increases with depth, reaching a maximum on upper continental slope depths (see below) and then decreasing out onto the abyssal plain.

Marine biogeography considers both these broad-scale patterns and the patterns and processes that affect such distributions in space. Some biogeographic regions such as the Mediterranean Sea have changed markedly in biodiversity with the entry of alien species from the Red Sea through the Suez Canal (**Lessepsian migration**), or other oceans. Moreover, the main theories on the evolution and maintenance of biodiversity consider equilibrium hypotheses (stability over time, spatial heterogeneity, available energy and productivity) and disequilibrium hypotheses (physical disturbance and biological disorder), and their combination. Here, we illustrate the different models to explain biodiversity patterns and potential relationships between biodiversity and ecosystem stability/functioning. In Chapters 18 and 19 we will consider how biodiversity loss may have different consequences in many different types of ecosystems, sometimes compromising ecosystem functioning in some, while producing unique and unpredictable (idiosyncratic) effects or no effects in others.

8.2 Processes Controlling the Distribution of Marine Biodiversity

8.2.1 Biodiversity Hotspots

Biodiversity hotspots refer to locations of exceptional biotic richness compared to adjacent locations of relatively low diversity. Many conservation programs have adopted this approach as a method of identifying areas to prioritize for conservation (Figure 8.1). Researchers have identified coral reefs (shallow and deep), hydrothermal vents, and cold seeps (which we will discuss in Chapter 25) as marine biodiversity hotspots, with corals reefs as particularly species rich habitats, and hydrothermal vents and seeps as habitats that support particularly unique species.

Total species number of both benthic and planktonic open ocean communities often exceeds that in coastal habitats, although the greater diversity of habitats in some coastal areas can result in higher regional diversity. In most cases, several interacting factors determine the diversity of a given ecosystem:

a) geological age, in that fewer species characterize "younger" communities;
b) habitat heterogeneity, where a greater variety of habitats offers more ecological niches, presumably lowering competition; both attributes tend to increase diversity;

Marine Biology: Comparative Ecology of Planet Ocean, First Edition. Roberto Danovaro and Paul Snelgrove.
© 2024 John Wiley & Sons Ltd. Published 2024 by John Wiley & Sons Ltd.
Companion Website: www.wiley.com/go/danovaro/marinebiology

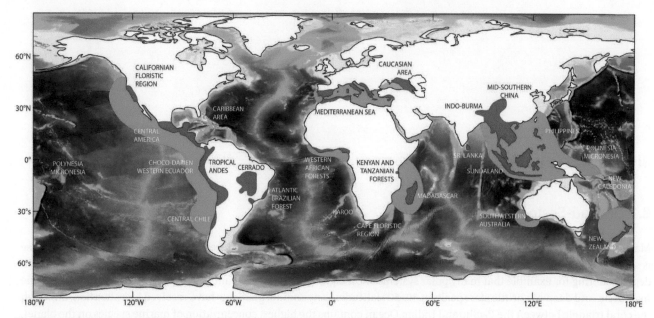

Figure 8.1 Hotspots of shallow-water biodiversity based on data on the wealth of marine species and the presence of endemic species.

c) climatic factors, where fewer species characterize environments with colder climates;
d) productivity, in which limited availability of trophic resources may limit diversity (e.g. abyssal plains); productive environments can support many individuals, though not always many species (e.g. hypoxic habitats under productive upwelling waters);
e) competition can enhance diversity by promoting shifts in ecological niches or promoting the movement of less competitive species into adjacent habitats;
f) predation, when intermediate in intensity and thus not resulting in extreme mortality, can decrease the abundance of dominant competitors that might otherwise decimate competitors or prey;
g) intermediate disturbance, based on a similar mechanism as predation, can increase biodiversity;
h) history of a region can be significant, in that centers of speciation such as the Indo-West Pacific typically support more species than recently colonized locations.

8.2.2 Latitudinal Gradient of Biodiversity

As noted earlier, global biodiversity generally decreases with increasing latitude, moving from the equator to the poles (Figure 8.2), representing the best-known spatial gradient in diversity. In the marine domain, biodiversity of many benthic taxa in particular indeed increase from the poles to the tropics, especially coastal species associated with hard substrata. Bivalves, for example, increase in diversity from the Arctic to Indo-Pacific tropics at the species, genus, and family level. High Antarctic diversity in many benthic taxa weakens a pole-tropic gradient in the southern hemisphere. The longer period of geographic isolation in Antarctica almost certainly played an important role in the development of its biodiversity.

Lower diversity with fewer endemic species characterizes comparatively young Arctic ecosystems, where glaciation and related effects reduced diversity significantly relative to the Antarctic.

Production also differs between the poles: Whereas many fish species (including some of commercial interest) dominate the Arctic, invertebrates (including krill and squid) dominate Antarctica and support bird and mammal communities and comparatively modest numbers of fishes. Macroalgal diversity, in contrast, peaks at temperate latitudes with lower species numbers in the tropics or at the poles, a pattern also seen in many oceanic taxa. Unusually, marine mammal diversity actually peaks at the poles, with lower numbers towards the equator (see Box 8.1).

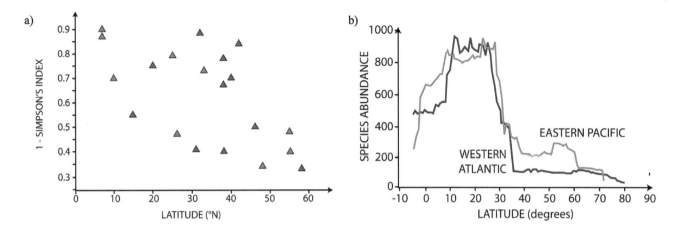

Figure 8.2 Examples of latitudinal gradients in biodiversity: (a) benthic biodiversity of estuaries, (b) latitudinal pattern of biodiversity in deep-sea gastropods in the Atlantic (brown line) and Pacific (blue line) Oceans. The latter image shows a biodiversity maximum at 10–30° of latitude.

Box 8.1   Explanations for Latitudinal Gradients: Eurythermal and energy theories

Researchers, often working in the terrestrial realm, have attributed latitudinal gradients on Earth to a variety of drivers including competition, mutualism, predation, epidemics, biotic spatial heterogeneity, population size, niche width, population growth rate, patchiness, epiphyte load, host diversity and harshness, environmental stability, environmental predictability, productivity, physical heterogeneity, abiotic filtering, sun angle, area, aridity, seasonality, numbers of habitats, and latitudinal ranges. In reviewing all of these possibilities, ecologist Klaus Rohde eliminated most explanations as circular or unsupported by evidence, and concluded that ecological and evolutionary time cannot provide a satisfactory explanation. Instead, he focused on energy and temperature. Considering seasonal climate oscillations at mid and high latitudes, the plants and animals that occupy these regions must necessarily tolerate temperature variation (eurythermal): higher tolerance of temperature differences allows them to occupy more ecological niches. Tropical species typically tolerate narrow temperature ranges because they rarely, if ever, experience substantial temperature changes, thanks to environmental stability of tropical environments over time. This strategy enables greater niche specialization. Warmer temperatures also favor evolutionary speed by shortening generation times, acceleration of selection, and increased mutation rates, all of which increase speciation potential.

Researchers have studied the distribution of biodiversity across latitudinal gradients for many years. Although general "rules" rarely exist in ecology, some have been proposed. Among them, **Rapoport's rule**, states that latitudinal ranges of species are generally narrower at lower latitudes than at higher latitudes; Eduardo H. Rapoport was the first to provide evidence for the phenomenon in 1975. This rule aligns with greater species diversity in the tropics than at the poles, noting higher latitudinal ranges (and thus available surface area) at higher latitudes. The rule motivated exploration of distributional patterns of plants and animals, and has since been extended to marine biodiversity. However, Rapoport's rule apparently does not perfectly apply to patterns observed in marine systems.

As species richness increases from the poles to the equator, the geographic range of species decreases. Decreased sea surface temperatures and available seabed at higher latitudes offers one hypothesis to explain reduced diversity at the poles. Research on the phylogeny of bivalves shows high speciation rates in inter-tropical environments, and particularly the Sino-Indonesian region, and that new species progressively adapt to life at higher latitudes. These studies also show that species that evolved in ancient times (old clades) often span large latitudinal gradients, whereas recent speciation (young clades) typically occur only in equatorial environments, not having had time to adapt to life at high latitudes. The presence of old and young clades in the tropics and largely older clades at the poles is therefore consistent with observed latitudinal gradients (Figure 8.3). However, more recent research on meiofauna shows increasing diversity with increasing latitude. These contrasting patterns among taxa suggest that variables associated with changes in latitude can affect biodiversity components differently. However, these differences in pattern may simply reflect differences in dispersal or other confounding factors.

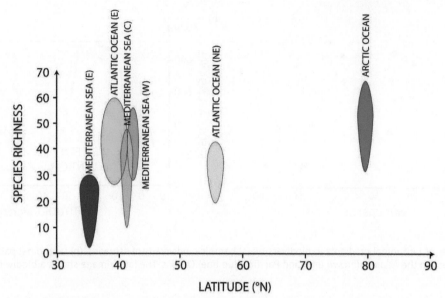

Figure 8.3 Trend in the number of species of marine nematodes with increasing latitude in different deep ocean regions.

8.2.3 Longitudinal Gradients in Tropical Biodiversity

Coral species and genera reflect the best-known longitudinal pattern in diversity, where highest diversity occurs in the Indonesian archipelago and decreases moving toward the west Pacific Ocean. Crossing the Indian Ocean, coral diversity decreases from the epicenter of highest diversity to the Red Sea, with similar patterns in mangroves and gastropods. The Indonesian archipelago seems to be the global epicenter for tropical marine biodiversity evolution, likely reflecting a long period of evolutionary stability and a great diversity in the types of islands and archipelagos. Presumably periods of isolation led to allopatric speciation, followed by periods of reunification that led to sympatric speciation. On a geological scale, rapid evolution and speciation often follow mass extinctions. Intriguingly, fossil and molecular data show that global biodiversity hotspots have "hopped" around Earth over the past 50 million years, demonstrating that environmental conditions and their changes must play a role in setting broad-scale biodiversity patterns.

8.2.4 Bathymetric Patterns in Marine Biodiversity

Diversity patterns as a function of depth have also received significant attention in marine environments. Researchers originally predicted that diversity would decrease with increasing depth because of the features of deep-sea environments, with limited food, high pressure, and the absence of light. Subsequent studies demonstrated relatively high biodiversity (expressed using the diversity measure "expected species") even at significant depth. The low abundance of individuals supports the idea of depauperate deep-sea environments, but with a truly impressive richness of species. Even more surprising, a "bell-shaped" pattern in diversity with increasing depth appears pervasive in the deep sea, contradicting any unidirectional or constant gradient in species richness with depth. This pattern arises as biodiversity progressively increases from the continental shelf to a depth of 1,000–2,000 m and then gradually decreases to abyssal depths (Figure 8.4).

This bell-shaped trend of diversity (lower diversity on the upper continental slope and abyssal plains, and higher at intermediate depths) seems to occur in many parts of the ocean, but with some variation (see Figure 8.5). For example, in upwelling regions, highest biodiversity and biomass occur at the surface and typically decrease at depths corresponding to where Oxygen Minimum Zones begin, or in particularly resource poor environments such as the deep eastern basin of the Mediterranean. To date, researchers do not fully agree on the drivers of this bell-shaped curve, though presumably marine biodiversity modulation plays some role. Researchers recently suggested that speciation processes may occur at intermediate (bathyal) depths, and new species then progressively adapt to spread to abyssal depths and to most surface environments. Alternatively, this pattern may reflect overlapping ranges for shallow and deep-water species.

8.3 Marine Biogeography

Biogeography, the study of organisms in space, depends on spatial distributions and combines geographical, ecological, and oceanographic studies. Biogeographic studies consider the spatial distribution of individuals, populations, species, and

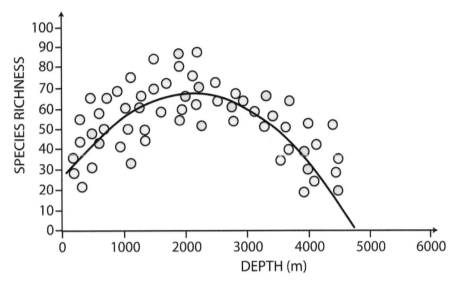

Figure 8.4 Pattern of macrofaunal biodiversity (mollusks, polychaetes, etc.) with depth. Highest values occur at intermediate depth, between 1,000 and 2,000 meters, with lower values at greater depth or on the shallower continental shelf.

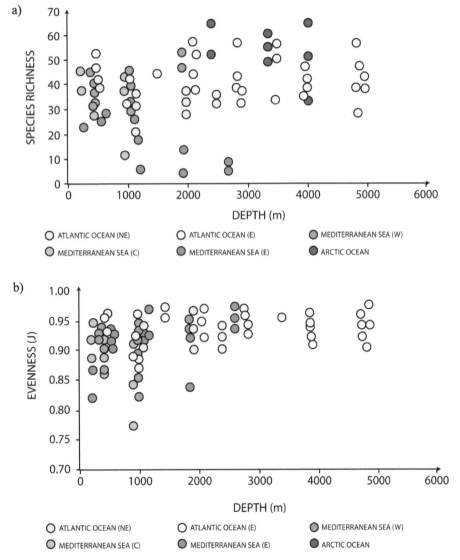

Figure 8.5 Depth-related patterns of meiofaunal biodiversity: (a) Biodiversity of nematodes; (b) Evenness.

habitats, taking into account both patterns and processes driving distributions. Patterns in biogeography, including range, density, diversity, and degree of isolation, define biogeographic areas, thus providing valuable classification tools. **Area**, in the context of biogeographical studies, refers to the portion of the geographic region in which an organism occurs and directly interacts with the ecosystem; experts more accurately refer to extent of occupancy.

By comparing different taxonomic groups within a geographic region, as well as climatic and topographic characteristics, scientists can identify distinct zones, each characterized by the presence of specific biota exclusive to that region, which can therefore define specific biogeographic regions, sub-regions, provinces, and sectors.

8.3.1 Terrestrial and Marine Biogeography

Most biogeography texts focus exclusively on the terrestrial domain, reflecting a widespread belief among many authors that continuity of marine ecosystems challenges efforts to define boundaries. The perceived lower speciation rate in the ocean than on land also implies that most marine species have been around long enough to disperse over vast marine areas and suitable habitats. In the ocean, ecological factors therefore play a greater role than historical factors in defining biogeographic limits. In fact, marine flora and fauna distinctly differ among geographical regions, with continents, ocean currents, and oceanographic fronts defining primary biogeographic barriers. Application of models developed in terrestrial environments to the ocean must consider the huge phyletic, functional, and ecological differences between terrestrial and marine ecosystems, and resulting differences in marine and terrestrial biogeographic patterns.

8.3.2 Biogeographical Regions

Scientists identified the major biogeographic zones on land in the mid-nineteenth century, in contrast to the first attempts to define marine biogeographical regions in the early decades of the twentieth century. The massive size and three-dimensional nature of the global ocean delayed efforts to define its biogeographical regions. Because of the proximity of the coastal zone to land and therefore easier access, knowledge of coastal biogeography has advanced more quickly than that of the deep sea and the pelagic environments. We now recognize a total of 23 coastal biogeographic regions and 29 pelagic regions (the latter shown in Figure 8.6). Despite many remarkable similarities between the polar regions, their biotas differ substantially. Two theories explain the truly **bipolar** species or genera (those limited numbers of species present at both poles but absent in between): relict distributions and migratory species. The first theory asserts that organisms

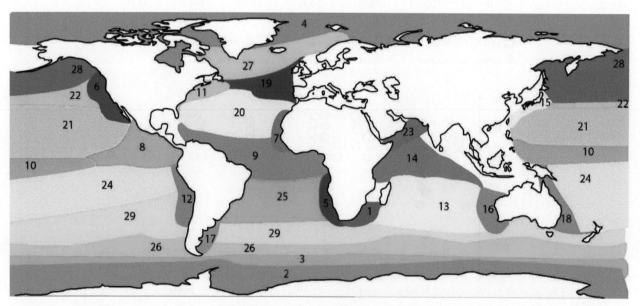

Figure 8.6 Pelagic ecoregions identified by the United Nations in 2007 (upper half): (1) the Agulhas Current, (2) Antarctica, (3) Antarctic Polar front, (4) Arctic, (5) Benguela Current, (6) California Current, (7) Canaries Current, (8) tropical Pacific Eastern Europe, (9) equatorial Atlantic, (10) equatorial Pacific, (11) Gulf of Stream, (12) Humboldt Current, (13) Tour of the Indian Ocean, (14) Tour of the Indian Ocean Monsoon, (15) Kuroshio front, (16) Current of Current, (17) Malvinas Islands Current, (18) Pacific Southwestern, (19) North East Atlantic, (20) Central North Atlantic gyre, (21) Central North Pacific gyre, (22) North Pacific, (23) Somali Current, (24) Tour of the central South Atlantic, (25) Tour of the South Central Pacific, (26) Sub-Antarctic, (27) Atlantic Subarctic, (28) Pacific Subarctic, (29) subtropical Convergence; and biogeographic regions identified by NRIC (lower half).

once widely distributed in the past went extinct at intermediate latitudes over time, whereas the second theory asserts that some organisms migrated over geological time between the two poles through deep, colder waters.

Biogeographic study of the deep-sea environment, a relatively recent discipline, requires constant revisions to explain some often-complex patterns, but some generalities exist. In the past, researchers believed that globally, all abyssal habitats harbored the same, somewhat depauperate fauna, which they grouped in a single biogeographic region. Today we recognize a diverse, though numerically and biomass sparse abyssal biodiversity. Even for the pelagic environment, a satisfactory biogeographical synthesis requires further investigation and assessment. In general, eurythermal species dominate the zooplankton, spanning wide longitudinal distributions that contrast the comparatively narrow longitudinal distributions of pelagic fishes.

8.3.3 Species Distributions Within Biogeographic Regions

Total numbers of marine species was estimated and censused across different geographic regions defined by the Census of Marine Life, an international collaboration that, for 10 years, conducted the largest worldwide study to document the diversity, distribution and abundance of ocean life (Figure 8.7). Nonetheless, a truly comprehensive analysis of ocean life remains incomplete, because many marine species remain undescribed or undiscovered and large portions of the ocean, particularly in the deep sea, remain badly undersampled.

In fact, recent estimates indicate that somewhere between 25 and 80% of marine biodiversity in Australia, Japan, the deep Mediterranean, New Zealand, and South Africa remains undescribed. Even higher proportions of undescribed Asian, Indian Ocean and tropical Pacific species are expected. Thus, the portion of marine species yet to be discovered could total 70–80% or more of global marine biodiversity. To date, scientists have described ~250–270,000 species (230,000 of which are already listed in the World Register of Marine Species, WoRMS database), and estimate that a total of ~2.2 million eukaryotic marine species live on Earth.

In many regions, the crustaceans, mollusks, and fishes that comprise the most (known) biodiverse-rich taxa account for more than half of known marine species richness (Table 8.1), almost certainly reflecting a taxonomic bias towards these

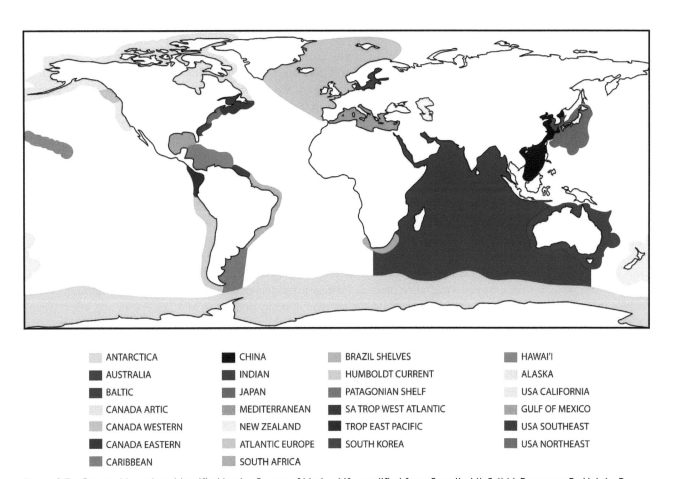

Figure 8.7 Geographic regions identified by the Census of Marine Life, modified from Costello MJ, Coll M, Danovaro, R., Halpin, P., Ojaveer, H., Miloslavich, P. (2010) A Census of marine biodiversity knowledge, resources, and future challenges. PLoS ONE 5(8): e12110.

groups, whereas protozoa and algae contribute ~10% each. The percentage by which each taxon contributes to regional species richness varies considerably among regions: Crustaceans, for example, account for 22–35% of the species in Alaska, Antarctica, Arctic, Brazil, California, Caribbean, Eastern Canada, and the Humboldt Current region, but only about 10% of Baltic species. Mollusks account for 26% of Australian and Japanese marine species, but for only 5–7% of Baltic, California, Arctic, and Canadian species. Fishes contribute from 18 to 32% in the Eastern United States, tropical Eastern Pacific, and tropical Eastern Atlantic, but only 3–6% of known Arctic, Antarctica, Baltic, and Mediterranean marine species.

"Plants and algae" (primarily algae) account for 28–38% of Baltic, Arctic, Atlantic European, and Eastern Canadian species, but only 5% of Antarctic, Caribbean, Chinese, Humboldt Current, tropical Eastern Pacific and tropical Eastern Atlantic species. The most variable groups reported for most regions are "plants and algae" and "other invertebrates," reflecting variability in their taxonomic classification in different regions. In contrast, relative proportions of crustaceans and mollusks vary less among regions. Experts used the percentage contribution of different taxa to total biodiversity in the best-known regions (such as Europe) to extrapolate how many unknown species of other taxa occur within less-studied habitats (Table 8.1). The high percentages of taxa other than fishes, crustaceans, and mollusks in some regions may also reflect a different classification of species or possibly taxonomic errors, which may explain greater variation in the "others" group, which varies more than individual taxa. It may also reflect sampling differences in that sampling in some regions may favor some taxa over others. Variation in the species-richest and best-known taxa such as fishes, which span 3% to over 20%, suggests that proportions of taxa that contribute to regional diversity actually vary among different parts of the world. To achieve similar species richness would require similar pattern of dispersion, speciation, and extinction, an unlikely scenario given the diversity of taxa and how they respond to their environment. Thus, corals and annelids in sediments, for example, occur in high diversity in the tropics but in lower diversity elsewhere. Species richness and size of biogeographic region correlate weakly in the ocean, indicating a poor relationship between species and habitat area, likely reflecting under-sampling of vast ocean regions. Good taxonomic knowledge exists for Australia, New Zealand, Atlantic Europe, China, Japan, the United States, and the Mediterranean Sea, with European seas as perhaps the best studied globally. Lack of specialists in some geographic regions may explain an apparent paucity of species.

8.3.4 Biogeography of the Mediterranean

Although the Mediterranean occupies only 0.82% of the world ocean surface and 0.32% of its volume, it represents a global hotspot of biodiversity. Species richness (about 8,500 species of **metazoans**, or multicellular organisms) accounts for 7.5% of all described marine species (between 4 and 18% depending on the taxonomic group in question): 67% of known Mediterranean species occur in the western Mediterranean, 38% in the Adriatic Sea, 35% in the central Mediterranean, 44% in the Aegean Sea, and 28% in the Levantine Sea. Several aspects of the Mediterranean Sea create great biogeographical interest. As the largest temperate-warm marine region on Earth that once occupied equatorial latitudes, its complex geological history contributes to ten distinct biogeographic regions (Figure 8.8). Mediterranean floral and faunal diversity concentrate in the central and western parts of the basin. Clear affinities between the Alboran Sea and Atlantic attest to continued penetration of flora and fauna through the Strait of Gibraltar. The northern Adriatic features significant peculiarities, with affinities in flora and fauna with the Black Sea and the North Sea (characterized by lower salinity and temperature, nutrient-rich waters, greater tidal range). The import of many Lessepsian species, referring to the steady flow of "alien" species that began with the opening of the Suez Canal in 1869, into the Eastern Mediterranean Sea has led to a proposal to consider the southeastern part of the Mediterranean Sea as an individual biogeographic province. Lessepsian takes its name from Ferdinand de Lesseps, who directed the excavation of the canal. The high salinity of the Bitter Lakes, located along the Suez Canal, created a migration obstacle for migrating species, however, towards the middle of the twentieth century, decreasing salinity in the lakes enabled the survival of many organisms and Lessepsian migration began. This ongoing migration drives the biggest ongoing marine biogeographic change; the opening of the Suez Canal has enabled the introduction of about 600 species from the Red Sea over the last 140 years, including many that compete with Mediterranean local and/or endemic species.

Past geological and climatic events over the past 10 million years have profoundly influenced the evolution of the biota of the Mediterranean Sea benthos and the basin's current biogeography. No other basin on Earth has experienced such intense geologic, oceanographic, and climatic events over such a limited geological time interval. Until the Miocene, the Mediterranean connected to the Indo-Pacific Ocean in the east and to the Atlantic Ocean in the west. In the late Miocene, the Atlantic was the sole connection with the Mediterranean. The Mediterranean sits between 30° and 45°N, and therefore experiences strong climatic oscillations that, over time, have played a major role in extinction events and geographic expansion of biota. During the Mesozoic Era (245–65 million years ago), the Tethys Ocean supported a rich biota of warm water species similar to that found today in the tropical Indo-Pacific region, including coral reefs. During the Cenozoic, the

Table 8.1 Percentage of known species of each taxon for the biogeographical regions illustrated in Figure 8.7.

	Total eukaryotes	Crusta- ceans	Mollusks	Fishes	Protozoans	Plants and algae	Anellids	Cnidarians	Other invertebrates	Platyhel- minthes	Echinoderms	Poriferans	Bryozoans	Other vertebrates	Tunicates
% area >10%		81	58	58	35	29	23	0	13	3	0	0	0	0	0
Australia	32889	19	26	16	2	6	5	5	3	2	5	5	3	1	3
Japan	32777	19	26	12	14	7	3	6	4	1	3	2	1	0	1
China	22365	19	18	14	21	5	5	6	2	2	3	1	3	1	1
Mediterranean	16848	13	13	4	24	7	7	4	13	6	1	4	2	0	1
Gulf of Mexico	15374	17	16	10	14	13	6	5	4	5	3	2	2	3	1
New Zealand	12780	17	18	10	12	11	4	6	4	2	4	4	5	1	1
South Africa	12715	18	24	15	2	7	6	7	5	3	3	3	2	2	2
Atlantic Ocean	12270	18	11	9	4	28	13	4	0	2	2	4	3	2	1
Caribbean	12046	24	25	11	7	5	5	8	3	1	4	4	1	0	1
Current of Humboldt	10186	31	12	11	7	5	6	5	8	2	4	2	4	2	1
USA California	10160	26	7	9	9	9	8	4	7	14	3	1	1	1	1
Chorea	9900	14	19	11	3	9	5	3	25	1	2	3	1	2	1
Brazil	9101	22	20	14	3	9	11	6	3	0	3	4	1	2	1
Hawaii	8244	16	16	15	10	12	4	6	3	8	4	2	2	1	1
Antarctica	8200	35	9	4	8	4	7	6	7	2	7	3	4	3	1
East Tropical Pacific	6696	13	13	18	14	5	28	2	1	0	3	1	1	1	0
Alaska	5925	26	8	7	13	7	9	4	10	2	3	3	6	2	1
Baltic	5865	10	5	3	20	30	7	2	13	5	1	0	1	2	0
USA NE	5045	16	17	19	1	12	14	4	3	2	3	1	3	4	1
USA NE	4229	16	17	28	4	8	9	9	1	0	0	3	2	2	1
Platform of Patagonia	7776	16	22	14	0	7	5	7	5	1	5	7	4	5	1
East Canada	7160	23	7	17	19	12	14	3	2	0	2	0	0	1	0
Arctic Canada	7038	24	5	6	12	36	11	2	2	0	1	0	0	1	0
West Tropical Atlantic	2743	19	16	32	2	5	6	5	2	0	4	1	0	8	1
West Canada	2636	18	7	14	4	38	14	0	2	0	1	0	0	1	0
Average	10759	19	17	12	10	10	7	5	5	3	3	3	2	2	1
Variation Coefficient		−0.29	−0.38	−0.54	−0.67	−1	−0.74	−0.39	−1.02	−0.97	−0.49	−0.65	−0.65	−1.04	−0.51

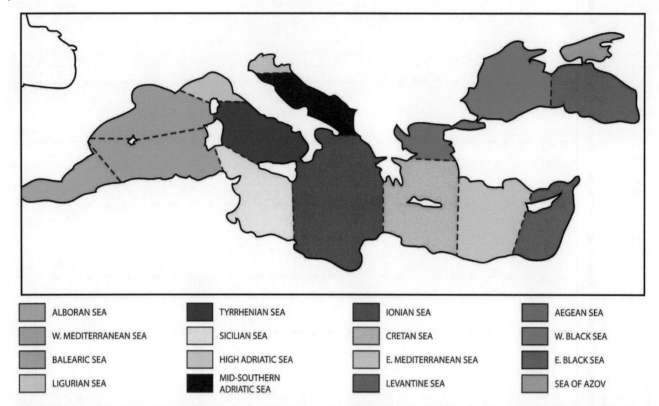

Figure 8.8 Biogeographic regions of the Mediterranean and Black Sea. The black lines delimit the approximate boundaries between the different sub-Mediterranean regions.

Tethys was largely tropical, and 30 million years ago, during the Oligocene, it gradually cooled. Starting in the late Miocene, the climate of the basin shifted from a tropical to cold boreal regime, with mainly temperate conditions over the last two million years. These oscillations, between tropical conditions in the Miocene glacially tempered conditions during the Quaternary led to radical biological changes in the Mediterranean basin.

The major steps of this change are as follows:

- **From 21 to 11 million years ago (from the Middle to the Late Miocene),** the warm climate supported Mediterranean biota including many species with tropical and subtropical affinity; towards the end of the period, subtropical conditions dominated, and only a fraction of the tropical marine biota survived.
- **From 11 to 7.4 million years ago (from the Late Tortonian to the Messinian),** the subtropical benthic marine biota supported tropical coral reef species. 10 million years ago, the rise of the Suez Isthmus separated the nascent Mediterranean from the Indo-Pacific. In the Messinian (about 6 million years ago) the Strait of Gibraltar connection with the Atlantic also closed, fully isolating the Mediterranean from adjacent oceans. As a result of a **negative water balance** (where evaporation losses exceed water inputs from rivers and rainfall) the Mediterranean almost completely dried up, forming large evaporite lakes and giant salt deposits on the seabed (the "salinity crisis" of the Messinian). The Messinian crisis lasted less than a million years, and in the Pliocene, 5 million years ago, the Strait of Gibraltar re-opened, and Atlantic waters refilled the basin, repopulating it with Atlantic species. The Mediterranean then became an Atlantic province.
- **From 7.4 to 5.3 million years ago (early Pliocene-middle Pliocene),** subtropical conditions of the Late Miocene persisted with progressive extinction of stenothermal species adapted to hot climates.
- **From 5.3 to 1.8 million years ago (from the Middle to the Late Pliocene),** the climate was considerably warmer than that present until the Middle Pliocene. In the late Pliocene, the continental climate shifted towards temperate conditions.
- **From 1.8 million years ago (Quaternary) to the Present,** the cooling trend marked the transition between the warm-temperate Pliocene to the cold-temperate Pleistocene. Mediterranean marine ecosystems profoundly changed as a result of waves of immigration of boreal "guests". Benthic coastal organisms that were important during glacial periods disappeared from the basin when interglacial conditions periodically recurred in the Mediterranean Sea, but other organisms failed to colonize the basin.

The current marine biota of the Mediterranean includes species with different biogeographic affinities, including some widely distributed species (although there are no truly cosmopolitan marine species that are present in all oceans at all latitudes). The biota also includes **panoceanic species** (circumtropical species that tolerate the warm southern Mediterranean) and, less commonly, **circumboreal species** that span the cold-temperate seas of the northern hemisphere and the northern Mediterranean. Atlantic-Mediterranean species dominate the current Mediterranean biota, including some subspecies that differ between the Mediterranean and Atlantic.

The Mediterranean also hosts numerous **endemic species** (those that live exclusively in one region). These include: (1) **paleoendemic species**, probably of Tethys Sea origin, (2) **neoendemic species**, especially of Pliocene origin that often include pairs of similar species, one Atlantic and the other Mediterranean. The Atlantic subtropical species represent what remains of the Senegalian biota, which penetrated from the Atlantic to the Mediterranean during interglacial phases (especially in the Tyrrhenian). Circumtropical species live mainly on the south coasts of the Mediterranean basin. The Atlantic boreal species (**Celtic biota**), which penetrated during glacial phases (Würm), as well as circumboreal species, occur together along the Mediterranean north shores.

Several regions demonstrate the northward movement of species with greater affinity for warm waters. The Ligurian Sea, one of the coldest areas of the Mediterranean, supports the lowest number of subtropical species and the highest abundance of species characteristic of cold-temperate waters. The warming of Ligurian waters facilitated the penetration of warm water species such as the ornate wrasse, *Thalassoma pavo*, which has established a large and stable population since 1985.

However, the increase in tropical Atlantic species in the northern Mediterranean may reflect a combination of climatic and anthropogenic factors. Recent studies correlate the **North Atlantic Oscillation** (NAO), a North Atlantic weather phenomenon in which atmospheric pressure at sea level varies over time periods of approximately decades, and climate variability of the North-Western Mediterranean.

A mass mortality event occurred in 1999, when a positive temperature anomaly during the summer, combined with an increase in the mixed layer below 40 m depth caused significant mortality of at least 28 species of invertebrates. This climate anomaly spanned much of the coasts of France to Italy, with a reduced impact on the coast of Corsica. The most strongly impacted benthic fauna included sponges and gorgonians, such as *Paramuricea clavata*, *Eunicella singularis*, *Lophogorgia ceratophyta* and *Eunicella cavolini* (Figure 8.9). Evidently, even short-lived temperature anomalies can dramatically alter Mediterranean faunal diversity, particularly when exacerbated by increases in opportunistic pathogens (including some fungi and protozoa). Once a species disappears, other species, pre-adapted to the new conditions, can replace it, so that the ecosystem cannot return to pre-impact conditions. Strong thermal anomalies can also impact cave faunas, leading to the replacement of endemic species with warmer water species.

Increased temperatures can also increase abundances of viruses. In the Adriatic Sea, the proportion of thermophilic fish species increased over the last 30 years, and species of previously rare fishes and zooplankton have become abundant, including species new to the region. These observations relate to intensified climate change effects after 1988, increasing temperature and salinity variation in the Adriatic Sea. These changes have altered climate in many ways, including increased frequency of storms and heavy rain, and altered wind speed and direction, which, in turn, affect the entire Adriatic ecosystem. These oceanographic changes have triggered blooms of jellyfish (*Pelagia noctiluca*) and thaliaceans (salps and doliolids), harmful algal blooms, and red tides (caused by blooms of several species of dinoflagellates; Figure 8.10).

Figure 8.9 Necrosis and detachment in gorgonians: (a) *Paramuricea clavata;* (b) *Eunicella cavolinii. Source:* Carlo Cerrano.

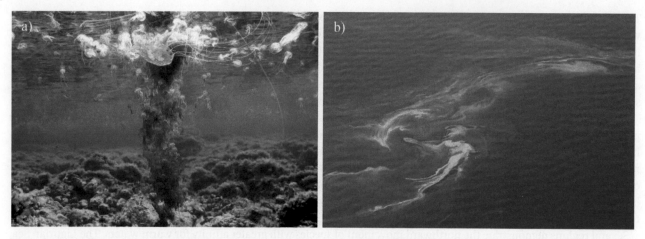

Figure 8.10 (a) The sudden and massive growth of marine organisms is called a bloom. The images show a bloom of jellyfish (*Pelagia noctiluca*), and (b) a red tide generated by dinoflagellates in the Adriatic, offshore of the coast of Conero (Adriatic Sea, Mediterranean). *Source:* Francesco Di Trapani.

8.4 Theories on Evolution and Maintenance of Biodiversity: The Deep-Sea as Examplar

The high diversity that characterizes deep-sea and other marine environments depends on complex biotic interactions with physical and geochemical variables that determine the number of coexisting species and on regional and global historical processes that determine the number of species in a community. Building on our earlier discussion of drivers of biodiversity patterns, and drawing on our own expertise in deep-sea ecosystems as an exemplar of the debate, we present two main groups of theories address deep-sea biodiversity "maintenance":

(1) Theories based on equilibrium hypotheses and (2) Theories based on disequilibrium hypotheses. The first group includes the stability-time hypothesis, and the habitat heterogeneity hypothesis. The second group includes the biological disorder hypothesis, the intermediate disturbance hypothesis, and the patch mosaic theory.

The **stability-time hypothesis** was developed by the deep-sea biologist Howard Sanders, who hypothesized in 1968 that the physical stability of marine environments (e.g., deep or tropical) allows species to adapt and specialize, minimizing competitive interactions between species. Relatively constant food resources contribute to high diversity. In **oligotrophic**, or low production regimes, diversity exceeds that in eutrophic areas, because individuals often specialize in order to acquire enough food. However, some researchers, including Robert Aller in 1989, argued that the deep-sea environment is not as stable as previously believed, in that various events such as benthic storms disrupt communities.

The **habitat heterogeneity hypothesis** attributes diversification to spatial heterogeneity rather than stability. In 1982, Peter Jumars hypothesized that sediment stability in most deep environments allows organisms to specialize on and exploit microhabitats, noting that some deep-sea environments experience current regimes.

The **predator cropping**, or **biological-disorder hypothesis**, theorizes that predators control or crop back the number of organisms present in a location. Predators may include **benthopelagic** organisms that move between the water column and sea seafloor such as grenadier fishes, or benthic macrofauna on meiofauna. This predation, unless excessive, adds a degree of "disorder" that reduces and controls numbers of opportunistic and dominant organisms that might otherwise outcompete other species for resources.

Physical disorder or **intermediate disturbance hypothesis:** episodic natural phenomena can disturb communities, resulting in mortality (Figure 8.11). If phenomena, such as seabed landslides (or **turbidity currents**), benthic storms, cold "waterfalls," occur at an intermediate frequency, then climax communities never form, and dominant competitors do not drive inferior competitors to extinction. Disturbance events reduce diversity and reset the re-colonization sequence. With highly frequent disturbances, only the most opportunistic species can persist, but intermediate disturbance also prevents the attainment of equilibrium and the associated dominance of one or a few species. In fact, low densities of organisms reduce competitive interactions, thus allowing potential competitors to coexist.

The **patch mosaic theory** correlates small-scale heterogeneity and disturbance (Figure 8.12). Fred Grassle and Howard Sanders hypothesized in 1972 that habitats with limiting food resources depend on external inputs of food; abundant resources create greater spatial heterogeneity (Figure 8.12a). However, when species reach a climax community, K-strategy species tend to dominate the system, excluding r-strategists and decreasing diversity. If a catastrophic phenomenon (major predator disturbance, major sediment resuspension event) defaunates an area of seabed (Figure 8.12b), colonization

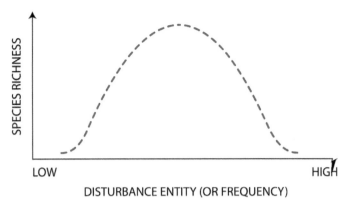

Figure 8.11 Example of the relationship between biodiversity and disturbance. With very infrequent disturbance, competitive dominants increase in abundance with a consequent reduction in biodiversity. With excessive disturbance, biodiversity decreases because most species cannot cope for extended periods. Intermediate levels of disturbance allow more species to coexist.

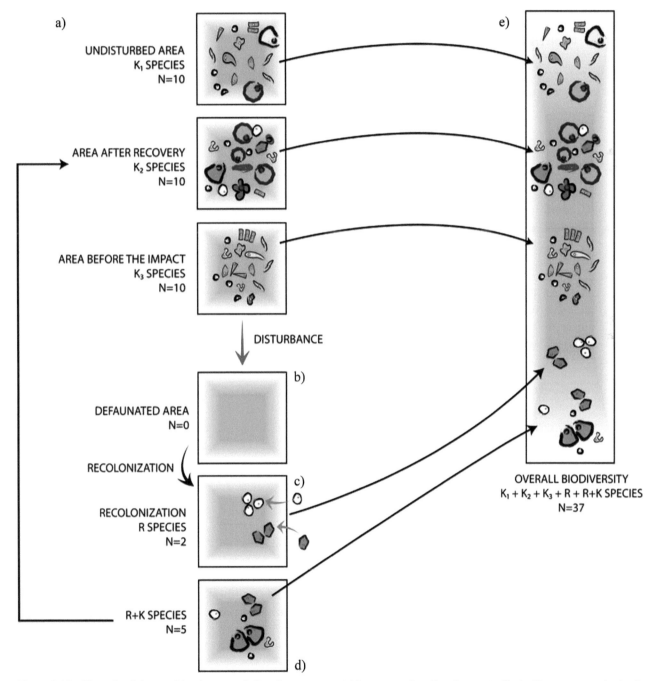

Figure 8.12 The role of the combined stages of disturbance on spatial heterogeneity: disorder creates "holes" in a system colonized over time by opportunistic and pioneer species. Other K strategists then follow sequentially. In summary, a mosaic created by different r-colonizers following disturbance events creates a heterogeneous habitat, in which K strategists begin to arrive and contribute to high biodiversity. The mosaic allows r-strategists and K-strategists to coexist.

begins, favoring r-strategists (Figure 8.12c). These species, though not numerous, differ from K-strategists (Figure 8.12d), and once K species also colonize the system, overall diversity increases. However, in the absence of small-scale disturbance, K strategists outcompete r strategists over time, and approach equilibrium with reduced diversity (Figure 8.12a). If a **mosaic**, referring to stages of heterogeneity and disturbance that includes different successional stages of K species and r species, then the overall diversity of the mosaic created by these microenvironments will greatly exceed that present a single area (Figure 8.12e).

Questions

1) How many phyla occur in marine environments and how many of them are endemic to the ocean?
2) How does biodiversity change with latitude?
3) How does biodiversity change with longitude?
4) How does biodiversity change with increasing depth?
5) What factors drive marine biodiversity patterns?
6) What are alien species and what are their implications for ecosystem functioning?
7) What are Lessepsian migrations and where do they take place?

Suggested Reading

Beaugrand, G., Luczak, C., Goberville, E., and Kirby, R.R. (2018) Marine biodiversity and the chessboard of life. *PLOS One*, 13(3), p. e0194006.

Brandão, M.C., Benedetti, F., Martini, S. et al. (2021) Macroscale patterns of oceanic zooplankton composition and size structure. *Scientific Reports*, 11, p. 15714.

Briggs, J.C. (1974) *Marine Zoogeography*. McGraw-Hill.

Brown, A. and Thatje, S. (2014) Explaining bathymetric diversity patterns in marine benthic invertebrates and demersal fishes: Physiological contributions to adaptation of life at depth. *Biological Reviews*, 89, pp. 406–426.

Chaudhary, C., Saeedi, H., and Costello, M.J. (2016) Bimodality of latitudinal gradients in marine species richness. *Trends in Ecology and Evolution*, 31, pp. 670–676.

Clark, M., Althaus, F., Schlacher, T. et al. (2015) The impacts of deep-sea fisheries on benthic communities: A review. *ICES Journal of Marine Science*, 73, pp. i51–i69.

Compaire, J., Gomez-Enri, J., Cama, M.C. et al. (2019) Micro- and macroscale factors affecting fish assemblage structure in the rocky intertidal zone. *Marine Ecology Progress Series*, 610, pp. 175–189.

Connell, J.H. (1978) Diversity in tropical rain forests and coral reefs: High diversity of trees and corals is maintained only in a nonequilibrium state. *Science*, 199, pp. 1302–1310.

Danovaro, R., Bianchelli, S., Gambi C. et al. (2009) α-, β-, γ-, δ- and ε-diversity of deep-sea nematodes in canyons and open slopes of Northeast Atlantic and Mediterranean margins. *Marine Ecology Progress Series*, 396, pp. 197–209.

Eddy, T.D., Bernhardt, J.R., Blanchard, J.L. et al. (2021) Energy flow through marine ecosystems: Confronting transfer efficiency. *Trends in Ecology and Evolution*, 36, pp. 76–86.

Ekman, S. (1953) *Zoogeography of the Sea*. Sidgwick and Jackson.

Fuhrman, J.A., Steele, J.A., Hewson, I. et al. (2008) A latitudinal diversity gradient in planktonic marine bacteria. *Proceedings of the National Academy of Sciences of the United States of America*, 105, pp. 7774–7778.

Jablonski, D., Roy, K., and Valentine, J.W. (2006) Out of the tropics: Evolutionary dynamics of the latitudinal diversity gradient. *Science*, 314, pp. 102–106.

Levin L.A., Etter R.J., Rex M.A. et al. (2001) Environmental influences on regional deep sea species diversity. *Annual Review of Ecology and Systematics*, 32, pp. 51–93.

Petsas, P., Doxa, A., Almpanidou, V. et al. (2022) Global patterns of sea surface climate connectivity for marine species. *Communications Earth and Environment*, 3, p. 240.

Pommier, T., Canback, B., Riemann, L. et al. (2007) Global patterns of diversity and community structure in marine bacterioplankton. *Molecular Ecology*, 16, pp. 867–880.

Renema, W., Bellwood, D.R., Braga, J.C. et al. (2008) Hopping hotspots: Global shifts in marine biodiversity. *Science*, 321, pp. 654–657.

Rex, M.A., Etter, R.J., Morris, J.S. et al. (2006) Global bathymetric patterns of standing stock and body size in the deep-sea benthos. *Marine Ecology Progress Series*, 317, pp. 1–8.

Rex, M.A., McClain, C.R., Johnson, N.A. et al. (2005) A source-sink hypothesis for abyssal biodiversity. *American Naturalist*, 165, pp. 163–178.

Rex, M.A., Stuart, C.T., Hessler, R.R. et al. (1993) Global-scale patterns of species diversity in the deep-sea benthos. *Nature*, 365, pp. 636–639.

Rex, M.A. (1983). Geographic patterns of species diversity in the deep-sea benthos, in Rowe, G.T. (ed.), *The Sea*, Vol. 8. John Wiley & Sons, pp. 453–472.

Rohde, K. (1992) Latitudinal gradients in species diversity: The search for the primary cause. *Oikos*, 1, pp. 514–527.

Roy, K., Jablonski, D., Valentine, J.W., and Rosenberg, G. (1998) Marine latitudinal diversity gradients: Tests of causal hypotheses. *Proceedings of the National Academy of Sciences of the United States of America*, 95, pp. 3699–3702.

Roy, K., Jablonski, D. and Valentine, J.W. (1994) Eastern Pacific molluscan provinces and latitudinal diversity gradient: No evidence for "Rapoport's rule". *Proceedings of the National Academy of Sciences USA*, 91, pp. 8871–8874.

Stehli, F.G., Douglas, R.G. and Newell, N.D. (1969) Generation and maintenance of gradients in taxonomic diversity. *Science*, 164, pp. 947–949.

Tittensor, D.P., Mora, C., Jetz, W. et al. (2010) Global patterns and predictors of marine biodiversity across taxa. *Nature*, 466, pp. 1098–1101.

Webb, T.J., Vanden Berghe, E., and O'Dor, R. (2010) Biodiversity's big wet secret: The global distribution of marine biological records reveals chronic under-exploration of the deep pelagic ocean. *PLOS One*, 5(8), p. e10223.

Witman, J.D., Etter, R.J., and Smith, F. (2004) The relationship between regional and local species diversity in marine benthic communities: A global perspective. *Proceedings of the National Academy of Sciences of the United States of America*, 101, pp. 15664–15669.

Worm, B., Sandow, M., Oschlies, A. et al. (2005) Global patterns of predator diversity in the open oceans. *Science*, 309, pp. 1365–1369.

Yasuhara, M., Wei, C.-L., Kucera, C.-L. et al. (2020) Past and future decline of tropical pelagic biodiversity. *Proceedings of the National Academy of Sciences of the United States of America*, 117, pp. 12891–12896.

Zinger, L., Amaral-Zettler, L.A., Fuhrman, J.A. et al. (2011) Global patterns of bacterial beta-diversity in seafloor and seawater ecosystems. *PLOS One*, 6(9), p. e24570.

9

Biodiversity of the Benthos

9.1 Introduction: Benthos and Plankton

Marine benthos derives from the Greek term "benthos" that means "depth of the sea" and refers to the assemblage of organisms of all size and life styles that live on, in, or near the seabed. The biotope that hosts benthic organisms is also known as the benthic zone. The benthos encompasses a complex of plant and plant-like photo-synthesizers, animals, and microbes associated with the seafloor. Three attributes distinguish benthic from pelagic domains: the time scale of change, the role of history, and their respective energy budgets.

1) Time scale: Typically, slower renewal (turnover) processes and longer life cycles in benthic organisms than their pelagic counterparts.
2) History: Because organisms persist for a longer period on the seafloor than in the plankton, their composition and abundance reflect past and present interactions with their environment (e.g., pollutants, temperatures) and with other species. Many benthic organisms therefore record their individual histories (e.g., within shells, scales and root structure of the seagrass, *Posidonia* spp.; Figure 9.1a), and affect the history of other organisms.
3) Energy balance: Overall, benthos utilize more energy than they produce, resulting in a net energy deficit. This energy deficit occurs because most seabed habitats (and especially those at depths >150 m) lack sufficient light to perform enough photosynthesis to offset their respiration needs, thus limiting net primary production. Most benthos, in fact, utilize **allochthonous** or non-local production (Figure 9.2), depending on production from other systems with higher turnover (plankton, microorganisms, debris, and suspended or deposited particles). Benthic environments often accumulate materials, including organic matter in the form of starches, lipids, and proteins.

9.2 Benthic Biota

The presence of substrate, whether rock or sediment, gives seafloor ecosystems much greater stability and persistence than the dynamic water column in which plankton live. Benthos includes all bottom-living animals (zoobenthos), photosynthetic flora (phytobenthos), and microbes (microbenthos), irrespective of whether all or part of their life cycle occurs on the seafloor. However, few researchers still use this classification, and most now use a system based on organism size. Benthos vary in size from 0.1 microns (viruses) to a few meters (megafaunal invertebrates and fishes that associate with bottom habitats (Table 9.1)). In all marine sediments, the abundances of these organisms increase exponentially with decreasing size of individuals (Figure 9.3). Abundances of benthic animals (expressed per m^2) typically vary from a few individuals of megabenthos, to 1000s of macrobenthos, to 10^6 meiofauna, 10^{12} bacteria, and 10^{13} viruses. The diversity of many benthic components remains poorly known, and almost totally unknown for viruses, prokaryotes, and nanobenthos. Estimates place prokaryotic diversity on the order of 500,000 taxa, and protistans at perhaps 60,000 species. Meiobenthos or meiofauna (organisms between microbenthos and macrobenthos in size) includes both large protistans and 22 of the 34 known phyla of metazoans, and several taxa adapted to living in extreme conditions (Nematoda, Gnatostomulida, and Loricifera). Meiofauna, once fully described, may include more than 100,000 species. Macrobenthic organisms include 18 animal phyla, potentially exceeding 500,000 species.

Marine Biology: Comparative Ecology of Planet Ocean, First Edition. Roberto Danovaro and Paul Snelgrove.
© 2024 John Wiley & Sons Ltd. Published 2024 by John Wiley & Sons Ltd.
Companion Website: www.wiley.com/go/danovaro/marinebiology

Figure 9.1 Example of benthic organisms belonging to the: (a) phytobenthos; and (b) zoobenthos that, through their shells or rhizomes, influence the nature of the substrate and affect sediment colonization by other organisms.

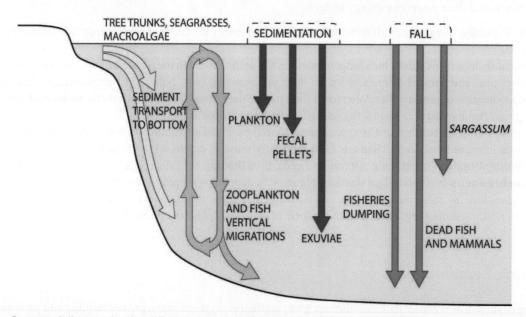

Figure 9.2 Conceptual diagram showing different types of allochthonous input that settles to the benthic compartment: these inputs include exported macrophytes (seagrass and macroalgae) and terrestrial material from the coastal zone; vertical migration of plankton with depth and the production of feces (fecal pellets); sedimentation of senescent or dead plankton (phytodetritus or dead zooplankton); sinking of pieces of dead organisms; fisheries discards; dead large fishes or whales carcasses; and sedimentation of large algae such as the seaweed, *Sargassum* from surface waters.

Some large megafauna, such as corals, play a key ecological structuring function (bioengineers and bioconstructors), providing substrate and shelter for many other organisms. Macrophytes (macroalgae and seagrasses), the photosynthetic counterparts of macro and megafauna, also play a key ecological role, forming submerged forests and beds. Researchers classify benthic consumers (heterotrophs) into different modes based on: (1) the size of the resource; (2) the type of the resource; (3) the environmental compartment in which the resource occurs; (4) foraging mode and associated mobility; and (5) the structures involved in ingesting food. Hard and soft surfaces support very different fauna and flora in composition and overall appearance. Benthic organisms exhibit different adaptations to gradients in: (a) latitude; (b) bathymetry; (c) distance from the coast. These adaptations reflect the locations to which each specific, characterizing community (**biocenosis**, referring to a group of interdependent organisms living at a given location) respond. Sometimes these communities form zones that change in response to some dominant environment variable, typically depth.

Researchers sometimes use the Pérès and Picard model (see below), to distinguish the **phytal zone**, the 6–8% of the seabed where sufficient light reaches the bottom to support photosynthetic seafloor flora. This zone potentially extends out to the edge of the continental shelf, from the **aphytal zone**, the deeper, 92–94% of the seabed that extends

Table 9.1 Classes of size and subdivision into autotrophic and heterotrophic components of the benthic organisms.

Group	Size (µm)	Hetero- / autotrophic	Present organisms
Femtobenthos	0.02–0.2		Viruses of sediments, or virio-benthos
Picobenthos	0.2–2.0	Autotrophic pico-benthos	Benthic prokaryotes (Bacteria, Archaea) and small eukaryotes
		Heterotrophic pico-benthos	Cyanobacteria, Archaea, autotrophic picoeukaryotes
Nanobenthos	2.0–20	Autotrophic nanobenthos	Protistans: unicellular eukaryotes, including heterotrophic nanoflagellates and other small heterotrophs dimensions
		Heterotrophic nanobenthos	Unicellular autotrophic eukaryotes, small benthic diatoms of small dimensions
Microbenthos	20–200	Microphytobenthos	Benthic microalgae (microphytobenthos)
		Benthic protozoans	Small protozoans such as amoebae, tintinnids, ciliates, micrometazoans (rotifers)
Meiobenthos	20–500 (0.5 mm)	Meiofauna	Practically, meiobenthos includes both unicellular (protozoans, foraminiferans included) and 22 phyla of metazoans (among which nematodes, copepods harpacticoids, tardigrades, loriciferans, gnatostomulida, etc.).
		Autotrophic Meiobenthos	Definition not used because this component is typically included in the microphytobenthos
Macrobenthos	>0.35–0.5 mm	Macrofauna	Polychaetes, amphipods, and organisms visible with naked eye
		Macroalgae	Multicellular algae with thallus visible with naked eye
Megabenthos	>1–2 cm	Megazoobenthos	Large crustaceans, echinoderms, poriferans, gorgonians, etc. visible with naked eye
		Macrophytes	Large algae (e.g., *Sargassum*), large *Laminaria*, marine phanerogams

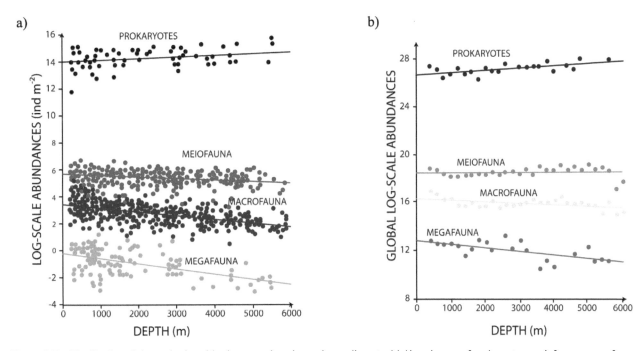

Figure 9.3 Distribution of the major benthic size groupings in marine sediments. (a) Abundances of prokaryotes, meiofauna, macrofauna, and megafauna per m^2 at different depths, and (b) total abundances at different bathymetric intervals (200 m) averaged globally.

roughly from the edge of the continental margin to the major ocean trenches, and where insufficient light penetrates to support bottom living photosynthetic organisms. The model also distinguishes subcategories of seafloor habitats (littoral, intertidal, infralittoral, and circalittoral) that host communities adapted to live in specific environmental conditions.

9.3 Classification of Benthos Based on Size

Abundance of different size components of the benthos vary with depth (Figure 9.3) and typically correlate with biomass (Figure 9.4) of the different benthic size groupings, which increase exponentially with decreasing size. Similarly, marine organism size relates inversely with abundance (Figure 9.5).

Differences in proportions of size classes as a function of depth leads to changes in relative importance of the megafauna, macrofauna, meiofauna, and prokaryotes at different depths (Figure 9.4). Specifically, whereas megafauna and macrofauna dominate shallow environments in total biomass (Figure 9.4a), meiofauna and prokaryotes become increasingly prevalent with increasing depth (Figure 9.4b), with near complete dominance of prokaryotes in abyssal environments (Figure 9.4c).

9.3.1 Femtobenthos

The **femtobenthos** (or **viriobenthos**) refers to all viruses that can be extracted from marine sediments and other seafloor environments, or from organisms living in those environments. **Viruses** include biological entities, sub-microscopic in size, that cannot live or reproduce outside the host cell, because they lack metabolic activity and biosynthetic functions. Viruses infect all forms of life and are grouped according to their hosts, as viruses of animals, plants, or prokaryotes. Each virus contains only one of the two types of nucleic acids (DNA or RNA, never both) that may occur as either single-stranded (ssRNA, ssDNA) or double-stranded forms (dsRNA, dsDNA). Differences in nucleic acids also define the taxonomic affiliation of viruses. Where there is life, viruses occur, defining the **viriosphere,** which spans every environment on Earth, from the atmosphere to the deep marine biosphere. Generally, a protein coat (**capsid**) composed of identical repeating units (**capsomeres**) surrounds the viral genome, followed by a protein matrix, and finally by an exterior glycoprotein-lipid coating (**pericapsid**). This structure protects the viral genome and allows it to penetrate into the host. Some viruses, such as T4 **bacteriophages**, referring to viruses that infect bacteria, also have accessory structures such as a tail; the fibers of the tail and the basal plate help in recognizing the host and injecting the nucleic acid into the host cell. Benthic viruses vary in morphology depending on the size of their genome and on the shape of the capsid. Aquatic phages include multiple morphotypes such as tailed, spherical, oval hook shaped, and cube shaped. However, phages with tails seem to dominate, with concentrations sometimes exceeding 90% of the entire viruses in a transmission electron microscopy sample.

Viruses use several strategies for replication, primarily lytic cycles and lysogenic cycles. **Lytic viruses** infect a cell, replicate, and release once the infected cell dies. **Lysogenic (or temperate) viruses** infect the cell and integrate their DNA into the host genome, replicating their genome until some other factor (e.g., UV, high temperatures, pollutants) induce the lytic cycle.

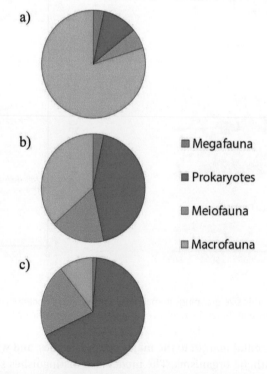

Figure 9.4 Relative importance of different size classes to the total biomass present in marine soft sediments. The components shown are: Prokaryotes (in orange), meiofauna (in grey), macrofauna (in yellow) and megafauna (in blue): (a) sediments from 200 to 2,000 m depth; (b) sediments from 2,000 to 4,000 m depth; (c) sediments from 4,000 to 6,000 m depth.

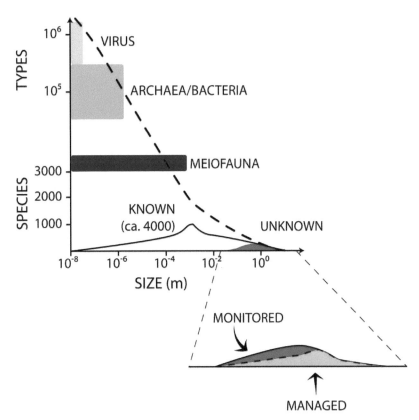

Figure 9.5 Much like the inverse relationship between size and abundance of marine organisms, organism biodiversity, and size also appear inversely related, acknowledging significant gaps in knowledge, particularly for smaller taxa. The graph also shows that scientists monitor, study, and manage (through marine protected areas or similar closures) only a small number of species, almost all of which are large in size.

Aquatic viruses, in the past (appropriately) called "virus-like particle" (VLP, because of some uncertainty about whether the particles observed under the microscope were actually viruses), are primarily bacteriophages. Marine viruses typically contain DNA (thus "DNA viruses") although recent **metagenomic** studies, those that consider all the genes of all organisms in a sample, demonstrate previously unknown RNA viruses in the ocean. Viral dimensions vary between 20 and 200 nm (0.02–0.2 μm), but most aquatic viruses range between 30 and 60 nm. However, some surface water contains viruses greater than 400 nm in size; given that this size exceeds that of most prokaryotes, they likely infect phytoplankton cells. **Giant viruses** of 750 nm have been reported in radiolarian vacuoles (0.75 μm). Viral community composition varies geographically, likely reflecting selective pressures. Acknowledging some endemic and other ubiquitous species, the vast majority of viruses occur widely and span different geographic areas, with greater diversity at lower than at higher latitudes.

Abundance of benthic viruses: Viruses are the most abundant biological entities in the global ocean, with particularly high concentrations in marine sediments that generally greatly exceed values typical of planktonic environments by up to several orders of magnitude. For example, 10^7–10^{10} particles · g^{-1} of sediment exceed the 10^4–10^8 particles · ml^{-1} typical of the water column Hypothetically, aligning all viruses in the ocean ($>10^{29}$) would span a distance of 42 million light-years (1 light year = 9,460,000,000,000 km $>10^{30}$) and form a biomass equivalent to 75 million whales (12,000 megatons), exceeding the abundance of prokaryotes by an order of magnitude.

Viral abundance can vary in space and time within a given ecosystem as a result of changes in biotic and abiotic factors. Many studies report a close relationship between prokaryotic and viral abundance in pelagic and benthic systems. In general, viral abundance increases with ecosystem productivity, as indicated by both chlorophyll-a concentrations in pelagic systems and organic matter in marine sediments. Higher abundances of viruses therefore characterize **eutrophic** (excess nutrients) systems relative to **oligotrophic** (nutrient-poor) systems. Consequently, viral abundance decreases with increasing depth and when moving from coastal to offshore environments (Figure 9.6). However, high concentrations of viruses have been reported from 1 m deep in sediments (10^8 viruses · g^{-1}) and meso-bathypelagic waters may contain many undiscovered viruses. Factors such as temperature and ultraviolet rays directly affect viral abundance, inducing the lytic cycle in bacteria through lysogenic infections, Nutrient concentrations and sediment organic load may also indirectly affect viral abundance. Sediments, in fact, offer an excellent environment for viral proliferation in that they typically provide: (1) relatively high concentrations of organic matter; (2) higher bacterial abundances than the water column; (3)

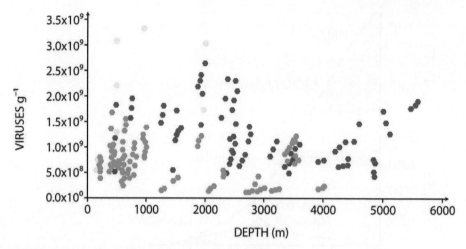

Figure 9.6 Example of bathymetric distribution of viruses in marine sediments.

higher probability of virus-host encounter. Viruses vary even on small spatial (cm) and time (minutes) scales, indicating their highly dynamic nature within pelagic and benthic communities.

Diversity of benthic viruses: Scientists have documented only a very small portion of viruses in the ocean and, of the ~5,000 types of bacteriophages isolated to date, only a small portion are marine. Scientists have sequenced only about a dozen marine viral genomes, but knowledge on viral richness in the ocean has rapidly increased in the last decade. Methodological issues have limited progress; cultivation of viruses has proven challenging, hardly surprising given that only 1% of marine prokaryotes have been cultivated. A comparison between the sizes of isolated phages and those of broader viral communities show that isolated phages are generally larger, reinforcing the idea that natural communities are much more diverse. Furthermore, we lack a common molecular marker for viruses akin to ribosomal RNA genes used for organisms other than viruses.

Culturing methods indicate that viruses lack universal genes, although some genes are specific subsets of the viral community and researchers have identified several conserved (consistently shared) regions for particular genes such as the DNA polymerase gene for Phycodnaviridae (algal viruses) or the capsid protein gene g20. Fingerprinting using Denaturing Gradient Gel Electrophoresis (DGGE), following Polymerase Chain Reaction (PCR) analysis, has demonstrated the genetic richness of viral communities without requiring cultivation of viruses. However, few studies have documented viral diversity in marine sediments, and transmission electron microscopy (TEM) indicates different morphologies and sizes of viruses in water and sediment samples, and higher morphological diversity in sedimentary habitats. RNA and single-stranded DNA phages, though rarely observed in the water column, appear relatively common in sediments. Given the limited number of viral morphotypes, TEM likely captures only a small fraction of total viral diversity (Figure 9.7).

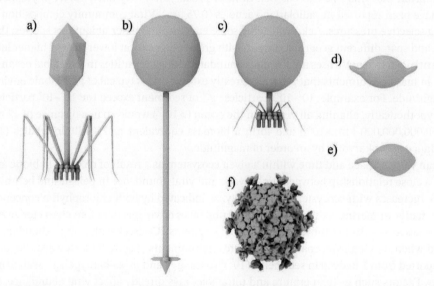

Figure 9.7 Examples of different types of phages, highlighting morphologies with tails (a-e), and others more spherical in form (f).

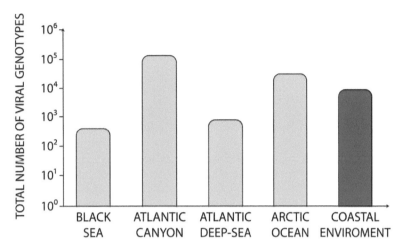

Figure 9.8 Examples of viral abundances in different sedimentary environments. Bars show numbers in 10 grams of sediment (in yellow) or kg of sediment (in red).

Metagenomic approaches bypass this limitation and point to enormous genetic diversity; indeed 1 kg of coastal sediment yielded 10^4–10^6 viral types (Figure 9.8).

Very rare (0.01–0.1% of the total viral community) genotypes apparently dominate sediments, with phages (44%) as the most common group (dominated by the family Siphoviridae, followed by Myoviridae and Podoviridae); eukaryote viruses appear less common (3%). Comparison of viral sequences in water and sediment samples indicates completely different composition, and that the majority of viruses do not originate from the water column. A recent study reported that about 75% of the 1,156 analyzed sequences were unknown, suggesting that viruses in marine sediments might represent the largest reservoir of genetic diversity on Earth, most of which remains uncharacterized.

9.3.2 Picobenthos

Picobenthos, by definition, includes all living organisms, whether prokaryotes or eukaryotes, between 0.2 and 2.0 microns in size. In the past, this group consisted largely of just bacteria, but recently, the advent of new molecular techniques enabled the discovery of novel and abundant eukaryotes (picoeukaryotes) and giant viruses. **Microorganisms**, a term that refers to forms of life invisible to the naked human eye and visible only through microscopic examination, includes picobenthos. This definition includes both **eukaryotes** (organisms whose cells have nuclei) and **prokaryotes** (i.e., single-celled organisms without a nuclear membrane). In terms of size, extreme variability characterizes these marine microorganisms, with dimensions varying between 0.2 microns (the smallest prokaryotic cells) and 30 microns (the largest sulfur bacteria that live in hydrogen sulfide-rich portions of upwelling areas). By comparison, this variability is proportionally similar to that between krill (~4 cm in length) and a 20-m whale.

Abundance of picobenthos: Phylogenetic classification divides benthic prokaryotes into two major cellular domains: Bacteria and Archaea. Benthic prokaryotes, the dominant microbial component (together with the viruses) generally occur in much higher concentrations (up to several orders of magnitude) in sediments, with 10^8–10^9 cells $\cdot$ g^{-1} compared with 10^5–10^6 prokaryotic cells $\cdot$ ml^{-1} in the water column. Prokaryotes generally dominate microbial abundance and biomass, often playing a major ecological role. Recent studies have documented the dominance of this biotic component within all ecosystems in the global biosphere. Therefore, understanding marine ecosystem functioning, Earth's biogeochemical cycles, and the flow of energy and material through food webs requires studies of Bacteria and Archaea.

Cyanobacteria, another component of the picoplankton, form an additional important and extremely widespread group. Cyanobacteria play a critical role in primary production by fixing atmospheric nitrogen, or transforming elemental N_2 into forms other photo-synthesizers can use, such as nitrate; cyanobacterial mats on sediments in some coastal environments also help to bind sediment grains together, and thus stabilize the substrate.

Diversity of benthic prokaryotes: Marine prokaryotes vary greatly in taxonomic affiliation and metabolism. Prokaryotes fall into three large groups based on nutritional types, energy demands, and substrates utilized: (1) "**photoautotrophs**", which obtain energy from photosynthesis and sunlight; (2) "**chemoautotrophs**", which obtain energy from chemical reactions, particularly by oxidizing reduced and/or inorganic compounds; (3) "heterotrophs", which can only obtain energy by breaking down organic compounds. Heterotrophs, which include both Bacteria and Archaea, are the

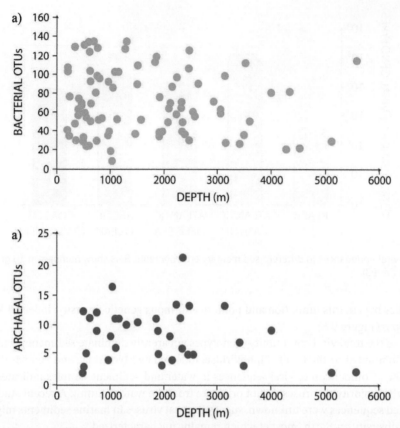

Figure 9.9 Diversity pattern of benthic prokaryotes along a bathymetric gradient. The values refer to OTU (Operational Taxonomic Units) determined by molecular fingerprinting techniques. Microbiologists consider OTUs as approximately equivalent to species because the standard definition of species does not work for microbes. Patterns related with benthic Bacteria were determined with the ARISA technique (a) and benthic Archaea were determined by the T-RFLP technique (b).

main "actors" in the biological transformation of organic matter and in CO_2 production, dominating prokaryotic communities in aerobic marine sediments. These organisms decompose comparatively large organic molecules by releasing hydrolytic enzymes outside the cell, which can degrade organic polymers. The products obtained by hydrolysis subsequently yield carbon and energy. In addition to organic carbon, these bacteria also require inorganic nutrients such as phosphorus, nitrogen, iron, and others in order to grow and reproduce.

Quantifying the role of bacteria in aquatic ecosystem functioning therefore requires studying these nutritional processes. Although molecular methods revealed extraordinary numbers of different gene sequences in benthic prokaryotes, we still lack an accurate estimate of the total number of benthic prokaryote species. Current estimates suggest a number between 10^5 and 10^6 species. Although we know little about the factors determining biodiversity patterns in prokaryotes, we do know that not all species live everywhere, and that environmental conditions select the forms that can colonize (often transported as spores in ocean currents), grow, and reproduce. In particular, depth strongly influences distribution and abundance of marine prokaryotes species, likely relating to availability of organic matter (Figure 9.9).

9.3.3 Nanobenthos

Nanobenthos includes all organisms between 2.0 and 20 μm in size, irrespective of whether autotrophic or heterotrophic. Their abundances in sediments (10^3–10^4 cells · g^{-1} of sediment) typically exceeds that in the water column (10–10^2 cells · ml^{-1}), but logistics have constrained studies to date. Nanobenthos includes an extraordinary diversity of organisms, but the imperfect correspondence between size class and taxonomic category means that the nanobenthos encompasses both large prokaryotes (some sulfur-bacteria and filamentous bacteria exceed 30 μm in size) and small metazoans. The smaller components of the nanobenthos typically form part of the microbenthos. Nanoflagellates (Figure 9.10) numerically dominate the nanobenthos, along with amoebae, ciliates, other unicellular eukaryotes and, occasionally, small metazoans.

Within the North Pacific nanobenthos, the dominant nanoflagellates prey on benthic bacteria. In some cases, the nanobenthos may be **mixotrophic**, meaning they are both primary producers and consumers of organic matter, depending on environmental and ecological conditions.

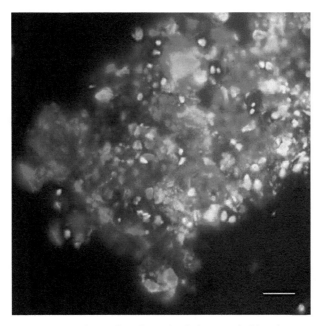

Figure 9.10   Epifluorescence microscopic image of nanoflagellates (scale bar equals 20 μm).

Diversity of nanobenthos: Protistans encompass a wide variety of organisms of different shapes and ecological habits. They are unicellular, but sometimes form colonies. These organisms occur widely in marine environments, including extreme habitats. Protistans include both autotrophic and heterotrophic organisms. A non-exhaustive list includes: (1) Microscopic unicellular algae such as **dinoflagellates** (photosynthetic organisms that play a similar ecological role as plants or primary producers); (2) unicellular eukaryotes, consisting of flagellates, amoebas, sarcodines, eliozoa, foraminifera, ciliates and (3) a group of oomycetes (heterotrophic protistans, closely related to the algae) and fungi (phyla Myxomycota and Dictyosteliomycota). Protistan biodiversity includes more than 200,000 species of microalgae and more than 60,000 other species. Many protistans form symbioses with other organisms. In the ocean, they also constitute a significant fraction of the total benthos in terms of biomass.

9.3.4   Microbenthos

Microbenthos encompasses all organisms between 20 and 200 microns in size. This forced classification, used for convenience, has little taxonomic basis. Indeed, for historical reasons, benthic ecologists have dealt only recently (in the last 20 years) with microbial components, and historically identified microbenthos as "all biota smaller than meiobenthos". In fact, because this size classification does not correspond to specific taxonomic groups (and has no operational advantages), many consider it artificial. However, the microbenthos compartment can be useful in studies that clarify the specific component, such as protistans (ciliates, amoebae, flagellates, etc.) or microphytobenthos (the autotrophs) greater than 20 microns in size. Microphytobenthos experts, for convenience, rarely separate the nano- and microbenthic size groupings of autotrophic eukaryotes. Microphytobenthos encompasses several groups of benthic unicellular algae, including the most important and different groups, the diatoms, and cyanobacteria (Figure 9.11).

Abundance and distribution of microbenthos: Microbenthos abundances typically vary between 10^2 and $10^4 \cdot cm^{-2}$ of sediment. The shell surface of some mollusks can support a density of 9,000–12,000 cells $\cdot cm^{-2}$. Diatoms attached to a substrate (whether using filaments, peduncles, or gravity) may be less prone to burial but more exposed to predator grazing. The ability of some benthic diatoms to migrate up and down in sediments likely reflects adaptations to different forcings, including surface feeder predation, re-suspension, excessive exposure to light, and burial resulting from seabed turbulence. Even in apparently homogenous sediments, patchy horizontal distributions typically characterize microphytobenthos at large (km), medium (m) and small (cm) scales; this patchiness likely reflects subtle patchiness in sedimentary variables such as nutrients, surface light, or grain size.

The vertical distributions of microphytobenthos span from a few tenths of a millimeter to several millimeters in depth, even though high biomasses may also occur several cm deep in the sediment. Migration of mobile species or physical transport (by waves and currents, or by benthic meiofaunal bioturbation) explain high subsurface biomass. Complex dinoflagellate life cycles often include benthic stages (in the form of cysts), but only a few species occur permanently within benthic

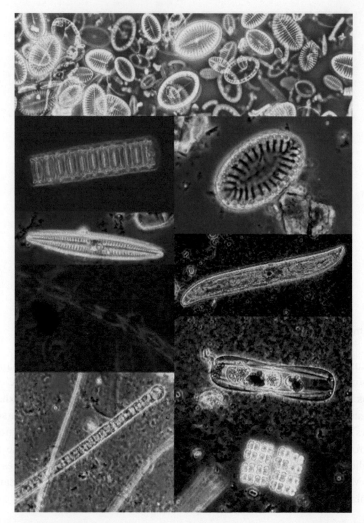

Figure 9.11 Top: mixed populations of epiphytic benthic diatoms, which in a marine context refers to organisms that grow on marine macroalgae. From the top to the bottom and from the left to the right: soft bottom microphytobenthos. *Paralia sulcate*, a common centric diatom on sandy bottoms. The pennate diatoms *Surirella fastuosa*, *Navicula directa*, and *Gyrosigma balticum* move actively above and inside the sediment; *Parlibellus* sp., a tubicular diatom, produces and lives within slime tubes. *Plagiotropis lepidoptera*, a common pinnate diatom in sediments, also occurs in the plankton; filamentous cyanobacteria (*Oscillatoriales* group) and colonial cyanobacteria (genus *Merismopedia*) occur commonly on shallow-water soft sediments.

habitats. At shallow depths (<1 m) microphytobenthic production may exceed phytoplankton production. In shallow water, excess light may limit phytoplankton production through photo-inhibition where microphytobenthos remain productive. Here, microphytobentic biomass exceeds that of phytoplankton.

Diversity of microphytobenthos. Of the microphytobenthos, diatoms generally dominate marine sediments numerically, in contrast to lake sediments where the most abundant taxa often include Cyanophyceae and some flagellates. In contrast, dinoflagellates typically dominate tropical seabeds. Benthic and planktonic forms generally differ significantly in composition. Life in sediments requires special adaptations that differ from those required to live in the plankton. In general, greater diversity and life forms characterize microphytobenthos compared to the plankton. In fact, this group includes forms living on sediment (moving on or within the sediment) that can move at speeds up to 240 μm · sec^{-1}, forms that attach to a substrate, (whether stone, rocks, or animals; see below), and planktonic cells that originate in the water column and settle on the bottom where they remain photosynthetically active, depending on light availability (Figure 9.11). Some species adhere to the substrate, despite strong motility (*Mastogloia* spp., *Amphora* spp., *Rhopalodia* spp.). Benthic dinoflagellates exhibit adaptations to a wide variety of habitats from tropical to boreal seas, and include the following groups:

1) **Epipelon** communities live above the sediment and occur in all waters where sediment deposits and light penetrates; associations above the sediment usually vary among the different habitats. These rich and diverse communities take advantage of abundant light, modest bottom currents, and flushing by tides.

2) **Epipsammon** communities attach to sand grains, and although this group has not been well studied, angularity of grains seems to influence likelihood of attachment; diatoms dominate this group, and larger forms can bind sediment grains together: Some chlorophytes and cyanobacteria also attach to sand grains.
3) **Endopelon** or **endopsammon** communities live within the sediment; some diatoms live and move within mucilage tubes that penetrate within the sediment, whereas other microalgae secrete mucus that binds sand grains together, potentially forming rocks through carbonate deposition.
4) **Epilithon** communities live on hard substrates, and establish themselves based on substrate, grazing rate, and wave exposure, particularly in the intertidal zone. The few available data indicate mixed epilithic/epipelic complex communities at high densities. These communities form the primary film (together with prokaryotes) that "prepares" the substrate for subsequent colonization by animals and plants, thus filling a specific ecological role.
5) **Endolithon** communities, comprising microalgae, live within hard substrates, including carbonate rocks, shells, corals, calcareous algae, etc.; Cyanobacteria represent the largest group of micro-endolithon, and grow in a wide range of environments (from rocky deserts to Antarctica, rock outcrops, animals with calcareous shells, algal reefs and rocks along ocean coastal regions);
6) **Epiphyton** communities live on other plants, and diatoms colonize almost all macrophytes (macroalgae and seagrasses), although with different intensity; for the most part the microepiphytic community colonize older, notched portions or macrophytes. Large diatoms can actually accommodate smaller epiphytes;
7) **Epizoon** live on animals, and benthic microphytes most commonly colonize metazoans with hard shells (mollusks and crustaceans).

9.3.5 Meiobenthos (Meiofauna)

The term "meiobenthos," first introduced by Molly Mare in 1942, refers to organisms intermediate in size compared to the smaller microbenthos, and the larger macrobenthos. Although meiobenthos refers to both animal and photosynthetic components, the term **meiofauna** refers only to animals between 20–30 μm and 0.5–1 mm in size. Although this size category could include both large protistans and metazoans, most researchers use this term to refer to metazoans. Meiofauna include small benthic metazoans of biomass typically ranging from 0.1 to 50 μg (dry weight), with an evolutionary history and feeding characteristics that clearly differentiate them from the larger macrofauna. Meiobenthic taxa include representatives of both interstitial and surface living organisms, including nematodes, copepods, and flatworms. Meiofaunal species living in sand and mud differ morphologically (Figure 9.12). Thinner, elongate bodies characterize taxa living in sand, facilitating movement between sand grains, whereas larger size and more variable morphologies characterize species living in mud. Permanent meiofauna refers both to multicellular individuals (noting that this group includes single-celled foraminifera, which represent a special case) and metazoans that remain meiofaunal throughout their life, such as nematodes, harpacticoid copepods, loriciferans, gastrotrichs, kinorhynchs, gnatostomulids as well as temporary meiofauna, referring to metazoans that remain meiofaunal in size for early life stages before becoming macrofaunal or even megafaunal as adults. These temporary meiofauna include many juvenile polychaetes, oligochaetes, cumaceans, tanaidaceans, and other crustaceans.

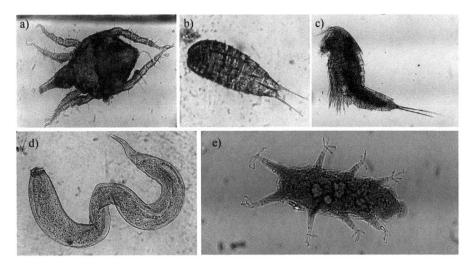

Figure 9.12 Examples of meiofaunal organisms stained with Rose Bengal to increase visibility: (a) mites; (b) kinorhynchs; (c) copepods; (d) nematodes; (e) tardigrades.

Abundance of meiofauna: As the most abundant group of metazoans in marine benthos, meiofaunal densities average between 10^5 and 10^6 individuals $\cdot$ m^{-2} (10–100 ind. $\cdot$ cm^2) and biomass averages 1–2 g DW (Dry Weight) $\cdot$ m^{-2} in coastal waters, less than 100 m. Of course, abundance and biomass vary with season, latitude, depth, tides, and sediment grain size. Highest abundances and biomass occur in muddy estuaries whereas lowest values generally occur in deep-sea sediments (noting some exceptionally high values reported for hadal depths). Box 9.1 briefly describes meiofaunal sampling.

The depth of oxygen penetration into sediments, measured as the **redox potential discontinuity depth**, or RPD, generally limits the depths within sediments at which meiobenthic taxa occur. Most meiobenthic species occur within the upper 2 cm of sediment, where oxygenation results in a redox potential greater than 400 mV; abundances decrease at redox potential values below 200 mV. In muddy environments rich in detritus, meiofauna occur only within the upper few oxidized millimeters or centimeter. Copepods, one of the most sensitive taxa to oxygen depletion, occur only within the oxygenated layer of the sediment. However, some meiofauna apparently tolerate hypoxic or even anoxic conditions, and may occur below the reduced layer. These organisms, such as gnathostomulids, form the **thiobios**, organisms that have apparently adapted to anoxic conditions (and the presence of hydrogen sulfide).

> **Box 9.1 Meiofaunal sampling**
>
> Scuba divers offer one of the most effective sampling approaches for coastal seafloor habitats because they can accurately target specific locations of habitats, a particular advantage for studies that require repeat sampling. Moreover, this type of sampling can document distributions at small spatial scales (microscale distribution) and link organisms to seabed features such as sediment morphology. Scuba divers can sample meiofauna using handheld **corers**, a Plexiglas tube ~15–20 cm in length and 2–10 cm in diameter that the divers push into the sediment and extract with a plug of sediment inside. Clear plexiglass cores enable visual measurement of the discontinuity depth of the redox potential, where sediments change rapidly in color from brown/gray to black. Statistical comparisons require at least 3–5 replicates (i.e., 3–5 cores) from each sampling site.

9.3.6 Macrobenthos

Macrobenthos, defined as organisms retained on a 0.5 or 1.0 mm sieve, primarily refers to macrofauna, which include the majority (18 of 34) of known phyla (Figure 9.13). Macrobenthos live in all marine environments, whether sediment covered or hard substrate, marine or estuarine, or at lower or higher latitudes and depths. In coastal environments, macrofauna dominate communities in terms of abundance, with polychaetes often (though not always) comprising more than 50% of abundance, diversity, and biomass, followed by crustaceans, bivalve mollusks, and echinoderms. Multiple factors influence macrofaunal spatial distributions including sediment grain size and sedimentation, light, oxygen, temperature, salinity, primary productivity, depth (e.g., pressure), hydrodynamics (tides and waves), physical disturbances (waves, currents), and history (e.g., glaciation).

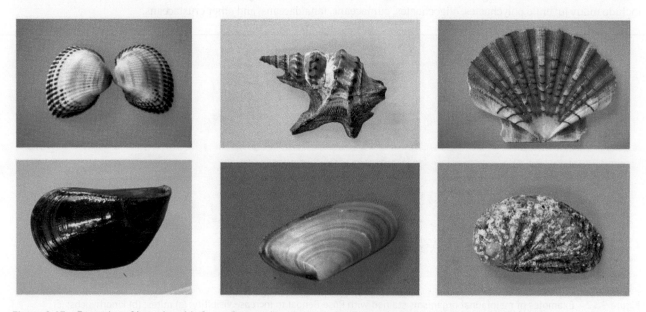

Figure 9.13 Examples of large benthic fauna. Reported are images of molluscs, including bivalves and gastropods.

> **Box 9.2 Macrofaunal sampling**
>
> Sampling tools vary depending on environment. Most commonly, researchers lower grabs, box-corers, and multi-corers on wire or rope to the bottom from a surface vessel. Fitting these devices with cameras (and thus requiring video cables) improves precision, but the high cost and complexity of this approach means that most studies do not avail of video and sampling generally occurs blindly, potentially attempting to sample over rocky or cobble bottom where these samplers are ineffective. At shallow depths, push cores deployed by scuba divers (similar to those described for meiofauna) offer a more precise and reliable sampling tool. Mixed sediments containing cobble or gravel support diverse macrofauna but create a challenge for these sampling devices; however, suction samplers offer one solution. Generally, a scuba diver deploys the suction sampler, which consists of a PVC pipe 1–2 m in length connected to a cylinder of compressed air with a mesh filtration bag (~0.5 mm mesh). The diver effectively "vacuums" a fixed area (often marked with a ~0.25 m^2 quadrat) to remove all animals. Statistical comparisons typically require five or more replicates from each sampling site.

All of these variables may act simultaneously, creating physical and ecological constraints on species distributions. Polar and temperate environments, with their strongly seasonal conditions, typically support lower diversity than tropical ecosystems. Macrofaunal abundance and diversity also vary with depth. In the distant past, ocean life was thought to disappear at great depth but the Challenger expedition in 1872–1876 proved this idea incorrect, and work in the 1960s by Howard Sanders and Robert Hessler completely revolutionized our views on the immense diversity of life in the "extreme" conditions of deep-sea environments (absence of light, low temperatures, high pressure, and low food availability).

Their semi-quantitative sampling of deep environments revealed diversity levels comparable to or greater than those in most shallow-water ecosystems, including the tropics (Box 9.2). Conservative estimates of potential macrofaunal species in deep-sea environments ranges between 500,000 and 5,000,000.

9.3.7 Megabenthos

The megabenthos includes the largest benthic organisms, and operationally refers to animals identifiable in bottom photographs (Box 9.3). These organisms often play important ecological roles, providing substrate and shelter to many epi- and endobionts. Megafauna includes either truly benthic species (both sessile and vagile fauna) and bentho-nekton forms. Sponges, corals, and other **gregarious** organisms (locally abundant species that form dense colonies) create important habitats for plant and animal communities (Figure 9.14). Some megabenthos accommodate symbiotic organisms (small crustaceans, larvae of insects, unicellular algae, cyanobacteria, etc.) within their tissue, providing a potential food source for other animals (fish, annelids, mollusks, echinoderms). The megafauna include some large crustaceans, such as spider crabs (genus *Maja*) that reach sizes close to a meter. Cnidarians, and specifically anthozoans, can dominate some benthic environments, sometimes forming more than half of benthic biomass. The anthozoans include the stony corals that form coral reefs, as well as the solitary sea anemones and soft corals that dominate many seascapes. Mollusks, another important phylum that include megafaunal taxa, include species that attach to the substrate (*Mytilus galloprovincialis*, *Tridacna gigas*, *Lima scabra*, *Ostrea edulis*, and *Pinna nobilis*), others that dig in hard substrates (shipworm, *Litophaga Litophaga*) and some that dig into sediments (*Tapes philippinarum*, the Venus clams). *Pinna nobilis*, the fan mussel, and *Tridacna gigas*, the giant clam, can reach sizes of one meter or more, with weights that exceed one ton. Some megabenthic gastropods, such as tritons, can weigh several kilograms

Many large marine organisms build special structures such as **exoskeletons** (an exterior skeleton such as in crab or lobster) or shells (mollusks) composed of inorganic substances to protect and support their body. Many sessile animals produce supporting structures composed of various spicules embedded in protein matrixes. Examples include sponges

> **Box 9.3 Megafaunal sampling**
>
> Because megafauna are, by definition, organisms visible in bottom photographs, they can often be sampled non-destructively through visual surveys by scuba divers, submersibles, or various types of drop cameras or towed camera/video. Prior to the development of these tools, scientists used a variety of trawls and dredges that they dragged through the water to sample enough volume of seabed or water column to enumerate abundances; these samplers work largely the way that commercial fisheries use fishing nets. Scientists today try to limit the use of these tools because they often damage habitat and often result in the death of the sampled biota. However, because taxonomy of some organisms can be difficult, researchers sometimes require physical specimens (either for morphological or genetics-based identification). Thus, a mix of tools is often necessary.

Figure 9.14 Examples of megafaunal organisms commonly encountered on coastal and deep seabed. *Source:* NOAA / Flickr / CC BY 2.0.

(calcite and sometimes aragonite spicules in Calciospongiae, siliceous spicules in *Hexactinella*, siliceous spicules and spongin structure in Demospongiae). Echinoderms possess an exoskeleton consisting of calcareous plates embedded in the dermis. Many cnidarians produce an exoskeleton, secreted by the epidermis. *Millepora* spp. stony corals produce a calcareous skeleton consisting primarily of aragonite. Even in octocoral colonies the mesoglea cells secrete a calcareous or horny skeleton. For example, the skeleton of the soft coral *Alcyonium* spp. consists of spicules containing an organic axis on which they deposit calcite crystals. Other species, including most gorgonian corals, produce axial skeletons with an elastic consistency resulting from "gorgonin", a protein similar to collagen with high thyroxine content. Similarly, hard and calcareous skeletons, such as in red coral *Corallium rubrum*, result from calcareous spicules secreted from ectodermic cells and cemented together by calcium carbonate. Among the Hexacorallia, or stony corals, the epidermis secretes a calcareous skeleton comprising mainly calcium carbonate in the form of aragonite. Crustaceans secrete a chitinous cuticle from the epidermis that forms a protective skeleton over soft body parts. The cuticle contains chitin, a nitrogenous polysaccharide that becomes impermeable and rigid once impregnated with calcium and proteins. In other crustaceans, such as barnacles, limestone impregnates the exoskeleton. Many benthic organisms also produce shells. Flexibility in some sessile animals reduces the drag force of bottom currents, much like a sapling bending in the wind. For example, the anemone *Metridium senile* can grow up to 1 m in size, but its flexibility allows it to bend in strong currents. When bent over in currents, its tentacles may collect food in the animal's wake, although in especially strong currents the anemone may retract its crown of tentacles.

Genomic approaches offer a pathway to investigate all components of the benthos (Box 9.4) that now enables scientists and technicians to identify all different benthic taxa, whether for assessing microbial, meiofaunal, macrofaunal, or megafaunal diversity.

Box 9.4   Metagenomic and Metagenetic Analyses of Benthic Biodiversity

Molecular and genetic analysis of the benthos

Metagenomics and metabarcoding analyses offer major advances in genomic analysis of single organisms because they contextualize associated species. These molecular analyses use next-generation technologies that provide massive de novo sequencing (e.g., Illumina) of the DNA extracted from the organisms. Bioinformatics analysis of the origin of the sequences using special software (Figure 9.15) produces a detailed genetic mapping of benthic species diversity (e.g., meiofaunal taxa that are difficult to identify at the species level using morphological taxonomy). Metagenomics and metabarcoding represent useful tools to analyze rare or unknown taxa. **Metagenomic analysis** considers all the genes of all organisms in a sample, whereas **metabarcoding** considers a single gene marker for all organisms in a sample. The extraction of organisms from sediments prior to DNA extraction, purification, quantification, and then sequencing allows recovery of a sufficient amount of high-quality DNA.

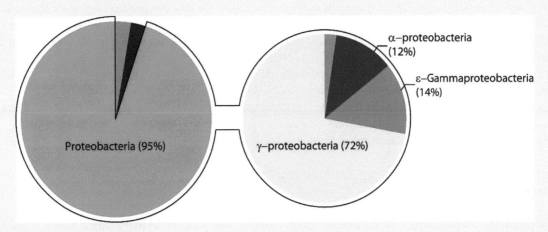

Figure 9.15 Example of the composition of the benthic community of prokaryotes resulting from analysis of metagenomic DNA extracted from marine sediments.

Questions

1) What are the main differences between plankton and benthos?
2) What are the main food sources utilized by benthic heterotrophs?
3) How are benthic organisms classified by size?
4) Describe the bathymetric patterns of different size classes of benthic organisms.
5) Discuss the importance of size classifications from a functional point of view.
6) Define marine meiofauna.
7) Is picobenthos composed of autotrophic and heterotrophic organisms? Provide examples.
8) Are viruses present in marine sediments?
9) Define the term megafauna.
10) Are marine sponges part of the benthos, and to what size class do they generally belong?

Suggested Reading

Clark, M., Althaus, F., Schlacher, T. et al. (2015) The impacts of deep-sea fisheries on benthic communities: A review. *ICES Journal of Marine Science*, 73, pp. i51–i69.

Coolen, J., van der Weide, B.E., Cuperus, J. et al. (2018) Benthic biodiversity on old platforms, young wind farms, and rocky reefs. *ICES Journal of Marine Science*, 77, pp. 1250–1265.

Corinaldesi, C., Barucca, M., Luna, G.M., and Dell'Anno, A. (2011) Preservation, origin and genetic imprint of extracellular DNA in permanently anoxic deep-sea sediments. *Molecular Ecology*, 20, pp. 642–654.

Corinaldesi, C., Beolchini, F., and Dell'Anno, A. (2008) Damage and degradation rates of extracellular DNA in marine sediments: Implications for the preservation of gene sequences. *Molecular Ecology*, 17, pp. 3939–3951.

Dayton, P. and Hessler, R.R. (1972) Role of biological disturbance in maintaining diversity in the deep sea. *Deep-Sea Research*, 19, pp. 199–208.

Dell'Anno, A. and Danovaro, R. (2005) Extracellular DNA plays a key role in deep-sea ecosystem functioning. *Science*, 309(5744), p. 2179.

Gooday, A.J., Bett, B.J., Shires, R., and Lambshead, P.J.D. (1998) Deep-sea benthic foraminiferal diversity in the NE Atlantic and NW Arabian sea: A synthesis. *Deep-Sea Research II*, 45, pp. 165–201.

Grassle, J.F. and Maciolek, N.J. (1992) Deep-sea species richness: Regional and local diversity estimates from quantitative bottom samples. *The American Naturalist*, 139, pp. 313–341.

Grassle, J.F. and Sanders, H.L. (1973) Life histories and the role of disturbance. *Deep Sea Research and Oceanographic Abstracts*, 20, pp. 643–659.

Gray, J.S. (2000) The measurement of marine species diversity, with an application to the benthic fauna of the Norwegian continental shelf. *Journal of Experimental Marine Biology and Ecology*, 250, pp. 23–49.

Gray, J.S. (2001) Marine diversity: The paradigms in patterns of species richness examined. *Scientia Marina*, 65, pp. 41–56.

Gray, J.S. (2002) Species richness of marine soft sediments. *Marine Ecology Progress Series*, 244, pp. 285–297.

Hessler, R.R. and Sanders, H.L. (1967) Faunal diversity in the deep-sea. *Deep Sea Research and Oceanographic Abstracts*, 14, pp. 65–78.

Hoshino, T., Doi, H., Uramoto, G.I. et al. (2020) Global diversity of microbial communities in marine sediment. *Proceedings of the National Academy of Sciences of the United States of America*, 117, pp. 27587–27597.

Maugeri, T.L., Lentini, V., Gugliandolo, C. et al. (2010) Microbial diversity at a hot, shallow-sea hydrothermal vent in the southern Tyrrhenian Sea (Italy). *Geomicrobiology Journal*, 27, pp. 380–390.

McClain, C.R., Allen, A.P., Tittensor, D.P., and Rex, M.A. (2012) Energetics of life on the deep seafloor. *Proceedings of the National Academy of Sciences of the United States of America*, 109, pp. 15366–15371.

Mokievsky, V. and Azovsky, A. (2002) Re-evaluation of species diversity patterns of free-living marine nematodes. *Marine Ecology Progress Series*, 238, pp. 101–108.

Piacenza, S.E., Barner, A.K., Benkwitt, C.E. et al. (2015) Patterns and variation in benthic biodiversity in a large marine ecosystem. *PLOS One*, 10(8), p. e0135135.

Poore, G.C. and Wilson, G.D. (1993) Marine species richness. *Nature*, 361, pp. 597–598.

Rossi, S., Coppari, M., and Viladrich, N. (2016) Benthic-pelagic coupling: New perspectives in the animal forests, in Rossi, S., Bramanti, L., Gori, A., and Orejas Saco del Valle, C. (eds.), *Marine Animal Forests*. Springer, Cham. pp. 855–885.

Roy, K., Jablonski, D., and Valentine, J.W. (2000) Dissecting latitudinal diversity gradients: Functional groups and clades of marine bivalves. *Proceedings of the Royal Society of London B: Biological Sciences*, 267, pp. 293–299.

Sanders, H.L. (1969) Benthic marine diversity and the stability-time hypothesis. *Brookhaven Symposia in Biology*, 22, pp. 71–80.

Sanders, H.L. and Hessler, R.R. (1969) Ecology of the deep-sea benthos. *Science*, 163, pp. 1419–1424.

Snelgrove, P.V.R. and Smith, C.R. (2002) A riot of species in an environmental calm: The paradox of the species-rich deep-sea floor. *Oceanography and Marine Biology: An Annual Review*, 40, pp. 311–342.

Soetaert, K. and Heip, C. (1995) Nematode assemblage of deep-sea and shelf break sites in the North Atlantic and Mediterranean Sea. *Marine Ecology Progress Series*, 125, pp. 171–183.

Soetaert, K., Muthumbi, A., and Heip, C. (2002) Size and shape of ocean margin nematodes: Morphological diversity and depth-related patterns. *Marine Ecology Progress Series*, 242, pp. 179–193.

Strugnell, J.M., Rogers, A.D., Prodöhl, P.A. et al. (2008) The thermohaline expressway: The Southern Ocean as a centre of origin for deep-sea octopuses. *Cladistics*, 24, pp. 853–860.

Stuart-Smith, R.D., Bates, A.E., Lefcheck, J.S. et al. (2013) Integrating abundance and functional traits reveals new global hotspots of fish diversity. *Nature*, 501, pp. 539–542.

Wilson, G.D. and Hessler, R.R. (1987) Speciation in the deep sea. *Annual Review of Ecology and Systematics*, 18, pp. 185–207.

Woolley, S.N., Tittensor, D.P., Dunstan, P.K. et al. (2016) Deep-sea diversity patterns are shaped by energy availability. *Nature*, 533, pp. 393–396.

Yasuhara, M., Hunt, G., Cronin, T.M., and Okahashi, H. (2009) Temporal latitudinal-gradient dynamics and tropical instability of deep-sea species diversity. *Proceedings of the National Academy of Sciences of the United States of America*, 106, pp. 21717–21720.

Part IIC

Life in Seas and Oceans: Fundamentals of Marine Biology

10

Ecology of Benthos

Marine benthic organisms have adapted to the specific conditions of the habitat in which they live, and include some species able to penetrate into the sediments, some that bore into calcareous hard bottoms, and some that simply affix to the bottom. In this chapter we examine the various size components of the benthos and provide information on their ecology, distribution, and contribution to ecosystem functioning.

10.1 Ecology of the Benthos: From Microbes to Megafauna

Benthic ecology of viruses: Viruses not only dominate our planet as biological entities, but they also represent the major cause of mortality on Earth. Indeed, about 10^{23} viral infections occur every second in the ocean, decreasing total living biomass by some 20% daily. Because **bacteriophages**, referring to viruses that target bacteria, dominate ocean environments, prokaryotes are the primary hosts of viruses. Prokaryote mortality from viral infection ranges 10–40%, but can reach 100% in benthic and pelagic environments. Because ocean ecosystems cover > 70% of the globe, viruses can affect overall functioning of the global biosphere.

Some theoretical models suggest that, if protozoan grazing primarily controls prokaryotic abundance, then most carbon would channel to the highest trophic levels in the food chain. Conversely, if viral infection primarily drives prokaryotic losses, carbon flow would deflect far below the largest organisms in the food web, accelerating transformation of nutrients from a dissolved to particulate state. Viral infection can then stimulate prokaryotic production and respiration, increasing nutrient regeneration through cell lysis and release of cytoplasmic components and cellular material, extracellular DNA, and nutrients. This release, in turn, can strongly influence ecological and biogeochemical processes, and therefore ocean productivity and its response to climate change.

The microbial community reuses organic detritus (both dissolved and particulate) transformed by viral lysis. This viral short circuit (**viral shunt**) results in increased respiration and decreased carbon transfer to higher trophic levels. This shunting not only affects global biogeochemical cycles, but also alters microbial communities. In some environments, viral infection causes bacterial mortality comparable to losses from protistan grazing on bacteria. However, in systems with limited protozoan abundance, such as in anoxic environments, viral lysis represents the major source of prokaryote mortality. Although lesser in impact, viruses also infect phytoplankton. The first viruses to be isolated that infect algae belong to a family of large double-stranded DNA viruses (the *Phycodnaviridea*) and include viruses that infect primary producers and algae that produce toxic blooms. Experimental addition of viral concentrates to natural samples can decrease primary production, suggesting that viruses could cause significant phytoplankton losses, especially during blooms.

One recent model suggests that about a quarter of organic carbon flows through the viral shunt. Both in marine sediments and in the water column, the lytic cycle appears to be the dominant strategy for producing new viruses, perhaps because temperate viruses can survive periods of reduced abundances of host cells. In sediments, abundant resources for heterotrophic prokaryote growth also favor lytic cycles.

Ecology of benthic prokaryotes: Marine microbial ecological research, a relatively young discipline, has recently exploded through the development of new genetic tools. Historically, microbial research followed an "auto-ecological" approach focused on studying individual prokaryotic taxa and specific metabolic reactions associated with them, or on a "synecological" approach in studies that treated bacterial communities as "functional units" and measured the processes

Marine Biology: Comparative Ecology of Planet Ocean, First Edition. Roberto Danovaro and Paul Snelgrove.
© 2024 John Wiley & Sons Ltd. Published 2024 by John Wiley & Sons Ltd.
Companion Website: www.wiley.com/go/danovaro/marinebiology

associated with that unit. This "black-box" approach arose because researchers typically could not identify the individual taxonomic components of these communities. Marine bacterial studies typically began with isolating a specific microbial strain in culture, in order to examine its metabolic properties. Once isolated, the potential role of the bacterium in a given ecosystem can be elucidated. However, traditional microbiology (i.e., using culture media) successfully grows just one cell out of 10,000, and isolates perhaps only 0.0001% of the species present in a sample. The application of methodologies such as epi-fluorescence microscopy to marine samples in recent decades revolutionized our understanding of the marine microbial world, allowing us to recognize bacteria as the most abundant component of this biota.

In 1979, Larry Pomeroy concluded that "...*our vision of food webs in the sea is gradually changing, from a vision of linear trophic chains consist of diatoms, copepods and nekton, to a complex network of ecological interactions in which an enormous variety of microorganisms plays a significant role in the production and in the consumption of organic matter....*" A few years later (1983), Farooq Azam led a group of researchers who coined the term **microbial loop**, a trophic pathway through which dissolved organic matter (DOM) reenters the food chain through its transformation into bacterial biomass. This concept recognizes the key role bacteria play in marine ecosystem functioning and in "rescuing" organic detrital matter produced by various trophic levels and converting it into biomass. Protistans and other organisms then feed on prokaryotic cells, which in turn, become prey for larger organisms, thus returning the dissolved organic matter to the food web. The introduction of the microbial loop concept revolutionized the field of marine ecology, focusing attention for the first time on a component that had been largely ignored, namely organic detritus.

Verification of many key theories in quantitative ecology, including those related to ecological succession, adaptation, biodiversity, competition, and predator-prey relationships require understanding the rules governing the spatial distribution of populations and communities. Prokaryotes in the marine environment are typically highly heterogeneous spatially, noting that a variety of bacteria, living both freely (dispersed in the water) or aggregated onto organic or inorganic particles. This "patchy," complex mosaic of organisms exhibits sophisticated behavioral strategies, such as the ability to move towards organic matter using chemotaxis. **Permeases**, or specialized transport enzymes, in the membranes of bacterial cells selectively transport substances through the membrane, including substances at very low concentrations. All of these abilities reflect adaptation to living in a heterogeneous environment with patchy resources. Ecological theory also suggests that such resource distribution might ultimately favor high bacterial biodiversity by creating a large number of ecological niches. Marine sediments offer an extremely complex environment for benthic bacteria, from both a physical and a chemical point of view, including a complex matrix of organic and inorganic particles of varying size.

Multiple factors control the spatial distribution of benthic bacteria: oxygen concentration, sediment particle size, physical disturbance (e.g., turbulence, re-suspension), availability of substrates, and predation pressure from protists, meiofauna, and other organisms. Often, the simultaneous effects of biological and physical forcing determine changes in **bottom-up** (i.e., determined by the resources) and **top-down** (determined by predators) controls on bacterial communities. In particular, oxygen strongly influences prokaryote distributions in sediments. Oxygen arrives on the seabed through the transport of water masses; penetration into the sediment occurs either (very slowly) by diffusion or (much more rapidly) by other mechanisms such as bottom mixing by waves, currents, or bioturbation by larger biota. Oxygen concentrations in sediment differentiate **"oxic,"** referring to oxygenated habitat, from **"anoxic"** or non-oxygenated habitat. Typically, highest oxygen concentrations in sediments occur at the water-sediment interface, with declining concentrations with increasing depth in the sediment. The scale of penetration varies greatly, but generally extends to 10s of centimeters at most and typically to just a few centimeters. In this oxygenated layer, aerobic respiration represents the primary process for organic matter decomposition, using oxygen as oxidizing agent.

In contrast, benthic bacteria in the anoxic zone use other available oxidants such as nitrate, sulfates, and iron and manganese oxides to break down organic matter anaerobically. In general, most bacterial biomass and metabolic activity occurs in sediment surface layers and decreases with depth in the sediment.

Ecology of nanobenthos: The small size of protistans means they perceive the environment both as a spatial and as a temporal mosaic; an apparently homogeneous environment for such small organisms might actually be highly heterogeneous, with environmental gradients that large organisms would not perceive. Typically, nanoflagellates move independently and perform all the functions of a unicellular organism but at much smaller scales. Nanoflagellates live in **interstitial** (between sediment grains) spaces, which can be mechanically unstable, particularly in coastal waters. Free oxygen occurs only in the uppermost sediment layers, thereby confining most nanobenthic organisms to this layer. In addition to this vertical sediment zonation gradient, horizontal patterns occur; for example, tiny aggregates of organic matter may create microhabitats < 100 μm in diameter. **Micropatches**, or micro-aggregates of organic matter or bacteria, strongly influence protists. Imagine a single-celled organism moving slowly through the sediment in search of food, experiencing sudden changes in environmental conditions, searching for a mate, or escaping a predator. Although we generally understand how environmental factors influence the most common eukaryotic organisms in sediments, such as flagellates and ciliates, we know far less about the nude amoebae with lobose pseudopodia (Order Gymnamoebae) that inhabit marine sediments, despite abundances >15,000 cm^{-3} of sediment. Although heterotrophic nanoflagellates consume more prokaryotes in marine sediments than other predators, meiofauna and macrofauna can also prey intensely on protists.

Ecology of meiofauna: Meiofaunal populations typically reproduce annually. However, some studies show reproductive cycles in meiobenthic species lasting more than three years, in some cases using resting cysts that remain inactive for years. Moreover, many species offset their reproductive periods relative to other species, presumably reducing competition. Deposit feeding dominates among meiobenthic organisms as a feeding strategy, with many grazing on diatoms and, especially in deep environments, on bacteria. Nematodes dominate most meiobenthic populations, accounting for >90% of total meiofaunal abundance. Harpacticoid copepods are generally the second most abundant group, followed by polychaetes, flatworms, and gastrotrichs. Meiofauna generally vary greatly in abundance over time.

Anatomical structures in nematodes, and particular oral cavity morphology, allow identification of their different food sources. This identification facilitates division of nematodes into trophic groups, which in turn, enables studies on trophic and functional biodiversity. Nematodes largely separate into four trophic groups based on the scheme proposed by Wolfgang Wieser in 1953: selective deposit feeders (or bacteriovores, known as group 1A) feed primarily on bacteria; non-selective deposit feeders (group 1B) feed on organic detritus; epistrate-feeders (group 2A) feed on benthic diatoms; and predators/omnivores (group 2B) feed on other organisms or dead bodies of meiofauna and some macrofaunal juveniles.

In 1987, Danish scientist Preben Jensen revised this classification by unifying the deposit feeders into a single group, retaining epistrate-feeders, and splitting group 2B into predators and omnivores. In 1997, Belgian scientists Magda Vincx and Tom Moens revised Wieser's classification, further subdividing group 1A into nematodes that feed primarily on ciliates as **ciliate feeders** and those that feed on bacteria as **microvores**. They also divided trophic predators (2B) into those feeding exclusively on other meiofauna as predators and those that feed occasionally as **facultative predators**. Some genera of nematodes, in addition to feeding on organic debris and bacteria, secrete and deposit mucus on the sediment. Bacterial populations rapidly colonize the mucus, converting the mucopolysaccharides into microbial biomass, which the same nematodes then ingest in a process known as **microbial gardening**. Many harpacticoids use a similar technique known as **mucus-trap feeding**. Studies of other meiobenthic taxa based on stomach content analysis indicate that flatworms are predators that feed mainly on other meiobenthos.

Meiobenthic grazing strongly impacts benthic diatoms and bacteria, ingesting an estimated 3% of bacterial biomass and 1% of diatom biomass every hour. This predatory activity stimulates microbial growth, maintaining it in a logarithmic phase. Meiofauna may actually consume the equivalent of their own weight in microbial biomass every day. Despite the small size of individual meiofauna, their secondary production plays an important role in benthic energetics. As a result of their small size, meiofaunal metabolic activity turnover (i.e., ratio of production: biomass, P:B) occurs five times faster, on average, than that of the macrofauna, matching macrofaunal production even in systems dominated by macrofauna. The P:B of meiofauna varies among taxonomic groups and even among species within each taxon. This variation complicates efforts to assign a precise overall value, but researchers conventionally assume a value of 9–10. Meiofauna plays an essential ecological role in benthic environments by linking organic matter (and thus energy transfer) to higher trophic levels. Harpacticoid copepods and nematodes comprise an integral part of the diet of macrobenthos and demersal fishes. In addition, meiobenthic predators feeding on macrobenthic larvae and juveniles (i.e., temporary meiofauna) can control the composition and structure of the adult macrobenthic community (the **bottleneck hypothesis**).

In recent years, researchers have identified meiofauna as a potential indicator of collective alteration of marine ecosystem functioning. Because the trophic characteristics of various species within a given taxonomic family typically remain constant, analysis at the level of genus or family may be sufficient in monitoring meiobenthic response to environmental changes. High sensitivity to environmental perturbations, large numbers of individuals, lack of planktonic larval forms, and a short life cycle all contribute to the utility of meiofauna in evaluating disturbance and re-colonization in marine environments. In particular, meiofauna can help in assessing impacts on benthic environments from disturbances ranging from pollution, presence of artificial reefs, mussel aquaculture, and fish farms. Furthermore, researchers use meiofauna to study benthic responses to material sedimenting from the water column, as in river plumes.

The short generation times and large number of juveniles produced by meiobenthic fauna qualify them as **opportunists**, or species that can quickly exploit new resources as they arise. In fact, although the lack of a pelagic dispersal stage for the permanent meiofauna necessarily retards re-colonization of disturbed substrates, limited dispersal adds an advantage in studies designed to identify the effects of pollution because larval immigration from other areas does not complicate interpretation. However, because of their small size, meiobenthos require considerable effort in sorting and identification. Only organisms with a cuticular coating or a chitinous casing (the "hard-bodied" meiofauna) are easily identified once preserved, whereas other groups (such as turbellarians) require taxonomic identification when still alive.

Ecology of macrofauna: Macrofaunal organisms living within the substrate, known as infauna, include annelids, bivalve mollusks, amphipod crustaceans, and anthozoans, among others. In contrast to many sedentary species of infauna, others move freely, tunneling through soft sediments. Some species, such as the clam *Macoma balthica*, burrow up to 10s of centimeters below the surface of the sediment in order to avoid predation. However, even organisms that dig into deep sediment layers must maintain a connection with the overlying water to obtain the oxygen necessary for respiration. Epifauna, in contrast to infauna, live on or just above the seabed. Sessile epifaunal organisms firmly attach to the substrate, whereas relatively sedentary or highly vagile (mobile) epifauna retain varying degrees of mobility. Common epifauna include barnacles, oysters, sponges, tunicates, gastropods, anthozoans, crabs and some species of amphipods. The juvenile stages of many sessile epifaunal species may be vulnerable to predators, but, as adults, they develop protective strategies such as hard shells (e.g., oysters, barnacles), antipredator compounds in their tissues that make them toxic or unpalatable (e.g., tunicates), or stinging cells to deter predators (e.g., anthozoans). Mobile epibenthic species (e.g., the polychaete *Neanthes succinea*) emerge from the interstices of the substrate to feed on detritus, algae, or small invertebrates, and then quickly retreat to avoid predation. Macrofauna living in muddy sediments feed on associated micro-algae, organic detritus, and prokaryotes.

Some filter feeders (suspension feeders) can ingest organic particles (including microphytoplankton) suspended in water. Most suspension feeders are relatively sedentary and depend on bottom currents to transport food to them. Environmental characteristics strongly influence filtering organisms, in that excessive organic load or re-suspension of sediments can increase turbidity to the point that it interferes with feeding, thereby limiting their development. However, some taxa, such as annelids, tunicates, and hydroids, can also occur in abundance in turbid estuaries and coastal marine habitats. Some groups of benthic macrofauna use microalgae as a food source whereas others are carnivores.

Macrofauna represent an important component of estuarine and coastal ecosystems, and link organic matter (e.g., produced by phytoplankton, benthic microalgae, or macroalgae) to higher trophic levels (e.g., fish), with significant ecological and economic importance for the ecosystem. For example, ~50% of fish production in highly productive systems link directly to the benthic food web generated by macrofauna.

Macrofauna includes some commercially important species, such as the hard clam *Mercenaria mercenaria* and the oyster *Crassostrea virginica*. Macrofauna also provide different ecosystem functions, including processing of organic matter and associated nutrient recycling, dispersing and burying pollutants, and contributing to secondary production. Filter feeders also improve water clarity by removing particles from the water column, thereby increasing light penetration, which benefits submerged aquatic plants and benthic microalgae. Bioturbation of sediments by macrofauna can stimulate microbial activity, enhancing degradation of some pollutants. Finally, macrofauna can influence transformation of dissolved nitrogen into elemental nitrogen gas, which diffuses into the atmosphere, reducing dissolved nitrogen load in the ecosystem.

Marine macroalgae mainly include Chlorophyta, or green algae, Phaeophyta, or brown algae, and Rhodophyta, or red algae. Although separation of algal groups by color may appear superficial, with some exceptions, fundamental morphological and biochemical differences support this subdivision (Figure 10.1).

All of these macroalgae contain chlorophylls but brown, blue, and red accessory pigments in brown and red algae (Phylum Chlorophyta and Rhodophyta, respectively) mask the green pigment to varying degrees – and thus confer distinct colors. Macroalgae occur along the coasts in relatively well-defined vertical zonation. These distributions overlap considerably but, with some exceptions, green algae occur most commonly in the upper intertidal zone, brown algae in the lower intertidal, and red algae from the lower intertidal to deeper zones.

Four key factors limit primary production in the benthic domain: (a) concentration of nutrients (nitrates, nitrites, phosphates, and silicates); (b) quantity and quality of light energy, which typically becomes limiting at ~15–20 m depth, influencing zonation and composition of photosynthetic pigments; (c) auxiliary energy, in the form of hydrodynamics and waves that may dislodge some algal forms, influence light penetration, and promote diffusion of nutrients and CO_2; (d) air and particularly water temperature, the former influencing desiccation and heating, the latter strongly affecting metabolic respiration/photosynthesis balance. These factors define zones particularly relevant to energy and primary producers (Figure 10.2): (a) supralittoral zone, (b) intertidal zone, (c) infralittoral subtidal "optimal" production zone, (d) seagrass or macroalgal zone (>15–20 m), and (e) light "limited" zone.

The **supralittoral zone**, which occurs above the high tide mark, creates extreme conditions for macroalgae through excessive light (high UV) and occasional wetting, with dominance by cyanobacteria, diatoms, and ephemeral macroalgae

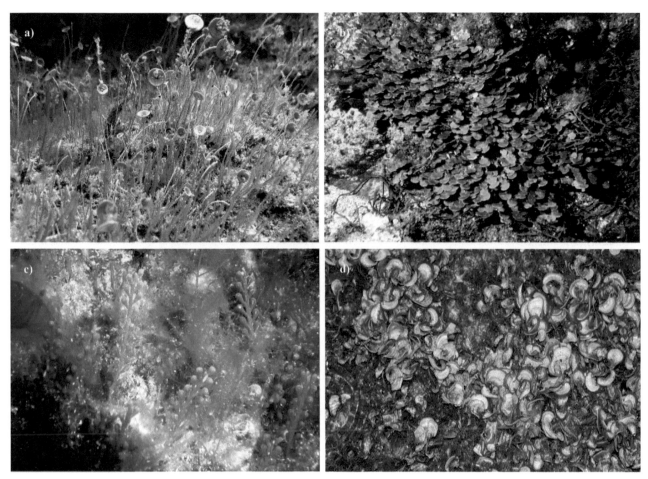

Figure 10.1 Examples of macroalgae in the optimal production zone: (a) *Acetabularia* sp.; (b) *Halimeda* sp.; (c) *Caulerpa racemosa*; (d) *Padina pavonica*.

that collectively result in low production (2 g C m^{-2} y^{-1}). The **intertidal zone** between low and high tide levels, with its excessive light, and regular wetting create conditions most organisms would find stressful, however, some seasonal micro and macroalgal species thrive with high turnover and moderate production (300 g C m^{-2} y^{-1}). The **infralittoral zone**, exposed only during spring tides and slightly deeper, has the highest production. This zone includes the high infralittoral, with abundant light but low UV, strong mixing, high macroalgal coverage (e.g., *Cystoseira* spp.) and production of 2000 g C m^{-2} y^{-1}. In regions with seagrass or macroalgae (>15–20 m), production decreases with depth and with light attenuation, which varies depending on dominant species and on local conditions (500–1,500 g C m^{-2} y^{-1}). With increasing depth, seagrasses begin to disappear, and calcareous coralline formations appear (coralligenous habitat) as light penetration decreases markedly (potentially to < 1%) and wave energy diminishes. Production here varies greatly, over an order of magnitude of around 40 g C m^{-2} y^{-1}. The average production of phytobenthos in a hypothetical transect 1 m wide × 20 m deep roughly equals the production of surface phytoplankton in a transect 1 m × 700 m in length. Nonetheless, the limited area covered by habitat suitable for phytobenthos (effectively the broad intertidal zone) concentrates this production and results in very modest contributions to overall ocean production. However, phytobenthos contributes fundamentally to the functioning of coastal ecosystems. Furthermore, given the comparatively small fraction of benthic primary production directly consumed by herbivores, most plant biomass undergoes both physical (mechanical) and biological (microorganisms) degradation, entering the pool of dissolved or particulate organic matter in the water column, potentially providing food for other filter feeders, both benthic and pelagic. Moreover, export of this biomass to other ecosystems, including the deep sea (depending on depth and distance from the coast), provides a vital source of food to some seafloor communities.

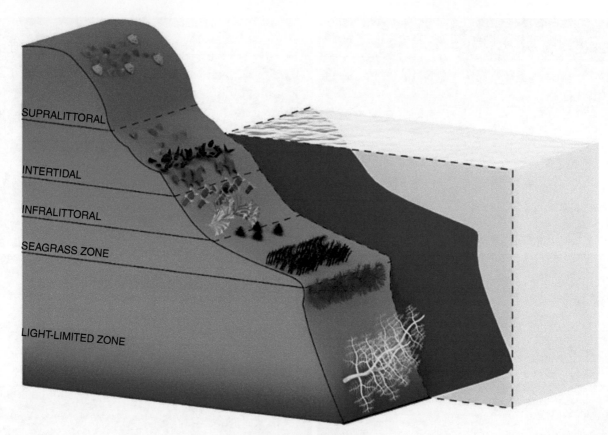

Figure 10.2 Benthic zonation based on the characteristics of each bathymetric range.

10.2 Trophic Groups: Classification of Benthos Based on Diet

Trophic needs fundamentally influence animal communities, particularly in the benthos. Benthic environments support five different feeding modes of heterotrophic organisms that vary with: (1) the size of available resource; (2) the type of resource; (3) the environmental compartment containing the resource; (4) the foraging mode and associated mobility; (5) the feeding structures involved in ingesting the food resource.

1) Resource size, which relates to the size of the food compared to the consumer, determines whether microphages or macrophages dominate. Sponges, ascidians, and other filter feeders capture and ingest particles only a few microns in size, and are therefore **microphages**, as are sea urchins that scrape diatom films or plant propagules, typically from hard surfaces. **Macrophages** such as decapod crustaceans and pennatulaceans (sea pens) feed on larger particles including other organisms.

2) Resource type: Benthic communities include carnivores, herbivores, omnivores, and scavengers. **Carnivores** refer to predators that eat animal tissues, whether parts or the entire animal, and also include parasitic and saprobic forms. **Herbivores** feed on plant tissues, whereas **scavengers** feed on often abundant, low quality organic detritus. Food sources may be animal (feeding on carrion, as scavengers) or vegetable (feeding on plant debris). **Omnivores** feed on a mixed diet of plants and animals; some scavengers are also omnivorous.

3) The resource environmental compartment: Benthic organisms feed on (a) particles suspended in the water column or (b) particles on or in the substrate. Some filter feeders or suspension feeders feed on suspended particles passively (when currents transport food particles to sessile animals, as occurs with coral polyps that intercept food transported by currents), whereas others feed actively, typically using some mechanism to pump water. For organisms scavenging organic particles on the seafloor, superficial scavengers feed on sediment surface layers whereas deposit-feeding species ingest particles in deeper layers. Organism size often influences the depths within the sediment at which they feed. The small gastropod *Hydrobia* sp. can dig a few millimeters into the sediment, whereas small polychaete worms may burrow

to several centimeters, the amphipod *Corophium volutator* digs "U" shaped dens and the polychaete *Arenicola marina* creates burrows to 25 cm deep. Some bivalves such as cockles and soft-shelled clams can move to even greater depth.
4) Feeding mode and mobility: most carnivores are vagile or mobile (with some exceptions such as sessile gorgonians and sea anemones that feed on suspended prey). In contrast, most sedentary organisms and those with limited motility often scavenge on surface or sub-surface particles. Feeding behaviors also vary among species. Some species feed with their head up, whereas others feed with their head down. Some predators ambush their prey, moving across the sediment surface in search of burrow holes that indicate the presence of potential prey. Other predators, such as the parchment worm, *Chaetopterus* sp., dig pits that organisms such as crustaceans crawling across the bottom may fall into, facilitating capture. Some benthic organisms build a kind of "garden" in which they grow/raise their food. For example, nematodes produce and then leave mucus on the surface of sediment particles; bacteria adhere and grow on this film of exopolysaccharides. Nematodes later return along the same path, feeding on the immobilized bacteria. By feeding on protein-rich bacterial cell, the nematodes exchange low nutritional value mucus for high nutritional value food (bacteria). The reduced digestive tract of some nematodes, such as *Astomonema* spp., reflects their close association with symbiotic bacteria.
5) Feeding structures: The mouthparts or oral structures of benthic fauna reflect their dietary preferences. Jaws or piercing structures, and structures for capturing or cutting characterize predators, some of which also use specialized structures to inject neurotoxins that paralyze, stun, or kill their prey; some species use tentacles, palps, or adhesive structures to capture and transport food. In contrast, suspension feeders or scavengers typically use finer structures such as flagellae or cilia to filter the water. Because deposit feeders do not need to attack prey, many possess a soft proboscis to ingest and swallow organic particles attached to or between sediment grains. Other structures such as the tongue-like, toothed **radula** of gastropods, allow them to scrape the food adhering onto hard surfaces.

All benthic organisms can be divided into **trophic groups**, referring to different species with similar dietary strategies that target similar food sources from a given trophic level. Thus, trophic groups refer not to numbers of species or individuals but to their functionality, and specifically their feeding mode. Other important functional aspects such as growth rates, biomass, and food supply complement structural information such as abundance and species richness but the latter alone tell us little about how a system works and we therefore need to understand what species actually do, as well as other critical aspects of their biology such as life cycles.

In the past, researchers argued for the importance of **amensalism** in sediments. Amensalism refers to an association between two different species in which one is inhibited or excluded from access to available resources while the other remains unaffected. Amensalism includes two basic modes: (1) competition in which a larger or stronger organism excludes a smaller or weaker one from living space or deprives it of food, and (2) **antibiosis**, in which one organism is unaffected but the other is damaged or killed by a chemical secretion. In sediments, competition between different species can result in the exclusion of one species from the environment. In particular, in sandy sediments we sometimes observe that high abundances of suspension feeders can prevent colonization by deposit feeders. This prevention occurs either by ingesting settling larvae, or by forming dense tube habitat that limits the burrowing of deposit feeders. Abundances of deposit feeders typically increase with depth as bottom currents weaken, resulting in accumulation of finer silt and clays on the seafloor. As deposit feeders burrow, they move and resuspend these sediments, which can clog the delicate filtering structures used by suspension feeders. Thus, deposit feeders and suspension feeders prefer different sedimentary conditions, although this preference does not result from direct interaction between species but rather through indirect interference; habitat and food preference likely play a major role in creating this pattern.

10.3 Comparison Between Hard and Soft Bottom Benthos

The two main kinds of substrate: hard and soft bottoms, host very different floras and faunas. Hard substrate favors attached species such as algae and sessile invertebrates such as sponges, gorgonians, stony corals, and tunicates, whereas soft substrate grading from course sands and gravel to fine muds hosts a variety of deposit-feeding invertebrates such as polychaetes, bivalves, and gastropods. In terms of photosynthetic organisms, mats of unicellular algae and seagrasses sometimes cover the shallowest coastal sediments helping to stabilize the bottom, much like plants on land reduce erosion and landslides. The substrate and the water immediately above it define benthic environments. Settlement onto soft or hard bottoms differs considerably (Figure 10.3).

The main differences between organisms inhabiting hard and soft bottoms can be summarized as follows (Table 10.1):

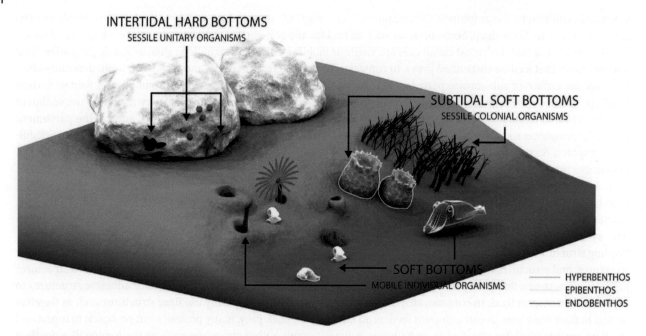

Figure 10.3 Differences between rocky and sedimentary bottoms. Note position of organisms with respect to substrate: hyperbenthos (blue line) living just above bottom, epibenthos (yellow line) living on sediment surface, and endobenthos (red line) living within the sediment.

Table 10.1 Types of benthic organisms present in different substrates.

Intertidal hard bottoms	Subtidal soft bottoms	Soft bottoms
Sessile organisms dominate	Sessile organisms dominate	Mobile organisms dominate
Individual (unitary) organisms dominate	Colonial organisms dominate	Individual organisms dominate
Larval recruitment dominates	Fragmentation/cloning augments recruitment	Movement of adults increases recruitment

Soft bottoms cover some 60% of Earth's surface and sedimentary communities thus broadly characterize our planet. These communities include benthic organisms from a few nanometers (virus) to several m in size, spanning similar orders of magnitude equivalent to the difference between a grain of sand and the moon. Not surprisingly, these organisms encompass quite a broad range of ecological niches and functions. Soft bottoms vary greatly in mineralogical composition, morphology, and structure. Soft sediments occur in estuarine, coastal, and deep-sea environments.

Benthic organisms live as **hyperbenthos** just above the substrate, **endobenthos** or **infauna** that live within the substrate, and as **epibenthos** that live upon the substrate and extend above the bottom into the water column to obtain oxygen and food. Intermediate forms also occur, in that many mollusks and polychaetes live partially buried in the substrate but extend siphons or tentacles up into the overlying water. Some crustaceans alternate periods of time buried in the sediment with periods in which they swim in the overlying water. Generally, epifauna dominate hard surfaces, whereas endobenthos or infauna dominate soft bottoms.

Researchers in the past described hyperbenthos by a variety of names, including "demersal zooplankton" in the tropics and as "benthopelagic plankton" in deep-sea environments. The highly mobile component of the hyperbenthos, and mysids in particular, are often referred to as **micronekton** or nekto-benthos. Hyperbenthos occur in greater densities in coastal waters and particularly estuaries; they also occur in many deep-sea environments. Many hyperbenthic taxa (decapods, copepods, amphipods, polychaetes) change in distribution over **diel cycles**, referring to migrations over 24-h time periods, or in relation to tidal phase. In deep-sea environments, however, distributions vary little with time of day. Hyperbenthic research really only started with F. Beyer's pioneer publication in 1958. At the moment, however, the subjective inclusion and exclusion of species and use of different sampling methods (such as volume of water filtered water versus bottom area swept) complicates direct comparison among different datasets.

Epibenthos vary in their capacity to move, and include: (1) sessile fauna that remains fixed to the substrate surface for almost their entire life, moving only as eggs or larvae. This group includes sponges, anthozoans, bryozoans, ascidians, barnacles, and others, some of which are **epibionts** that specifically attach to other organisms such as bivalve shells. **Epiphytic species** specifically attach to macroalgae, seagrass, or kelp. Some organisms such as barnacles "cement" themselves on hard surfaces, whereas others such as mussels use filaments and algae as adhesion structures; (2) sedentary species adhere to the substrate but can move short distances, as in the case of limpets; (3) mobile species move freely, often crawling across the substrate; (4) drifting organisms detach from the bottom and drift with currents until they sink again at another suitable location (as occurs with sea pens). **Demersal** benthos refers to fishes, crustaceans, mollusks, and other species that swim across and sometimes rest on the seafloor but depend on the seafloor as a habitat for feeding, shelter, or reproduction.

10.4 Ecology of Benthos Inhabiting Soft Bottoms

Soft sediments contain a mixture of inorganic and organic particles and interstitial water that strongly influence benthic organisms. The size of the sediment particles, which defines their **grain size**, generally indicates hydrodynamic energy, in that strong currents and waves resuspend and transport finer sediments such as silts and clays, whereas these fine sediments settle onto the bottom and accumulate in weak flows. Differences in morphology, including feeding structures, reflect species adaptations to different sediment grain sizes and current regimes.

Grain size, in fact, strongly influences the lifestyle of benthic organisms. For example, gravel sediments lack the fine particles required by organisms that ingest sediment for feeding. At the same time, sediments composed of only fine particles may lack the stability to support large, dense animals. On the one hand, small animals in habitats with strong currents may experience continuous resuspension and removal from the substrate unless they can burrow deep into the sediment or resettle rapidly to re-establish their position. On the other hand, animals that feed on organic matter can ingest fine sediments that characterize seafloor habitats in areas with weak currents, especially given increased organic material and small organisms compared with coarser sediments.

Precisely because of substrate instability, sessile forms occur much less frequently on soft bottoms and, when they do occur, they possess suitable anchor structures. Ascidians such as *Molgula* spp. attach to the seabed using branching rhizoids. Taproots form an anchor to the substrate with the lower part of their body or with peduncles that penetrate into the sediment. This strategy characterizes some sponges, pennatulaceans (*Pennatula phosphorea*), alcyonarians (*Alcyonium palmatum*), sea anemones, some algae, and, most notably, seagrasses. Seagrasses depend heavily on the presence of roots, because just like other vascular plants, and in contrast to macroalgae that draw their nutrients from the water, seagrasses absorb nutrients through roots embedded in the muddy substrate.

Living within the seafloor provides an important element of protection against predators or unfavorable environmental conditions, which explains the dominance of infauna in soft sediments. To be able to move through sediment, infauna must be able to move sediment particles. Initial displacement of the granules requires thrusting a rigid structure within the sediment with sufficient force. Many burrowing organisms, such as worms or some bivalve mollusks burrow using a soft structure they can modify in shape by varying internal hydrostatic pressure. To allow the operation of this organ a fluid stiffens the structure, as seen with the powerful foot that bivalve mollusks use to penetrate the sediment. They insert the distal part of the foot into the sediment and use hydrostatic pressure to widen the foot to become a real anchor. By contracting the remaining part of the body, the bivalve can move toward the base of the foot. Thus, a series of repeated expansions and extensions moves them through the sediment, with impressive speed in some species. Other burrowing invertebrates such as worms, sea cucumbers, and burrowing sipunculids apply this same general principle. Other species use muscular action of appendages that act as oars or paddles. For example, the crab *Emerita talpoida* uses a sword-shaped posterior appendix to dig. Polychaetes such as *Glycera* spp. undulate their body and move their parapodia to move through sediment. Regular echinoids (sea urchins) and asteroids such as *Astropecten* spp. dig by moving their spines. For burrowing infauna, sediments act as sandpaper. Therefore, the integument of relatively few species comes in direct contact with the sediment, and even those that do only burrow sporadically (e.g., synaptid holothurians or naticid gastropods). Real burrowers use a tunnel that communicates with the surface of the sediment, and typically consolidate the walls, for example with mucus, maintaining the integrity of the burrow by using cilia and appendages to create a water current.

The lugworm *Arenicola* sp., for example, which lives on sandy bottoms of the tidal zone, digs a U-shaped tubular gallery, cementing its walls by secreting mucous that hardens on contact with seawater. Only the shell of burrowing bivalves comes into direct contact with the sediment and they live buried within the sand or mud with only their two siphons extending outside the sediment. Their inhalant siphon brings water inside the shell, from which they obtain oxygen and filter food particles. The exhalent siphon then pumps water out (Figure 10.4).

Figure 10.4 Examples of infauna in soft sediments. Various types of macrofauna have different abilities to burrow in sediment including: (a) ophiuroids (brittle stars); (b) annelids; (c–e) bivalves.

Some species spend almost their entire lives within the same burrow, whereas others emerge sporadically to create a new burrow. In extremely unstable sediment, many infaunal species continually reposition themselves within the seabed. Waves in the surf zone frequently move species around, such as the surf clam, *Spisula solidissima*, but they quickly reposition themselves. Mobile species that crawl or walk on the bottom can colonize shifting sediment as long as it has some degree of compaction. The shapes of some organisms reflect an adaptation to avoid sinking into the mud. For example, the pencil urchin, *Cidaris cidaris*, uses its long quills as stilts; the long radii of the fins of the piper gurnard, *Trigla lyra* serve a similar function.

Organisms burrow within sediments for different reasons: (1) to use the three-dimensional environment; (2) to achieve greater stability in unstable sediment; (3) avoidance of and protection from predators; (4) to protect itself or its eggs from disturbance; (5) to facilitate oxygen exchange and waste removal from tubes or burrows; or (6) to enhance food supply in tubes or burrows. Benthic organisms have developed various modes of burrowing through sediment (Figure 10.5): (1) pushing themselves through the sediment; (2) movement of sediment laterally; (3) excavation, where they transport sediment toward the posterior part of the body, thus backfilling.

The mode by which these organisms burrow has profound implications for the sedimentary environment. An organism can move between grains of sand as meiofauna do, or burrow, or create a mucus-lined tube that oxygenates sediments below. All of these activities significantly impact the substrate. Benthic species depend on the substrate as a habitat and food source but, in turn, modify the substrate in a species-specific way. Much as plants modify the terrestrial landscape, barnacles, sponges, hydroids, bryozoans, gorgonians, ascidians, and tunicates structure and modify the seabed. Even in calcareous rocks, clionid sponges can erode the rock using acidic secretions.

As organisms, move through the sediment, they mix it and simultaneously increase oxygen exchange with the surrounding water, thus causing "**bioturbation**" (Figure 10.6). Digging, feeding, and movement of macrofauna, either at the water-sediment interface or within the sediment, play a particularly significant role in this process. Meiofauna augment this process through **crypto-bioturbation**, referring to small-scale and subtle bioturbation. Bioturbation may also decrease sediment stability, increase rates of remineralization, and transport oxygen and organic matter deeper into the sediment.

Without bioturbators the sediment would largely consist of anoxic, species-poor habitats made up of layers or **laminae** of sediment deposited in parallel thicknesses varying from 1 mm to 1 cm per year, depending on location. Instead, living organisms that bioturbate and thus break up this laminar structure create a heterogeneous environment that supports a wide diversity of species. Some bioturbators leave behind skeletons and shells that can stabilize the substrate.

Some organisms, by burrowing to greater depth, significantly alter sediment structure, sometimes oxygenating it to depths of 1 m or more. For example, the crustacean *Jaxea nocturna* (Figure 10.7), digs tunnels 1–1.5 m in length below the sediment surface, despite its small size (<10 cm). Bioturbation adds significant three-dimensional structure and, by

Figure 10.5 Examples of methods of excavation by organisms (a) intrusion: deep-sea tellinid bivalves, *Abra nitida* and *A. longicallus* show in light blue; (b) excavation by the lugworm, *Arenicola marina*; (c) backfill by the heart urchin, *Echinocardium cordatum*.

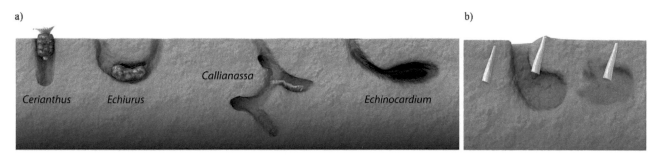

Figure 10.6 Bioturbation caused by movements of different organisms: (a) anemone, *Cerianthus* sp., the spoon worm, *Echiurus* sp., mud shrimp, *Callianassa* sp. and heart urchin, *Echinocardium cordatum* that burrows 15–20 cm below the sediment surface. These types of organisms play a key role in transferring organic matter from the water column and in transforming it into biomass in sediments; (b) scaphopod mollusk *Dentalium* sp.

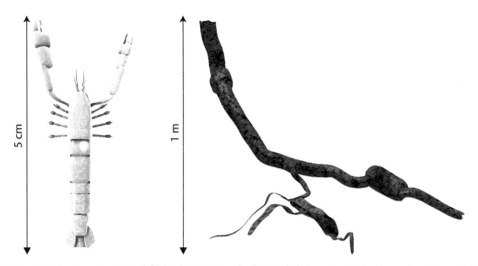

Figure 10.7 (a) crustacean *Jaxea nocturna*, and (b) intricate network of tunnels it burrows to depths up to a 1 m within the sediment.

enhancing water and therefore oxygen penetration, has important consequences for biogeochemical cycles. This penetration also affects organic matter cycling, influencing the relationship between carbon arrival, burial, respiration, and rates of microbial activity. Indeed, oxygen transport influences transformation, burial, and organic matter mineralization rates.

Many organisms use tubes, in some instances occupying those already produced by other organisms. Some amphipods, for example, live in tubes as juveniles until the tubes become too small, at which point they abandon them and look for other habitats of an appropriate size. This process contributes to the colonization process and how species use benthic substrates. The **mixing depth** refers to the depth to which organisms bioturbate.

10.5 Changes in Benthos in Space and Time

Benthic organisms adapt to specific abiotic environmental factors that act at various scales and define many environmental gradients. These include: (a) latitude, which results in large-scale changes in light radiation (intensity, variability) and temperature (mean and range); (b) depth or bathymetry, which results in medium to small-scale changes in hydrodynamics, light, temperature, and pressure; (c) proximity to land results in medium scale changes in input of terrestrial material, hydrodynamics, salinity, and pollutants, among others. These gradients in abiotic parameters contribute significantly to heterogeneous and often patchy distributions of benthic organisms. The interaction between these large- to medium and medium- to small-scale variables delineates the two major groups of factors that determine the composition of seafloor communities:

1) **Abiotic factors** group roughly into those related to the surrounding environment (precipitation, intensity and quality of solar radiation, temperature, pressure) and those related to habitat (organic and inorganic content, salinity, hydrodynamics – including turbulence, waves and currents, food supply) and substrate, which strongly influences choices made by settling larvae of hard and soft substrate species.

In a general sense, sedimentary communities form three latitudinal bands: intertropical (0–20/30°N/S), temperate (30–60°N/S) and polar (above 60°N/S). Temperature and productivity (linking to nutrient supply) fundamentally differentiate these environments. Tropical sedimentary environments support macrofaunal abundances of ~3,000 individuals m^{-2}, dominated by polychaetes, decapod crustaceans, isopods, and bivalves, and >1 million meiofaunal individuals m^{-2}. Similar groups dominate temperate zones, though **keystone species**, referring to a species that exerts a significant and disproportionate impact on its community or ecosystem relative to its abundance, as well as other species, differ. Structural or habitat-creating species, sometimes from the same genus, may play similar roles for communities in different environments. In contrast to keystone species, which are functionally important but few in number, biomass and function may both be significant for structuring species. Structuring species support a range of associated species that often influence the stability of the substrate and enhance food transfer to the seafloor. Some polar environments support high abundances of individuals (up to 130,000 macrofauna m^{-2} and 1–3 million meiofauna m^{-2}); high biomass (>1 kg wet weight m^{-2}) may also occur, for example with large densities of epifauna on the valves of the scallop *Adamussium colbecki*. As noted earlier, food supply (quality, quantity, and predictability) fundamentally influence seabed communities, their biomass, and how they function.

Depending on the nature and the characteristics of the habitat, the relative importance of structuring factors varies (Table 10.2). Many studies demonstrate the importance of environmental heterogeneity in controlling and generating biodiversity in terrestrial ecosystems, with greater biodiversity associated with higher environmental heterogeneity. This same relationship appears true for marine benthos. Furthermore, the fractal dimension (a measure of habitat complexity) adds an element of complexity that can affect abundance and benthic species richness more than other indices of habitat complexity. Fractal geometry can also account for differences in area when comparing estimates of species richness. One hypothesis to explain differences in diversity postulates that larger areas support more species than smaller areas, so that including additional habitats in a sampled area increases diversity. But expanding the spatial scale and heterogeneity of a habitat can also promote or facilitate speciation, and distinguishing the two effects is difficult. A special case occurs when complexity generates topographic diversity of habitats. Often, species richness associated with topographic variation links to an increased range of ecological niches, as one of the hypotheses of habitat diversity asserts. Recent studies demonstrate a topographic effect on species richness independent of area, suggesting that large-scale gradients in heterogeneity promote biodiversity. In addition, habitat complexity and structural components can separately affect communities. Environmental heterogeneity links to habitat complexity and the fractal dimension of spatial heterogeneity potentially connects habitat and species diversity through **facilitation**, where one species modifies a habitat in such a way that it facilitates establishment of subsequent species, supporting an even greater number of species.

Table 10.2 Main environmental factors (right) that influence benthic communities in relation to habitat type (left).

Nature of Substrate and Habitat	Greater Influence of Environmental Factors
Primary hard bottom (rocks)	Space, slope, roughness, mineralogy
Secondary hard bottom (biogenic structures, other organisms)	Morphology, biology, host dynamic
Unconsolidated or soft bottom (sands, muds)	Sedimentary regime, grain size
Mixed bottom	System characteristics (e.g., density of tubes or eelgrass blades)

The presence of a secondary substrate (a substrate created by other organisms) can greatly enhance environmental and habitat heterogeneity. Large sessile or epibenthic organisms can increase biodiversity by increasing available space, complexity, and fractal dimension by producing metabolites other organisms could potentially use, by altering chemical and physical conditions in the surrounding space, by facilitating co-evolutionary processes, or by attracting potential predators. Finally, the effect of heterogeneity and spatial scale links to the size of the organisms that colonize the system and to their mobility. The perception of space and its heterogeneity changes from prokaryotes to protozoa to megafauna. Furthermore, size classes that partially overlap, such as meio- and macrofauna, perceive the environment differently, given that the smallest organisms typically move into interstitial spaces between sediment grains, whereas larger individuals occupy a larger space within the sediment.

All of these organisms vary in life history strategies such as asexual reproduction versus sexual reproduction, or presence/absence of **meroplanktonic** larvae, referring to organisms that spend only part of their life cycle in the plankton. Differences in food resources in space and time can also contribute to spatial and temporal heterogeneity, creating a mosaic of conditions. Spatial and temporal heterogeneity in physical conditions help competing species coexist.

Sedimentation also strongly influences the establishment and growth of benthic organisms. For example, corals produce mucus to protect themselves from "rains" of sediment and particles, whereas sponges (particularly *Chondrosia reniformis*) incorporate the sediments, capturing suspended siliceous components using specialized cells called **exopinacocytes**. One key factor, the percentage of occupied substrate, reflects organism morphology, growth mode, organism needs with respect to fluctuations in different physical and chemical parameters, and protection from potential predators. These parameters strongly influence stratification of organisms, which influences exchange of matter or energy and protection. Even the influence of hydrodynamics on distributions of species depends on their morphology and consistency: for example, the most exposed hard substrates, which experience strong wave action, only support sessile and sedentary organisms that can firmly attach to the surface, in contrast to the more branched or delicate forms that dominate quiescent habitats. Living organisms can also change abiotic conditions, particularly in the case of habitat-forming species, which can change the nature of the substrate (e.g., calcareous algae that "pave" the seafloor, or lithophagous organisms that dissolve rock), or modify the composition, timing, or rate of larval settlement of both mobile and sessile species, potentially changing the predator-prey balance.

2) **Biotic factors** relate to different activities of living organisms that influence communities: (a) **competition**, defined as the simultaneous need of two or more organisms (or species) for an essential resource in limited supply (**exploitative competition**), or as the negative interaction of organisms or species as they seek an essential but non-limiting resource (**interference competition**). A **resource** refers to any aspect of the environment used to satisfy a need; b) **predation**, the consumption of an organism (the prey) by another (the predator) can drive community dynamics. Once benthic organisms, and particularly sessile forms, arrive and become established in a given habitat, they may face predation risk and competition for space and resources; (c) **symbiosis**, defined in several different ways including: (1) a close and sustained relationship between two organisms living together, whether both benefit (**mutualistic**), one benefits with no negative effect on the other (**commensalism**), or one benefits and the other suffers (**parasitic**); (2) the relationship between two interacting populations of a given species; (3) the relationship between members of two different species.

Among abiotic factors, the availability of space often limits sessile benthic organisms living on hard substrate. Organisms must therefore choose and then exploit suitable substrate; larvae respond to stimuli (chemical and physical) and inhibitory or attractive signals from the substrate or from previously settled individuals or conspecifics (**gregarious settlement**) or other species.

10.6 Organization of Benthic Assemblages

The term **biocoenosis** refers to a group of interdependent organisms of relatively consistent composition, number of species, and abundance that occurs in specific environmental conditions, and persists in a geographic area. For example, a seabed covered by the seagrass *Posidonia oceanica*, or by an algal forest, or a muddy deposit, or relict sand; each of these examples illustrate a habitat with specific characteristics and inhabited by a specific fauna. **Biotope** refers to the geographic area occupied by a biocenosis. Biocenosis reflects a relationship, typically trophic, whereas a "community" indicates a relationship between groups of organisms with interlinked life histories, not necessarily related to trophic type. **Population** refers to a group of individuals belonging to a given species that live in an environment and interact with one another more than with individuals from other population. A biocoenosis can include **characteristic species** that live only in that kind of biocoenosis, **preferential species** that live almost exclusively or preferentially in that kind of biocoenosis, but sometimes occur in similar biocoenosis, **accompanying species** that typically dominate numerically, and

accidental species that only occur in that habitat by accident. For example, the heart urchin, *Echinocardium cordatum* characterizes fine sands. Accompanying species dominate numerically and can: (a) characterize a group that occurs widely, for example, everywhere in the Mediterranean; or (b) indicate a specific trophic condition. For example, suspension feeding species live only in habitats with abundant suspended organic matter.

10.7 Zonation of Benthic Organisms

Zonation refers to bands of fauna and flora in an area characterized by specific communities. Zonation can occur on a large (biogeographic) or smaller (ecological) scale. Spatial zonation can form the basis of habitat classification. Well-defined zonation occurs in some types of environments such as the rocky intertidal, with its clearly distinguishable upper, middle, and lower intertidal zones. The lower intertidal rarely experiences exposure to air, the intermediate intertidal experiences frequent, periodic exposure, and near-continuous exposure characterizes the upper intertidal. Exposure to air strongly affects species distribution. Algae, gastropods, limpets, chitons, and barnacles characterize upper intertidal habitat, which is submerged only during high tide. In the intermediate zone, populations consist of mussels, anemones, sea stars, and, to a lesser extent, macroalgae. Hydroids, macroalgae, sea urchins, octopus, crustaceans, and gastropods occupy the lower intertidal. Communities can change over time as a result of different stages of **succession**, whereby different species colonize a given environment over time following disturbance. This "zonation in time" helps in understanding dynamics of populations that can be sequential in the case of a succession or following colonization of bare substrate or in relation to seasonal cycles.

Different models describe zonation, though they do not necessarily describe reality and aspects of how ecosystems function. However, these working tools can help to formulate useful hypotheses. Two main schools or lines of thought address zonation:

a) The European School considers a historically complex geographical context, emphasizing the importance of physico-chemical factors in controlling the presence and distribution of benthic species: particle size, currents, wave exposure, temperature, and salinity create conditions that favor a suite of species adapted to living in different zones;

b) The American School emphasizes the key role that biotic factors such as predation and competition play in establishing zonation. Highlighting the key role of biotic factors requires models based on experimental evidence.

The Pérès and Picard Model – The European School, promoted by Pérès and Picard, dominated marine biology from the sixties until the late eighties, building from Danish fisheries scientist Johannes Petersen's 1930s description of sedimentary communities based on characteristic species. Petersen's approach focused on the most abundant species, including the bivalves *Venus* sp., and *Macoma* sp., and the urchins *Brissopsis* sp., and *Echinocardium* sp.; researchers elsewhere recognized similar communities. This model has limitations, such as the presence of transient (**vicarious**) species or substitution of congeneric species, for example *Macoma balthica* replaces *Macoma americana* in some locations. The Danish ecologist Gunner Thorson advanced the model by adding details. *M. balthica* communities, for example, often include the bivalves *Mya* sp., *Cerastoderma* sp., and the polychaete *Arenicola* sp. In *Tellina* sp. communities, characteristic species include the bivalves *Donax* sp., *Dosinia* sp., and *Astropecten* sp. Characteristic species in *Venus* communities include the mollusks *Spisula subtruncata* and *Natica* sp., and the urchins *Echinocardium* sp. and *Spatangus* sp. Depending on the communities, several key species can strongly influence the composition of the remainder of the community, even though those key species may not dominate numerically. For example, the surf clam *S. subtruncata* plays a key role in transferring particulate matter from the water column to the sediment, thus creating conditions necessary for other species to live there.

This model offers an obvious advantage of clear biological criteria for classification, namely the species and typical environmental parameters associated with the biocenosis, irrespective of specific local differences such as precise bathymetry. It also offers utility across a large geographic scale, clarity, and simplicity of terminology. These models are limited in that their generality constrains formulating clear hypotheses about the ecosystems. Their practicality and ease of use (convenient classification), though advantageous, often lack accurate correspondence in reality. The Pérès and Picard model assumes primarily abiotic influences on benthic zonation, and distinction between the phytal and littoral system that extends to the edge of the continental shelf and supports photosynthetic organisms, and the aphytal system that extends from the edge of the shelf to the bottom of the ocean trenches. The specific ecological conditions vary depending on sea level and topography, thus precluding any sharp boundary edges of these zones. A margin or transition zone has mixed characteristics of the adjacent zones, thus forming an **ecotone**. Gradients in environmental factors determine the spatial extent of this transition. Phytal systems include the following zones: (1) **supralittoral,** the zone at the land-sea interface affected by the spray of the waves; (2) **mesolittoral** (or intertidal), which sits between the highest and lowest tidal limits, with frequent, periodic exposure; (3) **infralittoral**, which extends from just below the low tide mark to the limits of light penetration, and varies in water transparency; (4) **circalittoral**, which extends from the lower infralittoral limit (~30–40 m) to the edge of the continental shelf (~200 m). In contrast, aphytal systems include 3 zones: 1) **bathyal** (200–4000 m), 2) **abyssal** (4000–6000 m), and 3) **hadal** (6000–10,000 m) (Figure 10.8).

The Riedl Model – The Riedl model divides benthic system into three zones: a) spray or splash zone; b) intertidal zone; and c) subtidal zone (Figure 10.9).

The **Supralittoral zone** encompasses the transition between a fully terrestrial environment and a fully marine environment (Figure 10.10).

The **Intertidal zone (mediolittoral or mesolittoral):** this coastal zone between the highest and lowest tide levels (Figure 10.11) varies greatly in width, depending on the associated tidal range, which varies from <30 cm (e.g., the Mediterranean) to ~13 m (e.g., Bay of Fundy, Canada). This zone experiences frequent and periodic exposure and immersion, and the width of the zone can also be significant, large enough to be divided into subzones. In the Mediterranean, the action of waves defines these zones more so than tidal action, in that modest waves wet the highest horizon, whereas tides wet organisms in the lower subzone. Platforms, vermetid reefs (*Dendropoma petreum*), and coralline algae beds (e.g., *Spongites notarisii*, *Neogoniolithon notarisii*) also typically occur along high-energy shores. They exist as a result of the close association between vermetid species and coralline algae.

The variable conditions that characterize intertidal habitats include rapid changes in temperature, exposure periods, decreased oxygen concentration, and reduced feeding opportunities. Intertidal organisms have evolved several strategies: (a) reduced size and flattened bodies; for organisms such as limpets, their flattened foot plays a critical role in withstanding wave action; (b) shell color, which often varies in relation to typical substrate color to provide camouflage (e.g., chitons) and the light colors of shells of many tropical species reflect light and thus reduce heat absorption; (c) capacity to vary metabolic rate during low tide; (d) behavioral response to changing conditions. For example, the polychaete, *Arenicola* sp., retreats deep into sand or mud, where there is sufficient moisture to minimize **desiccation**, or drying out. In contrast, the mussel *Mytilus* spp. seals its valves during low tide to retain water.

Sandy and muddy intertidal habitats, with moderate and low energy respectively, as well as low productivity, contrast with high energy and comparatively productive rocky habitats. From a community ecology perspective, we know less about intertidal sedimentary communities than about those in the rocky intertidal. Caging experiments that helped provide major insights into rocky intertidal communities create flow artifacts in sedimentary environments, and the complex nature of organic matter in sediments, especially for deposit-feeders (refractory substances vs. organic matter actually available) complicates understanding. Furthermore, the three-dimensional nature of soft bottoms and the complex nature of sediment biogeochemistry severely limit the utility of manipulative experiments, in that trying to change one variable often changes many others.

Within this zone, epiphytic and endolithic species sometimes play a central role in structuring the seabed, including filling cavities. In the Mediterranean, the coralline red alga *Lithophyllum byssoides* builds cornices, an important biogenic component of the mesolittoral, where waves break. The cornices, with their numerous cracks and crevices, create habitat niches, in which different species can specialize, including sponges, cnidarians, bryozoans, polychaetes, and other species including terrestrial forms such as springtails and spiders that avoid submergence. Calcareous algae often reinforce the basal structure that may become quite impressive, with thicknesses up to 1 m. During calm conditions or during low tide, the upper 20–30 cm of these platforms may protrude above the ocean surface, sustained by wetting from wave motion and the porosity of the structure. Morphology may vary from a simple algal coating to a thick and complex step structure that can grow in thickness 2–3 cm y^{-1}. These structures also occur in some locations with moderate wave energy, although the most extensive structures occur in areas with strong wave exposure.

Three distinct layers occur in well-developed cornices (Figure 10.12), including: (1) an inner pink or beige porous layer several centimeters thick with living **thalli**, referring to a vegetative body in plants and fungi that is not differentiated into roots, stems, or leaves; (2) dead thalli that resemble rock, and may eventually become rock through **diagenesis**, referring to the physical and chemical processes that affect sedimentary materials after deposition and before metamorphism, and collective organic matter transformations; (3) the outer layer, dead, covered by **sciaphilous** ("shade loving") and drilling organisms, such as sponges, bivalves, and polychaetes, which are bioeroders.

Subtidal zone (infralittoral and circalittoral): Classification schemes outside the Mediterranean often combine infralittoral and circalittoral zones in the subtidal zone. Subtidal zones demarcate the boundary between phytal and aphytal systems. Rocky subtidal environments support a broad spectrum of organism morphologies, including many relatively erect forms compared to sediment-covered bottoms and, especially, the presence of **epilithic** organisms,

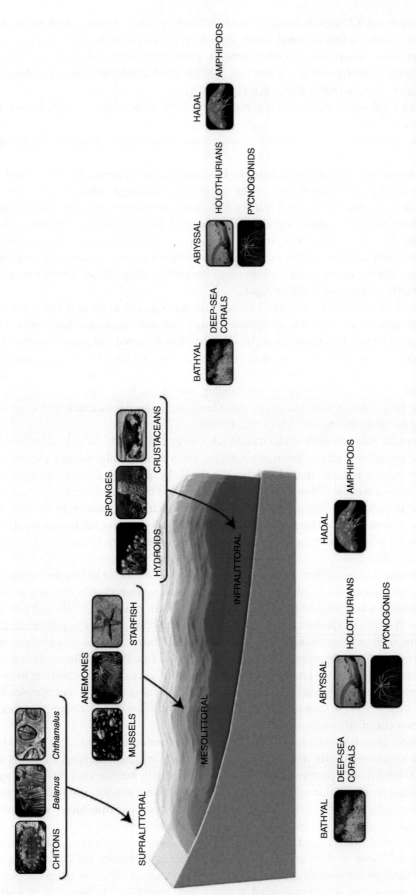

Figure 10.8 Subdivision of marine benthic environment in zones from the supralittoral to sublittoral (left) and from bathyal to oceanic trenches (right). *Source:* Kirt L. Onthank / Wikimedia Commons / CC BY-SA 3.0, Auguste Le Roux / Wikimedia Commons / CC BY-SA 4.0, Mark A. Wilson / Wikimedia Commons / Public Domain, Nhobgood / Wikimedia Commons / CC BY-SA 3.0.

10.7 Zonation of Benthic Organisms

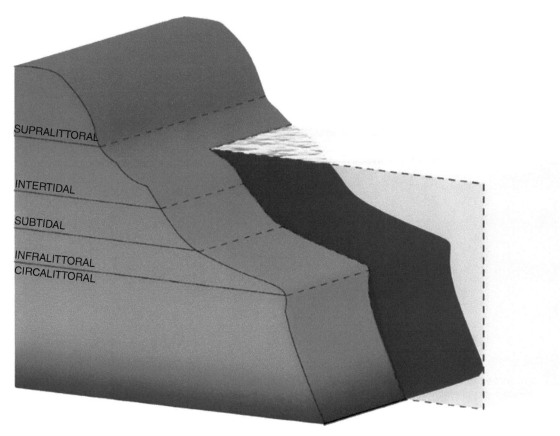

Figure 10.9 Diagram of benthic vertical zonation of hard surfaces based on the Riedl model.

Figure 10.10 Typical organisms that inhabit intertidal depths: (a) chitons; (b) a gastropod; (c) a gastropod mollusk photographed in the Mediterranean intertidal; (d) sea anemone; (e) Grapsid crab photographed along the coast of Chile.

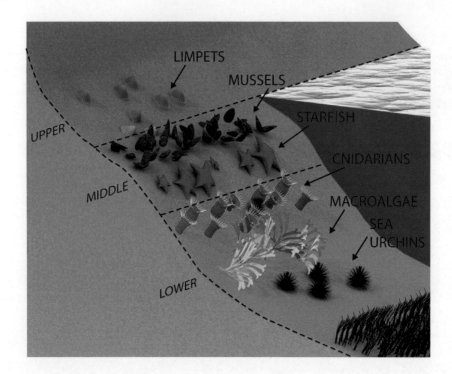

Figure 10.11 The three intertidal zones, characterized by considerable tidal range, spans upper, middle, and lower subzones, which each contain different types of organisms.

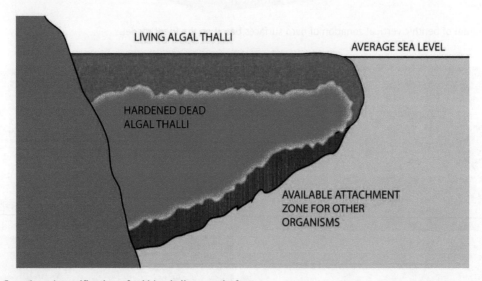

Figure 10.12 Growth and stratification of a *Lithophyllum* sp. platform.

referring to those that grow on rock. In the higher part of the infralittoral, plants structure the landscape, whereas animals dominate the lower part of the environmental architecture. The hard-bottom communities dominated by vegetation include photophilic algae; those dominated by animals mainly consist of coral formations and the most developed forms occur in tropical seas. The circalittoral zone extends from the lower limit of the infralittoral (about 30–40 m deep) out to the edge of the continental shelf. The infralittoral zone extends from a few centimeters below the limit of low tide to a depth that varies with light penetration and thus water transparency (~ 35 m in the Mediterranean but considerably less in more productive temperate environments). Biologically photophilic algae and seagrasses dominate this zone. Platforms and vermetid and coralline algae reefs, which are almost always submerged, occur in wave-battered coastlines.

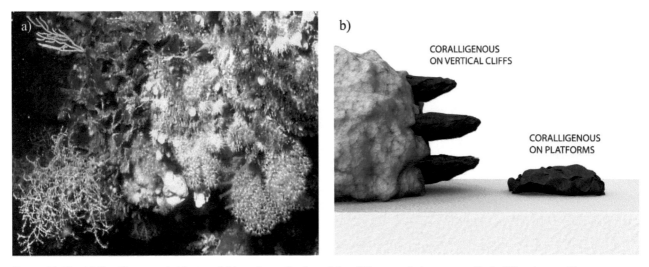

Figure 10.13 (a) Coralligenous habitat, and (b) a schematic view of the differences between a vertical cliff and a coralligenous platform.

Vermetid snails and oysters create submerged structures 0.5–1 m thick, with vermetid skeletons (*Dendropoma* sp.) forming the upper surface below and between which a dense layer of small oysters occurs. The hard-lower part of the cornice consists of coralline algae such as *Corallina elongata* that likely settle following pioneering oysters that pack tightly together, with a non-calcified sandy matrix filling space. After oysters form a rim, the vermetid layer develops, covering all the structure. The coralline algae layer spreads below the oysters forming terraces so they can capture required light.

Sedentary polychaetes of the genus *Sabellaria* live in rigid tubes they build by agglomerating sand and fragments of organic material, eventually forming barriers. Organic material may actually exceed sand within the tubes, likely because waves easily resuspend flat shells, increasing their accessibility to polychaetes. The barriers can form platforms perpendicular to the main current that can help stabilize shorelines and reduce coastal erosion. The high environmental heterogeneity created by these structures supports a high diversity associated biota.

Circalittoral zone: Differences in temperature and salinity conditions, as well as phytoplankton abundance, result in changes in this zone with latitude. Coralline algae often dominate this zone, forming cliffs or platform banks. The term "coralline algae" was coined in the late nineteenth century to indicate calcareous algal formations. Actually, the term "coralline" refers to the secondary hard substrate formed by algal thalli concretion and, to a lesser degree, animal skeletons. The depths of coralline algae banks vary with water clarity and substrate topography, penetrating to depths up to 130 m (Eastern Mediterranean). In the Western Mediterranean, most coralline algae banks occur at 20–60 m depth. This biocenosis requires specific conditions, including dim light, cool temperatures, and relatively constant and modest sedimentation that support typically slow growth rates of <1 mm y^{-1}, and high mortality with excessive sedimentation. The families *Corallinaceae* and *Peysonelliaceae* exemplify coralline algal reefs, whose thalli form structures several meters thick. A wide variety of animals with carbonate skeletons play important roles in structuring seabeds, including foraminiferans, bryozoans, serpulids, scleractinians, sponges, crustaceans, and gorgonians (Figure 10.13). Pre-coralline algal habitat occurs in locations with insufficient light and other environmental factors that influence coralline algae.

Shallow formations that develop on pre-existing rock (cornices of coralline algae, and on the lower horizon of the littoral rock) can form cliffs, whereas platforms (or plateaus) develop at different depths, beginning with the stabilization of soft bottom. Cornices of coralline algae that form parallel planes facing towards the sun support coralline algal growth (particularly *Mesophyllum lichenoides* and *Lithophyllum frondosum*), or towards the shade, on which a rich community of calcium carbonate producers settle (*Corallium rubrum*, *Leptopsammia pruvoti*, *Smittina cervicornis*, *Myriapora truncata*, *Pentapora fascialis*). In addition to bioconstructors, numerous other sessile organisms bind the sediment. Coralline algae reefs create enormous structural and ecological complexity through numerous crevices and cavities of different sizes, light, sedimentation and hydrodynamics that offer myriad habitats that support a diverse cryptic fauna. Boring organisms, such as sponges or mollusks, amplify the complexity and add to the biodiversity by creating cavities and holes. For example, one study showed that 333 cavities created by *Pholas dactylus* supported 32 species of mollusks, crustaceans, and polychaetes. Species characteristic of soft sediments (especially mollusks) live within crevices filled with fine sediment.

Questions

1) What are the differences between plankton and benthos?
2) What are the main differences between benthos on hard surfaces and benthos on soft bottoms?
3) What criteria are used to classify the benthos?
4) What are accompanying species?
5) Describe the zonation model of Pérès and Picard for the Mediterranean.
6) What are the factors that influence benthic zonation?
7) What factors influence meiofaunal distributions?
8) What is bioturbation and what ecological role does it play?
9) What does the bottleneck hypothesis affirm?

Suggested Reading

Bartoli, M., Nizzoli, D., and Viaroli, P. (2003) Microphytobenthos activity and fluxes at the sediment-water interface: Interactions and spatial variability. *Aquatic Ecology*, 37, pp. 341–349.

Bik, H.M., Sung, W., De Ley, P. et al. (2012) Metagenetic community analysis of microbial eukaryotes illuminates biogeographic patterns in deep-sea and shallow water sediments. *Molecular Ecology*, 21, pp. 1048–1059.

Bolam, S., Fernandes, T., and Huxham, M. (2002) Diversity, biomass, and ecosystem processes in the marine benthos. *Ecological Monographs*, 72, pp. 599–615.

Braeckman, U., Provoost, P., Moens, T. et al. (2011) Biological vs. physical mixing effects on benthic food web dynamics. *PLOS One*, 6(3), p. e18078.

Danovaro, R., Gambi, C., Mirto, S. et al. (2004) Mediterranean marine benthos: A manual of methods for its sampling and study, in Gambi, M. and Dappiano, E. (eds.), *Biologia Marina Mediterranea*, Vol. 11, SIBM, Genova, Italy, pp. 55–97.

Emmerson, M.C., Solan, M., Emes, C. et al. (2001) Consistent patterns and the idiosyncratic effects of biodiversity in marine ecosystems. *Nature*, 411, pp. 73–77.

Fabiano, M., Danovaro, R., and Fraschetti, S. (1995) A three-year time series of elemental and biochemical composition of organic matter in subtidal sandy sediments of the Ligurian Sea (Northwestern Mediterranean). *Continental Shelf Research*, 15, pp. 1453–1469.

Fresi, E., Gambi, M.C., Focardi, S. et al. (1983) Benthic community and sediment types: A structural analysis. *Marine Ecology*, 4, pp. 101–121.

Giere, O. (2008) *Meiobenthology: The Microscopic Motile Fauna of Aquatic Sediments*. Springer.

Godson, P.S., Vincent, S.G.T., and Krishnakumar, S. (2022) *Ecology and Biodiversity of Benthos*. Elsevier.

Jensen, P. (1987) Feeding ecology of free-living aquatic nematodes. *Marine Ecology Progress Series*, 35, pp. 187–196.

Mees, J. and Jones, M.B. (1997) The hyperbenthos. *Oceanography and Marine Biology: An Annual Review*, 35, pp. 221–255.

Meysman, F.J.R., Middelburg, J.J., and Heip, C.H.R. (2006) Bioturbation: A fresh look at Darwin's last idea. *Trends in Ecology and Evolution*, 21, pp. 688–695.

Puglisi, M.P., Sneed, J.M., Ritson-Williams, R., and Young, R. (2019) Marine chemical ecology in benthic environments. *Natural Products Reports*, 36, pp. 410–429.

Snelgrove, P.V.R. (1999) Getting to the bottom of marine biodiversity: Sedimentary habitats. *BioScience*, 49, pp. 129–138.

Walag, A.M. (2022) Understanding the world of benthos: An introduction to benthology, in Godson, P.S., Vincent, S.G.T., and Krishnakumar, S. (eds.), *Ecology and Biodiversity of Benthos*. Elsevier, pp. 1–19.

11

Biodiversity of the Plankton

11.1 Introduction to the Plankton

Plankton refers to all organisms, whether microbial, macroalgal, or animal, that live suspended in the water column and cannot make significant headway against currents. "Plankton" derives from the Greek and means "rambling". Victor Hensen, a marine biologist, proposed this term in 1889 to indicate "the complex of organic particles in water", but without specifying their nature. Plankton include a wide diversity of biota from viruses, bacteria, protistans, and microalgae (Figure 11.1), to small metazoans such as crustaceans (Figure 11.2a-c), to large gelatinous animals such as jellyfish that can grow to lengths of 8–20 m (Figure 11.2d). The best-known components, the phytoplankton (autotrophic photosynthetic organisms) and zooplankton, consist of chromistans, protistans, cnidarians, ctenophores, annelids, chaetognaths, tunicates, and some crustaceans. The zooplankton include most marine phyla, noting that almost all benthic macrofaunal and megafaunal species spend a period of their lives (typically egg and/or larval stages) in the plankton. Among the zooplankton, copepods dominate numerically, followed by cladocerans and euphausiids. Plankton abundance decreases exponentially with increasing size of individual organisms. Virioplankton and bacterioplankton concentrations of 10^8–10^{11} viruses or cells l^{-1} contrast abundances between 10^3–10^5 cells l^{-1} for phytoplankton, such as diatoms and dinoflagellates; mesozooplankton rarely exceed abundances of 10^2 cells l^{-1}.

Over large spatial scales of kilometers to thousands of meters, physical variables determine plankton distributions, whereas biological factors dominate at smaller scales. Macroplankton and micronekton undertake the largest daily migrations of animal biomass on Earth, moving to great depths in the ocean to limit energy use or to escape predation, but then moving upward to shallower waters in search of food. Reproductive rates of different planktonic species vary from a few hours for picoplankton, days for phytoplankton, weeks for protozoans, months for mesozooplankton, and up to more than one year for macrozooplankton (e.g., Antarctic euphausiids). Seasonality of phytoplankton in temperate and polar waters plays a dominant role in the temporal dynamics of all planktonic components; for example, the zooplankton that link the base of the food web to higher trophic levels peak in total abundance in late spring and early summer with lowest values in winter. Phytoplankton have developed sophisticated mechanisms to coexist, despite similar basic needs (light, nutrients and vitamins) that should presumably lead to competition among species. By specializing on particular light conditions and nutrient concentrations, these species can coexist.

Most zooplankton feed on micro-phytoplankton and on nano- and picoplankton, generally by filtering large volumes of water; a copepod only a few millimeters in size filters a few hundred milliliters each day. Size primarily determines which organisms zooplankton graze (phytoplankton) or prey (animals) upon. Optimal foraging theory predicts that predators should favor larger and more abundant prey that provide more calories. Some phytoplankton reduce risk to grazers using bioluminescence, protective structures, or chemical substances that inhibit zooplankton reproduction. As noted earlier, grazers rarely consume all of the phytoplankton they graze on, typically releasing substantial particulate organic material in the water through a process known as sloppy feeding that leaves behind an estimated 20 – 40% of grazed biomass. This released material, along with production of feces, has important consequences for the functioning of the planktonic domain and for the export of organic matter to the benthos.

Marine Biology: Comparative Ecology of Planet Ocean, First Edition. Roberto Danovaro and Paul Snelgrove.
© 2024 John Wiley & Sons Ltd. Published 2024 by John Wiley & Sons Ltd.
Companion Website: www.wiley.com/go/danovaro/marinebiology

Figure 11.1 Examples of planktonic organisms: Viruses, bacteria, and diatoms in a sample treated with fluorescent dyes.

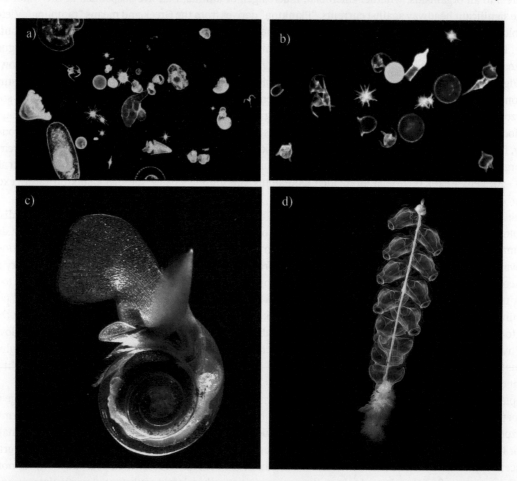

Figure 11.2 Examples of planktonic organisms: (a) a Mixture of crustacean larvae, small jellies, protists, dinoflagellates, and diatoms; (b) single-celled organisms (diatoms and dinoflagellates); (c) a sea butterfly with small copepods and an ostracod; (d) large gelatinous plankton. *Sources:* Kevin Raskoff / Wikimedia Commons / Public Domain, NOAA's National Ocean Service / Wikimedia Commons / CC BY 2.0.

11.2 Planktonic Organisms

All planktonic organisms share a common inability to make significant headway against currents, generally because of their small size and lack of efficient locomotory systems. Nonetheless, many plankton are capable of moving small distances and undertaking vertical migrations. Because many such organisms tend to sink towards the bottom, movement largely helps them maintain or adjust their position vertically within the water column. The plankton includes an extraordinary diversity of organisms, such as viruses, bacteria (Figure 11.1), protozoans, microalgae, and a wide range of zooplankton (Figure 11.2).

Researchers classify plankton according to their functional characteristics, size, distribution, and life cycle. Functional groups include:

1) Biological particles such as viruses (virioplankton); 2) prokaryotes (bacterioplankton and archaeo-plankton), including cyanobacteria and autotrophic prochlorophytes; 3) unicellular eukaryotes (nanoplankton), autotrophic diatoms (phytoplankton), **mixotrophic** species capable of photosynthesis and heterotrophic feeding, such as dinoflagellates, and some protists; 4) metazoans (zooplankton), a diverse mix of heterotrophic organisms, such as copepods, amphipods, and mysids.

Prior to scientific recognition of the importance of prokaryotes, the field of **planktology** (or planktonology) referring to the study of plankton, focused primarily on phytoplankton and zooplankton. Phytoplankton consists of photosynthesizing, autotrophic organisms, able to transform inorganic nutrients into organic compounds. Because phytoplankton form the basis of the food web, zooplankton, nekton, and benthos all depend on them, whether directly through grazing or indirectly through the food web. Highest phytoplankton abundances typically occur along coastlines, and especially in enclosed seas and gulfs with a steady supply of nutrients from rivers and terrestrial runoff, or in regions where circulation carries cold, nutrient-rich water to the surface through upwelling.

11.3 Planktonic Classification Based on Water Column Distributions

Organism distributions within the water column reflect biological adaptations and their position within trophic pyramids. Indeed, we can distinguish different categories of planktonic organisms according to the depth at which they occur; this vertical distribution largely depends on the movements of water masses. **Pleuston** live at the atmosphere-ocean interface in the upper 10 cm and include floating organisms with parts of their bodies emerging from the water, moving passively with drifting currents and wind, such as the cnidarians *Physalia physalis*, *Velella velella*, and *Porpita porpita*, and the gastropod *Janthina janthina* (organisms discussed in Chapter 5 on adaptations).

Neuston refers to organisms adapted to living in contact with the sea surface film (produced by surface water tension). This group, which includes micro- and macro-organisms, can be subdivided into organisms that live immediately above (**epineuston**) or below (**iponeuston**) the sea surface film. Typical epineuston includes insects from the family Gerridae and the genus *Halobates* (Figure 11.3a), a group of >40 species. The iponeuston includes hydroids, mollusks, copepods, isopods, decapod crustaceans, fishes, and algae (genus *Sargassum* spp.; Figure 11.3b).

The distribution of plankton through the water column below the surface includes several relatively distinct layers (Figure 11.4).

- The **epiplankton** includes organisms that live in the epipelagic zone between the ocean surface to 100–200 m depth; this zone includes the most abundant and most diverse phytoplankton and zooplankton, dominated by copepods. Some species only live in the epiplankton for part of the day, in that vertical migrating species may move thousands of meters over a 24-h cycle.
- The **mesoplankton** includes organisms that live in the comparatively **oligophotic** (food poor) mesopelagic zone (200–1,000 m) where sunlight penetrates only weakly or not at all; phytoplankton living here will survive only if upward

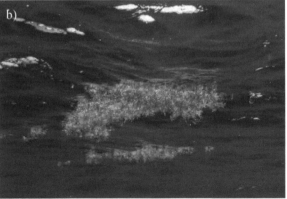

Figure 11.3 (a) Marine insect of the family Gerridae, genus *Halobates* sp.; (b) brown alga *Sargassum* sp. *Source:* Cory Campora / Wikimedia Commons / CC BY-SA 2.0.

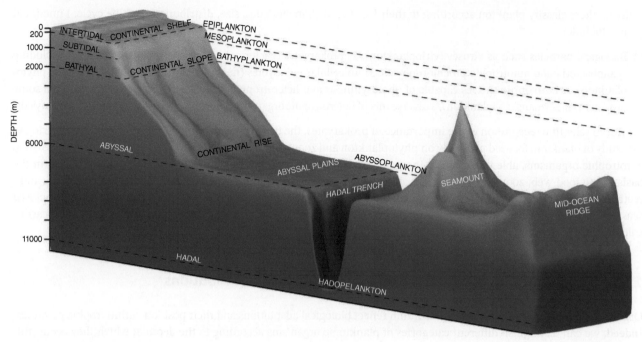

Figure 11.4 Ocean profile showing plankton categories divided by depth.

movements bring them quickly in the upper photic layer. In contrast, the zooplankton in this layer closely resemble those in the epipelagic zone. **Infraplankton** refers to zooplankton that live permanently at depths of 500–600 m.
- The **bathyplankton** includes organisms that live in the truly deep bathypelagic layer, totally devoid of light (1,000–3,000/4,000 m). Copepods dominate the bathyplankton, both in number of species and in biomass, however, species composition differs greatly from that in surface layers. Other groups typical of deep plankton, including species of jellies and decapods, also occur in this layer;
- The **abyssoplankton** consists of plankton living at depths of 4,000–6,000 m, dominated by chaetognaths, amphipods, and mysids.
- The **adoplankton** or **hadoplankton** consists of the plankton in the great ocean trenches from 6,000 to 11,000 m, consisting primarily of amphipods, copepods, and ostracods.

11.4 Plankton Classification Based on Life Cycles

Depending on their **life cycle**, referring to the developmental stages of organisms, plankton can be divided into holoplankton and meroplankton. **Holoplankton** consists of all microorganisms (Figure 11.5a), metazoans (Figure 11.5b), and photosynthetic organisms/plants that live their entire lives in suspension in the water column (Table 11.1). Other species, known as **meroplankton** (or temporary plankton) spend only part of their life cycle in the plankton. These taxa include eggs and larval stages of many animals that, once they develop sufficiently, abandon their planktonic habitat and metamorphose and settle onto the seafloor to become benthic; this group also includes spores and gametes released by macroalgae and plants. In fact, most macro- and megabenthos that spend much of their life associated with the seafloor, including mobile forms, move comparatively little once settled on the seafloor, so much of their dispersal occurs during the planktonic egg and larval stages that can transport these stages over large distances. Currents vary from average speeds of just a few cm s^{-1} (5–10 cm s^{-1}) to >100 cm s^{-1} in extreme cases of jet currents. Consider a medium current speed of 10 cm s^{-1}; a larva can travel an impressive distance (10 cm × 60 s min^{-1} × 60 min h^{-1} × 24 h d^{-1} × 7 d) of 60 km in about one week, an unimaginable distance for most benthic organisms during their lifespan.

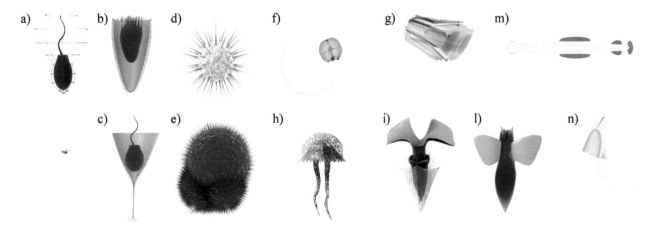

Figure 11.5 Representative holoplankton. Chromistans, and protozoans: (a) choanozoan, *Stephanoeca* sp.; (b) ciliophoran, *Parafavella* sp.; (c) chromistans, *Bicoeca* sp.; metazoans: cnidarians; (d) radiozoan, *Acanthometron pellucida*; (e) foraminiferan, *Globigerina*; (f) the ctenophore, *Pleurobrachia* sp.; (g) the ctenophore, *Beroe* sp.; (h) the scypohozoan, *Pelagia* sp.; (i) the gastropod, *Euclio pyramidata*; (l) the gastropod, *Clione limacina*; (m) the chaetognath, *Sagitta setosa*; (n) the siphonophore *Muggiaea* sp. *Source:* Richard Lampitt / Wikimedia Commons / CC BY 2.5.

Table 11.1 Main taxonomic groups of holoplanktonic zooplankton.

Phylum	Subphylum	Common genera
Protozoa	Dinoflagellata	*Noctiluca*
	Zooflagellata	*Bodo*
	Foraminifera	*Globigerina*
	Radiozoans	*Aulacantha*
	Ciliophora	*Strombidium; Favella*
Cnidaria	Medusae	*Aglantha; Cyanea*
	Siphonophora	*Physalia; Nanomia*
Ctenophora	Tentaculata	*Pleurobrachia*
	Nuda	*Boroe*
Chaetognatha		*Sagitta*
Annelida	Polychaeta	*Tomopteris*
Mollusca	Heteropoda	*Atlanta*
	Thecosomata	*Limacina; Clio*
	Gymnosomata	*Clione*
Arthropoda		
	Cladocera	*Evadne; Podon*
	Ostracoda	*Conchoecia*
	Copepoda	*Calanus; Oithona*
	Mysidacea	*Neomysis*
	Amphipoda	*Parathemisto*
	Euphausiacea	*Euphausia*
	Decapoda	*Sergestes; Lucifer*
Urochordata	Appendicularia	*Oikopleura*
	Thaliacea	*Salpa; Pyrosoma*

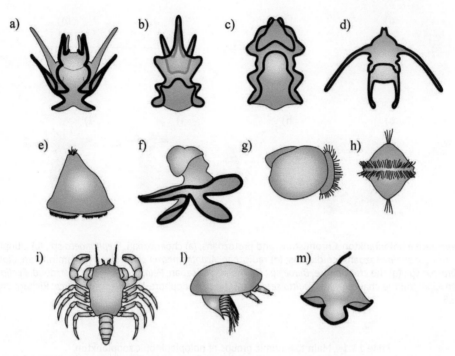

Figure 11.6 The main larval forms that comprise the meroplankton and their corresponding taxa include: (a) echinopluteus (echinoids); (b) bipinnaria (asteroidean echinoderms); (c) auricolaria (holothuroideans); (d) actinotroch (brittle stars); (e) ciphonauta (bryozoans); (f) pediveliger (mollusks); (g) veliger (mollusks); (h) trochophore (polychaetes); (i) megalopa-zoea (decapod crustaceans); (j) naupliocypris or cyprid (decapod crustaceans); (k) pilidium (nemertines).

Considering that larval durations can range from a few days up to 2 y, depending on the species, dispersal can clearly influence colonization of new habitats. Given the limited dispersal ability of most benthic organisms, the evolutionary "choice" to produce larvae that develop in the water column and disperse in the environment allows species to colonize distant habitats. A life cycle with planktonic larval development also adds serious risks; larvae, because of their small size and lack of defenses, experience high mortality rates.

Researchers name larvae differently depending on their morphology and their specific Phyla. Some higher-level taxa include markedly different larval forms (Figure 11.6): within echinoderms, for example, **echinopluteus** larvae in echinoids (sea urchins), differ from **bipinnaria** and **brachiolaria** in asteroids (seastar larvae), and the **auricolaria** in holothurians. In contrast, **cyphonaute** larvae occur in bryozoans and **actinotrophs** occur in brittle stars. Mollusks produce **veliger** larvae, which become **pediveligers** once they develop and encounter a suitable substrate. **Trochophore** larvae in polychaetes and **megalopa-zoea** in decapod crustaceans add to the larval mix, along with **pilidium** larva in nemertines (Figure 11.6). Although specific larval types characterize many phyla, some polychaetes may have a veliger larval stage that resembles veliger larvae in mollusks.

The crown of cilia that characterizes many larvae not only provides them with some minimum capacity for spatial orientation, but also to perceive water movement around them, influencing their capacity to intercept food particles. Despite the limited larval ability to move, most species undertake vertical migration and move up and down within the water column and carry out small-scale excursions above the bottom. For example, the trochophore larva of polychaetes, characterized by a "crown" of cilia (Figure 11.7) moves in response to its physico-chemical needs: salinity, depth, light, pressure, and the presence of a suitable substrate on which to settle, metamorphose, and grow into an adult form. As noted earlier, **planktotrophic larvae**, referring to larvae that feed on plankton, contrast **lecithotrophic larvae**, referring to larvae whose yolk sac contains sufficient energy reserves to support them for several days. In addition to the risk of currents transporting them into unfavorable habitats, these larvae risk being preyed upon by other organisms. (Figure 11.7).

11.5 Plankton Size Classes

Within the last 10–15 years the development of new molecular tools has greatly expanded the study of smaller marine organisms. With increasing depth in the ocean, the relative importance of microbial components increases relative to that of larger organisms: the more oligotrophic an environment and the more physically removed from primary production, the greater the relative importance of microbial processes (Table 11.2).

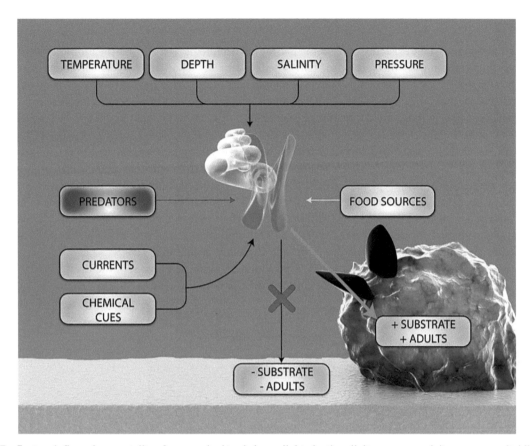

Figure 11.7 Factors influencing mortality of a meroplanktonic larva: light, depth, salinity, pressure, adult presence, suitability of available substrate (such as soft bottoms vs hard surfaces), surface and bottom currents (direction, strength), chemical stimuli, food availability (e.g. phytoplankton), predator abundance and, finally, microbial biofilm on the substratum, which may provide a larval settlement cue.

Table 11.2 Size groupings and functions of autotrophic and heterotrophic planktonic organisms.

Size Grouping	Size Range (μm)	Component/function Hetero- / autotrophic	Types of Organisms
Femtoplankton	0.02–0.2	Virioplankton	Viruses in the water column
Picoplankton	0.2–2.0	Autotrophic picoplankton	Planktonic prokaryotes (Bacteria, Archaea) and small eukaryotes
		Heterotrophic picoplankton	Cyanobacteria, Archaea, autotrophic picoeukaryotes
Nanoplankton	2.0–20	Autotrophic nanoplankton	Unicellular autotrophic eukaryotes, benthic diatoms of small dimensions
		Heterotrophic nanoplankton	Protozoans and Chromista: unicellular eukaryotes, including heterotrophic nanoflagellates and other small heterotrophs
Microplankton	20–200	Microphytoplankton	Planktonic microalgae
		Microzooplankton planktonic Protozoans and Chromista (Protozooplankton)	Small protozoans and Chromista such as amoebae, tintinnids, foraminiferans and ciliates, and micrometazoans (rotifers)
Mesoplankton	0.2–2.0 cm	Mesozooplankton	Multicellular organisms (crustaceans dominated by copepods)
Macroplankton	2.0–20 cm	Macrozooplankton	Polychaetes, amphipods, and organisms visible with the naked eye
Megaplankton	>20 cm (to 200 cm)	Megazooplankton	Large jellies, siphonophores, and ctenophores

11.5.1 Virioplankton

Although scientists recognized the existence of viruses in marine environments by the 1990s, their importance as functional members of microbial communities was recognized only when researchers reported viral concentrations of 10^4–10^8 virus ml^{-1} in seawater samples. Studies on the impact of infection and viral lysis on aquatic microorganisms quickly followed, placing importance on the previously underestimated planktonic component known as virioplankton (Box 11.1). Subsequent studies showed 10 times more numerous virioplankton than bacterioplankton, the group previously considered the most abundant (Figure 11.8).

Bacteriophages, which attack bacteria, dominate marine viruses but planktonic viruses can also infect many other organisms, from algae to marine mammals. Marine viruses can infect eukaryotic phytoplankton cells such as diatoms, Chrysophyceae, Primnesiophyiceae, Haptophyiceae, Raphidophyceae, and Cryptophyceae, and almost certainly all phytoplankton species (even those for which pathogenic viruses have not yet been identified). The dimensions of most aquatic viruses span 30–60 nm (Figure 11.9), but "giant" viruses in some surface waters may be 200–400 nm. Viral infections may have important implications for nutrient cycling, evolution of infected species, and potentially effects on climate change and biodiversity through mortality effects on hosts or through virus-host gene transfer.

Although viruses may comprise 94% of biological particles in the plankton and represent the most abundant biological entities in the ocean, they represent only 5% of the total biomass because of their small size (Figure 11.10). Although

Figure 11.8 Bacterioplankton observed under a scanning electron microscope.

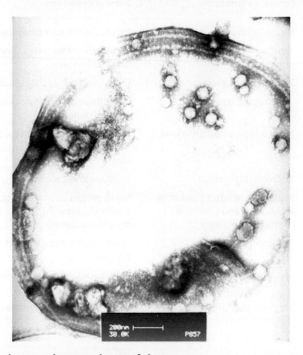

Figure 11.9 Transmission electron microscope image of viruses.

prokaryotes reach only 10% of viral concentrations, they represent >90% of microbial biomass. Chromista and Protozoa, the Kingdoms that include single-celled organisms such as unicellular algae, can represent more than a half of the biomass in ocean surface waters, but at meso- and bathypelagic depths they may contribute only for a small percentage to total biomass. Consequently, as an overall percentage, they represent an even lower percentage than viruses.

Virioplankton, as obligate parasites, must infect other marine plankton or other species for their development. Viruses can control bacterial populations, limiting their concentrations and causing bacterial mortality rates of 10–50% in surface waters over short periods (days–weeks) of time, and 50–100% in environments hostile for bacterial life.

Viruses can also exert direct control on phytoplankton populations. Microcosm experiments showed decreases of up to 50% in phytoplankton biomass and primary production in virus-enriched seawater samples. Viral infection of bacterioplankton and phytoplankton can also play a major role in controlling bacterial secondary production (potentially reducing it by >70%), contributing to the viral shunt, described earlier (Figure 11.11). Viral infection may therefore significantly alter marine food webs, affecting the lowest trophic levels with potential cascading effects on upper trophic levels.

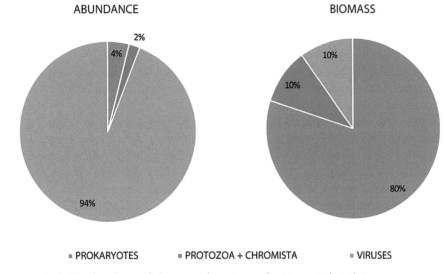

Figure 11.10 Biomass and relative abundance of viruses, prokaryotes, and protozoa + chromista.

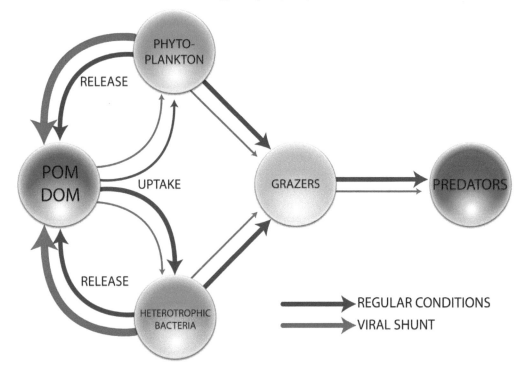

Figure 11.11 Example of the viral shunt, a pathway of the microbial food, which plays a key role in cycling marine organic matter. In the viral shunt, viruses infect phytoplankton and bacteria, thus diverting carbon and nutrients from the main food web and limiting the transfer of energy and material to higher trophic levels. In this way, viruses increase the conversion of living particulate organic matter (POM) into non-living dissolved organic matter (DOM).

Among the most susceptible phytoplankton to viral control, the alga, *Aureococcus anophagefferens*, causes devastating brown tides on the northeast coast of the United States. *Heterosigma akashiwo* causes fish kills, and the ubiquitous coccolithophore, *Emiliana huxleyi*, causes major blooms at mid-latitudes. Viruses can infect over 50% of the cells of *E. huxleyi*, during the middle phase of blooms, and up to 100% during the final phase, thus directly controlling phytoplankton blooms. Bacterial and viral abundances strongly correlate; the higher the number of host cells present in an environment the greater the abundance of viruses. Viral abundances often vary with changes in water masses, such as fronts or pycnoclines. These discontinuities represent ecotones, characterized by elevated abundances and increased organism diversity relative to adjacent systems.

Numerous studies document variation in total viral abundance on monthly, weekly, daily and hourly temporal scales. In the Northern Adriatic Sea, for example, viral abundance increases by an order of magnitude between winter ($<10^6$ ml^{-1}) and late summer (10^7 ml^{-1}). Bacterial abundance varies seasonally by only a factor of 5.

Changes in chemical and physical properties with depth strongly influence planktonic microorganisms. Within the euphotic zone, the intensity and wavelength of light, along with nutrient concentrations, collectively determine distribution patterns of photoautotrophic organisms. In open ocean waters, virioplankton abundance rapidly decreases below the euphotic zone to relatively constant, low values (10^6 ml^{-1}). Cells that experience nutrient limitations or become mixed below depths with sufficient light intensity may be more susceptible to viral infection.

The place where a population lives does not define its **ecological niche**, which instead refers to the multidimensional space (hypervolume) an organism occupies, in which each dimension represents an environmental factor (ecological status or resource) that determines population survival. Predation and competition may restrict this "**fundamental niche.**" The niche an organism could potentially occupy until it becomes a "**realized niche,**" the habitat it actually occupies. For example, a given virus may tolerate a wide temperature range (a dimension of the fundamental niche), but it may live only in a restricted habitat characterized by minor changes in temperature (a dimension of the realized niche).

Only fragmentary information exists regarding virus ecological niches, but these variables presumably include water temperature (marine viruses are more sensitive to heat than other viruses), hydrostatic pressure, light radiation, ion concentrations, oxygen, pH, organic matter and "availability" (abundance) of potential hosts. Solar radiation, in particular UV-B (ultraviolet radiation) rays, represents one factor primarily responsible for decreases in viral abundance. Viral sensitivity to UV-B rays varies greatly, with decay (viral "death" and decay) rates of 5–80% h^{-1}. The importance of this variable decreases with depth as a result of attenuation of light radiation as it passes through the water. In the presence of light, bacterial host cells can often repair any slightly damaged portion of viral DNA, restoring its infectious power. Thus, viruses in the photic zone can be infectious.

Studies on viral decay show that viral turnover time in water varies from ~1 h to a few days. Viral turnover depends strictly on the trophic levels of a given system and, in fact, faster turnover occurs in nutrient-rich coastal waters than in oligotrophic waters (e.g. offshore). In addition to turnover, even viral abundance in the ocean correlates with productivity and, indeed greatest viral abundances occurs with high concentrations of bacteria and chlorophyll, changing with seasonal cycles of primary and secondary production. Viral abundances may increase by as much as an order of magnitude going from oligotrophic (nutrient-poor) to eutrophic (nutrient-rich) systems. Lower abundances occur in deep environments (10^4–10^5 indiv. ml^{-1}), intermediate abundances in open sea surface waters (10^5–10^6 because of their small size and lack of defenses), and highest concentrations in coastal habitats (10^6–10^7 indiv. ml^{-1}). Physical forcing, such as wind, waves, and currents, likely influences viral distributions within the water column, for example by transferring them from water to air. The small size (and density) of viruses means they do not sink in the water column and are thus unaffected directly by turbulence. Pollutants can also influence viruses and may act synergistically with eutrophication to create conditions that favor viral development.

The ocean hosts a huge diversity of prokaryotic communities, although their ecological role remains largely unknown. Some researchers argue that the specificity of the host involved in a viral infection means that viruses could control prokaryote community composition. The "**killing the winner" model** postulates that viral infection controls microbial community diversity whereas non-specific grazing controls microbial abundance; viral infections focus on the dominant species, infecting abundant species more so than less abundant species. Lysis of a cell transforms it into cellular debris composed of dissolved molecules (monomers, oligomers, polymers), colloidal substances, and cell fragments.

Research in the 1990s showed that bacteria use almost the totality of the dissolved organic matter derived from viral lysis. If other bacteria use the lysate from lysed bacteria, this transfer results in a kind of trophic loop in which bacteria degrade bacterial cellular content. The release of dissolved organic matter by phytoplankton and grazers further supplements this loop, with the regeneration of inorganic nutrients as the end result. On the one hand, viral infections decrease bacterial abundance and increase availability of dissolved organic matter that the same bacteria use to increase their production. On the other hand, this loop decreases grazer production. In the final analysis, viral lysis serves the important function of converting biomass into dissolved and particulate forms. Similarly, lysis of phytoplankton results in decreased grazing by heterotrophic organisms by transforming the phytoplankton into smaller forms of biomass (Figure 11.12).

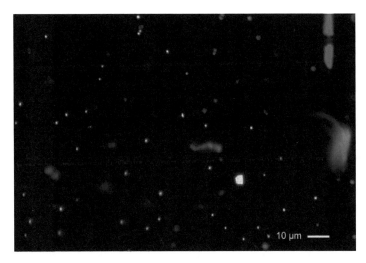

Figure 11.12 Picophytoplankton observed with an epifluorescence microscope; orange dots correspond to the cyanobacterium, *Synechococcus* sp., and the red dots to picoeukaryotes. Largest cells in the upper right are diatoms. *Source:* Daniel Vaulot / Wikimedia Commons / CC BY-SA 2.5.

Sequestration of organic substances in a system with abundant viruses into viruses, bacteria, and dissolved material retains nutrients in the euphotic zone for a longer time because such small forms do not tend to sink. This retention is important, especially for nutrients such as nitrogen, phosphorus, and iron that potentially limit primary production and occur in relatively high concentrations within bacterial cells. Reduced viral activity leads to sinking of organic and inorganic nutrients with phytodetritus and living cells out of the photic zone and into deeper waters.

Box 11.1 Methods for studying virioplankton
Currently, epifluorescence microscopy and flow cytometry represent the most common methodologies for determining viral abundances in the water column and in marine sediments. Sample collection typically uses instruments such as Niskin bottles/rosette samplers for water or box corers and multi-corers for sediments, depending on the sampling depth. Analyses should begin immediately to minimize viral decay, although samples can be stored at −80 °C for later analysis. Researchers use similar procedures for determining abundances of sedimentary and water column viruses. Both analyses use specific fluorochrome viral strains, with *SYBR Green I-II* and *SYBR Gold* as the most common choices currently. However, fluorochrome *SYBR Gold* produces greater DNA fluorescence (both double and single-stranded) and RNA viral particles. Because these fluorochromes bind nucleic acids, fragments of extracellular DNA could be counted as viruses, thus overestimating viral abundance (in particular within sediments). Addition of low concentrations of DNase to the sample can eliminate this problem.

11.5.2 Picoplankton

The history of microbial evolution in the oceans corresponds to the history of life itself. In contrast to terrestrial ecosystems, microorganisms dominate marine biomass, representing one of the largest living populations on the planet. In addition to cyanobacteria (mainly prochlorophytes, the dominant types of bacterioplankton), picoplankton include some eukaryotic organisms <2.0 μm in diameter (Figure 11.12). Bacterioplankton occupy a wide range of ecological niches in marine ecosystems. The total mass of marine prokaryotes exceeds the combined mass of zooplankton and fish, representing the largest biomass on Earth.

Bacterioplankton play an important role in nitrogen fixation, nitrification, denitrification, remineralization, and methanogenesis. Many microbes are saprophytic, obtaining energy by consuming dead organic matter. In some cases, this material occurs in a dissolved form that microbes absorb directly; often bacteria live and grow in association with particulate forms of organic matter, leading to the formation of "marine snow". Many other species of bacterioplankton are autotrophic, drawing energy from photosynthetic and chemosynthetic processes. Researchers typically include the former with picophytoplankton, along with cyanobacterial groups such as *Prochlorococcus* spp. and *Synechococcus* spp.

The taxonomic clade known as SAR11 (*Pelagibacter ubique*) belongs to the α-proteobacteria and represents one of the most common planktonic prokaryotic organisms; recent estimates suggest that these bacteria represent >50% of the total bacteria within the surface waters of the Sargasso Sea.

Other common components of α-proteobacteria resemble *Roseobacter* spp. and typify coastal and open ocean communities. Seasonally they represent between 10 and 25% of the picoplankton.

Proteobacteria likely represent the single most important phylum of bacteria. This large and diverse group of **gram-negative bacteria**, referring to bacteria that do not retain the crystal violet stain used in the Gram staining method of differentiating bacteria, includes many pathogens, with highly plastic metabolisms that include nitrogen-fixers and photosynthesizing organisms.

Marine picoplankton abundances oscillate between 10^3–10^6 ml^{-1} (averaging 10^5–10^6 ml^{-1}), with a total abundance in the ocean of 12×10^{28} cells, a number that exceeds the number of stars in the universe (estimated at 10^{21}). Despite the ubiquity of picoplankton in the ocean, they dominate oligotrophic waters in particular. Typically, 1 µl of superficial seawater can contain 10,000 viruses, 1,000 bacteria, 100 *Prochlorococcus* sp. (cyanobacteria), 10 *Synechococcus* sp. (cyanobacteria), 10 eukaryotic algae, and 10 protists; these numbers can vary greatly in time and space.

Scientists recently discovered that Archaea, which were initially considered typical of high salinity, high temperature, and anaerobic environments, constitute a significant fraction of picoplankton in most marine environments. In fact, Archaea can comprise >30% of coastal picoplankton in Antarctic waters, and 40–60% of total bacteria in temperate waters (at depths of 200–600 m). Archaea (mainly Crenarchaeota) comprise a large fraction of prokaryotes in the deepest waters of the open oceans.

As critical components in marine ecosystems, bacterial abundances in the water column vary between 10^5 and 10^6 ml^{-1}. On average, a liter of seawater typically contains 100–1,000 different types of bacteria. Prokaryotes represent "an ocean of biodiversity". Recent estimates suggest > 1 million different types of prokaryotes in the biosphere, of which only 6,000 species have been formally described to date. This "known diversity" represents those taxa currently grown in pure culture. We now explore unknown diversity using molecular techniques, particularly cloning and sequencing. Although a relatively small number of microbial groups (~50 species) apparently dominate the planktonic prokaryotes, these populations exhibit high genetic variability. As soon as researchers developed genetic markers for ecological studies, they immediately discovered vertical stratification of some of the dominant clades of microbial plankton. The first indications of these patterns emerged in comparing the distribution of clones of rRNA genes among sample libraries and among different depths. The most obvious interpretation is that many of these groups specialize on exploiting different aspects of physical, chemical, and biological environments. For example, as obligate phototrophs, cyanobacteria occur only in the photic zone. Depth differences also occur in many microbial groups for which we have yet to identify metabolic strategies. A clear boundary exists between the photic and the dark mesopelagic zone; below the photic zone the abundance of the pico-phytoplanktonic group SAR86 sharply decreases and Archaea I, SAR406, and SAR324 become dominant (Figure 11.13).

Anaerobic phototrophic bacteria occur in relatively low abundance in pelagic marine ecosystems, in contrast to the ubiquitous distributions of two groups of oxygenic (oxygen producing) photosynthetic bacteria (coccoid cyanobacteria and prochlorophytes) in marine pelagic ecosystems that can contribute significantly to phytoplankton biomass and production. Most coccoid cyanobacteria belong to the genus *Synechococcus* (Figure 11.14a), which occurs in high abundance in the euphotic zone of coastal waters and open ocean (with concentrations ranging from 10^2–10^5 cells ml^{-1}). These cyanobacteria appear unimportant only in polar ecosystems. Two other species, *Trichodesmium* spp. (Figure 11.14b) and *Richelia intracellularis*, both filamentous forms, play a significant role in pelagic marine ecosystems because they fix nitrogen. **Nitrogen fixation** refers to conversion of N_2, which most organisms cannot use, into ammonia, which many organisms can, a trick few organisms can achieve. *Richelia intracellularis*, an intracellular symbiont of a large diatom (*Rhizosolenia* sp.), obtains energy for nitrogen fixation from organic carbon produced by the host diatom and simultaneously provides additional nitrogen to the diatom, which often finds itself in nitrogen-limited environments. Pelagic prochlorophytes of the genus *Prochlorococcus* are slightly smaller phototrophs than those of the genus *Synechococcus*, with different pigments than Cyanobacteria. Numerous studies on *Prochlorococcus* spp., particularly in the Sargasso Sea and in the equatorial Pacific, demonstrate their fundamental role in terms of biomass and productivity in these regions.

Sulfur oxidizing and nitrifying methanogenic bacteria comprise most of the chemoautotrophic prokaryotes. These microbes include anaerobic (methanogens) as well as oxygen-tolerant forms that thrive in sub-oxic environments with chemically reduced substrates such as ammonia. The ocean surface can be supersaturated in methane, demonstrating the ubiquitous distribution of methanogenic bacteria. Researchers believe that methanogens may occupy anaerobic microniches within suspended particles in which respiration consumes any available oxygen. Methanogenesis occurs in regions of intense upwelling where deposition and subsequent oxidation of organic matter reduces oxygen concentrations.

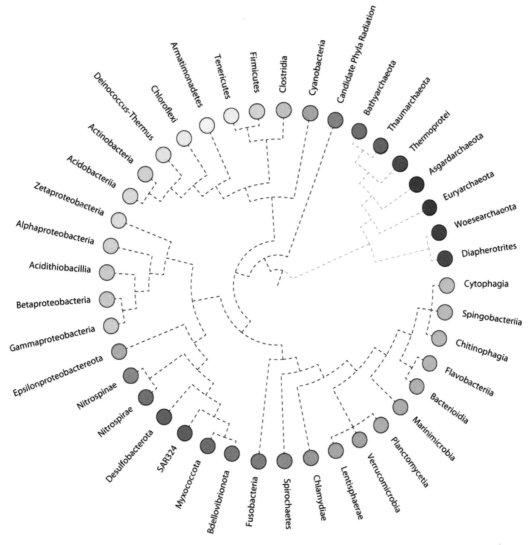

Figure 11.13 Phylogeny of the major clades of planktonic prokaryotes. Orange indicates dominant groups in the photic zone, blue indicates groups confined to mesopelagic depths and surface waters during polar winters. Green indicates microbial groups associated with coastal ecosystems.

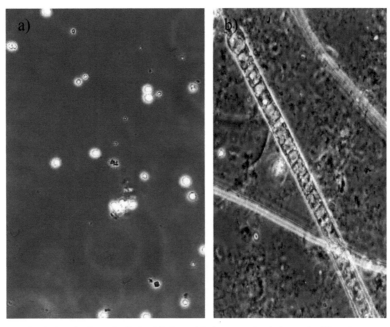

Figure 11.14 Examples of typical picoplankton include: (a) Coccoid photosynthetic bacteria (*Synechococcus* sp.); and (b) filamentous cyanobacteria (Order *Oscillatoriales*).

A variety of gram-negative heterotrophic bacteria that oxidize organic substrates numerically dominate the bacterioplankton community. In oxygen-rich waters, microbes grow by catabolizing organic molecules aerobically, with oxygen as the ultimate electron acceptor. In environments with low oxygen availability, microbes may use nitrogen and sulfur as alternative electron acceptors during anaerobic respiration. Fermentation becomes important in the absence or near-absence of oxygen.

In the early years of the last century, researchers proposed that the majority of planktonic microorganisms were cosmopolitan, with few endemic species and low global diversity. All microbes potentially occur everywhere. Despite the absence of obvious geographic barriers not all taxa occur everywhere in significant numbers. Microbial biogeography differs from other taxa in that historical factors appear less important than biological and physical variables. Three main features distinguish microbes from multicellular organisms: large populations, short generation time between one generation and the next, and large potential for passive dispersal over time. These features lead to fundamental differences between bacteria and other organisms, especially in comparing the time scales of life cycles. Autotrophic microorganism can form blooms and disappear in just a few days, whereas terrestrial autotrophic organisms in the form of woody plants require decades to reach maturity.

In the past, researchers believed that the high abundances and adaptability of planktonic marine organisms to different environmental conditions obliterated biogeographic patterns observed in macro-organisms, such as latitudinal gradients (increased species richness from polar to equatorial regions). The small size of individuals, the enormous spatial dimension of populations, the few ecological barriers, and mixing of surface waters by wind, waves, and currents, presumably facilitates the dispersion of marine bacterioplankton, potentially on a global scale. However, recent studies show geographic latitudinal patterns that resemble other taxa, with increasing diversity from the poles towards the equator. The exact nature of this geographic gradient in bacteria remains unknown but current explanations for this pattern encompass three classes: historical, ecological, and evolutionary drivers. The high abundance and dispersion ability of bacteria may reduce the influence of climatic and tectonic events on diversity patterns, leaving ecological and evolutionary factors as primary causes.

Therefore, while "everything is everywhere" (referring to bacterial species) may prove true, environmental conditions select those species that survive and develop. Of the numerous ecological and evolutionary hypotheses formulated to date, current data support two ecological mechanisms: (1) diversity increases with increased productivity because abundant resources can support large numbers and more specialized types of organisms; (2) diversity increases with increased water temperature because of increased kinetics of biological processes, including reproduction and dispersion rates, specific interactions, mutations, adaptive evolution, and speciation.

11.5.3 Nanoplankton

Nanoplankton, which includes organisms 2–20 μm in size, consists primarily of Chromista and Protozoa of autotrophic species (Classes Cryptophyceae, Prymnesiophyceae, Euglenoidea and Chrysophyceae, small forms of Dinophyceae and Diatoms) and heterotrophic species. This size class also includes microzooplankton juvenile stages. Flagellates (especially nanoflagellates) dominate the heterotrophic component and represent the main predators of picoplankton and a potential food resource for larger zooplankton. They can therefore transfer a significant portion of bacterial production to higher trophic levels. Heterotrophic nanoflagellates form a heterogeneous group of unrelated eukaryotes, including choanoflagellates, facultative or non-pigmented autotrophic chrysomonads, non-pigmented euglenoids, dinoflagellates, and elioflagellates. The small-sized nanoplankton has had limited in-depth study until recently, when researchers demonstrated their major contribution to marine primary production of organic matter, and as critically important prey for larvae of organisms at higher trophic levels. Nanoplankton dominate epipelagic habitats in many temperate and tropical seas, both numerically and in terms of their contribution to primary production; for example, tropical nanoplankton may account for 80% of open ocean photosynthesis. In neritic locations, nanoplankton play a lesser role.

Nanoplankton total abundances may exceed 10–100,000 individuals ml^{-1}. We know relatively little about the taxonomic composition of nanophytoplankton in the ocean, although studies to date indicate that phytoflagellates and small coccoids dominate. Coccolithophores, one of the most important groups of nanoplankton (Figure 11.15a) consist of unicellular organisms ranging 5–20 μm in diameter. Thin calcareous plates cover the entire surface, and cells contain brown chromatophores. Their morphology differs markedly from the common flagellate microalgae (Figure 11.15b). Nanoplanktonic flagellates are the most efficient bacterial grazers. Larger Chromista (ciliates and dinoflagellates) and larval forms of most marine metazoans prey on them, providing an important link in the planktonic food web. The autotrophs contribute to primary productivity, whereas heterotroph predation controls the abundance of other planktonic fractions, particularly picoplankton.

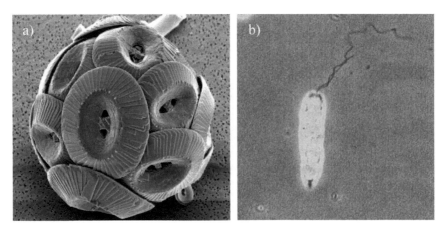

Figure 11.15 Examples of nanoplankton: (a) coccolithophore, *Coccolithus pelagicus*; (b) euglenoid, *Eutreptiella gymnastica*. Source: Plankton Net.

11.5.4 Microplankton

The microplankton include both autotrophic and heterotrophic organisms ranging 20–200 μm. Microphytoplankton includes diatoms, dinoflagellates, coccolithophores, silicoflagellate, Chlorophyceae, Euglenoids and Cryptophyceae (Figure 11.16). Prymnesiophyceae, Prasinophyceae, Chrysophyceae, Cryptophyceae, Dinophyceae, and small diatoms also contribute to the autotrophic fraction.

Some microplankton are heterotrophic or mixotrophic, including the most common marine forms, the "chrysonomads" such as *Paraphysomonas* sp. – Chrysophyceae – and *Ochromonas* sp. Both groups have two flagella, one short and smooth and one long and hairy. They use the longer flagellum for cell motility, and the shorter flagellum to trap bacteria and other particles. New technological advances have resulted in constant revision of microalgal species of continuously growing numbers: initial classification was based on morphology with identification of microalgae using optical microscopes. In the 1940s, application of scanning electronic microscopy improved the identification and description of new species. Today we

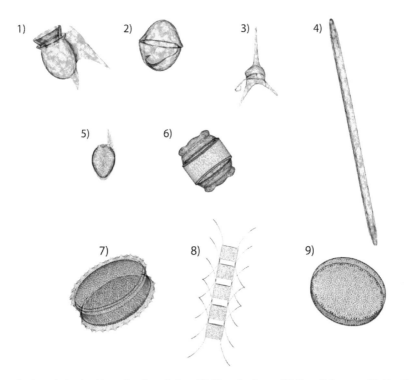

Figure 11.16 Examples of microphytoplankton: dinoflagellates: (1) *Dinophysis* sp.; (2) *Gyrodinium* sp.; (3) *Ceratium* sp.; (4) *Nitzschia* sp. (5) *Prorocentrum* sp.; and the diatoms; (6) *Biddulphia* sp.; (7) *Thalassiosira* sp.; (8) *Chaetoceros* sp.; (9), *Coscinodiscus* sp.

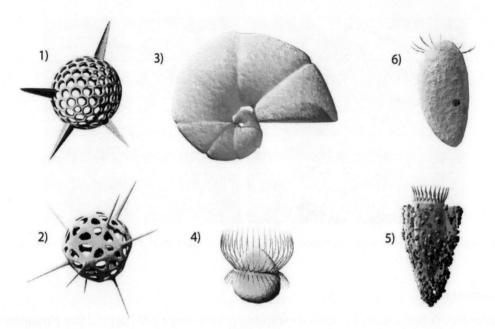

Figure 11.17 Examples of microzooplankton: (1) radiozoans (*Hexastylus* sp.); (2) *Plectacantha* sp., (3) foraminifera (*Eponides* sp.); (4) ciliates (*Mesodinium* sp.); (5) *Tintinnopsis* sp.; (6) *Amphisia* sp.

know ~12,000 species of diatoms, but experts estimate that 100,000–10 million diatom species await discovery. Typical phytoplankton abundances vary from 10^3–10^7 cells liter^{-1}, and in eutrophic areas, concentrations average about 10^6 cells l^{-1}. Typically, 1 ml of water contains 1–10 micro-planktonic organisms. Phytoplankton concentrations vary temporally by order of magnitude over periods of days, meaning that planktonic communities change very quickly.

Microzooplankton (Figure 11.17), which includes all heterotrophic organisms (Protozoa, Chromista, and larval stages of metazoans) 20–200 μm in size, provide a critical link in the "classic" food web by preying on micro-phytoplankton. Several groups of micro-planktonic organisms, primarily ciliates and heterotrophic dinoflagellates, consume bacteria, nanoplankton, and micro-phytoplankton. Coccolithophores, the largest microplankton in the ocean, represent one of the most important primary producers in open ocean ecosystems. Some phytoplankton may become bacterivorous in order to acquire nutrients during periods of low food availability.

Over the last few years, researchers recognized the key role microzooplankton play in the food web by demonstrating the effectiveness of predation on both autotrophic and heterotrophic nanoplankton and, to a lesser extent, picoplankton. As already noted, heterotrophic nanoplankton prey on the bacterial fraction, which in turn uses dissolved organic matter to increase in biomass.

Microzooplankton therefore provide the main vehicle for energy transfer from the microbial loop to the classical food web, including mesozooplankton (copepods and larval fishes). Often, microzooplankton comprise a major portion of the total zooplankton biomass in oceanic environments, with abundances on the order of 100 individuals ml^{-1}. The high growth rates of microzooplankton populations allow rapid response to increased food availability, thus closely linking production and consumption in the euphotic zone. Protists comprise the most widespread and abundant taxon in this plankton fraction, including ciliates (Figure 11.18a), radiozoans, foraminifera, and heterotrophic dinoflagellates. The first stage larvae of metazoan and cladocerans, copepods, and ostracods (Figure 11.18b) add additional important components.

11.5.5 Mesozooplankton

Mesozooplankton include heterotrophic organisms varying in size from 0.2–20 mm, which swim vertical distances in the water column of 100s of meters but cannot make significant headway against currents. Crustaceans dominate those organisms >200 μm in size, with abundances of 10^2 individuals ml^{-1} that represent ~90% of ocean plankton biomass.

Copepods comprise the main group of meso-zooplanktonic crustaceans (Figure 11.19) in terms of biomass, abundance (about 80%), and diversity (~10,000 known species) of zooplankton. They play a critical role in the ocean, providing the first link in the classic pelagic food web as the major primary consumers. Nonetheless, their abundant fecal pellets provide an important energy source for scavengers and benthic scavengers in particular. The larval stages of many fishes feed directly on copepods.

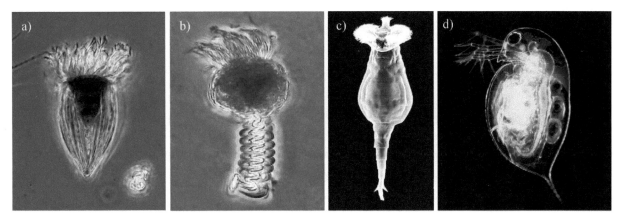

Figure 11.18 Examples of microplankton: (a) and (b) Protistans (tintinnid on left, peritrichous on right); (c) and (d) metazoans (rotifer on left, and cladoceran on right). *Source:* Liza Gross, 2007 / PUBLIC LIBRARY OF SCIENCE (PLOS) / CC BY 2.5.

Figure 11.19 Copepods represent the most important mesozooplankton component. Long appendages (antennae and tails) facilitate their suspension in water, and characterize pelagic calanoid copepods shown here.

Beyond copepods, amphipods such as *Praunus* spp. also contribute to the mesozooplankton, though in lower abundances in pelagic environments than in the benthos. Planktonic mollusks also occur throughout the ocean. These mollusks include heteropods (small pelagic "sea elephants") that use well-developed eyes for visual predation and a modified foot as a wing or fin for swimming. Pteropods or "sea butterflies" refer to filter (suspension feeding) gastropods with thin carbonate shells, in which their foot has evolved into a pair of "wings" they use to "fly" through the water and produce mucus nets to catch their prey. The mesozooplankton include early developmental stages of amphipods, euphausiacean, decapods, and planktonic mollusks, whereas the macrozooplankton include the mature stages. The sections on macro and megazooplankton treat these taxa in greater detail.

11.5.6 Macro and Mega-zooplankton

Macrozooplankton consists of a wide range of organisms (Figure 11.20) from multiple major taxonomic categories.

This group includes the Euphausiacea (common krill), predators ranging 15–100 mm that swim quickly but also represent important prey for many species of commercial fish (herring, mackerel, salmon, and tuna), penguins, and whales (in Antarctica). Their life cycle of 2+ years is long compared to many groups of plankton, and they may occur from the surface to abyssal depths. Crustaceans, and particularly calanoid and cyclopoid copepods, comprise a large part of the macroplankton. Shrimp and meroplanktonic crabs dominate the decapod members of this group; the larvae zoëa and megalopa of brachyuran crabs, which grow quite large even in their planktonic stages, can undertake significant vertical migrations. *Tomopteris* spp. represent the only important genus of planktonic polychaetes. Planktonic tunicates include Thaliacea (doliolids and salps) and Appendicularia, a group of chordates that live in a gelatinous coating they periodically abandon.

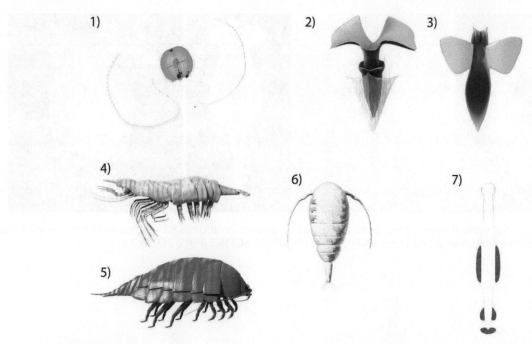

Figure 11.20 Examples of macrozooplankton: (1) ctenophore; (2–3) pteropod mollusks; (4) Euphausiacea; (5) isopod; (6) large copepods; and (7) Chaetognatha.

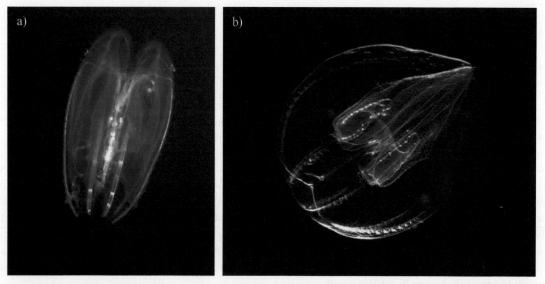

Figure 11.21 Examples of gelatinous zooplankton, which includes ctenophores: (a) *Beroe* sp.; (b) *Bolinopsis* sp. *Source:* NOAA / Wikimedia Commons / Public Domain.

The macrozooplankton also include carnivorous chaetognaths, which can be important predators. Researchers often refer to jellies, ctenophores, tunicates, and siphonophores collectively as "gelatinous zooplankton" (Figure 11.21).

Micronekton refers to free-living animals, such as juvenile fishes, sufficiently large and mobile enough to avoid capture in plankton nets.

11.6 Abundance Comparisons Among Different Planktonic Components

Generally, the abundance of plankton decreases exponentially with increasing size. Virioplankton represent, by far, the most abundant component in the ocean, averaging 2–4 to 10–100 times more individuals than planktonic prokaryotes, with abundances up to 10^{11} viruses l^{-1} of water or, in other words, 100 billion viruses l^{-1}! Viruses include many host-specific and obligate parasites that can result in very high abundances within a given large host. Imagine, if you accidentally swallow about 50 ml of seawater, you could ingest 100 million viruses (bacteriophages)!

Abundances decrease exponentially moving from viruses to prokaryotes, and that trend continues from prokaryotes to unicellular eukaryotes (Figure 11.22). Bacterioplankton represent the second most abundant component (10^8–10^9 cells l^{-1}), followed by cyanobacteria and prochlorophytes, which reach total abundances of 10^7 cells l^{-1}. Total prokaryotes reach abundances of 10^8–10^9 cells l^{-1}, with autotrophic (i.e. Cyanobacteria, Prochlorophyta) abundances apparently 1–2 orders of magnitude lower than heterotrophic (i.e. Archaea and Eubacteria) abundances. Even lower abundances characterize unicellular eukaryotes at ~10^5–10^6 cells l^{-1} whether autotrophic eukaryotic picoplankton, heterotrophic nanoplankton, nanoflagellates, or protistans (~10^4–10^5 cells l^{-1}). Abundances of the larger components of microzooplankton (ciliated protozoans, tintinnids) average 10^3–10^4 cells l^{-1}, or as few as 10^2 cells l^{-1}.

Importantly, abundances vary greatly in space and in time, meaning that individual samples may yield much higher or lower numbers than the averages described here.

The unicellular eukaryotes include all phytoplankton components, including diatoms and dinoflagellates that occur in concentrations of 10^2–10^3 cells l^{-1}, and up to 10^6 cells l^{-1} during blooms that turn surface waters green or red. These phenomena, known since ancient times, occur seasonally at localized sites greatly influenced not only by abundant nutrients, but also by weather conditions and particularly temperature, hydrodynamics, and stratification of surface waters that strongly influence ecological processes (Box 11.2).

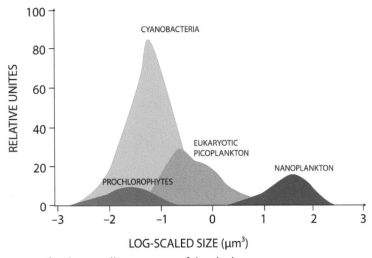

Figure 11.22 Relative importance of various small components of the plankton.

Box 11.2 What are the consequences of a bloom for water masses?

Bloom: a phytoplanktonic bloom refers to the intense proliferation of one or a few algal species that rapidly reach high concentrations of 10^6–10^7 cells per liter.
- absorption of CO_2 during photosynthesis increases the pH of surface waters by a few tenths
- oxygen release by photosynthesis results in over-saturation of oxygen (up to 130%)
- phytoplankton uptake reduces concentrations of inorganic phosphate and nitrate.

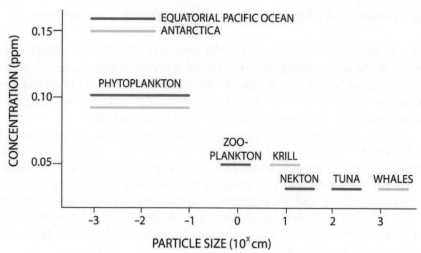

Figure 11.23 Concentrations of various organisms in relation to size.

A similar pattern of reduced concentrations occurs when comparing phytoplankton to zooplankton to nekton (Figure 11.23). The patchy distribution of some species also means that that in 100 samples, a given species might be absent in most but relatively abundant in some.

Questions

1) Define plankton.
2) How are plankton classified by size?
3) How are plankton classified based on their distribution within the water column?
4) Describe the main life cycle of zooplankton.
5) What are the virioplankton?
6) What are the main mesozooplankton taxa?
7) How many bacteria typically live in a liter of seawater?
8) What are the consequences of a phytoplankton bloom?

Suggested Reading

Amaral-Zettler, L., Artigas, L.F., Baross, J. et al. (2010) A global census of marine microbes, in McIntyre, A.D. (ed.), *Life in the World's Oceans: Diversity, Distribution and Abundance*. Wiley-Blackwell, pp. 223–245.

Behrenfeld, M.J., O'Malley, R., Boss, E. et al. (2021) Phytoplankton biodiversity and the inverted paradox. *ISME Communications*, 1, p. 52.

Breitbart, M., Thompson, L.R., Suttle, C.A., and Sullivan, M.B. (2007) Exploring the vast diversity of marine viruses. *Oceanography*, 20, pp. 135–139.

Bucklin, A., Nishida, S., Schnack-Schiel, S. et al. (2010) A census of zooplankton of the global ocean, in McIntyre, A.D. (ed.), *Life in the World's Oceans: Diversity, Distribution, and Abundance*. Wiley-Blackwell, pp. 247–265.

DeLong, E.F. (1992) Archaea in coastal marine environments. *Proceedings of the National Academy of Sciences of the United States of America*, 89, pp. 5685–5689.

Di Camillo, C.G., Luna, G.M., Bo, M. et al. (2012) Biodiversity of prokaryotic communities associated with the ectoderm of *Ectopleura crocea* (Cnidaria, Hydrozoa). *PLOS One*, 7, p. e39926.

Dolan, J.R. (2005) An introduction to the biogeography of aquatic microbes. *Aquatic Microbial Ecology*, 41, pp. 39–48.

Fuhrman, J.A., McCallum, K., and Davis, A.A. (1992) Novel major archaebacterial group from marine plankton. *Nature*, 356, pp. 148–149.

Gili, J.M., Bouillon, J., Pages, E. et al. (1998) Origin and biogeography of the deep-water Mediterranean Hydromedusae including the description of two new species collected in submarine canyons of Northwestern Mediterranean. *Scientia Marina*, 62, pp. 113–134.

Giovannoni, S.J., Britschgi, T.B., Moyer, C.L., and Field, K.G. (1990) Genetic diversity in Sargasso Sea bacterioplankton. *Nature*, 345, pp. 60–63.

Hutchinson, G.E. (1961) The paradox of the plankton. *The American Naturalist*, 95, pp. 137–145.

Righetti, D., Vogt, M., Gruber, N. et al. (2019) Global patterns phytoplankton diversity driven by temperature and environmental variability. *Science Advances*, 5, p. eaau6253.

Suttle, C.A. (2005) Viruses in the sea. *Nature*, 437, pp. 356–361.

Suttle, C.A. (2007) Marine viruses – Major players in the global ecosystem. *Nature Reviews Microbiology*, 5, pp. 801–812.

Wommack, K.E. and Colwell, R.R. (2000) Virioplankton: Viruses in aquatic ecosystems. *Microbiology and Molecular Biology Reviews*, 64, pp. 69–114.

12

Ecology of the Plankton

12.1 Plankton Distribution

Large gyres associated with higher or lower overall productivity of water masses influence broad-scale plankton distributions in the ocean. Generally, lower plankton concentrations characterize the large regions in the middle of the oceans; surface waters typically depend on inputs from terrestrial sources, and the great distance of the middle of the ocean from the continents (and therefore from nutrient sources) supports lower productivity than in coastal regions. Plankton distributions also vary locally, depending primarily on depth and season; light, temperature, salinity, and nutrient concentration primarily drive plankton ecology (Figure 12.1).

Physical and biological factors influence plankton distributions. Specifically, physical factors largely determine patterns at large spatial scales, whereas biological factors dominate at smaller spatial scales (Table 12.1). For example, at the largest scale, major circulation features such as gyres and the Antarctic Convergence determine patterns over megascales of >3,000 km.

At macro-scales of 1,000–3,000 km, biomass gradients occur with high biomass along continental margins and lower biomass toward the open ocean. Environmental gradients also occur between stenothermal species such as *Calanus finmarchicus* that live in cold regions and most eurythermal species that occur at mid-latitudes. At meso-scales of 100–1,000 km, surface currents, tides, and upwelling form local eddies and other circulation features that influence phytoplankton biomass and primary production. At smaller scales of 0.1–100 km, river plumes and upwelling around coastal headlands or submarine canyons alter biomass patterns. At fine-scales of 1–100 m, biological factors (vertical movement, grazing, and predation) dominate, while at micro-scales of centimeters to meters physiological and behavioral aspects (e.g., chemotaxis) influence organism movement and position. Even at the smallest spatial scales, plankton distribution is patchy: coastal transects often reveal large changes in concentrations over distances of just a few meters, from low concentrations or complete absence of zooplankton to aggregated patches.

12.1.1 Physical Control of Macro-scale Plankton Distribution: The Case of El Niño

El Niño, an episodic climatic event, creates anomalies in the distribution of plankton on macro- and mega-scales. ENSO (**El Niño-Southern Oscillation**) refers to the cyclical phenomenon of extreme weather events in the south-central Pacific Ocean that alternates between El Niño, La Niña, and intermediate conditions.

During El Niño years, the waters in the equatorial Indo-Pacific region gradually warm, progressively extending toward Central America along the equatorial belt. Surface water temperatures in the eastern Pacific Ocean increase ~0.5–3.0 °C, persisting for periods of five months or longer. These conditions lead to a slowing of the Humboldt Current and reduce upwelling of deep water along the continental margin between Chile and Peru with a lowering of the thermocline depth. This reduction in upwelling stops the nutrient supply from deep environments and reduces rates of primary and secondary production, and therefore the abundance of fish (Figure 12.2). A drastic reduction of water oxygen content also occurs as an additional consequence of warming of water masses. The most intensive phase of El Niño events (usually around December), confines pelagic fish to the surface, often causing beach strandings that become a real boon for local communities who collect them by hand. El Niño translates to "the child" and refers to the proximity of this event to the birth of Jesus. The opposite phenomenon, La Niña, occurs when surface water temperatures drop sharply.

Marine Biology: Comparative Ecology of Planet Ocean, First Edition. Roberto Danovaro and Paul Snelgrove.
© 2024 John Wiley & Sons Ltd. Published 2024 by John Wiley & Sons Ltd.
Companion Website: www.wiley.com/go/danovaro/marinebiology

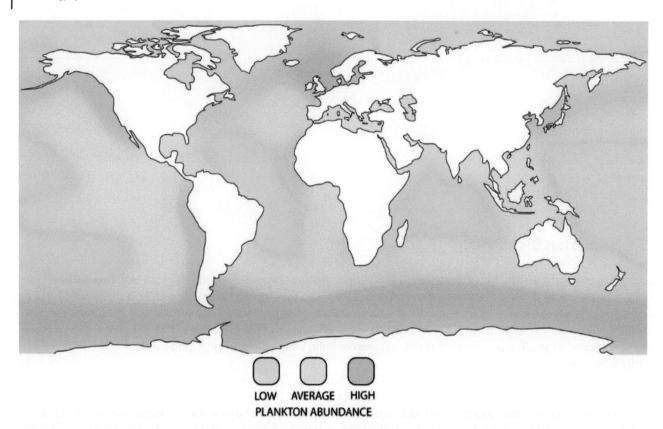

Figure 12.1 Global distribution of plankton with respect to large oceanic gyres, highlighting low plankton concentrations in the center of gyres and high concentrations at mid latitudes and near continents.

Table 12.1 Physical and biological factors influencing the distribution of zooplankton at different spatial scales.

Range (km)	Physical	Biological	Time (days)
>1.000	Gyres, Antarctic Convergence	–	>1.000
100	• Fronts • Warm/cold waters • Upwelling	Seasonal growth	>100
10	Turbulence	–	–
1	–	Vertical migrations	1
0.1	–	• Physiological adaptations-flotation-light	0.1
0.01	Langmuir circulation	Behavioral adaptation	0.01

12.1.2 Small-scale Plankton Distribution: The Vertical Migration Example

The *Challenger* expedition in the nineteenth century first reported vertical movement in plankton, documenting the presence of copepods primarily during their night samples. Subsequent research demonstrated that copepods descend during the day and ascend to the surface during the night.

Sonar, which use sound waves to detect particles, demonstrated that the depth distribution of layers of high zooplankton concentrations (called the **Deep Scattering Layer** – DSL or reflective deep layer; Figure 12.3) changed over a 24-h cycle. The DSL includes groups of marine organisms that move to different depths at different times of day. The zooplankton that undergo vertical migrations include a broad range of taxa, dominated by euphausiids, followed by shrimp, copepods, siphonophores, cnidarians, pterotracheoids and pteropods; larger predators such as squid and small fishes prey on these

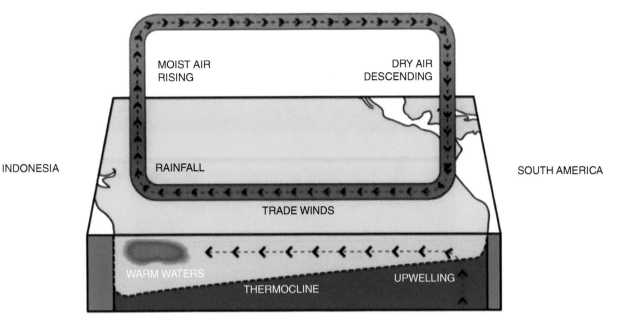
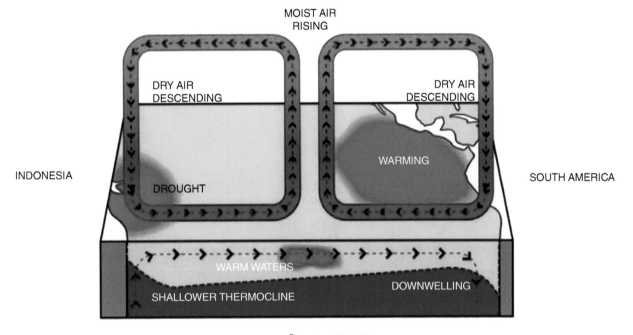

Figure 12.2 Dynamics of ENSO (El Niño Southern Oscillation). Upper panel shows normal conditions in which the Humboldt Current flows northerly along the continental margins of South and Central America, transporting cold, nutrient-rich waters. Bottom panel illustrates El Niño effect and weakening of the Humboldt Current.

taxa. Although the organisms that comprise the plankton cannot make significant headway against currents, they can nonetheless carry out large vertical migrations.

The phenomenon of daytime vertical migration (or diurnal migrations; Figure 12.4) differs between phytoplankton and zooplankton. Phytoplankton generally move towards the surface during the day, but some species favor different depths as a result of excessive sunlight. During nocturnal migration, zooplankton move to the surface at sunset and remain in surface waters during the night. Before sunrise, they begin to migrate down to greater depths, where they remain during the day.

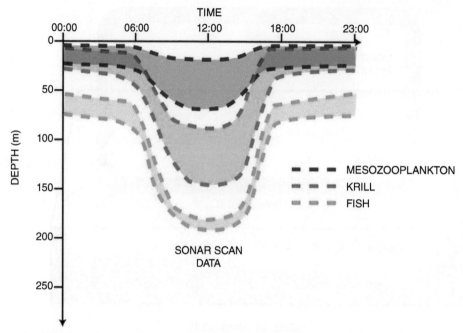

Figure 12.3 Schematic of the Deep Scattering Layer showing changes in zooplankton distributions over a 24-h period.

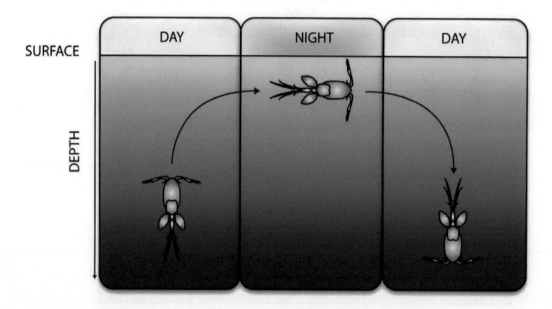

Figure 12.4 Typical diel vertical migration of copepods with nocturnal ascent and diurnal descent.

During diurnal (as opposed to nocturnal) vertical migrations, animals begin the ascent to surface waters at sunrise and begin to return to depth at sunset. Such migrations (both diurnal and nocturnal) commonly occur in surface waters, but also occur to depths > 1,000 m. Double migrations occur, though less commonly, where two ascents and two descents occur over 24 h: a short period in deeper waters follows an upward migration at dusk, followed by a descent to the bottom during midday. Later in the night they move back up to the surface before descending again at sunrise. The extent of migration varies among species and diel migrations occur in the epipelagic zone (upper 200 m), in the mesopelagic zone (up to 1,000 m depth), or in the bathypelagic zone with vertical migrations of >1 km (Figure 12.5). Even within the same species, however, the extent of vertical migrations varies substantially, depending on the life phase of the organism. For example, juveniles typically complete minor excursions compared to adults (Figure 12.6).

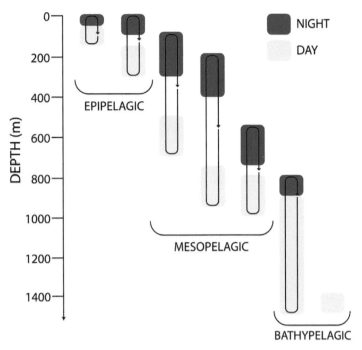

Figure 12.5 Examples of plankton vertical migrations that differ in depth range and position within the water column.

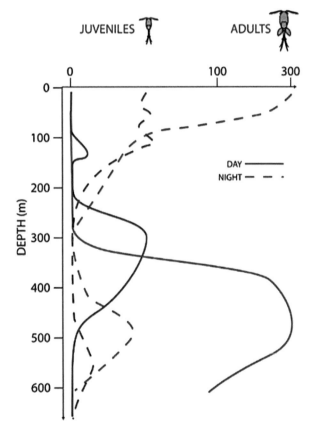

Figure 12.6 Example of the difference between vertical migration of juvenile forms (blue lines) and adults (red lines) where zooplankton juveniles (e.g., copepodites) peak in abundance during the day at ~300 m depth whereas adults generally migrate to 400–500 m depth.

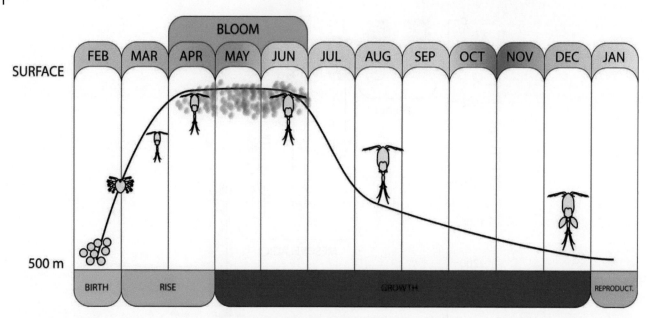

Figure 12.7 Migration associated with copepod life cycles, illustrating seasonal vertical migration in which gamete release results in late winter nauplii production. Nauplier ascent to the surface in spring corresponds with the periods of phytoplankton blooms, and also the subsequent juvenile descent to depth (with pre-adult stages that mature to begin a new reproductive event). Note that gamete release and juvenile production occur at relatively deeper depths.

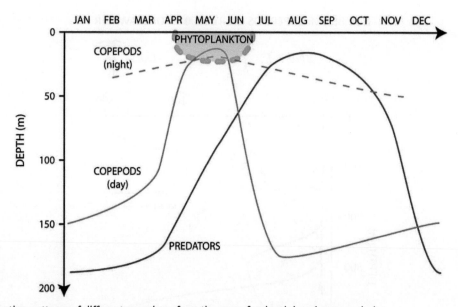

Figure 12.8 Migration patterns of different organisms from the same food web in a 1-year period.

III–IV. Later copepodite stages (V and VI) move back down to greater depths, where they mature and become fully developed adults during autumn-winter, ready to reproduce (spawn) in late winter-early spring of the following year.

Vertical migration also occurs in macro- and micro-nektonic organisms (Figure 12.8), producing the largest animal migration, in terms of biomass, on Earth. The widespread nature of this migration across taxa and geographic locations suggests some common forcing factor. For example, migration presumably limits the damage caused by ultraviolet radiation, but this explanation cannot account for vertical migrations of 10s, 100s, or even 1,000 m depth.

Most migration patterns involve populations moving from deeper layers during the day, to shallower depths at twilight. One hypothesis suggests these organisms follow constant lines of luminosity. When the sun rises, the luminosity line moves to greater depth and animals migrate to follow it. With sunset, this line shoals again, along with the

Figure 12.9 Example of organisms of different sizes able to move at different speeds: (a) *Calanus* sp., can move 15 m in 1 h; (b) the larger copepod, *Centropages* sp., can swim 100 m in 1 h; and (c) the euphausiacean, *Meganyctiphanes norvegica*, can travel up to 900 m in 1 h. *Source:* Anita Slotwinski – IMOS/CSIRO, NOAA / Public Domain, Øystein Paulsen / Wikimedia Commons / CC BY-SA 3.0.

organisms. Some species of copepods and jellyfish migrate 400–800 m in a single day, with maximum upward vertical migrations in the copepod *Calanus* spp. of about 15 m h^{-1} and descent rates of 100 m h^{-1}. Euphausiids move vertically at 100–200 m h^{-1}. Compared to the sizes of the migrating organisms (only a few mm, Figure 12.9), these migrations represent impressive distances. For most marine organisms, the speed of movement is proportional to the square of the length according to:

$$V = KL^2$$

where V = velocity m h^{-1}, K = constant depending on taxon and characteristics of the animal swimming, and L = length of the organism.

The main hypotheses to explain vertical migrations encompass two broad categories. The first category suggests that migration reduces energy expenditure, in that residing in cold waters during the day provides substantial advantages by reducing metabolism; metabolism in cold-blooded organisms depends on surrounding water temperature. At night, they move into warm waters to feed. However, few existing data support these hypotheses and substantial evidence suggests just the opposite – that daily vertical migration of this kind may be energetically disadvantageous. The second category starts with the assumption of "predator avoidance", where vertical migrations reduce the risk of mortality from visual predators.

This hypothesis is simple and intuitive; if mesozooplankton remain close to the surface during the day, visual predators (mainly fishes) would have a greater chance of seeing their prey. Mesozooplankton therefore move deeper around sunrise to the refuge of deep ocean darkness, reducing the likelihood of attracting the attention of predators. Basically, the benefit of reduced likelihood of becoming prey outweighs the cost of reduced feeding opportunities during the day. In other words, better hungry than dead! Experiments with some planktonic species demonstrate that chemical exudates emitted by predators trigger the migratory behavior.

Migration adds the additional benefit of increased gene exchange among populations. Indeed, when plankton ascend and descend, some individuals fall behind their original migrating swarm and they subsequently join a different swarm of conspecifics with which they can reproduce, thus reducing **inbreeding**, referring to limited mixing with other populations during reproduction.

Among the taxonomic groups of macroplankton and micronekton (individuals >4.5 mm in size), the largest day-night disparity occurs in euphausiids (> 40%), followed by fishes (10–20%), and decapods (10–20%). Differences in vertical migration distance width, species number, and migratory species composition can affect ecological processes in the overlying water column. Changes in composition of the migratory group between day and night can substantially impact grazing, predation pressure, and interspecific interactions. Indeed, zooplankton play an important role in transferring organic material (and thus food) from the euphotic zone to depth both as food supply (i.e., predators and prey) and in producing fecal pellets that sink toward the bottom (Figure 12.10). Through their intense daily and seasonal vertical migration, zooplankton also transfer food energy from the euphotic zone to deeper layers. Copepods, and particularly adults, play a particularly important role in this process. At high latitudes, for example, a single species of copepod, *Calanus finmarchicus*, can comprise more than half of zooplankton biomass.

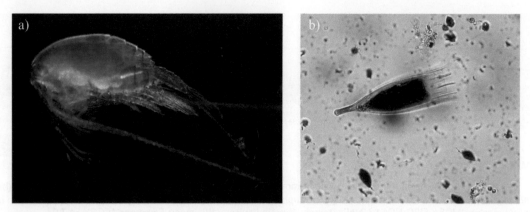

Figure 12.10 (a) Example of planktonic copepod producing fecal pellets. (b) Zooplankton surrounded by fecal pellets.

12.2 Ecology of Plankton

The large variety of life histories, phylogenetic origin, size, and function all add to plankton complexity. Nonetheless, the importance of some environmental variables transcends all plankton. Euphotic zone depth and surface mixed layer depth, together with solar light intensity and nutrients, primarily control phytoplankton community composition, and their variation (seasonality in particular) (Figure 12.11). Short-term temporal variation in planktonic communities depends on

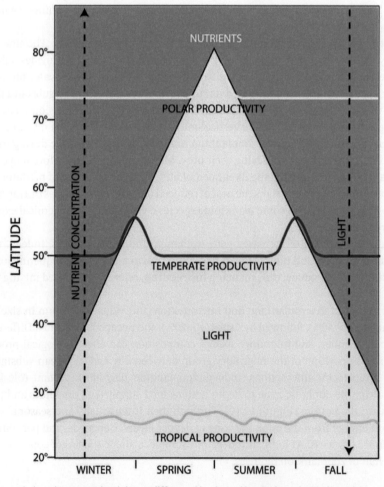

Figure 12.11 Temporal trends in primary productivity at different Northern Hemisphere latitudes, with a general pattern of no peaks at low latitudes, two phytoplankton blooms at mid-latitudes, and a single bloom event at high latitudes.

Table 12.2 Rates of reproduction/replication of various plankton components in relation to size.

Plankton size	Life cycle/replication time
Picoplankton	3–24 h
Heterotrophic nanoplankton (nanoflagellates)	1–3 d
Microphytoplankton	1–7 d
Protozooplankton (ciliates)	1–2 weeks
Mesozooplankton (copepods)	2–3 months
Macrozooplankton (Antarctic Euphausiacea)	3 months–2 y

species turnover and rates of growth and mortality. The descent of plankton to depth, migration (diel and circadian), and predator activity also cause short-term changes. Below, we provide examples of representative reproductive rates of various plankton categories (Table 12.2):

In Arctic polar waters, sufficient light for phytoplankton production exists only for a single summer algal bloom, followed by an increase in zooplankton biomass. In North Pacific, North Atlantic, and Mediterranean waters, a considerable phytoplankton bloom (high rates of primary productivity and potentially large numbers) and high biomass of standing primary production typically occurs during spring followed by a "minor" event in autumn. In contrast, stratified tropical surface waters at low latitudes support large numbers of species but comparatively few individuals, greatly reducing the likelihood of phytoplankton blooms; primary production therefore remains uniformly low throughout the year. Seasonal variation in limiting nutrients strongly influences plankton seasonal dynamics, along with water temperature and production of exudates that may inhibit other species. Several species typically dominate at a given latitude, changing over time in a successional sequence that varies greatly among different geographic regions (Figure 12.12).

Phytoplankton seasonality plays a critical role in the temporal dynamics of all plankton components (Figure 12.13). For example, temporal patterns in zooplankton communities at mid-latitudes closely track those in the phytoplankton community but delayed a few weeks. This phytoplankton–zooplankton time course resembles predator–prey dynamics, in

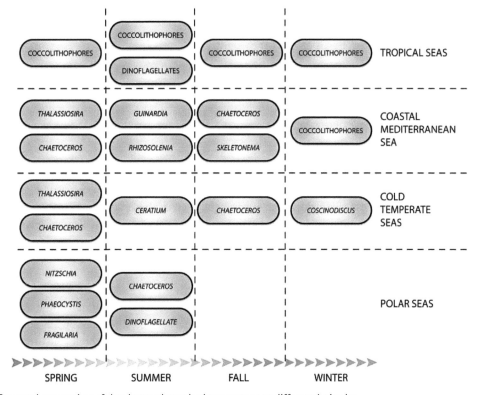

Figure 12.12 Seasonal succession of dominant phytoplankton genera at different latitudes.

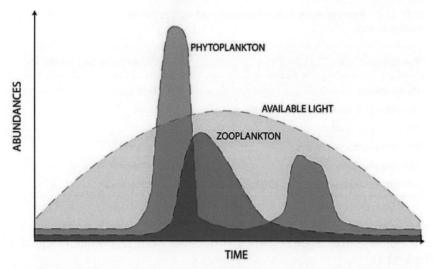

Figure 12.13 Example of temporal dynamics of phytoplankton (green) and zooplankton (orange). Zooplankton abundances follow those of phytoplankton, with a temporal lag related to their longer life cycle.

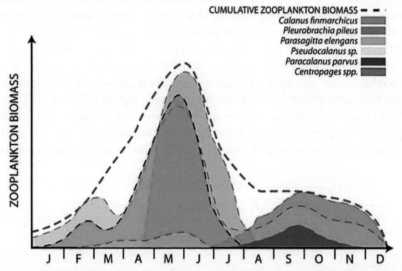

Figure 12.14 Temporal zooplankton dynamics highlighting two periods of zooplankton bloom, one in late spring and one in late summer-autumn (dotted line). These dynamics integrate the dynamics of many species that alternate in peak abundance during different months (colored portions).

that increased grazing associated with higher abundances of mesozooplankton can decrease phytoplankton abundance. However, some researchers suggest that these dynamics reflect "betting" by zooplankton, because whereas phytoplankton reproduce very quickly (hours–days), especially during an exponential growth phase, zooplankton require many weeks (or even months) to develop from eggs to nauplii, and then through copepodite stages. Given that a phytoplankton bloom usually lasts 2–4 weeks, zooplankton cannot wait for bloom onset to begin breeding because they would miss the window of maximum food availability. Zooplankton therefore begin to breed in winter before bloom onset, so they are ready to take full advantage once it happens. This "anticipation" adds significant risk for reproductive failure if the bloom begins early or later than usual, potentially resulting in high mortality in zooplankton.

As with phytoplankton, succession also occurs in zooplankton within the two periods of maximum abundance (Figure 12.14), resulting from species interactions that limit overlap between some species. Zooplankton abundance peaks in late spring and summer, with lowest numbers in winter, typically dominated by copepods, which constitute 70–80% of total zooplankton. Scientists currently recognize ~11,500 species of marine copepods globally, 10 of which account for ~50% of total zooplankton abundance. Indeed, zooplankton ecologists argue that the copepod *Calanus finmarchicus* may be the single most abundant animal in the world!

Structurally, zooplankton communities include several common types:

1) open water communities, characterized by low primary production rates and dominated by micro-phytoplankton and zooplankton dominated by tintinnids, herbivorous copepods, and carnivorous species. High stability, high diversity, and typically top-down control characterize these communities.
2) intermediate communities with moderate primary production rates and biomass, with bottom-up control of primary production.
3) coastal communities, characterized by low diversity, high primary production, and high biomass. Ciliates and tintinnids consume abundant nanoplanktonic primary producers, which provide a critical food resource for abundant filter feeders. Here, top-down grazing cannot control phytoplankton dynamics.

12.3 How Many Phytoplankton Species Coexist in a Volume of Water? "Homage to Santa Rosalia"

The **principle of competitive exclusion** states that: two species cannot occupy the same niche, because the strongest will persist and the weakest will disappear with time. This "paradox of the plankton," highlighted by the famous ecologist G. Evelyn Hutchinson, notes that many species of phytoplankton apparently occupy the same volume of water and use the same resources (nutrients, light, etc.). How do these multiple species of phytoplankton compete for the same resources (nutrients) and live within the same niche (the same volume of water)? Hutchinson explained this paradox in a classic study published in 1961, "The Paradox of the Plankton" based on the following observations:

1) different species vary in their ability to use nutrients, depending on their size and characteristics. In general, the greater surface area – volume ratio in smaller species, enables them to use nutrients efficiently, even at low concentrations. In contrast, larger-sized species can grow more quickly, but only with high concentrations of nutrients;
2) nutrient concentrations vary significantly over small spatial scales (with highest concentrations around dissolved organic particles);
3) nutrient concentrations vary in time, in large part because of phytoplankton exploitation.

These conditions create such a dynamic scenario that no single species can dominate others for a sufficiently long time to drive them to competitive extinction, despite their overlapping trophic niche. For example, species S_1 dominates over species S_2 at high nutrient concentrations (because it reaches a higher maximum growth rate μ_{max}) (Figure 12.15a); if high nutrient concentrations persist over time, species S_1 would outcompete and exclude species S_2. However, if a similar μ_{max} characterizes the two species but they differ in their affinity for the substrate (K_n), the two species would reach half of the μ_{max} at different nutrient concentrations (Figure 12.15b). In this instance, some species grow better than others at low nutrient concentrations, with a greater substrate affinity in species S_1 compared to species S_2 and thus a competitive advantage at low nutrient concentrations. For this reason, in prolonged oligotrophic conditions, species S_1 would increasingly dominate species S_2. Alternatively (Figure 12.15c) species S_1 might grow well at high nutrient concentrations, whereas S_2 shows great substrate affinity and superior growth at low nutrient concentrations. In this case, the environment might gradually transition from one characterized by high nutrient concentrations to one with low concentrations of nutrients resulting from phytoplankton uptake. These conditions, compounded by spatial variation in the distribution of light, nutrients, and phytoplankton, allow large numbers of species to coexist within the same volume of water.

12.4 Zooplankton Nutritional Mode

Meso- and macrozooplankton typically consist of: (1) herbivores that feed on phytoplankton; (2) carnivores that feed on other zooplankton or fish larvae; (3) scavengers that feed on suspended organic debris in the form of organic particulate matter; and (4) omnivores sustained on a mixed diet of plants and animals or debris. Evaluating the role of zooplankton in marine ecosystem functioning requires investigating the associated energy budget, which requires analysis and quantification of nutrition (filtration and predation), assimilation, respiration, and excretion processes.

The nutritional sources for zooplankton can vary greatly. Some species use bacteria (1–2 μm) as a food source, and many use flagellates, micro-metazoans (microzooplankton), and/or phytoplankton (2–50 μm). Copepods dominate mesozooplankton feeding using filtration (or often through predation). Mesozooplankton filtration uses maxillae with long

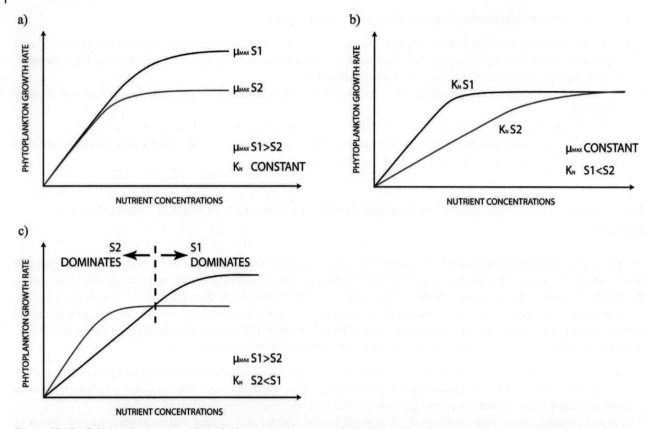

Figure 12.15 Relationship between phytoplankton growth rates (vertical axis) and nutrient concentration (horizontal axis), which resembles the Michaelis-Menten equation for enzymatic dynamics of substrate degradation. S_1 and S_2 denote the two species of interest, μ_{max} represents maximum growth rate of each species, K_n denotes the semi-saturation constant, referring to the concentration at which each species reaches 50% maximum growth rate.

appendages and "networks" of bristles equipped with mucus to remove the finest particles (<1 μm). The following simple equation calculates filtration rate:

$$\delta C / \delta t = -C(nG/V + S)$$

Where: C = particle or cell concentration; G = filtration rate (volume filtered x units of time), n = number of individuals; V = volume of water; S = sedimentation rate (of phytoplankton sinking to the bottom and thus unavailable as suspended food). Integrating this equation yields a filtration rate G: $C_T = C_0 - (nG/V + S) t$ that, following log transformation, produces:

$$G = V/nt(\log C_0 - \log C_t - S_t)$$

Where: C_0 = cells' concentrations of cells at time 0; C_t = cell's concentration at time t. This equation yields an estimate of filtered water of a few hundred $cm^3 \, d^{-1}$. Filtration rate for a given species increases with individual weights and generally decreases with increasing concentrations of phytoplankton. Surface area of *Calanus finmarchicus* maxillae of 0.3–0.4 mm^2 corresponds to a volume of filtered water of ~100 $cm^3 \, d^{-1}$. Therefore, zooplankton grazing can significantly reduce phytoplankton stocks.

Phytoplankton use multiple mechanisms to reduce grazing, including bioluminescence, as seen in some dinoflagellates to discourage predatory copepods, and protective skeletal structures such as silica thecae of some diatom species. Some species also produce inhibitory or stimulant substances for grazing. The unsophisticated appendages used for capturing and manipulating phytoplankton results in "sucking" or crushing of prey rather than fully ingesting it, resulting in release of much of the cytoplasmic material into the water – a mechanism known as **sloppy feeding**. This process releases dissolved material (DOM) and detritus into the water column in amounts equivalent to 20–40% of the grazed biomass.

Prey size primarily determines particle selection in zooplankton, whether referring to grazing in herbivores or predation in carnivores. In the strict sense, **optimal foraging theory** describes predation, in that predators select larger prey when prey items occur in equal concentrations because larger food items provide more calories. In contrast, when prey concentrations differ, predators generally select the most abundant species because rates of contact between predator and prey increase, along with probability of predator success. The ratio between predator and prey size typically falls somewhere between 14:1 and 20:1.

Food selection may also change with age and predator size, and behavior may also be important. Some predators require living and mobile prey to provoke an attack. Calculation of rate of assimilation of ingested organic material uses the following equation:

$$A = \frac{C-F}{C} \times 100$$

Where: A = assimilation efficiency, C = amount of food consumed, F = amount excreted with feces.

Assimilation rate varies from 6 to 99% and depends on quality of food ingested and processed by proteases, glucosidases, nucleases, and lipases enzymes. Calculation of zooplankton energy balance considers the following variables.

$$Q_{food} - Q_{feces} = Q_{growth} + Q_{respiration}$$

Where Q_{food} = energetic value of food consumed; Q_{feces} = energetic value of substances excreted within feces and those assimilated but not metabolized; Q_{growth} = total investment of energy in body materials (growth); $Q_{respiration}$ = energy required for respiration.

Respiration represents the minimum energy required for an organism to survive in nature. Respiration varies with environmental variables that include: (a) temperature; (b) chemical composition of food; as well as biological factors that include: (c) mass and (d) species. The dependence of respiration on organism body mass is well known and quantified by:

$$R = a\, M^b$$

R = respiration rate; M = body mass; a = species-specific constant; b = species-specific constant.
Respiration also depends on other variables:

$$Q_{respiration} = Q_{methabolism} + Q_{degradation} + Q_{activity}$$

$Q_{methabolism}$ = energy required for basic metabolism;
$Q_{degradation}$ = energy required for deamination of amino acids not used in the production of proteins and digestion;
$Q_{activity}$ = energy required for movement.

$$Q_{feces} = Q_{feces} + Q_{nitrogen}$$

Q_{feces} = caloric value of feces; $Q_{nitrogen}$ = caloric value of nitrogenous substances that are not metabolized.

After digestion, copepods expel feces as elongate fecal pellets surrounded by a film. Copepods excrete nitrogen as ammonium ion (ammoniotelic excretion) or in the form of amino acids. In proportion to body size, they excrete about 2% in winter to 10% in spring, with slightly higher proportions of phosphate excretion (5–25%). Of the nitrogen digested by *Calanus*, they excrete 36% in soluble form, eliminate 37% in fecal pellets, and invest 27% in growth.

Excretion increases with temperature and with size as follows:

$$Q_{food} - Q_{feces} - Q_{nitrogen} = Q_{growth} + Q_{methabolism} + Q_{degradation} + Q_{activity}$$

Questions

1) What factors control plankton distributions?
2) What is the Deep Scattering Layer?
3) What are the differences between phytoplankton and zooplankton vertical migrations?
4) What is the extent of vertical migration?

5) Describe how copepod seasonal migration links to their life cycles.
6) What factors control phytoplankton blooms?
7) How does seasonal succession vary with latitude?
8) How do phytoplankton dynamics influence zooplankton communities?
9) How do multiple species of phytoplankton compete for the same resources and live within the same niche?
10) What are the primary food sources for zooplankton?

Suggested Reading

Azam, F. and Malfatti, F. (2007) Microbial structuring of marine ecosystems. *Nature Reviews Microbiology*, 5, pp. 782–791.

Ban, S., Burns, C., Castel, J. et al. (1997) The paradox of diatom-copepod interactions. *Marine Ecology Progress Series*, 157, pp. 287–293.

Boero, F. and Bouillon, J. (1993) Zoogeography and life cycle patterns of Mediterranean hydromedusae (Cnidaria). *Biological Journal of the Linnean Society*, 48, pp. 239–266.

Boero, F., Bouillon, J., Gravili, C. et al. (2008) *Gelatinous plankton*: Irregularities rule the world (sometimes). *Marine Ecology Progress Series*, 356, pp. 299–310.

Bouillon, J., Medel, M.D., Pagès, F. et al. (2004) Fauna of the Mediterranean hydrozoa. *Scientia Marina*, 68, pp. 5–438.

Brotz, L., Cheung, W.W.L., Kleisner, K. et al. (2012) Increasing jellyfish populations: Trends in Large Marine Ecosystems. *Hydrobiologia*, 690, pp. 3–20.

Brown, E.R., Cepeda, M.R., Mascuch, S.J. et al. (2019) Chemical ecology of the marine plankton. *Natural Products Reports*, 36, pp. 1093–1116.

Castellani, C. and Edwards, M. (2017) *Marine Plankton: A Practical Guide to Ecology, Methodology, and Taxonomy*. Oxford Academic.

Fuhrman, J.A. (1999) Marine viruses and their biogeochemical and ecological effects. *Nature*, 399, pp. 541–548.

Garcés, E., Vila, M., Reñé, A. et al. (2007) Natural bacterioplankton assemblage composition during blooms of Alexandrium spp (Dinophyceae) in NW Mediterranean coastal waters. *Aquatic Microbial Ecology*, 46, pp. 55–70.

Garrabou, J. and Flos, J. (1995) A simple diffusion-sedimentation model to explain planktonic gradients within a NW Mediterranean submarine cave. *Marine Ecology Progress Series*, 123, pp. 273–280.

Gasol, J.M. and Kirchman, D.L. eds. (2000) *Microbial Ecology of the Oceans*. Wiley.

Haury, L., Fey, C., Newland, C., and Genin, A. (2000) Zooplankton distribution around four eastern North Pacific seamounts. *Progress in Oceanography*, 45, pp. 69–105.

Hay, S. (2006) Marine ecology: Gelatinous bells may ring change in marine ecosystems. *Current Biology*, 16, pp. 680–682.

Kiørboe, T. (2009) *A Mechanistic Approach to Plankton Ecology*. Princeton University Press.

Lang, A.S., Rise, M.L., Culley, A.I., and Steward, G.F. (2009) RNA viruses in the sea. *FEMS Microbiology Reviews*, 33, pp. 295–323.

Lovejoy, C., Legendre, L., Therriault, J.C. et al. (2000) Growth and distribution of marine bacteria in relation to nanoplankton community structure. *Deep-Sea Research II*, 47, pp. 461–487.

Mangialajo, L., Bertolotto, R., Cattaneo-Vietti, R. et al. (2008) The toxic benthic dinoflagellate *Ostreopsis ovata*: Quantification of prolife-ration along the coastline of Genoa, Italy. *Marine Pollution Bullelin*, 56, pp. 1209–1214.

Menge, D.N.L. and Weitz, J.S. (2009) Dangerous nutrients: Evolution of phytoplankton resource uptake subject to virus attack. *Journal of Theoretical Biology*, 257, pp. 104–115.

Olson, D.B., Hitchcock, G.L., Mariano, A.J. et al. (1994) Life on the edge: Marine life and front. *Oceanography*, 7, pp. 52–60.

Ortmann, A.C., Lawrence, J.E., and Suttle, C.A. (2002) Lysogeny and lytic viral production during a bloom of the cyanobacterium *Synechococcus spp. Microbial Ecology*, 43, pp. 225–231.

Pomeroy, L.R., leB. Williams, P.J., Azam, F., and Hobbie, J.E. (2007) The microbial loop. *Oceanography*, 20, pp. 29–33.

Sarno, D., Zingone, A., Saggiomo, V., and Carrada, G.C. (1993) Phytoplankton biomass and species composition in a Mediterranean coastal lagoon. *Hydrobiologia*, 271, pp. 27–40.

Suthers, I.M., Rissik, D., and Richardson, A.J. eds. (2019) *Plankton: A Guide to Their Ecology and Monitoring for Water Quality*. 2nd Edition. CSIRO Publishing.

Part IID

Life in Seas and Oceans: Fundamentals of Marine Biology

13

Biodiversity of the Nekton

The term "**nekton**" (Greek for "swimming things") refers to organisms that can make significant headway against ocean currents through active swimming. Nekton includes fish, cephalopods, mammals, reptiles, and some birds, and a subset of nektonic taxa (reptiles, birds, and mammals), the **xeronekton**, referring to organisms that must maintain a connection with the ocean surface in order to obtain oxygen. The true nekton, or **eunekton**, spend their entire lives below the surface. The fishes comprise almost the totality of nekton, whereas the cephalopods (the only invertebrates, other than some crustaceans with strong swimming capacity) include squids and octopuses, some of which can be quite large. Marine mammals, reptiles, and seabirds comprise the other major nektonic groups.

13.1 Species Contributing to the Nekton

Nekton biodiversity generally decreases from the equator to the poles, peaking at depths of 100–300 m and then declining to abyssal depths. Fishes, with more than 20,000 known species, represent Earth's most diverse vertebrates. Cartilaginous fishes, which include the sharks, skates, and rays, contribute some 1,000 species and encompass the majority of marine top predators. The ~1,000 species of cephalopod mollusks include squid, cuttlefish, octopus, and nautilus, including some species that can exceed 15 m in length and live to depths of 5,000 m. These mollusks can capture and handle prey effectively, but are also prey for tunas, marlins, whales, and elephant seals. Marine reptiles include 110 species, mainly sea serpents, 8 turtle species, 2 crocodile species, and 1 species of marine iguana. Adult reptiles live in the ocean but move onto land for breeding.

For some species of nekton that must breathe air, diving plays a key role as they chase fish, small plankton, gelatinous zooplankton, and other organisms to depths >100 m. Seabirds include diverse forms, some flightless (e.g., penguins) that move most efficiently when swimming, and others (petrels and albatrosses, pelicans, gulls, terns, and auks) that fly easily through the air, and dive into the ocean to feed on zooplankton and fish. Other seabirds, such as puffins, can fly in air but move much more gracefully in water. Marine mammals must breathe at the surface, and generally dive to feed. Seals, sea lions, walruses, dugongs, manatees, and cetaceans all exhibit significant adaptations for swimming and diving. They include other mammals, such as bears and otters that live primarily on land but use marine food resources along coastlines. The four extant marine species of Sirenia, the herbivorous dugongs and manatees, occur only in the tropics, where they graze on seagrasses. Some of the 33 extant species of pinnipeds, which include walruses and seals, can dive to 1,000 m depth; these voracious predators can strongly influence the composition and abundance of pelagic fishes. Cetaceans, the mammals that fully transitioned to an oceanic life, include the parvorder Mysticeti (baleen whales), which feeds on krill and other zooplankton, and the predatory parvorder Odontoceti (toothed whales) that feeds on fishes and other whales. The ~44 species of dolphins dominate the Odontoceti, which produce ultrasonic sound waves to locate their prey (**echolocation**) and to communicate. Many nektonic species undertake major migrations, both for foraging and for reproduction. Gray whales migrate ~9,000 km, in contrast to 8,000 km in some Odontoceti, and 4,000 km for sea lions and seals. Some whales migrate to escape predation from orcas. Some large sharks normally feed on mammals and marine reptiles but significant global declines in large sharks has led to increases in their prey, including seabirds, turtles, seals, and reef sharks, with cascading effects on overall abundance, biodiversity, and ecosystem functioning.

Marine Biology: Comparative Ecology of Planet Ocean, First Edition. Roberto Danovaro and Paul Snelgrove.
© 2024 John Wiley & Sons Ltd. Published 2024 by John Wiley & Sons Ltd.
Companion Website: www.wiley.com/go/danovaro/marinebiology

13.2 Main Organisms and Characteristics of Nekton

The diverse phyla of marine organisms that comprise the nekton illustrate a wide range of substantial and often convergent adaptations (Figure 13.1). Physiologically, nektonic life generally reflects adaptations to facilitate buoyancy and thus minimize the energetic cost of living suspended within, and swimming through, water. For example, the swim bladder present in some fishes allows them to vary their buoyancy by filling or emptying their bladder with an oxygen, nitrogen, and carbon dioxide gas mixture. Other fishes use rich fat deposits to reduce their specific weight relative to the surrounding water. Mammals, reptiles, and seabirds use air sacs or lungs to adjust buoyancy by adding or removing air. Sharks lack a swim bladder, meaning most are heavier than water and must therefore swim continuously to oxygenate their gills, unless they live in particularly well-oxygenated water where they can rest on the bottom.

Most nektonic animals are structurally and functionally adapted to swim actively and relatively quickly in water (Figure 13.2). Because of their generally large size and rapid movement, viscosity and the turbulent flow generated by their movement become important factors in water. Recall from Chapter 5 that Reynolds number (Re) refers to a dimensionless ratio of inertial and viscous forces that quantifies the relative importance of these two forces for given flow conditions.

Depending on the value of the Reynolds number, we can distinguish between:

a) **Planktonekton**, referring to small, fast organisms, and large organisms with relatively modest locomotor apparatus ($5 \times 10^3 < Re < 5 \times 10^5$); this group includes small fishes that have not yet developed the muscles that enable them to swim against currents once they become adults;

b) **Eunekton**, referring to the true nekton, consists of those species that live a fully nektonic existence for the duration of their life ($Re > 5 \times 10^5$). Morphologically, hydrodynamically efficient shapes characterize most eunekton, as illustrated by the fusiform appearance of tunas and dolphins.

In addition to reduction or loss of hind limbs, many of these species have adapted to the threat of predation with coloration that helps them escape from visual predators. Thus, a light-colored ventral side provides camouflage against the surface of the water when viewed from below and a dark dorsal side provides camouflage against the progressively darker water when viewed from above. Reduced scales in some species minimize frictional resistance as they move through water.

c) **nekto-benthos** or **bentho-nekton** refers to species such as flatfish that actively swim close to the bottom, without actually settling onto the seabed for extensive periods. These species live near the bottom where they usually feed, sometimes hiding in the sediment or in bottom crevices. Some species, especially fish and crustaceans, can be nekto-benthic during the day and strictly nektonic at night.

d) Xeronekton, as discussed earlier, refers to organisms that occasionally occupy the marine environment, such as reptiles, seabirds, and coastal mammals with swimming abilities.

13.3 Fishes

Fishes, a ubiquitous component of the marine community, all belong to the phylum Chordata (Table 13.1), subphylum Vertebrata, and are subdivided into Agnatha (an infraphylum containing jawless fish), Chondrichthyes (a class containing cartilaginous fish) and Osteichthyes (a superclass containing bony fishes). Fishes include more species than all other vertebrates combined (~58% of known species of vertebrates). Fishes continue to yield many new species and genera: the World Register of Marine Species (WoRMS) reported the discovery of 340 new species and 25 new genera of fish in 2011 alone. The Census of Marine Life, which ran from 2000 to 2010, coordinated biodiversity research around the world, reported the discovery of 400 new species within just the first two years of the project. Using submersibles and other tools, scientists discovered 30 new species of fishes near the Galapagos Islands, a relatively well-studied environment. *The Catalogue of Fishes* publishes an update on fish species every 6–8 weeks. The total number of known species of fishes (including the Agnatha, or jawless fishes) was 16,733 in 2012. Accounting for synonyms (multiple names applied to the same species), experts estimate that ~500 new fish species remain undescribed, with an estimated 200–300 cryptic species that molecular methods will soon resolve. Model extrapolations estimate that scientists have described only ~77% of fish species on Earth, noting that experts described 1,577 new species between 1999 and 2008 alone. Despite these discoveries, efforts to evaluate the conservation status of marine fishes lags far behind that of other classes of vertebrates. The IUCN (International Union for the Conservation of Nature) has assessed the global conservation status of all birds, mammals, and amphibians, but has examined only 6% of all fishes globally (including freshwater species).

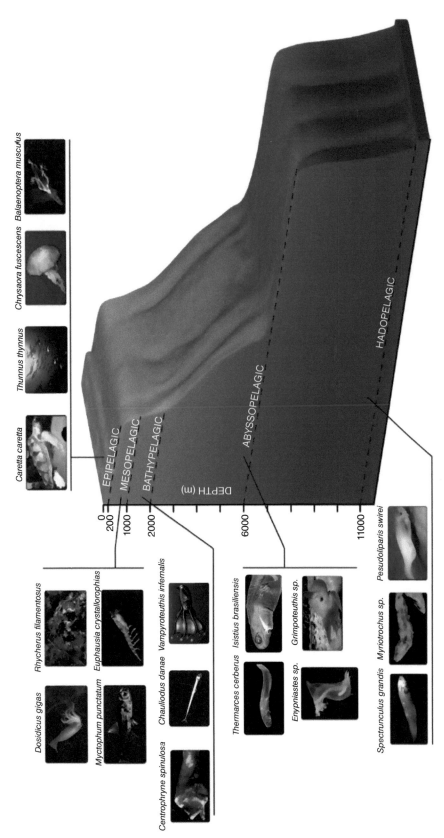

Figure 13.1 Examples of nektonic organisms, which include a wide variety of animal phyla representing diverse life cycle, feeding habits, and ecology but with a shared ability to move actively in the ocean. *Source:* NOAA/MBARI / Wikimedia Commons / Public domain and Scubagirl85 / Wikimedia Commons / CC BY-SA 3.0. Strobilomyces / Wikimedia Commons / CC BY-SA 3.0, Dan90266 / Wikimedia Commons / CC BY-SA 2.0, NOAA Photo Library / Wikimedia Commons / Public Domain, Chiswick / Wikimedia Commons / CC BY-SA 3.0, Uwe Kits / Wikimedia Commons / CC BY 3.0, Theodore W. Pietsch / Wikimedia Commons / CC BY 3.0, Citron / Wikimedia Commons / CC BY-SA 3.0, Karsten Hartel / Wikimedia Commons / Public Domain, National Oceanic and Atmospheric Administration (NOAA), Trang Nguyen / Wikimedia Commons / CC0 1.0, Dr Thom Linley and Professor Alan Jamieson.

Figure 13.2 Nekton includes phylogenetically diverse biota, including: (a) bony fishes, (b) elasmobranchs, (c) cetaceans, (d) mollusks, (e) reptiles, and (f) birds.

Table 13.1 Classification and biodiversity of fishes.

Category	Classification				
Kingdom	Animalia				
Phylum	Chordata				
Class	Myxini	Petromyzontida	Chondrichthyes	Actinopterygii	Sarcopterygii

13.3.1 Osteichthyes – The Bony Fishes

Osteichthyes contains 96% of living fishes and include the ray-finned fishes (class Actinopterygii) and the lobe-finned fishes (class Sarcopterygii). The class Actinopterygii includes the most species-rich group of living vertebrates: the modern bony fishes. Lungfishes and coelacanths represent the second class, the Sarcopterygii. Morphology and ecology form the primary basis for fish classification. A long, tapered body with fins spaced along the body enhances maneuverability in large, predatory bony fishes such as tunas and marlins. Upward-facing mouths designed to catch prey from the very surface characterize surface-feeding pelagic species. In contrast, bottom fishes encompass diverse body shapes. For example, the flattened bodies of plaice and sole bear some semblance to the dorso-ventrally flattened head and part of the body of some bentho-nektonic sharks. In contrast, modified pelvic fins in the order Scorpaeniformes (redfish, scorpaanids) allow them to sit on the bottom. The laterally flattened shape of many deep-sea fishes enables efficient swimming, albeit slowly. In contrast, the pelagic sunfish (*Mola mola*), one of the largest species of teleosts (bony fishes) may reach 3 m in length, but with a laterally compressed shape (Figure 13.3).

Swimming encompasses three functional components: (1) acceleration, maximized by strong caudal fin propulsion, allows rapid movements, such as escape predators, in contrast to; (2) cruising, which involves continuous undulating body movements; the skipjack tuna (*Katsuwonus pelamis*), with a very simple body shape, specializes in this type of swimming; 3) maneuverability, best illustrated by discoid-shaped fishes such as butterfly fishes on coral reefs, that allows the body to flex and suddenly change direction. Swimming ability and characteristics also depend on body shape (Figure 13.4).

The fins play an important role in locomotion. In fishes that must maneuver in tight spaces, the pectoral fins usually sit near the midline, but for fishes that move rapidly the fins sit more towards the ventral surface. Caudal fin shape also reflects

Figure 13.3 A sunfish (*Mola mola*) approaching the surface.

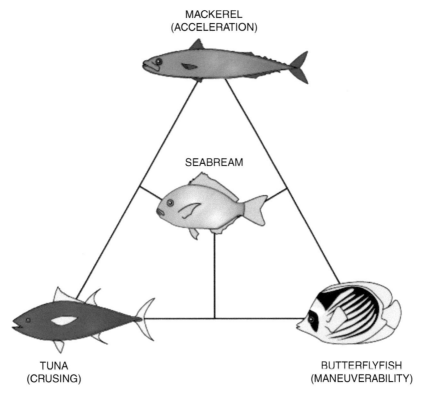

Figure 13.4 Body shape of the most common fishes.

fish swimming speed. A caudal fin shaped like a crescent moon characterizes fast swimmers such as tunas, connected to the body by a narrow peduncle that reduces frictional resistance. A forked tail occurs in actively swimming fishes such as different species of sharks that can increase speed rapidly. Although basking sharks swim slowly while feeding on zooplankton, their forked tail helps accelerate when actively swimming or leaping out of the water.

A well-developed sensory system characterizes most fishes: eyes, a lateral line system that senses water movement, and, in many species, apparatus for smelling and hearing. Visual development varies among fishes; in well-lit surface waters, many species perceive color, noting that some species use bright coloration as part of mating. The mechanoreceptors that form the lateral line system resemble that in cartilaginous fishes. The inner ear of fishes provides hearing and balance: otoliths in suspension encounter hairy fibers, providing information on orientation in space.

The 96% of all living fishes represented by teleosts likely demonstrate the greatest diversity of all vertebrates in eating behavior and diet. Their behavior and ecology, mode of locomotion, body shape, sensory physiology, and oral morphology reflect this variation. The diet of many fishes typically includes a subset of available prey within the environment. A series of largely behavioral filters eliminates many potential prey, and determines what an individual will eat (Figure 13.5).

Many fishes feed by suction, rapidly opening their mouth to suck water inside their oral cavity, dragging along prey at the same time. This versatile mechanism of feeding can capture fairly large and also smaller prey, which they filter using gill extensions (or gill lamellae). Large carnivorous fish can act as a sort of battering ram, moving forward through the water with their jaws open to "collide" directly with their prey. Some fishes, such as parrot fish and surgeon fish, feed by scraping algae off solid surfaces, or crushing shells or skeletons of prey such as sea urchins, filtering zooplankton (menhaden, anchovies, herring), scavenging carcasses as in bream, or actively capturing prey (tuna, mackerel). These different feeding modes require a wide diversity of mouthpart morphologies.

13.3.2 Chondricthyes – The Cartilaginous Fishes

The presence of a cartilaginous skeleton (aside from minor calcification in some shallow-water forms) defines the Chondrichthyes (sharks, skates, rays, and chimaeras). Chondrichthyes arose from gnathostomes (jawed fishes) and flourished particularly during the Devonian and Carboniferous. The Chondrichthyes are divided into the orders Holocephali (chimaeras, ratfishes) and Elasmobranchii (sharks, skates and rays). A long and thin tail, large pectoral fins, and non-toothed jaws characterize Holocephali. Their jaws bear large flat plates with the upper plate fused to the cranium. They live close to the bottom feeding on seaweeds, mollusks, echinoderms, crustaceans, and fishes.

Elasmobranchs contain two subdivisions: the Selachii (modern sharks) and Batoidea (rays, skates, and sawfish). The morphological and physiological characteristics of these organisms adapt them particularly well for predation. Teeth grow in parallel rows akin to a conveyor belt ready for rapid replacement, given that every attack brings tooth loss. The Selachii differ from bony fish in combining ventral mouth position with effective (and impressive!) teeth. The jaw is not fused with the skull and varies in shape, size, and orientation depending on feeding mode. Particularly strong jaw muscles ensure strong bite pressure. The jaw opens to different degrees to increase bite penetration, teeth rotation, and removal of prey body parts. Placoid scales (dermal denticles) comprise a critical part of the **integumentary system**, referring to the set of organs forming the outermost layer of an animal's body. This protective layer also decreases friction as the animal moves through the water and smooths forward movement. The body form optimizes forward thrust and speed.

The pectoral fins provide stability whereas thrusts of the caudal fin generate propulsion. The digestive system varies depending on feeding mode, but coiling of the intestine generally increases digestive capacity with pancreatic enzyme secretions similar to those in mammals. The large liver (about a quarter of total body mass) secretes squalene and other oils that increase buoyancy and provides an important energy reserve during fasting periods. The advanced sensory system has evolved a strong sense of smell that can detect scent molecules at extraordinarily high dilution levels. Thus, sharks can

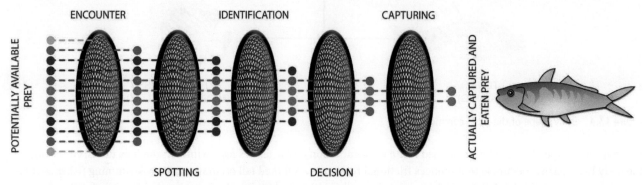

Figure 13.5 Representation of factors determining what an organism feeds on (left): encounter, hold, recognize, decide to attack, and then capture prey.

perceive blood drops at distances of several km aided by a connection between their epithelium and olfactory bulb that supports a remarkably sensitive chemoreceptor and olfactory system. Their sense of smell uses apical sensory cells with high sensitivity and specificity that allows them to identify conspecific individuals or other species. The visual system consists of eyes placed laterally on their head and equipped with lids or membranes (except in abyssal species). Species that live in environments with low light have retinas with rod to cone ratios of 1–100, facilitating perception of bioluminescence at depth. The lateral line is also part of the sensory system (Figure 13.6), and, in Elasmobranchii, the mechano-receptors sit within interconnected tubes and pores located along the sides of the body. The association of these receptors with other sense organs allows sharks to detect low-frequency vibrations of prey from long distances. The ear of sharks can sense different frequencies, even quite low ones, to distances of up to 250 m, suggesting that Elasmobranchii do not communicate with each other (dolphins communicate but use very high frequencies).

All animals produce electric fields, whose intensity and characteristics vary among species. Sharks and rays exploit this property to locate their prey. Sensitivity to weak electric fields in fishes represents a primitive character that occurred in groups now extinct: the first agnates (jawless fishes), the acanthodians (jawed fishes), and the Chondrostei. Of living groups, this sensitivity persists only in Selachii, in Dipnoi, in lampreys, in *Latimeria* spp. and in a few other species of teleosts. The electrosensory system of Selachii consists of the **ampoules of Lorenzini** (specialized tubules innervated with nerve endings in sharks), an extension of the lateral line system. However, rather than sensing vibration this system detects weak electric fields and responds to local reversals of polarity. Electroreception forms the basis of prey identification as demonstrated in aquarium experiments in which dogfish (*Scyliorhinus canicula*) identified electrodes hidden under the sand. Importantly, electroreception also links to orientation and movement. Field studies on hammerhead shark, *Sphyrna lewini*, demonstrated its ability to follow "electromagnetic tracks" along the seabed to move between feeding and social networking sites.

Elasmobranchs are almost exclusively marine organisms, but some forms of tiger shark use estuaries, and some others live in lakes, but few freshwater species exist. They produce urea which also occurs in high concentrations in the blood, giving shark meat its characteristic ammonia taste. The copulatory organ consists of modified pelvic fins, called **pterigopods**, which are long, flexible structures with partly calcified tips (depending on sexual maturity). Reproduction is sexual and takes place through internal fertilization. Sperm delivered in aggregates reduces dilution and dispersal in the water and maximizes fertilization after copulation. We lack detailed information on reproduction for most shark species, although microsatellite DNA analysis has revealed some information of kinships.

Elasmobranchs include **oviparous** (eggs hatch after parents lay them), **ovoviviparous** (eggs hatch inside the parent), and **viviparous** (live birth of young that developed inside the parent) species. Viviparity may be placental, where eggs remain in the oviducts until embryos develop sufficiently from nourishment from the placenta; ~10% of known species use this strategy).

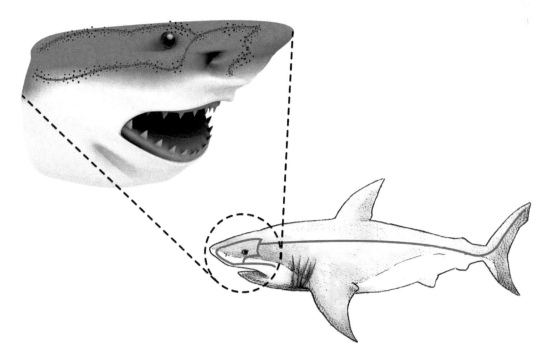

Figure 13.6 Sensory organs in sharks, showing: (a) lateral line and (b) detail of nose with pores of ampullae of Lorenzini.

In aplacental species, eggs remain in the oviducts until embryos develop, feeding through the yolk sac. In some species, nourishment comes from feeding on other eggs produced by the mother (**oophagia**) or other embryos present in the oviduct (**adelphophagy**). The long gestation period in elasmobranchs can last up to two years, increasing population vulnerability. Better management of this resource requires knowledge of rates of reproduction and age data for different shark species.

13.3.3 Agnatha – The Jawless Fishes

Agnathans (superclass Agnatha), the primitive, jawless fishes, include the lampreys (order Petromyzontiformes), hagfishes (order Myxiniformes), and several extinct groups. Hagfishes and lampreys, because of their parasitic nature, can play a significant ecological role. The group is of great evolutionary interest because it includes the oldest known **craniate** (skulled) fossils and because living agnathans retain many primitive characteristics. Hagfish bodies are soft-skinned, scaleless, and nearly cylindrical, with a single nostril at the anterior end, overlying the mouth, and a low caudal fin around the tail. Skin covers the vestigial eyes. All ~70 known species are restricted to cold, marine bottom waters at depths ranging from 10 m in high latitudes to >1,300 m in equatorial oceans. Adults are 40–80 cm long. All species are superficially similar except in the number and position of the gill apertures. Hagfishes locate their food by scent, scavenging and preying upon annelids, mollusks, crustaceans, and dead fishes. The best-studied species, *Myxine glutinosa*, normally feeds on soft-bodied invertebrates and larger dead animals. *M. glutinosa* burrows into soft marine sediments with only part of their head visible. When stimulated by the scent of dead fish, *M. glutinosa* leaves its burrow and swims against the current. To eat the fish, *M. glutinosa* coils around it and bites into it by protruding and retracting the comb-like horny tooth plates on the floor of its mouth. A row of prominent glands long each side of their body produces a gelatinous slime as a defence mechanism to escape predation. Recently, researchers proved that the slime can clog fish gills, forcing potential predatory fish to release the hagfish to avoid suffocation.

Lampreys, from Latin *lampetra*, "stone licker" (*lambere* "to lick," and *petra* "stone"), which number about 43 species and occur in cool, fresh, and coastal waters of all continents except Africa. The smooth, scaleless, and eel-shaped body has well-developed dorsal and caudal fins; a suctorial oral disk bearing horny teeth surrounds the mouth. The eyes are well developed, and the single nostril sits atop the head. Lampreys possess seven pairs of external gills. Adults range 15–100 cm long. Although hagfish gonads usually include both ovary and testis, there is no evidence of either hermaphroditism or self-fertilization. Marine lampreys are anadromous, so adults leave the ocean and ascend freshwater streams to reproduce. Females produce tough-skinned yolk-filled eggs that are immediately fertilized after spawning and quickly covered with sand. After two weeks the eggs hatch and the larval lampreys, called **ammocoetes**, leave the nest and drift downstream to burrow into soft sediments where they live as slow-growing suspension feeders for 3–7 years, depending on species. After that period, the ammocoetes metamorphose into adults. Metamorphosis can last 3–4 months, depending on species. During this process they lose their eyes, replace the hood with an oral disc and keratinized teeth, enlarge their fins, modify their gill openings, and develop gonads. Adults of marine species migrate to the ocean where they live for up to four years before migrating to freshwater for reproduction.

Marine lampreys are parasitic/carnivores that attach their sucker-like mouth to large fish, feeding on their host's body fluids. Using their sharp teeth to rasp through the fish's flesh, they obtain liquid nourishment. Some species eat fishes captured by fishing nets and scavenge on cadavers of large vertebrates. Body form varies considerably among extinct agnathans (Figure 13.7).

13.4 Marine Mammals

The first recognizable marine mammal was a cetacean that dates back 50 million years, when amphibious forms of the (now extinct) order Archaeocetes began the move into the ocean that would eventually produce the Odontoceti and Mysticeti. Ancestral whales lived in arid environments and in rivers, only occasionally diving into the water; many were predominantly terrestrial and thus comparatively poor swimmers. The genus *Squalodon*, also now extinct, lived 22 million years ago and resembled modern dolphins. The genus *Basilosaurus*, or "lizard king" was completely aquatic and lived in tropical and subtropical environments, reaching 25 m in length. Their smaller relative *Dorudon* sp. resembled modern dolphins whereas *Cetotedium* sp., which occurred million years ago, resembled modern baleen whales.

Marine mammals form three orders: Cetacea, Sirenia, and Carnivora (including the suborder Pinnipedia, family Mustelidae, and family Ursidae). They include seals, sea lions, walruses, manatees and dugongs (represented by very few species), and cetaceans (which include numerous species). Other mammals, such as some bears and otters, though lacking substantial adaptation to the marine environment, depend on the sea for food. Marine mammals include all mammals fully adapted to the marine environment (Table 13.2).

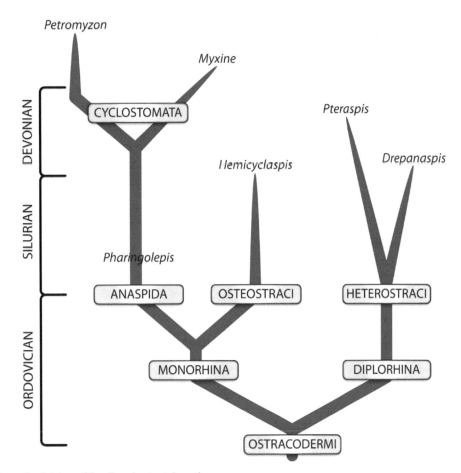

Figure 13.7 Systematic division of fossil and extant Agnatha.

Table 13.2 Classification and biodiversity of marine mammals.

Categories	Classification		
Kingdom	Animalia		
Phylum	Chordata		
Class	Mammalia		
Order	Cetartiodactyla	Sirenia	Carnivora

At present there are ~125 marine mammal species, among the 6,513 species of mammals on Earth (~2.1%; source ASM Mammal Diversity Database). These species inhabit a diverse range of habitats from riverine, brackish, mangrove, and estuarine habitats, to coastal shallows and pelagic seas, with some even foraging at the edge of the abyssal plains. They also use a diverse range of food items, from seagrass or zooplankton, through to fish, penguins, and other marine mammals. Their diverse niches result in a wide range of behaviors.

All marine mammals are viviparous, with mothers feeding their milk to their offspring. As warm-blooded organisms, they maintain a constant body temperature, which increases energy expenditure, and thus the need for large amounts of food.

Living in the ocean requires that marine mammals develop significant adaptations for diving. Marine mammals can exchange up to 90% of the air in their lungs with each breath (humans, by comparison, can exchange up to 20%), and their blood has higher oxygen-carrying capacity. They can decrease their heart rate during dives and control blood flow, reducing supply to non-essential areas and concentrating blood and oxygen flow to vital organs such as the brain. This adaptation further reduces energy expenditure and oxygen consumption. During dives, the excessive pressure of the surrounding

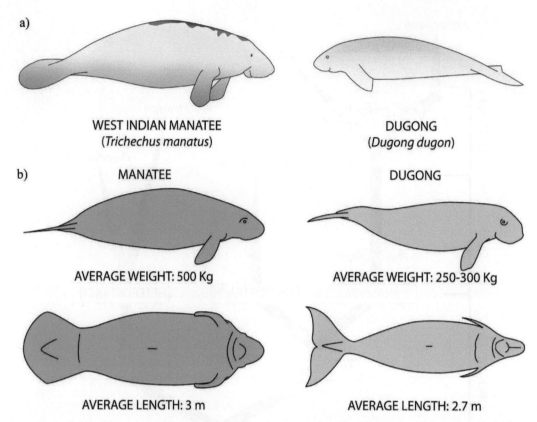

Figure 13.8 (a) Simplified diagrams of sirenians, illustrating size differences between manatees and dugongs; (b) Comparison of dugongs and manatees illustrating some key morphological differences. Note paddle-like tail of manatees relative to dolphin-like tail of dugongs.

water collapses their lungs, reducing air contact with their blood vessels. This effect also greatly reduces dissolved nitrogen in their blood, which might otherwise cause embolisms if nitrogen bubbles expanded during ascent.

Cetacea – The order Cetacea comprises two extant sub-orders, Mysticeti (baleen whales) and Odontoceti (toothed whales), and an extinct sub-order. Both Mysticetes and Odontocetes are thought to be descendants of Archaeoceti, the extinct suborder of ancient whales. The 103 known living species of cetaceans include 46 genera in 14 families. Cetaceans inhabit all of the world's oceans. Some species, such as killer whales (*Orcinus orca*) occur in all of the world's oceans, whereas others are limited to one hemisphere (Antarctic minke whales) or a single ocean (Pacific white-sided dolphins). These large animals range in size from 20 kg–180 metric tons and from 1.2 to >30 m. Blue whales (*Balaenoptera musculus*) are the largest animals that have ever existed.

Sirenians (Figure 13.8a) include only four species of marine herbivores, all limited to the tropics. Sirenians and cetaceans have almost completely transitioned from terrestrial to marine life. Phylogenetically, Sirenia most closely link to elephants, and historically they included many more species than the four that remain today (Figure 13.8b).

Carnivora – Fissipeds, the "split-footed" members of the order Carnivora, align more closely with terrestrial carnivores such as weasels, than seals or whales. Evolutionary newcomers to the marine environment, these species lack many of the physiological adaptations to marine life seen in pinnipeds and cetaceans. Both species are considered marine mammals because of the roles they play in the marine environment, but they spend most of their time on land and only part of their time in the ocean, mainly to hunt for food.

The Fissipedia include sea otters, of the weasel family (Mustelidae), which live a primarily marine life: they rest, mate, give birth, and suckle their young in the water. Their webbed hind limbs facilitate swimming, and contrast their padded front paws with separate, clawed digits. They lack blubber, but air trapped in their thick fur, the densest fur among all mammals, provides insulation. Mustelidae and Ursidae are phylogenetically close to Felidae (cats) and Canidae (dogs), respectively. Polar bears, of the bear family (Ursidae), spend most of their lives associated with marine ice and waters. Although competent swimmers, they are the marine mammal least adapted to aquatic existence. They rest, mate, give birth, and suckle their young on the ice and, as such, are vulnerable to reduced extent and duration of sea ice.

13.5 Cephalopods

Cephalopods belong to the phylum Mollusca and include squid, cuttlefish, nautilus, and octopus; they also include some of the largest invertebrates (Table 13.3). The giant squid *Architeuthis dux*, a deep-ocean species of the family Architeuthidae, may exceed 18 m in length, including its two long tentacles (Figure 13.9). It can grow to a tremendous size, illustrating deep-sea gigantism. Only the colossal squid *Mesonychoteuthis hamiltoni* has a similar length (~13 m). The mantle of the giant squid measures about 2 m long (more for females, less for males), and squid length excluding its tentacles (but including head and arms) rarely exceeds 5 m. Scientists have not documented claims of specimens measuring >20 m. Cephalopods are exclusively marine and stenohaline, rarely living at salinities <17.5. One key exception, the Gulf of Mexico squid *Lolliguncula brevis*, tolerates salinities as low as 16. Cephalopods occur in most marine habitats.

Ink glands in squids and octopus distract predators. The mantle of these species ejects a volume of water through a siphon, which they can point in all directions as they expel water to provide propulsion. The well-organized nervous system in many cephalopods includes well-developed eyes with similarities to vertebrate eyes. Their nervous system supports skilled motility. Known cephalopod species currently number slightly <1,000, with molecular methods greatly aiding taxonomy over the last 25 years.

Squids (order Teuthida) encompass a large and diverse group of families that vary greatly in habitat use and spatial distribution. The two extant suborders include commercially important species. The order Myopsida mostly comprises squids of the genus *Loligo* that typically occupy coastal environments and may vary in number greatly seasonally. The order Oegopsida encompasses many families, including some almost completely unknown species that occupy abyssal environments.

Table 13.3 Classification and biodiversity of Cephalopods.

Categories	Classification
Kingdom	Animalia
Phylum	Mollusca
Class	Cephalopoda

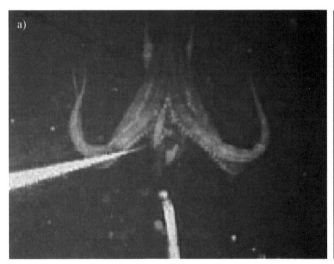 

Figure 13.9 Giant squid: (a) *Architeuthis* sp. feeding on bait; (b) a tentacle near the hull of a boat. *Source:* Associated Press / Alamy Stock Photo.

13.6 Reptiles – The "Land-based" Marine Species

Marine reptiles (Table 13.4), though biologically and ecologically interesting, contribute minimally to marine biodiversity totaling only ~110 known species to date. With some 60 known species, sea snakes dominate this group in terms of diversity, followed by turtles (8 species), crocodiles (2 species) and only 1 species of marine iguana (*Amblyrhynchus cristatus*). The Order Testudines includes 3 suborders of turtles, the Antichelidyiae (extinct), the Pleurodira (tortoises, or species that retract the neck horizontally in an S-fold), the Cryptodira (species that retract their neck into the shell with a vertical fold). This latter group includes all living species of sea turtles. The family Chelonidae includes species covered by horny plates (scutes), whereas the family Dermochelyidae lack bony plates, and thick skin covers the shell. Figure 13.10 illustrates extant marine species.

13.7 Seabirds

Seabirds (Table 13.5) include four major groups (Figure 13.10):

- Penguins (Figure 13.11a) have lost the ability to fly by evolving wings better adapted to swimming. They live within the Antarctic and sub-Antarctic cold waters and at mid-latitudes (near Central America) in colonies that vary in size from a few pairs to hundreds of individuals. They swim near the surface preying on small fishes, utilizing a layer of fat and a top layer of feathers to insulate them from the cold. Protection from the cold influences other aspects of their biology; royal penguins, for example, cluster together in colonies to minimize heat loss.

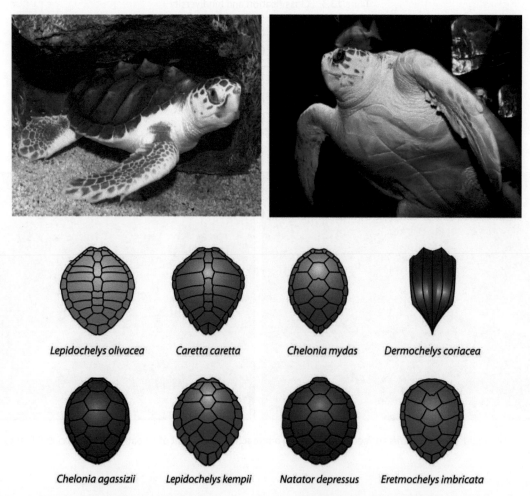

Figure 13.10 (a) Schematic representation of distinctive characteristics of the 8 known species of sea turtles (top image), along with images of *Caretta caretta* (bottom). *Source:* Brian Gratwicke / Wikimedia Commons / CC BY 2.0, Becky & Tim Gregory / B+T Photography.

Figure 13.11 The four main groups of marine birds include: (a) Antarctic penguins, (b) petrels, (c) pelicans, and (d) gulls. This highly heterogeneous group includes individuals that have completely lost the ability to fly and others that can travel great distances. They all share the common trait of close interaction with the marine environment as their primary food source.

- Petrels and albatrosses (Figure 13.11b) use large external nostrils to help sense their prey. The massive 3-m wingspan of albatrosses makes them excellent gliders capable of long foraging ranges and migrations to breeding colonies.
- Pelicans (Figure 13.11c) include some of the largest and heaviest species of seabirds, some with bright colors and ornamentation. Though mostly tropical, some species nest in the Arctic and Antarctic, and others, such as frigates, stay relatively close to land with brief forays to the ocean.
- Gulls (Figure 13.11d), terns, and auks include the most varied seabirds by far. They live in colonies with hundreds to thousands of individuals and feed primarily on small fish and zooplankton.

13.7 Patterns of Biodiversity in Nekton

Many marine taxa, including teleost and elasmobranchs, exhibit non-linear patterns of species richness, with biodiversity generally decreasing from the equator to the poles but peaking at intermediate depths (Figure 13.12). High diversity of benthic fishes at mid-bathyal depths in the North-West Atlantic contrasts the higher diversity of pelagic fish in surface waters, including numerous species limited to that depth range (Figure 13.13), with slightly higher numbers from 100–200 m compared to the very surface layer. This pattern likely reflects the influence of the environment on depth distributions of pelagic fishes (Table 13.6).

Many factors affect species distribution at different depths, including intense predation in well-lit environments or the need to adapt physiologically to high or variable temperatures and high light levels in surface waters. The apparent inverse relationship between pelagic fish species diversity and depth let to a hypothesis that low temperature and food limitation play an important role in explaining species richness and distribution (Figure 13.14). Pelagic fish abundance and plankton biomass decreases exponentially with increasing depth, before stabilizing at abyssal depths (4,000–6,000 m), a pattern also generally seen in temperature.

A more careful inspection suggests that species diversity remains relatively constant from the surface to 300 m, whereas temperature and plankton biomass decrease more rapidly. This pattern of distribution in fish depth distribution contradicts the hypothesis that high pelagic fish diversity correlates with habitat size given that deep environments are the most extensive in the world. Thus, the causes of diversity patterns in pelagic fishes with depth remain unclear.

Table 13.4 Classification and biodiversity of reptiles.

Categories	Classification		
Kingdom	Animalia		
Phylum	Chordata		
Class	Reptilia		
Order	Crocodilia	Testudines	Squamata

Table 13.5 Classification and biodiversity of seabirds.

Categories	Classification
Kingdom	Animalia
Phylum	Chordata
Class	Aves

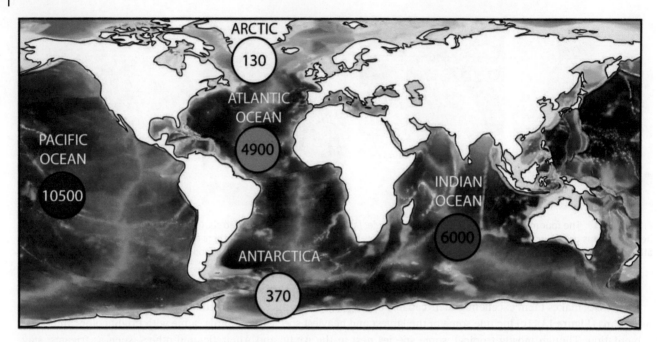

Figure 13.12 Distribution of numbers of fish species in different ocean regions. Numbers of species of fish decreases moving from the equatorial belt to the poles, illustrating differences in species number in different oceans, from richest (Pacific) to poorest (Arctic).

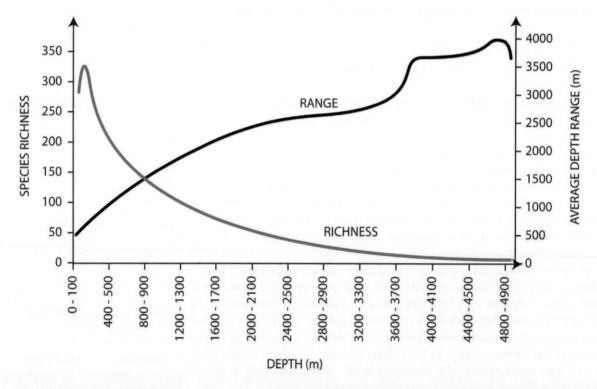

Figure 13.13 Species richness and average depth range of pelagic fishes plotted as a function of depth, suggesting pelagic fish ranges increase along a depth gradient.

Teleosts and elasmobranchs differ in several important ways. With the evolution of internal fertilization, elasmobranchs have developed a reproductive strategy that generally (except in oviparous, or egg bearing species) favors the production of numerically few small offspring, which the maternal body protects and nourishes internally for a period of time. This strategy requires significant investment of material and energy for each offspring. Because resource availability decreases with depth, this type of investment in offspring becomes more and more "expensive". Selection therefore favors few offspring in food-limited environments such as the deep sea.

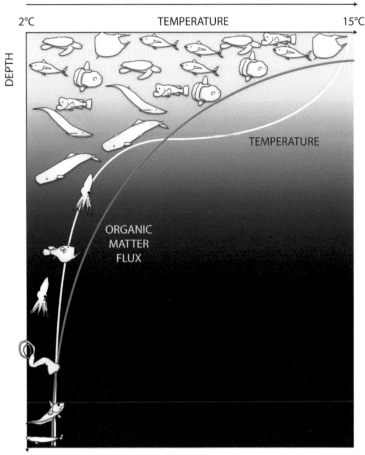

Figure 13.14 Changes in biodiversity and composition of taxa with increasing depth. Note the association with decreasing temperature and organic carbon flux to the bottom (orange line) with ocean depth.

Table 13.6 Number of marine pelagic fish in different bathymetric zones between 40°N and 50°S in the North East Pacific.

Zone	Depth (m)	Species number
Euphotic	0–200	69
Euphotic-Mesopelagic	0–1,000	180
Euphotic-Mesopelagic-Bathypelagic	0–3000/4000	52
Euphotic-Mesopelagic-Bathypelagic-Abyssal	0–6,000	4
Mesopelagic	200–1,000	30
Mesopelagic-Bathypelagic	200–3,000/4,000	49
Mesopelagic-Bathypelagic- Abyssal	200–6,000	5
Bathypelagic	1,000–3,000/4,000	12
Bathypelagic-Abyssal	1,000–6,000	6
Abyssal	3,000/4,000–6,000	2

Several general patterns characterize fish distribution:

1) diversity decreases rapidly with depth, peaking at ~200 m and declining to very low values at abyssal depths;
2) pelagic fish species depth distributions reflect available habitat, and ranges thus increase with depth;
3) depth distributions in pelagic species correlate with food availability and temperature;
4) the relationship between body size and depth apparently differs between teleost and elasmobranchs.

Questions

1) Discuss the classification of nekton.
2) Differentiate between nektobenthos and benthos.
3) Define xeronekton.
4) Which are the differences between Teleostei and Agnatha?
5) How many extant species of marine mammals have been censused?
6) Which mollusks belong to the nekton?
7) How many known species of marine reptiles are there?
8) What are the main groups of seabirds?
9) Which are the differences between teleosts and elasmobranchs?
10) Describe the general patterns of fish distribution.

Suggested Reading

Arai, T., Limbong, D., Otake, T. and Tsukamoto, K. (1999) Metamorphosis and inshore migration of tropical eels Anguilla spp in the Indo-Pacific. *Marine Ecology Progress Series*, 182, pp. 283–293.

BirdLife International (2021) *The BirdLife Checklist of the Birds of the World, with Conservation Status and Taxonomic Sources*. Version 3. www.birdlife.org.

Brodeur, R.D. and Pakhomov, E.A. (2019) Nekton. In Cochran, J.K., Bokuneiwicz, and Yager, P.L. (eds.), *Encyclopedia of Ocean Sciences*. 3rd Edition. Academic Press, pp. 582–587.

Clarke, M.R. (1996) The role of cephalopods in the world's oceans: General conclusions and the future. *Philosophical Transactions of the Royal Society B: Biological Sciences*, 351, pp. 1105–1112.

Grady, J.M., Maitner, B.S., Winter, A.S. et al. (2019) Metabolic asymmetry and the global diversity of marine predators. *Science*, 363(6425), p. eaat4220.

Klimley, A.P. (2013) *The Biology of Sharks and Rays*. University of Chicago Press.

Levi, C., Stone, G., Schubel, J.R. (1999) Censuing non-fish nekton. *Oceanography*, 12, pp. 15–18.

Matkin, C.O., Ellis, G., Olesuk, P. and Saulitis, E. (1999) Association patterns and inferred genealogies of resident killer whales *Orcinus orca*, in Prince William Sound, Alaska. *Fishery Bulletin*, 97, pp. 900–919.

Menegotto, A. and Rangel, T.F. (2018) Mapping knowledge gaps in marine diversity reveals a latitudinal gradient of missing species richness. *Nature Communications*, 9, p. 4713.

Mora, C., Tittensor, D.P. and Myers, R.A. (2008) The completeness of taxonomic inventories for describing the global diversity and distribution of marine fishes. *Proceedings of the Royal Society B: Biological Sciences*, 275, pp. 149–155.

Rasmussen, A.R., Murphy, J.C., Ompi, M. et al. (2011) Marine reptiles. *PLOS One*, 6(11), p. e27373.

Rosa, R., Dierssen, H.M., Gonzalez, L. et al. (2008) Ecological biogeography of cephalopod molluscs in the Atlantic Ocean: Historical and contemporary causes of coastal diversity patterns. *Global Ecology and Biogeography*, 17, pp. 600–610.

Schipper, J., Chanson, J.S., Chiozza, F. et al. (2008) The status of the world's land and marine mammals: Diversity, threat, and knowledge. *Science*, 322, pp. 225–230.

Whalen, C.D. and Briggs, D.E.G. (2018) The Palaeozoic colonization of the water column and the rise of global nekton. *Proceedings of the Royal Society B: Biological Sciences*, 285(1883), p. 20180883.

14

Ecology of the Nekton

14.1 Introduction

The nekton encompass a wide range of taxa from cephalopods to fishes to seabirds to marine mammals, and each of these taxa have developed solutions to living in their specific worlds, whether that means fully submerged to species that move from land, air, or ice, into marine habitats. These adaptations include behavioral, physiological, morphological and trophic/ecological specializations that all merit discussion.

14.2 Fishes and Formation of Fish Shoals

Many fishes display gregariousness behavior in that pelagic species, and particularly nekton, gather and move together in the same direction (Figure 14.1). These gatherings may be temporary or persistent. A shoal made up of individuals of different sizes cannot persist for long periods because smaller individuals cannot swim as quickly as large individuals. Much as cyclists draft behind one another during races, fishes in the front of a school must work harder, necessitating a continuous exchange of individuals between the inside and the outside of the shoal. Fishes may move in schools or in shoals. **School** refers to a group of fish that swim in synchrony at the same speed and in the same direction. Schools maintain a constant and regular NND (**Nearest Neighbor Distance**), the minimum distance between two "points" that are simultaneously occupied), meaning that individuals maintain the same distance from adjacent fish.

In a **shoal**, however, fish orient randomly in a group with variable NND. Shoals of fish in motion almost always form schools. Many fish that form schools tend to maintain a NDD 0.5–1 time the length of their body. Only fish in the front and sides of the shoal see the "external world," whereas others see only their neighbors. The shoal can change direction and speed rapidly when it perceives a threat, moving like a single individual as they use their senses to "maintain" the shoal. Visual stimuli help them recognize and relate to other members of the group. Fish recognize the markings, spots, or stripes that resemble those on their body and can see and identify colors and pattern. Even small changes in orientation and frequency of colors and patterns induce changes in fish movement. Changes perceived by neighbor fish allow them to maintain a NND or move in the same direction as others.

In addition to visual stimuli, some fishes can sense pressure, using their lateral line to sense pressure waves caused by fish motion. The sense organs react quickly, transmitting information rapidly across the shoal, ensuring synchronized movement. The sensitivity of the lateral line to pressure waves explains why some fishes react to tapping on the glass of an aquarium.

Whether or not fish choose to form shoals or not depends on several factors. First, fish form shoals to protect themselves from predators. Imagine two shoals, one containing 1,000 fish and another with 10 fish; the corresponding probability of being eaten increases from 1 in 1,000 in the large shoal to 1 in 10 in the small shoal. Shoal formation also helps avoid predators. Shoaling fish use their senses to perceive not only movements of other members of the shoal, but also to sense predators. The number of fish within the shoal obviously increases the number of "sensors" that simultaneously work to detect predators. Indeed, the shoal can act like one big fish, whereby the rotation and synchronized movement in different directions inside the shoal can confuse an attacker. Presumably the predator takes more time to choose an individual prey item as shoal size increases, thus increasing escape time. However, shoal formation may increase the likelihood that a predator

Marine Biology: Comparative Ecology of Planet Ocean, First Edition. Roberto Danovaro and Paul Snelgrove.
© 2024 John Wiley & Sons Ltd. Published 2024 by John Wiley & Sons Ltd.
Companion Website: www.wiley.com/go/danovaro/marinebiology

Figure 14.1 Example of a school of fish. Individuals creating the school move in the same direction with the same spatial orientation. *Source:* Unknown author / Wikimedia Commons / Public Domain.

may notice the fish. Indeed, many predators "associate" with their prey, such as barracuda that often accompany shoals of pelagic fish, their natural prey. Shoals in shallow water also attract aerial predators (e.g., seabirds) by increasing visibility from above. Shoal members also use interesting defensive tactics. Predators recognize that choosing and capturing an individual is easier than capturing an individual from the group. Therefore, remaining inside the shoal is in the best interest of the fish for all the reasons listed above (reduced probability of capture, better warning system). During an attack, the shoals communicate a defined number of maneuvers to escape. Some individuals escape and make a U-turn on both sides of the aggressor to create confusion. In other species, the shoals divide into smaller shoals, forcing the predator to choose one group to attack. Other species employ both tactics (visual confusion and forced choice); the shoal literally explodes in every direction. This evasion technique has one disadvantage; if the shoal fails to reform itself quickly, the probability of capture increases for scattered individuals.

Fishes in the open ocean often adapt a different tactic, whereby they form shoals in which individuals remain very close to one other. They can also form spherical groups to minimize the number of individuals exposed to predation (Figure 14.1). Predators also gather in shoals (or schools) to attack; a simultaneous attack by a group of predators breaks up shoals more effectively, facilitating capture of individuals.

Unfortunately, shoals limit availability of resources, leading to stronger competition for food. Research shows that well-fed fish generally associate with larger shoals, whereas hungrier fish associate with smaller shoals with weaker competition. The largest shoals can prioritize predator avoidance whereas the smallest shoals must sacrifice the security of a large group to meet their food requirements. Shoal formation brings other benefits such as finding food in less time, because more individuals are searching, and shoaling helps conserve energy just as it does for flocks of birds. Fish at the back of the shoal follow in the wake of the fish leading the shoal, reducing frictional resistance. Fish may also benefit from more precise navigation. A shoal that lacks a single leader may increase migration accuracy by averaging the group. Despite a general assumption that only conspecific fish (fish of the same species) form shoals and, although most shoals tend to be monospecific, multi-species shoals often form to defend themselves from predators. In some cases, this multi-species shoaling may produce a feeding benefit in that adding species could diversity dietary preferences and needs.

14.3 Ecology of Chondrichthyes

The Chondrichthyes include many large species (i.e., great white shark, Figure 14.2). The widespread commercially fished species *Centroscymnus coelolepis*, a deep-sea shark (also known as Portuguese dogfish), occur in all the oceans, typically at depths >800 m. As the deepest living shark known to date, it occurs at depths up to 3,750 m. These top predators use large nostrils and jade-colored eyes to scour the seabed in search of crustaceans and other prey such as cephalopods and decapods. In the Mediterranean, dwarf forms of these sharks occur with much smaller (~17 cm) juveniles than those in Atlantic conspecifics (generally >30 cm). The Mediterranean form likely reflects differences in food availability compared to larger individuals in the Atlantic.

Elasmobranchs are top predators par excellence but many feed on diverse food sources (Figure 14.3). The basking shark, *Cetorhinus maximus*, for example, feeds by using its mouth as a sort of plankton net that retains smaller organisms by filtering ~1000 metric tons of water through its gills in one hour while swimming at a speed of two knots. Similarly, rays capture plankton that swim vertically from the bottom to the surface. Some species of elasmobranchs feed on large pelagic species such as octopus, marlin, and other sharks (cannibalism often occurs). Rays feed on benthic crustaceans, but studies on shark stomach contents demonstrate that they can ingest almost anything, including surfboards, books, armor, and other objects.

Different species use different predatory strategies (such as attacking their prey as individual predators or in a group); great white sharks attack their prey from behind, relying on surprise. For example, in South African waters they jump out of the water to capture seals. Some sharks enter "feeding frenzies" in which some animals bite each other as they frantically feed.

Figure 14.2 Great white shark, *Carcharodon carcharias*. *Source:* Terry Gross / Wikimedia Commons / CC BY 2.5.

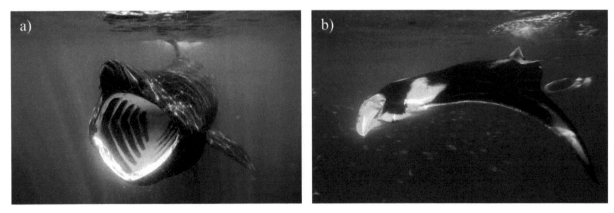

Figure 14.3 Large pelagic elasmobranchs: (a) *Cetorhinus maximus*, a planktotrophic shark feeding with its mouth open; this species occupies a wide range of boreal and warm waters, (b) a manta ray on Fiji's Astrolabe Reef. *Source:* jidanchaomian / Flickr / CC BY-SA 2.0.

14.4 Sharks at Risk of Extinction from Indiscriminate Hunting

14.4.1 Measuring Abundance of Sharks

Humans, the single greatest predators on Earth, kill an estimated 30–100 million sharks annually (~50% as bycatch), whether for food (shark fin soup in China), for pharmacological products (many para-pharmaceuticals use squalene), for commercial products (use of skin for bags and belts) or largely recreational (sport fishing) purposes. Uses can vary; fishers cut off fins and release the immobilized shark into the sea alive, then dry the fins to fetch market prices of 100 euros kg^{-1}. Fishing nets deployed for other species often entangle and kill sharks. In future, surgeons could use shark corneas for human transplants.

Biologists and engineers are developing low-frequency acoustic systems to keep sharks and dolphins away from nets, but this approach risks a boomerang effect: sharks and dolphins may associate the specific wavelengths with the presence of nets loaded with fish and therefore approach rather than avoid them. According to the FAO (Food and Agriculture Organization of the United Nations), in 2004 alone EU countries recorded shark catches (including rays and chimaeras) of 109,121 metric tons (t), including 51,000 t by Spain, (46% of the total EU), followed by France at 21,300 t (19.5%), United Kingdom with 16,000 t (14.6%) and Portugal at 7,200 t (6.5%). In recent decades, shark catches in Europe have declined a precipitous 66% despite unchanged fishing effort, indicating major population decline. The exploitation of sharks >2 m in length has intensified in recent decades, largely in response to increased demand for shark fins and meat, as well as by-catch. Given the role of sharks as apex predators in many marine ecosystems, this decrease has inevitable repercussions for the entire food web.

14.4.2 Shark Attacks – How Often Do Sharks Attack Humans?

Despite widespread fears, shark attacks on humans rarely occur (Figure 14.4a), noting that the last fatal attack in the Mediterranean occurred in 1989 in the Tyrrhenian Sea), and few attacks prove fatal (Figure 14.4b). In Australian waters, 25 attacks occurred 1974–1998, and sometimes death resulted from bleeding from victims' arms and legs, many of whom were surfers. The attacks depend on the activity taking place in the water, the area, and the time of the day, and their severity depends on whether damage to vital organs occurs. Often, deaths result from infections associated with microorganisms present in shark mouths.

14.5 Ecology of Cephalopods

Cephalopods occur at depths 0–5,000 m and many oceanic species migrate vertically daily, moving from depths of 1,000–4,000 m during the day to depths as shallow as 200 m during the night. Cephalopod abundances vary depending on group, habitat, and season, from isolated individuals such as octopus, to small schools of a few dozen individuals to millions of squid that form large oceanic shoals.

In addition to major differences in anatomy and physiology, cephalopods differ from other nektonic organisms in their biology and ecology. Cephalopod populations fluctuate greatly in abundance for two main reasons: (1) they reach maturity in one or two years, preventing overlapping generations, (2) their migratory patterns make them particularly susceptible to changes in oceanography. Nonetheless, their ability to vary their growth rates, extend their breeding season, and vary spawning depth "territories," give them remarkable resilience to fluctuating conditions. As active predators, cephalopods hunt using versatile and effective strategies. Despite high metabolic activity, they cope with seasonal fluctuations in available food using specialized structures adapted to facilitate prey capture and handling, and absorb sufficient nutrition to support their growth. As opportunists, they hunt, ingest, and digest animals of many phyla. Their unique feeding apparatus uses arms with adhesive structures, offering a major advantage compared to fish that capture food using only their mouths. Cephalopod populations respond quickly to natural and anthropogenic environmental changes in marine ecosystems.

Cephalopods display a remarkable ability to alter their coloration; using a combination of three types of colored cells, cuttlefish and squids can dramatically change color almost instantaneously. Typically, they change color to mimic their background in order to camouflage themselves from visual predators.

Genera such as *Todarodes* and *Illex* (family Ommastrephidae) characterize locations with strong currents and upwelling. The Ommastrephidae migrate long distances, releasing large masses of neutrally buoyant eggs, completing their life cycle in open ocean waters. As active predators, squids catch fish and crustaceans both in the water column and on the seabed. This flexible feeding allows large populations to spread beyond the continental shelf during annual migrations.

Nautilus spp. (order Nautilida) fill the chambers of their shell with gas and water in varying proportions to stabilize their body at a specific depth. The few species of *Nautilus* spp. occur in the Indo-Pacific region, typically in deep water outside

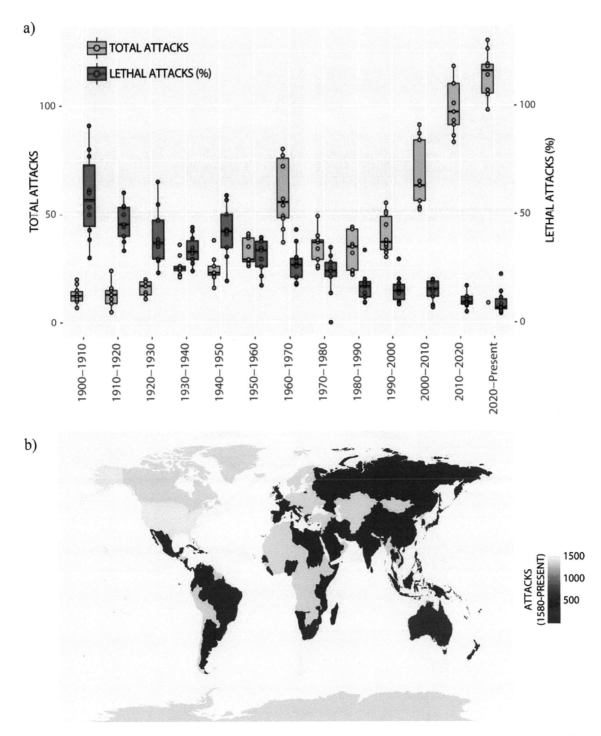

Figure 14.4 Number of: (a) Shark attacks in the last century and; (b) total and fatal reported attacks globally (data: International Shark Attack File, University of Florida).

reefs. Although most common in waters between 60 and 500 m, they have been captured at depths >800 m. These relatively slow animals use their strong shell to defend themselves. They anchor their eggs to hard substrate and adults usually feed on benthic crustaceans. Both observations suggest a strong relationship with the seabed at some stage of their life cycle. *Nautilus* spp. illustrate a largely ecologically extinct species that aquariums can now successfully breed in captivity.

Octopuses (order Octopoda) include multiple families, only one of which lives on the seabed. This family (Octopididae) includes all species commonly used for studies in captivity. Octopuses that live in coastal environments occur globally in epi-benthic habitats closely associated with the seabed. They usually occupy hard or rocky substrates where they find both refuge and food (mostly invertebrates). In fact, many octopus species occur widely in complex substrates where they can use their flexible shape and movement to hide in cavities and fissures in rocks. Octopuses of the suborder Cirrina (named for the fine cirri on their suckers) occupy deep environments, and specimens of dead individuals form much of the basis of our knowledge. Some other species (Figure 14.5), were only recently discovered. Cuttlefish (order Sepioidea, families Sepiolidae and Sepiidae)

Figure 14.5 Two cuttlefish (*Sepia officinalis*) of the class Cephalopoda.

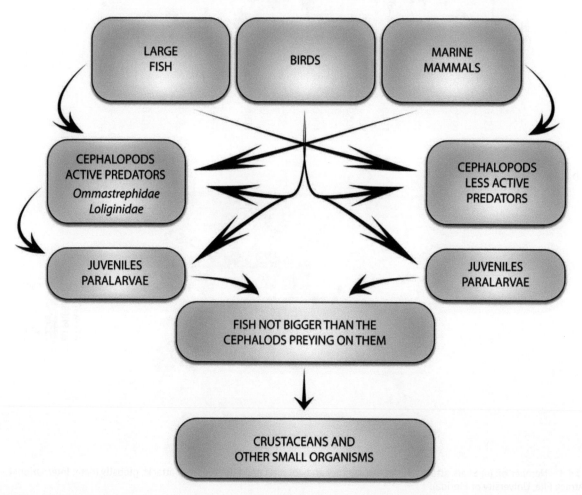

Figure 14.6 Ecological role of cephalopods in the ocean, which play a key role in benthic energy transfer, both in coastal and especially in deep-sea ecosystems.

characterize temperate and tropical coastal waters. As with octopus, these active predators associate with the seabed but instead of hiding in stones and rocks, they create hollows in sediment they can exit quickly from using lateral "fins."

To understand the role of cephalopods in marine environments, consider what would happen if their abundance and biomass declined drastically. Cephalopods feed mainly on crustaceans during the first phase of their life and subsequently on fish, although **cannibalism** (feeding on individuals of the same species) or predation on other cephalopods (Figure 14.6) also occurs, likely when food becomes scarce. The main predators of squids include fishes such as tuna and marlin, marine mammals, and birds, which feed mainly on adult individuals; some seabirds such as petrels prefer larvae and juveniles.

Cephalopods represent high quality food in that they concentrate and convert food resources into tissue with high nutritional value for large oceanic predators. Removing cephalopods in coastal habitats may strongly influence their prey (crustaceans and fish), potentially increasing prey populations, whereas populations of cephalopod predators may decline with the disappearance of this important energy source. The equilibrium of these species may therefore vary, although cephalopods in coastal environments represent only a small part of the community, in contrast to deep environments where their removal could substantially reduce food availability for many toothed whales, pinnipeds, and large fishes. In this case, the considerable impact on top predators would result in a parallel decrease in their abundance.

14.6 Marine Reptile Ecology

Turtles use the terrestrial environment for spawning following many weeks of courtship in the water during which two or more males court a single female. Copulation occurs within the water, with internal fertilization. Adults then return to land to lay their eggs in the sand and, once eggs hatch, juveniles immediately return to the ocean. Most juvenile turtles remain at sea for a year or more, returning to land only once they reach a considerable size. The duration of incubation time for eggs varies among species from 45 to 70 d, generally depending on temperature and humidity. Many individuals build nests at considerable distance from the shoreline in locations with optimal humidity and temperature. Environmental conditions, and temperature in particular, determine sex. Newborns use their teeth to break the shell apart, emerging several days later. Nocturnal hatching occurs asynchronously to minimize predation, which nonetheless affects approximately 90% of newborns. Individuals that survive this early mortality can live to 80 y, reaching sexual maturity at 20–50 y, noting challenges in determining age of turtles. The habits of different species vary, with some preferring coastal versus reef waters, depending on diet. Trophic aspects vary from species to species: they can be carnivores, herbivores, or omnivores. The adults of many herbivorous species usually feed on seagrasses and algae. Historically, sea turtles played a significant role in the functioning of seagrass ecosystems, noting that much larger individuals continuously grazed on seagrass blades. Indeed, experts estimate that Caribbean turtles once exceeded 60 million individuals.

Carnivorous species feed mainly on jellies, salps, sponges, and pelagic organisms. The generally carnivorous loggerhead turtle, *Carretta carretta* eats crabs, mollusks, and tunicates, but also plants, using their beak to split the hard portions of the bodies of their prey. However, diet can also vary with age; carnivorous juveniles of some species become herbivorous as adults.

Adult turtles have few natural predators other than large sharks such as tiger sharks, as well as orcas. Birds prey on small turtles, whereas crabs and lizards prey on eggs. Nesting locations illustrate the clear impacts of humans on this group of organisms where beachfront hotels and lights encroach on nature. Turtles also exemplify the tremendous impact of habitat fragmentation and reduction of spawning habitat. Many adults also die from ingesting trash, or entanglement in commercial fishing gear, whether as bycatch or as capture for food.

We cannot accurately estimate the total number of living turtles, with current projections derived from estimates of numbers of females that come to the land – often based on the density of stripes they create as they haul up onto beach sand. Experts estimate that <100,000 individuals remain of some species. Over the past 30 years, the number of nests in some sampling locations has decreased from >100 to 14. Some species once represented by >600,000 individuals have declined drastically in number.

Green turtle, *Chelonia mydas*, the only almost exclusively herbivorous species, occurs around the globe in tropical and subtropical waters. Juveniles live pelagically and move to the coasts only when 20–25 cm in length, but carapace length can exceed 1.5 m. Breeding takes place March–October in the Atlantic and Caribbean. *C. mydas* migrates along the coast to nest and can travel >2,000 km to cross the Atlantic and nest in Brazil. Many males may fertilize a single female, a pattern typical of turtles. Females can store sperm from one or more males in their oviducts and fertilize the eggs produced in subsequent cycles. They nest on beaches and in one night a female can lay > 100 eggs. In many cases egg deposition cycles can occur every 2–3 years rather than annually, a pattern with strong implications for population dynamics.

Loggerheads, *Caretta caretta*, the most abundant turtle in the Mediterranean Sea, spawn along the Turkish coasts, in Lampedusa, and in quieter areas with less human activity. Historically, turtles also spawned in the center and lower Adriatic. *C. caretta* migrates from its feeding grounds to travel thousands of kilometers to breed. More limited distributions and, in some cases, lower abundances characterize other species. Leatherback turtles, *Dermochelys coriacea*, the largest sea turtle, attain carapace lengths of up to almost 2 m and a body weight up to almost a metric ton. Members of the genus *Dermochelys* can travel 5,000 km across different seas and hold the migration record for turtles. In 1990, some 1,500 leatherback turtles nested on the beach of Playa Grande. Now only 30–40 nesting females use this beach each year. Projections suggest that the population of this species could decrease by 7% every ten years, dropping by 75% by 2100.

Figure 14.7 Sea snakes, *Laticauda colubrine*, return to the surface to breathe (Bunaken Park, Indonesia).

Few scientists have studied diving ability in turtles, but preliminary data indicate that turtles can dive to depths 300–1,220 m. Because turtles are heterothermic their descent to depth and thus low temperature corresponds to a lowering of their metabolism. They can remain immersed for time periods of 35 min to 7 h and, in some cases, reduce their heart rate to one beat every 9 min. Turtles hibernate during the winter and the high concentration of red blood cells and myoglobin in their muscle tissue supports more efficient oxygen transport. With each breath, they exchange ~50% of the air in their lungs. They do not need fresh water because glands for salt secretion (located close to the eyes and giving the impression of moist eyes) allow them to use seawater. Generally, turtles live as solitary animals and do not exhibit social behavior, although they can often aggregate and form groups in the open ocean, especially for breeding. Some species follow the deep scattering layer, which is rich in fish, plankton, jellies, and other organisms.

A flattened tail commonly characterizes marine sea snakes to facilitate swimming. Some species (such as the yellow-bellied sea snake, *Pelamis platurus*) can be gregarious. Keeping in mind that snakes are air breathers (Figure 14.7), valves in their nose close during dives and a salt excretory gland requires continuous cleaning to prevent fouling. Carnivorous feeding mainly targets fish eggs. Some species return to land to spawn, even if mating occurs within the water after a highly competitive courtship. Sea snakes pose minimal risk to humans, and only 5% of attacks prove fatal, although they become particularly aggressive during breeding season.

The one extant species of marine iguana (*Amblyrhynchus cristatus*; Figure 14.8), ranges 70 cm – 1 m or more, and spends long periods on land in the tropics. These herbivores feed underwater on macro-algae on rocky bottoms.

Figure 14.8 Marine iguanas, *Amblyrhynchus cristatus*, from a and b) Mexico; and c and d) Galapagos Islands. *Source:* Wragge / Wikimedia Commons / CC BY-SA 2.5.

Figure 14.9 Marine crocodile, *Crocodilus porosus*. *Source:* Robin W. Baird.

Crocodiles include two marine species. *Crocodilus porosus* (Figure 14.9), the largest (>7 m in length) and more aggressive species that consumes mainly crustaceans, birds, and fishes, lives in estuarine and mangrove habitats, but also occurs in open marine waters of the Pacific and Indian Oceans. The second species, *Alligator mississippiensis*, is not a truly marine species, in that it lives mainly in estuaries of the Gulf of Mexico and the Caribbean Sea and prefers freshwater environments. Even estuarine crocodiles are carnivores, with juveniles consuming crustaceans, small fish, and birds whereas adults hunt large fish, snakes, and any terrestrial vertebrate that comes too close to the water, including dogs and (occasionally) humans!

14.7 Ecology of Seabirds

Seabirds live in all oceans and encompass many forms, from cormorants to frigates, including birds that have lost the ability to fly, such as penguins, to others such as albatross that migrate thousands of kilometers. Prey size varies from zooplankton to large fish (as in pelicans). Salt glands secrete excess salts absorbed from seawater and food. Some species consistently use nearby locations for nesting and feeding, whereas others migrate hundreds of kilometers. Seabirds are long-lived organisms, and albatrosses quite often live for 50 years. Many species live in colonies of several hundred birds; marine swallow colonies may consist of millions of pairs.

Birds can comprise an important faunal component of steep rocky coasts because nesting colonies use vertical rock faces as ideal sites for reproduction. The inaccessibility of cliffs provides a level of protection for nests from predation by many species, including humans. Large, species-rich colonies occur in high abundance along the coasts of North Atlantic. These rocky shore species encompass the Orders Procellariiformes, Pelecaniformes, Charadriiformes, Falconiformes, Columbiformes, Apodiformes, and Passeriformes. Procellariiformes, the true seabirds, rarely visit land and, depending on the species, feed on small fish or marine invertebrates carried in surface currents, such as small jellyfish, planktonic mollusks, small squids, and shrimps. In some cases, they follow great whales and great white sharks as they hunt, taking advantage of scraps. Normally they sleep floating on the ocean surface, and come ashore only during the spawning season to nest in colonies on rocky coasts, sometimes producing only one egg per brood. During this period, they are most active nocturnally and emit particular raucous or guttural calls.

Albatrosses fly 100s of kilometers across the open ocean in search of prey. New tracking technologies demonstrate that albatrosses actively follow killer whales (*Orcinus orca*), presumably to feed on any surplus food left behind by the killer whales.

Cory's shearwater (*Calonectris diomedea*), Manx shearwater (*Puffinus puffinus*), and storm petrel (*Hydrobates pelagicus*) all nest on Italian rocky shores. Cory's shearwater, the only species of Procellariiformes in the Mediterranean, flies high above the water like gulls and albatrosses. In contrast, an irregular and flickering flight next to the water surface characterizes storm petrels, and with rough seas these little Procellariiformes become particularly active because water turbulence brings up debris and small marine organisms on which they feed. They nest on rocks in isolated and inaccessible coastlines, often building nests in cavities at low elevations near sea level.

Nesting sometimes occurs asynchronously, meaning that pairs do not nest together at the same time, but instead alternate in occupying the site. In the Marettimo Island (Aegadian Archipelago, Sicily), a colony of about 1,000 pairs of Cory's shearwaters nest within a complex system of caves with a single access point to the sea, where breeding persists for ~6 months each year. This pattern results from the birds abandoning breeding sites when excessive tourist activity, and recreational boating in particular, drives them away, illustrating the vulnerability of this species to human impact.

Two species of cormorants (Order Pelecaniformes, the cormorant *Phalacrocorax carbo*) and the shag (*Phalacrocorax aristotelis*) nest in Italy. Cormorants commonly occur throughout mid-latitudes in winter, but rarely breed at that time. Individuals that live near the coast often roost on the cliffs and dive in search of fish, ingesting up to 15% of their body weight in a day. Populations of royal Mediterranean gull (*Larus cachinnans*) have shown impressive population growth, despite human disturbance. Puffins, parrot-like birds that occur in cooler waters of the North Atlantic and North Pacific, feature a wide and brightly laterally compressed and colored beak, and use their vibrant colors to indicate sexual maturity and vitality. The large internal capacity of the beak allows them to bring more prey to the nest (up to a dozen capelin or sand lance). Given the awkward flight of the parents and the need to transport food to their fledglings, this is a non-trivial advantage. Like other auks, females lay only one egg per brood in nests often formed from digging holes in sediment along the shoreline or in nests abandoned by shearwaters or wild rabbits.

14.8 Ecology of Marine Mammals

Carnivores – Polar bears, *Ursus maritimus* (Figure 14.10), illustrate excellent adaptations to marine life and the polar cold. *U. maritimus* is one of eight species of brown bears, from which it differentiated 200,000 years ago. It occurs almost exclusively north of the Arctic Circle (noting that polar bears and penguins only live together in comic strips!) It swims surprisingly well given its bear-like morphology, and feeds exclusively on marine organisms (fish, seals, walruses, and occasionally whales).

The sea otter, *Enhydra lutris* (Figure 14.11), spends most of its time swimming on the ocean surface, living primarily in North Pacific kelp forests. Its dense fur provides excellent thermal insulation as it feeds on benthic invertebrates, such as mollusks and echinoderms (using stones to break the shells of urchins), but also on fishes. It plays a major role as a keystone species in kelp ecosystems, as shown when human hunting drastically reduced their numbers and led to major increases in their sea urchin prey. The increased numbers of urchins grazed back macroalgae to the point of creating **urchin barrens**, referring to seafloor habitat devoid of kelp as a result of urchin overgrazing, leading to a form of desertification (see Chapter 18 on Ecosystem functioning).

Figure 14.10 *Ursus maritimus* (polar bear), a species listed by the IUCN (*International Union for Conservation of Nature*) as vulnerable: Experts estimate that populations declined more than 30% over just three generations (~ 45 years) resulting from decreased ice and decline in habitats quality. *Source:* Lukas Riebling / Wikimedia Commons / CC BY-SA 3.0.

Figure 14.11 *Enhydra lutris* (sea otter) feeding and resting at sea. A species once targeted by intensive hunting, it has rebounded well through legal protection and continues to re-colonize the Pacific Northeast coast. *Sources:* a) David Menke / Wikimedia Commons / Public domain. b) Mikebaird / Wikimedia Commons / CC BY 2.0.

Figure 14.12 Seals, sea lions, and walruses (clockwise from left to right). True seals, in contrast to sea lions, lack outer ears. Walruses use their characteristic tusks to dig into the ice and haul themselves out of the water. *Sources:* c) Ansgar Walk/Wikimedia Commons. d) Ansgar Walk / Wikimedia Commons / CC BY-SA 3.0.

Despite excellent adaptations to life in the ocean, pinnipeds (which include seals, sea lions, and walruses; Figure 14.12) must return to land or solid ice in order to give birth. Most species live in cold environments, insulated by a thick layer of fat that also acts as an energy reserve; this fat may comprise up to 30% of the animal's weight. Pinnipeds include three large groups, the Phocidae (true earless seals), the Otariidae (eared seals and sea lions) and the Odobenidae (walruses). The Phocidae are better swimmers, whereas the placement of the limbs in the otariids reflects adaptations for skillful

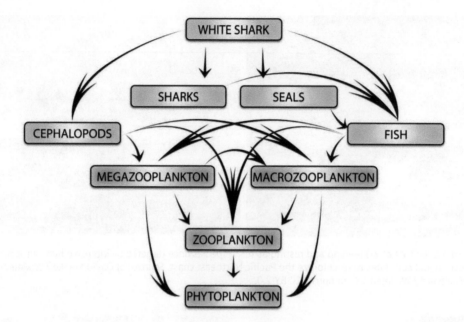

Figure 14.13 Pelagic food web, which involves sharks, pinnipeds, and cephalopods.

locomotion on land. The Phocidae group (the subfamilies Phocinae, seals – and Cystophorinae, elephant seals, in particular) contain the largest number of pinnipeds with 19 species. These excellent swimmers use their pelvic fins to produce thrust, but also move reasonably well on land. They can dive to 1,000 m depth and their weights may exceed 2 metric tons. Sea lions are somewhat quiet and social animals that spend most of their time on land. Walruses spend a third of their lives on land but also dive to depths up to 100 m and remain underwater for up to 30 min. Walruses feed on bivalves and other sea-floor invertebrates, with polar bears as their primary predators. Pinnipeds breed on land or on ice, generally once a year. Parental care by seals typically lasts a few weeks, with mothers rarely leaving their pups during that time. Seal pups remain either on land or on ice during this time whereas pups of other species follow their mothers into the ocean.

Almost all pinniped species reinitiate breeding-related activity immediately after weaning or nursing pups, sometimes even immediately after whelping (giving birth). Males typically leave females as they give birth, thus giving females their own feeding area to access sufficient resources to feed their pups.

Elephant seals with numerous females, known as a harem, reproduce with a single alpha male, and only a small number of males therefore reproduce successfully. Seals generally feed on slower prey (cephalopods, squid, cuttlefish), often at depth. Despite impressive maneuverability and agility, seals are not fast swimmers. Some species feed almost exclusively on krill. Seals with outer ears feed on fish, but generally not at great depth. Prey can vary depending on specific environment, but most pinnipeds can adapt to available prey as they and their prey migrate. Harp seals, for example, prefer to feed on Arctic cod from coastal waters but they also feed on the most abundant species. Leopard seals eat other seals or penguins but when food becomes scarce, they also feed on krill. As warm-blooded mammals, all pinnipeds prey voraciously, often controlling numbers of shoaling coastal pelagic fishes. Potential competition of seals with fishing activities has created significant debate. Many researchers attributed the collapse of northwest Atlantic cod stocks in the Atlantic to overfishing and, to a lesser degree, unfavorable environmental conditions; but seal predation may play a role in the lack of any substantive recovery of cod stocks decades after declaration of a moratorium on cod fishing (Box 14.1). Figure 14.13 illustrates a pelagic food web involving sharks, pinnipeds, and cephalopods.

Cetaceans – Cetaceans occur in all climates, including regions where sea water approaches freezing temperatures. Small cetaceans can cope with cold temperatures because they combine high metabolic rates, and thus heat generation, with thick insulation. Some cetaceans are considered the most intelligent non-primates. Cetaceans use ~12% of the oxygen that they inhale, compared to 4% by terrestrial mammals. They also have at least twice as many erythrocytes and myoglobin molecules in their blood. Cetaceans (Figure 14.15) are often sexually dimorphic in size. For example, female blue whales are larger than males, and male bottlenose dolphins (*Tursiops truncatus*) are larger than female bottlenose dolphins. In a few species, such as narwhals (*Monodon monoceros*) and beaked whales (Ziphiidae), males may use their enlarged, protruding teeth in aggressive male-male encounters.

Box 14.1 Role of Seals and Walruses in the Arctic Food Web

Many pinniped species help connect different ecosystem components. In fact, whereas walruses prey on invertebrates and benthic megafauna, seals prey on bento-nekton and pelagic fishes, thus affecting pelagic and benthic food webs (Figure 14.14).

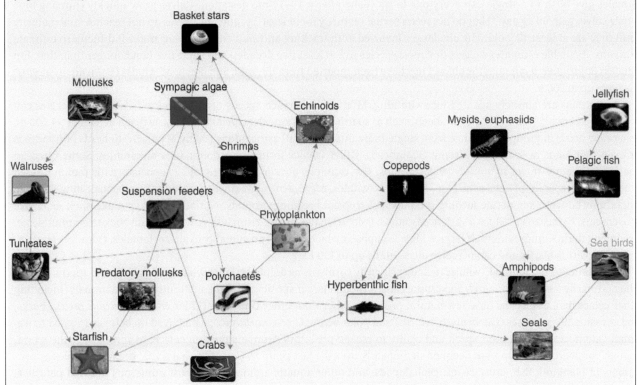

Figure 14.14 Bentho-nektonic food webs in the Bering Sea that connect benthic organisms (brown), pelagic organisms (blue), sympagic algae (gray) and sea birds (light blue). *Source:* Visviva / Wikimedia Commons / public_domain, Steve Clabuesch / Wikimedia Commons / Public domain, Joel Garlich-Miller / Wikimedia Commons / Public domain, U.S. Department of the Interior, NASA / Public domain, Nhobgood / Wikimedia Commons / CC BY-SA 3.0, Graham Curran / Wikimedia Commons / CC BY-SA 3.0, Flickr/Harry Rose, NOAA Fisheries, Unknown author / Wikimedia Commons / Public domain.

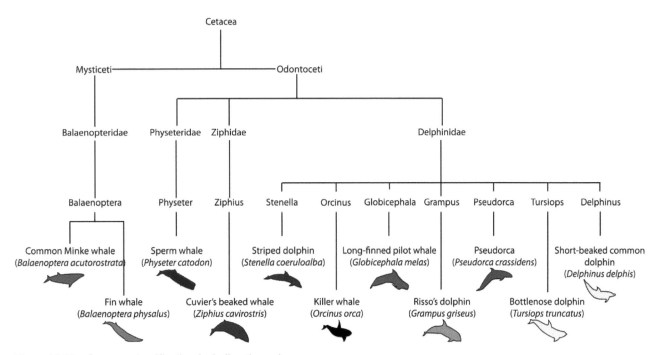

Figure 14.15 Cetacean classification, including the main genera.

Cetaceans are difficult creatures to study both logistically and ethically and, for this reason, the behavior and biology of many species remains a mystery. Most species that have been studied exhibit polyandry, polygyny, or polygynandry. With **polyandry**, the female of some species may mate with several males in succession, or even with two at the same time, in contrast to **polygyny** where a male mates with multiple females. **Polygynandry** involves multiple matings for both sexes. Females give birth to a single calf every one to six years, after a 10–17 mo gestation. Calves grow quickly (nursing blue whale calves gain 90 kg day^{-1}) but do not reach sexual maturity for at least 2 y, and mysticetes do not reach sexual maturity until they are at least 10 y old. The challenges involved with tracking and studying cetaceans make it difficult to estimate lifespans. Available estimates suggest that most species live at least two decades, and some live much longer than that. Fin whales (*Balaenoptera physalus*) as old as 116 y have been reported from the wild, and bowhead whales (*Balaena mysticetus*) may live up to 200 y.

All cetaceans are strongly adapted for swimming. Mysticeti can reach speeds of up to 26 km h^{-1} and odontocetes can swim at >30 km h^{-1}. Some of the Odontoceti, such as sperm whales (*Physeter catodon*), regularly reach depths of >1,500 m. Cetaceans occur in groups that range from single individuals to small associations (many Mysticeti) to herds of hundreds or even thousands of individuals (some Odontoceti). Killer whales form stable dominance hierarchies. Some cetacean species frequently travel in mixed-species groups. For example, Fraser's dolphins often associate with melon-headed whales. Some species of cetaceans (e.g., humpback whales, *Megaptera novaeangliae*) undertake seasonal, long-distance migrations between temperate feeding grounds and tropical breeding grounds. Not all cetaceans make such extensive movements. Cetaceans produce a variety of sounds. Baleen whales moan, grunt, chirp, whistle, and click to communicate, producing these sounds with their larynx. Male humpback whales "sing," presumably to attract females. Odontocetes communicate with whistles and can perceive ultrasounds up to 120 kHz.

The Odontoceti or "toothed" whales include primarily carnivorous dolphins, killer whales, narwhals, and sperm whales. They primarily feed on fish, squid, and crustaceans, although larger species also eat aquatic birds and mammals (including other cetaceans). Cetaceans have few natural predators other than other cetaceans (killer whales, *Orcinus orca*), sharks, and occasionally walruses (*Odobenus rosmarus*) and polar bears (*Ursus maritimus*), which feed on belugas trapped in ice. Small Odontoceti rely on their speed and agility to escape predators. Humans prey on cetaceans throughout the world, though commercial whaling has declined significantly over the last century. Cetaceans have vital ecosystem roles as consumers of plankton, fish, crustaceans, cephalopods, and other aquatic animals. They host numerous internal parasites, including cestodes in their intestines (*Tetrabothrium* sp. and *Diplogonoporus* sp.), plerocercoids in their blubber and peritonea (*Phyllobothrium* sp. and *Monorygma* sp.), trematodes in their stomachs, livers, intestines, and sinuses, acanthocephalans in their intestines (*Bolbosoma* sp. and *Corynosoma* sp.), and nematodes in their stomachs and urinogenital tracts (*Anisakis* sp., *Crassicauda* sp., and *Placentonema* sp.). In addition, cetacean lungworms (family Pseudaliidae) parasitize the Odontoceti.

Mysticeti, the "toothless" whales, use baleen to collect their planktonic food, and includes several species of whales, such as humpback, fin whales, minke whales, endangered North Atlantic right whales, and bowhead whales. Mysticeti feed by ingesting large volumes of water and filtering out planktonic organisms. This filter-feeding mechanism means that baleen whales are secondary consumers (planktivores) that feed at lower food web levels rather than as apex species, despite their large size. Gray whales feed both on benthic invertebrates and opportunistically on plankton. They open their mouth and scoop sediment (and associated invertebrates) into their mouth. Atlantic gray whales were hunted to extinction in the 18th century, but Pacific gray whales have rebounded slightly from historical lows caused by whaling. Gray whales undertake spectacular 8,000–11,000 km migrations (each way!) along the coast between Mexico and Alaska

The connections in humpback jaws allow then to open their mouths between 30° and 90° to aid in plankton capture. As they open their mouth, water quickly fills their mouth depending on their swimming speed, optimizing prey capture. They then discharge water from the corners of their mouth and pharynx, which expand to increase mouth capacity. They feed with a gulping system and also produce a circle of ascending air bubbles that create a sort of barrier (known as **bubble-netting**) to isolate shoals of herring and other species inside the bubble curtain. The whale then swims up from below with its jaws wide open to swallow their prey.

Mysticeti comprise an important component of many marine communities involving complex interactions with other species: the drastic decline of Antarctic cetaceans (whaling reduced whale biomass from 45 to 9 million metric tons) allowed krill biomass to increase to about 150 million metric tons. This release from whale predation could explain the enormous increase in Antarctic seals (*Arctocephalus pusillus*), penguins, and perhaps the lesser increases in fin whales (*Balaenoptera bonaerensis*), whose smaller size attracted less interest from whalers. This increased number reduces phytoplankton biomass, with potential cascading effects on ecosystems functioning and equilibrium.

The Odontoceti (or toothed whales) include numerous species and families. Swimming ability varies among species in that some can dive for a few minutes at shallow depths, while others (such as sperm whales and beaked whales) can dive to depths 1,000–1,500 m or more for longer than one hour. The Odontoceti produce ultrasonic sound for echolocation; using whistles and "clicks" and, depending on the frequency used and, on the response, these organisms cannot only estimate their distance from objects but also assess their size and shape. Groups of individuals of these species often produce "signature whistles" unique to their extended family; indeed, language develops in dolphins from birth to adulthood almost like humans. Several families of dolphins identify specific objects with specific sounds, suggesting that each family group develops its own specific "language." Females with parental relationships generally live in stable groups; males, however, may live in groups with no family relationships or they may be solitary. Males of sperm whales are larger (< 15 m long) than females (< 11 m). These carnivores feed mainly on cephalopods and fish.

The family Monodontidae contains two species. *Delphinapterus leucas* or beluga whales may reach 3–4.5 m in length. These "sea canaries" emit a continuous chirping sound. They live in coastal environments but also undertake considerable migrations. Usually, they occur at depths < 20 m, but some dives reach 400–600 m. *Monodon monoceros*, or narwhal males use their characteristic hypertrophied incisor tooth (similar to a real horn) to fight other males and impress females. The ~50 species that comprise the family Delphinidae include the most common Odontoceti. Some species live near the coast in small groups, whereas other species form groups of 1,000s of individuals. These intelligent animals prey upon large mollusks, fishes, and mammals, and some species use tools or other organisms as tools (e.g sponges) to defend themselves. When they sleep, only part of their brain rests.

Sirenids – Sirenians are exclusively marine and swim using vertical movements of their thick, fat-covered tail. Valves on their nostrils prevent water from entering. Females produce just one offspring every 3 y, which results in slow replacement. The larger size of manatees relative to dugongs means fewer potential predators in the ocean, but tiger sharks place dugongs at risk. Unlike dugongs, manatees often spend time in fresh water, feeding almost exclusively on marine plants by stirring up the sediment surface and extracting rhizomes. Manatees even graze on seagrass beds using their prehensile lips. As warm-blooded animals, they need substantial food to sustain their high metabolism, and can therefore significantly impact food webs. For example, dugong grazing can alter biomass and species composition of seagrasses, maintaining a maximum growth rate in seagrasses – not so different from mowing your lawn!

14.9 Great Migrations of Nekton

Migration refers to movement from one location to another one. Ecologists define three types of migration:

1) **Local and seasonal movements** refer to occupation of different geographical locations during different times of year, typically involving movements over varying distances. Feeding behavior often influences this category, where individuals interrupt their movement to feed upon a specific resource. These movements may vary in duration, from entire seasons to as little as daily excursions, and spatially over vertical and/or horizontal planes. In fact, animals move for many reasons, but mostly to pursue resources, whether food, shelter, or mates. Usually these resources occur within an individual "**home range**" referring to the habitat area that provides the requirements for breeding, feeding, or survival. Movements within this area, sometimes also called **station keeping**, include foraging and reproduction over short time scales and small spatial scales. Generally, round-trip movements (often daily), known as **commuting**, include daily vertical migrations by nekton chasing plankton, or albatross (*Diomedea* spp.) and other seabirds that breed on land masses and move out to coastal waters to feed.
2) **Dispersal** implies departure from preferred breeding habitat, with distances spanning from a few centimeters to basin scales.
3) Strictly speaking, **migration** involves movements between clearly separated and well-defined locations. Migration confers many advantages, including increased feeding opportunities, escape from adverse environmental conditions, and improved reproductive success. Obviously, migration brings cost, including energy costs required to move and adapt to the new environment, as well as increased predation risk. Factors influencing migrations include:
    a) Foraging movements driven by spatial and temporal distribution of food and its predictability. Marine mammals often associate with specific habitats, such as canyons that can concentrate large numbers of prey. Size, sex and reproduction influence movement to their feeding grounds. In seals, foraging ranges link to body size in that males average 26 km, compared to distances of 15 km in females. Although less common, some species move for foraging in response to tides. Certain fish species, including scorpion fish, move close to the shoreline during flood tides occasionally entering mangroves, and then move out to sea as the tide recedes. Some species migrate across latitude in search of food: seals, for example, move north during summer and south during winter. Seasonal movements also occur between surface coastal waters and the deep waters of the open ocean, as in Hector's dolphin (*Cephalorhynchus hectori*).

b) Reproduction movements differ among marine mammals, depending on the social system or species in question. Many marine mammals aggregate together during the breeding season, often traveling long distances, providing some protection for vulnerable offspring. Whereas cetaceans and sirenians breed in water, other marine mammals use land or ice as breeding locations. Scientists often do not understand why animals use specific breeding locations. The degree of separation between breeding and foraging locations varies considerably. For pinnipeds, the need to give birth and raise their pups on land or ice leads them to move towards sites far removed from the feeding locations they use during the remainder of the year. Thus, many Phocidae do not feed during weaning. These movements resemble whales migrating from high latitudes to whelp at low latitudes, though often reversing seasons relative to whales. In contrast, some pinnipeds breed in colonies located adjacent to foraging habitats. In fact, the offspring of otariids, the eared seals, remain on land for extended periods, so females must continue to feed, making proximity essential. Pelagic dolphins use a similar strategy; they can reproduce throughout the year and hence their foraging and reproduction sites do not differ.

"Loyalty" to foraging locations occurs commonly among animals but also to sites that confer reproductive advantages; however, loyalty varies depending on species' reproductive systems and sex of the individual. Humpbacks and gray whales, for example, aggregate to breed and wean offspring, and many individuals generally return to the same breeding area year after year, noting that small numbers of humpback whales shift reproductive locations.

c) Molting: Phocids molt their fur every year; otariids also molt but more continuously than phocids. Growth of new hair requires increased blood supply to the skin. If the animal remained within the water, increased blood flow could increase heat loss and seals therefore spend much of their time on land during molting. Cetaceans molt excess skin continuously, often producing unusual behaviors such as rubbing against rocks.

d) Avoidance: some animals migrate to leave a region where environmental conditions become hostile. We know little about predation and disease influence on migration, but evidence points to a selective advantage in moving long distances to avoid disease. One study suggested that the long period of fasting during gray whale migration contributes to their lower parasite load compared to other Mysticetes. Despite impressive adaptations to withstand thermal stress, marine mammals must sometimes abandon otherwise suitable habitats if temperatures shift beyond their optimum range. For example, manatees (*Trichechus* spp.) can move hundreds of miles between marine, brackish, and fresh waters throughout the year, but they withdraw from locations where industries discharge water used for heating/cooling.

Bony fishes, sharks, crustaceans, turtles, and marine mammals often migrate between spawning/ reproduction and foraging locations. Food availability largely drives migration (in bony fish in particular), often to pursue shoaling prey. Movement of both older juveniles and adults after the vulnerable reproductive period reflects double adaptations: that populations ensure reproduction in a favorable environment, and that all life cycle stages access high concentrations of food resources. Thus, migration characteristics in a species clearly depend on both water temperature fluctuations as well as trophic potential. Juvenile movement from nursery areas to adult foraging areas may reduce competition between juveniles and adults for limited food resources, or it may result from different food needs of each life cycle stage. In some species, the goal may be to avoid cannibalism! Migratory patterns are classified based on spawning and foraging location.

Gray whales migrate > 9,000 km in the Eastern Pacific, from foraging regions at high latitudes to breeding regions at lower latitudes. Other Mysticeti travel ~8,000 km, whereas sperm whales cover distances of ~6,500 km. Elephant seals can travel 3,000 km, and some seals migrate up to 4,000 km. Within a species, migration distances can vary greatly between individuals and populations. Humpback whales migrate at least 1,200 km between California and Mexico, this distance pales in comparison with their 5,000 km seasonal migrations in the North Atlantic and 8,000 km between South America and Antarctica. Some non-migratory groups of whales live in tropical systems, and in highly productive upwelling areas.

Animal "**cruising speeds**," referring to speeds that organisms can maintain for extended periods, depends on their hydrodynamics and physiology, as illustrated by comparing different marine mammal species (Table 14.1). Cruise speed range varies considerably within a species. Not only the distances, but also the patterns can differ within a species depending on sex, age, and reproductive status. This variation may result from different energy demands by animals of different sizes, and sexual dimorphism in particular.

In some species, only one of the two sexes migrate. In sperm whales, adult males regularly migrate to polar waters, feeding near the productive ice edge during summer and returning to the tropics in winter. In contrast, females and juveniles travel widely but remain in tropical and temperate waters. In other cases, males and females migrate to different locations. Beluga males in the Canadian Arctic migrate during the summer to specific locations such as Viscount Melville Sound, whereas females and juveniles migrate to Amundsen Gulf. Sexual segregation in migratory patterns of elephant seals relates to their sexual dimorphism. Males can weigh 10x more than females, and therefore reach larger foraging areas. Males migrate towards foraging areas beyond the continental shelf, whereas females migrate into deeper water or to the Antarctic polar front.

Table 14.1 Cruising speeds of several taxa of marine mammals.

Taxa	Speed km h^{-1}	Taxa	Speed km h^{-1}
Mysticeti	2.3–6.4	Otariidae	3.6
	2.7–7.9		4.7–5.7
	0.7–12.6		3.6–5.4
	0.8–4.6		1.5–1.7
	1.5–3.9	Phocidae	3.1–4.2
	1.3–6		3.0–4.0
Odontoceti	2.5–5.2		0.7
	1.1–6.0		3.1–4.2
	2.3–3.9		3.5
	1.5–4.5	Sirenidae	0.9–2.1
	0.6–2.3		

Migration timing often varies with age, sex, and reproductive status. In Mysticeti, females tend to migrate earlier than males, followed by juveniles. Pregnant females usually arrive first to foraging areas, followed by adult males. Females, together with their calves, usually arrive last. The return to breeding locations occurs in the reverse sequence. Birds and whales perform the most extensive migrations of all animals, with some species moving from pole to pole.

Green turtles (*Chelonia midas*; Figure 14.16) are a real scientific enigma: they move thousands of kilometers between foraging areas and spawning beaches where females lay their eggs. Tagging experiments show that many females tend to return to the same beach to spawn, but whether this behavior begins at birth when hatchlings move into the ocean or if females simply find beaches on which to spawn and stick with them remains unclear.

Many researchers consider migration a mechanism to avoid unfavorable environments by moving to favorable locations. In other words, migrations may represent seasonal movements of organisms from one place to another under various environmental conditions. The energy required to migrate to a site different than that of their origin factors in the significant possibility of death as a result of this movement. For some species, ideal habitats for reproduction may be unfavorable for foraging, or vice versa. Thus, some species have evolved a strategy of separating foraging and reproduction in time and space with the ability to navigate along appropriate distances and, in some cases, to tolerate the physical and energy needs necessary to traverse long distances.

Different species respond to environmental changes in contrasting ways that reflect their individual adaptations and their range of tolerance. Comparison of marine and terrestrial organisms helps illustrate this phenomenon. Terrestrial

Figure 14.16 Migration pattern of green sea turtle, *Chelonia midas*, in the Coral Triangle between Pacific and Indian Oceans.

organisms expend significant energy to move, whereas the density of marine species closely approximates the seawater that surrounds them, leading to less energy expenditure to move or maintain a given position. Finally, the constant motion of ocean waters influences patterns of productivity to a much greater extent than the wind on land, both by transporting essential nutrients and gases, and by transporting organisms on ocean currents.

Migrations often develop in response to environmental changes, whereas adaptation provides a long-term, but less plastic solution. Some organisms possess genes that enable survival in a particular environment. Thus, some of the **alleles** (referring to different forms of a given gene) of populations of a particular species that live in cold waters typically differ from those in warm water; this variation has significant ramifications for restoration of depleted populations.

Within the ocean, water and seabed form distinct (though linked) domains with distinctive features that support different communities. Benthic organisms critically depend on the sediment or seabed properties that they live within, or close to, in contrast to characteristics of seawater (temperature, light, nutrients salinity, clarity) that determine water column biota. As conditions change, organisms can react or move, with some species distributed widely and others geographically confined to a narrow range. Fish migration patterns differ within and among species and correlate with numerous biotic and abiotic factors. Many species migrate to find a suitable spawning location. Obviously, each species requires specific environmental characteristics that determine their migration patterns, but some of these characteristics may change from year to year and potentially alter migratory routes or destinations. Prey and predator abundance vary, and migrations to pursue food or avoid predators may change concurrently.

Because migration patterns differ among and within species, categorizing migrations can prove difficult. However, fish biologist George Myers managed to group fishes broadly into **diadromous** species that migrate from fresh water to the sea and vice versa. A subset of diadromy, **anadromous** fishes, refers to those species that spend a significant portion of their life cycle in the ocean, before returning to fresh water to spawn. Examples of anadromous species belong to the following groups: salmon, trout, brook trout, smelt, whitefish, shad, herring, stickleback, and lamprey (Figure 14.17). A second subset of diadromy, the **catadromous** fishes, spend most of their life in freshwater but migrate to the sea to spawn (e.g., American and European eels, Figure 14.18).

Myers also described a third type of diadromy, **amphidromous**, referring to species that regularly migrate throughout their life cycle between environments of freshwater and sea but not specifically driven by reproduction (e.g., some gobies). **Potamodromous** species refers to those that spend their entire life in fresh water with well-defined migration patterns between lakes, rivers, and streams. In contrast to these freshwater species, **oceanodromous** species refers to those that live and migrate exclusively within marine environments. Researchers consider the large-scale migrations of oceanodromous species as closed journeys rather than as round-trip migrations. These migratory patterns may include specific locations that meet the demand for resources of the migrants at specific times of their life cycle, whether for spawning, nurseries, foraging, overwintering, oversummering, migration pathways, etc. Individuals, small groups of individuals, or populations visit these different habitats, and each location may overlap both space and time.

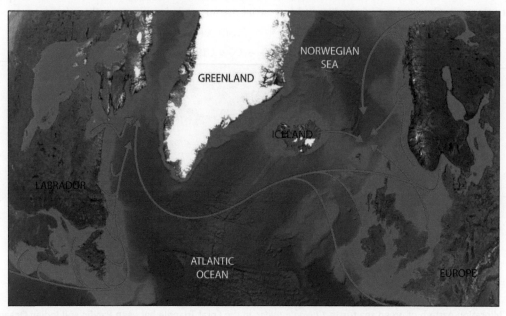

Figure 14.17 Map of migration routes of Atlantic salmon to their main foraging areas offshore of Greenland and Labrador, and north of Faroe Islands. *Source:* Google LLC.

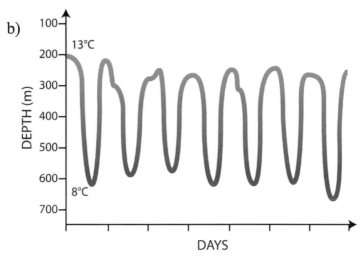

Figure 14.18 Migration of eels. (a) Migratory patterns during larval (leptocephalus) stage of European eel spans from Europe to the Sargasso Sea; (b) daily migration during which leptocephali move between 200 to >600 m depth. *Source:* Michael Tangerlini.

In summary, for marine species we can distinguish between oceanodromous species (which migrate within the ocean) and diadromous species (migratory species that pass from seawater to fresh water or brackish water, and vice versa) based on the mode of migration. Within diadromous species we can further discriminate between anadromous species, such as salmonids, that move to freshwater to spawn (now under intense anthropogenic pressure from the construction of dams and dykes) and catadromous species that move from fresh water into the sea to reproduce. Amphidromous species, however, move from marine to fresh water with no specific link to reproduction.

The European eel (*Anguilla anguilla*) migrates ~5,000 km from Europe to the Sargasso Sea. The Gulf Stream and North Atlantic Drift transport larvae. Tagging of 1,300 individuals enabled elucidation of the complete migration to the Sargasso Sea.

Migration represents a critical part of the life history of many marine vertebrates. Over the last decade, new tracking technologies have enabled a rapid increase in studies on migration of both sea turtles and other marine vertebrates such as birds, mammals, and fish. These new approaches include non-invasive tracking that increase knowledge on the frequency and extent of movements of individuals. For example, satellite tags that measure sunrise and sunset, light intensity, and time of day, provide spatial resolution of a few km. Alternatively, researchers use GPS (Global Positioning System) tags in species that come to the ocean surface long enough for tags to obtain a satellite fix and relay positional information to satellites with a resolution of 10s of m (Figure 14.19). The *Tagging of Pacific Predators* (TOPP) project, one of the 17 projects of the Census of Marine Life, dedicated their research to understanding better the migratory patterns of pelagic organisms.

Several variables, direct and indirect, guide migrations; some species apparently respond to moon cycle changes, which they perceive either visually or by detecting tidal change. *Leuresthes tenuis*, a small pelagic fish (family Atherinidae) migrate to spawn only at night from March to August during full or new moons, when tides reach their highest levels. These fish lay their eggs in the sand where they remain for two weeks, until tidal height increases. Salinity can also significantly influence migration; river currents and changes in osmolarity can stimulate migration. Adult Atlantic salmon (*Salmo*

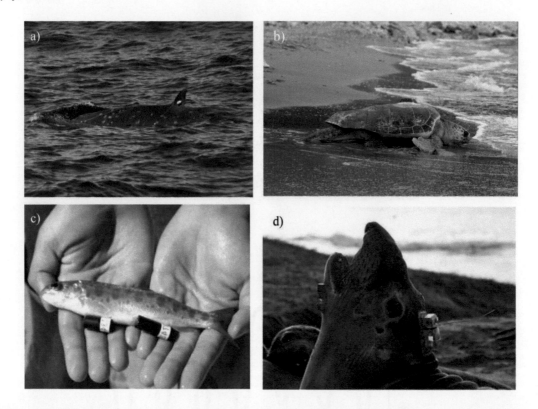

Figure 14.19 Example of species with sensors and tags, and tracked with satellites: (a) Note tag attached near the dorsal fin base; this "popup tag" records positional data for about two weeks, and then floats to the ocean surface to transmit data to satellites; (b) sea turtles; (c) salmon, in which acoustic tags may be surgically implanted within the body cavity; (d) male elephant seal with an electronic tag and sensor temporarily fastened to its head. *Source:* Daniel Costa, ulvio Maffucci.

salar) use currents, magnetic signals, or celestial bodies while migrating in the open sea, but switch to olfactory stimuli once in freshwater. Little evidence supports the hypothesis that photoperiod and temperature represent important stimuli in initiating migration. For example, Atlantic salmon begin their migration at different sites and speeds during periods with quite different photoperiods but similar water temperatures. The transition from light to dark seems more important with respect to the length of photoperiod. Once an individual attains a certain physiological or behavioral state, many species of fish migrate, but only at a certain time of the day; many crepuscular species migrate at dawn (*Salmo trutta*), at sunset (juvenile *Clupea harengus*) or both (*Alosa pseudoharengus*). Salmonid juvenile migration in the Pacific takes place mainly at night, likely to avoid predation.

The sardine *Sardina pilchardus* in the Adriatic Sea (Figure 14.20) further illustrates migration from spawning areas to locations that maximize growth. Scientists have identified two spawning areas in the northern and southern eastern Adriatic Sea along the Croatian coast. In the northern location, in winter, adult sardine populations migrate from northern nutrient-rich but hydrologically unstable shallow Adriatic waters to the south in search of habitat more favorable for larval development.

Following reproduction in spring, spawners move in the opposite direction toward the productive waters of the northern Adriatic Sea in search of food. In contrast, in the south Adriatic Sea, sexually mature sardines migrate offshore away from coastal temperature and salinity gradients. At the beginning of spring, adults, larvae, and post-larvae migrate to more productive coastal waters. The first migration (in winter) coincides with initial sexual maturation, in which critical lipid reserves allow adults to leave foraging areas in search of conditions more favorable for reproduction. The search for food seems to induce the spring migration to replace the enormous energy expended during the intense reproductive activity.

Many species of fish that undertake long or short migrations return to the locations from which they started. Other species, however, migrate from one place to another and may not return to their place of origin, a pattern known as one-way or non-return migration. A loop or circular pattern, in contrast, occurs when, after the start of the migration, the animals moves from location to location but never retrace their path while somehow returning to the starting point. For many species, oceanic movements cover impressive distances (Figure 14.21). For example, as noted earlier for elephant seals (*Mirounga angustirostris*), leatherback turtles (*Dermochelis coriacea*), tuna (*Thunnus thynnus*), and shearwaters (*Puffinus* spp.), move distances of many hundreds and even many thousands of kilometers.

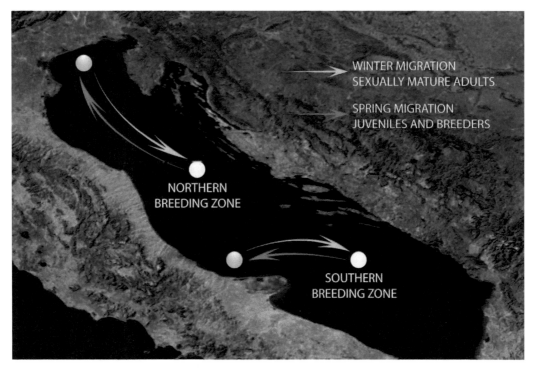

Figure 14.20 Migration of sardine, *Sardina pilchardus*, in the Eastern Adriatic, along the Croatian coast. Spawning occurs in the northern sector and southern regions. Adult, larval, and post-larval migration patterns of *S. pilchardus* depend on spawning location. *Source:* Sansculotte / Wikimedia Commons / Public Domain.

Figure 14.21 Migration pathways of various marine species obtained by satellite and acoustic tagging. In the Atlantic: tuna (blue line), turtles (green line), and petrels (pink line). In the Pacific: turtles (green line), salmon species (blue line), seabirds (pink line), humpback whales, and elephant seals (orange line).

These studies identify "areas of high use" that represent habitat/hotspot locations critical to the biology and ecology of the species involved. Identifying the rules that the animals follow for "finding" these hotspots is therefore fundamentally important not only in understanding foraging ecology, but also in protecting these organisms from by-catch or other conservation threats.

Sea turtles spend their entire life in marine or estuarine environments, occupying terrestrial habitat only during nesting, and, in some rare cases, to use heating by the sun. As a result, these animals have evolved different physiological, anatomical, and behavioral adaptations to their aquatic life, and sea turtles actually share many elements with large fish and cetaceans in how they use habitat and in how they migrate. Juvenile sea turtles move immediately from the beach to the sea after hatching, usually in the dark, and actively swim out to deeper water. Then, after a few years the larger and older juveniles move into neritic habitats of tropical and temperate environments where they develop and mature. The populations of juveniles in some temperate zones migrate seasonally to foraging areas at higher latitudes in summer and lower latitudes in winter, whereas tropical species tend to limit their movement to a particular region. The late juvenile stages of some species, such as loggerhead (*Caretta caretta*) and green sea (*Chelonia midas*) turtles migrate seasonally between summer and winter habitats.

As they approach maturity, sea turtles move to other foraging habitats that may be geographically distinct from the habitats used by juveniles, whereas others may overlap or directly coincide. As soon as they reach maturity, and spawning season approaches, adults migrate to the nesting beaches. Mating often occurs in "courtship" locations near or distant from the nesting beaches, depending on the population, far from distant feeding grounds. During spawning season, females usually become residents of nesting habitat, near the beach where they eventually lay their eggs. Many species of sea turtles migrate intermittently throughout their life.

Other species, such as leatherback turtles, *Dermochelys coriacea*, typically remain in the open ocean. After a few years migrating in the open ocean feeding and growing, juveniles of several species leave the oceanic environment and take up "residence" in foraging habitats in coastal areas. The juveniles of loggerhead and green sea turtles in these neritic habitats often show fidelity to specific foraging locations, returning after long seasonal migration. Adults, like juveniles, may settle in small, isolated, foraging locations. Once mature, the turtles of almost all species migrate from their foraging habitats to specific areas for mating and nesting, before moving back to foraging locations. The nomadic lifestyle of sea turtles depends on their ability to navigate through the ocean.

Relative predictability defines an important characteristic of environmental influences on movement patterns. The dynamic nature of the ocean means that patches of favorable or unfavorable habitat may change over time and space. Movements can then reflect adaptation of the species to a fragmented and variable environment. The forces that drive animals to migrate include the need to carry out activities in specific types of habitat at specific times, and the need to optimize resource use in those habitats – whether food, shelter, or other attributes – that maximize fitness. These drivers may also include a need to move away from high-risk locations, including escape from predation or disease, or to escape harsh environmental conditions.

Migration has produced parallel co-evolution in some migratory species. In polar region, migration might reflect the need to avoid getting trapped by ice. Researchers assumed that whale migrations to tropical regions reflect a response to relatively high risk of predation on vulnerable offspring by killer whales at high latitudes, but the need to maintain a neutral or positive net energy balance in warm waters for part of the year may offer a more likely explanation for migration. Two factors compete in determining energy gain and loss in an animal: distribution and abundance of energy resources (food) in the ocean and thermal characteristics of the ocean, noting that warm endothermic animals should maintain a body temperature of 40 °C, above that of surrounding water. Marine mammals therefore maximize their fitness by evolving phenotypic characteristics that include body size (small surface area/volume ratio to minimize heat loss), migratory behavior, longevity, and low reproductive rates. Cetaceans exemplify some of the most extreme applications of these adaptations. The impressive annual migrations of whales still lack a truly definitive explanation.

The hypothesis that pregnant female whales migrate to reduce predation risk by killer whales (*Orcinus orca*) on their newborns by moving to low latitude waters merits some discussion. In fact, killer whales occur in significantly higher abundances at high than at low latitudes, and killer whales generally do not migrate to low latitudes. In addition, pinnipeds, the primary prey of killer whales, largely determine killer whale distributions. What makes whale migrations particularly interesting is that, at least for some species, food resources do not appear to offer the primary benefits for undertaking a long migration. In contrast, such resources (food, water, or special nutrients) play a primary role in determining seasonal migrations in terrestrial mammals. In general, whale migrations represent seasonal movements between high latitude foraging habitats (their foraging areas) and low latitude mating locations. Mysticeti seem to follow migration routes they learned during their first year of life. Mysticeti display remarkable **philopatry**, the tendency of some species of migratory animals to return to a particular place to feed or reproduce. Maternal communication seems to convey loyalty to these post-weaning sites.

In contrast to birds and fish, evidence suggests that whales often use ocean currents to facilitate migration. However, fin whales (*Balaenoptera* spp.) pass through the West Indies in winter and follow the North Atlantic gyre, proceeding west with the Equatorial Current and then north with the Gulf Stream. Even for those species that do not normally associate in groups, staying close to conspecifics may provide advantages. Vocalizations can help maintain contact, even over long distances, and maintain cohesion of groups during migrations. Because studies demonstrate particular sensitivity in some marine mammals to chemical signals in urine and feces, these organisms may also detect and follow "traces" left by conspecifics, further helping to maintain group cohesion during migration. Evidence also suggests segregation along migration routes based on age and sex. Gray whale females and their offspring travel along the coastline closer to each other than other individuals. In many species of marine mammals, offspring follow their mothers to learn locations of feeding and mating grounds.

The ability to undertake large-scale migrations and return to specific locations indicates that marine mammals have evolved sophisticated navigation modes. Several pinniped species show excellent homing behavior after moving distances of 100s of kilometers. We lack direct evidence on how marine mammals find their "way" in the ocean; however, we can reasonably assume they use multiple senses and information sources. Other animals with wide-ranging migrations, such as birds, use different information sources. Signals potentially available for marine mammals include celestial bodies, bottom topography, terrestrial reference points, currents, temperature and salinity gradients, smell and taste, sound, and geomagnetism. Hearing and sight can help in detecting physical characteristics, such as seabed features, coastline, and ice cover, thus providing information, both general and specific, on location. For example, some porpoises follow the isobath at ~92 m during large-scale migrations, suggesting they use seabed topography to navigate. Cetaceans living in coastal environments use prominent terrestrial structures to aid navigation, although evidence remains limited. The sun provides information on location given its location in the sky and photoperiod. Some researchers propose that whales perceive day length in polar region as a stimulus to begin their migration. Patterns of daily movements for seals and cetaceans suggest they use the sun for navigation, but these movements may also reflect responses to changes in prey distribution.

Water masses, currents, and frontal systems can provide important information for navigation and orientation of marine mammals because they often encompass gradients of temperature, salinity, turbidity, and other factors. In many species water temperature strongly influences seasonal distributions, indicating an ability to detect and respond to thermal gradients. Marine mammals reported outside their normal distribution limits may represent failure of the navigation system. Alternatively, they might represent animals exploring new areas. Cetacean strandings, which may represent navigational failures, have fascinated people since Aristotle. But living animal strandings, particularly mass strandings, raise questions regarding why animals that spend their entire lives at sea swim onto beaches. Repeated strandings at particular sites suggest some biological conditions in these locations linked to events such as parasite outbreaks that cause neurological diseases. Several factors likely contribute to these events. The loss (or lack) of a specific sense, or confusion regarding one source of information seems inadequate in explaining the loss of navigation abilities given many other potential sources of information. In birds, individual experiments that remove specific cues show some ability to maintain migration patterns.

14.10 Role of Top Predators in Food Webs

In the past, predators constituted the majority of fish biomass in coastal environments; sharks comprised some 74% of this biomass (*Triaenodon obesus, Carcharhinus amblyrhinchos, Carcharhinus limbatus*). The proportion and biomass of sharks gradually decreased with increasing human presence, leading to 0–13% predator biomasses today relative to historical numbers. In the North Atlantic, a 94% decline has occurred since 1930, and in South Africa, >99% decline since 1961. **"Fishing down the food web"** refers to the process by which fishing has selectively removed large species leading to increases in smaller species (Figure 14.22), thereby altering food webs. Indeed, decreases in large fishes correspond to the top of the food web and fishing effort gradually shifts to lower tropic levels.

These studies suggest that fishing pressure, even when managed, strongly compromises survival of top predators (Figure 14.23). The situation in demersal species appears no better. Over 90% of elasmobranch species that inhabit demersal ecosystems globally are vulnerable to trawling. Large sharks typically disappear first as a result of bycatch, and small elasmobranchs (mesopredators) soon dominate the community. Some shallow demersal species such as the stingray, *Dasyatis dire* and the butterfly ray, *Gymnura micrura* decreased by 60 and 99%, whereas largely pelagic species (angelsharks, *Squatina* sp., and smooth-hound sharks, *Mustelus canis*) increased by 6 and 13x, respectively. In Australia, trawling in the open sea in the 1970s yielded fish catches that consisted of almost 50% shark biomass. After 20 years, catch rates of elasmobranchs were already reduced by 80%. A century of trawling led to the near elimination of 16 of 31 species of elasmobranchs in the Tyrrhenian Sea, 6 of 33 species in the Adriatic Sea, and half of elasmobranch species reported by trawlers in the Gulf of Lyons since 1950. North Sea pelagic populations have changed radically from a relatively diverse group of top predators to one currently dominated by small elasmobranchs of productive species, such as dogfish (*Scyliorhinus canicula*) and rays (*Raja naevus* and *R. montaguy*). Industrial fishing in the open sea beginning in 1950 focused mainly on tuna and swordfish but led to bycatch of sharks for each yellow-fin tuna (*Thunnus albacares*) captured, as well as 2–3 sharks per swordfish.

In recent years, the large pelagic silky shark, *Carcharhinus falciformis*, decreased by 92%, and the largely temperate and oceanic whitetip shark, *C. longimanus*, declined >99%. At the same time, in the Pacific and in the Southern Atlantic, the biomass of *Prionace glauca* (blue shark) apparently increased by 20% and this species is now considered the most abundant deep-sea shark. Mako sharks appear to have decreased less than other large species, but intense exploitation continues (Box 14.2).

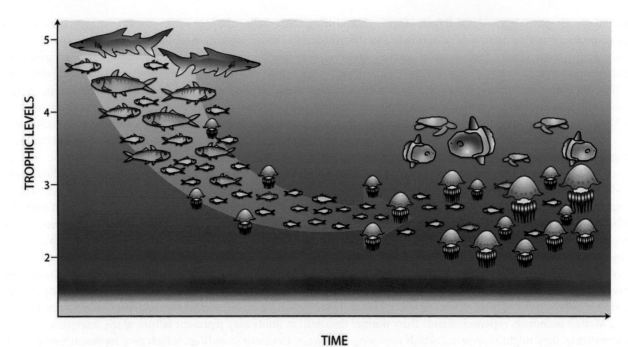

Figure 14.22 The concept of Fishing down the food web; the disappearance of large sharks leads to increased numbers of jellies and their predators (turtles and sunfish).

Figure 14.23 (a) Tuna; (b) barracuda; (c) swordfish; and (d) sailfish (marlin) number among the primary pelagic predators that often compete with sharks in hunting pelagic prey. *Sources:* a) National Oceanic and Atmospheric Administration (NOAA). c) Citron/Wikimedia Commons / CC BY-SA 3.0. d) Flickr/Dominic Sherony.

Box 14.2 The Loss of Large Predatory Sharks

Global assessments of 881 of the 1,159 known species of Chondrichthyes, report 14% threatened, 11% vulnerable, more than 4% endangered, and 2.4% critically endangered; thus >31% of species face significant pressure. Status varies from region to region, but the Mediterranean and North-East Atlantic contain the highest proportion of threatened species (see box "Shark extinction resulting from indiscriminate hunting by humans"). In the Mediterranean ~10% of known shark species occupy <1% of ocean surface area, clearly defining a hotspot of elasmobranch diversity. Widespread species include *Carcharhinus plumbeus* (sandbar shark), *Prionace glauca* (blue shark), *Sphyrna zygaena* (hammerhead shark), *Scyliorhinus canicula* (dogfish), Alopias *vulpinus* (thresher sharks), *Cetorhinus maximus* (elephant shark), *Squalus acanthias* (spiny dogfish), *Carcharodon carcharias* (Great White Shark), *Isurus paucus* (mako), and *Lamna nasus* (porbeagle). The Mediterranean's long history of intensive fishing offers a unique perspective on fish population declines. Records of landings of fish dating from the mid-nineteenth century enable comparisons in temporal trends of 5 species: *Sphyrna* spp., *Prionace glauca*, *Isurus oxyrinchus*, *Lamna nasus*, and *Alopias vulpinus* (Figure 14.24). Declines in abundances of these species between 96 and 99.99% relative to historical values meet the criteria of the *International Union for the Conservation of Nature* (IUCN) to consider these species in danger of extinction.

Figure 14.24 Some of the shark species that have declined most precipitously in the Mediterranean Sea: (a) Blue shark, *Prionace glauca*; (b) Smooth hammerhead, *Sphyrna zygaena*; (c) Bignose shark, *Carcharhinus altimus*; (d) Shortfin mako, *Isurus oxyrinchus*; € school shark, *Galeorhinus galeus*; (f) Sandbar shark, *Carcharhinus plumbeus*. Sources: a) Mark Conlin/NMFS / Wikimedia Commons / Public domain. b) suneko / Wikimedia Commons / CC BY 2.0. c) Apex Predators Program, NOAA/NEFSC / Wikimedia Commons / Public domain. d) Mark Conlin / Wikimedia Commons / Public domain. e)Taylor Hand /Wikimedia Commons / CC BY-SA 2.0. f) Brian Gratwicke / Wikimedia Commons / CC BY 2.0.

We are only now beginning to understand the potential broader ecological consequences of declines in sharks, a problem exacerbated by the challenges of studying sharks and their prey in the natural environment. The lack of empirical studies makes it difficult to draw conclusions regarding how removal of top predators will alter their constituent ecosystems. We generally lack information on the effects of shark predation on population dynamics of teleost fishes and cephalopods, which constitute a large part of their diet. Presumably larger, long-lived species respond most rapidly to reduced shark predation.

Consumption mechanisms (direct predation) and behavioral mechanisms (perceived risk) drive these effects. In general, behavioral mechanisms should produce greater effects on large and long-lived prey, which could invest more resources in predator avoidance than species with short life spans. A study on trophic interactions between 249 species in the Caribbean that included 10 shark species highlighted how shark removal induced trophic cascades that increased community vulnerability to disturbances and reduced the degree of omnivory. This shift contributed indirectly to the alteration of reefs from a coral dominated to algae-dominated system by increasing fish consumers (e.g., groupers) that reduced herbivore abundances (e.g parrot fishes) (Figure 14.25). However, sharks prey on both groupers and herbivorous parrot fish, potentially reducing the indirect effects of shark predation on coral reef ecosystems.

Some models suggest that the effects of shark removal depend on the species and context, with strongest effects predicted to occur for large sharks in coastal environments, where a decrease of tiger sharks would cause increased numbers of prey

Figure 14.25 Trophic cascades in coral reefs: declines in sharks lead to increases in groupers and meso-predators that can reduce abundances of herbivorous fishes such as parrot fish, resulting in increased algal abundance. *Source:* Stormy Dog / Wikimedia Commons / CC BY 2.0.

including seabirds, turtles, monk seals, and reef sharks, which in turn would lead to rapid declines in tuna and trevally. In contrast, shark removal would produce little effect on coral reefs. By way of explanation, reef sharks feed on fish and invertebrates that have relatively high turnover rates compared to birds, turtles, seals, and sharks consumed by tiger sharks. In addition, tiger sharks represent the only major predator of turtles. The loss of all sharks in the Galapagos might lead to increased abundances of toothed whales and sea lions, but not of commercially fished reef predators that would otherwise decrease reef fish and increase small invertebrates through four levels of trophic cascades.

In Alaska, depletion of large pelagic fish and salmon shark (*Lamna ditropis*) had transient effects on pinnipeds and large fish. In other cases, in the North Pacific, sharks do not represent keystone predators in the pelagic community. Predatory fish, such as tuna and sailfish, characterized by comparatively rapid biomass turnover, could possibly replace sharks without significantly affecting other species dynamics.

Killer whales (*Orcinus orca*) near the coast and sleeper shark *Somnius pacificus* in deeper waters prey upon seals (*Phoca vitulina richardsi*). Tagging experiments reveal that killer whales and sharks both affect diving behavior of seals and how they use resources. However, sharks elicit a stronger behavioral response in seals than killer whales, although they are less specific predators.

Sharks consume large numbers of small elasmobranchs, and several studies of coastal habitats demonstrate increases in these meso-predators in response to decreases in large sharks, a phenomenon rarely observed in pelagic ecosystems. Dogfish (*Squalus* spp.) have increased markedly in many regions: *S. acanthias* increased 17–20x in the Gulf of Alaska and *S. megalops* increased by 5x in Australia. Substantial increases in pelagic rays (*Pteroplatytrygon violacea*) and small teleost fishes can also reflect changes in habitat use. The elimination of large sharks may allow small species to move more easily into epipelagic waters during daytime, a habitat once dominated by large predators. Many large sharks normally feed on marine mammals and reptiles, exerting both direct predation and perceived risk effects. The preferential distribution of predatory sharks in tropical and temperate latitudes could be a key factor in limiting pinniped and bird expansion in these regions. For example, tiger sharks are primary predators of some sirenians, dolphins, turtles, sea snakes, and cormorants. Detailed studies in Australia show that tiger shark presence influences seasonal distribution, habitat use, and feeding behavior of their prey

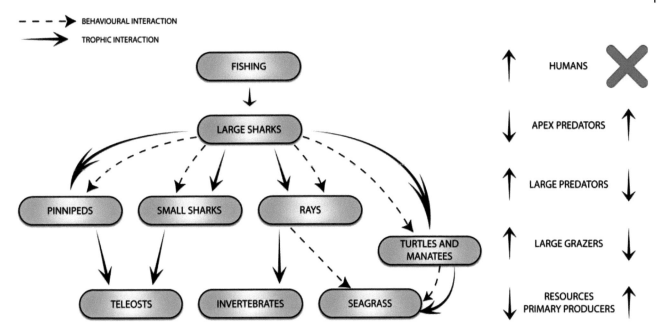

Figure 14.26 Cascading ecosystem effects of removing large sharks, showing trophic interactions (full arrows) and behavioral interactions (dashed arrows) between humans, large predators (sharks), meso-predators (smaller sharks), and prey species. Arrows to the right indicate general trends within various trophic groups.

(green turtles, *Chelonia mydas*, dugongs, *Dugong dugon*, and bottlenose dolphin, *Tursiops aduncus*). Decreased shark predation may have partly contributed to recovery of some populations of megafauna and gray seals (*Halichoerus grypus*) with a 6% increase in annual pup birth during the 1980s prior to a collapse caused by viral infections. Decreased abundances of predatory tiger sharks have led to some recovery in endangered Hawaiian monk seals (*Monachus schauinslandi*).

The decline in large sharks, in particular, has contributed to increased abundances of marine mammals, reptiles, and **mesopredators** (mid-level predators) such as elasmobranchs, inducing cascading effects down the food web in some marine ecosystems. Increased numbers of rays, in turn, have reduced scallop (*Agropecten irradians*) abundances, a key prey item. *Rhinoptera bonasus* (cownose ray) illustrate trophic cascades nicely. Overall, hyper-abundant populations of cownose rays consume large numbers of bivalves and potentially inhibit recovery of oysters and other species of bivalves, which also face pressures of overexploitation, disease, habitat destruction, and pollution. Green turtles and dugongs affect spatial distributions and species composition of seagrasses through their searching and digging for food. Some models suggest that green turtles and dugongs alter their foraging behavior and locations in response to tiger shark presence, with cascading effects on seagrasses. Where tiger shark populations have declined, green turtles have begun to recover, with a consequent reduction in abundance of seagrass beds. The presence of strong interactions between species, mediated by direct consumption and risk avoidance behaviors, lead to important trophic cascades (Figure 14.26). Thus, reduced predation may partly explain increases in mesopredators, which sometimes lead to increased pressure on basal prey species, such as invertebrates, teleost fishes, or even seagrass (Figure 14.26).

Questions

1) What are the most commonly used methods for sampling nekton?
2) Describe the characteristics of nekton migrations.
3) What role do reptiles play in marine environments?
4) Describe diving adaptations in marine mammals.
5) What is the ecological role of large mammals (pinnipeds, cetaceans) in marine food webs?
6) Define trophic cascades.
7) What is the ecological effect of the collapse of sharks in marine environments?

Suggested Reading

Alonso, M.K., Crespo, E.A., Pedraza, S.N. et al. (1999) Food habits of the South American sea lion, Otaria flavescens, off Patagonia, Argentina. *Fishery Bulletin*, 98, pp. 250–263.

André, M., Johansson, T., Delory, E., and van der Schaar, M. (2007) Foraging on squid: The sperm whale mid-range sonar. *Journal of the Marine Biological Association of the United Kingdom*, 87, pp. 59–67.

Bailey, D.M., Jamieson, A.J., Bagley, P.M. et al. (2002) Measurement of in situ oxygen consumption of deep-sea fish using an autonomous lander vehicle. *Deep-Sea Research*, 49, pp. 1519–1529.

Beaugrand, G., Brander, K.M., and Lindley, J.A. (2003) Plankton effect on cod recruitment in the North Sea. *Nature*, 426, pp. 661–664.

Bossart, G.D. (2006) CASE STUDY marine mammals as sentinel species for oceans and human health. *Oceanography*, 19, pp. 134–137.

Boyle, P.R. and Rodhouse, P.G. (2005). *Cephalopods: Ecology and Fisheries*. Blackwell Publishing.

Brodziak, J. and Hendrickson, L. (1999) An analysis of environmental effects on survey catches of squids Loligo pealei and Illex illecebrosus in the northwest Atlantic. *Fishery Bulletin*, 97, pp. 9–24.

Carlson, J.K. and Brusher, J.H. (1999) An index of abundance for aoastal species of juvenile sharks from the Northeast Gulf of Mexico. *Marine Fisheries Review*, 61, pp. 37–45.

Costa, D.P., Breed, G.A., and Robinson, P.W. (2012) New insights into pelagic migrations: Implications for ecology and conservation. *Annual Reviews of Ecology, Evolution and Systematics*, 243, pp. 73–96.

Dudley, S.F.J., Anderson-Reade, M.D., Thompson, G.S., and McMullen, P.B. (2000) Concurrent scavenging off a whale carcass by great white sharks, *Carcharodon carcharias*, and tiger sharks, *Galeocerdo cuvier*. *Fishery Bulletin*, 98, pp. 646–649.

Fanelli, E., Bianchelli, S., and Danovaro, R. (2018) Deep-sea mobile megafauna of Mediterranean submarine canyons and open slopes: Analysis of spatial and bathymetric gradients. *Progress in Oceanography*, 168, pp. 23–24.

Fano, E.A., Mistri, M., and Rossi, R. (2003) The ecofunctional quality index (EQI): A new tool for assessing lagoonal ecosystem impairment. *Estuarine and Coastal Shelf Science*, 56, pp. 709–716.

Ferretti, F., Worm, B., Britten, G.L. et al. (2010) Patterns and ecosystem consequences of shark declines in the ocean. *Ecology Letters*, 13, pp. 1055–1071.

Hays, G.C., Bailey, H., Bogard, S.J. et al. (2019) Translating marine animal tracking data into conservation policy and management. *Trends in Ecology and Evolution*, 34, pp. 459–473.

Hays, G.C., Ferreira, L.C., Sequeira, M.M. et al. (2016) Key questions in marine megafauna movement ecology. *Trends in Ecology and Evolution*, 31, pp. 463–475.

Heckel, G., Murphy, K.E., and Compean Jimerez, G.A. (2000) Evasive behavior of spotted and spinner dolphins (*Stenella attenuata* and *S. longirostris*) during fishing for yellowfin tuna *(Thunnus albacares)* in the eastern Pacific Ocean. *Fishery Bulletin*, 98, pp. 692–703.

Johnson, C.K., Tinker, M.T., Estes, J.A. et al. (2009) Prey choice and habitat use drive sea otter pathogen exposure in a resource-limited coastal system. *Proceedings of the National Academy of Sciences of the United States of America*, 106, pp. 2242–2247.

Morell, V. (2011) Killer whales earn their name. *Science*, 331, pp. 274–276.

Natason, L.J., Casey, J.G., Kohier, N.E., and Colket, T. (1999) Growth of the tiger shark, *Galeocerdo cuvier*, in the western north Atlantic based on tag returns and length frequencies; and a note on the effects of tagging. *Fishery Bulletin*, 97, pp. 944–953.

Poole, D., Givens, G.H., and Raftery, A.E. (1999) A proposed stock assessment method and its application to bowhead whales, *Balaena mysticetus*. *Fishery Bulletin*, 97, pp. 144–152.

Snelgrove, P.V.R. and Haedrich, R.L. (1983) Structure of the deep demersal fish fauna off Newfoundland. *Marine Ecology Progress Series*, 27, pp. 99–107.

Tzanatos, E., Moukas, C., and Koutsidi, M. (2020) Mediterranean nekton traits: Distribution, relationships and significance for marine ecology monitoring and management. *PeerJ*, 8, p. e8494.

Witzeil, W.N. (1999) Distribution and relative abundance of sea turtles caught incidentally by the U.S. pelagic longline fleet in the western North Atlantic Ocean, 1992–1995. *Fishery Bulletin*, 97, pp. 200–211.

Yung, M.S. and Page, B.N. (1999) Diet of Pacific sleeper shark, *Somniosus pacificus*, in the Gulf of Alaska. *Fishery Bulletin*, 97, pp. 406–409.

15

Life Cycles and Larval Ecology

In this chapter, we consider life cycles, reproductive processes, and larval ecology of marine species, the latter of which often involves a planktonic phase. Unlike nekton, early life history stages of most marine organisms lack the swimming capacity typical of mature individuals, and thus depend very much on passive transport by ocean currents. We consider larval supply strategies and the key role larvae play in colonizing habitats and supporting marine ecosystem functioning. Complex life cycles characterize many marine species, and survival of recruits (progeny) strongly influences population dynamics. Larval biology and ecology, their type of development, and their settlement dynamics form the basic biological characteristics of species and community dynamics. These features, together with analysis of factors that determine survival of offspring, strategies of survival in unfavorable conditions, and production of cysts or other resistant forms, form the basis of this chapter.

15.1 Life Cycles and Reproduction

The life cycles of marine organisms represent critical basic knowledge for marine biologists. The role of a given species in its environment, how an ecosystem functions, the dynamics of that species, how that species interacts with its environment and other species, the ecosystem components on which it feeds, and almost every kind of interaction change during the different phases of its life. To date, we have some level of knowledge regarding the life cycles of <1% of marine species.

An organism's **life cycle** refers to the different developmental stages an organism goes through during its life, spanning all the **ontogenetic** (or developmental) stages of a species, from the **zygote** or fertilized egg stage, through sexual maturity. These stages include spawning, embryonic development, larval development, post-larval growth, and maturation with the production of new gametes. **Larva** refers to a stage of the life cycle that occurs in many marine species and that differs morphologically, physiologically, and ecologically from adult forms.

Cnidarians, organisms that appeared early on Earth and sit at the base of metazoan evolution, illustrate the importance of understanding of species' life cycles. Completely different morphologies characterize different life history stages of cnidarians, creating major confusion regarding which species is which (Figure 15.1). Zoologists in the past sometimes gave two different names to the polyp and medusa ("jellyfish") stages of the same species, believing they represented two distinct species rather than two different stages of the same species. The classes that comprise this phylum (Anthozoa, Hydrozoa, Scyphozoa, Cubozoa, Staurozoa, and possibly Polypodiozoa, although the phylogenetic relationship remains unclear) can include three distinct types of biological cycle: (1) a completely pelagic life cycle (medusa stage), (2) an alternating pelagic and benthic cycle (polyp and medusa stages), and (3) a completely benthic life cycle (polyp stage) (see Figure 15.2 for examples).

The life cycle of an organism progresses through stages of initial differentiation: development, growth, and reproduction. These four components correspond to the main life history phases: differentiation and development occur during the embryonic and larval stages, rapid growth occurs during the juvenile stage, and reproduction occurs once an organism matures sexually.

Life history analysis refers to the study of a species' life cycle based on different parameters, such as fecundity, fertility, birth rate, and recruitment. For example, life history analysis considers factors including numbers of eggs an organism produces during a single reproductive event, how many of those eggs hatch, how many reproductive cycles occur each year, embryonic duration, embryo survival rate, larval duration, larval ecology, larval diet, larval mortality, larval settlement rate, post-larval or juvenile duration, post-larval life or juvenile mortality, typical organism lifespan, and other factors determining mortality or recruitment success.

Marine Biology: Comparative Ecology of Planet Ocean, First Edition. Roberto Danovaro and Paul Snelgrove.
© 2024 John Wiley & Sons Ltd. Published 2024 by John Wiley & Sons Ltd.
Companion Website: www.wiley.com/go/danovaro/marinebiology

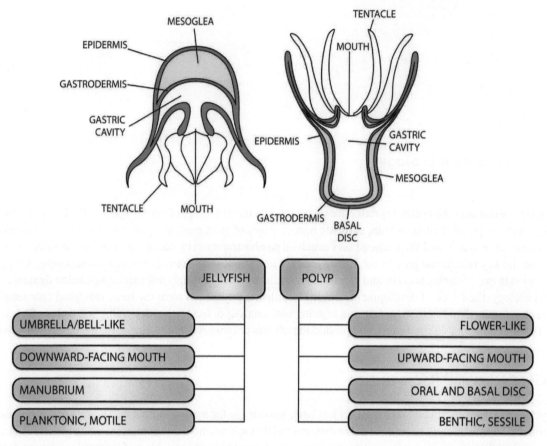

Figure 15.1 Schematic representation of two stages in life cycles of the same species of cnidarian: planktonic jellyfish (left) and benthic polyp (right).

Different species of marine organisms reproduce sexually by producing two gametes that fuse into a zygote, or asexually through mechanisms such as budding and fragmentation that give rise to new individuals without genetic recombination. Many marine organisms use both reproductive strategies, depending on specific circumstances.

Sexual reproduction increases genetic diversity in a species by exchanging genes between two individuals and simultaneously reducing the possibility of a "**genetic bottleneck**" caused by low intraspecific genetic diversity that increases the risk that a species cannot cope with atypical environmental conditions. Sexual reproduction dominates stable environments characterized by highly competitive species and adds considerable energy cost given the need to find a partner or engage in courtship. In more extreme cases, such as some crabs and elephant seals, males compete for females in ritualized fights and clashes with other males that can lead to injury or even death for one or more of the contenders. In addition, during courtship or fighting bouts, male vulnerability to predators may increase, thus elevating mortality rates. Female presence often results in intrusion of other males into a territory already colonized by a resident male, thus interrupting courtship. In this situation, a male must then divide its time between courting the female and defending its territory from rivals. Not surprisingly, studies show that males invest less time in courtship in the presence of a rival. The reduced courtship time suggests that animals must decide periodically how much time to allocate to each behavior, but we cannot say whether all individuals respond similarly. Indeed, individuals from the same population and even of the same age, sex, and reproductive stage may respond differently to such situations.

If individuals behave consistently over time and in different contexts, differences among individuals may originate from genetic variation. However, given the dynamic nature of behavior and influences of social context and motivational state, genetics, experience, environment, or context of the stimuli could all influence consistency and differences in behaviors.

Asexual reproduction requires less time than sexual reproduction and, in circumstances with a temporarily abundant resource, organisms can reproduce quickly and expand population size over a short time in order to exploit the resource efficiently. However, with asexual reproduction genetic mutation, which occurs at low frequencies, offers the only source of genetic diversity.

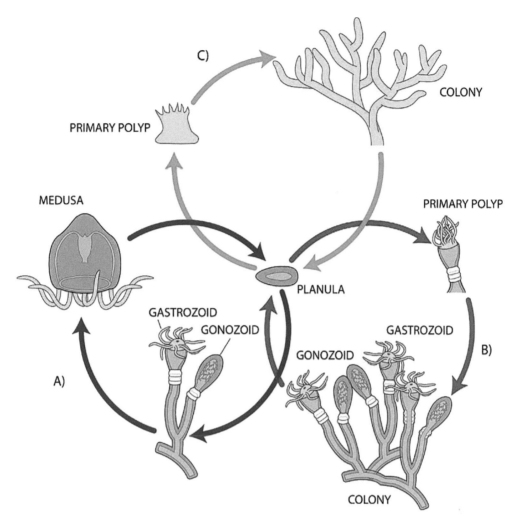

Figure 15.2 Examples of cnidarian life cycles: (A) life cycle of the hydrozoan, *Podocoryne carnea*, showing typical alternation between medusoid and polyp phases. In sessile colonial forms, some polyps differentiate into gonozoids (specialized reproductive polyps), from which medusae develop and grow and, after a series of developmental stages, are released into the water where they eventually reproduce. Embryos transform into planulae that settle and become polyps; (B) life cycle of *Hydractinia illepor*, a colonial hydrozoan that typically adheres to crab shells: the fertilized egg develops into a planula that adheres to shells and metamorphoses into a primary polyp. Through asexual reproduction, the polyp develops into a mature colony of male and female polyps; (C) life cycle of the anthozoan, *Acropora millepora*, where female polyps release hundreds of eggs and male polyps release sperm into the water, where fertilization occurs. The embryo develops into a ciliated larva (planula) that metamorphoses and settles to become a sessile polyp.

Depending on their reproductive mode, we can divide species into **gonochoric** (referring to species that can produce sperm or eggs, but not both) and **hermaphroditic** (which can produce gametes of both sexes) forms. Gonochoric species dominate the animal kingdom. Hermaphroditic species may display both sexes simultaneously, or they may metamorphose from one sex to another. In this case, we differentiate hermaphrodites as **proterogenic** if they begin as females and metamorphose into males, and **protandric** if they begin as males and then become females. In contrast to their terrestrial or freshwater counterparts, gonochoric strategies dominate marine annelids and mollusks (Figure 15.3).

Organisms may undergo **direct** or **indirect development**, depending on whether they go through a larval stage. In direct development, organisms develop through embryo, juvenile, and adult (sexually mature) stages, whereas with indirect development, the embryo develops into a larval stage but subsequently metamorphoses directly into an effective adult stage. Gamete fertilization and fusion may occur externally (releasing gametes into the water) or internally, where males directly or indirectly introduce gametes into the female reproductive apparatus. External fertilization occurs widely in the marine environment, especially in sessile or less mobile species. Significant energy cost and high risk of failure represent key disadvantages. Some species lay eggs in large numbers, attaching them to the substrate in the case of mollusks, or as individual egg cases in the case of sharks (Figure 15.4).

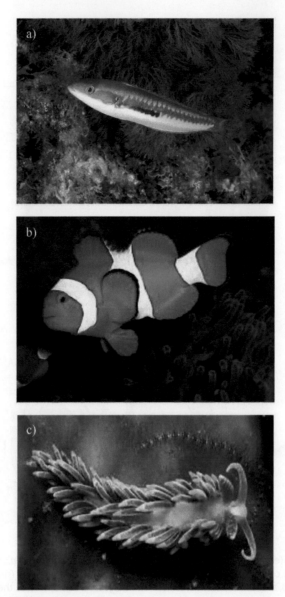

Figure 15.3 Examples of species that are: (a) proterogynic (*Coris iulis*); (b) protandric (clownfish); and (c) hermaphroditic (nudibranchs). *Sources:* Etrusko25 / Wikimedia Commons / Public domain., Nhobgood / Wikimedia Commons / CC BY-SA 3.0, and Parent Géry / Wikimedia Commons / Public Domain.

15.2 Larval Ecology

Larval ecology, development, and settlement dynamics fundamentally drive the biology of species and ecological communities. Some species evolve a reproductive strategy (**gregarious settlement**) that ensures larvae and juveniles settle near conspecific adults and thus increase the likelihood of finding favorable environmental conditions for survival. In some cases, currents transport larvae into new environments, creating risk of not finding suitable habitat, but also creating opportunities to settle in different communities or spaces than their conspecific adults. In **sessile species**, referring to those organisms with limited ability to move for most of their life cycle, egg and larval stages provide the main opportunity for dispersal. Life cycles can be complex and often use one or more intermediate larval stages that occupy an environment substantially different from that of the adults. Biologists distinguish four main types of life cycles (Figure 15.5):

1) **meroplanktonic** species, in which larval stages occur in pelagic environments;
2) **merobenthic** species, in which larval stages occur in benthic environments;
3) **holoplanktonic** species that complete their entire life cycle in pelagic environments;
4) **holobenthic** species that complete their entire life cycle within the benthic compartment.

Dispersal, settlement, and larval survival represent fundamental aspects of understanding the ecology of marine species. Larval dispersal in some species, for example, spans large distances of 100s and even 1000s of kilometers whereas other species totally lack a larval dispersal phase. A merobenthic cycle characterizes many species, especially those that form resting stages such as cysts, which typically occur in sediments.

Figure 15.4 Timing of gamete release in marine organisms fundamentally alters probability of fertilization. Sessile organisms on coral reefs illustrate this phenomenon: (a) coral *Montastraea cavernosa* releases gametes into the water; (b) spawning during night; (c) cephalopod eggs; (d) shark egg case. *Sources:* Haplochromis / Wikimedia Commons / Public domain and NOAA / Wikipedia Commons / Public Domain.

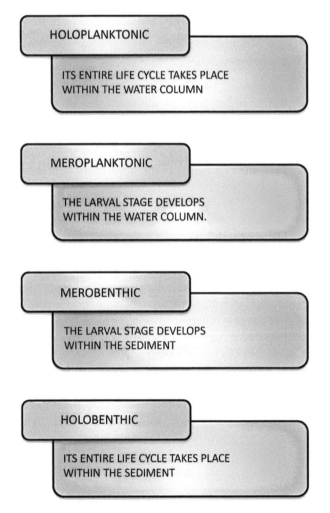

Figure 15.5 Summary of different types of life history strategies: holoplanktonic (cycle entirely within the water); meroplanktonic (planktonic as larval stage, benthic as adults); merobenthic (benthic larvae, planktonic or pelagic adults); holobenthic (larvae develop in sediment).

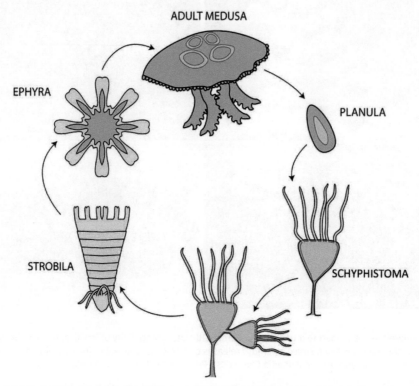

Figure 15.6 Example of merobenthic life cycle illustrating markedly different morphologies and alternation of sexual and asexual reproduction. Separate sexes of adults of the scyphozoan, *Aurelia aurita*, release gametes into the water, and zygotes either develop on the oral arms of females or swim freely, eventually giving rise to a ciliated planula larva that settles onto the substrate and develops into polypoid larvae or scyphostomae. Through strobilation (transverse splitting), scyphostomae produce pelagic medusoid larvae known as ephyrae that grow into adult scyphomedusae.

Cnidaria (class Scyphozoa), which reproduce both sexually and asexually, are considered merobenthic because their developmental (~larval) polyp stage lives as a sessile form within the benthic environment, propagating asexually through strobilation and giving rise to adult planktonic medusae that reproduce sexually. **Strobilation** refers to the process of segmentation and detachment of a portion of the polyp that becomes the body of the planktonic medusa. In contrast, pelagic medusae reproduce sexually by releasing gametes that fuse in the water column, developing into larvae (**planulae**) that settle on the bottom to form a new colony (Figure 15.6).

Most polychaetes, in contrast, produce meroplanktonic larvae (Figure 15.7), referred to as **trochophores**, that develop from the fusion of gametes. This larva metamorphoses into another pelagic larval stage called a **nectochaeta** that lives in the water column until settling into a benthic habitat, metamorphosing, and developing into an adult. Some holobenthic species, such as *Nereis diversicolor*, live their life cycle in the benthos. The scyphozoan and polychaete examples above illustrate how meroplankton differ from merobenthos in that the two distinct phases of their life cycles occur in two very different environments (benthic and planktonic). The benthic larval stage in Scyphozoa (merobenthic) and planktonic medusoid stage, contrast the meroplanktonic larval stage and adult benthic stage of polychaetes.

Marine invertebrates encompass a wide range of different larval types in addition to those listed above. Lophophorates produce **actinotroch larvae**, bryozoans produce **cyphonautes**, nemertines produce **pilidium larvae**, and turbellarians produce **Müller's larvae**. However, some taxa can produce similar types of larvae. For example, some bivalves and some annelids produce larval stages known as **veligers**. Within the same phylum, different taxa sometimes produce different types of larvae. For example, among crustaceans, copepods produce **nauplii** with different developmental stages that precede a series of **copepodite** developmental stages. In contrast, barnacles develop from a nauplkiar stage into a larval **cyprid**, and decapods develop through **zoeal** stages followed by a **megalopa** stage prior to settlement. Among echinoderms, echinoids produce **echinopluteus** larvae, and asteroids produce **diplanula** larvae. These diverse larval forms often lead different lifestyles (Figure 15.8).

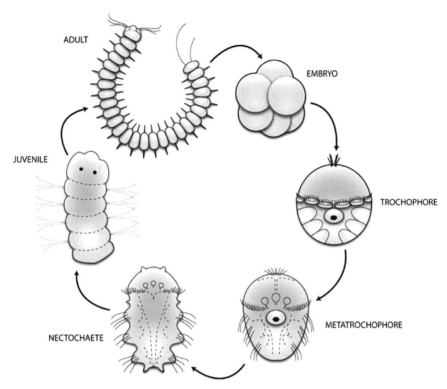

Figure 15.7 Example of life cycle of a typical macrofaunal organism, a polychaete, showing the different stages of life cycle. The embryo formed by fusion of gametes during sexual reproduction gives rise to trochophore larvae, which develop through a metatrochophore stage, a post larval form (juvenile), and then pre-adult and adult (sexually mature) stages.

15.3 Life History Strategies

Morphological variation in a species, resulting from "interaction" with the environment, known as **phenotypic plasticity**, influences important aspects of reproduction. As a result, a given species does not necessarily reproduce synchronously or produce the same numbers of eggs or larvae and may differ in reproduction in many other respects. The affinity (or differences) between various aspects of life histories of different species points to recurring patterns of life history strategies known as r and K selection theory.

In general, this theory divides **vital** (life history) strategies into two major categories: K and r strategies (general characteristics in Table 15.1), where k links to the **carrying capacity**, the maximum value of population growth in a location described by a sigmoidal curve:

$$dN/dt = rN(k - Nk)$$

If a population overcomes the k threshold, intraspecific competition reduces biotic and reproductive potential r. In contrast, populations living in non-limiting conditions may increase rapidly in number through exponential growth described by:

$$dN/dt = rN$$

Obviously, not all species fall neatly into one of these two groups. In fact, some taxonomic groups include both r and K strategy species. These two extreme strategies represent a simplification in that many intermediate forms also exist.

Different species exhibit remarkable variability in reproductive strategy. **Semelparous** species, such as Pacific salmon and catadromous eels, reproduce just once during their life cycle, in contrast to **iteroparous** species, such as Atlantic salmon and cod, that reproduce multiple times during their lives. Sometimes, highly evolved species can show plasticity with different reproductive events during their life cycle rather than just one. Different life cycles and larval stages characterize K and r strategies. Larvae group into two broad categories (Table 15.1): **lecithotrophic** larvae use food reserves in the egg and can therefore survive without external food; **planktotrophic** larvae feed on phytoplankton and/or small zooplankton immediately or shortly after hatching.

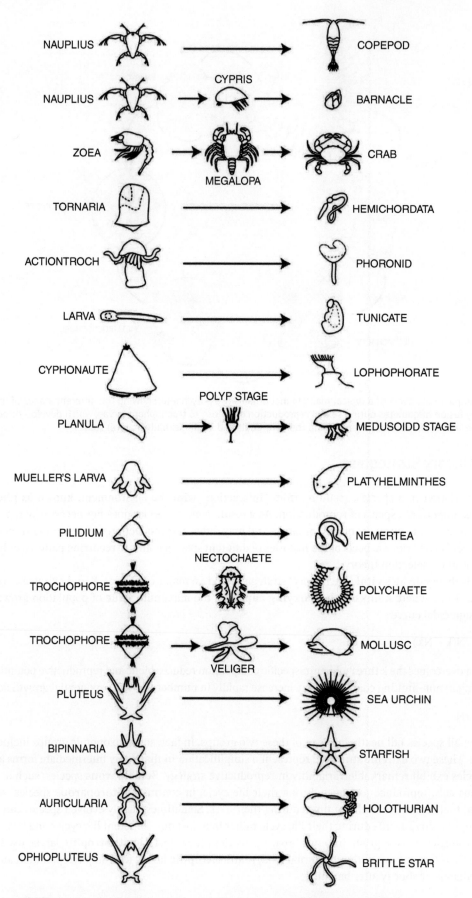

Figure 15.8 Larvae of different groups of marine organisms, indicating corresponding adult organisms.

Table 15.1 Key differences between K-strategy species (hereafter referred to as K species) and r-strategy species (hereafter referred to as r species): (1) K species live in relatively predictable and stable environments, in contrast to the unpredictable environments characteristic of r species; (2) K species reproduce relatively infrequently compared to r species; (3) K species grow at slow rates in contrast to rapid growth and often small size in r species; (4) K species invest large amounts of resources (energy, parental care) per offspring and thus produce fewer individuals, whereas r species invest little energy per offspring and produce many individuals; (5) generally stable populations in K species contrast highly variable population size for r species; (6) K species strongly compete with other species for resources, whereas r species are weak competitors.

Ecological/Biological or Environmental features	K Strategists	r Strategists
Environment	Predictable – stable	Unpredictable – unstable
Number of offspring	Few	Many
Growth	Slow	Fast
Reproductive investment/ offspring	High	Low
Mortality	Low	High
Development	Slow	Rapid
Adult Size	Large	Small
Juvenile Size	Large	Small
Parental Care	Present	Absent
Population Size	Stable	Variable
Competitive ability	High	Low

Lecithotrophic larvae rely on their stored energy reserves, resulting in a relatively short pelagic life that varies from minutes in some species to a few days at most. These species produce large, but relatively few eggs rich in energy reserves. A short larval stage also means limited dispersal potential. Indeed, lecithotrophic larvae often settle close to parental habitats. Planktotrophic larvae, in contrast, are produced in large numbers, live longer, and may remain planktonic for several weeks (or longer), thus increasing dispersal potential, but they also experience high mortality (<1% typically reach metamorphosis). Some planktotrophic larvae, known as **teleplanic larvae** may remain planktonic for up to three months, greatly increasing their dispersal potential, despite high mortality rates.

Benthic larvae (Table 15.2), which are typically lecithotrophic and short-lived, occur close to the bottom (demersal species swim actively near the sediment-water interface). In these species, most dispersal occurs during the adult phase, for example in many species of cnidarians and the mollusk, *Adamussium colbecki*, whose veligers settle on the shells of the adults that produced them. This strategy also occurs in many abyssal species, whose larval stages often cannot survive in the water column. With direct development, however, individuals bypass the first life history stages and develop from an egg into a juvenile form ready to settle, even if not yet sexually mature. Nematodes exemplify a benthic species lacking a larval stage; the zygote develops through several juvenile stages (moults) similar to adults but prior to sexual maturity.

Table 15.2 Larval classification based on life history strategies reflecting different reproductive effort, mortality rate, and dispersal potential (+ + + = high, + + = medium, + = low, – = null).

Category	Resource' use	Reproductive effort	Number of Eggs Produced	Dispersal Potential	Eggs' Mortality
Teleplanic	Very high	Low	Very high	Very high	Very high
Pelagic Planktotrophic	High	Low	Very high	High	High
Pelagic Lecithotrophic	Very low	High	Low	Low	Medium
Benthic	Low	Very high	Very low	Very low	Very low
Direct development	Low	Very high	Very low	Very low	Very low

15.4 Supply Side Ecology

Dispersal plays a crucial role in the ecology and evolution of organisms, especially in marine systems where life cycles of different species encompass both benthic and planktonic stages. At the end of the dispersal phase, larval abundance and larval settlement patterns often differ from those at the point of origin. Recent studies attribute benthic composition to variation in larval mortality, transport mechanisms, and larval behavior during settlement. Supply-side ecology (Figure 15.9) recognizes the key role that larval supply plays in determining local adult population size: each community, in fact, reflects larval recruits from elsewhere. It is not just studies of adult ecology that lead to understanding of species dynamics, but rather studies on larval biology, larval contributions from other locations, settlement cues, and related factors that advance understanding of development of biotic communities. A given community results from a continuous history of colonization following mortality of adults. In marine benthic communities, pelagic larvae (most macrobenthos possess pelagic larvae) contribute to most colonization. These larvae rarely come from resident adults from that site; more often, these propagules were produced upstream of the location and then transported by currents. In many cases, abundances of adults in a given location depend primarily on larval recruitment, which, in turn, reflects physical transport processes, particularly currents, and benthic boundary layer flow, and subsequent biological interactions (food, predation, juvenile mortality).

15.5 Forms of Resistance and Benthic-pelagic Coupling

Some species can respond to adverse environmental conditions by producing resting stages and cysts as a physiological response to adverse environmental conditions. Many planktonic organisms, including some holoplanktonic species, produce benthic resting stages. Despite limited information on the life cycles of these organisms, resting stages occur widely in temperate coastal species, from microalgae to crustaceans and polychaetes. This strategy suspends life activity, enabling organisms to survive harsh conditions.

In nature, organisms deal with these conditions with two different processes:

a) **quiescence**, or a developmental delay in immediate response to environmental stressors. In quiescent organisms, reactivation depends on when environmental conditions become favorable again, at which point development immediately continues;

b) **diapause**, a biological development disruption "programmed" by the genome that suspends the life cycle of an organism, typically in advance of unfavorable environmental conditions. In this case, development resumes only after completion of a fixed period in diapause and when environmental conditions become suitable.

Resting stages and cysts are typically spherical in shape and highly robust, often covered by a resistant wall and spines and protuberances to discourage ingestion by predators (Figure 15.9). Identifying many forms of resting cysts can be challenging given relatively few morphological characters, and the difficulties in extracting and amplifying DNA contained within cysts hampers molecular analysis. Identification often becomes certain only after a cyst hatches and the organism develops (Figure 15.10).

Figure 15.9 Supply-side ecology refers to a mechanism by which larval stages move from areas or origin (source areas) to other locations. If juveniles can grow within this new location and become reproductive adults, these areas become new source locations. In contrast, if larvae arrive in locations where they cannot grow and reproduce, we refer to these locations as "sinks" or "loss areas." Individuals in these locations do not contribute to future generations.

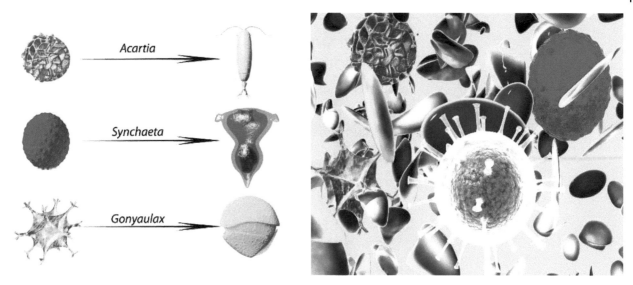

Figure 15.10 Examples of cysts of various marine organisms, including copepods (upper), polychaetes (middle) and dinoflagellates (lower), noting that similarity in morphologies often makes identification difficult. Cysts similar in shape develop into organisms spanning different phyla or kingdoms. In these examples, copepods (*Acartia* sp.), annelid worms (*Synchaeta* sp.), and dinoflagellates (*Gonyaulax* sp.) produce cysts similar in size.

In adverse environmental conditions, many active marine organisms encyst or develop other resistant forms, but some species also produce resistant eggs that, when released, develop only after a resting period (Figure 15.11).

Many planktonic organisms produce cysts and resting stages that reside within the benthos and represent merobenthos (in contrast to meroplankton, typically referring to the planktonic larvae of benthic organisms). The inert resistant forms behave much like sediment particles on the seabed. These inert particles preferentially accumulate at sites with particles with similar characteristics. In sediments, "banks" of cysts and resistant forms can accumulate, and quickly develop once environmental conditions become favorable. Isotope dating of copepod eggs in an anoxic coastal basin in the northeastern United States indicated eggs >40 years old were still viable.

Resting stages contribute to benthic-pelagic coupling, which refers to biological linkages between the two realms, such as the release of resting stages from the benthos into the water column, sometimes leading to rapid and generally unpredictable increases (blooms) of planktonic species such as toxic dinoflagellates. This rapid change occurs because these organisms were

$$a = d + eld$$

a: number of unborn resting stages
d: number of dead resting stages
eld: number of resting stages entering extra-long diapause

$$N = A - a + held$$

N: number of individuals starting a new cohort
A: number of produced cysts
a: number of unborn resting stages
held: number of resting stages entering extra-long diapause (derived from distant cohorts)

Figure 15.11 Number of individuals that produce a new cohort depends on cysts produced (A) which must be subtracted from the total because they may die or enter an extended diapause (a), and previous generations that emerge from extended diapause (held) must be added.

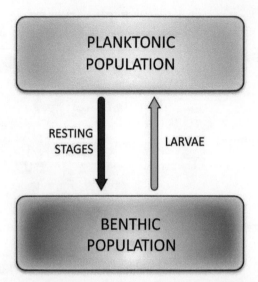

Figure 15.12 Exchange between planktonic populations producing resting stages and cysts and benthos that periodically release larvae, cysts, and hatched resting stages.

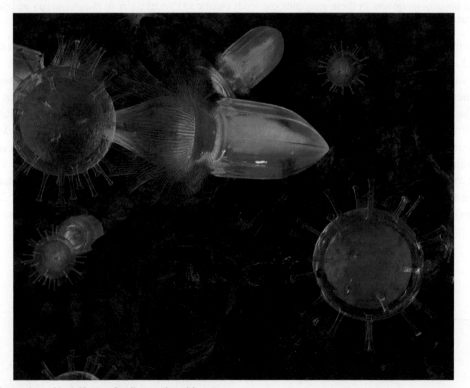

Figure 15.13 Illustration of loriciferans feeding on benthic cysts.

absent from the water column prior to their release from seafloor sediments. This phenomenon also represents one of the main reasons why planktonic species seemingly disappear and reappear from marine environments. In many cases, they simply encyst and settle to the seafloor, suspending their activity, periodically replenishing populations with propagules and cysts hatched from the seafloor. This process, which depends on exchanges between plankton and benthos, helps marine organisms persist over time (Figure 15.12).

Despite limited knowledge on the physiology of resistant forms or even on their distribution and abundance in different marine sediments, evidence indicates that multiple factors play an important role in controlling planktonic abundance. For example, bottom currents determine the spatial distribution of resistant forms within sediment depths. Benthic predation on merobenthos also plays an important role in that some organisms with specific mouthparts or feeding appendages (such as loriciferans and tardigrades) can feed on these cysts (Figure 15.13).

Nematodes, tardigrades, and other organisms can pierce the cells and suck out their contents much like using straws to drink from (miniature) "coconuts." Once in sediments, benthic deposit feeders also feed on resting stages. Predation on these banks of resting stages may affect future structure of planktonic communities: predation on cysts reduces their contribution to plankton populations. Furthermore, selective predation may favor some species over others.

Resting stages may also have important implications for gene exchange. For example, crustacean cysts produced hundreds of years ago can successfully hatch, despite an active life cycle phase that lasts just a few weeks. At the time of hatching these individuals encounter conspecifics produced hundreds of generations after them. This encounter can be important for genetic drift, particularly, where genetic bottlenecks occur. The "sawtooth" trend model describes temporal dynamics in these species, with periods during which the population experiences strong natural selection that coincides with dramatic decreases in abundance, followed by an increase in population size (explosion) and then a sudden decrease (crash). Surviving organisms then become the founders of successive populations. These dynamics also depend on complex interactions between resistant forms in normal diapause and those in extended diapause.

Questions

1) Summarize the key differences between meroplankton and merobenthos.
2) Provide examples of K and r strategists in marine systems.
3) Describe the life cycle of a marine polychaete.
4) What are the differences between life cycles and life histories?
5) Differentiate between lecithotrophic and planktotrophic larvae.
6) Describe the main types of larvae.
7) What is supply-side ecology?
8) Contrast diapause and quiescence.
9) What are resting stages?
10) Which marine organisms produce resting stages?

Suggested Reading

Carrier, T., Reitzel, A., and Heyland, A. eds. (2017) *Evolutionary Ecology of Marine Invertebrate Larvae*. Oxford Academic.

Casagrandi, R. and Gatto, M. (2006) The intermediate dispersal principle in spatially explicit metapopulations. *Journal of Theorerical Biology*, 239, pp. 22–32.

Cowen, R.K., Gawaekiewicz, G., Pineda, J. et al. (2007) Population connectivity in marine systems: An overview. *Oceanography*, 20, pp. 14–21.

D'Aloia, C.C., Bogdanowicz, S.M., Francis, R.K. et al. (2015) Patterns, causes, and consequences of marine larval dispersal. *Proceedings of the National Academy of Sciences of the United States of America*, 112, pp. 13940–13945.

Franco, A., Elliott, M., Franzoi, P., and Torricelli, P. (2008) Life strategies of fishes in European estuaries: The functional guild approach. *Marine Ecology Progress Series*, 354, pp. 219–228.

Lee, S.C. and Bruno, J.F. (2009) Propagule supply controls grazer community structure and primary production in a benthic marine ecosystem. *Proceedings of the National Academy of Sciences of the United States of America*, 106, pp. 7052–7057.

Lutjeharms, J.R.E. and Heydorn, A.E.F. (1981) The rock lobster *(Jasus tristani)* on Vema Seamount: Drifting buoys suggest a possible recruiting mechanism. *Deep-Sea Research*, 28, pp. 631–636.

MacKenzie, B.R., Hinrichsen, H.H., Plikshs, M. et al. (2000) Quantifying environmental heterogeneity: Habitat size necessary for successful development of cod *Gadus morhua* eggs in the Baltic Sea. *Marine Ecology Progress Series*, 193, pp. 143–156.

Marcus, N.H. and Boero, F. (1998) Minireview: The importance of benthic-pelagic coupling and the forgotten role of life cycles in coastal aquatic systems. *Limnology and Oceanography*, 43, pp. 763–768.

McEdward, L. (1995) *Ecology of Marine Invertebrate Larvae*. CRC Press.

Mollet, H.F., Cliff, G., Pratt, H.L., Jr, and Stevens, J.D. (2000) Reproductive biology of the female shortfin mako, *Isurus oxyrinchus* Rafinesque, 1810, with comments on the embryonic development of lamnoids. *Fishery Bulletin*, 98, pp. 299–318.

Montresor, M. and Tomas, C.R. (1988) Growth and probable gamete formation in the marine dinoflagellate *Ceratium schrankii*. *Journal of Phycology*, 24, pp. 495–502.

Montresor, M., Zingone, A., and Sarno, D. (1998) Dinoflagellate cyst production at a coastal Mediterranean site. *Journal of Plankton Research*, 20, pp. 2291–2312.

Mullineaux, L.S. and Mills, S.W. (1997) A test of the larval retention hypothesis in seamount-generated flows. *Deep-Sea Research*, 44, pp. 745–770.

Neuheimer, A.B., Hartvig, M., Heuschele, J. et al. (2015) Adult and offspring size in the ocean over 17 orders of magnitude follows two life history strategies. *Ecology*, 96, pp. 3303–3311.

Palumbi, S.R. (1994) Genetic divergence, reproductive isolation, and marine speciation. *Annual Review of Ecology and Systematics*, 25, pp. 547–572.

Privitera, D., Noli, M., Falugi, C., and Chiantore, M. (2011) Benthic assemblages and temperature effects on *Paracentrotus lividus* and *Arbacia lixula* larvae and settlement. *Journal of Experimental Marine Biology and Ecology*, 407, pp. 6–11.

Swearer, S.E., Treml, E.A., and Shima, J.S. (2019) A review of biophysical models of marine larval dispersal. *Oceanography and Marine Biology: An Annual Review*, 57, pp. 325–356.

Part III

Comparative Marine Ecology: Habitat Types, Their Biodiversity, and Their Functioning

The five chapters on *Ecosystem Functioning* represent the roles played by the different actors, and specifically their activities, how they interact, and their importance to the overall "plot" of our play or functioning of the ecosystem. These functions or processes contribute to the overall health of the ocean, and thus the health of the planet. In the 1970s, a chemist named James Lovelock and a microbiologist named Lynn Margulis proposed the Gaia Hypothesis, which posits that interacting living organisms and their surroundings on Earth form a sort of superorganism that sustains the conditions necessary for life on our planet. Although this hypothesis has many critics, most ecologists agree that living organisms play vital roles in the various processes that go on in the ocean, and loss of organisms compromises those roles. Building on the earlier sections on diversity and composition of species in different environments, this section asks "so what?", and in answering the question illustrates the importance of biodiversity for critical ocean processes and discusses whether loss of biodiversity will compromise those processes.

Following descriptions of the roles that biota play we must place those processes into a spatial context, namely the specific habitats in which they occur. Much like the greater visibility of the setting of a play at the front of the stage, most humans are more familiar with the *Coastal Ecosystems* described in this section than any other aspect of the ocean. From the beaches and rocky shorelines that many of us visit, to the estuaries on which many of our great cities (New York, Cairo, San Francisco, Montreal, Kolkata (formerly Calcutta)) have grown, to the continental shelves where recreational and commercial fisheries concentrate their efforts, proximity breeds familiarity. But these coastal ecosystems also encompass some of the most productive and dynamic ocean environments, as well as some of the best known and harshest environments that experience not only natural variability but also the most intense human-induced change. From the harsh rocky intertidal to shifting sand- and mudflats, to the dense vegetation of mangroves and salt marshes that transition land to ocean and the waters above them, these habitats also support specialized faunas adapted to their unique environmental characteristics. The chapters on *Coastal Ecosystems* take us from the shoreline to the edge of the continental shelf, from tropical waters to the poles, and describe the many habitats and processes that characterize these ecosystems. *Open Ocean Ecosystems*, where total depths typically range from 200 to 10,000 meters, represent the least visible part of our ocean stage, and the least sampled. Far from the shoreline, sometimes 100s of kilometers from the nearest land mass, most humans only see the surface of these dominant features of our planet from the window of a plane, or perhaps from the deck of a cruise ship. The margins or "sides" of the ocean basins support a diverse and unique variety of habitats, spanning from the canyons that resemble a drowned version of the Grand Canyon, to superheated hydrothermal vent oases supported by toxic chemicals bubbling from the seafloor. The seawater that fills these basins forms the largest habitat on the planet, supporting living organisms from its surface to its greatest depths. Seamounts and mid-ocean ridges interrupt the rolling plains on the bottom of the basin and, though invisible from the ocean surface, rival the largest mountains and expansive mountain ranges we see on land. The three chapters that comprise this subsection describe the biota and processes that characterize open ocean ecosystems.

16

Ecosystem Functioning I: Primary and Secondary Production

16.1 Introduction

Marine ecosystem functioning depends on a complex network of interactions among organisms and between them and the environment. **Primary production** results from a series of processes through which photosynthetic organisms synthesize organic matter from inorganic carbon (mainly H_2O and CO_2), thus representing one of the most important examples of ecosystem functioning. This chapter begins by examining the main factors affecting marine primary production, such as (1) light intensity; (2) nutrients; (3) hydrodynamics of the system; and (4) the type of primary producers that are present in a system. In marine environments, and particularly in the open ocean, the smallest photoautotroph fraction, the phytoplankton, dominates primary production (picoplankton, in particular, can contribute up to 90% of primary production).

We then consider **secondary production**, referring to tissues and cells formed by primary consumers; secondary producers or heterotrophic organisms grow by ingesting organic matter produced by plants, algae, and photosynthetic and chemosynthetic bacteria (primary producers). Only a small fraction of organic matter accumulates in ocean sediments, because respiration ("breathing") **remineralizes** or decomposes the remaining material (>99%) into its raw constituents (carbon dioxide, nutrients, and minerals). In this chapter, we also consider extra-specific processes, which cover aspects of organic matter cycling that do not involve direct relationships between species, with particular reference to different types of organic matter (and related classifications) and to the use of nonliving organic matter (detritus) by organisms. The functioning of all ecosystems critically depends on quality and quantity of food resources.

16.2 Primary Production

All cells require energy to sustain their metabolism. They can obtain this energy through three pathways: **chemorganotrophy**, where organisms obtain energy from organic compounds, **chemolithotrophy**, where organisms use inorganic compounds, and **phototrophy**, in which light triggers electron flow (Figure 16.1).

As discussed in an earlier chapter, primary production refers to the complex processes through which living organisms transform inorganic carbon into organic material. The chemical reactions through which they convert carbon from inorganic to organic forms represent one of the fundamental mechanisms driving ecosystem functioning. Primary production can follow two pathways: (1) **photoautotrophic** (organisms capture light as an energy source) or (2) **chemoautotrophic** (organisms use potential energy of chemical bonds as an energy source).

- **Photoautotrophic** organisms use light energy to synthesize organic compounds starting with simple inorganic molecules (photosynthesis). Photosynthetic pigments capture the energy of photons, either inside chloroplasts, in the case of plants and unicellular algae, or scattered in the cytoplasm in the case of cyanobacteria and photobacteria.
- **Chemoautotrophic** organisms synthesize organic substances, oxidizing inorganic molecules. Prokaryotes exclusively perform chemosynthesis, especially archaea that use inorganic compounds and carbon dioxide to produce energy.

In both cases, the conversion of carbon dioxide forms the basis for organic matter production. Just a few decades ago, scientists believed that photoautotrophic production (through photosynthesis) provided the engine for all ecosystems, a

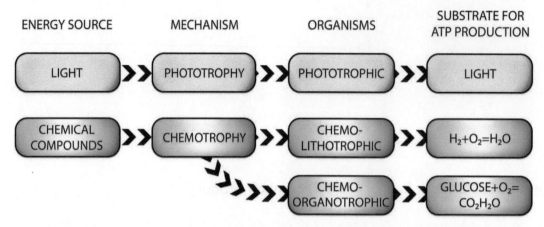

Figure 16.1 Metabolic options to obtain energy. Oxidation of organic and inorganic compounds by chemoautotrophic organisms yields ATP, whereas photoautotrophic organisms convert solar energy into chemical energy (in the form of ATP).

natural transfer of concepts derived from the terrestrial realm to marine environments; life on land depends fully on energy derived from sunlight, without which plants, the basis of all terrestrial food webs, could not exist. Historically, scientists largely ignored the importance of chemoautotrophic production except in specialized environments (e.g., hydrothermal vents, seeps), but recent evidence suggests a need to reevaluate this assumption because of the common, though less intense, occurrence of primary production other than photoautotrophic production across large swaths of the ocean.

Photosynthesis accounts for some 99.9% of primary production in terrestrial environments, but new evidence suggests this percentage may drop to as little as 50–60% in marine habitats as a result of chemosynthetic contributions.

Scientists quantify primary production as the total amount of organic matter (or organic carbon) per unit area or volume expressed as units of mass (grams, moles) of carbon fixed by primary producers per unit area (e.g., m^2), or unit volume (e.g., m^3) and primary productivity refers to primary production per unit of time (hour, day, year).

Terrestrial and ocean ecosystems recycle energy and matter efficiently. Terrestrial ecosystems produce about 56 Pg C yr^{-1} net primary production, exporting a small portion of the carbon fixed and recycled on land (1–4 Pg C yr^{-1}) to the ocean through rivers (0.4–0.5 Pg C yr^{-1}). Globally, marine primary production yields an estimated 49–54 Pg C yr^{-1} (or Gt, Giga tons of carbon y^{-1}), or about half of the primary production on Earth. Almost all of net marine primary production is recycled, with about 1% burial in marine sediments. Although heterotrophic organisms recycle organic matter efficiently, they cannot use all of the material because they must divert some into metabolites such as ammonium, and, in anoxic conditions, as reduced compounds such as sulfide.

Chemoautotrophic organisms obtain energy for inorganic carbon fixation from reduced inorganic compounds to yield global chemoautotrophic primary production of about 0.77 Pg C yr^{-1}, including 0.3 Pg C yr^{-1} in coastal sediments, 0.3 Pg C yr^{-1} from nitrification in the euphotic zone, and ~0.1 Pg C yr^{-1} in the deep sea. This input of new organic carbon to the ocean approximately equals the amount delivered to the ocean by the rivers of the world, and eventually it becomes buried in marine sediments. Chemoautotrophic production derived from the recycling of organic matter exceeds the production fueled by deep hydrothermal vents globally.

16.3 Chemosynthetic Primary Production in the Ocean

The famous Russian microbiologist Sergei Winogradsky developed the concept of **chemolithoautotrophy** (referring to organisms that derive energy from reduced inorganic compounds) toward the middle of the last century. Studying populations of bacteria of the genus *Beggiatoa*, he demonstrated how an organism oxidizes and uses inorganic substrates (in this case hydrogen sulfide, H_2S) as their sole source of energy. Chemolithoautotrophic metabolism, which typically involves aerobic respiration using an inorganic rather than an organic energy source, occurs only in prokaryotes that exhibit high metabolic plasticity, especially regarding energy use.

We currently know of two mechanisms for generating energy (i.e., metabolic processes): respiration and fermentation. Metabolic diversity in prokaryotes lies in their ability to use different electron donors and acceptors. Microorganisms further subdivide into those capable of using NADH (nicotinamide adenine dinucleotide + hydrogen) to reduce the CO_2 in organic compounds, and those unable to do so. Thus, chemoautotrophic microorganisms couple chemolithotrophic metabolism with CO_2 fixation. These organisms use a wide spectrum of inorganic compounds (e.g., hydrogen sulfide, iron, and ammonia; Table 16.1), even if a specific prokaryote may specialize on related inorganic compounds. Chemolithotrophic metabolism

Table 16.1 Examples and sources of electrons and electron acceptors of chemoautotrophs.

Name	Examples	Source of electrons	Respiration electron acceptor
Iron bacteria	*Acidithiobacillus ferrooxidans*	Fe^{2+} (ferrous iron) $\rightarrow Fe^{3+}$ (ferric iron) $+ e^{-}$ [21]	O_2 (oxygen) $+ 4H^{+} + 4e^{-} \rightarrow 2H_2O$ [21]
Nitrosifying bacteria	*Nitrosomonas*	NH_3 (ammonia) $+ 2H_2O \rightarrow NO_2^{-}$ (nitrite) $+ 7H^{+} + 6e^{-}$ [22]	O_2 (oxygen) $+ 4H^{+} + 4e^{-} \rightarrow 2H_2O$ [22]
Nitrifying bacteria	*Nitrobacter*	NO_2^{-} (nitrite) $+ H_2O \rightarrow NO_3^{-}$ (nitrate) $+ 2H^{+} + 2e^{-}$ [23]	O_2 (oxygen) $+ 4H^{+} + 4e^{-} \rightarrow 2H_2O$ [23]
Chemotrophic purple sulfur bacteria	Halothiobacillaceae	S^{2-} (sulfide) $\rightarrow S^{0}$ (sulfur) $+ 2e^{-}$	O_2 (oxygen) $+ 4H^{+} + 4e^{-} \rightarrow 2H_2O$
Sulfur-oxidizing bacteria	Chemotrophic Rhodobacteraceae and Thiotrichaceae	S^{0} (sulfur) $+ 4H_2O \rightarrow SO_4^{2-}$ (sulfate) $+ 8H^{+} + 6e^{-}$	O_2 (oxygen) $+ 4H^{+} + 4e^{-} \rightarrow 2H_2O$
Aerobic hydrogen bacteria	*Cupriavidus metallidurans*	H_2 (hydrogen) $\rightarrow 2H^{+} + 2e^{-}$ [24]	O_2 (oxygen) $+ 4H^{+} + 4e^{-} \rightarrow 2H_2O$ [24]
Anammox bacteria	Planctomycetes	NH_4^{+} (ammonium) $\rightarrow 1/2N_2$ (nitrogen) $+ 4H^{+} + 3e^{-}$ [25]	NO_2^{-} (nitrite) $+ 4H^{+} + 3e^{-} \rightarrow 1/2N_2$ (nitrogen) $+ 2H_2O$ [25]
Thiobacillus denitrificans	*Thiobacillus denitrificans*	S^{0} (sulfur) $+ 4H_2O \rightarrow SO_4^{2-} + 8H^{+} + 6e^{-}$ [26]	NO_3^{-} (nitrite) $+ 6H^{+} + 5e^{-} \rightarrow 1/2N_2$ (nitrogen) $+ 3H_2O$ [26]
Sulfate-reducing bacteria: Hydrogen bacteria	*Desulfovibrio paquesii*	H_2 (hydrogen) $\rightarrow 2H^{+} + 2e^{-}$ [24]	$SO_4^{2-} + 8H^{+} + 6e^{-} \rightarrow S^{0} + 4H_2O$ [24]
Sulfate-reducing bacteria: Phosphite bacteria	*Desulfotignum phosphitoxidans*	PO_3^{3-} (phosphite) $+ H_2O \rightarrow PO_4^{3-}$ (phosphate) $+ 2H^{+} + 2e^{-}$	SO_4^{2-} (sulfate) $+ 8H^{+} + 6e^{-} \rightarrow S^{0}$ (sulfate) $+ 4H_2O$
Methanogens	Archaea	H_2 (hydrogen) $\rightarrow 2H^{+} + 2e^{-}$	$CO_2 + 8H^{+} + 8e^{-} \rightarrow CH_4$ (methane) $+ 2H_2O$
Carboxydotrophic bacteria	*Carboxydothermus hydrogenoformans*	CO (carbon monoxide) $+ H_2O \rightarrow CO_2 + 2H^{+} + 2e^{-}$	$2H^{+} + 2e^{-} \rightarrow H_2$ (hydrogen)

adds an additional advantage in that it avoids competition with chemoorganotrophic organisms, which gain energy by metabolizing organic compounds. In addition, chemoorganotrophic metabolism produces many oxidized inorganic compounds; these organisms have therefore developed several strategies to exploit resources that many organisms cannot use.

Proteobacteria that constitute the largest and most physiologically heterogeneous bacterial group show the greatest diversity in energy production mechanisms among bacteria. In fact, this group includes phototrophic, chemolithotrophic, and chemoorganotrophic taxa. This phylum includes purple phototrophic bacteria (which may or may not use hydrogen sulfide as an electron donor), nitrifying bacteria (chemolithotrophically capable of utilizing reduced nitrogen compounds), sulfide- and iron-oxidizing bacteria with chemolithotrophic metabolism (divided into taxa that live at neutral pH and those living at acidic pH), hydrogen-oxidizing bacteria, methanotrophic and methylotrophic bacteria (using methane and a few other compounds and carbon as electron donors to produce energy), acetic acid bacteria (which can oxidize alcohol and sugars into organic acids), aerobic nitrogen-fixing bacteria, and sulfate- and sulfur-reducer bacteria (which decompose >90% of organic matter in marine coastal sediments).

16.4 Photosynthetic Primary Production

Most primary production occurs within the euphotic zone, the depth range in which >1% surface irradiance occurs (as discussed in Chapter 3, light attenuates exponentially with depth). Photosynthetic pigments in primary producers (seagrasses, macroalgae, phytoplankton, including photosynthetic eukaryotes and prokaryotes) enable their autotrophic capacity that forms the base of most food webs (hence the name primary producers). Through photosynthesis, organisms can fix inorganic carbon to synthesize organic compounds rich in energy (carbohydrates) using light as an energy source. Photosynthesis involves several complex steps that encompass a: "light phase" and a "dark phase." During the light phase, photosynthetic pigments capture photons to produce energy in the form of adenosine triphosphate (ATP), the energy storage molecule so critical for living organisms. During the dark phase that follows they reduce carbon dioxide into complex organic molecules. However, **anabolic metabolism**, referring to the process by which living organisms synthesize proteins, nucleic acids, and many other organic molecules, requires other elements from the environment, including nitrogen and phosphorus.

As noted earlier, the major factors that affect primary production include light availability and intensity, nutrient availability, turbulence and mixing of water, and primary producer composition, but we must also add biological control of grazing by zooplankton and viral infections.

Intensity of light radiation and duration of light availability strongly influence marine primary production. Of the total light energy reaching the surface of the ocean, reflection and diffusion caused by particles in the ocean surface result in a 15% loss. The water column selectively absorbs different wavelengths of light as it passes through to greater depths, creating microclines in different potential absorption spectra. The photosynthetic process does not use all of the solar radiation, in that water almost immediately absorbs the ultraviolet component (10–400 nm) of incident light, whereas light useful for photosynthesis corresponds to wavelengths between 400 and 720 nm (visible light), defined as **PAR** (referring to photosynthetically available radiation). Light penetration results in vertical zonation in phytoplankton, so that different groups of phytoplankton use specific wavelengths and photosynthesize most effectively at specific light intensities, corresponding to specific depths within the water column. Some phytoplanktonic organisms migrate vertically in response to changing PAR (for more details see plankton vertical migration in Chapter 12, Ecology of the Plankton).

Three factors determine the amount of light absorbed by the ocean: (1) latitude (lower light intensities occur at higher latitudes), (2) light penetration (influenced by the amount of water vapor/cloud cover/fog and particles suspended in surface waters, such as sediments, phytoplankton and organic matter, all of which absorb light, decreasing the intensity of light that reaches the phytoplankton), (3) seasonality (which changes the photoperiod, especially with increasing distance from the equator).

Light intensity (I) and rate of photosynthesis (P) are not directly proportional; excessive light intensity can lead to photoinhibition, where physiological changes, such as wrinkling of the chloroplasts reduce photosynthetic rates. The photosynthesis/light intensity curve, known as the **PI curve** (Figure 16.2) illustrates how different intensities alter photosynthesis. Photosynthesis increases with increasing light intensity but reaches an asymptote where maximum rate of photosynthesis (P_{max}) and light saturation occur. The maximum rate of photosynthesis P_{max} and $\Delta P/\Delta I$ (α line, which relates variation in photosynthesis to variation in the amount of incident light) define the most important characteristics of the curve describing photosynthetic production and light intensity.

The initial slope of the curve (the α straight line portion in Figure 16.2) represents **photosynthetic efficiency**, the efficiency of light utilization. The initial steepness of the curve is a function of the light reaction of photosynthesis (as opposed to environmental factors). In contrast, environmental factors (low temperature, nutrient concentrations) strongly

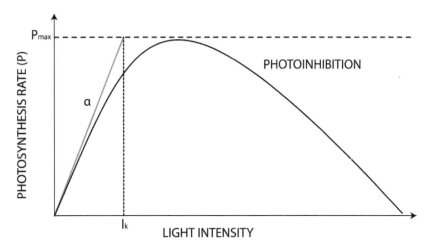

Figure 16.2 Relationship between light intensity (I) and photosynthesis rate (P). An increase in light intensity results in an exponential increase in photosynthesis, up to a point at which photosynthesis reaches an asymptote, after which increasing light intensity provides no additional benefit and typically results in photoinhibition.

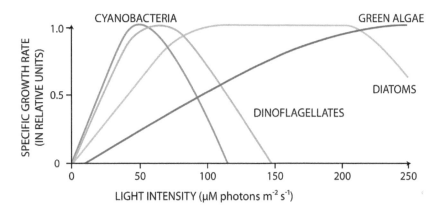

Figure 16.3 Relationship between light intensity and rate of photosynthesis in different groups of phytoplankton and specific growth rate (1.0 = 100% of the potential growth rate of each group). After growth rate rapidly increases initially with increasing light intensity, photoinhibition occurs once maximum specific growth is exceeded, though the intensity at which this occurs varies greatly among species.

affect P_{max}. At low light intensity, P increases linearly with increases in I. The I_k value, the optimum light intensity for production and photosynthetic activity of an organism, corresponds to the point where the straight line $\Delta P/\Delta I$ intersects P_{max}.

In the open ocean and in coastal phytoplankton communities, high I_k values characterize the summer months, whereas low I_k characterize winter and deep-water communities. P_{max} values change with algal physiological state, even under constant conditions; experiments with *Skeletonema costatum* showed increasing P_{max} in young cells that reached a maximum prior to cell division. Depending on species, algae may respond differently to light because maximum photosynthetic activity in each species corresponds to specific wavelengths, and unsuitable wavelengths inhibit, or at least significantly slow, this process (Figure 16.3). Seasonal variation in light intensity also influences growth rates of different groups of primary producers. Growth rate reflects the photosynthetic ability of a specific group: maximum production for cyanobacteria occurs at low light intensity (in fact, at light intensity of 1%, only picoplankton can photosynthesize).

Photosynthetic rate decreases with depth, but photoinhibition in the upper layers often results in a sub-surface maximum in primary production (Figure 16.4). Below the sub-surface photosynthetic maximum, productivity values typically decrease exponentially with depth and decreased light intensity, declining to a depth, known as the **compensation depth**, where productivity and respiration rates for a given phytoplankton cell balance when averaged over >24 h (Figure 16.5). Below this light level, photosynthesis can still take place, though at insufficient levels to counterbalance respiration needs. The compensation depth marks the bottom of the photic zone where net positive photosynthesis occurs. At the **critical depth**, the total amount of productivity (or oxygen produced) integrated for phytoplankton over that depth range balances

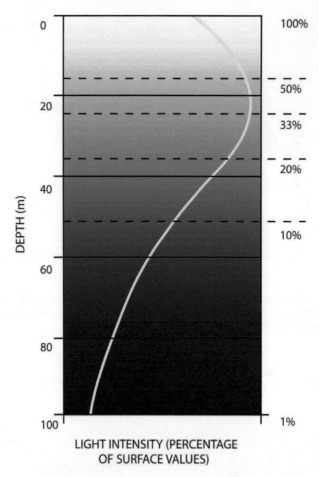

Figure 16.4 Variation in primary production with depth. Maximum production typically occurs at subsurface depths because of photoinhibition in the upper layers of the water column. The figure also shows exponential decline in the proportion of surface light intensity penetrating to different depths.

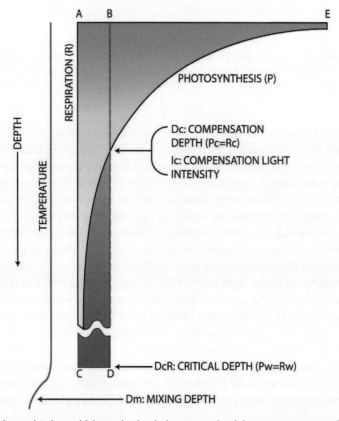

Figure 16.5 Red dotted line shows depth at which respiration balances productivity, on average, over 24 h, known as compensation depth. Critical depth denotes depth at which mixing carries phytoplankton below where photosynthesis offsets respiration needs.

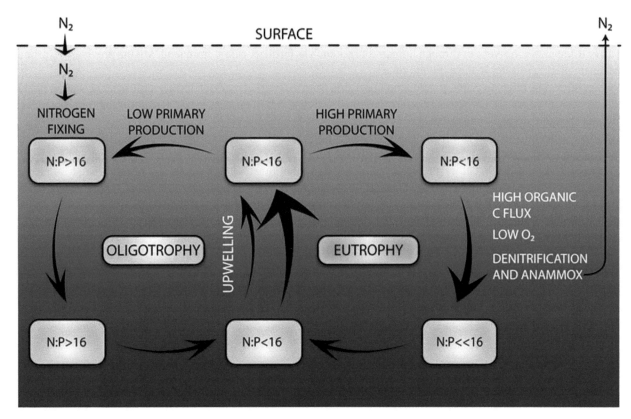

Figure 16.6 Primary production processes showing the major macronutrients. Numbers represent stoichiometric ratios between nitrogen and phosphorus, the denitrification process, anaerobic ammonium oxidation (anammox), and nitrogen fixation.

total respiration (or oxygen consumed) (Figure 16.6). If wind mixing results in a critical depth that exceeds the compensation depth, the population declines, whereas mixing conditions that result in a compensation depth shallower than the critical depth, result in population increases, sometimes quite rapidly in the form of a **phytoplankton bloom**.

In addition to light intensity and photoperiod, primary production depends on temperature and pH, which affect phytoplankton metabolism. Species typically experience a rapid growth increase at their optimal temperature, suggesting a specific genetic adaptation to thermal conditions, a characteristic seen in algal species. Respiration and catabolic processes in general, which influence ecosystem production, link strongly to temperature. However, despite its buffering capacity, pH in coastal environments and beyond may vary, and phytoplankton may lack the capacity to cope with this variation. In addition to these physicochemical factors, which act in concert with one another, biological variables also play a role: in other words, how organisms within a community interact, whether through predation, competition, or other processes. For example, zooplankton grazing in late spring reduces phytoplankton biomass to levels often lower than those found during the winter, a situation exacerbated by reduced nutrient concentrations.

Cell growth and photosynthesis critically depend on nutrients. Production and growth require nitrogen (N, in the form of ammonia and nitrate) and phosphorus (P, in the form of orthophosphate), as well as water and carbon dioxide. The Redfield ratio "$C_{106}H_{263}O_{110}N_{16}P_1$," named for the famous oceanographer Alfred Redfield, indicates that producing 1 mole of organic material requires 1 mole of phosphorus, 16 of nitrogen, and 106 of carbon. Clearly, algal cells must live in environments with sufficient quantities of these elements in a usable form, even if their concentrations vary. The Redfield ratio suggests that phosphorus limits primary production, given the atomic ratio for nitrogen:phosphorus of 16:1, meaning that 1 atom of phosphorus is required for 16 atoms of nitrogen. Therefore, in situations with large quantities of available nitrogen, phosphorus deficiency limits photosynthesis and formation of organic molecules. In surface layers, as phytoplankton assimilate nutrients, those nutrients may become limiting during periods of rapid algal growth. When ambient nutrient concentrations, and particularly N and P, deviate significantly from the Redfield ratio (N: P = 16), limiting conditions for phytoplankton growth may occur. Nitrogen limits growth when N: P $\ll$ 16, whereas phosphorus becomes limiting at N: P $\gg$ 16 (e.g., ratios of 30 or 60) (Figure 16.6).

Ammonia (NH_4^+) and nitrate (NO_3^-) comprise the primary N sources: in the euphotic zone, organism excretion provides ammonia (reduced form of nitrogen); nitrate (oxidized form of nitrogen) occurs deeper in the water column, where remineralization of nutrients by bacteria regenerates nutrients. Thus, phytoplankton may use autochthonous (locally produced) nitrogen (usually in the form of reduced NH_4^+) or allochthonous nitrogen (usually in the oxidized form of NO_3^-). Given the richness of the deep ocean in inorganic nutrients, phenomena such as upwelling can transport nitrates to the surface from depth, or it can arrive in rivers, or through human sources such as fertilizer runoff from land. Mixing provides the primary mechanism placing nitrates in the euphotic zone.

New production refers to primary production resulting from delivery of oxidized forms of nitrogen as well as fixation of atmospheric nitrogen (N_2), whereas **regenerated production** refers to primary production resulting from the reduced form (and therefore derived from nutrient recycling) (Figure 16.7).

In surface waters with low N and P availability (**oligotrophic**, referring to nutrient-poor regions), reduced primary production occurs and these conditions often favor **nitrogen fixation** (N_2) by cyanobacteria and archaea, which oxidize dinitrogen (N_2) into forms that other phytoplankton can use (ammonium and then nitrite and nitrate). The fixation of N_2 increases the N:P ratio of organic substance to values above the Redfield ratio. Global rates of N_2 fixation, however, are not sufficient to balance losses of N fixed by processes such as **anammox**, a process only discovered in 1999, in which specialized bacteria convert nitrite and ammonium into N_2, and **denitrification**, a process in which specialized bacteria reduce nitrate to nitrite and N_2. Consequently, denitrification in the ocean decreases the N:P ratio below the Redfield ratio.

In regions such as the open ocean where production depends strongly on nitrogen fixation, low efficiency in using N_2 (breaking triple covalent bond requires large amounts of energy) results in low production rates, whereas high primary production occurs in regions with high nutrient concentrations (such as with upwelling). As a result, upwelling regions account for approximately 50% of global marine primary production (Figure 16.8).

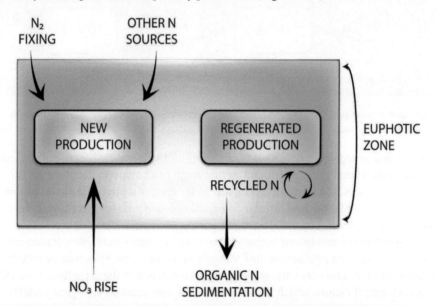

Figure 16.7 New production, which depends on delivery of allochthonous nutrients from depth (oxidized through remineralization) and regenerated production (which depends on euphotic zone nutrients provided by organism excretion).

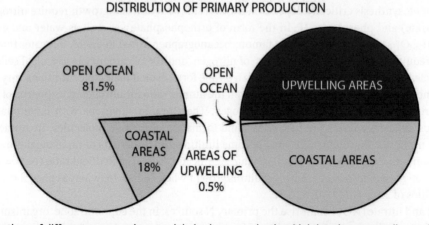

Figure 16.8 Contributions of different ocean regions to global primary production (right) and corresponding surface areas they represent (left). The open ocean represents 80% of ocean surface area but produces a small fraction of total global primary production. Upwelling areas, in contrast, comprise a small portion of the ocean, but produce ~50% of global primary production.

Estuarine environments play a particularly important role in ecosystem function, not only because they often support comparatively high productivity, but also because they provide nutrients that support new production. Atmospheric deposition can also enhance primary productivity in oligotrophic marine environments, where limited availability of inorganic nutrients leads to lower primary production. In particular, researchers have linked atmospheric deposition of both wet and dry fractions of dust (such as Saharan sand in the Atlantic and Mediterranean), with significant increases in primary productivity. Recent estimates indicate that Saharan sands contributes 10–50% (20–100 mmol m^{-2} yr^{-1}) of total external sources of nitrogen and about one third of total organic phosphorus in the Mediterranean.

The input of nitrogen from rivers into the ocean increases primary production. Despite the seemingly minor contribution of atmospheric deposition (resulting in 2–4 g C m^{-2} yr^{-1} new production for nitrogen, <1 g C m^{-2} yr^{-1} for phosphorus) to total Mediterranean production (estimated at 50–100 g C m^{-2} yr^{-1}), atmospheric deposition can trigger some episodic phytoplankton blooms, not only related to nitrogen and phosphorus inputs, but also iron, which sometimes limits primary production. Scientists also hypothesize a causal link (direct link) between episodic blooms and specific atmospheric (e.g., winds) and oceanographic forcing (e.g., mixing).

The hydrology of a system also influences photosynthetic rates. For example, upwelling and downwelling regions (discussed in Chapter 1) lead to enrichment and depletion of oxygen and nutrients, respectively. Riverine input causes major horizontal differences in nutrient distributions along coastlines. In the vertical dimension, strong turbulence influences phytoplankton distribution within the water column, transporting phytoplankton to depths beyond the compensation point, resulting in decreased primary production. Thus, seasonal stabilization and destabilization of surface water, linked to formation of the spring thermocline and its breakdown during fall surface cooling and windstorms, strongly influence primary production. In spring, warming air temperatures and increased light intensity and duration begin to stabilize the water column and create conditions favorable for photosynthesis (Figure 16.9).

However, by summer, strong stabilization of the water column and reduced mixing above and below the thermocline lead to progressive depletion of nutrients by phytoplankton, limiting delivery of new nutrients from depth to support primary production. With autumn cooling and associated storms, water column stratification breaks down, mixing deeper

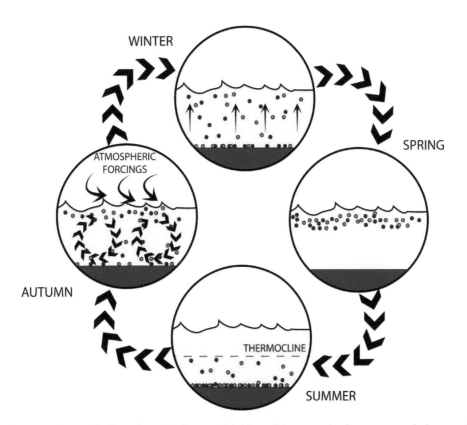

Figure 16.9 Classic perspective on plankton dynamics in coastal habitats. Primary production ramps up during a spring phytoplankton bloom: (right) High production of organic matter that is gradually depleted during summer, (bottom) increasing water column stratification limits nutrient exchange; nutrients needed for photosynthesis become depleted and primary production becomes low (left) fall mixing reinjects nutrients into surface waters but low light in winter (top), limits nutrient uptake.

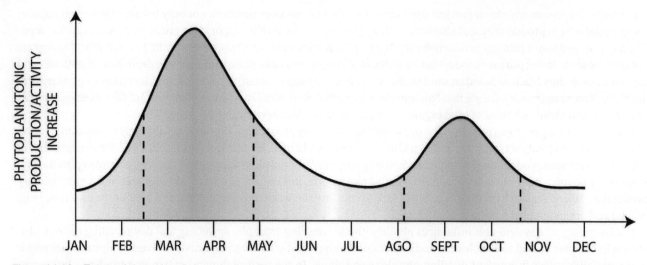

Figure 16.10 Temporal dynamics of phytoplankton blooms in temperate latitudes. After the spring bloom, a summer period of low productivity follows, prior to a new autumn bloom typically of lesser magnitude and duration. This cycle depends on water column stratification, nutrient and light availability, and mixing depth.

water and surface water to provide new inputs of nutrients re-mineralized in deeper layers during the summer; this process often results in a fall phytoplankton bloom. This seasonal cycle, typical of mid-latitudes, results in two blooms annually, a major spring bloom and an often smaller bloom in autumn (Figure 16.10).

Biological processes, and particularly competition (inter- and intraspecific) and zooplankton grazing, also contribute to variation in primary production. In many regions, grazing by copepods and other zooplankton may exert a **top-down effect**, where higher trophic levels control abundances of lower trophic levels, in this case phytoplankton. In some ecosystems, zooplankton can consume 30–50% of phytoplankton daily.

More recent studies demonstrate the importance of phytoplankton viral infections in ending phytoplankton blooms. High concentrations of phytoplankton cells (>1 million l^{-1}) create ideal conditions for viruses specializing on infecting phytoplankton, sometimes triggering die offs of >50% of phytoplankton cells daily.

Simultaneously, sinking phytodetritus, fecal pellets from zooplankton, and other organic matter originating from phytoplankton production sink and accumulate on the seafloor, where breakdown occurs through reingestion and decomposition (an oxygen-consuming process). During autumn, bottom mixing may release nutrients produced by decomposition and accumulation in sediments. Assuming sufficient light, this injection of nutrients may give rise to a second bloom followed by a winter of intense mixing and low production.

16.4.1 Primary Producers

In the marine environment, phytoplankton account for most of primary production, especially in the open ocean where the pico- and nanoplankton fraction (sizes 0.2–2.0 μm and 2–20 μm, respectively) dominate numerically, in large part because they have the highest photosynthetic efficiency. In coastal waters, however, a micro-phytoplankton community (sizes 20–200 μm) typically dominates, and abundances can vary significantly throughout the year with values ranging from 10 to 100 cells cm^{-3} depending on season and system, but intense blooms can also reach 500 cells cm^{-3}. Picoplankton (sizes 0.2–2.0 μm), a ubiquitous component of marine communities, vary in their contribution to marine primary production, ranging 1–90% (Box 16.1).

In addition to the planktonic fraction, **phytobenthos**, or benthic photosynthetic organisms, sometimes play a significant role in coastal systems: seagrasses, for example, contribute up to 12% of net global oceanic primary production. Recent estimates of global primary production of seagrasses indicate average annual production of 10^{12} gDW m^{-2} yr^{-1}. Microphytobenthos can also contribute significantly to primary production, especially in shallow waters and intertidal environments, where they often dominate primary producers, with primary productivity rates that can equal or exceed those for phytoplankton. Even macroalgae play a major role in coastal primary production, especially in shallow rocky bottom habitats where they may dominate primary production at 10–20 m depth.

Box 16.1 Methods to measure primary production and productivity in the sea
Researchers measure primary production in milligrams of carbon per cubic meter per day [mg C m^{-3} d^{-1}] or in micrograms of carbon per liter per hour [μg C l^{-1} h^{-1}]. The most widely used methods to estimate primary production on broad (e.g., satellites) and small scales (e.g., water samples) take advantage of pigments and, specifically, concentrations of chlorophyll-a. Using the natural auto-fluorescence of photosynthetic pigments, satellites can quantify chlorophyll, although with significant caveats. Still, different environmental conditions and cell physiological state can change greatly, along with concentration of chlorophyll-a per unit biomass. For example, chlorophyll a content differs in nitrogen-deficient environments or in relation to light intensity. Evaluating biomass based on chlorophyll-a content in seawater samples requires a conversion factor between biomass and chlorophyll content. With appropriate conversion factors, measurements of chlorophyll-a can be converted to biomass (mg of carbon), though recognizing that chlorophyll to organic carbon ratios (by weight) range 22–154, depending on metabolic state and phytoplankton community composition. Counting phytoplankton cells and determining their relative biovolume enable estimates of phytoplankton biomass. Specifically, following concentration of water samples by letting cells settle to the bottom through sedimentation, microscopic counts can produce estimates of biomass based on the total volume of phytoplankton cells in a given sample volume. This approach also allows calibration of satellite data. Estimates of cells based on direct counts with epifluorescence microscopy of autofluorescent cells offer the most reliable method for estimating phytoplankton biomass. Up until the 1950s, researchers measured productivity rates based on oxygen concentrations, by comparing dissolved oxygen before and after exposing samples to light. Specifically, this method compares oxygen concentrations in clear bottles (light available) and dark bottles (no light available). In the presence of light, both photosynthesis and respiration occur (therefore producing and consuming oxygen), whereas in the absence of light, only respiration occurs (oxygen consumption). Importantly, respiration occurs not only in phytoplankton, but also in the heterotrophic compartment within the sample. Zooplankton presence reduces accuracy, given that zooplankton may feed on primary producers and also add to total respiration. Moreover, biofouling on the sides of the bottles may affect both photosynthesis and respiration. Finally, because the bottles are closed and nutrient depletion can also occur, decreased primary production may result. Another method for calculating primary production uses concentrations of phosphates that algae could potentially use, assuming a C:P ratio = 106 to calculate carbon produced for each mole of P removed. However, the method does not account for variations in concentration associated with transport of water masses and removal of nutrients by heterotrophic prokaryotes. Finally, the ^{14}**C method** involves adding known amounts of NaH^{14}CO$_3$ to water samples from dark bottles and clear bottles (in which light can penetrate), followed by exposure to sunlight. Through photosynthesis, phytoplankton assimilate the inorganic (and radioactive) ^{14}C from the water sample (just as they would for the non-radioactive ^{12}C naturally present in seawater). After 1–12 h incubations, filtering separates phytoplankton from seawater prior to measuring the radiation incorporated by phytoplankton captured on the filter. Knowing the initial concentration of radioactive material in the sample and the amount retained on the filter the quantity of C produced by photosynthesis during the exposure period can be calculated based on: ^{14}C$_{inorg}$ added / ^{14}C assimilated = total inorganic ^{12}C present / ^{12}C assimilated, with P measured in mg C · m^{-3} · hour^{-1}. Incubations must last long enough to allow measurable concentrations of uptake whereas incubations in excess of 12 h risk recirculating ^{14}C in dead phytoplankton. However, because phytoplankton can also use ^{12}C naturally present in the sample and filtration can damage cells, accurate estimates of primary production require appropriate correction factors.

16.5 Secondary Production

Herbivores and bacterivores use a portion of organic matter produced by plants and photosynthetic and chemosynthetic bacteria for tissue formation. **Gross secondary production** refers to energy channeled into heterotrophic organism biomass, whereas **net secondary production** takes into account losses to heterotroph respiration and excretion. Therefore, secondary production always relates per heterotrophic consumers. As with primary production, secondary production provides a measure of mass per unit of volume in unit of time. Calculating secondary production as a change in biomass over time is complicated, especially when considering smaller components such as bacteria. Secondary production of organisms depends on: (1) reproduction and growth rates of the organisms; (2) metabolism; (3) food availability and **Liebig's Law** ("law of the minimum," which states that the scarcest resources rather than total resource availability limits growth); (4) spatial-temporal coupling between food availability and recruitment. Although the first factor links to intrinsic aspects of each species or population, others depend on the external environment and primarily determine ocean secondary production.

Spatial and temporal variation in living organism dynamics in marine ecosystems mainly reflects specific metabolic needs. Wherever they live, organism activities require energy transformation. They convert inorganic matter into organic forms (cells and tissues), use this stored energy to survive, grow, and reproduce, and, finally, eliminate and excrete metabolites back into the environment in altered forms. Variability among ecosystems, including changes in chemical composition, matter and energy flow, populations, and diversity of species, depends on the metabolic characteristics of the resident biota. Most variability among different organisms, including differences in life cycles and ecological roles, depends on body size, temperature, and chemical composition.

16.5.1 Methods of Measurement of Secondary Production in the Sea

16.5.1.1 Cohort Analysis

A **cohort** refers to a group of individuals belonging to the same generation.

Within cohort analysis, secondary production refers to variation in average biomass multiplied by the number of individuals within the cohort of a given population. The size distribution of individuals from a single cohort or generation forms a hump that, over time, decreases in abundance because of mortality of recruits, whereas biomass of individuals within the cohort increases (Figure 16.11). This method has significant limitations, because identifying which individuals belong to a given cohort can be quite difficult, mainly because of differences in growth and development of individuals. Nonetheless, researchers often use this method to study zooplankton, macrobenthos, and fishes.

A method to estimate production based on identifiable cohorts uses:

$$P = [(N_2 - N_1) \times (W_1 + W_2)/2] + (B_1 - B_2)$$

Where:
P is secondary production
N is number of individuals of a specific cohort,
B is Biomass,
W is individual weight,
2 and 1 refer to time 2 and 1 of sampling.
Annual production of each species is calculated as the sum of the production of each species for each time interval.

$$P = P_{t1} + P_{t2} + P_{t3} + P_{t4} + \ldots + P_{tn}$$

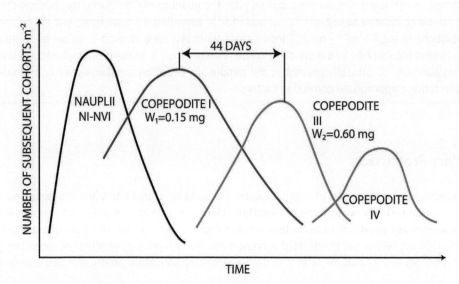

Figure 16.11 Change in a cohort over time.

The unidentifiable cohorts' method

$$P = \sum n_i g_i$$

n = individual number
g = increase in weight per size class
i = the various populations/species present in the sampling unit

This method, which also can be difficult to apply, uses the sum of the product of the number of individuals in a population and average biomass increase over time.

The cumulative growth method

This method was proposed by Wolf Arntz and is used empirically in numerous applications because of its simplicity. It considers increases in biomass in time (Δt), without considering predation or variation in biomass produced. These analyses report biomasses as dry weight or C content.

$$P = [\Delta B1 + \Delta B2 + \Delta B3 + \Delta B4]/\Delta t$$

Where:
P is production,
ΔB is increase in biomass of the organisms/assemblage in time interval Δt.

Overall production is calculated as the sum of the increases of biomass (considering biomass decrease as a loss to predation or other mortality events).

The production/biomass method

$$P = P/B \times B$$

This popular method calculates annual secondary production by multiplying the ratio between production and biomass (turnover = time for biomass to replace itself) by biomass. This model considers the time required for an individual to replace itself. In some case, we know these relationships (e.g., P/B ratio = 9–10 for meiofauna), calculated simply by adding together the production of each individual component.

Metabolism encompasses a complex network of biochemical reactions catalyzed by enzymes that regulate concentrations of substrates, products, and reaction rates. **Metabolic rate** (the energy necessary to carry out vital life functions) represents the fundamental biological rate, because it encompasses assimilation, transformation, and allocation of energy within organisms. Metabolic rate varies greatly among organisms, whether terrestrial or marine, and also links to organism size (Figure 16.12). These differences reflect environmental constraints and limitation of resources, although some theories attribute this variability to the diversity of ecological roles of organisms and their energy needs.

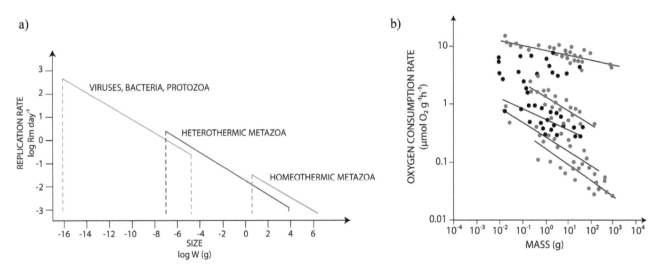

Figure 16.12 Relationship between body mass and metabolism in marine organisms: (a) inverse relation between size and rate of replication/reproduction of marine organisms, from phages (viruses), with biomass measured in femtograms (10–15 fg), to the largest cetaceans (10^6 g or tons); (b) inverse relationship between metabolic rates of different marine taxa as a function of their body weight. Light blue symbols represent coastal taxa and black symbols represent deep-sea taxa.

Temperature plays an important role in metabolism, especially for poikilothermic organisms. For a heterotrophic organism, metabolic rate equals respiration rate (noting some energy may be put into growth and reproduction), given that heterotrophs obtain their energy by oxidizing organic compounds as described by the reaction:

$$CH_2O + O_2 = energy + CO_2 + H_2O$$

For a photoautotroph, metabolic rate mirrors photosynthetic rate because the same reaction takes place in reverse order, using energy provided by the sun to fix carbon. Metabolic rates in animals vary greatly. Some researchers argue that all organisms converge at a common metabolic rate, once corrected for mass and temperature. Another theory suggests that metabolic patterns vary with depth, reflecting energy demands for predator-prey interactions and that these interactions often depend on visual acuity and light availability. Therefore, metabolism in some groups (for example, cephalopods), but not others, declines with temperature. Although researchers have successfully captured and maintained a wide range of organisms in captivity, the logistical challenges of keeping organisms from >1,000 m depth alive limit knowledge of deep-sea metabolic processes.

Many studies demonstrate metabolic rates 10–200x lower in pelagic fishes, cephalopods, and crustaceans in deep-sea environments compared to shallow depths, a trend that differences in temperature or size cannot fully explain. This pattern is less evident in benthic species. Patterns of enzymatic activity, buffer capacity, and protein content in locomotor muscles reflect this decrease. Most measures of metabolic consumption and aerobic enzymes in deep benthic species, including echinoderms, meiobenthic taxa, crabs, shrimps, and sponges, indicate rates 1–5x lower than surface species and many of these taxonomic groups do not show any decrease. Published values indicate a significant decline in anaerobic metabolic enzymes (e.g., lactate dehydrogenase) with depth in both pelagic and benthic taxa.

Surface pelagic predators that orient themselves using visual cues (such as epipelagic squid, fishes, and crustaceans) spend at least part of their day actively pursuing prey in subsurface waters, thereby consuming significant energy. The "visually limited" pelagic groups, including representatives of taxa that live in water depths > 100 m, often use tactile sensory mechanisms (e.g., vibration from prey movement) and ambush strategies, requiring low, but variable, rates of energy consumption at all depths. Energy consumption in benthic organisms generally varies little with depth. These patterns indicate that environmental parameters do not strictly determine metabolic rates in deep environments. However, a strong evolutionary convergence among pelagic predators with visual skills in surface waters and between pelagic predators with (or without) visual skills in deep environments points to a common selection regime, mediated by the presence or absence of visual predator–prey interactions, operating across depths. Recent studies suggest that differences in metabolic rates between shallow and deep pelagic organisms may relate to different locomotor abilities.

Food availability plays a key role in limiting metabolism and growth rate of animals. Reduced metabolic rates in deep marine environments link to environmental factors, and limited food supply in particular. However, this hypothesis requires experimental verification. The presence of animal groups that span a wide range of metabolic rates, independent of depth, contradicts a simple theory of food limitation alone. In addition, the presence of highly active animals in oligotrophic surface waters with even greater food limitation suggests that scant availability of food does not preclude high metabolic rates.

16.6 Respiration

Only a small fraction of organic matter produced accumulates permanently in ocean sediments, with remineralization through respiration accounting for the vast majority of produced organic matter (>99%). Respiration by prokaryotes, and heterotrophic bacteria in particular, accounts for a large fraction of total water column respiration, especially in less productive environments. Although dissolved organic carbon (DOC) supports this microbial respiration, researchers have traditionally focused on particulate organic matter flow, noting that this source represents only part of total ocean carbon flux. Despite an overall positive relationship between respiration and production levels of ecosystems on large spatial scales, the two processes often decouple considerably over smaller scales. The typically high respiration rates in ocean surface layers, with mean values of ~1.2 g C m^{-2} d^{-1}, if extrapolated over the 3.7 × 10^8 km^2 of the open ocean, represent a global respiration rate of 143 Gt C yr^{-1}. This value is about 3x the accepted estimate for ocean production (about 50 Gt C yr^{-1}), because of bias in available estimates of respiration in the photic layer toward more productive regions and periods of elevated metabolic activity.

Table 16.2 Estimates of respiration and organic matter input.

Components	Minimum estimate (Gt C yr^{-1})	Maximum estimate (Gt C yr^{-1})	Average estimate (Gt C yr^{-1})
Respiration			
Photic Zone	32	42	37
Mesopelagic Zone	21	28	24.5
Deep Environments	1.3	1.6	1.5
Mesozooplankton	1.5	4.5	3
Vertebrates			0.01
Total Respiration	**55.8**	**76.1**	**66**
Organic matter input			
Measured primary production (based on ^{14}C)	28	52	40
Non-measured gross production	13.4	25	19.2
Total production	41.4	77	59.2
Coastal areas input	6	6	6
Atmospheric input	3	3	3
Old organic matter	0.5	0.5	0.5
Total input	**50.9**	**86.5**	**68.7**

Respiration rates abruptly decrease below the photic layer, and often remain low through the thermocline, typically 1–30% of upper photic layer rates. Nevertheless, microplankton respiration integrated over mesopelagic depths rivals that within the photic layer. Low rates of oxygen consumption below 1,000 m depth create challenges in obtaining good measurements. Combining deep benthos respiration estimates with estimates of organic matter flux through the water column yields a total ocean respiration estimate of 1.2×10^{14} mol O$_2$ yr^{-1}, of which benthic respiration accounts for 45%; this approach yields estimates consistent with other independent calculations. The latest estimates (Table 16.2) place oceanic respiration at 55–76 Gt C yr^{-1}, therefore representing one of the largest CO$_2$ sources in the biosphere.

Questions

1) What factors control primary production in marine environments?
2) What are the compensation depth and the critical depth?
3) How do we measure primary production?
4) What are POM and DOM?
5) In what ways can we classify organic matter?

Suggested Reading

Behrenfeld, M.J. and Falkowski, P.G. (1997) Photosynthetic rates derived from satellite-based chlorophyll concentration. *Limnology and Oceanography*, 42, pp. 1–20.

Behrenfeld, M.J., Worthington, K., Sherrell, R.M. et al. (2006) Controls on tropical Pacific Ocean productivity revealed through nutrient stress diagnostics. *Nature*, 442, pp. 1025–1028.

Berg, I.A., Kockelkorn, D., Buckel, W., and Fuchs, G. (2007) A 3-hydroxypropionate/4-hydroxybutyrate autotrophic carbon dioxide assimilation pathway in Archaea. *Science*, 318, pp. 1782–1786.

Carr, M.E., Friedrichs, M.A.M., Schmeltz, M. et al. (2006) A comparison of global estimates of marine primary production from ocean color. *Deep-Sea Research II*, 53, pp. 741–770.

Falkowski, P. (2012) Ocean science: The power of plankton. *Nature*, 483, pp. S17–S20.

Guidetti, P. (2000) Leaf primary production in *Posidonia oceanica*: Two reconstructive aging techniques give similar results. *Aquatic Botany*, 68, pp. 337–343.

Guidetti, P. (2004) Consumers of sea urchins, *Paracentrotus lividus* and *Arbacia lixula*, in shallow Mediterranean rocky reefs Helgoland. *Marine Research*, 58, pp. 110–116.

Laufkötter, C., Vogt, M., Gruber, N. et al. (2015) Drivers and uncertainties of future global marine primary production in marine ecosystem models. *Biogeosciences Discussions*, 12, pp. 3731–3824.

Loreau, M. (2001) Microbial diversity, producer-decomposer interactions and ecosystem processes: A theoretical model. *Proceedings of the Royal Society of London B: Biological Sciences*, 268, pp. 303–309.

Mistri, M. and Ceccherelli, V.U. (1994) Growth and secondary production of the Mediterranean gorgonian *Paramuricea clavata*. *Marine Ecology Progress Series*, 103, pp. 291–296.

Pauly, D. and Christensen, V. (1995) Primary production required to sustain global fisheries. *Nature*, 374, pp. 255–257.

Robinson, C. (2000) Plankton gross production and respiration in the shallow water hydrothermal systems of Milos, Aegean Sea. *Journal of Plankton Research*, 22, pp. 887–906.

Sarmiento, J.L., Gruber, N., Brzezinski, M.A., and Dunne, J.P. (2004) High latitude controls of thermocline nutrients and low latitude biological productivity. *Nature*, 427, pp. 56–60.

Soliveres, S., van der Plas, F., Manning, P. et al. (2016) Biodiversity at multiple trophic levels is needed for ecosystem multifunctionality. *Nature*, 536, pp. 456–459.

Thingstad, T.F. and Rassoulzadegan, F. (1995) Nutrient limitations, microbial food webs, and 'biological C-pumps': Suggested interactions in a P-limited Mediterranean. *Marine Ecology Progress Series*, 117, pp. 299–306.

Unsworth, R.K., Nordlund, L.M., and Cullen-Unsworth, L.C. (2019) Seagrass meadows support global fisheries production. *Conservation Letters*, 12, p. e12566.

Voss, M., Bange, H.W., Dippner, J.W. et al. (2013) The marine nitrogen cycle: Recent discoveries, uncertainties and the potential relevance of climate change. *Philosophical Transactions of the Royal Society B: Biological Sciences*, 368, p. 20130121.

17

Ecosystem Functioning II: Organic Matter Recycling

17.1 Introduction – Extra-Specific Processes

When considering energy flows through ecosystems, we tend to think of living organisms such as herbivores and predators, but significant food and nutrients move through nonliving compartments as well as through living organisms. **Extra-specific organic matter processes** refer to processes in organic matter cycling that do not involve direct relationships between living organisms but instead depend on the relationship between species (whether vertebrates, invertebrates, protists, or prokaryotes) and organic detrital matter (nonliving matter that was once living) in the environment.

17.2 Organic Matter and Detritus in the Ocean

Biological activity produces most organic matter in the ocean. Phytoplankton exudates represent one major source of (nonliving) organic matter, accounting for ~20% of primary production that phytoplankton release into the environment rather than investing it in somatic growth. This release might create a protective structure against potentially harmful environmental conditions. Cell **lysis** provides another source of organic material through rupture or fragmentation of unicellular organisms (bacteria and phytoplankton) after their death from viral infection; viruses can regulate phytoplankton blooms and decrease primary production by 50% or even up to 100%. This fraction of primary production, rather than passing through the grazing food web, circulates into the ocean as dissolved or particulate material. Grazing itself also produces organic matter through **sloppy feeding** where zooplankton crush phytoplankton rather than swallowing them whole, resulting in leakage of cell content. Calculations show that zooplankton can lose up to 20–40% of captured phytoplankton production in the form of dissolved material. Degradation of waste products or carcasses releases another 20–40% of organic material into the environment. Organisms often reuse this material, which forms a dynamic equilibrium with particulate organic matter (POM) to fuel additional heterotrophic production, thus avoiding loss of this huge pool of nutrients and energy.

Robert Wetzel and co-workers coined the term **detritus** in 1972, referring to the sum of non-predatory losses of organic carbon (including egestion, excretion, secretion, etc.) from each trophic level, or the sum of the input of sources external to the ecosystem that enter and circulate in the system (allochthonous organic carbon). We could define detritus, more generally, as all organic, nonliving material, irrespective of its size, composition, or origin (Figure 17.1). Organic detritus plays a major role in biogeochemical cycles, but biogeochemists downplay its importance in marine food webs, likely reflecting a historical focus on grazing food webs. Organic matter contributes extensively to remineralization, nutrient, and production cycles, thus defining a critical component for marine ecosystem functioning. Organic detritus also moves via passive transport both vertically through sinking and horizontally through tides and currents, thereby connecting ecosystems. Organisms actively exploit this pool or material through selective filtration and passive capture.

The definition of organic detritus has important implications for understanding system processes and defining trophic groups. Detritus encompasses all nonliving organic matter and therefore excludes the microbial components (prokaryotes, protozoa) that play a central role in both water column and seafloor organic matter decomposition. These microbial components generally contribute negligibly to organic detritus biomass; calculations indicate that microbial biomass represents <3% (often <1%) of total organic carbon in sediments. However, because microbes contribute a greater percentage of nitrogen

Marine Biology: Comparative Ecology of Planet Ocean, First Edition. Roberto Danovaro and Paul Snelgrove.
© 2024 John Wiley & Sons Ltd. Published 2024 by John Wiley & Sons Ltd.
Companion Website: www.wiley.com/go/danovaro/marinebiology

Figure 17.1 Rain of detrital particles deposited on sediments includes dead microalgae, remains of organisms that includes exoskeletons, fish bones, and other forms of organic matter, which constitute the macroscopic forms of organic debris in the sea. Rachel Carson, in her book *The Sea around Us*, said, "When I think of the floor of the deep sea, the single, overwhelming fact that possesses my imagination is the accumulation of sediments. I see always the steady, unremitting, downward drift of materials from above, flake upon flake, layer upon layer – a drift that has continued for hundreds of millions of years, that will go on as long as there are seas and continents... For the sediments are the materials of the most stupendous snowfall the Earth has ever seen".

and organic phosphorus, they provide high quality food (and sometimes palatability) for organisms feeding on detritus that may select for detritus with higher microbial content.

This microbial aspect of detritus, spanning bacteria, archaea, protozoa, to fungi, strongly influences ecosystem functioning, noting the dual role that microbes can play: (1) Microbes can contribute directly to sediment nutritional value; (2) microbes can act as catalysts for remineralization. Researchers in the 1980s first proposed that microbial components contribute to nutritional value of sediments, and subsequent studies provided some support. Nonetheless, evidence in recent years suggests a modest role for microbes in filling overall nutritional demands of benthic organisms except for some specific components, such as larvae and the smallest metazoans. In the latter case, microbial components act as catalysts because, as bacteria process organic matter in sediments, they alter qualitative characteristics, making it easier for deposit feeders to digest and exploit detritus. In this case, the microbial component plays a vital role in ecosystem functioning by mediating detritus access for planktonic and benthic consumers.

Classification of organic matter in the sea uses five different approaches: nature, size, origin, composition, functional characteristics, and trophic value (Table 17.1).

Nature: The living components of organic matter in marine ecosystems represent only 10% of total organic matter, with detritus accounting for the other 90%. Prokaryotes, protozoans, phytoplankton and microzooplankton comprise the major

Table 17.1 Classification of organic matter according to nature, size, sources, composition, and trophic value.

Typology	Origin	Size	Examples	Quantity
Nature	Living		Organisms, feces, molting, excretions	10%
	Detrital			90%
Size	DOC	<0.20–0.45 µm	Bacterial fragments, colloidal material, viruses, cellular debris, phytoplankton	
	POC	0.45–200 µm		
Sources	Autochthonous		Produced *in situ*	10–90%
	Allochthonous		erosive, aeolian, terrigenous origin	0–90%
Composition	Biopolymeric			5–15%
	Geopolymeric			85–95%
Trophic value	Bioavailable			1–10%
	Refractory			90–99%

living component of marine organic matter. Generally, the largest size components (macro and megabiota) contribute negligible amounts because of their comparatively low concentrations. Of course, classification provides a convenient, but imperfect, organizational convention; a whale carcass, for example, changes dramatically in size as it decomposes, eventually becoming fine detritus and DOM.

Size: A threshold or 0.2 microns operationally demarcates particulate organic matter (POM) from dissolved organic matter (DOM). From a practical perspective, size classification enables marine biogeochemical studies (such as analysis of organic chemistry or biochemical components), based on the concept of progressive fragmentation of detrital particles, from their beginnings as a dead animal or plant through a continuum of POM that eventually transforms into the tiniest DOM. Within the water column, the dissolved components typically occur in concentrations 10–20x higher than particulate components in terms of carbon. In contrast to the water column, the dissolved organic carbon and particulate organic carbon (DOC/POC) ratio in sedimentary environments favors the particulate fraction, with only 5% of organic carbon in the dissolved fraction. Within the particulate organic matter component, particle sizes 0.2–200 μm support an important living fraction consisting of bacterioplankton (0.2–2 μm), autotrophic nanoplankton (2–20 μm; diatoms, dinoflagellates, and coccolithophores) and microplankton (20–200 μm), which sometimes contribute significantly to the total organic matter pool. Therefore, the heterotrophic fraction living on particulate organic matter also includes proto-zooplankton, which generally feed on bacteria. This size class includes small nanoflagellates (genera such as *Bodo*, *Monas*, *Rhynchomonas*, and choanoflagellates and euglenoids) and ciliates (e.g., *Euplotes* sp., *Holosticha* sp., *Huronema* sp., and *Cyclidium* sp.). Particulate organic matter suspended within the water column typically varies in concentrations 7–20x lower than dissolved organic matter, representing 40–95% of total suspended matter. Particulate organic matter (POM) occurs ubiquitously in all marine systems, both in the water column and in sediments. Collectively, particles suspended in water that include living forms and organic aggregates form the **seston** (with **tripton** referring to the collective nonliving fraction).

Origin: External sources provide an important fraction of oceanic organic matter. Organic matter may therefore derive from allochthonous sources outside the system, or autochthonous sources produced within the system. Neither of these two fractions consistently dominates a given environment, because allochthonous contributions vary greatly both in space and time. Nonetheless, allochthonous contributions from terrestrial sources gradually decrease from the coast to the open ocean and, non-local terrestrial components therefore contribute little to open ocean environments. Freshwater sources, and river inputs in particular, deliver an important fraction of allochthonous organic matter to the ocean. These inputs fundamentally enrich not only inorganic nutrients (that support primary production), but also deliver large amounts of organic matter to the ocean, and especially detritus, which undergoes processing and transformation within marine environments. Particularly large freshwater and detrital inputs occur in estuarine and delta coastal environments, where modest average depth and rapid exchange of material with the seafloor enhance enrichment. In river deltas with sedimentation accumulation of m y^{-1} a considerable amount of this input may be buried before processing, thereby effectively removing it from the system permanently.

Trophic value refers to the quality of organic matter as a food source for consumers. The POM fraction includes proteins, carbohydrates, lipids, and nucleic acids. In contrast, a wide range of molecular components of variable weight (from a few to >100,000 Daltons) comprises DOM, and includes viruses, gelatinous matrices, colloidal organic matter, dissolved sugars and aromatic compounds (such as phenols and lignin), free amino acids, dissolved RNA and DNA, humic acids, and vitamins. Historically, classification based on trophic value focused on the stoichiometric ratio between organic carbon and organic nitrogen (C/N ratio) which ranges from 3.5 for bacteria, 7–9 for phytoplankton, 7–10 for zooplankton, >10 for detrital matter and >15 for refractory, low-value material that few organisms can digest. Thus, a C/N ratio < 10 indicates organic material with a higher trophic value, and organic matter with a C/N ratio >15 indicates refractory material. Some studies have reported trophic values > 60, suggesting very low trophic value.

The **biopolymeric fraction** refers to the bioavailable fraction (smallest triangle in Figure 17.2) formed by organic matter produced both by autotrophic and heterotrophic processes from biopolymers (proteins or nucleic acids) and other compounds (such as non-structural and relatively simple carbohydrates or lipids). The geopolymeric fraction refers to inorganic polymers. Biopolymeric organic matter may become geopolymeric and vice versa; microbial components in a biopolymeric fraction may mobilize a part of the geopolymeric matter. In fact, geopolymeric and biopolymeric components link to the concept of refractory (or **recalcitrant**, difficult to digest) material versus bioavailable (or labile, readily digestible by organisms) organic matter. A strong link between organic matter and mineralogical sedimentary fraction contributes to the geopolymeric fraction following formation of an organic matter complex that becomes progressively buried and subsequently transformed by decomposition under anaerobic conditions. This process eliminates nutritional value from organic matter as it becomes increasingly refractory.

Generally, the geopolymeric fraction represents a greater proportion than the biopolymeric fraction, which represents 10–30% of total organic matter. Scientists evaluated this fraction by simulating systems of bioreactors represented by

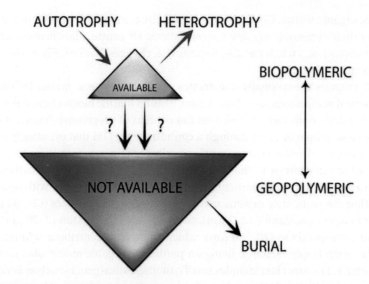

Figure 17.2 Flows between different fractions of organic matter.

digestion in polychaete worms in which they added sedimentary organic matter, a cocktail containing protease, lipase, and glucosidase that simulates digestion within the organism. These enzymes mobilize the digestible fraction, thereby increasing its bioavailability, as opposed to the unavailable geopolymeric fraction. This approach makes it possible to distinguish the fraction of organic matter that actually contributes to polychaete diets, or those of other deposit feeders that ingest sediment, from the fraction organism cannot digest as it passes through their gut and thus cannot contribute significantly to their diet. However, as organic matter passes through the intestine of deposit feeders, its very nature changes. On the one hand, the digestive process reduces the nutritional quality of the remaining food because the animal retains and assimilates the digestible (and typically higher quality) portion. On the other hand, this "reworking" of organic debris facilitates microbial processing, which makes the food easier to digest for another consumer.

Therefore, bulk measurements of total organic load that do not consider food quality do not represent actual availability of food for marine organisms (the labile portion of organic matter; Figure 17.3). In optimal foraging theory, deposit feeders could potentially ingest large amounts of organic matter of which they can digest only a small fraction, whereas an organism

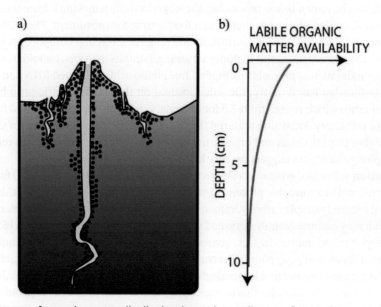

Figure 17.3 Vertical patterns of organic matter distribution in marine sediments: Reported are: (a) Representation of particle concentration and benthic activity (in terms of frequency, number and depth of burrows excavated by scavenging organisms); (b) concentration profile of labile organic matter in sediments: clearly organic matter content decreases exponentially with depth in sediments and disappears almost completely below 10 cm.

that ingests sediment with less organic matter could benefit from a high percentage of more readily available matter for digestion. Given the energy required for foraging, the difference in energy expenditure between nutrition in oligotrophic and eutrophic environments may be less than expected based on total concentrations of organic matter.

High loads of organic debris in an environment lead to rapid uptake of available oxygen in bottom water by microbes as they decompose organic matter, creating micro-zones that can lead to **hypoxia** (low oxygen concentration, often defined as <2–3 mg O_2 l^{-1}) or even **anoxia** (absence of oxygen, O_2< 0 ml l^{-1}). This link means that a high organic detrital load may not always be good for an ecosystem. Imagine optimal conditions for benthic organism food supply described by a uni-modal curve in which the abscissa (x-axis) shows organic matter concentration and the ordinate (y-axis) shows nutritional benefit to the community, in arbitrary units. This overall benefit increases with increases in organic matter supply to a maximum point, after which the benefit decreases and eventually disappears. In fact, excessive increases in organic matter concentration may create conditions unsuitable for organism growth and survival within that environment. **Eutrophication** occurs when a water body becomes overly enriched with nutrients, leading to plentiful growth of photosynthetic organisms, often followed by decomposition of that material and potential hypoxia (Figure 17.4).

The trophic value of detrital organic matter depends on concentration and bioavailability of proteins, carbohydrates, lipids, and nucleic acids. Of particular importance, protein concentration increases detrital nutritional quality because it elevates concentrations of organic nitrogen, a nutrient necessary to support growth of consumers. Carbohydrates also play an important role in benthic metabolism when comprised of simple sugars and easily used energy sources (especially ribose and deoxyribose, fundamental sugars associated with nucleic acids). Phytoplankton release approximately 60% of the glucose present in the water column, with the remainder arising from degradation of often refractory and complex structural carbohydrates. Overall, the heterogeneous pool of carbohydrates includes easily assimilated molecules (free sugars), and highly refractory structural components (which can pass through intestinal tracts of consumers unaltered). Lipids represent the third important organic chemical compound extractable in organic solvents and include fatty acids, carotenoids, and phospholipids. These compounds can act as energy reserves accumulated during periods of food availability that enable coping during periods of scarce food. Therefore, lipids fulfill multiple roles. For example, growth of sea cucumbers, and potentially many other benthic organisms, requires carotenoids, and phospholipids represent an important dietary component by providing phosphorus and PUFA (polyunsaturated fatty acids produced primarily by

Figure 17.4 Eutrophication in coastal ecosystem can lead to hypoxia, particularly near the seafloor. The Pearson-Rosenberg model summarized above shows the progression from a healthy, well-oxygenated seafloor environment with rich diversity on the left to a hypoxic or even anoxic environment on the right. Reproduced from Solan, M., Bennett, E. M., Mumby, P. J., Leyland, J., & Godbold, J. A. (2020). Benthic-based contributions to climate change mitigation and adaptation. Philosophical Transactions of the Royal Society B, 375(1794), 20190107.

phytoplankton and fungi). Larval development, metamorphosis, and growth of many species require PUFAs because they cannot produce them themselves. Some lipids, however, are difficult to digest and species that use them require specific strategies.

The **diagenesis of organic matter** refers to transformation processes that occur both in the water column and in sediments, but stronger intensity, gradients, and rate processes characterize the sedimentary compartment relative to the water column. The dynamic nature of biopolymeric and geopolymeric processes complicates differentiation between labile and refractory components. In fact, decomposition or interaction with the sedimentary matrix can alter labile components so they become refractory (Figure 17.2). Diagenesis refers to these collective organic matter transformations; some of the most interesting phases of diagenesis occur during initial degradation of organic matter, which begins with release of organic matter into the environment.

Starting from the definition of labile organic matter based on the main biochemical classes (proteins, carbohydrates, lipids, and nucleic acids), one approach to evaluating the labile fraction begins with simulating the digestion process, and then measuring the hydrolysable fraction, the enzymatically digestible components (aminopeptidase, glucosidase, lipase and DNAse). This portion provides a reasonable representation of what fraction organisms can actually use and subsequently assimilate.

Detritus, when strongly dominated by refractory material, results in a low turnover rate of organic matter, a change in oxygen concentrations, and alteration of biological processes (production) within benthic communities. Availability of labile organic matter creates the opposite situation by promoting increased activity and elevated biomass of benthic components. At the same time, availability of labile components for benthos can result in negative feedbacks. Increased organic matter results in increased benthic abundance and biomass, which, in turn, increases organic matter consumption that becomes more and more refractory, leading to reduced growth in benthos. Two processes can counteract this effect, either through mobilizing the geopolymeric pool to the biopolymeric pool, or de novo rebalancing through contributions of new labile organic matter (Box 17.1). Over the long term this issue becomes increasing important, particularly with accumulation of material from land and river runoff.

Box 17.1   Methods of Study of Organic Matter in Marine Environments

Researchers use multiple approaches in studying organic matter in marine ecosystems depending on organic matter origin, size, availability, and biochemical composition. Study techniques vary from primary and biochemical analysis, to stable isotopes, lipid biomarkers, and analysis of nuclear magnetic resonance. Studies on amount and biochemical composition of organic matter from a trophic perspective generally quantify concentrations of photosynthetic pigments (chlorophyll-a, phaeopigments), proteins, total carbohydrates and lipids, bioavailable proteins, and carbohydrates (enzymatically hydrolysable) within the matrices exploited by marine organisms (water, sediments, or tissues of organisms).Historically, researchers have mostly used gas chromatography and gravimetric determination after combustion to quantify total organic matter. Gas chromatography enables direct measurement of total organic carbon by means of elemental analysis of C, H and N. Instrumental measurements of organic C with CHN analyzers have several limitations: (1) thermal separation between organic and inorganic C; (2) efficient removal of inorganic C from the sedimentary matrix. Determination of total organic matter content (TOM) consists of dehydration of wet sediment samples in an oven until it reaches a constant weight to remove water associated with inorganic salts. Dry sediment is then weighed, placed in an oven and burned at 450 °C for 2 h, and reweighed. Sample organic matter content is obtained by comparing the weights of dried sediment and burned sediment. The content must then be standardized to dry sediment weigh to yield carbon per gram of sediment. Usually, researchers also determine the amount of organic carbon easily available to higher trophic levels by measuring phytopigments (chlorophyll-a and phaeopigments), which reflect concentrations of phytodetritus and autotrophic benthos.

The extraction of chlorophyll-a and phaeopigments from sediment consists of adding magnesium carbonate to wet or frozen sediment to prevent rapid degradation of chlorophyll-a. To facilitate extraction of pigments from the sediment, following treatment with ultrasound, the addition of 90% acetone and storage at 4 °C for >12 hours in the dark dissolves cells and thus extracts pigments. Following measurement of pigments in the acetone extract on a spectrofluorometer and, multiplying its relative concentrations by appropriate conversion factors, analyses can determine total chlorophyll present in the sample. Measuring chlorophyll in water samples uses a similar approach, though most samples must be concentrated on a filter to produce a sufficiently strong signal for the spectrophotometer.

In recent years, researchers have given greater attention to nutritional value of organic matter in marine sediments by determining its biochemical composition in terms of proteins, carbohydrates, and lipids. The three major classes of

Box 17.1 (Continued)
biochemical compounds (biopolymeric organic carbon) collectively represent 10–70% of total organic carbon in marine sediments. The most widely used method for determining protein in sediments uses a colorimetric method, which exploits the reactive properties of protein that determine color in proportion to protein content in the reaction mixture. These samples are also read with a spectrophotometer. Carbohydrate determination also uses a colorimetric method that exploits the reaction between sugars and phenol in the presence of concentrated sulfuric acid (causing carbonization of all phenol-carbohydrate compounds). The strongly exothermic reaction further catalyzes total carbonization of the sample. Normally, organic solvents (chloroform and methyl alcohol) are used to extract total lipids, following colorimetric determination after carbonization at high temperatures with concentrated sulfuric acid. Consumers can only use a relatively small fraction of sediment total protein and carbohydrate pools, which are assayed using the protocols described above. For this reason, determining the most bioavailable fraction of these two classes of organic compounds becomes important in trophodynamic studies. Researchers normally use an approach based on commercial hydrolytic enzymes, whose actions mimic that of enzymes present in the digestive tract of consumers or in microbes.

17.3 Dissolved Organic Matter in the Ocean (DOM)

Dissolved organic matter (DOM) represents one of the most extensive reserves of organic carbon in marine environments. This dynamic carbon pool links components of the geosphere, hydrosphere, and biosphere that enter the global carbon cycle. DOM may include gelatinous matrices, transparent exopolymeric particles (TEP), colloidal organic matter (COM), dissolved carbohydrates (e.g., glucose), aromatic compounds (phenol, lignin, lipids), dissolved amino acids (DFA-dissolved free aminoacids), dissolved DNA and RNA, vitamins, and humic and fulvic acids. In marine ecosystems, autochthonous sources dominate the DOM pool, largely because phytoplankton excretion and exudates represent the primary sources of DOM, whereas terrigenous (allochthonous) contributions represent only 10% of DOM inputs on a global scale. DOM can change easily through biological activity (in particular enzymatic action), but also through physical and mechanical processes (disintegration of particle aggregates during turbulent ocean conditions). Ultraviolet light can also transform DOM; in the upper water column ultraviolet rays cleave compounds with high molecular weight into smaller, more refractory molecules of lower molecular weight and more labile molecules of higher molecular weight. Therefore, ultraviolet rays help process DOM, increasing its bioavailability for bacterial heterotrophs, but at the same time potentially inhibiting bacterial activity. These rays decrease secondary production and enzymatic activity and, because bacteria lack repair systems to fix the damage to DNA caused by UV-B radiation (despite some capacity to repair damage from UV-A), exposure to sunlight might create irreparable damage to the bacterial genome.

Consider exchange of organic matter between the ocean surface and subsurface depths. Ultraviolet light breaks down organic matter at the surface, increasing availability and, therefore, utilization in subsurface as well as surface layers. Other analyses on consumption and respiration processes in the ocean indicate that POC in the disphotic layer can support only a small fraction of respiration. In the Mediterranean, for example, POC supports only 20% of respiration, meaning that dissolved organic matter must supply the remaining 80%, thus defining the dominant mechanism supporting heterotrophic ocean metabolism.

DOM contains no nitrogenous compounds, consisting mainly of carbohydrates or products derived from decomposition of polymers and monomers, as well as aromatic compounds (phenols, low molecular weight quinones, lignin deriving from aromatic alcohols). DOM compounds that contain organic nitrogen include proteins and their derivatives, nucleic acids, vitamins, and two other refractory compounds, cartilage and chitin. Cartilage, derived from vertebrates, contains highly refractory mucoproteins, whereas chitin, produced by invertebrates and particularly crustaceans, is important in terms of abundance and biomass at all depths. Lipids, a third category that includes fats, generally include esters of fatty acids that, through hydrolysis, can produce corresponding acids, glycerol, aliphatic acids, sterols, etc. Generally, these compounds strongly resist decomposition and, therefore, comprise a considerable portion of refractory DOM. The fourth category consists of complex substances produced from protein decomposition and complexation, along with products derived from carbohydrates decomposition. These complex substances supply the geopolymeric component of a system including humic or fulvic substances composed of 50–60% carbon, 5–10% nitrogen, and 35% oxygen. Complex substances typically resist enzymatic degradation (i.e., highly refractory); high organic nitrogen and a relatively low C/N ratio (9–12) may therefore sometimes provide an ambiguous measure of organic matter quality.

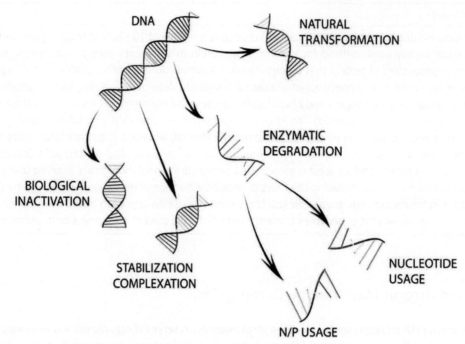

Figure 17.5 Transformation of DNA released in the marine environment (environmental DNA).

DNA, deoxyribonucleic acid, long considered the "molecule of life," typically associated with synthesis and transmission of genetic information, contributes to four major processes linked to cycling or organic matter and biogeochemical processes (Figure 17.5):

1) **Enzymatic degradation or mechanical fragmentation**, which progressively liberates nitrogen, phosphorus, and nucleotides. The rich nitrogen and even richer phosphorus content of DNA means that degradation in deep environments can satisfy up to 50% of microbial metabolic demand for organic phosphorus and can also support a significant portion of organic nitrogen and phosphorus requirements in coastal environments.
2) **Inactivation**, which occurs through biological activity, temporarily denatures or transforms DNA, but does not fully degrade it; inactivation can lead to preservation of DNA within sediments, thereby creating potential markers to support paleoecological reconstructions of oceans past.
3) **Stabilization**, through weak electrostatic forces, can bind DNA to inorganic matrices and thereby stabilize it, reducing susceptibility to extracellular enzymatic attack and preserving the DNA over time. Indeed, this process helps preserve ancient DNA.
4) **Amplification** (or potential natural transformation), where extracellular DNA can preserve part of the genetic information that bacterial cells can use. These cells import genetic information that can bind stably to their chromosome or amplify using a vector to increase their functional plasticity, thereby enabling expression of the incorporated gene.

17.4 Pelagic-Benthic Coupling

Organisms in the vast majority of marine environments live in the complete absence of sunlight, preventing any form of photosynthetic primary production. This lack of primary production means a negative energy budget (i.e., the balance between amount of primary production and amount of organic matter consumed by respiration). In general, positive energy budgets in coastal systems contrast negative budgets in deep-sea environments. However, negative energy budgets can also occur in transitional and coastal environments where respiration exceeds production. How can a system that lives by consuming more than that it produces persist over time? This scenario can occur because a large proportion of the respired organic matter was not produced in situ but imported from another system with a positive balance.

The term **pelagic-benthic coupling** (or pelago-benthic coupling) refers to the process by which energy and matter transfers from the water column to the seafloor. The photic zone (or coastal zone) produces most food resources that arrive on the seafloor, following primary production and transfer of primary production to the benthos through particle sedimentation.

Export of particulate organic matter from the photic zone, produced primarily by photosynthesis, often reaches the seafloor as detrital particles. This exported organic matter represents the major food source for benthos and for the ocean's most extensive marine ecosystems, deep-sea environments. Historically, researchers believed this transfer occurred slowly, taking weeks or even months, but recent studies demonstrated that sedimentation can occur rapidly and that organic matter sinking from the surface can sometimes reach the abyssal seafloor (5,000 m depth) in just a few days. Organic matter sedimentation typically occurs episodically, following strong fluctuations in primary production over time. Therefore, the "rain" of detritus to the benthos varies in intensity both in time and space. In some cases, deposition of living, visible "mats" of prochlorophytes and cyanobacteria (prokaryotic phytoplankton) occurs, however, these organisms soon die because of lack of light. Benthic organisms then quickly graze this fresh layer of organic material. This rain of particles to the seafloor fundamentally influences benthic metabolic processes and survival of seafloor living components from bacteria on up through the food web.

Phytodetrital contributions to benthos represents a critical food source for seafloor biota not only from a nutritional point of view, but also in terms of growth and development. Carotenoids, for example, contribute to overall diets of organisms but also provide contributions necessary for normal development and reproduction. In fact, some species directly incorporate the antioxidants contained in carotenoids into the yolk sac of their offspring, supporting larval development. Holothurians use carotenoid pigments contained in phytoplankton cells as precursors of hormones. Changes in phytoplankton species contributing to primary production in the North Atlantic led to rapid movement into the area by the holothurian, *Amperima rosea*, and 50,000x increases in abundance over the previous 15 years, coincident with the disappearance of another holothurian species.

Pelagic-benthic coupling illustrates the need to study ecosystem functioning in different ways. The diet of many organisms, for example holothurians, that often dominate deep-sea megafauna, strongly depends on pigments and lipids delivered from the water column, because without those resources they will not reproduce successfully. This example illustrates how a remote connection between organic matter transfer to the seafloor and benthic response influences not only growth but also reproduction. This response was reported for large organisms, but meiofaunal turnover rates on the order of a few days means they reproduce much more frequently. Thus, episodic contributions of organic matter to the seafloor may be more important for macro and megabenthic fauna, whereas many smaller organisms can also use low concentrations of organic matter within sediment subsurface layers.

Sedimentation rate depends on particle size, shape, and density (Figure 17.6), Organic matter preferentially binds to small particles, but degree of cohesion and floc formation can create increasingly larger flocs. The largest fecal pellets produced by euphausiids reach sinking rates of 1 km d^{-1}. Lower sinking rates of perhaps 50 m d^{-1} characterize the smallest fecal pellets. Thinking of the average ocean depth of ~4,000 m and ignoring lateral transport in an environment where vertical transport dominates, a particle produced by a copepod would require >78 d to reach the seafloor, undergoing further processing during descent.

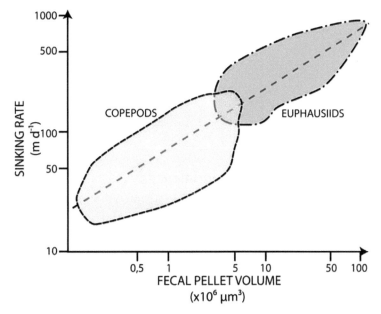

Figure 17.6 Relationship between fecal particle volume and sedimentation rates.

17.4.1 Organic Aggregates in the Ocean

Researchers classify the rain of sinking detrital particles, including larger particles aptly called **marine snow**, based on particle size. Larger particles sometimes form flakes of, thin, laminated layers with rough edges, containing carbohydrates as well as proteins that strongly resist enzymatic degradation with half-lives of months. In surface waters, these formations become little "submarines" – gelatinous masses that host a mixture of organisms, releasing oxygen during the day through phytoplankton photosynthesis.

Researchers have developed several hypotheses regarding mucilage formation, attributing the origin of this material to alteration of available nutrients or their stoichiometric ratio, leading to over-production of mucopolysaccharides. Mucilage could also reflect an excess of primary production in the absence of significant zooplankton grazing (when dissolved material and particles exceed a certain concentration threshold, they tend to "snowball" by aggregating material they encounter as they sink). A third theory posits that phytoplankton produce mucus as a mechanism to cope with attacks from bacteria and exoenzymes. Other possibilities include viral infection, which could lyse phytoplankton or bacteria and decouple the microbial circuit: excess dissolved material not used by prokaryotes forms aggregates.

Fecal pellets: refers to fecal waste particles produced by zooplankton. They consist of a bag of polysaccharides containing partially digested phytoplankton (such as frustules of diatoms) and remains of other partially degraded organic particles.

The larger aggregates, known as marine snow, form from highly heterogeneous and amorphous matrices at all depths and they cycle with interannual production and phytoplankton blooms. Different types of organic particles form these amorphous aggregates (Figure 17.7), including fecal pellets, masses of organic substance, and breakdown products (Box 17.2). The typically irregular shapes vary in size from microscopic particles to aggregates up to a few meters in diameter. The most extreme aggregates accumulate considerable amounts of material, sometimes producing gelatinous masses of exopolysaccharides that can form stripes, clouds, and sheets that may extend over several kilometers. Rich communities of phytoplankton, zooplankton, and bacteria form the mucilage core. The continental shelf exports an important fraction of organic nutrients that reach the seabed, combining terrigenous inputs delivered in rivers and land runoff as well as organic matter produced in situ. Detrital transport includes not only particulate organic matter (POM), but also dissolved organic matter (DOM). Transport to deep environments occurs through sedimentation and water mass downwelling. If productive surface waters release a large amount of organic detritus in the form of exudates (dissolved material) downwelling can transport both POM and DOM to greater depths together with water masses that would otherwise contribute little to sedimentation rates. In addition to sedimentation or transport by downwelling currents, zooplankton migration can move particulate

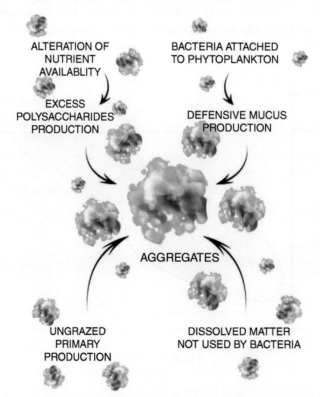

Figure 17.7 Primary biological components that colonize mucilaginous aggregates and processes that lead to their formation and dissolution.

organic matter and release organic detritus at different depths. During daylight hours, when most zooplankton live at depth, metabolic activities lead to the production of larger quantities of debris at depth but based on organic material ingested in shallow waters at night. These two mechanisms are important both in transporting organic matter within water masses, and in influencing oxygen availability and thus enabling more rapid decomposition processes.

Box 17.2   A Whale of a Carbon Story

How do whales affect carbon cycling?

Researchers connect three carbon transport pathways to whales: (i) the "whale pump" in which excretion of nutrients in urine and feces may stimulate primary production; (ii) export of whale carcasses to the deep ocean; (iii) export of fecal carbon to the deep ocean, thus contributing to C sequestration (Figure 17.8).

1. **Whale pump.** The whale pump theory posits that dissolved inorganic nutrients from whale fecal plumes stimulate primary production and hence contribute to carbon sequestration. Phytoplankton require light, dissolved inorganic macronutrients (nitrogen, N; phosphorus, P; and silica, Si), as well as micronutrients such as iron and other trace metals, for growth and development. However, open ocean surface waters often limited in dissolved macronutrients contrast micronutrient deficiency in oligotrophic waters and in the Southern Ocean. Dissolved macronutrients and micronutrients in whale feces may increase primary production, though no studies have fully quantified their significance for open ocean primary production. Whales also excrete key nutrients in their urine. In fact, minke whale feces contain low concentrations of N, compared to P, potentially excreting most N in urine as urea (CH_4N_2O)

2. **Deadfalls.** In contrast to carcasses of large terrestrial organisms, most carcasses of whales and large marine fishes sink and sequester carbon from surface waters into the deep ocean. Before industrial whaling, sinking of carcasses of great whales (baleen whales) and sperm whales would have transported 1.9–19 million metric tons of carbon into the deep ocean each year, which corresponds to ~10% of global marine organic carbon burial in sediment. Despite the relatively small total biomass of whales compared to total overall biomass of marine biota, they could represent significant carbon export because it occurs in large parcels.

3. **Feces.** Feces sinking from the euphotic zone likely export organic carbon to deep waters. Whale feces form visible plumes that apparently dissolve quickly in surface waters, but include some recalcitrant, carbon-rich, high-density particles that sink to the seafloor.

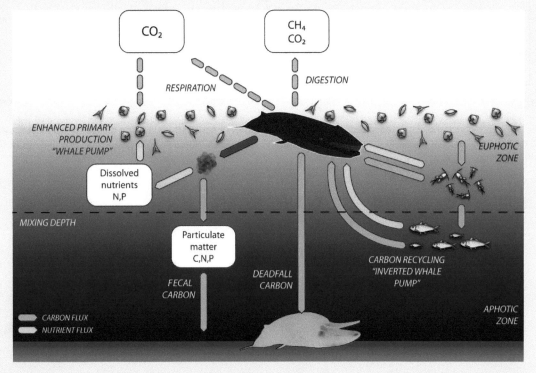

Figure 17.8   Carbon fluxes associated with large cetaceans: (1) Enhancement of primary production ("whale pump"); (2) sinking feces; and (3) carcass export. Orange circles (2, 5 and 6) indicate fluxes.

(Continued)

> **Box 17.2 (Continued)**
>
> **Carbon emission fluxes.** Baleen whales and their terrestrial relatives, the ruminants, have multichambered stomachs, which include a large non-glandular forestomach. The forestomach of whales harbors microorganisms capable of fermenting the chitin-rich exoskeletons of krill. In addition to respiration, which directly returns CO_2 to the atmosphere, forestomach digestion in whales leads to carbon release in the form of methane (CH_4). This "inverted whale pump" may be significant in that deep diving species that forage below the mixed-layer depth may "recover" carbon that would otherwise be "sequestered" at these depths. Whales store part of carbon consumed as biomass and eventually return it to depth as carcasses. However, they excrete most of the carbon they gain from prey either at the surface or release it into the atmosphere as CO_2 or methane, thus representing a carbon recycling pathway from the deep ocean to surface waters and the atmosphere.

Inputs of organic matter to the seafloor and pelagic-benthic coupling critically influence the functioning of most marine ecosystems on Earth, but periodicity and variability of these events results in non-continuous input of organic matter to the seafloor. Concentrations of organic matter within the water column in coastal and oceanic environments vary significantly over time (Figure 17.9). This mechanism can deliver massive amounts of organic matter to benthic ecosystems, where significant amounts of this material become integrated in superficial sediment layers, some of which may become permanently buried.

Variability of organic matter delivery extends all the way to the abyssal seabed (Figure 17.10), as illustrated by ROV images at 5,000 m depth during contrasting seasons; the greenish-brown color of phytodetritus contrasts more homogenous substrate where perforations indicate burrowing organisms that collect material on the surface and move that material into their protective burrows within the sediment.

Long-term studies conducted using imagery (Figure 17.10) and sediment trap analysis (Figure 17.11) demonstrate that massive deposition occurs in some years, in contrast to other years with no obvious deposition. Thus, although pelagic-benthic

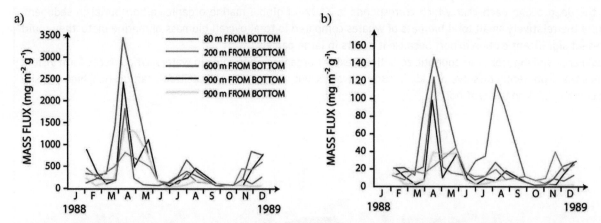

Figure 17.9 Amounts of material collected in sediment traps during different periods, highlighting differences in amounts arriving at contrasting depths (less material arrives at greater depths, expressed both as total mass or organic content) over time (for example, peaking during spring).

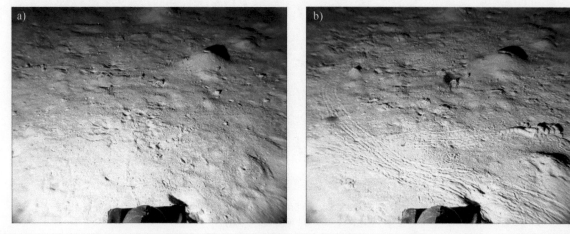

Figure 17.10 Photographs taken by a ROV on the abyssal plains of the Atlantic Ocean at 5,000 m depth. Phytodetritus creates the green haze (left), increasing biological activity (right) that depletes the phytodetritus. *Source:* National Oceanography Centre.

Figure 17.11 Sediment trap samples at different times of year; the coarsest material occurs as currents transport material off the continental shelf into the open ocean, or during periods with abundant carbonate from foraminiferans.

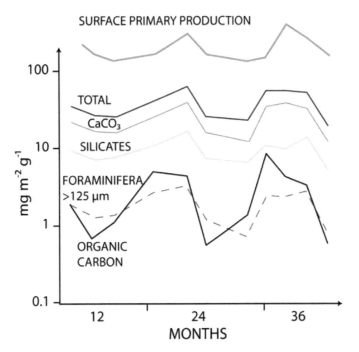

Figure 17.12 Variation in surface primary production over time as well as supply of organic matter to the seafloor. Upper line denotes primary production, followed (from top to bottom) by concentrations of different types of material collected in traps: calcium carbonate, silicate, planktonic foraminifera, and particulate organic carbon.

coupling can deliver substantial organic matter to the seafloor in some locations, the accumulation of this material provides something of a buffer to compensate for periods (sometimes lasting several years) with minimal injections of phytodetritus.

Complex processes determine the amount of organic matter that reaches the seabed. Analysis of temporal variation in primary production in surface waters and organic matter delivery to the seafloor (Figure 17.12) shows similar patterns, illustrating the fundamental role of primary production for organic matter export to the seabed. However, some peaks in organic matter do not closely coincide with peaks in primary production, and instead reflect calcium carbonate likely associated with the presence of planktonic foraminiferans. This example illustrates not only the importance of primary production for organic matter flow to the seabed, but also the importance of secondary production by foraminiferans. Clearly, seasonality in organic carbon may differ greatly from that for superficial primary production, leading scientists to hypothesize that the time for phytodetritus to sink to depth results in a time lag (delay) between primary production and arrival of organic material on the seafloor. This lag adds additional challenge for the many species that time their reproduction to coincide with the arrival of this material.

Remains of plankton, fecal material (fecal pellets), fragments of organisms left behind by predators, cell aggregates, complex aggregates (marine snow), and other forms such as fragments of leaves or carcasses of large organisms (small and large fish or marine mammal carcasses) sink to the seafloor producing a non-homogeneous distribution of organic matter in time and space. Terrestrial material provides a substantial portion of detrital material that reaches the ocean, including leaves, logs, and material eroded from the coast, transported by rivers, eroded from beaches or other shoreline substrate, and in some cases blown by wind; rivers alone discharge $13–20 \times 10^9$ metric tons of material y^{-1}. Fragmentation of detrital matter accelerates its aging and degradation. This material adds to autochthonous primary production and detrital material produced in coastal environments. Planktonic and benthic scavengers living in continental shelf environments use much of these two pools of detritus (autochthonous coastal material and allochthonous terrestrial material), but currents export some debris outside of this system. Transport of material beyond the continental shelf follows simple processes. The gentle slope of the continental shelf and relatively shallow depths facilitate re-suspension of deposited material off the bottom by waves and currents, transporting fragmented (smaller and less dense) particles seaward. Currents associated with downwelling (where winds blowing toward the coast carry surface water shoreward) result in subsurface counterflows out to sea. On the steeper upper continental slope, seismic events or instability in sediments that have built up over time can slump suddenly, forming intense and rapid turbidity currents that flow down the slope in submarine canyons, carrying detrital material and sediments to bathyal and abyssal depths. These mechanisms of **lateral transport** of particulate organic matter represent fundamental food supply sources to organisms living in the **twilight zone** of dim light and the **dark zone**, of no light.

Transport of organic matter through the water column can require long time periods, during which the prokaryotic component and nutritional value of organic matter may change substantially. Recent research demonstrates intense hydrolytic activity within sedimented particles and refute a key trophodynamic assumption about deep-sea environments, namely that low nutritional quality of organic matter transported to depth results from degradation of labile components in surface layers during descent. This scenario would result in a nutritionally poor deep-sea environment, but despite the comparatively low production available in these environments, more recent studies demonstrate relatively high percentages of labile organic matter at depth in some instances. Thus, although degradation occurs through the water column, transport of organic matter to the bottom occurs much more rapidly than settlement rates calculated based on particle size and density.

Organic matter concentration tends to decrease exponentially with increased depth in oceanic environments. On continental shelves this transfer from surface waters to the seafloor normally occurs quickly (from hours to a week). If the relationship between carbon supply at depth and surface carbon flow/production (Figure 17.13) approaches 1, this would indicate export of all surface production. With a relationship of ~0.1 only 10% of production is exported and so on. At a depth of ~100 m, some 30–60% of surface primary production reaches the bottom. At ~1,000 m depth this percentage drops to ~10%, and only ~1% of the material arrives at 5,000 m.

New sediment trap technologies, combined with coupling of measurements from fixed and drifting traps (which help differentiate between vertical and lateral transport), demonstrate that much of the organic matter flow reaching bottom environments descends laterally rather than just vertically. In fact, bottom currents responsible for lateral transport (called **lateral advection**) transport significant amounts of organic material (Box 17.3).

This organic matter transport mechanism has important consequences for ecosystem functioning because lateral advection can strongly affect particle concentration and composition, so that material reaching the bottom along the continental margins can differ greatly from that reaching the seafloor in the open ocean (Figure 17.15). For example, in continental margin environments the biogenic fraction rarely exceeds 30–40%, with a substantial fraction of fecal pellets (up to 10%), but a consistently dominant lithogenic fraction (60–70%). Flux can also differ greatly among canyons (Figure 17.16): Lacaze-Duthiers, an **active canyon** (thus subject to strong turbidity currents and mass transport) differs from **passive** (subject to largely vertical flow, much like the open ocean) Grand Rhone Canyon. Sediment traps collect much more material at depth in an active compared to a passive canyon, thus demonstrating the fundamental importance of hydrodynamics for particle transfer from the water column to the seafloor.

17.5 Consequences of Organic Matter Export to the Seabed

Except in the shallowest environments, survival and growth of benthic communities depends on food supplied from the water column, emphasizing the importance of organic matter export. In the Mediterranean, for example, difference in organic matter supply to the seafloor result in significant differences in benthic biomass between the western and eastern basins. Similarly, organic matter flow strongly influences abundances of benthic organisms (Figure 17.17). Low supply of organic matter to the seafloor and more rapid turnover in warmer surface layers also characterize tropical environments, but

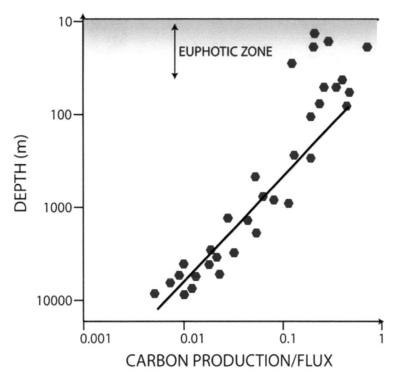

Figure 17.13 Carbon flow exported as settling particles from photic zone as a function of depth.

Box 17.3 Measuring Organic Matter Flow

Researchers often measure material flux using sediment traps anchored above the seafloor on a **mooring**. A mooring consists of a wire cable held vertical by buoyant floats, with a heavy weight (such as a train wheel) to anchor it to the seabed; by attaching instruments such as sediment traps to the mooring, researchers can sample at a specific depth over time (Figure 17.14). A device called an **acoustic release** disconnects the wire from the anchor weight when triggered remotely by an acoustic signal through an electronic mechanism. Upon release, the upward thrust of the buoys (Figure 17.14) floats the wire and instrument to the surface. A single mooring can contain multiple sediment traps or current meters (Figure 17.14) placed at specific depths from either the bottom or the surface in order to measure variation in particle flux or other variables.

Because of costs, researchers launch and recover moorings infrequently; a mooring can collect data for very long periods, of months to a year or more. The high aspect ratio of the sediment traps minimizes resuspension. By analyzing variation among traps at different depths, and analyzing composition of trap material, researchers can determine type and origin of material, degradation and transformation processes of organic matter during descent, and contributions of lateral transport. Some studies that reported higher amounts of material in traps at greater depths than nearer the surface strongly suggest lateral input. More recently, researchers have used ocean gliders for collecting flux measurements.

(Continued)

Box 17.3 (Continued)

Figure 17.14 Different phases of sampling with sediment traps; top, from left to right: launching of a sediment trap; buoys that enable the mooring to return to the surface; bottom, from left to right: external and internal structure of sediment traps, which typically feature a high aspect ratio (height to diameter ratio) to reduce resuspension of trapped particles A funnel-like structure in the lower part of the trap further helps to retain material, and sometimes helps contain a fixative such as formalin that preserves the collected material so it does not decompose over time.

even a small increase in supply leads to rapid increases in benthic abundance and biomass. Benthic communities in environments normally rich in organic matter, however, respond much less dramatically to increased carbon flux to the seafloor.

Food limitation means that the most oligotrophic systems normally accommodate fewer organisms but can often increase rapidly with increased organic matter supply. Higher abundances characterize food richer environments, but these environments respond less rapidly (and sometimes negatively) to increased carbon flux from the photic zone. Thus, different systems may not respond in the same way to similar contributions of organic matter. Because of competition among organisms, systems with strong food limitations often use organic matter much more efficiently, thus compensating to some degree for reduced organic matter availability.

The contrasting flux rates of organic matter to the seafloor in oligotrophic and eutrophic basins produces cascading effects on different benthic compartments (Figure 17.18). As just seen, increased flux of organic matter flow to the seafloor increases organic carbon available on the sediment surface, even at great depths, illustrating critical coupling between primary production in the water column and productivity in marine sediments.

With increased food resources in both basins of the Mediterranean, concentrations that reach the sediment differ between basins. The benthic prokaryote compartment (particularly heterotrophic bacteria) responds rapidly to increased organic matter in both systems, but abundances increase faster in the richer environment. In contrast, the benthic meiofaunal compartment clearly increases in the richest environment, but remains largely constant in the most oligotrophic environments. This pattern suggests ongoing competition between bacteria and meiofauna for organic material that reaches the

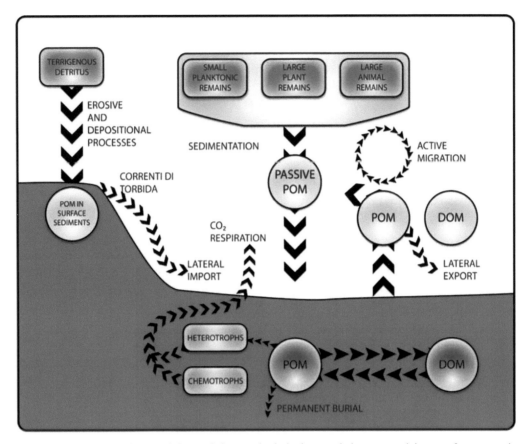

Figure 17.15 Detrital pools, their different origins, and changes in their characteristics as material moves from coastal environments to deep environments, from the water column to the seafloor or vice versa, or within the sedimentary matrix.

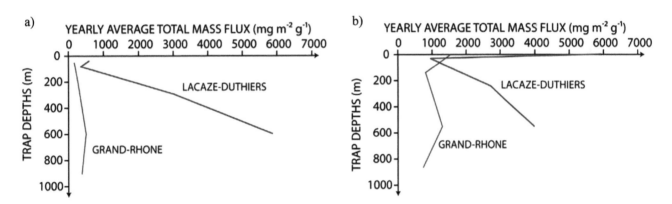

Figure 17.16 Flows of organic C and total mass average within Grand Rhone and Lacaze-Duthiers Canyons, illustrating how living at depth can potentially mean severe food limitation, depending on canyon location and morphology. Reported are: (A) the total mass flux in the two canyons in winter; (B) the patters observed in summer.

bottom. Only a sufficiently strong input of organic matter addresses food demands of both the prokaryotic and meiofaunal compartments, triggering potentially rapid increases in total benthic biomass.

An inverse process to pelagic-benthic coupling occurs, appropriately termed **bentho-pelagic coupling**, where material released from sediments (nutrients, sediments, detritus, larvae) flows from the benthic to the pelagic compartment. For example, resuspension events within the benthic boundary layer can mobilize significant amounts of both mineral and organic material as well as benthic organisms. Resuspension plays a critical role in energy supply for suspension feeders.

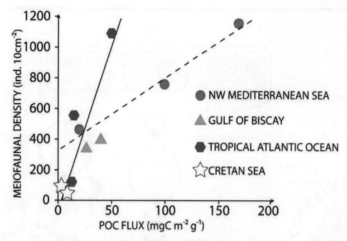

Figure 17.17 Relationship between meiofaunal abundance and supply of particulate organic carbon (POC) to the bottom; increased carbon supply to the bottom tends to be associated with increased meiofaunal abundance, though the specific relationship varies among environments.

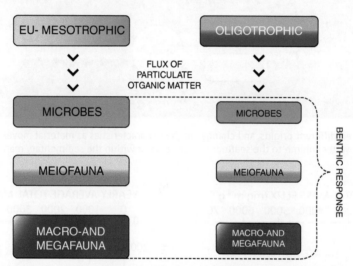

Figure 17.18 Changes in the inputs of organic matter (such as in the case of winter and summer) can have profound effects of the response of benthic components. In this example greater biomass of all major groups of biota coincides with mesotrophic compared with oligotrophic conditions.

In fact, resuspension remobilizes a portion of particulate organic matter otherwise "trapped" in sediments, making it newly available to consumers. Detritus dominates resuspended organic matter because degradation has already begun, both during its descent through the water column, and after mixing into sediments through activities of benthic suspension feeders, deposit feeders, and scavengers (Figure 17.19).

Bentho-pelagic coupling also plays an important role in life cycles of marine organisms with resting cysts or eggs that remain dormant for long periods of time prior to release from the benthic compartment into the water column. In other words, sediments can be the starting point for both merobenthic organisms (copepods that produce resting cysts, for example), and benthic organisms such as macrobenthos and megabenthos that produce meroplanktonic larvae. Both examples illustrate bentho-pelagic coupling, through movement of matter or organisms from the benthic environment into the planktonic environment. These dynamics play an important role not only for recruitment and survival of the species but also for ecosystem functioning of planktonic systems from a biogeochemical cycling and nutrient supply point of view, among other functions. In short, the interaction between the water column and seafloor works in two directions: the water column provides organic matter to the benthos, and the seafloor provides organic matter, nutrients, and organisms to the water column. This process has been important over geological timescales and correlates with the diversification of life and productivity of the ocean.

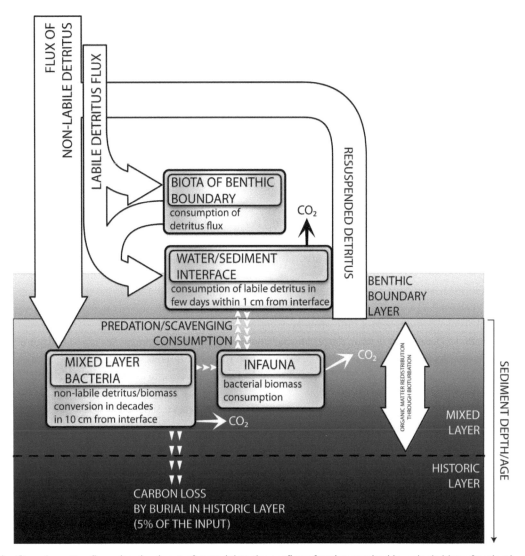

Figure 17.19 Organic matter flow, showing input of material to the seafloor, fraction respired by pelagic biota, fraction buried deep inside the sediments, and fraction respired in sediments. Flux lines to the right indicate those occurring in sediments and at the sediment-water interface. Organic matter that reaches the seafloor does not remain there, but is respired (and thus transformed into CO_2), and remineralized. Nutrients remineralized in surface layers of sediments return to the water column through benthic fluxes as nitrates, phosphates, or silicates. Some material, however, becomes buried and sequesters organic carbon permanently or semi-permanently in sediment subsurface layers.

Questions

1) Explain the difference between labile and refractory organic matter.
2) Discuss sedimentation rates of fecal material and the variables that contribute to it.
3) What does the DOM-POM equilibrium refer to?
4) Define marine snow.
5) Is flux of organic material to the seafloor constant?
6) Discuss the composition of phytodetritus.
7) What are the ecological consequences of organic matter input to the seabed?
8) What is pelagic-benthic coupling?
9) What is bentho-pelagic coupling?
10) How do we measure flux of organic matter to the sea floor?

Suggested Reading

Accornero, A., Picon, P., de Bovée, F. et al. (2003) Organic carbon budget at the sediment-water interface on the Gulf of Lions continental margin. *Continental Shelf Research*, 23, pp. 79–92.

Albertelli, G., Covazzi-Harriague, A., Danovaro, R. et al. (1999) Differential responses of bacteria, meiofauna and macrofauna in a shelf area (Ligurian Sea, NW Mediterranean): Role of food availability. *Journal of Sea Research*, 42, pp. 11–26.

Chase, J.M. (2000) Are there real differences among aquatic and terrestrial food webs? *Trends in Ecology and Evolution*, 15, pp. 408–412.

Coma, R., Ribes, M., Gili, J.M., and Hughes, R.N. (2001) The ultimate opportunists: Consumers of seston. *Marine Ecoogy Progress Series*, 219, pp. 305–308.

Danovaro, R., Della Croce, N., Dell'Anno, A. et al. (2000) Seasonal changes and biochemical composition of the labile organic matter flux in the Cretan Sea. *Progress in Oceanography*, 46, pp. 259–278.

Danovaro, R., Gambi, C., Dell'Anno, A. et al. (2008) Exponential decline of deep-sea ecosystem functioning linked to benthic biodiversity loss. *Current Biology*, 18, pp. 1–8.

Hansell, D.A. (2013) Recalcitrant dissolved organic carbon fractions. *Annual Review of Marine Science*, 5, pp. 421–445.

Hewson, I., Govil, S.R., Capone, D.G. et al. (2004) Evidence of *Trichodesmium* viral lysis and potential significance for biogeochemical cycling in the oligotrophic ocean. *Aquatic Microbial Ecology*, 36, pp. 1–8.

Jiao, N., Herndl, G.J., Hansell, D.A. et al. (2010) Microbial production of recalcitrant dissolved organic matter: Long-term carbon storage in the global ocean. *Nature Reviews Microbiology*, 8, pp. 593–599.

Luna, G.M., Bianchelli, S., Decembrini, F. et al. (2012) The dark portion of the Mediterranean Sea is a bioreactor of organic matter cycling. *Global Biogeochemical Cycles*, 26(2), p. GB2017. doi: 10.1029/2011GB004168.

Maldonado, M., Navarro, L., Grasa, A. et al. (2011) Silicon uptake by sponges: A twist to understanding nutrient cycling on continental margins. *Scientific Reports*, 1, p. 30.

Middelburg, J.J. (2011) Chemoautotrophy in the ocean. *Geophysical Research Letters*, 38(24), p. L24604. doi: 10.1029/2011GL049725.

Middelburg, J.J. (2018) Reviews and syntheses: To the bottom of carbon processing at the seafloor. *Biogeosciences*, 15, pp. 413–427.

Miquel, J.C., Fowler, S.W., La Rosa, S. et al. (1998) Particulate and organic fluxes in a coastal hydrothermal area off Milos, Aegean Sea. *Rapport et Proces Verbaux de la Commission Internationale pour l'Exploration Scientifique de la Mer Mediterranée*, 35, pp. 276–277.

Misic, C., Schiaparelli, S., and Harriague, A.C. (2011) Organic matter recycling during a mucilage event and its influence on the surrounding environment (Ligurian Sea, NW Mediterranean). *Continental Shelf Research*, 31, pp. 631–643.

Mopper, K., Kieber, D.J., and Stubbins, A. (2015) *Marine Photochemistry of Organic Matter: Processes and Impacts. Biogeochemistry of Marine Dissolved Organic Matter*. 2nd Edition. Academic Press.

Ngai, J.T. and Srivastava, D.S. (2006) Predators accelerate nutrient cycling in a Bromeliad ecosystem. *Science*, 314, p. 963.

Pearson, H.C., Savoca, S.C., Costa, D.P. et al. (2023) Whales in the carbon cycle: Can recovery remove carbon dioxide? *Trends in Ecology and Evolution*, 38, pp. 238–249.

Repeta, D.J. (2015) Chemical characterization and cycling of dissolved organic matter, in Hansell, D.A. and Carlson, C.A. (eds.), *Biogeochemistry of Marine Dissolved Organic Matter*. 2nd Edition. Academic Press, pp. 21–63.

Sarà, G. (2009) Variation of suspended and sedimentary organic matter with depth in shallow coastal waters. *Wetlands*, 29, pp. 1234–1242.

Smith, Jr., K.L., Ruhl, H.A., Huffard, C.L. et al. (2018) Episodic organic carbon fluxes from surface ocean to abyssal depths during long-term monitoring in NE Pacific. *Proceedings of the National Academy of Sciences of the United States of America*, 115, pp. 12235–12240.

Snelgrove, P.V.R., Blackburn, T.H., Hutchings, P. et al. (1997) The importance of marine sediment biodiversity in ecosystem processes. *Ambio*, 26, pp. 578–583.

Solan, M., Aspden, R.J., and Paterson, D.M. eds. (2012) *Marine Biodiversity and Ecosystem Functioning: Frameworks, Methodologies, and Integration*. Oxford University Press.

18

Interspecific Interactions and Trophic Cascades

The previous chapters summarized the roles of primary and secondary production in ecosystem functioning, but biota support ecosystem functioning in many other ways relating to how they interact with one another. Below we consider the relationship between biodiversity, ecosystem functioning, and ecosystem services, and then examine how species interactions affect the major functions of trophic (food web) support and provisioning of habitat.

The *Millennium Ecosystem Assessment* (MA, 2005), which gathered ecological experts from around the world, defined **ecosystem services** as "the multiple benefits provided by the ecosystems to the human race," and grouped these services into four major categories: (1) support services such as nutrient cycling, (2) regulating services such as coastline protection, (3) provisioning services such as biotic habitat structures for other species, and (4) cultural services such as artistic inspiration. The *Group on Earth Observations Ecosystem Biodiversity Observation Network* (GEOBON) defines **ecosystem functioning** as "the collective life activities of plants, animals, and microbes and the effects these activities (e.g., feeding, growing, moving, excreting waste) have on the physical and chemical conditions of their environment." *The Convention on Biological Diversity* defines **natural capital** as "the world's stocks of natural assets, which include geology, soil, air, water and all living things."

18.1 Biodiversity and Ecosystem Functioning

Scientists worldwide worry about the consequences of biodiversity loss in marine ecosystems. Many marine environments have seen a decrease and/or replacement of species locally, and the arrival of a growing number of nonnative (alien) species. Despite far fewer documented global extinctions in the marine environment than in terrestrial or freshwater system, numerous studies demonstrate regional declines in species richness and in biodiversity indices, sometimes over short periods of time. This biodiversity loss often links to habitat destruction, not only along the coast but also at greater depths. In 2019, the *Intergovernmental Panel on Biodiversity and Ecosystem Services* summarized and ranked the key threats as changing use of sea and land, direct exploitation of organisms, climate change, pollution, and invasive nonnative species; all of these threats contribute to marine biodiversity loss. Because ecosystem services and natural capital of marine ecosystems depend on their biodiversity, changes in biodiversity can potentially compromise ecosystem functioning, and thus potentially alter the goods and services they offer (Figure 18.1).

Although experts debate the numbers, socio-economic calculations suggest that coastal environments and continental shelf wetlands provide ~43% of the global value of ecosystem services, and that >50% of these environments already show evidence of degradation and biodiversity loss. Therefore, effective management of marine environmental use requires understanding the relationship between diversity and ecosystem functioning.

Increasing evidence support the conclusion that:

1) ecosystem functioning depends on the number and composition of species present;
2) ecosystem functioning depends on species interaction;
3) ecosystem functioning depends on the roles that these key species play.

In turn, ecosystem functioning links tightly to the provision of good and services (ES), on which human wellbeing depends (Figure 18.1).

Marine Biology: Comparative Ecology of Planet Ocean, First Edition. Roberto Danovaro and Paul Snelgrove.
© 2024 John Wiley & Sons Ltd. Published 2024 by John Wiley & Sons Ltd.
Companion Website: www.wiley.com/go/danovaro/marinebiology

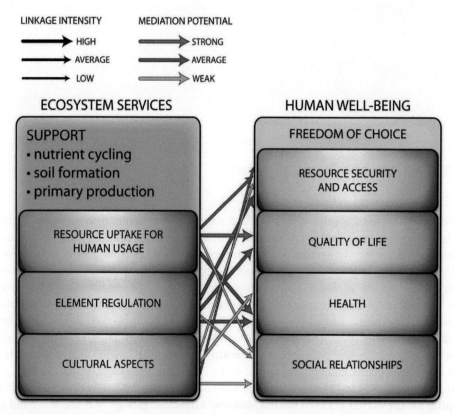

Figure 18.1 Relationship between ecosystem services (left) and components of human well-being (right).

Understanding the nature of any relationship between biodiversity and ecosystem functioning requires understanding how biodiversity affects two key processes: **resistance,** referring to the ability of a system to withstand disturbances without altering its community structure, biodiversity, and functioning, and **resilience**, referring to the capacity of a system to return to its previous state once a disturbance ceases. However, assessing resistance and resilience is not easy: although resilience can be assessed as the time that a system requires to return to its original condition after a perturbation, that time could span many years. Therefore, studying the relationship between biodiversity and functioning requires practical and measurable indices of ecosystem functioning, as well as appropriate metrics for measuring biodiversity. Studies on biodiversity changes, whether loss or replacement of species, represent a recent and expanding field of marine ecological research because of the potential implications for critical processes within ecosystems, and predicting changes in such processes with biodiversity loss.

In understanding how changing biodiversity can affect ecosystem functioning, consider why some environments harbor more species than others. For example, one hectare of tropical forest can accommodate > 300 species of trees, a number 10x greater than in temperate forests. In the ocean, a hectare of coral reef supports many more species than a similar area of coastal sediment. Availability of energy in each system offers one possible explanation for such differences, in that systems with higher energy availability can potentially accommodate more species related to a given function, and sometimes in greater abundance. However, some highly productive systems, such as mangroves, support low species richness despite high production.

Scientists have long pondered the relationship between biodiversity and stability with no obvious answer. One school of thought argues that high biodiversity ensures greater stability and a high level of functioning in marine ecosystem, but some really diverse systems such as coral reefs also appear highly vulnerable, with strong sensitivity to disturbance. In contrast, lower diversity systems may exhibit strong resistance to changes because sustaining fewer species may be simpler than sustaining many. In either case, numerous rare species may occur in low abundance. These examples illustrate potentially complex relationships between biodiversity and ecosystem stability, and that the role of specific species may matter, as occurs with some keystone species that sometimes play a disproportionate role in ecosystem functioning (see Section 18.5).

The remainder of this chapter considers positive interspecific processes (those processes involving interactions of multiple species: competition, predation, and facilitation) as well as complex interactions such as symbiosis and trophic cascades. These processes form the basis for understanding the primary mechanisms that regulate marine species abundance.

Knowledge of the main primary and secondary producers provides a basis to study the main characteristics of food web structure and functioning in the ocean. Below, we discuss several examples of trophic networks and trophic pyramids in marine environments. Starting from the concept of the classical food web (grazing), we define sources and characteristics of living and detrital organic matter, and of different food webs. We also consider functioning of the microbial food web, microbial loop, and viral shunt, as well as their implications for marine ecosystem functioning. In the ocean, only a small subset of the ecosystem exploits photosynthetic primary production directly. Many systems depend either on chemoautotrophic primary production or, in most cases, on organic detritus. Moreover, complex processes drive important changes in marine ecosystem characteristics, both in terms of production and cycling of organic matter, including bottom-up, top-down, and mixed effects. Some trophic cascades explain complex marine ecosystem functioning and suggest, for example, how removing a key species (keystone species) from a particular marine habitat can significantly change biota several trophic levels lower in the food web.

Numerous studies on **trophic cascades** demonstrate how changes to a specific component of the system or to a trophic level may influence species and trophic levels that do not interact directly with that specific component. Given that ecosystem services and marine ecosystem natural capital depend on biodiversity, changes in biodiversity can potentially alter ecosystem functioning, and thus the services they offer.

18.2 Facilitation and Cooperation – Positive Interactions

Positive (or "**facilitative**") interactions between two organisms refer to those interactions that benefit at least one of the participants and do no harm to either of them. These positive interactions include "**mutualistic**" interactions where both species benefit. Positive interactions can occur when an organism creates an environment more favorable for another, either directly (for example by reducing environmental stress) or indirectly (by removing competitors or predators).

Facilitation includes very close, **obligate** (required) relationships, as well as **facultative** relationships, referring to a potential but not required relationship. Facilitation occurs everywhere, from the origin of eukaryotic cells to the spreading of seagrasses, and the productivity of coral reefs. Some species modify the environment and facilitate other species living in that same habitat simply through their presence. For example, corals form complex structures and barriers that provide microhabitats for numerous other species. Remove the corals and the associated species disappear.

The inclusion of facilitation in classic ecological models leads to the paradox that **realized niches** (the complex of functions actually expressed by a species) can potentially exceed the **fundamental niche** (the functions potentially expressed by a species). In intertidal environments, macroalgae form "canopies" (akin to large trees in terrestrial forests) that reduce thermal and desiccation stress, thereby extending the environmental niche of some organisms that might otherwise dry out. Mutualistic relationships between corals and symbiotic zooxanthellae allow corals to occupy a wider range of physical environments than would be possible without their symbionts.

Consider four ecological models, with and without facilitation (Figure 18.2).

a) With facilitation: (a1), the realized niche (the blue circle) can exceed the fundamental niche (light-blue outer circle). Incorporating facilitation in the concept of niche; (a2) illustrates the amount of space needed to satisfy the demands of the fundamental niche, and potential mitigation factors reducing the niche.
b) Interactions between species often include facilitative and competitive components. This fact complicates efforts to measure competition because many experiments quantify only net interaction and assume minimal or constant facilitation. However, variation in net effect along an environmental gradient may reflect interaction of the two components: (b1) weak and constant facilitation, (b2) strong and variable facilitation.
c) The relationship between diversity and invasive species success (Figure 18.2): (c1) becomes negative, (c2) unimodal, or (c3) positive when it considers facilitation in promoting colonization and post-colonization survival.
d) The current concept of dominant organisms (**community dominants**) sometimes ignores their role as facilitators of the entire community that create habitat and thereby increase diversity. When considering facilitation: (d2), predictions of the intermediate disturbance hypothesis differ for primary (red line) and secondary (black line) exploiters of space. Specifically, small organisms often depend on habitat complexity, which can be greater when habitat-forming species dominate and with low disturbance or predation frequency or intensity.

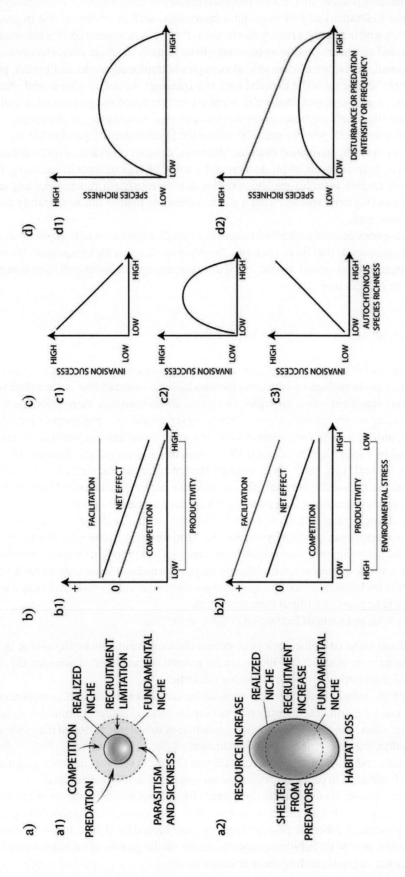

Figure 18.2 Fundamental models of ecological relations, with and without facilitation. (a) When excluding facilitation, fundamental niches can exceed realized niches (a1). Incorporating facilitation into the niche concept, (a2) illustrates how different processes expand the niche. (b) Species interactions often include both competitive and facilitative components, complicating efforts to measure competition effects, either with constant facilitation (b1), or variable facilitation (b2). (c) relationship between species diversity and invasion success becomes negative (c1), unimodal (c2) or positive (c3) when including facilitation by increasing colonization and post-colonization survival. (d) Facilitation may play a major role in "dominant communities" (d1). When including facilitation (d2), effects of disorder differ because intermediate facilitators can generate habitat, greatly increasing species diversity. Small organisms, in particular, often depend on complex habitat, which can be greater when habitat-forming species dominate and re-phase equally of disturbance or predation.

In subtidal environments, abundances of grazer gastropods depend on abundances of herbivorous urchins (e.g., *Evechinus chloroticus*). Urchins, in fact, remove macroalgae (or limit their settlement), creating a surface favorable for gastropod grazing. Similarly, the purple urchin *Strongylocentrotus purpuratus* promotes the chiton, *Tonicella* sp., and the limpet, *Acmaea mitra*, which feed on encrusting coralline algae; once again urchins eliminate leafy algae, enabling growth of coralline forms and facilitating chiton and limpet grazing. Similarly, the limpet, *Cellana tramoserica*, prevents macroalgal development, providing a grazing surface suitable for barnacles and the seastar, *Patiriella exigua*.

Links between biodiversity and stability/functioning of marine ecosystems has inspired research on the functional role of biodiversity in marine ecosystem functioning, particularly since the *Convention on Biological Diversity* (CBD) developed at the Earth Summit in Rio de Janeiro in 1992. Research on how biodiversity relates to ecosystem characteristics, determining stability and capacity of communities to respond to disturbance phenomena (resistance, resilience) has led to the development of several theoretical models to represent different experimental and field results. Recent studies have focused on marine benthic fauna as a model system to explore relationships between biodiversity and functioning. Benthic ecosystems offer the opportunity to conduct replicated, manipulative experiments with multiple species of macrofauna, meiofauna, and microbes and measuring different functional variables within a system. The following major models consider interactions between biodiversity and functioning:

1) **Complementarity model:** This model, initially proposed by Charles Darwin in 1859, suggests that numbers of species correspond to the complexity of their interactions, where complexity refers to the different ways in which energy can pass through a community. This model proposes a direct, positive relationship between species richness and system stability (Figure 18.3a). It assumes that a greater number of species corresponds to greater complexity of interactions, and thus a greater number of pathways along which matter and energy can transfer from one trophic level to another. In this scenario, availability of an alternative pathway of transfer of matter and energy to higher trophic levels offers a "plan B" if, for any reason, biodiversity loss interrupts some connections. This model implies a relatively open ecological space where progressively adding species to the community does not result in saturation. Studies carried out in subtidal environments support this model, showing a positive effect on stability and functioning of the system with increased diversity of macrobenthic species (and particularly of spatangid sea urchins). These species, in fact, determine levels of bioturbation within sediments, promoting re-mineralization of organic matter and nutrient release of nutrients at the

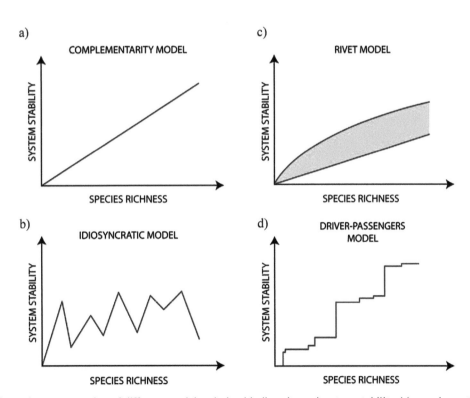

Figure 18.3 Schematic representation of different models relating biodiversity and system stability: (a) complementarity model; (b) idiosyncratic model; (c) rivet model (or functional redundancy); (d) driver and passengers models.

water-sediment interface. These nutrients stimulate primary production (phytoplankton and phytobenthos) resulting in a positive effect on functional processes. This model illustrates how higher species richness increases functioning (in this specific example, primary and secondary production, and biomass) and stability of the system.

2) **Idiosyncratic model:** In this model, increased biodiversity can positively or negatively affect ecosystem functioning, depending on specific characteristics of the species added to (or subtracted from) the system (Figure 18.3b). The effects of species addition or loss therefore depend on the nature of the added species and their relationship with species already present in the system. Studies conducted so far show neither a positive nor negative effect of biodiversity in absolute terms. However, experiments related to this model manipulated only a few species of macrobenthos in a species-poor coastal environment. Still, the introduction of alien species in a system may cause idiosyncratic changes, depending on how alien species interact with native species. Introducing a species with no evolutionary history in that community to enable functional integration could produce negative effects by altering the historical balance among different species, removing resources from the system, etc. This model hypothesizes that interactions among species strongly influence the contribution of each species to system functionality, and studies on relationships between stability and diversity should thus consider findings case by case.

3) **Rivet or functional redundancy hypothesis:** Recent studies show that growth of stability or functions of a system does not follow a linear trend with increasing species richness, but instead initially grows and then flattens onto a plateau (Figure 18.3c). This pattern likely indicates a phenomenon defined as **ecological** or **functional redundancy**, where species introductions or removals can produce major, minor, or zero effects, depending on the type of system. In other words, high diversity does not further increase system functionality because some species become redundant. In this sense, redundant species allow the system to continue to function, even with the loss of some species, which are replaced by others with the same functional characteristics. In this model, failure of the last rivet (or the last species in a system performing a particular function), causes the system to become unstable. Ecological stability therefore increases with increased specific richness as ecological space becomes progressively crowded and resilience of ecological function increases thanks to redundancy. Several studies report this model in lagoon systems, where multiple species perform similar functions; for example, several different bivalve species stabilize the substratum.

4) **Drivers and passengers model:** This model hypothesizes that some species, in addition to performing a specific function, play a fundamentally important role in controlling overall ecosystem functioning. "Driver species" refers to those taxa with strong roles in essential ecological functions required for ecosystem stability, whereas "passenger species" play a minor role in ecological functioning and thus add accessory value (Figure 18.3d). The model invokes the idea of a bus with multiple passengers but noting that the bus cannot move without the driver. In light of this hypothesis, variation in the number of "driver" species (also called "keystone" species) clearly significantly affects ecosystems if species loss (or addition) occurs, relative to the role of passengers. The graphical representation of the driver and passenger model resembles that of the rivet model. The rivet model suggests significant overlap in ecological functions among species, whereas the "driver and passengers" model suggests many species have functional importance, but ecosystem functioning requires some species much more than others. Both models emphasize that ecological stability requires different types of ecological function.

5) **Inverse model:** Some microbial studies reported a negative relationship between biodiversity and functioning, where increased diversity led to decreased ecosystem functioning. Some research shows higher levels of ecosystem functioning in lagoon and coastal environments with greater microbial diversity. However, in instances where one or a few opportunistic species rapidly carry out biogeochemical (or other) functions relative to larger numbers of less active species, high biodiversity and reduced function may coincide.

6) **Facilitation model:** A study of deep-sea environments identified a new model for biodiversity and ecosystem functioning, which shows a positive, exponential relationship (Figure 18.4). Increased biodiversity coincides with an exponential increase in ecosystem functions, such as rates of biomass production, rates of re-mineralization of nutrients, or secondary production, etc. This exponential relationship implies that some species play a facilitation role, increasing the likelihood that additional species will co-occur, yielding a positive, exponential relationship.

Teleost fishes provide additional examples of facilitation. Sediment re-suspension by red mullet benefits sole by exposing prey. Other fish benefit from feeding by rays, which also resuspend organisms. Predators often chase small fish toward the surface, where birds eat them. In some cases, more indirect facilitation can occur. For example, *Haematopus moquini* (oystercatchers) reduce numbers of limpets, promoting algal growth that supports the cryptofauna on which small wading birds (typically birds with long legs, such as herons) can feed.

The sea stars, *Pisaster ochraceus* and *Pycnopodia helianthoides*, and the anemone, *Anthopleura xanthogrammica*, further illustrate faciliation. *P. ochraceus* feeds mainly on barnacles and mollusks whereas *P. helianthoides* preys on echinoids, so

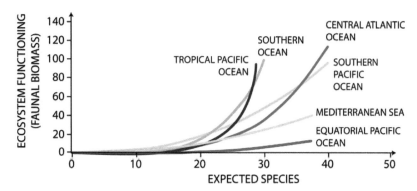

Figure 18.4 Relationship between biodiversity and functioning in different basins in benthic environments. (biomass) and diversity (*Expected species number*).

the two seastars do not compete for food resources. However, the anemone diet overlaps with that of both seastars. Increases in numbers of both seastars can limit the resources available to the anemone, decreasing number of prey this anemone shares with seastars. But a decrease in abundances of both seastars has the opposite effect. Manipulative experiments demonstrated that, when the three species coexist, the presence of both seastars facilitates anemone feeding, and seastars and anemone do not compete. Increasing abundances of *P. ochraceus*, increases the number of mussels and barnacles consumed by the anemone because seastars remove the mussels and barnacles from the substrate, which roll in the waves where the anemone can capture them. Similarly, the seastar, *P. helianthoides*, increases the number of urchins the anemone can consume by detaching the urchins from the substrate. The deep sea illustrates facilitation, as seen in the positive, exponential relationship between biodiversity and ecosystem functioning described earlier. These data suggest that higher biodiversity supports higher rates of ecosystem processes (Figure 18.4) and greater efficiency in performing these processes.

18.3 Symbiosis

The term **symbiosis** (from the Greek: συν = together; βιος = life) refers to a constant and intimate relationship between different species. A symbiont refers to an organism that lives or cooperates with an individual of a different species, irrespective of whether both individuals benefit (**mutualism**), only one benefits (**commensalism**), or one benefits to the detriment of the other (**parasitism**). Some symbionts are so interdependent they cannot live separately, whereas others may survive but marginally. Symbioses vary in: (1) mode of interaction; (2) ecology; (3) ethology (behavior); (4) levels of intimacy of structural association between organisms. The latter category includes: (a) ectosymbiosis or episymbiosis, where the symbiont adheres to the surface of the host (including inner surfaces and spaces such as the intestine or the excretory ducts of the endocrine glands); (b) intercellular ectosymbiosis, where the symbiont lives between cells; (c) intracellular endosymbiosis, where the symbiont lives within cells. Functional modes of symbiosis include:

1) As noted above, with mutualism all associated organisms receive benefits (relationship between species +/+). Coral reefs offer a classic example, through corals and their symbiotic zooxanthellae (see Chapter 21 on coral reefs). Pomacentridae fishes of the genera *Amphiprion* and *Premnas* protect themselves from predators by living among the poisonous tentacles of anemones (Figure 18.5a). A mucous coating protects the fish from cnidarian stinging cells (**nematocysts**). In fact, during an acclimatization period, the fish feed on the tentacles, assimilating the poison over time and thus becoming somewhat tolerant. The anemones also gain many advantages; the fish supplies nitrogen through excretion, which benefits the algae living within the anemone. Anemone fish defend the anemone from predators and parasites, and provide feeding scraps the anemone can use. This example of mutualism illustrates how both partners benefit from the symbiosis, a relationship also seen in some anemones and crabs. In some cases, however, the anemone (*Calliactis parasitica*) that adheres to the shells of pagurid crabs (*Pagurus bernhardus*) benefit from the mobility the crab adds by moving around on the seafloor, whereas the crab apparently receives no clear benefit. In many cases, however, especially with hermit crabs, *Dardanus* spp., the crab actively places an anemone on its shell. In these associations, the crab presumably gains some advantage, but whether that advantage is camouflage, aid in capturing prey, protection from predators, or shell reinforcement remains largely unknown, though some evidence points to predation protection.

Figure 18.5 Examples of symbiotic relationships between organisms: (a) The clownfish *Amphiprion* associated with an anemone on an Indonesian coral reef; (b) The porcelain crab (genus *Neopetrolisthes*) associated with the anemone *Stichodactyla* sp.; (c) a pair of gobies *Amblyeleotris guttata* associated with the snapping shrimp *Alpheus bellulus*; (d) The anemone, *Adamsia palliata*, attached to growing on the hermit crab, *Pagurus prideaux. Source:* MASSIMO BOYER and Hans Hillewaert / Wikimedia Commons / CC BY-SA 4.0.

When *Octopus vulgaris* was confined with two Mediterranean species of hermit crab, *Dardanus arrosor* and *D. calidus*, with and without their guest anemone, *Calliactis parasitica*, the octopus attacked in all cases. The octopus attacked and swallowed all hermit crabs that lacked anemones but could not successfully ingest crabs with the commensal anemone present. In the absence of an anemone, the octopus extracted the hermit crab by inserting an arm through its shell opening and pulling on the crab for up to 3–4 d. Hermit crabs with anemones successfully escaped, aided by the anemone's nematocysts and the octopus eventually gave up and left the crabs alone within experimental cages. *Neopetrolisthes* sp. crabs live between the tentacles of the sea anemones *Stichodactyla* spp. (Figure 18.5b), whereas others live at the anemone base (*Mithrax* sp. and *Lissocarcinus laevis*). *Alpheus* sp. shrimp build underground burrows and cannot see approaching threats, however, *Amblyeleotris* sp. fish that occupy the shrimp lair opening warn of danger (Figure 18.5c). The shrimp might otherwise emerge from their lair only to fall victim to a waiting predator. These species have developed a form of body communication that benefits them both. The fish gains a comfortable lair it could not build by itself, whereas the shrimp gains a guardian.

2) **Commensalism**, a relationship in which one organism benefits at no cost or benefit to the other (relationship between species +/0), typically involves a non-obligatory association. The "host" neither gains nor loses, whereas the commensal gains some advantage (often food). For example, guardian shrimp, *Pontonia custos*, spend their lives almost exclusively within fan mussels, *Pinna nobilis*, presumably utilizing the mussel as habitat. In rare cases this shrimp also lives in large sponges, where male and females cohabitate within the same host.

3) **Inquilinism** refers specifically to one organism that shares part of the living space of another, and neither "occupant" benefits or suffers (relationship between species 0/0). For example, hermit crabs often occupy empty gastropod shells, a space also used by polychaetes. Hermit crabs and annelids are inquiline because they share the same space, and commensal because the annelid feeds on hermit crab discards. Some species of hermit crabs, such as *Pagurus prideaux* live symbiotically with the anemone *Adamsia palliata* (Figure 18.5d), where the hermit crab moves the anemone onto its new shell as it grows and molts. Tiny crabs, *Porcellanella triloba*, live on Pennatulacean sea pens, along with other small shrimp and crabs.

4) **Amensalism** refers to an association disadvantageous for one species and neutral for the other (relationship between species -/0)
5) **Parasitism** occurs when one organism benefits and the other (typically a "host") suffers negative effects (relationship between species +/-). Moreover, we lack a clear understanding of the benefits/disadvantages of some cryptic symbioses such as the dinoflagellates *Mesodinium rubrum* and *Durinskia dybowskii*, which contain algal cells that have lost their independence and appear unable to live separately from their host.

Symbioses also differ in dependency; **facultative symbionts** can live independently of the other symbiont, in contrast to **obligate symbionts** that depend on another organism and cannot survive separately. Symbiosis classification can also consider the type of interacting organisms:

1) **Symbiosis between vertebrates and invertebrates.** These symbioses include forms of commensalism with **epizoic** or **endozoic organisms** that commonly live on bryozoans, crustaceans, and mollusks.
2) **Symbiosis between unicellular eukaryotes and prokaryotes.** Despite their high morphological diversity and complexity, eukaryotes possess limited metabolic capacity, in that almost all live aerobically, cannot photosynthesize (except for plants, algae, and some protistans), cannot fix nitrogen, and some have lost the ability to synthesize amino acids essential for basal metabolism. Prokaryotes, in contrast, possess high metabolic plasticity, particularly mechanisms for energy conservation. A recent hypothesis reduced the concept of symbiosis to associations in which at least one partner organism provides the other with new metabolic capabilities. Indeed, many symbiotic associations involve a eukaryotic organism with high structural complexity, but often limited metabolic capabilities. In this perspective, symbiosis reflects, and may continue to reflect, an evolutionary strategy through which eukaryotes access a wider range of metabolic resources.
3) **Symbiosis between vertebrates and prokaryotes**. This type of symbiotic relationship requires four essential characteristics: (1) It must involve at least two different species; (2) the symbiosis must transmit unaltered to the next generation; (3) the association must be intimate, meaning strong interaction between the species; (4) it must confer a new character to at least one of the partners, such as new metabolic capacity. For example, *Vibrio fischeri* occurs widely in all temperate and subtropical regions, both in free and saprophytic forms. This bacterium also occurs within the intestinal microbial communities of many marine mammals, or in the **luminous** (light producing) organs of many species of fish. Bacteria of the genus *Photobacterium* occur only within a body cavity near the animal's surface. These photobacteria require O_2 to produce bioluminescence, as well as a long aldehyde chain that increases photon production by 1,000x as the enzyme luciferase returns to its original state. Thus, the bacteria obtain food and shelter within the animal, and, in exchange, enable the animal to produce light for defense or to lure or locate prey. These bacteria occur in fishes or other organisms, in which the animal develops lenses to focus the light.
4) **Symbiosis between invertebrates and invertebrates**. Cases of endozoic symbionts occur in some turbellarians and their resident nematodes.
5) **Symbiosis between invertebrates and eukaryotes**. Zooxanthellae sometimes live in the margin of the mantle of mollusks that contain "hyaline bodies," whose function is to facilitate light penetration. Zooxanthellae crowd around these bodies to maximize photosynthesis. The alga *Tetraselmis convolutae* lives in the flatworm *Symsagittifera roscoffensis*, whose anterior region contains pigments and optical statocysts associated with the nervous system. This region facilitates three-dimensional orientation in response to light, required to feed the abundant symbionts. Some poriferans (sponges) contain symbiotic algae in their cytoplasm and aid the algae only by increasing the surface suitable for catching light. In cnidarians with symbiotic algae, modified endoderm and mesoderm result in decreased tentacle sizes and a reduced digestive system. Algal adaptations include loss of locomotor flagellum, decrease in cell wall thickness, and loss of the photosynthetic apparatus typical of free-living forms.
6) **Symbiosis between invertebrates and prokaryotes** – Symbioses between prokaryotes and metazoans occur in all marine environments and many different taxa. These include: chemolithotrophic proteobacteria that reside in the tissues of different species of invertebrates living near hydrothermal vents or in anoxic sediments; archaea within deep-sea sea cucumber intestines; bacteria and archaea symbionts of sponges; luminescent bacteria within squid light organs. In benthic marine environments, most symbiotic relationships involve chemoautotrophic Bacteria and macroinvertebrates (Table 18.1). Researchers have documented >200 species of metazoans with symbionts that span the Mollusca and Annelida, but major gaps remain on knowledge of associations between bacteria and smaller metazoans (such as the meiofauna). Some interactions of this type occur within the outer shell or epidermis, whereas others involve extracellular symbioses, such as obligate symbionts that lack a complete digestive system. For example, the nematode *Astomonema southwardorum* lives near methane seeps but lacks a digestive system, depending on extracellular

Table 18.1 Overview of primary symbioses between chemoautotrophic prokaryotes and marine invertebrates.

Phylum	Class	Family	Tissue containing symbionts	Localization	Habitat
Mollusca					
	Bivalvia		Gills	Intracellular or extracellular	Hydrothermal vents, cold seeps, anoxic sediments
		Solemyidae, Lucinidae, Vesicomyidae, Mytilidae, Mactridae	Gills	Intracellular	Hydrothermal vents
Annelida					
	Oligochaeta			Intracellular	Anoxic sediments
	Polychaeta	Siboglinidae	Trophosome	Intracellular	Hydrothermal vents, cold seeps
					Cold seeps / anoxic sediments
Nematoda					
		Halomonisteridae	Subcuticular space		Cold seeps
		Stilbonematinae	Subcuticular space	Extracellular	Coralline sands and anoxic sediments
			Cuticle	Epicuticular	Anoxic sediments

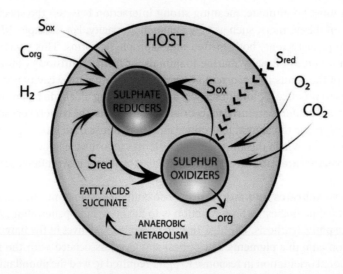

Figure 18.6 In conditions of low sulfide flow from sediment, electron donors use the sulfide produced internally by sulfate reducers to fix CO_2 autotrophically. The worms provide the electron donor to sulfate reducers as succinate and fatty acids for anaerobic metabolism. Organic carbon or H_2 taken from the sediment provides the external electron donor.

intestinal bacteria that represent 50% of the host's volume. Symbioses between chemosynthetic bacteria and marine invertebrates were first discovered in 1981 living within the giant tube worm, *Riftia pachyptila*, at a hydrothermal vent at 2,500 m. This vestimentiferan worm lacks a digestive system and has developed a specialized organ called a **trophosome**, in which they house chemoautotrophic symbiotic bacteria. The unusually high oxygen affinity of *R. pachyptila* hemoglobin enables it to capture oxygen in hypoxic conditions and accumulate oxygen for use during short periods of anoxia. Moreover, similar affinity for hydrogen sulfide allows this hemoglobin to transport gases in the trophosome and avoid chemical reactions within the blood. Therefore, the animal transports otherwise toxic hydrogen sulfide through its internal tissues, which symbiotic bacteria then use for their growth (Figure 18.6). Bacterial oxidation of sulfur compounds produces energy and simultaneously reduces their toxicity. Bacterial exudates then provide a source of dissolved organic matter for the host, such as sugars and amino acids, which support growth in the tube worm. Bacteria digest different compounds using alternative metabolic pathways.

The oligochaete worm *Olavius algarvensis* provides another example of sulfate-reducing endosymbionts. One bacteria and cephalopod symbiotic relationship offers a particularly interesting story. The symbiont *Vibrio fischeri* is a luminescent heterotrophic marine bacterium, in the family Vibrionaceae, a family with many other examples of symbiotic and pathogenic relationships. The nocturnal cephalopod, *Euprymna scolopes*, hunts for prey in shallow waters, contracting and expanding chromatophores in its skin to change colors and hide in the sand. But the unique aspect of this species is a light organ within its ink bag in the mantle cavity; *V. fischeri* lives within this light organ. The organ consists of several layers of host tissue, including epithelial layers containing symbiotic bacteria and a yellow filter lens to enhance the light produced by bacteria and select the wavelength of light. The cephalopods emit the light produced by the symbionts downward, manipulating light intensity to simulate moonlight intensity, or hide its silhouette to confuse predators. The invertebrate and bacterial symbiont partner to ensure that appropriate bacteria colonize the light organ. *Vibrio fischeri* comprise a very low percentage of the total bacteria in seawater (<0.1%), yet bacteria colonize the light organ within just a few hours after the cephalopod hatches. No bacteria other than *V. fischeri* can colonize this organ. Mutants of *V. fischeri* that lack the ability to produce bioluminescence can colonize the organ but non-mutants soon outcompete and eliminate them. The organ lasts the entire life of the host and no specimen of this cephalopod has ever been found without this organ. The host feeds the bacteria with carbon and organic nitrogen (peptides and proteins).

7) **Symbiosis between invertebrates and unicellular eukaryotes**. Many invertebrate species have symbiotic algae: sponges, corals, sea squirts, ctenophores, polychaetes, flatworms, mollusks, and spoon worms (echiurans). With many of these symbioses all of the algal cells live within the invertebrate, growing and reproducing, but the invertebrate does not digest the algae. In other symbioses, only the chloroplast lives within the invertebrate; when the organism ingests the algal cells, it transfers the chloroplast from the digestive tract into another tissue (Figure 18.6).

18.4 Complex Biotic Interactions: Trophic Networks and Cascades

18.4.1 Trophic Networks

Trophic networks regulate vital processes and ecosystem functioning. Classical ecology viewed the transfer of energy in an ecosystem as a linear process with linear ramifications, or specifically as a **food chain** (Figure 18.7) in which primary producers (autotrophic organisms) provide food for primary consumers (herbivores) that provide food for secondary consumers (carnivores), then to tertiary consumers, and so on. We now recognize many more types of trophic interactions between marine organisms than initially assumed. The concept of food webs (Figure 18.8) better illustrates possible interactions among organisms and shows that the same organism may be involved in multiple trophic levels. This model also considers the role of decomposers, which recover inorganic substances from the physical environment by decomposing carcasses and feces of organisms. As noted earlier, trophic level refers to a set of organisms spanning multiple species that exploit a common trophic resource; these organisms and their shared resources define functional typologies. Within a given trophic level, these trophic groups or guilds share the same resources and environmental compartment. Despite the generally high number of trophic levels in the ocean, numbers vary depending on the characteristics of the system. **Guild** refers to species that exploit a similar resource in a similar way, even if they do not occupy a similar ecological niche. The species of a guild at a single location use a shared resource and may therefore interact strongly.

Different models describe trophic networks depending on specific environment. Diagrams representing trophic networks may contain information on species, on energy flow direction, and on amounts of energy or matter transferred from one trophic level to the next. Trophic networks should always include producers, consumers, and decomposers, enabling researchers to follow energy transfer from one network node to another, with increasingly higher trophic levels with increasing distance from organic matter production. In the ocean, trophic relationships link various sectors of benthic and planktonic environments (the bentho-pelagic network; Figure 18.8) or involve only one of them.

Primary production forms the base of two trophic pathways: the grazing web and the detrital web. The grazing chain begins with photosynthetic autotrophic organisms, through herbivores, and ends with carnivorous consumers; organisms typically increase in size and decrease in abundance with increasing trophic level through to the apical consumer. Primary producers consist of phytoplankton and phytobenthos upon which herbivorous zooplankton feed, defining the second

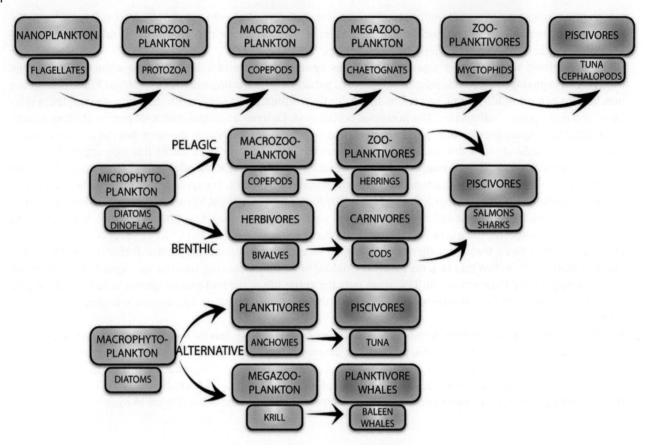

Figure 18.7 Different types of food webs in the ocean, including the possible different fates of primary production (phytoplankton, macro-phytobenthos) as it moves through different trophic levels.

trophic level. Carnivorous zooplankton and small fishes comprise the third level, fish, crustaceans (mainly cephalopods), and larger birds comprise the fourth, and mammals and large fishes (sharks, for example) the fifth. The detrital food web begins with detrital organic matter, on which microorganisms feed prior to passing on to detrital feeders (consumers of detritus), which return basic inorganic material from the physical environment to the food web, eventually ending with their predators.

Trophic pyramids can represent trophic relationships in terms of numbers of individuals, biomass, or energy. In pyramids denoting numbers (Figure 18.9), the area of the different levels crudely indicates the relative proportions of individuals within each level, with the horizontal rectangles corresponding to each of the different trophic levels. The height of a given pyramid depends on the number of trophic levels, so different ecosystems could differ greatly in trophic pyramid shape. With biomass pyramids (Figure 18.9), sizes of different rectangles represent the product of average weight and number of individuals. In the ocean, inverted biomass pyramids often exist because the biomass of the lower trophic levels (such as phytoplankton) replaces itself very quickly, even if the **standing crop** (biomass) of phytoplankton at any given time is not large.

Although inverted biomass periods exist, inverted productivity (energy) pyramids cannot because each trophic level of consumers depends on productivity from the trophic level on which they feed; basic energy laws require bottom heavy productivity pyramids (although their height and relative widths can vary among ecosystems) (Figure 18.10). The amount of energy transferred between trophic levels (**trophic efficiency**) averages ~10% but may vary 5–30% depending on the specific system. This wide variation arises from our inability to fully understand what material marine organisms actually consume in nature (in laboratory experiments, organisms must consume whatever food we give them). They lose much of the energy as heat, undigested food, and movement, and organisms at the next trophic level cannot use all the energy (efficiency always falls well below 100%). Furthermore, food pyramids "lose" some material at each level; some 50% of phytoplankton production may settle to the bottom (thus entering a different trophic pyramid in which subsequent mineralization returns that material to the water column as nutrients and carbon dioxide).

18.4 Complex Biotic Interactions: Trophic Networks and Cascades | 325

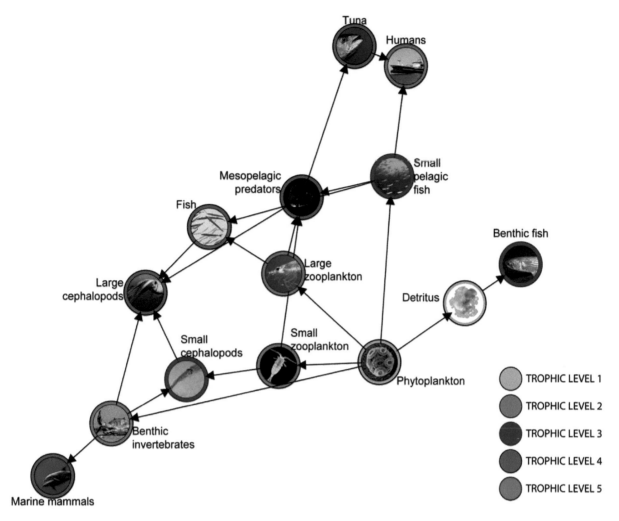

Figure 18.8 Example of a simplified bentho-pelagic food web, showing different network elements (functional groups), the flow of energy and matter between them, and trophic levels, which indicate each functional group location within the food web.

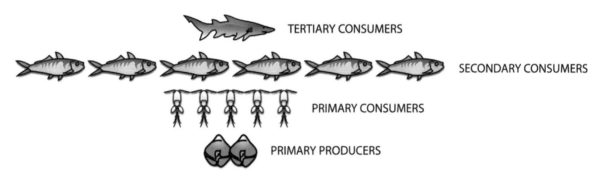

Figure 18.9 Trophic pyramids for numbers (or biomass); each trophic level consists of the number/biomass comprising each functional group (often expressed as total dry weight of the organisms). Most pyramids decrease in size from the base to the top, but a reverse pyramid occurs in some aquatic ecosystems. In this example, phytoplankton grow and reproduce quickly enough to support a large population of zooplankton, despite lower biomass of phytoplankton at a given point in time where zooplankton biomass exceeds that of phytoplankton.

Energy pyramids (Figure 18.10) provide the most informative description of ecosystem trophic structure, in that the succession of rectangle widths denote the proportional amount of energy accumulated at one level per unit time and volume. Globally, terrestrial ecosystems produce more organic material than marine ecosystems, but more efficient energy transfer from the first to second trophic level in marine ecosystems means that, despite less energy provided by primary production, marine systems apparently lose less energy during the first transfer.

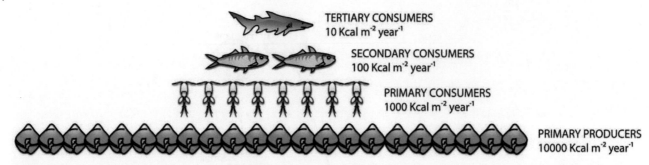

Figure 18.10 Energy trophic pyramid (productivity), where energy flows from producers (photosynthetic organisms) to consumers (herbivores, carnivores). Ecological pyramids typically represent energy as the amount of living material (or its equivalent energy) present within different trophic levels. Primary producers convert only 1% of solar energy. Only 10% of energy typically moves through the next trophic level on average relative to the previous trophic level because most organisms use much of the energy for maintenance, growth, and reproduction. These losses typically limit food pyramids to 3–4 steps.

18.4.2 Detrital Trophic Network

Historically, most studies on energy flow in the marine environment understandably focused on living biota and their abundance, biomass, and composition, a focus that also shaped thinking on ecosystem functioning. Until relatively recently, researchers largely ignored the importance of organic matter in marine food webs and in energy flow. Scientists now recognize the essential role organic matter plays in the maintenance and functioning of most marine organisms. This bias arose from the classic view of food webs based on grazing food webs reported in terrestrial ecology. The discovery of a food web not based on grazing but instead on the detrital circuit (Figure 18.11) has completely changed our understanding of marine ecosystem functioning in the last twenty years.

Each trophic level releases material into the environment from two pools that form a dynamic equilibrium with each other: DOM (**Dissolved Organic Matter**) and POM (**Particulate Organic Matter**) (Figure 18.11). Together, these sources contribute significantly to the detrital pool potentially exported from the system. Of all the material produced by primary producers, herbivores consume some portion, losing some through natural death of cells, senescence of phytoplankton, phytoplankton exudate, and waste material or feces of primary consumers. The previously held view that DOM and POM had no influence on ecosystem functioning changed when studies verified that many plankton and benthic organisms depend on particulate organic matter for food, filtering and ingesting particles and retaining those components with nutritional value. We also know that some biota use dissolved organic matter, such as sugars. Given compelling evidence that many organisms use these pools of organic matter, detritus must play a non-negligible role in the ocean. Indeed, that role becomes particularly important when considering the critical function performed by prokaryotes, in particular by bacteria, especially in decomposing these organic compounds and releasing inorganic nutrients as they respire carbon. Therefore, detrital organic matter provides an essential source of energy and nutrients that supports prokaryotic biomass, which constitute a substantial majority of living biota. In oligotrophic regions, bacteria may represent 40–70% of particulate organic carbon, demonstrating the importance of detritus in supporting ocean biomass.

The relatively low organic carbon:nitrogen (C/N) ratio of 3.5–5 that characterizes microbes means high nutritional value. As microbes break down organic matter, they convert it to material with higher nutritional value (by respiring and releasing

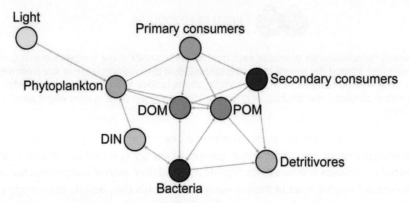

Figure 18.11 A marine food web, highlighting flows of organic matter.

carbon as they break down organic matter, they lower the C/N ratio). This transformation of detrital matter or "**protein enrichment**" refers to increased organic matter nutritional value and quality. The organisms that feed on detritus, known as **detritus feeders** or **deposit feeders,** fill a similar niche in seafloor sediments as earthworms on land. In fact, annelids, and polychaetes in particular, often dominate this trophic guild, ingesting large amounts of sedimentary matter and retaining the associated organic matter and presumably the microbes, which add substantial nutritional value. These combined food sources (microbes and detritus) and the difficulty in discriminating between detrital material and living organisms, complicate efforts to assign these organisms to a specific trophic level. Deposit feeders (detritivores) typically also feed on microbes.

18.4.3 Trophic Networks Based on Dissolved Organic Matter

The ocean contains many macroscopic and microscopic organisms that specialize on DOM. Bacteria, Archaea, and protozoa can all use DOM; urea and amino acids represent alternative but particularly important sources of nitrogen for phytoplankton. Phytoplankton cells produce phosphatases that release phosphorus from some compounds, as well as aminopeptidase that releases nitrogen to use in their synthesis activities. Use of these compounds requires active transport and often results in production of mucous substances. Recent studies demonstrated that zooplankton such as *Tigriopus brevicornis* and *Calanus finmarchicus* use mono- and polysaccharides. **Gelbstoff** ("yellow substance"), or optically measurable dissolved organic matter in water with an absorption peak at ~220 nm, varies in composition but represents an important product, especially in coastal environments. Typically, dissolved material delivered from freshwater habitats contributes to this DOM. Microscopic and multicellular organisms both use this DOM, although we lack information on how larger organisms use this dissolved material. Taxa ranging from pycnogonids (sea spiders) to many organisms lacking digestive tracts (roundworms and siboglinids) can incorporate DOM directly through their tissues.

18.4.4 Microbial Loop

Microorganisms play a fundamental role in transferring energy through marine ecosystems and in biogeochemical cycling of elements such as carbon, nitrogen, and phosphorus. Understanding the structure and function of these microbial ecosystems and their role in biogeochemical cycles links to the processes that govern the functioning of all marine ecosystems. In the ocean, different mechanisms convert a large fraction of primary production into dissolved organic matter, and only heterotrophic bacteria and archaea can access this component of primary production. As a result, the uptake of organic matter by bacteria represents a major pathway for carbon flow, and variability in this uptake can alter the pattern of carbon flux. The ability or inability of bacterial communities to grow on specific types of organic matter may influence how an ecosystem functions. **The microbial loop** in marine ecosystems refers to the microbial food web through which microbes reintroduce DOM produced by various trophic levels. Prokaryotes, in fact, can reuse organic detritus (nonliving) resulting from cell deaths, excretions, and exudates from marine organisms, and convert it into biomass. Protists such as ciliates and flagellates subsequently prey on prokaryotic cells, and larger organisms (i.e., mesozooplankton), in turn, prey upon protists and thus return organic matter to the grazing trophic chain. This process increases availability of a portion of organic matter derived from primary production that might otherwise be lost. This process has important implications for carbon, nitrogen, and phosphorus biogeochemical cycles, and on overall marine ecosystem functioning. Researchers introduced the microbial loop concept to describe a lateral network complementary to the classic food chain (Figure 18.12). Bacteria can use DOM that would otherwise be lost from the food web and transform it into new biomass, thus representing something of a keystone process in transferring prokaryotic biomass to higher trophic levels.

The second step involves heterotrophic Protozoa (mainly nanoflagellates) that regulate bacterial abundance through bacterial grazing, simultaneously providing a food source for higher trophic levels. Ciliates (microzooplankton) define the next trophic level, which, mesozooplankton, in turn, prey upon, although this latter group often directly consumes bacteria associated with particles of organic detritus. Bacterivorous microzooplankton play a key role in transferring bacterial production, and therefore energy, to upper trophic levels through mesozooplankton. In oligotrophic environments such as the open ocean, the microbial loop provides an essential mechanism for transferring matter and energy to higher trophic levels. In contrast, the microbial loop plays a much less important role in comparatively nutrient-rich systems such as upwelling regions. Moreover, the microbial loop interacts with gas production. Dimethyl sulfide (DMS), for example, a key gas in forming cloud condensation nuclei and thus climate regulation, interacts with the microbial component because, when released in the form of dimethyl-sulfoniopropionate (DMSP) by microalgae through viral infection or other mechanisms, bacteria can use it as a source of carbon and sulfur. Therefore, microbes influence two opposing processes: gas release associated with death of phytoplankton cells, and consumption of cells by bacteria. The viral shunt further works against the microbial loop through a viral short circuit described below.

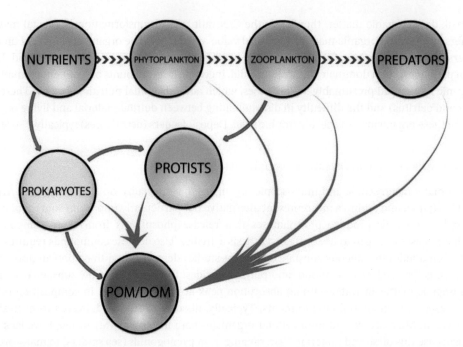

Figure 18.12 Microbial loops in the ocean recycle POM/DOM that would otherwise be lost because organisms at higher trophic levels within the classic food chain cannot access it. This process links to phytoplankton primary production in supporting higher trophic levels, including fisheries production.

Heterotrophic bacteria therefore fill a fundamental role in marine ecosystem that extends beyond "simple remineralizers" in that they support different functions: (a) decomposers of DOM and POM, (b) a food source for micro-organisms at the base of the food web, (c) competitors with phytoplankton for inorganic nutrients and with microorganisms for POM, and symbionts of other organisms such as protozoa, d) pre-digesters of refractory organic matter, thus releasing POM and nitrogen. The capacity of heterotrophic bacteria to decompose dissolved and particulate organic matter and absorb inorganic nutrients at very low concentrations links to their small size and large surface/volume ratio. This capacity contributes to their high adaptability to different environmental conditions and nutrient availability, including a "no growth" or dormancy phase within their life cycle. These sometimes "silent" bacteria do not participate in functioning of systems during adverse environmental and nutrients conditions.

18.4.5 Viral Shunt

Viruses can infect all organisms. The virus encounters a host cell, adheres to its wall, and finally penetrates inside or through host cell endocytosis prior to direct injection of its genetic material. The discovery of abundant concentrations of viral populations in aquatic environments has changed how we study some aspects of marine ecosystems. The idea of a **viral short circuit** arose from the study of phytoplankton and prokaryotic cell mortality from viral infection. This process occurs within the microbial loop where the virus, through infection and lysis of prokaryotic and phytoplankton cells, alters rates of transfer of living organic matter in the pool of dissolved and particulate organic debris. In fact, cell lysis products such as nutrients, DNA, cell debris, and virus particles contribute to the DOM and POM pools, diverting the flow of carbon fixed by phytoplankton and prokaryotes from higher trophic levels. This process is also known as the **viral shunt** (Figure 18.13).

This short circuit causes a negative feedback control on the microbial loop and prokaryotic community because viral lysis converts bacterial biomass into DOM. Dissolved organic matter (DOM) used by bacteria and converted into biomass potentially available to higher trophic levels can be re-converted into DOM because viruses can kill prokaryotes and other organisms; the microbial loop brings DOM to higher trophic levels whereas the viral shunt returns it to microbes. Every time viruses increase in abundance, export of materials to the seafloor declines because sedimentation rates of DOM approach zero. These events help in understanding the complexity of this process. Reduced transfer of organic material to the seafloor can result from predatory interactions between bacteria and viruses or between viruses and plankton and can therefore occur independent of changes in primary production. Although we lack thorough studies and models of this type of interaction, research on the viral shunt within sediments shows that this process promotes the breakdown of organic matter assimilated by bacteria that would otherwise have moved to higher trophic levels.

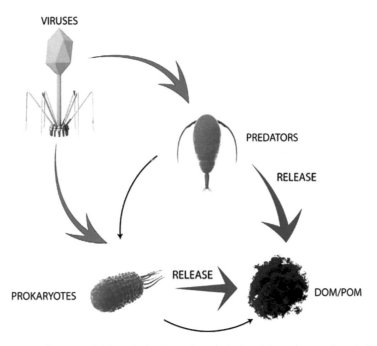

Figure 18.13 The viral shunt transfers material from heterotrophic and photoautotrophic organisms to POM and DOM. This process alters the chemical composition of POM and DOM pools relative to source organisms. Recycling of highly labile materials, such as amino acids and nucleic acids, typically occurs in the photic zone, whereas more recalcitrant or refractory material rich in carbon generally sinks to deeper waters. Thus, material exported to deep environments may be richer in carbon than its source. The viral short circuit occurs when viruses infecting bacteria and other organisms cause their death by bursting, thereby releasing DOM into the ocean. This short circuit therefore acts in opposition to the microbial loop, which tends to increase DOM cycling and energy transfer to higher trophic levels.

Viral lysis yields labile dissolved organic matter immediately available for bacteria. Consequently, the release of cellular content from cells killed by viral infection increases bacterial production and respiration as it reduces matter and energy transfer to higher trophic levels, and thus overall food web productivity. Viral lysis resulting in greater availability of DOM also increases efficiency of nutrient recycling. The small size and kinetics regarding use of inorganic nutrients favors bacteria compared to phytoplankton. Phytoplankton, in fact, dominate eutrophic environments, even if bacteria can use DOM to grow. In oligotrophic systems, increased viral abundance and the associated increased role of the viral shunt in the microbial loop leads to significant loss of production by higher trophic levels. The reverse scenario occurs in eutrophic or mesotrophic systems. Models of trophic networks that also include viruses demonstrate that the effects of the microbial circuit on ecosystem functioning also depend on the extent of viral infection.

Therefore, analysis of the implications of microbes on carbon and other nutrient cycling should consider the efficiency with which prokaryotes can transfer organic matter to higher trophic levels and the efficiency of viral lysis and associated diverting of organic matter from higher trophic levels. Viral infection can then modify the effect of the microbial loop with respect to the fate of carbon in the ocean. Therefore, viruses merit further study not only because of their potential effect on ecosystem trophic status, but also because of the extent of viral control on mortality, particularly that of prokaryotes. Because prokaryotic mortality from viral infection varies so greatly both in pelagic and benthic ecosystems (generally varying 10–40%, but potentially reaching 100%), their effect on efficiency of the biological pump could vary dramatically. Recent studies suggest that viral lysis could decelerate carbon export by transforming living organic matter into dissolved or particulate detritus, and promoting conversion into dissolved inorganic carbon through respiration and photolysis in the photic zone. However, some studies suggest that availability of nutrients such as N and P that limit primary production may control carbon export. Because organisms immediately assimilate the most labile products released through cell lysis, such as nucleic acids and amino acids, the net result would be greater retention of N and P in the photic zone. This retention would produce a positive effect on primary production and biological pump efficiency.

18.4.6 Bottom-Up Control of Trophic Food Webs

Bottom-up control (Figure 18.14) occurs where environmental conditions, and food resources in particular (inorganic nutrients that support growth of primary producers), strongly affect other components of the food web. Abundant

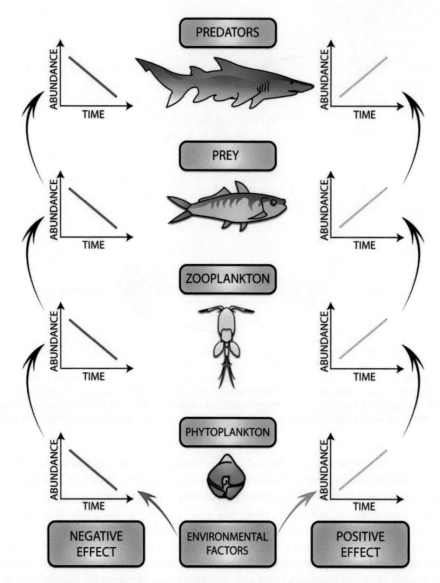

Figure 18.14 Bottom-up control in a simplified (four-level) food web in a marine ecosystem. A less favorable environment leads to decreased phytoplankton abundance, in turn leading to reduced zooplankton abundance. Zooplankton decline means fewer prey for small fishes, leading to fewer predators.

nutrients sustain phytoplankton, which support a large biomass of zooplankton grazers, which in turn support numerous small pelagic fishes, which can then support top predators such as marine mammals. Primary producers and nutrient input therefore regulate other food web components. Comparative analysis of twenty marine ecosystems indicates that increased nutrient concentrations generally lead to increased phytoplankton biomass. Although vascular plants dominate terrestrial ecosystems, the ocean contains <1% of photosynthetic biomass. Recognizing that marine animals feed directly or indirectly on a "soup" of small algae (phytoplankton), and that this resource limits overall ecosystem productivity, has motivated scientists to understand the dispersed and heterogeneous nature of the marine environment. Environmental variables thus drive the carrying capacity and fish biomass of ecosystems, but apparently in complex ways.

Species and individual distributions vary greatly in space, and marine populations, including some fishes, vary widely from year to year. Considerable evidence now demonstrates that natural variability in ocean circulation and mixing play a major role in marine productivity, as well as in population distributions. Scientists now regard availability of nutrients and physical constraints, such as retention, concentration, or enrichment processes associated with currents and turbulence, as important factors that influence larval survival, fish recruitment, and stock sizes.

A recent study on phosphorus limitation in the highly oligotrophic environment of the eastern Mediterranean demonstrated that addition of inorganic phosphates to surface water caused an unexpected response. Rather than stimulating primary production, phosphorus addition resulted in decreased chlorophyll and phytoplankton biomass, and increased

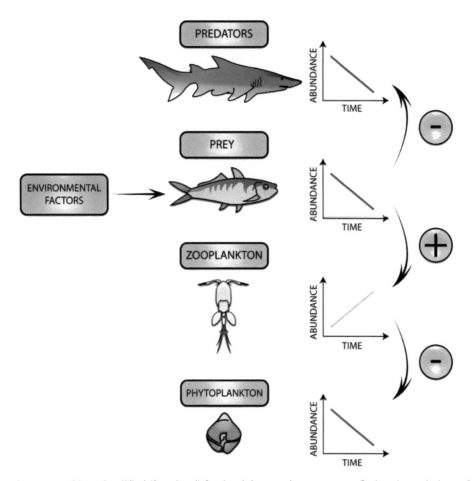

Figure 18.15 Top-down control in a simplified (four-level) food web in a marine ecosystem. Reduced populations of top predators (large fish) lead to reduced predation and increased abundances of small fishes. Increased numbers of small fishes result in increased predation on zooplankton, decreasing zooplankton population size. This decrease in zooplankton abundance reduces grazing pressure on phytoplankton, which consequently become more abundant.

bacterial production and copepod egg abundance. Although nitrogen and phosphorus can co-limit phytoplankton growth, this study suggests that phosphorus passes through the microbial food web to support copepods. This process can take two non-mutually exclusive major pathways: in the first pathway heterotrophic bacteria can also use inorganic nutrients for their growth and thus compete directly with phytoplankton; the second pathway involves rapid cycling of phosphate that quickly changes the stoichiometric composition of copepod prey. In this scenario, copepods (and therefore zooplankton) link to lower trophic levels through mechanisms not previously considered.

We do not know whether this type of interaction occurs widely in all marine ecosystems or specifically in highly oligotrophic environments such as the Cyprus gyres that lack inorganic nutrients. Studies show that nitrogen and phosphorus taken up by **phagotrophs** (organisms that eat other organisms) may first pass through **osmotrophs** (organisms that incorporate organic molecules obtained from organic particles through chemical breakdown such as phytoplankton or heterotrophic bacteria) that compete with them. This mechanism should help optimize transfer efficiency of essential elements for zooplankton growth without losing elements stored within phytoplankton, which may disperse more easily in the euphotic zone. When senescent phytoplankton die, if not preyed upon, their larger size means they sink relatively rapidly. In contrast, sinking bacteria usually adhere to particles. This mechanism requires a very complex conceptual model of pelagic food webs, which in some ways resemble deeper layers only to those developed for **high nutrient low chlorophyll (HNLC) water**, referring to locations with abundant nutrients but low primary productivity.

18.4.7 Top-Down Control on Trophic Food Webs

In contrast to bottom-up control, the **top-down control** model of food webs focuses on purely biological control and predator structuring of food webs. Predation pressure exerted by higher trophic levels can control abundances throughout the entire ecosystem. With bottom-up systems, increases in abundances at lower levels of food webs also lead to increases in higher trophic levels, collapsing only when nutrients become depleted. With top-down control, concentrations of available inorganic nutrients do not play the controlling role, which instead comes from the apical predators of the food web. These predators exert top-down control by preying on a large fraction of the prey and, in doing so, decreasing prey population size (Figure 18.15).

Of course, predation also occurs in bottom-up control systems but results in negligible effects that cannot control the dynamics of prey populations. Nonetheless, predators in these systems can influence community structure. A key predator can "remove" a dominant competitor and thus prevent competitive exclusion. The predator may therefore allow coexistence of other species competing for the same resource, thus increasing diversity. Predators disrupt competition and avoid dominance of single species.

18.4.8 Mixed Wasp-Waist Control

More recently, other models of interaction between different trophic levels such as the **wasp-waist model** (Figure 18.16) have emerged. In this model an environmental phenomenon can influence an intermediate trophic level. For example, small pelagic species decreased because of unusually low winter temperatures associated with Bora winds in the Adriatic Sea in 1899. This phenomenon caused bottom-up effects on top predators that were adversely affected by reduced food resources, and top-down effects on planktonic crustaceans that developed without strong predatory control. In turn, abundant zooplankton negatively affected phytoplankton.

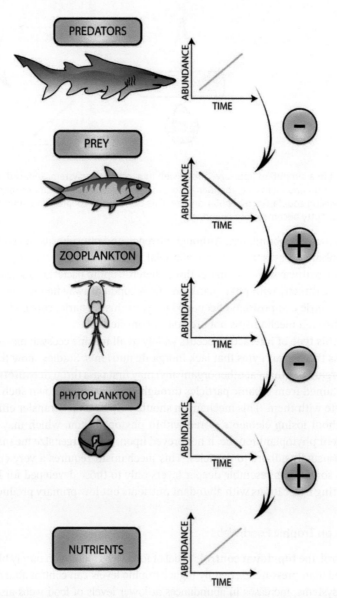

Figure 18.16 Wasp-waist environmental variable control applied to a simplified (four-level) food web in a marine ecosystem. Reduction of population size of small pelagic fishes negatively affects top predators (large fishes). Reduction of small fishes also decreases zooplankton predation pressure, allowing zooplankton populations to increase. This increase results in more grazing on phytoplankton, which consequently become less abundant.

18.5 Keystone Species

Keystone species generally refer to a species that exerts a significant and disproportionate impact on its community or ecosystem relative to its abundance. Keystone species typically occur in limited numerical abundance but play a crucial ecological role in maintaining ecosystem functionality. Several studies in aquatic systems (such as kelp forests, microcosms, microbial communities of marine invertebrates), showed a positive effect of species richness on community stability and illustrate higher community stability in species-rich communities that experience strong environmental variation, including stress and disturbance. However, not all species play an equally important role in overall system functioning. Keystone species fundamentally influence ecosystem functioning because changes in their abundance leads to an ecosystem shift. The absence of other species able to duplicate fully their role also characterizes keystone species.

Keystone species often (but not always) sit near the apex of the food web, though not necessarily at the highest trophic level. These species affect others through consumption, competition, etc., but also by modifying habitat characteristics (we refer to these keystone species as **ecosystem engineers**). Ecologists have devoted much attention to identifying keystone species. In fact, some researchers believe that conservation efforts should focus on maintaining keystone species rather than attempting to protect and manage all potentially important or vulnerable species. Examples of keystone species include:

a) sea otters, which support kelp forests and their resident biodiversity;
b) the seastar *Pisaster ochraceus*, which controls the abundance and biodiversity of rocky intertidal environments;
c) *Galeocerdo cuvier*, known as the tiger shark or "dustbin of the sea" because it eats practically everything. Although not a classic keystone species, it controls sea turtle and dugong populations that graze on Australian seagrass beds; removal of seagrass habitat leads to a drastic decrease in local biodiversity.

18.6 Trophic Cascades

Trophic cascades refer to changes in one trophic level that produces cascading effects on the abundance, biomass, or productivity of a community, population, or trophic level through multiple food web links. Cascading effects share a common structure: a predator, a species on which the predator feeds (species 1), a species on which species 1 feeds (species 2), and finally a species on which species 2 feeds (species 3). Predation pressure on species 1 inevitably reduces abundance of that prey. Reducing abundance of species 1 typically results in increased abundances of species 2 because of reduced predation mortality. However, greater numbers of species 2 can presumably reduce abundances of species 3 (Figure 18.17). This trophic cascade then alternates positively and negatively through successive trophic levels. This simplified food web results in just three effects; where numbers of effects in trophic cascades vary in direct proportion to food web complexity.

True trophic cascades often involve keystone species. These cascades were first described in lakes and intertidal environments. Initially, ecologists thought that trophic cascades were relatively unusual mechanisms linked to short food webs and restricted to particular types of marine ecosystems. However, recent examples from contrasting ecosystems suggest that cascading effects may occur in in a wide range of ecosystems, including the open ocean. The impacts of trophic cascades have strong impacts on ecosystems, potentially stabilizing them in alternative states.

The interaction between the mussel *Mytilus californianus* and seastar *Pisaster ochraceus* (Figure 18.18) provides a well-known example of trophic cascades. In 1966, Robert Paine reported the top-down effect of sea stars on rocky intertidal communities, showing for the first time how some species can play a disproportionately important role within their ecosystems by preventing any one species from monopolizing limited intertidal space. In the presence of the keystone species, *P. ochraceus*, a different set of algae, mollusks, barnacles, chitons, limpets, sponges and nudibranchs comprise the intertidal community in Mukkaw Bay (Washington). Experimental removal of the sea star allowed its most important prey, the mussel, *M. californianus*, to rapidly proliferate and become dominant, reducing species diversity, and Paine concluded that variation in this keystone interaction resulted in a mussel "monoculture." A subsequent study reported that reduced *Mytilus-Pisaster* interaction during environmental changes could dramatically impact the ecosystems.

Some researchers believe that predation on megafauna in the deep sea causes patchiness (fragmentation), which leads to decreased abundances of species that strongly compete for limiting resource, allowing weaker competitor species to coexist.

Many primary consumers exert top-down control. Several species of amphipods, grazing on seagrass epiphytic algae, can significantly alter ecosystem processes, such as epiphytic grazing, and *Zostera marina* biomass. In the northern Arabian

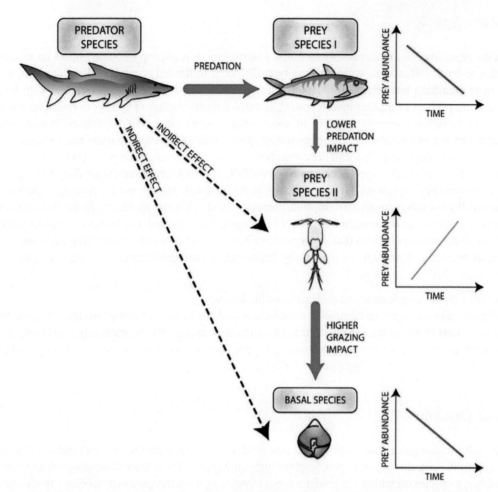

Figure 18.17 Conceptual diagram of trophic cascades and the effects of predation on subsequent trophic levels. In this case, increased predation results in decreased abundances of their major prey species and a minor impact on the second prey species. In this case predation on species 1 positively affects prey species 2. Increases in this species may then impact species at lower trophic levels.

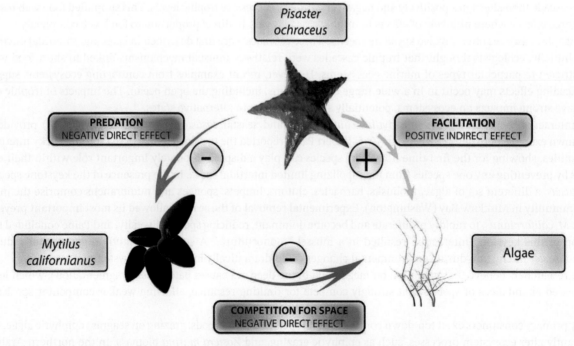

Figure 18.18 Schematic representation of the *Pisaster ochraceus* – *Mytilus californianus* interaction.

Sea, the primary consumer, *Pegea confoederata*, appears in large numbers at the end of winter, in February. This thaliacean, an intensive filter feeder that defecates quickly, produces fecal pellets ~0.25 cm² in diameter. Microscopic examination of these fecal pellets illustrates the enormous diversity of phytoplankton present in the water during this period. This major phytoplankton consumer grazes relatively uniformly on a wide range of local phytoplankton species, reducing competition and promoting biodiversity.

Diseases can also disrupt marine communities with profound effects on biodiversity. Several recent studies demonstrate major effects of viral diseases on phytoplankton diversity.

Kelp forests (*Laminaria* spp.: large brown algae forming the equivalent of underwater land forests) occur in shallow waters of rocky American coastlines with cool, nutrient rich waters and rank among the most productive ecosystems in the world. Historically, sea otters, a keystone species, occurred widely along the west coast of the United States. Kelp forests can persist for a long time; an individual kelp bed can last from 1 to 10 years. Typically, warming waters or grazing by urchins and fishes causes death of these adult macroalgae. During the "early" life history phase, in which embryonic kelp attach to the bottom by means of a rhizoid and begin to grow, young kelp are particularly vulnerable to grazing (or predation) by small sea urchins. Sometimes, however, sea urchins even attack adult forms. The interaction between otter-sea urchin-kelp (*Macrocystis*, *Nereocystis*, and *Laminariales*) that create real "marine forests" or kelp forests (Figure 18.19) provides a classic example of a trophic cascade.

Sea otters feed on urchins (*Paracentrotus* spp.) and bivalves. When abundant, sea otters stabilize kelp forest ecosystems by consuming sea urchins and thus limiting total grazing activity. Because they are a keystone species, reduced sea otter numbers can cause system shifts so that sea urchins increase in dominance, overgrazing and substantially reducing kelp cover and thus ecosystem productivity. These complex interactions can even cascade through to bald eagles (*Haliaeetus leucogaster*), apex predators at risk of extinction. Bald eagles prey on small fishes living near the kelp forest. Over time, in the absence of sea otters, sea urchin numbers increase greatly and graze down kelp to create seafloor barrens effectively devoid of vegetation cover. Rich kelp vegetation provides habitat for abundant small pelagic fishes on which bald eagles

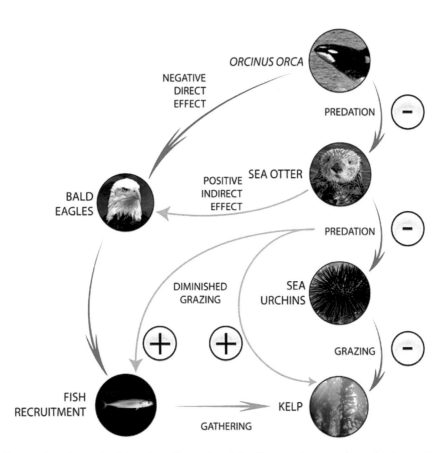

Figure 18.19 Sea otter-urchin-kelp trophic interaction. The text explains the complex cascade trophic interactions between different components. Plus signs indicate positive effects and negative signs indicate negative effects. *Source:* Robert Pittman/National Oceanic and Atmospheric Administration (NOAA), Marshal Hedin / Wikipedia Commons / CC BY-SA 2.0, AngMoKio / Wikimedia Commons / CC BY-SA 3.0, Hans Hillewaert / Wikimedia Commons / CC BY-SA 4.0.

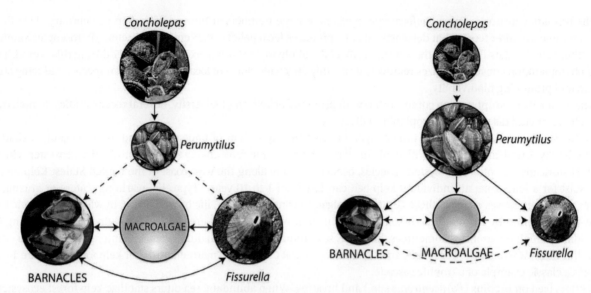

Figure 18.20 Interactions among *Concholepas concholepas*, *Perumytilus purpuratus*, barnacles, and macroalgae. Circle size is proportional to species abundances; top images indicate linkages prior to abalone removal whereas the lower images indicates a scenario with reduced abalone. *Source:* Auguste Le Roux / Wikimedia Commons / CC BY-SA 4.0.

feed, but the loss of kelp from the cascading effect of urchin overgrazing eliminates habitat for small fishes and a key food source for bald eagles disappears; the bald eagles then struggle to reproduce successfully. Clearly, trophic cascades can lead to surprising shifts both in ecosystem appearance and properties.

Initially, researchers attributed progressive declines in sea otter numbers to hunting and poaching, but they later established that predation by Alaskan killer whales (*Orcinus orca*) had played a major role in sea otter population decline. Killer whales may have recently increased their predation on sea otters as a result of the collapse of seal and sea lion populations, their preferred prey. This example illustrates how indiscriminate removal of some ecosystem components can dramatically alter how marine ecosystems function, as well as the complex linkages between terrestrial and marine ecosystems. This system also shows how grazers, such as sea urchins, can cause major shifts in ecosystem functioning. This dramatic change also emphasizes the value of monitoring sea urchin abundance, given their capacity to structure benthic communities, as well as seascapes. Moreover, these grazers can also cause mortality in settling meroplanktonic larvae, limiting their capacity to colonize hard surfaces.

Chilean abalone, *Concholepas concholepas* (Figure 18.20) provides a second example. These abalone live in rocky intertidal habitats in Chile, preying on small mussels (*Perumytilus purpuratus*) that compete for space with another mollusk, the limpet, *Fissurella* spp., as well as barnacles and macroalgae. When abundant, abalone predation on mussels limits their abundance, thereby freeing up space for other species. In the absence of abalone predation, these little mussels eventually occupy all available space in the rocky intertidal, where substrate represents a key limiting resource that results in intense spatial competition.

If human intervention selectively removes key species, especially abalone, it alters the balance of the whole ecosystem and intertidal biodiversity collapses. The Chilean abalone model illustrates the delicate balance of interactions among species and how humans, a multi-species predator, can profoundly alter the equilibrium of marine ecosystems. Human predation can alter all equilibria in marine systems (Box 18.1).

Bream fishing in Spain's Medes Islands provide a third example of keystone predators similar to that of sea otter. In Medes Islands, different species of bream (*Diplodus* spp.) prey on sea urchins, which graze on algae such as *Cystoseira* sp., a leafy seaweed characterized by high fractal complexity (i.e., highly branched); its three-dimensionality houses a rich diversity of meiofauna, mollusks, gastropods, and amphipods. Increasing *Cystoseira* sp. abundance leads to increases in associated species and thus to greater biodiversity. Decline in bream resulting from fishing activity results in an explosion of sea urchins. Only small encrusting algal species escape urchin grazing, so decreasing bream numbers turn the rocky environment into a two-dimensional barren with reduced heterogeneity and thus lower biodiversity and primary production (Figure 18.22). Creation of protected areas that ban bream fishing does not produce recovery until bream numbers rebound and individuals become sufficiently large to feed on urchins. Smaller bream feed on urchin juveniles, whereas adult forms continue to graze the algae.

Box 18.1 Impacts of Humans as an Apex Predator on Marine Food Webs

The potential direct effects of overfishing include:

1. Reduction of numbers of the target species, sometimes to extirpation. Interestingly, fishing cessation during the Second World War allowed multiple overfished populations to recover;
2. Impacts on habitats, in that some legal (such as bottom trawls) or illegal (dynamite fishing, use of poisons) practices affect bottom substrata, sometimes with irreversible effects. Reduced structural complexity of habitat reduces available refuges, decreasing recruitment of sessile species, potentially decreasing local productivity, and reducing overall diversity. Fishing activities can also yield (often unpredictable) effects on ecosystem functioning that may affect the overall community. Many fisheries exploit top predators, potentially resulting in altered food webs and trophic cascades triggered by reduction of functionally important taxa, particularly where target prey species modify entire communities. Increased resource exploitation has resulted in decreasing catches since the late 1980s. Fishing changes community structure and may impact (or disturb) timing of reproduction and favor species with shorter reproductive cycles.

As noted earlier, fishing can also impact organism size. "Fishing down the food web" refers to the removal and effective replacement of large, long-lived fishes with other smaller and short-lived species typically of lesser commercial value. This reduced value creates a need to catch more individuals of smaller fish in order to maintain the same profit level (Figure 18.21). See earlier discussion on fishing down food webs.

A similar situation occurred in the Ustica marine protected area in the Mediterranean, where ornate wrasse, *Thalassoma pavo*, compete with rainbow wrasse *Coris julis*, a common species that preys on adult and juvenile sea urchins. *C. julus* predation on sea urchins positively affects the seaweed *Cystoseira* sp., whose increased cover accommodates a whole associated community. Competition with *C. julis* by *T. pavo*, a species favored by climate change, forces *Coris julis* into deeper water (niche displacement) where no sea urchins occur. When this happens, sea urchins increase in number and reduce algal coverage and the biodiversity (and invertebrate biomass) associated with it.

Elimination of large groupers in the system through overfishing clearly illustrates a cascading effect in the system. Groupers and other **piscivores** (fish predators) affect *T. pavo* sp. more than *C. julis*. Based on optimal foraging theory, assuming equal predation cost, groupers prefer to feed on larger *T. pavo* than on smaller (and thus less energy rich) *Coris* sp. This process leads to increased abundance of *Coris* sp. that, by preying on sea urchins, promote *Cystoseira* sp. growth as well as associated zoobenthos. This increase provides more food resources to the multiple fish species preyed upon by groupers, thus increasing their food source (Figure 18.23).

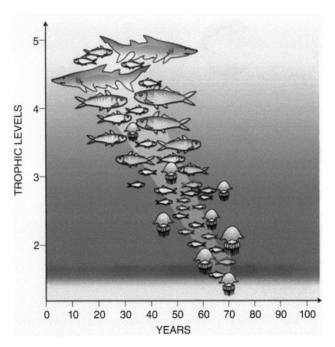

Figure 18.21 Systematic reduction in number of trophic levels.

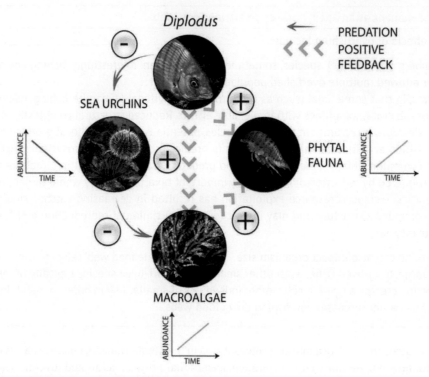

Figure 18.22 Cascading effect of bream predation (*Diplodus* spp.) on sea urchins on macroalgae and their associated fauna. *Source:* Nhobgood / Wikipedia Commons / CC BY-SA 3.0, Hans Hillewaert / Wikimedia Commons / CC BY-SA 4.0, Josh Pederson/NOAA MBNMS/Wikimedia Commons/Public Domain.

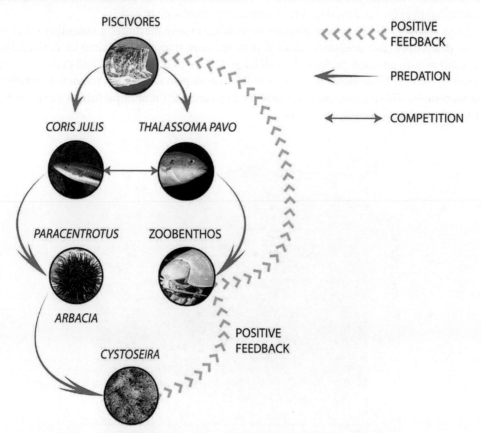

Figure 18.23 Example of a trophic cascade controlled by Mediterranean groupers, which selectively prey on *Thalassoma pavo*, and thus help *Coris julis*, which preys on sea urchins and thereby leads to increased algal coverage and increased abundance and diversity of associated small invertebrates.

Questions

1) Describe models of positive interactions amongst species.
2) Describe different models of species interactions.
3) Summarize the relationships between biodiversity and ecosystem functioning.
4) How do we classify marine symbioses? Provide examples.
5) Describe a detrital trophic network.
6) Describe a trophic network based on dissolved organic matter.
7) Explain the microbial loop and viral shunt.
8) Differentiate between "bottom-up" and "top-down" control.
9) Explain the concept of mixed wasp-waist control.
10) Define and provide examples of keystone species.
11) Describe the interaction between *Mytilus californianus* and *Pisaster ochraceus*.
12) Explain trophic cascades in kelp forests.
13) What is the impact of humans as an apex predator on marine food webs?
14) What does the example of *Concholepas* in the Chilean coasts illustrate?
15) Provide an example of a trophic cascade in the Mediterranean Sea.

Suggested Reading

Bell, G. (2000) The distribution on abundance in neutral communities. *The American Naturalist*, 155, pp. 606–617.

Bell, J. and Williamson, J. (2017) Positive indirect interactions in marine herbivores and algae, in Shields, V.D.C. (ed.), Chapter 6, *Herbivores*. InTech Open, pp. 135–153.

Bulleri, F. (2009) Facilitation research in marine systems: State of the art, emerging patterns and insights for future developments. *Journal of Ecology*, 97, pp. 1121–1130.

Cardinale, B.J., Palmer, M.A., and Collins, S.L. (2002) Species diversity enhances ecosystem functioning through interspecific facilitation. *Nature*, 415, pp. 426–429.

Cavanaugh, C.M. (1994) Microbial symbiosis: Patterns of diversity in marine environment. *American Zoologist*, 34, pp. 79–89.

Duffy, J.E., Cardinale, B.J., France, K.E. et al. (2007) The functional role of biodiversity in ecosystems: Incorporating trophic complexity. *Ecology Letters*, 10, pp. 522–538.

Eriksson, B.K., Ljunggren, L., Sandstrom, A. et al. (2009) Declines in predatory fish promote bloom-forming macroalgae. *Ecological Applications*, 19, pp. 1975–1988.

Ferreira, C.M., Connell, S.D., Goldenberg, S.U., and Nagelkerken, I. (2021) Positive species interactions strengthen in a high-CO_2 ocean. *Proceedings of the Royal Society of London B: Biological Sciences*, 1954, p. 20210475.

Filotas, E., Grant, M., Parrott, L., and Rikvold, P.A. (2010) The effect of positive interactions on community structure in a multi-species metacommunity model along an environmental gradient. *Ecological Modelling*, 221, pp. 885–894.

Goffredi, S.K., Orphan, V.J., Rouse, G.W. et al. (2005) Evolutionary innovation: A bone-eating marine symbiosis. *Environmental Microbiology*, 7, pp. 1369–1378.

Göltenboth, F. and Schoppe, S. (2006) Mangroves, in Goltenboth, F., Timotius, K., Milan, P., and Margraf, J. (eds.), *Ecology of Insular Southeast Asia*. Elsevier, pp. 187–214.

Gross, K. (2008) Positive interactions among competitors can produce species-rich communities. *Ecology Letters*, 11, pp. 929–936.

Lynam, C.P., Llope, M., Möllmann, C. et al. (2017) Interaction between top-down and bottom-up control in marine food webs. *Proceedings of the Royal Society of London B: Biological Sciences*, 114, pp. 1952–1957.

Mann, K.H. and Lazier, J.R.N. (2005) *Dynamics of Marine Ecosystems*. Wiley.

Post, E. (2013) *Ecology of Climate Change. The Importance of Biotic Interactions*. Princeton University Press.

Sarà, M., Bavestrello, G., Cattaneo-Vietti, R., and Cerrano, C. (1998) Endosymbiosis in sponges: Relevance for epigenesis and evolution. *Symbiosis*, 25, pp. 57–70.

Part IIIB

Comparative Marine Ecology: Habitat Types, Their Biodiversity, and Their Functioning

19

Interspecific Interactions II: Negative Interactions

Although some species interact positively, others can result in injury or death of one of the species. Below, we consider predation, competition, parasitism, and disease. In all cases, these relationships involve interactions between organisms in which one species gains an advantage at the expense of another. We explore these interactions, which researchers often test through experimental manipulations and hypothesis testing (Box 19.1), in greater detail.

> **Box 19.1   Hypothesis Testing and Statistical Errors**
>
> In ecology, manipulative experiments enable researchers to move beyond simple descriptions of ecological phenomena to understanding the underlying processes by testing specific hypotheses. For instance, studies investigating the effects of predation and competition require an experimental approach. Scientists develop strong hypotheses logically, on the basis of evidence derived from direct field observations. They subsequently perform hypothesis-driven experiments to explain why communities are structured and function in a certain way. These experiments require appropriate sampling designs to eliminate alternative explanations for the observed results. The effort begins with a description of the process and documentation of patterns and trends. Based on these observations, researchers formulate several alternate theories/hypotheses, including a null hypothesis of no relationship between an experimental variable and a response variable that may or may not change as a result of the manipulated variable.
>
> A good experimental design can accept or refute this null hypothesis and therefore "prove" one specific theory. The experiment includes "controls" in which the researcher can monitor the system under unmanipulated conditions. Importantly, the researcher must replicate both the control treatment and the experimental treatment(s) in order to ensure any difference between the two does not simply reflect random variation rather than a cause and effect relationship. Sufficient replication of control and experimental treatments minimizes the likelihood of making a **type 1 error** (resulting in the researcher concluding that a treatment effect exists where in reality it does not) or a **type 2 error** (ineffective control treatments resulting in the researcher concluding that no differences exist between the experimental treatment and controls when in reality they actually differ). To avoid these errors requires replicating the experimental system and increasing numbers of samples. The more variable the system, the greater the number of replicates required to avoid experimental error.

19.1   Predation

Predation represents one of the most important interspecific relationships in nature, usually referring to an interaction between organisms in which one feeds on the other. Nevertheless, some experts think of predation as an interaction between two species where one "gains" and the other "loses." This definition includes not only predator-prey relationships but also plant-herbivore and host-parasite interactions. All of these interactions play a central forcing role in energy transfers

Marine Biology: Comparative Ecology of Planet Ocean, First Edition. Roberto Danovaro and Paul Snelgrove.
© 2024 John Wiley & Sons Ltd. Published 2024 by John Wiley & Sons Ltd.
Companion Website: www.wiley.com/go/danovaro/marinebiology

through food webs. Predation fundamentally drives population ecology through influences on prey mortality and increases in predator abundance (Box 19.2). It also drives evolutionary changes, in that natural selection favors the most effective predators and the most elusive prey. Many biological characteristics of organisms represent adaptations to reduce the probability that predators will capture and consume them; these characteristics include protective armor, cryptic coloration, and chemical defenses, as well as migration patterns, sociality, and many others.

> **Box 19.2 Predator Exclusion Experiments**
>
> Experimental manipulations that exclude predators allow scientists to understand the role of predation in organizing benthic assemblages. Researchers exclude predators by placing cages made from 5×5 mm^2 plastic mesh, for example, or using physical barriers that exclude vagile organisms (Figure 19.1). Mesh size selection depends on the size of predators to exclude. All cage treatments are replicated and compared with the controls that do not exclude predators. At the end of the experiment, the researcher determines species identity, abundance, and other characteristics (native/cryptogenic changes versus the species of interest).
>
> Most manipulative experiments have focused on intertidal environments that offer relative simplicity of biota and ease of access. In this case researchers set up the following experimental conditions: (a) a control treatment that allows light and grazer access ("natural" system); (b) an experimental cage treatment that excludes predators/grazers but allows light for algal growth; (c) a shaded roof treatment to prevent algal growth, but with no cage to exclude grazers; (d) a full cage treatment that completely excludes both light and grazers. A procedural control comprised of a partial cage assesses effects of the cage itself (e.g., sedimentation) but allow predator access. This combination of manipulations allows researchers to evaluate the relative importance of light and grazing of limpets on algal populations. To evaluate the importance of light, researchers compare dark roof treatments and controls, whereas comparison of cage and control treatments tests the role of grazers. Comparison of all 4 treatments tests the combined effects of light and grazers.

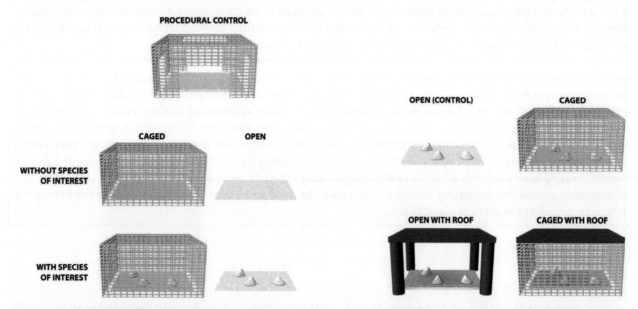

Figure 19.1 Set up of manipulative experiments: Left side: exclusion and partially open cages and experimental plate treatments used to manipulate biotic interactions of predation pressure and facilitation. Procedural control to test for caging effects; full cage with species of interest; full cage without species of interest; open seafloor with presence of species of interest; and open without presence of species of interest. Right side: manipulative field experiment in intertidal environments: open-control (unmanipulated seafloor), caged (cage that excludes predators/grazers). Below: open cage with roof to limit light penetration and full cage with roof that reduces light and also excludes predators and grazers.

Predator-prey interactions profoundly influence the dynamics of ecological systems. Many studies have focused on numbers of prey killed by predators (i.e., the direct effect of predation). But predation also promotes development of anti-predator behaviors that can lead to reduced energy assimilation, changes in patterns of habitat use, and reduced reproductive efficiency associated with increased risk. Many organisms exhibit anti-predator adaptations, including gastropods, most of which have developed increasingly durable "armors" over time. In parallel, their predators, such as crabs, have developed massive claws with increasingly greater crushing power. Even the mere presence of predators can influence prey habitat selection or prey activity, irrespective of whether predation actually occurs. Predation impacts prey both directly and through risk effects. Risk effects may be particularly pronounced in large prey with long life cycles, such as sea turtles, which generally experience low predation rates but deadly consequences result from some individual encounters.

The previous chapter provided multiple examples of hard substrate predators including keystone species, so to switch things up a bit we focus more on sediments to discuss predation. In soft sediment ecosystems, several experiments that excluded epibenthic predators showed increased abundances of infauna compared with non-manipulated systems. Major epibenthic predators on sediments and hard substrates include gastropods, crabs and other decapod crustaceans, fishes, and birds.

Researchers have used different strategies to exclude various predators, but the challenge has been to avoid changing the system in unintended ways, such as increased organic flux with mesh cages because of dampened bottom flow that results in particle accumulation. Researchers work around this problem by using partial cages that do not exclude predators but test for experimental artefacts. The use of different mesh sizes showed that small predators (crabs, shrimps, and gobies) affect infaunal abundances more than larger predators (for example, flatfish of the order Pleuronectiformes). The use of metal rods to exclude horseshoe crab (*Limulus carcharias*) from an experimental intertidal habitat separated the effects of crabs from those of horseshoe crabs. Using total exclusion cages and others with a movable wall (attached floats "opened" the door to fishes during high tide), researchers separated effects of fishes (excluded only by complete cages) from those of birds (which both types of cages exclude). These experiments suggest a less important role for fishes, but bird effects varied among habitats.

Fewer studies have focused on infaunal predators, mainly nemertean worms and polychaetes, than on epibenthic predators. However, scattered studies suggest that infaunal predators can strongly impact sedimentary community structure. One well-known example involves the polychaete *Glycera carcharias* preying on the polychaete *Alitta virens (*formerly *Nereis virens)*. Exclusion experiments (Box 19.2) that excluded *G. carcharias* resulted in increased survival of *A. virens*, which subsequently reduced abundances of the amphipod, *Corophium volutator*. The presence of *A. virens* in a community in which *C. volutator* is absent results in increased infaunal abundance rather than the expected decrease. Presumably *A. virens* removes an intermediate predator such as *Nephthys incisa*, which preys on infauna.

Some predators remove only part of their prey: for example, sole primarily target the palps of spionid polychaetes and bivalve siphons. Although many infaunal organisms can regenerate body parts, we do not fully understand the implications of this predation mode for prey. For example, removal of a bivalve's siphon forces the organism to move toward the sediment surface to maintain contact with bottom water. Closer proximity to the sediment surface increases vulnerability of the bivalve to other predators. Multiple levels of predation occur within infaunal communities, such as those just described above.

Predator size may also have important consequences for prey abundance. Using fences with different size openings in North Carolina salt marshes and estuaries, scientists observed greater infaunal abundance in the presence of large mummichog, *Fundulus heteroclitus*; these large fishes prey on grass shrimp, *Palaemonetes pugio*, an intermediate predator, which in turn preys on macroinfauna. Thus, by reducing shrimp abundance, mummichog support high infaunal density.

Some structures and environmental characteristics protect infauna from predators. Sediment type influences the ability of crabs to burrow and prey on bivalves. Seagrasses deter digging predators. Exclusion of predators in seagrass beds usually results in modest increases in infaunal abundance, demonstrating their effectiveness as a predator refuge, whereas faunal abundances inside cages in adjacent non-vegetated areas were similar to those in seagrass beds. Dense polychaete tubes on the sediment surface also provide a refuge from digging predators.

Numerous studies in recent decades have documented the role of cnidarians as predators; these organisms feed on other cnidarians, fishes, and arthropods, including macro- and microzooplankton. However, we know little about their importance as prey, largely because predators digest them quickly and quantification would require near instantaneous analysis of predator stomach contents. Scyphozoans (pelagic cnidarians), comprise the most important group of gelatinous predators; they prey mostly on other scyphomedusae, hydromedusae, and ctenophores. Benthic anthozoans, such as anemones, can prey on gelatinous zooplankton that swim near the bottom. Larvae of anemones, such as *Edwardsiella lineata* and *Peachia quinquecapitata*, ingest ctenophores and jellies feeding on the food ingested by the host or on its internal tissues. Most hydromedusae and scyphomedusae feeding on other cnidarians eat other prey as well, such as copepods. This feeding mode means that jellies may prey on members of the same guild, consuming species that could potentially compete with them for food.

Among the mollusks, gastropods exert the strongest predatory pressure on cnidarians. The nudibranch *Coryphella archaria*, largely regulates polyp numbers, and the release of *Aurelia* ephyrae. Among arthropod predators, malacostracan amphipods prey most strongly on pelagic cnidarians. Other predators include gammarids, euphausids, mysids, and decapods. Copepods, ostracods, barnacles, and pycnogonids also eat gelatinous zooplankton. Even fishes (mostly generalists) feed on gelatinous zooplankton. Specialists that feed exclusively on jellies belong to the suborder Stromateoidea, especially the fish families Centrolophidae and Stromateidae, as well as the commercial species *Peprilus triacanthus* (American butterfish).

Marine fishes experience high risk of predation. Among invertebrates, squids represent the major invertebrate predator of fishes, and fishes, in turn, prey upon squids. Eggs and larvae of fishes make up a major dietary component of many marine invertebrates, including many zooplankton: jellies, copepods, chaetognaths, amphipods, and euphausiids. Among vertebrates (other than humans), other fishes comprise the main predators on fishes, followed by marine mammals, especially whales and seals, and seabirds. Fishes, in particular, mainly feed on smaller fishes.

Even if a larva escapes from a predator, stress can reduce its chances of survival. Furthermore, predators can consume 40–60% of the total eggs produced by fishes. Cannibalism can represent a significant portion of egg mortality: in anchovies, for example, adults eat 17% of eggs d^{-1}, resulting in a 32% mortality rate from cannibalism alone. Often, cannibalism is **density dependent**, in that the number of eggs present in the stomach of an adult increases exponentially with the concentration of eggs present in the plankton; increased filtration rates likely occur when shoals of adult anchovies encounter high concentrations of eggs.

Jellies dominate the stomach contents of some species of sea turtles. Green sea turtles, *Chelonia mydas*, feed mainly on seagrasses and algae, but also on ctenophores, jellies, salps, and sponges. Loggerhead sea turtles, *Caretta caretta*, feed on the hydrozoan *Velella velella*, the siphonophore *Physalia* sp., scyphomedusae, and ctenophores.

Megafauna also face predation risk. Predators target all sea turtle adults, although rarely. In addition to vulnerable juveniles, females laying their eggs on beaches may also encounter predators. Even crocodiles feed on turtles, though at extremely low rates. In the ocean, sharks prey on adult turtles, especially bull sharks, *Carcharhinus leucas*, sand tiger sharks, *Carcharodon carcharias* and tiger sharks, *Galeocerdo cuvier*. Killer whales (*Orcinus orca*) also prey on turtles. Low predation rates, however, result in few observations of predator-prey interactions with sea turtles.

Sharks and killer whales, the main predators of cetaceans and pinnipeds, occur widely through the ocean. Predation levels on cetaceans vary depending on species, population, and region. Understandably, small cetaceans such as dolphins and porpoises are more vulnerable than larger species. Small cetaceans often respond to killer whale attacks by forming groups that move rapidly to surface waters.

Predator-prey relationships often develop into an evolutionary arms race as prey evolve to avoid capture and consumption while predators evolve to capture and kill prey more efficiently. The intensity of this selective force depends on the intensity of interactions between predators and prey. As predators, marine mammals feed mainly on fishes and zooplankton, which in turn feed on other fish species, zooplankton and phytoplankton. Marine mammals have evolved in order to capture their prey, using their senses (sight, hearing), morphology (dentition), and physiology (capacity to hold their breath for a long time). The evolution of behavioral strategies further aids in prey capture, including cooperation, bubble production (which forms an underwater "curtain"); humpback whales (*Megaptera novaeangliae*) use this mechanism to capture herring. Marine mammals have also evolved specific capture behaviors, including diel vertical migrations between the surface and depths of hundreds of meters, or capturing migrating prey. This strategy affects life history strategies not only for marine mammals, but also for their prey. For example, humpback whales feed for about six months at locations with abundant plankton, and then effectively fast for the rest of the year in warmer waters with scant plankton.

The different anti-predatory strategies seen across different organisms represent two main coexisting and coordinated models, one physiological and ethological, the other genetic. Both can occur during different stages of development from larval and juvenile to adult stages. The primary anti-predation mechanisms include:

1) Protection – exoskeletons, shells, hairs, stinging spines or quills, toxin secreting organs,
2) Posing (aggressive or threatening behavior).

3) Disguise – adapting colors and morphologies that perfectly match the background or specific environmental features that confuse or distract predators.
4) Escape or hide – many marine invertebrate species and some fishes bury themselves in sand or mud.

Fishes and other cold-blooded prey species have evolved a variety of strategies to increase their chances of survival. **Countershading**, with dark dorsal coloring and light ventral coloring helps fish blend into the dark background of the seafloor or deep water when viewed from above, and against sunlight when viewed from below. Many species of fishes, invertebrates, and zooplankton escape from predators by living in deep, dark waters during the day and migrating towards the surface at night. Marine mammal prey such as salmon and herring form large groups during spawning in order to overwhelm predators and thus reduce predation losses for populations. Schooling causes confusion by creating stimuli as the shoal moves to escape predators, thus making it difficult for marine mammals, for example, to select and pursue a single individual. Many species use rapid dispersion to escape predation.

As prey themselves, marine mammals must escape aquatic and terrestrial predators. Bears, killer whales (*Orcinus orca*), and sharks prey on some species of pinnipeds, for example. Thus, some pinniped species reduce predation risk by moving out of the water. Similarly, pinnipeds such as Steller sea lions, *Eumetopias jubatus*, and Northern fur seals, *Callorhinus ursinus*, reduce risk from terrestrial predators by raising juveniles between rocky intertidal environments or on islands devoid of predators. Pack ice offers a partial but imperfect refuge from orcas (Figure 19.2). Ringed seals, *Pusa hispida*, give birth inside caves formed between ice and snow to protect pups from polar bear (*Ursus maritimus*) predation.

When killer whales attack other species of whales, the threatened group responds by moving closer together and moving their tails in order to generate significant foam, thus confusing predators. Gray whales exemplify this behavior, which respond to attacks by grouping together and, in encounters within a few miles of the coast, swim shoreward, sometimes reaching where waves break, or where kelp beds occur. The mothers defend their young with rapid flicks of their tail towards the killer whales, or quickly surfacing to create considerable splashes. Researchers recently divided cetacean antipredatory strategies into two distinct categories: fight or escape. Active defense characterizes the first strategy, including techniques for individual defense as well as for coordinated defenses within groups of whales. Many fight strategies also include retreating toward refuges such as swimming quickly from killer whales.

The term "cryptic" refers to species that hide from their predators, often utilizing special attributes that make them difficult to detect in the environment (Figure 19.3a-l). These attributes may relate to color or morphology known as homochromia and homomorphia, respectively. In **homochromia**, the species matches its substrate colors by minimizing any silhouette effect. For example, many pelagic fish possess characteristic homochromic coloration, with blue dorsal coloration and silver ventral coloration. As a result, they appear similar in color to seawater (blue) when viewed from above, but silvery like the water surface when viewed from below. In **homomorphia**, the individual takes on a shape similar to the substrate or by means of "disruptive" colors that make its contour less visible. Some animals can even blend in by perfectly matching the color of the environment in which they live.

The difficulty in directly observing predator-prey interactions involving cetaceans contributes to our poor understanding of these relationships, especially compared to terrestrial predator–prey interactions. Nonetheless, cetaceans may experience significant predatory pressure and therefore develop various anti-predatory strategies. For example, some researchers believe that predation primarily drives group behavior in many toothed whales.

Figure 19.2 Seals often spend most of their time on land to escape predation by orcas or sharks. *Source:* Robert Pitman / Wikimedia Commons / Public domain.

Figure 19.3 Examples of cryptic species in marine environments: (a–c) gastropods (family Ovulidae, Indonesia) on tropical corals; (d) pygmy seahorse (*Hippocampus bargibanti*) on tropical gorgonian (Indonesia); (e–f) leaf fish within its habitat and in isolation (Lembeh Strait, Indonesia); (g) race (trigon) Red Sea; (h) the leafy seadragon (*Phycodurus eques*) in an aquarium; (i) seahorses between artificial seagrass blades in Los Angeles Aquarium; l) chitons on rock.

19.2 Methods to Escape Predation

The evolutionary history of predator-prey relationships has led to a great variety of both morphological and chromatic adaptations that ultimately increase a species' chances of survival. Mimicry uses two very different approaches:

1) **cryptic mimicry,** where some animals attempt to avoid being seen by adapting coloration (homochromia) and/or forms (homomorphism) that resemble the substrate on which they live. Others instead use colors to confuse predator or prey perception of their shape (disruptive coloration). On sandy bottoms, animals such as turbot and sole, that cannot completely bury themselves, can match the color of their bottom habitat perfectly. In contrast, animals such as scorpion fish and crustaceans living on rocky bottoms use color and appendages of various shapes that mimic the substratum (e.g., kelp) on which they live. Fish living on coral reefs often use disruptive coloration. In the Mediterranean, fishes such as gurnard (*Chelidonichthys lastoviza*), comber (*Serranus scriba*) and many wrasses use color in this way. The white nudibranch, *Peltodoris atromaculata*, looks conspicuous on the purple and orange sponges on which they feed, but the presence of large spots confuses perception of shape.

2) **faneric mimicry** uses color to provide information to the observer. Some animals cannot produce poisonous secretions or tastes or smells to deter predators. They instead advertise predators with a showy approach, **aposematic** warning coloration that (typically falsely) warn predators that already experienced negative encounters with organisms similar in appearance. Nudibranchs, for example, feeding on sponges and cnidarians, accumulate toxins and stinging cells in their tissues that make them unpalatable. Their showy colors remind would-be predators of this characteristic. Other species of nudibranch that lack these deterrents mimic warning coloration to discourage predators but avoid the energetic cost of producing deterrents. **Batesian mimicry** refers to instances where an organism imitates a defense mechanism present in another species but lacks the specific mechanism. **Mullerian mimicry** describes situations where both the original model and the mimic possess similar deterrents, so predators learn more quickly to avoid both species.

Generally, predators intercept their prey as they attempt to escape, and thus try to anticipate their direction and type of movement. Therefore, any attribute that confuses the predator on possible prey direction of escape could make the difference between life and death. To this end, some species feature "false eyes" or spots at the posterior of their body to create confusion. Fishes such as bream and white bream and the crustacean *Squilla mantis* illustrate this adaptation. "False eyes" force the predator to guess which end is the head, and thus the likely direction of attempted escape.

19.3 Competition

Competition refers to the exploitation of a common resource, where those individuals that can obtain and use the resource win. Competition dominates in marine habitats with some limiting resource such as food or space. In these limiting situations, as populations increase in abundance, competition for the limiting resource(s) increases. In contrast to predation, mutualism, and similar ecological interactions, competition in nature may be difficult to observe. Competition occurs in two forms, either through direct exploitative competition or through interference competition. **Exploitative competition**, where organisms reduce or consume the resource, occurs more frequently in animals than interference competition. **Interference competition**, occurs where the dominant individuals or species deny their competitors access to the resource. This latter type of interaction often implies some form of aggression or hindrance, including fights or similar injurious behavior. Territoriality in many fishes and invertebrates illustrate such behavior.

These two types of competition can occur simultaneously. The limpet *Scutellastra cochlear* (formerly genus *Patella*), for example, actively prevents other animals from invading its foraging territory. It preferentially feeds on the seaweed *Ralfsia* sp. within its foraging territory, allowing only encrusting algae to grow. Competition can touch upon multiple aspects of each species' life. Apparently distinct resources can, in fact, relate closely to each other For example, competition for food links closely to competition for space in species living in coastal rocky habitat.

In the coastal waters of the Pacific, the bryozoans *Onychocella alula* and *Akatopora tincta* live near one another. Feeding rates of the bryozoans do not vary within different parts of a colony but when the two species grow together on the same substrate, *A. tincta* feeding rates on the portion of the colony in contact with *O. alula* decrease noticeably. Growth rate depends on feeding rate, so individuals of *A. tincta* on the edge of the colony grow more slowly, and the *O. alula* colony eventually overgrows the *A. tincta* colony. This happens even though *A. tincta* in isolation grows faster than *O. alula*. The considerably longer tentacles in *O. alula* relative to *A. tincta* may deviate currents carrying food away from *A. tincta*, markedly decreasing food supply for this species. In this case, space and food do not represent independent resources.

Intraspecific competition between fish can occur in density-dependent and territoriality forms. Density-dependent growth is well documented in fish and presumably reflects competition for food. As an example, intraspecific competition in striped seaperch, *Embiotoca lateralis*, a micro-carnivorous teleost, results in reduced growth, fecundity, and feeding. This species lives along the coast, with some specialist individuals feeding on caprellid amphipods and other specialists feeding on gammarid amphipods. Specialists on caprellids grow larger and achieve higher fecundity than gammarid specialists, with generalists in between. Aggressive behavior of larger fish against smaller fish apparently maintains this pattern, in that larger fish occupy the foraging microhabitats occupied by caprellids.

Territoriality clearly represents a form of intraspecific competition in defending resources. Typically, individual territories remain relatively permanent, encompassing a specific, though usually limited, space and involving defense of shelter, food, and often nests (and eggs), especially from conspecifics. Individuals of mussel blenny, *Hypsoblennius jenkinsi*, occupy and defend abandoned crevices and burrows to feed on planktonic and benthic invertebrates. Following translocation of 42 individuals into a location already inhabited by 17 fish, the population returned to its original size 50 d later, with 27% loss of individuals in the population in the first 3 d, and 50% in the first 18 d. These declines point to the role of intraspecific competition in regulating territoriality. Fish constantly defend their territory from conspecifics that may represent a threat. Laboratory observations indicate that territorial individuals of mussel blennies aggressively dominate non-territorial congeneric individuals (*Hypsoblennius gentilis*), thus limiting their use of the habitat when the two species co-occur.

Intraspecific competition occurs in *Tridacna crocea*, the crocea clam that lives in coral reefs. Larvae aggregate when they settle (which seems counter intuitive given competition for space), orienting as far from the adjacent individual as possible. Adults reduce larval settlement by spreading their mantle over the substrate. Thus, intense intraspecific competition may lead to >40% of mortality. *Tridacna crocea* lives ~80 years, but intense competition contributes to high mortality during the first three years of life.

Competition for space commonly occurs on tropical coral reefs. A neighboring *Galaxea* sp. colony can attack a *Montipora* sp. colony, and the nematocysts cause tissue mortality (Figure 19.4).

Two species of California rockfish, *Sebastes carnatus* and *Sebastes chrysomelas*, dominate different locations: the first located in coastal areas and the second in deeper waters. The transition depth between the two species occurs at ~10–15 m. This transition depth correlates inversely with a steep drop off in depth that likely affects abundances of benthic invertebrates on which the two species of rockfish prey. These species exhibit remarkable overlap in food needs, so only space separates these congeners. Manipulative experiments removed all *S. chrysomelas* individuals from one site and all individuals of *S. carnatus* from another site while monitoring a third site as an unmanipulated control. Over the subsequent three years, researchers continued to remove individuals from the sites. The deeper species responded to this manipulation by moving into shallow waters where they had removed *S. chrysomelas*, and the opposite pattern emerged where researchers removed *S. carnatus*. Neither species changed in distribution at the control site. Therefore, competition between these

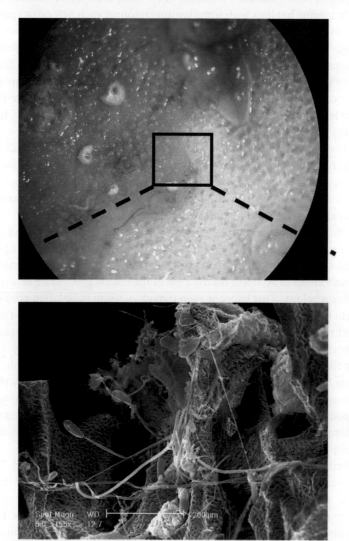

Figure 19.4 Competition for space in corals can result in whitening and loss of tissue. Here, severe tissue loss in a *Montipora* sp. colony (upper half), resulted from attack by a neighboring *Galaxea* sp. colony. SEM observations of the boundaries between healthy tissue and recently exposed skeleton revealed high abundances of *Galaxea*-like nematocysts (lower half).

Figure 19.5 *Coris julis* and *Thalassoma pavo*, two species that now compete for resources as a result of warming of Mediterranean surface waters. *Sources:* Parent Géry / Wikimedia Commons / CC BY-SA 3.0 and Matthieu Sontag / Wikimedia Commons / CC BY-SA 3.0.

species partly drives bathymetric segregation. Some studies suggest that this segregation begins with differences in larval settlement depth followed by reinforcement by territoriality. How do these species coexist without one eliminating the other? Evidence suggests that *S. chrysomelas*, the dominant species, occupies the shallowest, and food-rich depths. *S. carnatus* then finds refuge from competition in deeper, food-poorer environments.

A similar mechanism occurs in the Ligurian Sea, where damselfish (*Coris julis*, Figure 19.5a) and damsel Pavonina (*Thalassoma pavo*; Figure 19.5b) co-occur. *C. julis* lives in shallow coastal waters whereas *T. pavo* prefers warmer water. This difference segregated the two species until just a few years ago, where *C. julis* occurred in the northern Mediterranean and *T. pavo* in the south. However, the damsel Pavonina recently shifted to higher latitudes in response to increased surface temperatures. Its larger size and greater ability to compete for resources (i.e., food) compared to damselfish is progressively pushing damselfish into deeper water (>20 m).

The three-dimensional structure of soft-sediment habitat creates habitats both horizontally and vertically that reduce potential competitors. However, direct interactions commonly occur among conspecifics or closely related species. For example, the spionid *Pseudopolydora paucibranchiata*, competes aggressively with associated infauna, reducing foraging opportunities for these species. Direct interactions between the polychaete, *Hediste diversicolor*, and the amphipod, *Corophium volutator*, lead to declines in amphipod populations. Protobranch deposit feeder bivalves vary their position in sediment depending on densities of other species. The three-dimensional structure of the habitat minimizes direct competitive interactions, leading to expectations of indirect competition mechanisms.

Several experiments with bivalves that showed competition reduces growth and survival rates suggest an 80% reduction in growth of the clam, *Nuttallia nuttallii*, when confined with two deep species, whereas surface-living bivalves had no effect. These results suggest competition for space. Placing *Nuttallia nuttallii* with dead surrogates of the deeper living bivalves (shells placed in the sediment at depths normally occupied by living individuals) also reduced growth rate in *Nuttallia nuttallii*.

Environment can strongly influence intensity of competition. Density-dependent growth occurs in the bivalve, *Limecola balthica* (formerly *Macoma balthica*), maintained in muddy sediment, in contrast to density-independent growth when maintained in sand. Facultative feeding behavior in this bivalve may explain this difference; in muddy substrates, deposit feeders dominate and negatively affect *Limecola balthica* as they capture resources, whereas suspension feeders in sandy sediments capture suspended food particles transported by bottom currents. Migration and mortality rates of deposit feeders provide a metric of their responses to changes in density. Movement and feeding behavior in the gastropod, *Ecrobia ventrosa*, decreased with increased density.

19.4 Parasitism

The term **parasitism** refers to a close relationship between two organisms, in which one member (the parasite) depends on and benefits from the other (the host), typically to the host's detriment. The parasite often receives a trophic benefit, noting that metabolic dependence differentiates the nature of associations, and parasitism involves 100% dependence upon the host. Parasites include many bacteria, viruses, and fungi. A parasite, depending on circumstances, can be commensal, mutualistic, or a predator. Any marine organism could potentially host parasites, but choice of host depends on location, season, and specific parasite species; not all organisms are suitable hosts for all parasites. Parasites represent an integral part of all marine ecosystems, and a major component of global biodiversity. Host-parasite checklists suggest at least 3–4 metazoan parasites, on average, for each species of marine fish within a specific environment.

Parasites infect hosts at all trophic levels and play a primary role in regulation of marine biodiversity. Parasitism has proven to be a highly successful lifestyle, and all organisms are susceptible to parasitic infection during at least one phase of their life. Indeed, some scientists believe that numbers of parasite species at least equals that of non-parasitic species. Therefore, marine biologists tend to underestimate the great success of parasitism as a life history strategy, partly because of the considerable challenge in identifying many parasites and parasitic relationships.

Parasite life cycles can vary greatly, sometimes with remarkable complexity and crossing different environments and hosts. Parasites with a **direct life cycle** use a single host. For example, many adults of Monogenea (flatworms) parasitize fishes, but their free-living larvae are non-parasitic. Parasites with an **indirect life cycle** use multiple hosts. The adults of Digenea (a subclass of Trematoda), for example, infect multiple vertebrate species and their larvae infect at least one, and often several, hosts. Similarly, parasite hosts fall into different categories. **Definitive** or **final hosts** host the sexually mature stages of a parasite, whereas **intermediate hosts** host developing and immature stages and **paratenic hosts** support larval forms that cannot complete their development within the host. For parasites, larval dispersal is particularly critical because hosts can be difficult to find in an ocean. To spread, parasites could "follow" their dispersing hosts (for example following migrating fish and mammals), or spread through passive mechanisms dependent on ocean currents.

Mechanisms of infection also vary. For example, other parasites inject flagellates (such as *Trypanosoma* spp.) into the blood of fishes. Many Trematoda infect hosts that ingest them as encysted parasites, or live within intermediate host tissues, whereas other parasites actively penetrate the skin (seabirds, for example). Many parasites change their host's behavior, increasing the chance of infecting the next host during its life cycle. Many parasites have asexual and **parthenogenetic** reproduction mechanisms, where unfertilized eggs develop into a new individual. **Hermaphroditism** also occurs, which means that a single individual possesses both male and female reproductive organs. All species, whether parasitic or not, require specific habitats. In parasites, specific hosts represent a required habitat. No one parasite can universally infect all hosts, although parasite specificity can vary. Some trematodes, for example, infect a wide range of marine fish species, whereas some monogeneans exclusively infect a single host species and a small segment of the gills. Among >900 samples of monogeneans collected from 2,014 samples from 17 fish species, all but 2 of 18 monogenean species were highly host specific.

In terms of phylogeny, even predominantly free-living groups include lineages that gave rise to obligate parasitic forms. An evolutionary line, once parasitic, never evolves free-living forms. Thus, in evolutionary terms, parasitism seems irreversible. Most lineages that successfully adopted parasitism are species-rich or, at least, they interact with different hosts. Parasitism drives genetic diversity in hosts, which relates to the persistence of sexual reproduction – which promotes genetic diversity. The **Red Queen hypothesis** proposes that species (e.g., hosts) must constantly adapt, evolve, and proliferate in order to survive against ever-evolving opposing species (e.g., parasites). The same concept applies to predators and their prey.

Many researchers assume that, over evolutionary time, parasites have lost much of the complex behavior of free-living organisms through their dependence on the host for food and protection. Thus, many parasites do not need a complex nervous system or sophisticated organs for feeding. For example, the barnacle, *Rizocephalo sacculina*, which parasitizes marine crabs (Figure 19.6a) possesses all of the characteristics typical of a barnacle during its free living larval stage, but the adult

Figure 19.6 Examples of marine parasites: (a) barnacle (*Rizocephalo sacculina*); (b) copepod parasites showing host degeneration; (c) *Anisakis simplex*, a nematode parasite of anchovies. *Sources:* Hans Hillewaert / Wikimedia Commons / CC BY-SA 4.0 and Anilocra / Wikimedia Commons / Public Domain.

resembles a bag attached to the ventral surface of the crab, with extensive cytoplasmic extensions into various host tissues, thus bearing little resemblance to any typical crustacean. The ~250 species of parasites barnacles almost all belong to the Superorder Rhizocephala.

Although most parasites are much smaller in size than their hosts, some large parasites exist; the trematodes (family Didymozoidae), common parasites of many marine fishes, can reach considerable sizes. Some species of parasites can infect several hosts. For example, the platyhelminth, *Neobenedenia girellae*, reportedly infects greater yellowtail, *Seriola dumerili*, amberjack, *Seriola quinqueradiata* and plaice, *Paralichthys olivaceus*. The parasitic copepod *Caligus orientalis* occurs in 22 species that span 8 orders and 13 families of marine fishes in temperate Japanese waters.

The main groups of marine parasites include chromistans, protozoans, fungi, arthropods (especially copepods and barnacles), platyhelminthes and nematodes, as well as many less relevant groups in terms of number and distribution of species (Table 19.1). The chromistan, *Ulkenia amoeboidea*, parasitizes octopus, *Eledone cirrhosa*. The algae *Schizochytrium* spp. grows on nudibranchs, *Tritonia tetraquetra*, and on octopus, *Illex illecebrosus*, whereas the fungus, *Cytospora rhizophoae*, is an important parasite of red mangrove, *Rhizophora mangle*. Within the alga *Asterionella formosa*, decreased light reduces zoospore production by the fungal parasite *Rhizophydium planktonicum*. Many dinoflagellates and ciliates parasitize animals intracellularly, absorbing nutrients through osmotrophy. The ciliate, *Vampyrophrya pelagic*, parasitizes adults of 25 species of pelagic copepods in the Sea of Japan. Some parasites, such as amoebae of the genus *Amoebophrya*, limit blooms of toxic dinoflagellates, *Karlodinium veneficum*.

Copepods represent the most numerous parasitic groups; of ~11,500 known species, half live in symbiotic associations and most are probably parasitic. A Mediterranean study reported higher species richness of these crustaceans on average, compared to nektonic fish. The copepod, *Lernaeocera branchialis*, develops from a free-living naupliar larva with typical larval crustacean characteristics, but the parasitic adult stage inserts its mouthparts into a blood vessel of its host fish and completely transforms its body into a bag-like structure through a process known as **sacculinization** (Figure 19.6). Often, this degeneration is not obvious, and the complexity of parasitic species may exceed that of their free-living counterparts.

Table 19.1 Examples of marine parasites.

Protozoa – Fungi – Chromista	Platyhelminthes + other worms	Crustaceans	Other groups with parasitic forms
Sarcomastigophora	Turbellaria	Copepoda	Porifera
Labyrinthomorpha	Monogenea	Isopoda	Cnidaria
			Ctenophora
Haplosporidium	Digenea	Branchiura	Orthonectida
Apicomplexa (Phylum Myzozoa)	Amphilinidea	Tantulocarida	Dicyemida
Microsporidia	Gyrocotylidea	Ascothoracida	Myzostomida
Mikrocytos mackini	Eucestoda	Cirripedia	Polychaeta
Ciliophora	Nematoda	Amphipoda	Hirudinea
	Acanthocephala		Cycliophora
			Nemertea
			Rotifera and *Seison* spp.
			Nematomorpha
			Acarina
			Pycnogonida
			Insecta
			Tardigrada
			Pentastomida
			Mollusca
			Echiura
			Echinodermata
			Fish

The parasitic Platyhelminthes are particularly species rich, encompassing >5,000 species of Cestoda, 3,000–4,000 species of Monogenea and >25,000 species of Trematoda (which primarily infect fishes and invertebrates).

Nematodes also include many species of parasites. For the most part, marine nematode parasites cycle through four larval stages and four molts. After the fourth molt, adults emerge with developed gonads. Adult nematodes can continue to grow in size without further molting. *Anisakis simplex* (Figure 19.6), a parasitic nematode with a complex life cycle, commonly occurs in small pelagic fishes such as anchovy, and can infect humans who ingest raw fish. The nematode parasite, *Philometra lateolabracis*, occurs in the ovary of groupers, *Epinephelus marginatus*, and the yellowtail, *Seriola dumerili* where it sucks blood, causes atrophy in developing eggs, fibrosis in ovarian tissue, and increased granulocytes and hemorrhaging. Parasitic roundworms of the genus *Trichinella* span a circumpolar distribution within Arctic marine mammals, with reports of infections in muscle tissue of polar bears, *Ursus maritimus*, and walrus, *Odobenus rosmarus*, and less commonly in seals and whales. One species that infects the sunfish, *Mola mola* (which can weigh a metric ton) reaches lengths of 12 m, but with a very small diameter. Free-living species of nematode are often significantly smaller (from one to a few millimeters) than related parasitic species (sometimes growing to meters in length). Parasites persist by producing many offspring with reproductive capacity often exceeding that of free-living species.

Biotic variables such as host size and abiotic variables (temperature in particular), can influence host infection rates. Host size strongly influences parasitism processes. Therefore, different parasites can infect a host species at each of the host's different developmental stages. For example, different species of parasites may parasitize larvae compared to those that infect adults. Multi-cellular organisms tend to parasitize small metazoans. The largest living marine animals, the cetaceans, host the longest nematode parasite. Environment also plays a role. Moderately high temperatures seem to favor settlement and development of the parasite *Anguillicoloides crassus* in European eel *Anguilla anguilla*.

19.5 Diseases of Marine Organisms

The topic of diseases in marine organisms alone could fill books, and here we focus on examples of diseases that can alter the structure and functioning of marine ecosystems by directly causing sometimes massive mortality of marine vertebrates (mammals, turtles, and fishes), invertebrates (corals, crustaceans, and echinoderms), algae, and plants (seagrasses). Recent studies show that diseases in marine organisms increase over time. In recent decades, new diseases have emerged that affect marine mammals, often exhibiting complex pathogenesis and sometimes involving immunological dysfunction. "New" or re-emerging pathogens that infect marine mammals include various papilloma viruses, varicella viruses (a group of DNA viruses that infect through vesicular lesions) in dolphins, and other viral infections, various neoplasmic diseases, algal toxic blooms, cold stress syndromes in dugongs, and cardiomyopathy in sperm whales.

Cancer rates (especially malignant cancers) appear to be steadily increasing among marine mammals, but several other diseases also occur, including emerging infectious agents. For example, a recently described urogenital cancer associated with a herpesvirus or with the continuous exposure to chemical contaminants such as PCBs (polychlorinated biphenyls) and DDT (dichlorodiphenyltrichloroethane), has affected ~17% of California sea lions (*Zalophus californianus*). Scientists have discovered lingual and genital papilloma, and squamous cell carcinoma in the dolphin, *Tursiops truncatus*. An epidemic of lobomicosi caused by a type of yeast has recently been reported in dolphins on the Atlantic coast; humans are the only other known host. Despite a well-developed immune system in Florida manatees, *Trichechus manatus latirostris*, scientists recently described the first viral disease (a cutaneous papillomatosis); this disease adds to the challenges for one of the most endangered marine mammal species on Earth. Toxoplasmosis now causes major mortality rates in sea otters (*Enhydra lutris nereis*). Because they prey on certain types of food that humans also consume, sea otters represent real "sentinels" for pathogenic protozoa.

Since 1987, *Morbillivirus* has caused at least eight different epidemics that caused mortality in global populations of pinnipeds and cetaceans. Researchers characterized the viruses responsible for these outbreaks, caused by the genus *Morbillivirus* (family Paramyxoviridae), using antigenic and genomic biomolecular tools. Through this approach they distinguished among strains of distemper virus (*Canine Distemper Virus*, CDV), capable of infecting pinnipeds, as well as of three new morbilliviruses, which can infect pinnipeds and cetaceans. In 1988, ~189,000 common seals (*Phoca vitulina*) and several hundred gray seals (*Halichoerus grypus*) died in a serious episode along the northern Europe coast. The causative agent proved to be a new *Morbillivirus* sp., which they called *Phocine* (*Phocid*) *Distemper Virus* (PDV). They later attributed another major mortality event to a similar viral agent. Once again, this event affected North Sea seal populations during 2002, causing an estimated 21,000 deaths over a period of just six months. Thousands of seals (*Phoca sibirica*) in Lake Baikal, in Siberia, exhibited similar conditions to those reported in European seals.

We still cannot identify the origin of the morbilliviruses described in several species of aquatic mammals, despite the use of antigenic and genomic tools that have enabled researchers to classify them as entirely new viral agents. The effects of certain contaminants, such as organochlorines (PCBs, dioxins, 4-4′DDE, etc.) in particular and heavy metals (mercury, Hg, lead, Pb, cadmium, Cd, etc.) remain largely unknown. Nonetheless, these contaminants might play an important role in triggering the pathogenic action of morbillivirus in marine mammals.

19.5.1 Coral Diseases

Studies on coral diseases began relatively recently, with the first descriptive reports on coral tissue degradation in hard corals dating back to the 1970s. The first reported disease in corals in 1973 was "black band" disease (black banding), in which dark bands appear among apparently healthy coral tissue and recently exposed coral skeleton. In 1977, "white band" and "plague" diseases were reported in massive corals. Different types of bacteria appear to cause all three of these diseases, destroying coral tissue at a rate of several millimeters per day. Numbers and incidence of diseases and other syndromes in corals have increased in recent years, and studies have evolved from simple descriptions of their effects to microbiological studies focused on causative agents. Currently, researchers have documented 18 coral diseases but identified the causative agents for only a few. Known pathogenic bacteria appear to cause "white band II", white pox, aspergillosis (Figure 19.6a) and "white plague II" diseases. For the corals *Oculina patagonica* and *Pocillopora damicornis* (Figure 19.7), researchers have identified *Vibrio shiloi* and *V. coralliilyticus*, respectively, as the disease agents.

Single agents cause some diseases, whereas a consortium of microbes cause others. For example, "black band" disease in Caribbean and in the Indo-Pacific region, seems to contain at 50+ different types of bacteria within a lesion. Historically, coral disease research focused mainly on bacterial and fungal infections, but more recent studies have identified other sources, such as trematodes on *Acropora* sp. (Figure 19.7) and ciliates on *Porites* sp. The chromistan, *Halofolliculina corallasia*, causes skeleton-eroding band (SEB), a progressive disease spreading throughout the Indo-Pacific. Although ciliates cause brown band disease (*Brown Band*, BRB) by invading coral tissues, this disease arises after feeding by predators such as seastars, *Acanthastaer planci*, leaves open scars (Figure 19.8).

Figure 19.7 Examples of diseases in hard corals and gorgonians: (a) aspergillosis on a gorgonian; (b) white band in *Pocillopora damicornis*; (c) white band in *Acropora* sp.; d) white band in *Stylophora* sp.

Figure 19.8 Effects of predation on stony corals by the seastar, *Achanthaster* sp. The signs on the coral closely resembles – and can be confused with – those caused by an infectious "white syndrome."

The major known coral diseases include the following:

Black band disease (BBD), one of the best characterized diseases, has contributed significantly to global decline of coral reefs. It causes a characteristic ring shaped black band (0.1–7 cm) ring shaped, which kills healthy coral tissues at rates up to 1 cm d^{-1}. This disease preferentially affects massive corals such as *Orbicella annularis, Montastraea cavernosa*, and *Pseudodiploria strigosa*, but has been reported on 21 Caribbean species and on >40 species of coral on the Great Barrier Reef. The consortium of microorganisms responsible for this disease consists of filamentous cyanobacteria, sulfide-oxidizing bacteria (*Beggiatoa* spp.), bacteria sulfate reducers (*Desulfovibrio* spp.) and other bacteria producing marine toxins.

Yellow band disease (YBD) mainly affects large **hermatypic** (reef building) corals and expands at a rate of 0.5–1.0 cm mo^{-1} as a pale yellow color developing on coral tissue. No pathogenic species have yet been identified, but a recent study suggests a possible role for some *Vibrio* spp.

White band disease (WBD) causes degradation of coral tissues. Researchers recently divided the disease into type I and type II, based on the mode of tissue and zooxanthellae loss. *Acropora cervicornis* and *A. palmata* lose tissue slowly (a few millimeters per day), generally starting from the colony base, though sometimes it starts from the middle of a branch. In the Caribbean, this disease has destroyed up to 95% of *Acropora* spp. populations in the past few decades, possibly by a *Vibrio* pathogen similar to *V. harveyi*.

White plague (WP) affects more than 40 species of scleractinians and represents one of the most destructive diseases in the Caribbean, where its impact extends over 400 km of coral reef. The disease causes a strong and rapid progression of lesions, with necrotic tissue progressing at a rate of a few millimeters to 2 cm d^{-1}, typically leading to colony death. The bacterium *Aurantimonas coralicida*, the only member of the family of α-proteobacteria, causes type II infections.

White pox (WPO) affects only the coral *Acropora palmata*, though it has literally decimated populations with losses of up to 88%. It differs from "white band" in the presence of irregular shaped lesions that form white patches where tissue has disappeared. Lesions can range from a few square centimeters to areas > 80 cm^2, and they can appear simultaneously on different parts of the coral colony. *Serratia marcescens*, a Gammaproteobacteria, is the causative agent.

Growth anomaly (GA) indicates a wide range of injuries and coral skeletal anomalies, which arise from both tumors and altered growth (hyperplasia or accelerated growth) in response to an external stimulus (for example, the presence of parasites, fungi, or algae). The paucity of available scientific information limits any detailed discussion on this subject. Although contact between sick and healthy organisms does not transmit coral tumors, these growths can reduce overall reproductive potential of the colony and increase coral susceptibility to infectious diseases. Furthermore, *Vibrio* sp. may contribute to coral tumor development, but the relationship between cause and effect remains hazy. In contrast, direct contact from diseased colonies to healthy *P. lutea* and *P. lobata*, suggests an infectious agent contributes to GA development in scleractinian corals.

Skeleton erosion band (SEB), caused by the ciliate, *Halofolliculina corallasia*, results in significant damage to calcareous coral skeletons. Ciliate density reaches a maximum of 417 indiv. mm^{-2} forming a black-gray band on tissue. SEB affects 24 species of corals, including *Acropora* spp. and *Stylophora* spp.

White syndrome. For a long time, the term "white syndrome" was considered synonymous with "white plague" (WPD). However, in recent years several studies on wild corals demonstrated some key differences. Infected corals were found to lack bacteria, but molecular evidence points to **apoptosis** (programmed cell death) and a possible link to a phenomenon known as Rapid Tissue Necrosis (RTN), characterized by an extremely rapid loss of living tissue potentially leading to rapid death of the entire affected colony (*Acropora* sp., *Pocillopora* sp., *Seriatopora* sp., *Stylophora* sp.; Figure 19.9). Researchers identified a strain in *Pocillopora damicornis* close to *Vibrio harveyi* (99% 16S rDNA similarity).

Black disease causes the appearance of dark areas and/or lesions on the surface of the coral. It includes a family of syndromes caused by unknown agents that include: (1) dark spots syndrome, (2) brown band syndrome, (3) *Atramentous* sp. necrosis, (4) thin dark line, and (5) black fungus syndrome. This syndrome causes the production of a characteristic brown mucus.

Pink-line syndrome (PLS), caused by several species of fungi and a cyanobacterium, *Phormidium valderianum*, infects the coral, *Porites lutea*, in the Arabian Sea.

Ulcer white spots disease (PUWSD) produces round spots (3–5 mm diameter) that may regress or progress, sometimes resulting in colony death. This disease spreads from diseased to healthy colonies, but the responsible agents remain unknown.

Figure 19.9 Different kind of disease and injuries of tropical stony corals. Images show possible cases of: (a) skeleton erosion band, (b) possible pink syndrome in the coral *Fungia* sp.; (c) and (d) white syndrome.

Finally, the term **"bleaching"** refers to loss of color in symbiont dinoflagellates, *Symbiodinium* sp., and the host corals. The term "bleaching" describes the loss of typical coral color, which bleaches to become whitish. This process may result from: (1) the loss of zooxanthellae, or (2) the loss of photosynthetic pigments.

"Bleaching" refers to a condition and not a "real" disease. Although we do not fully understand the mechanisms of coral bleaching, several factors can trigger this condition, including: (1) temperature stress, (2) UV radiation, (3) salinity fluctuations, (4) bacterial infections, (5) cyanide, copper, and pesticides. Recently, researchers added sunscreens to the list of pollutants, because they can induce bleaching in *Acropora* sp.

Questions

1) What are manipulative experiments and why are they needed?
2) What are the impacts of predation, competition, and disturbance on populations of soft bottoms?
3) Describe the main strategies in marine organisms to avoid predation.
4) Provide some practical examples of positive and negative interactions between organisms.
5) Provide some examples of interspecific competition.
6) Describe the Red Queen hypothesis.
7) Is parasitism important in the ecology of marine species? Why?
8) What main diseases affect tropical stony corals?
9) What are the possible causes of bleaching?

Suggested Reading

Bavestrello, G., Arillo, A., Calcinai, B. et al. (2000) Parasitic diatoms inside Antarctic sponges. *Biological Bulletin*, 198, pp. 29–33.

Côté, I., Darling, E., and Brown, C. (2016) Interactions among ecosystem stressors and their importance in conservation. *Proceedings of the Royal Society of London B: Biological Sciences*, 283, p. 20152592.

Hewitt, J.E., Ellis, J.I., and Thrush, S.F. (2016) Multiple stressors, nonlinear effects and the implications of climate change impacts on marine coastal ecosystems. *Global Change Biology*, 22, pp. 2665–2675.

Lafferty, K.D., Allesina, S., Arim, M. et al. (2008) Parasites in food webs: The ultimate missing links. *Ecology Letters*, 11, pp. 533–546.

Leroux, S.J. and Loreau, M. (2009) Disentangling multiple predator effects in biodiversity and ecosystem functioning research. *Journal of Animal Ecology*, 78, pp. 695–698.

Mumby, P.J., Harborne, A.R., and Brumbaugh, D.R. (2011) Grouper as a natural biocontrol of invasive lionfish. *PLOS One*, 6, p. e21510.

Quijon, P.A. and Snelgrove, P.V.R. (2005) Predation regulation of sedimentary faunal structure: Potential effects of a fishery-induced switch in predators in a Newfoundland sub-arctic fjord. *Oecologia*, 144, pp. 125–136.

Russo, T., Pulcini, D., O'Leary, Á. et al. (2008) Relationship between body shape and trophic niche segregation in two closely related sympatric fishes. *Journal of Fish Biology*, 73, pp. 809–828.

Vinebrooke, R., Cottingham, K., Scheffer, J. et al. (2004) Impacts of multiple stressors on biodiversity and ecosystem functioning: The role of species co-tolerance. *Oikos*, 104, pp. 451–457.

20

Intertidal Ecosystems and Lagoons

The term **intertidal** (or **mesolittoral** or mediolittoral zone) refers to the stretch of coast at the land-sea interface between highest and lowest tide levels, and includes coastal lagoons, rocky intertidal environments, and sedimentary intertidal environments (mudflats and sandflats). Several other specialized intertidal environments (mangroves, salt marshes, and coral reefs) will form the focus for other chapters. We begin by discussing some of the general characteristics of these systems and follow with a more detailed treatment.

Lagoons occupy ~12% of the world seacoast and exhibit strong ecological heterogeneity, strong environmental gradients, high primary production (up to 3,500 g C m^{-3} y^{-1}), and high biomass. Lagoon ecosystems exhibit low biodiversity with strong dominance by plants, including multiple species of seagrasses such as *Zostera noltii*, *Cymodocea nodosa*, and *Zostera marina*. Assemblages of marine invertebrates, with relatively low diversity, have developed the ability to withstand strong environmental fluctuations typically associated with lagoons. Noting limited average depths of lagoons (often <1 m), these species also experience considerable temporal variability.

Like lagoons, rocky intertidal environments share low biodiversity as a result of: (a) strong mechanical stress, (b) strong temperature variations, (c) periodic drying, (d) decreased oxygen concentration, and (e) reduced feeding opportunities. Primary production varies among habitats but can be quite low. In many cases, the ability of organisms to withstand prolonged periods of exposure to air defines their upper distribution limit, whereas interactions with other organisms define lower limits. In these environments, keystone species such as sea stars strongly affect ecosystem structure and functioning. Organisms within intertidal zones often form horizontal bands divided into zones, each with its own characteristic species composition.

Intertidal sedimentary environments support very different biota compared to their rocky counterparts but the same environmental characteristics, albeit somewhat dampened by limited protection that sediments provide, results in generally low diversity and productivity.

20.1 Rocky Intertidal Habitats

Tides immerse most of the intertidal zone every day; high tide places almost all of this environment underwater, whereas low tide exposes most of the intertidal zone to air. Tides can move rapidly in the intertidal zone, carrying significant amounts of suspended material with it. Widths of this zone can vary greatly depending upon bottom topography and basin shape. For example, typical Mediterranean tidal ranges of ~30 cm (though up to 1 m in the Gulf of Trieste), pale compared to 11 m tidal ranges in the Bay of Fundy and Ungava Bay in Canada. These periodic emersions and submersions can expose substantial habitat (Figure 20.1), which researchers have divided into depth-related zones. Three zones (upper, mid-, and lower intertidal) characterize many marine regions with extensive tidal ranges, in contrast to two zone (upper and lower intertidal) in regions with minimum tidal ranges, such as the Mediterranean.

In the Mediterranean, wave action largely defines the zones, particularly the upper zone, where splashing waves wet biota, in contrast to tides that fully immerse lower intertidal organisms. In this system a fully terrestrial environment transitions to a completely marine environment without an obvious ecotone such as seen with lagoons (where communities intertwine). Because of ease of access, many key concepts in marine ecology (supply-side ecology, keystone species, and

Marine Biology: Comparative Ecology of Planet Ocean, First Edition. Roberto Danovaro and Paul Snelgrove.
© 2024 John Wiley & Sons Ltd. Published 2024 by John Wiley & Sons Ltd.
Companion Website: www.wiley.com/go/danovaro/marinebiology

Figure 20.1 Intertidal habitats vary greatly in tidal range, such as the 7-m tidal range in a Chilean fjord.

niche) arose from observations and experiments in this environment. Indeed, intertidal habitats provide a unique opportunity to study community structure in relatively simple and accessible marine environments.

Intertidal zonation (Figure 20.2) often appears quite obvious to the naked eye, driven by a range of environmental and biological conditions that influence interactions among organisms. These environments create significant environmental challenges for biota, including strong mechanical stress, significant temperature fluctuations, periodic drying out, decreased oxygen concentrations, hyper- and hypo-saline conditions, and reduced feeding opportunities. In fact, two dominant gradients in physical conditions characterize intertidal environments, the first driven by wave motion (typically higher in exposed headlands and decreasing in protected bays), and the second driven by tidal excursions. These two environmental gradients combine to produce a complex system of environmental conditions that characterize and differentiate intertidal habitats at very small spatial scales. For example, wave disturbance can limit mobile species and/or, feeding opportunities compared to more sheltered environments. Reduced wave motion can also lead to increased thermal stress with associated desiccation, although wave spray can help to wet intertidal habitats. Although erosion affects high-energy environments in particular, greater sedimentation rate and movement often occur in sheltered, low-energy environments, especially in waters with high suspended load of inorganic particles.

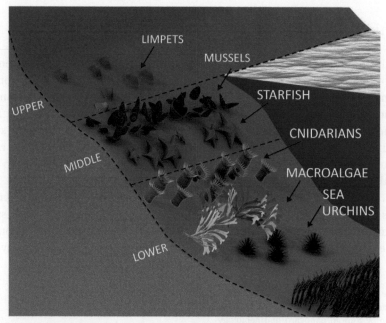

Figure 20.2 Rocky intertidal zonation, highlighting upper, mid- and low intertidal zones.

Salinity may vary greatly as a result of freshwater contributions from rain and land runoff. At the upper range of intertidal habitats, marine organisms may experience long periods of emersion, and evaporation of tide pools can result in hypersaline conditions. Air temperature generally varies much more than seawater temperature spatially but also temporally; air temperatures can vary by 10–20 °C over 24 h with associated changes in humidity. The combined effects of heat and low relative humidity create the primary source of stress, both in temperate and in tropical environments. However, in boreal and polar regions cold temperatures and ice create the biggest problems. Although seawater temperature never drops below –2 °C before it freezes, air temperatures may drop to –40 °C. The time available for respiration and feeding typically decrease in the upper intertidal. Filtering organisms, such as barnacles and many mollusks, can feed only when fully immersed, limiting food assimilation. Whatever the predominant factor (desiccation, low temperature, lack of nutrients or food), intertidal environments create vertical gradients in stress for many marine organisms. In the pelagic realm, light represents the only environmental variable that always decreases with depth, whereas in intertidal environments, wave drag on organisms can detach them from the substrate and wave energy can abrade organisms with suspended sediment, logs, or ice. Therefore, organisms must adapt to avoid detachment from intertidal substrate, and to withstand abrasion. In fact, species living in exposed environments differ greatly in morphology from those living in protected locations.

20.1.1 Survival Strategies for Rocky Intertidal Environments

To overcome the challenges of life in intertidal environments, species have evolved multiple strategies:

1) Body size and shape: Rounding of organisms reduces their surface area to volume ratio. Some organisms such as limpets and chitons use their expanded foot to remain attached, despite wave motion.
2) Shell color: Color varies with substratum in order to provide camouflage (e.g., chitons) and escape predators. Furthermore, light exterior coloration reflects sunlight more than dark colors, and thus help cool the animal.
3) Ability to change their metabolic rate as the tide ebbs. For example, as the tide recedes the polychaete *Arenicola* sp. moves to depths where sand or mud retain sufficient moisture and the blue mussel, *Mytilus* spp., retains water inside its valves to enable respiration even when the shell closes to reduce desiccation.

Some organisms that occupy intertidal environments, such as lichens, seagrasses, and arthropods, are essentially terrestrial biota that tolerate exposure to marine environments and largely occur in the uppermost intertidal. These species do not respond to a gradient of exposure, but rather to a gradient of increasing immersion and the associated flood of saltwater that produces a stress gradient opposite of that described above for truly marine organisms (Figure 20.3).

Habitat structure and complexity, biological productivity, species diversity, and intensity of biological interactions all increase with depth. The presence of dense seagrass mats, beds of bivalves and tunicates, or reefs comprised of polychaete tubes alter coastal environments. In many cases these features result in positive change, preventing drying or providing shelter from wave action. Sunlight blockage by seagrass and macroalgae can reduce light available to conspecifics or recently settled smaller individuals. Vegetation traps large amounts of sediment that can either suffocate other plants and animals or, alternatively, provide habitat for benthic species.

Behavioral strategies of intertidal invertebrates further enable them to survive in hostile environments, primarily through isophasic strategies and isospatial strategies. In **isophasic strategies,** mobility allows species adapted to either aquatic or land habitats to enter the intertidal zone during favorable periods. In this "hit and run" strategy, organisms only temporarily occupy intertidal habitat. Therefore, isophasic adaptation involves periodic migration between habitats. In contrast, with an **isospatial strategy,** animals do not use mobility to avoid unfavorable conditions, and instead remain in the intertidal irrespective of tidal phase and resist rather than avoid unfavorable conditions. In both cases, these species require a biological clock, an internal compass that regulate rhythms of activity, and specific orientation mechanisms.

20.1.2 Rocky Intertidal Zonation

Ecologists also group intertidal communities into supralittoral and eulittoral zones. The spray line from waves defines the upper limit of the **supralittoral zone**, so that exposed locations with stronger waves support wider supralittoral zones than sheltered locations. The high tide mark defines the lower limit of this zone. Lichens, cyanobacteria, and green microalgae dominate primary producers, with scant macroalgae. Gastropods typically dominate the herbivores (Figure 20.4),

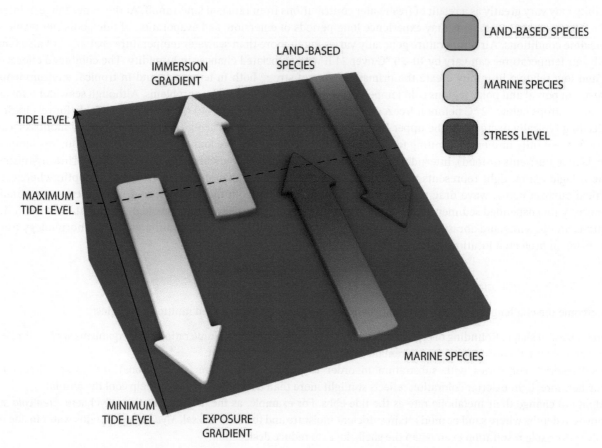

Figure 20.3 Primary factors, gradients, and processes that influence the presence and distribution of intertidal species.

particularly members of the family Littorinidae: the most common genera in the northern hemisphere are *Littorina* and *Nerita*, whereas the most common southern hemisphere genera include *Nodilittorina* and *Melanerita* (Figure 20.5).

The **eulittoral zone** extends from the lower limit of spring tides to the upper limit of complete immersion. The upper limit of the eulittoral zone of Arctic and tropical environments supports an impoverished algal flora consisting of ephemeral green algae and coralline algae. At temperate latitudes fucoid algae such as *Fucus* spp. and *Polvetia* spp. form a canopy that shades small red and coralline algae. Barnacles numerically dominate the animal fauna in the upper eulittoral zone globally. The lower eulittoral supports a much richer flora. At Arctic and temperate latitudes, fucoid brown algae often dominate, whereas kelps such as *Hedophyllum* spp. and *Polstelsia* spp. characterize the eulittoral North Pacific. Along tropical coasts, warm temperatures and dry climate limit algal abundance and diversity, and red calcareous algae dominate. In North Africa an algal turf sometimes dominates the tropical supralittoral, consisting of red and brown filamentous algae only a few millimeters in length. Aerial exposure defines dominant populations within different zones (Table 20.1 and Figure 20.6).

High tide covers the upper intertidal for daily periods of 2–3 h, once or twice daily depending on whether the location experiences diurnal (daily) or semidiurnal (twice daily) tides, in contrast to similar periods of exposure for the sublittoral fringe. Immersion and wave action apparently largely determine establishment and growth of intertidal organisms. Under these conditions, animals with protective shells (bivalves, limpets, gastropods, and barnacles) dominate (Figure 20.4).

Intertidal zonation depends on several complex and interlinked factors: (1) Larval settlement; (2) adult movements; (3) physiological tolerances; and (4) biological interactions (competition and predation). Higher biodiversity characterizes lower horizons because marine species comprise intertidal populations, and marine conditions characterize these zones. However, research also implies variation across depth in intraspecific (density dependent) and interspecific competition, as well as differences in predator abundance. Organisms that thrive in deeper zones cannot survive in shallower depths, whereas those that live at shallower depths could survive in deeper layers, but cannot compete with species adapted to

Figure 20.4 Examples of intertidal organisms including (a) mussels (*Mytilus californianus*) and (b) crustaceans (*Lepas* sp.) along the Eastern Pacific (California) coast. At the center, crustaceans (c) *Lepas* sp., and (d) hermit crabs living in the rocky intertidal near Indonesia; below, (e) encrusting algae and (f) *Littorina* spp., which tend to aggregate in the upper intertidal in rock pools or between rocks.

lower intertidal zones in terms of metabolic performance, growth, and ability to exploit resources. Space limitations result in strong competition on rocky hard intertidal bottoms.

20.1.3 Rocky Intertidal Primary Producers

Whereas epiphytic microalgae and macroalgae provide most of the primary production in intertidal environments on a larger scale, macroalgae, primarily from the orders *Fucales* (typical of the intertidal) and *Laminariales* (typical of subtidal environments where they create kelp forests that also penetrate into the intertidal), comprise the main primary producers. However, a significant proportion of animals in these environments derive most of their energy and nutrients from phytoplankton they obtain by filter feeding during high tide. Community composition varies greatly; in some locations a macroalgal canopy covers much of the intertidal zone, whereas sessile animals such as barnacles and bivalves dominate elsewhere (Figure 20.7). In these systems, organisms typically form horizontal bands that contribute to zonation. Estimates

Table 20.1 Dominant species within different intertidal zones, depending on tidal exposure.

Zone	Exposed	Less exposed
Upper intertidal Zone	Periwinkles, Barnacles	Periwinkles, barnacles (less abundant)
Mid-intertidal zone	Mussels	Mussels often replaced by macrophytes
Lower intertidal zone (Infralittoral fringe)	Invertebrates, Algae	Invertebrates, algae (kelp diminish, "turf" algae increase)

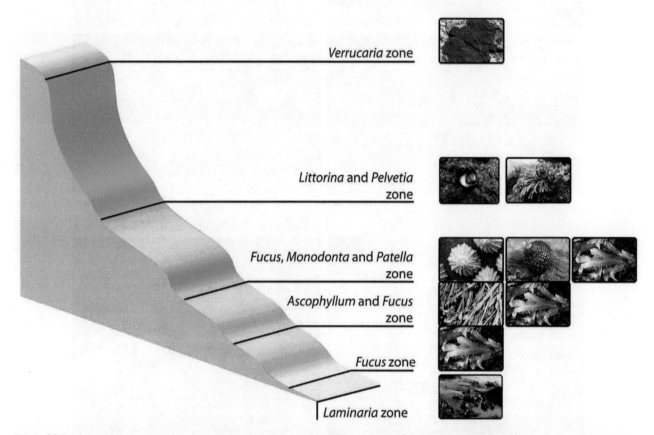

Figure 20.5 Intertidal zonation showing species that characterize different zones. *Sources:* Jerzy Opioła / Wikimedia Commons / CC BY-SA 4.0, Guttorm Flatabø / Wikimedia Commons / CC BY-SA 3.0, Stemonitis / Wikimedia Commons / CC BY-SA 2.5, Tango22 / Wikimedia Commons / CC BY-SA 3.0, ふうけ / Wikimedia Commons / Public domain, Stemonitis / Wikimedia Commons / CC BY-SA 2.5, Dozens / Wikimedia Commons / CC BY 2.5, Stemonitis / Wikimedia Commons / CC BY-SA 2.5, Stemonitis / Wikimedia Commons / CC BY-SA 2.5, Jerry Kirkhart / CC BY 2.0.

of daily primary production range from 8–17 g C m^{-2}, noting strong seasonality in some locations, but regions may differ greatly, depending on coverage of different primary producers. Plants generally experience short periods of aerial exposure, during which limited growth and photosynthesis occur. Near negligible rates of primary production characterize many intertidal locations.

20.1.4 Rocky Intertidal Consumers

Because desiccation poses such a major risk for biota in intertidal environments subject to tidal rhythms, the dominance of shelled mollusks is hardly surprising. As described above, limpets, *Patella* spp., and gastropods, *Littorina* spp. dominate the upper intertidal zone, whereas mussels, *Mytilus* spp., often dominate the intermediate intertidal zone. Barnacles, and the genus *Balanus* in particular, are sessile crustaceans that use their calcareous plates to protect themselves from desiccation and maintain high abundances in the intermediate intertidal zone. These animals frequently occur in different parts

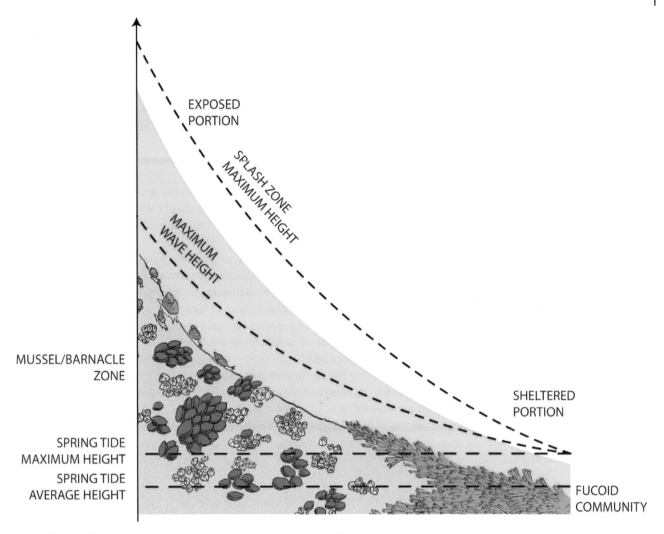

Figure 20.6 Effect of exposure on zonation patterns in rocky intertidal environments.

Figure 20.7 Direct observation of intertidal populations (left), and use of quadrats to estimate macroalgal cover in lower intertidal environments.

of the world. Three groups of barnacles, *Chthamalus* spp., *Octomeris* spp. and *Tetraclita* spp., occur in high abundances in the lower intertidal zone, thus explaining the alternate name for this band, the "barnacle zone." Other species include two kinds of limpets, *Patella* spp. and *Cellana* spp., mussels *Perna* spp. and bivalves *Saccostrea* spp. Mussels, oysters, and barnacles feed on suspended particles by filtering water, and their main interaction with plants and macroalgae involves competition for space, even though limpets and gastropods actively feed on algae. Small filter feeders such as bryozoans and hydroids live attached to the algae. Amphipods live within *Fucus* spp. and *Ascophyllum* spp. beds, where algal branches shelter them from wave action and desiccation.

20.1.5 Competition in Rocky Intertidal Environments

Among marine environments, some of the highest intra- and interspecific competition rates occur in rocky intertidal environments. Competition intensity and its impacts depend on: (1) larval distribution in the water column and their ability to settle on the substrate; (2) presence and abundance of conspecifics; (3) presence of chemical stimuli; (4) interspecific interactions with other species.

The classic manipulative caging experiment on competition in rocky intertidal communities by ecologist Joseph Connell examined the barnacles *Semibalanus balanoides* and *Chthamalus stellatus*, by selectively removing the predatory gastropod *Nucella lapillus*, which normally mediates direct competition between the two barnacle species. These experiments focused on understanding factors that influence species distributions and thus determine intertidal zonation. By selectively removing species from the system and observing effects on remaining species, he could observe structural and functional changes associated with each species. He discovered that larvae of *S. balanoides* could settle throughout the intertidal, whereas *C. stellatus* settled only in the upper intertidal (Figure 20.8).

Subsequently, post-settlement selection factors, including light, temperature, and desiccation altered species distributions (Figure 20.8). These factors had no obvious effect on *C. stellatus* but reduced the upper limit of *Semibalanus balanoides* to the medium high tide line. Finally, where the two species came into direct contact, the larger size of *S. balanoides* helped to drive *C. stellatus* to competitive exclusion (Figure 20.8). Thus, intertidal zonation reflects the combined effects of physical and biological factors and different species of barnacles arrive in the same rocky intertidal habitat, but environmental selection and competitive interactions alter their distributions.

20.1.6 Predation in Intertidal Habitats and the Intermediate Disturbance Hypothesis

Predation may play an important role in limiting competitive interactions between species and in limiting settlement and use of spatial resources. The major predators in intertidal communities include few species (Figure 20.9) such as sea stars, *Pisaster* spp., which feed on bivalves by forcing their valves apart, and gastropods such as *Nucella* spp. that pierce the shells of barnacles or bivalves with their radula and suck out soft tissue. In contrast to these relatively slow-moving predators, some predatory decapod crustaceans such as brachyuran crabs move quickly (e.g., *Pachygrapsus marmoratus* commonly occur in the Mediterranean, Figure 20.9c). In locations with less risk of desiccation, such as in algal habitat, soft-bodied predators such as nudibranchs and anemones exert strong predation pressure. When the tide ebbs, seabirds, such as gulls, ducks, and oyster-catchers prey on intertidal mollusks. However, during flood tides, many species of small fishes swim into the intertidal zone and prey on invertebrates.

As noted in Chapter 18, Robert Paine removed *Pisaster ochraceus* from the intertidal zone and also prevented them from recolonizing, resulting in a remarkable change in the community as it shifted from one of ~30 species to one largely dominated by bivalves, *Mytilus californianus*. These results suggest sea stars somehow help to maintain biodiversity in the community. He hypothesized that bivalves were adept competitors for space that can eliminate all other species if left unchecked. In this context, the keystone predator *P. ochraceus* reduces mussel coverage, thereby providing space that other species may colonize and use.

The intermediate disturbance hypothesis suggests that intermediate levels of disturbance favor high diversity by limiting the dominance of a superior competitor or predator but not with such intensity that the disturbance eliminates most species. Different authors have presented this hypothesis in multiple forms, and it even aligns with the keystone predator concept. Specifically, predation (considered a form of disturbance because it leads to the death of individuals of a dominant species), if moderate in intensity, leads to greater numbers of species by limiting dominant species that might otherwise exclude all other species through competition. Excessive predation removes large numbers of individuals and potentially eliminates biodiversity but excessively weak (or no) predation leaves the dominant species unchecked and able to exclude all the other species through competition (Figures 20.10, 20.11).

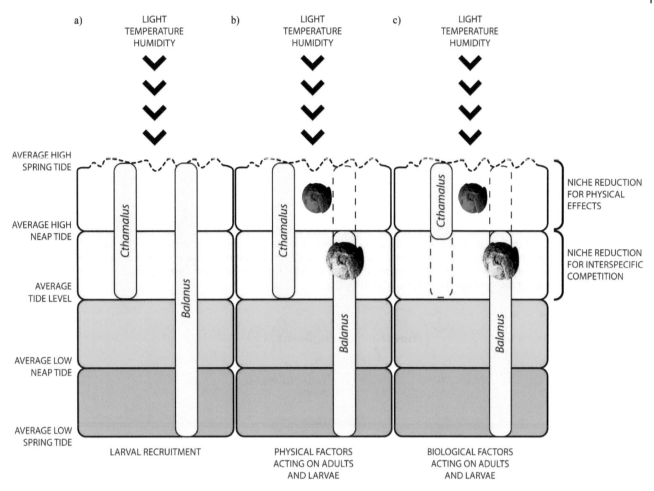

Figure 20.8 Effects of abiotic variables and competition on distributions of *Chthamalus stellatus* and *Semibalanus balanoides* distributions in the rocky intertidal.

Figure 20.9 Some key intertidal predators: (a) seastar, *Pisaster* sp.; (b) gastropod, *Nucella* sp.; (c) crab, *Pachygrapsus marmoratus*. *Sources:* Franco Folini / Flickr / CC BY-SA 2.0, Minette Layne / Wikimedia Commons / CC BY-SA 2.0, George Chernilevsky / Wikimedia Commons / Public domain.

Experiments examining predation interactions between *Pisaster* sp., *Nucella* sp., and mussels (Figure 20.12) verify the long-term effects (over time scales of 10 years) of *Pisaster* on mussels. The presence of *Pisaster* sp., an intertidal predator, almost totally eliminates mussels, resulting in a strong increase in algae and invertebrates, but in its absence, mussels increase in abundance and biodiversity decreases. A model of cascading interaction between predator, prey, and other species illustrates the positive effect of predation on ecosystem structure (Figure 20.13).

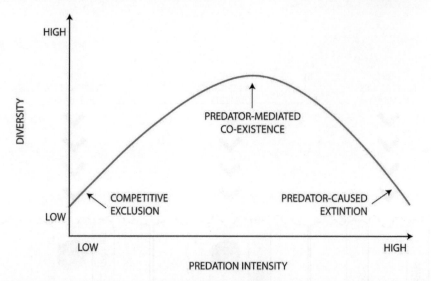

Figure 20.10 The predation hypothesis and its effect on biodiversity. Predation produces a unimodal trend where weak predation and minimal biological disturbance result in low biodiversity. With intermediate disturbance (i.e., with the "right" number of predators), the predators effectively remove dominant species and allow weaker competitors to persist, albeit at modest abundances. This relationship creates a new equilibrium with a positive overall effect on system biodiversity. With intense predation, many species decline and may disappear locally, leading to decreased biodiversity.

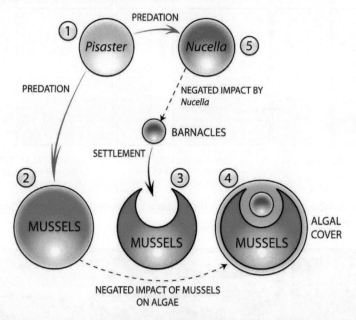

Figure 20.11 In rocky intertidal communities, competition for space often results in a hierarchy; the sea star *Pisaster* sp. (1) preys on mussels (2), freeing up space (3) in which young barnacles can settle and grow (4). The gastropod *Nucella* sp. (5) does not prey upon the barnacles, because *Pisaster* sp. prey on *Nucella* sp. Algal cover increases because *Pisaster* sp. prey on mussels. The band of mussels covers almost the entire substrate, leaving little space for other species of invertebrates and algae to colonize.

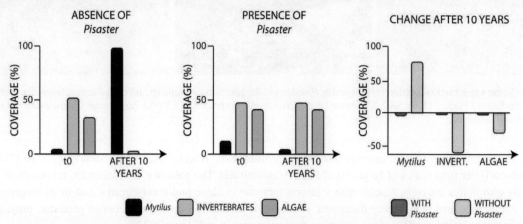

Figure 20.12 Long-term effect of presence/absence of intertidal predator *Pisaster* sp. By removing this species, after 10 y, *Mytilus* spp. increase to near total dominance, eliminating invertebrates and algae. By reducing *Mytilus* spp., *Pisaster* sp., leaves room for algae and invertebrates to sharply increase, thus increasing overall biodiversity. Removal of *Pisaster* sp. destabilizes the system and leads to reduced biodiversity.

Figure 20.13 Diagram illustrating the effect of a predator (*Pisaster* sp.) on rocky intertidal communities also colonized by algae, mussels, and other sessile benthic invertebrates. The symbols "+" and "-" indicate positive and negative effects respectively.

We close out this discussion on intertidal predation by noting that recent sea star epidemics on the United States coasts clearly show that the keystone specie concept lacks predictive power. Mussel dominance did not follow mass mortality of sea stars, either because other predators filled the gap, mussel recruitment failed, or for other reasons. These findings illustrate that ecological generalizations often require caveats and contingencies that new generations of researchers must address!

20.1.7 Keystone Species

As discussed earlier, keystone species strongly influence community structure and overall ecosystem functioning disproportionately relative to their abundance (indeed they are typically numerically rare; Figure 20.14). These species exhibit high "functionality" in that their removal from a community alters the diversity (and often functioning) of the community. *Pisaster* sp. and many other infralittoral species apparently play a key role in defining community structure, thus illustrating the keystone species concept.

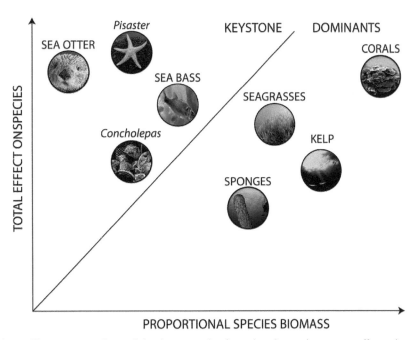

Figure 20.14 Comparison of keystone species and dominant species. Low abundances but strong effects characterize keystone species (left), whereas dominant species (abundance and biomass, right) often play important roles in ecosystem functioning. *Sources:* Citron / Wikimedia Commons / CC BY-SA 3.0, Toby Hudson / Wikimedia Commons / CC BY-SA 3.0, Frédéric Ducarme / Wikimedia Commons / CC BY-SA 4.0, National Oceanic and Atmospheric Administration (NOAA), Nhobgood / Wikimedia Commons / CC BY-SA 3.0.

20.1.8 Rocky Intertidal Trophic Food Webs

The intertidal zone relies on several primary trophic sources: (1) algal and bacterial microfilm; (2) photophilic algae; (3) planktonic (and benthic) prey and 4) suspended matter. Exposed locations (Figure 20.15a) export significant macroalgal production in the form of detritus, whereas herbivores such as limpets and gastropods generally consume microalgal production locally. Phytoplankton communities support abundant populations of suspension feeders. Crabs and gastropods prey on both herbivores and suspension feeders, which in turn become prey for birds, the top predators in this system. This trophic structure forms a classic energy pyramid, with drastic decreases in production with each trophic level.

The trophic network of sheltered locations (Figure 20.15b) differs in that little export of algal detritus occurs, and detritus instead feeds a local scavenger pathway. Here, carnivores find more diverse prey to feed on, including scavengers, herbivores, and suspension feeders. Grazers feed directly on available macroalgae, resulting in primarily local use of macroalgal production, with less importance for microalgal grazing. Unfortunately, estimating the influence of organisms present only during low tides or seasonally can be challenging; many fishes migrate into the intertidal zone only briefly during high tides, whereas birds represent the main predators, especially in winter.

20.1.9 Comparison Between Soft and Hard Bottom Intertidal Environments

Mobile sediments dominate ~75% of global coastlines. Forests of seagrass, macroalgae, and mangroves populate many of these environments. Corals typically colonize rocky submerged substrates in tropical and equatorial latitudes; however, sometimes during low tides corals may emerge and survive aerial exposure, temporarily forming another type of intertidal environment (Figure 20.16); many tropical environments lack major freshwater inputs to provide sources of sediments. These systems also support numerous algae able to colonize spaces between sparse stones or rocks, and seagrasses that can withstand temporary air exposures.

From a community ecology perspective, fewer studies have focused on intertidal soft bottom environments compared to the rocky intertidal. But even soft bottom intertidal communities experience highly variable conditions that cause physiological and mechanical stress (exposure to waves and tides). Nonetheless, sediments dampen these effects compared to the rocky intertidal because organisms can take refuge in subsurface sediments that can hold moisture and that vary less in temperature (Figure 20.17). The three-dimensional nature of sediment defines the main difference between this type of habitat and that of hard bottoms, but also complicates efforts to observe and understand soft bottoms. In fact, most manipulative caging experiments in sedimentary habitats tend to alter near bottom flow and therefore sedimentation, organic

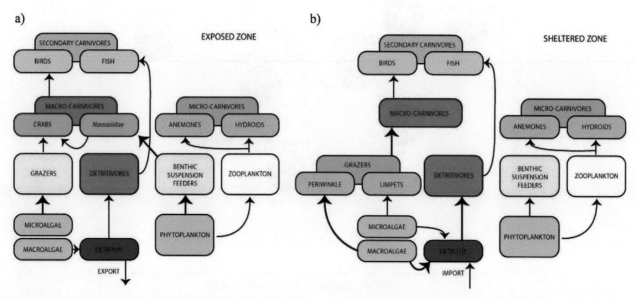

Figure 20.15 Two models of trophic networks specific to rocky intertidal habitat. (a) Food web in an environment exposed to strong hydrodynamic energy. (b) Food web in a sheltered environment.

Figure 20.16 Parts of coral reefs may emerge during low tide, thus forming another type of intertidal environment. *Source:* Roberto Danovaro (Book author).

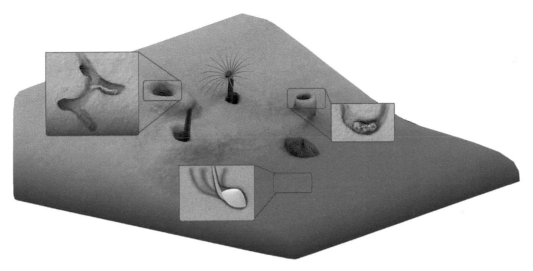

Figure 20.17 Zonation of vagile (mobile) fauna in intertidal soft sediments. Some organisms can dig deep tunnels to take refuge inside wetted sediment during low tide.

matter supply, and even larval supply. Moreover, aspects related to nutrition, especially deposit feeders, complicate understanding of these systems.

20.1.10 Niche Displacement to Reduce Competition

When similar species coexist, they tend to differ in resource exploitation and morphology more so than when spatially distant. Behavioral observations on two species of Gastropoda (*Peringia ulvae* and *Ecrobia ventrosa*, both formerly genus *Hydrobia*) formed the basis of the theory of limiting similarity. If the two species live in spatially distinct environments, competition does not occur and the species do not evolve differences in size distribution; if the species co-occur in the same environment (mudflats or mud plains) one of the two species grows larger in size (upper part of graph in Figure 20.18). Competition "displaces" characters, thereby minimizing competitive effects between the two species. The importance of competition for substratum space increases with intensity of environmental stress, until conditions become so extreme, they become dominant structuring forces.

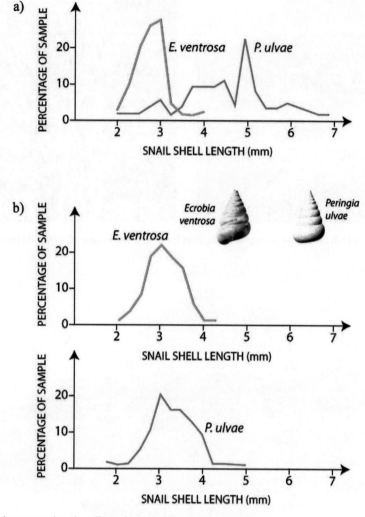

Figure 20.18 Experiment demonstrating the effect of competition between *Peringia ulvae* and *Ecrobia ventrosa*: (A) When they co-occur they differ in size (*P. ulva* becomes larger), thus limiting competition for resources; (B) when the two species live separately, they can occupy the same niche and therefore do not differ in size.

20.2 Transitional Environments Between Land and Ocean

Transitional aquatic environments refer to typically coastal environments characterized by a more or less pronounced mixture of inland waters and seawater. Depending on geomorphological characteristics and geological history, transitional environments include estuaries, lagoons, coastal lagoons, and salt ponds (Figure 20.19), among others that we will consider elsewhere. Estuaries, in particular, represent the terminal part of a river in which periodic tidal flows can strongly affect the system to some distance upstream. For example, the St. Lawrence River in Canada, the largest estuary in the world, links the St. Lawrence River with the Atlantic Ocean. "**Delta**" refers to a particular mouth shape of a river near which the river deposits sedimentary material originating from continents, resulting in multiple branches of flow (e.g., the deltas of the Po River, Northern Adriatic Sea). "**Lagoons**," in contrast refer to fully or largely enclosed coastal basins, that connect with the ocean but also receive continental waters (e.g., Lesina Lagoon, Southern Adriatic Sea); a "saltpond" refers to a basin in which exchange with the sea no longer (or rarely) occurs (e.g., Molentargius Pond, Tyrrhenian Sea, off-shore of Southern Sardinia). This chapter considers coastal lagoons as surface water bodies separated from the ocean by a barrier, usually oriented parallel to the coast, and at least intermittently connected with the adjacent open ocean through one or more channels.

20.2.1 Lagoons

Among transitional aquatic environments, coastal lagoons (Figure 20.20), occur in highest abundance and frequency along coasts of closed or semi-enclosed seas such as the Mediterranean and Baltic Seas, both characterized by **microtides**, or

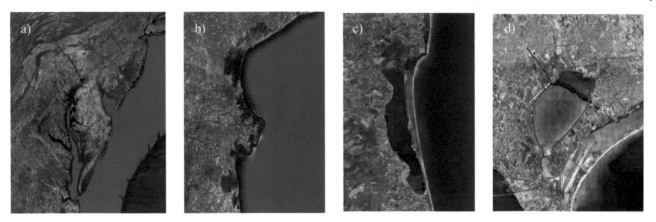

Figure 20.19 Four types of transitional aquatic environments. a = estuary (Chesapeake Bay); b = delta (Po River Delta); c = lagoon (Lesina Lagoon); d = pond (Molentargius Pond). *Source:* Michael Tangerlini.

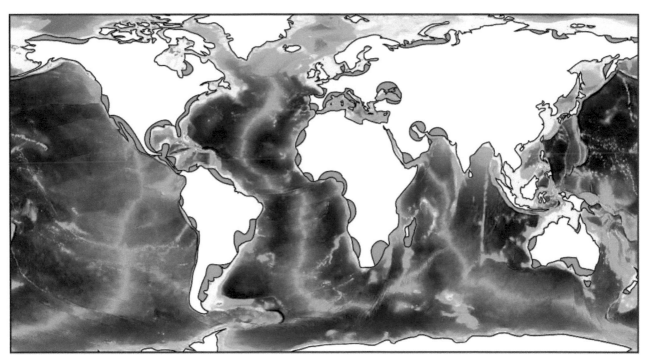

Figure 20.20 Worldwide distribution of lagoons.

very small tidal excursions. The Convention of Ramsar on Wetlands of International Importance defines wetlands as "areas of marsh, bogs, natural and artificial water, static or flowing, brackish or salty, including marine areas, whose depth at low tide does not exceed 6 meters." If we extend the concept of "transitional environments" to the broader concept of "wetland," the Mediterranean alone contains ~28,500 km^2 of wetlands, of which natural lakes and ponds comprise 12,000 km^2, wetlands comprise 10,000 km^2, and coastal lagoons represent ~6,500 km^2.

The structural and functional heterogeneity of transitional environments as a whole limits generalizations regarding their ecology and biology. For this reason, we focus on coastal lagoons, one of the more transitional environments on some coasts; we then cover mangroves and salt marshes.

Coastal lagoons occur frequently in low-lying coastal regions affected by tides, and are characterized by proximity of mouths of large rivers. From a geological perspective, coastal lagoons are relatively temporary, and their persistence depends on raising and lowering of sea level, or on **subsidence** or geological sinking. Lagoon environments, in fact, form through deposition of sediment supplies from riverine (fluvial) inflow or as a result of coastal erosion, where deposition of material forms beach ridges that separate the lagoon basin from the ocean. Frequently, lagoons contain brackish water enclosed by beach ridges that communicate directly with the ocean through one or more "mouths" at which a network of meandering "lagoon channels" converge; seawater enters through these channels during flood tide and exits during ebb tide (Figure 20.21).

Linear barrier islands, produced by accumulation of coastal sands, separate the lagoon from the ocean, and they generally support both a "real" beach on the ocean side and backshore dunes on the lagoon side. The role of fluvial sediment transport in forming coastal lagoons suggests dividing these environments into two broad categories: lagoons associated with floodplains (including estuaries and deltas), and lagoons no longer maintained by fluvial intake (or lacking them altogether) (Figure 20.22).

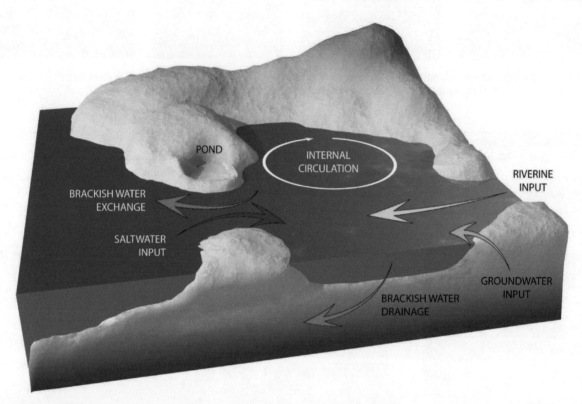

Figure 20.21 Lagoon system, highlighting the dominant hydrodynamic processes.

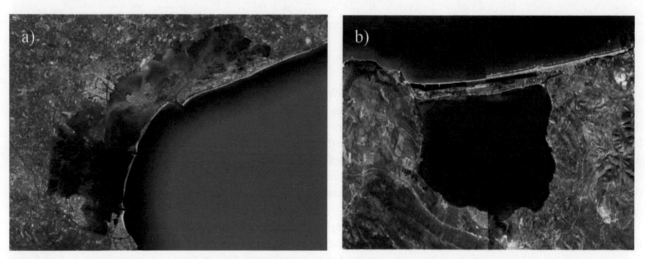

Figure 20.22 Examples of coastal lagoons associated with a floodplain (left the Venice Lagoon) and those not associated with a floodplain (right Varano Lagoon, Apulia). *Source:* Michael Tangerlini.

20.2.2 Ecology of Coastal Lagoons

Considerable structural and functional complexity characterizes transitional environments, and coastal lagoons in particular, as a direct result of the influences of marine-coastal and inland waters. For the same reason, lagoons experience wide and unpredictable variability in environmental conditions. This variability not only helps to define transitional environments but also represents their major source of vulnerability. In addition to this intrinsic variability, each lagoon uniquely differs from every other, precluding any general model of lagoon structure and functioning. But lagoons nonetheless share some general characteristics such as strong salinity gradients where freshwater and seawater input meet, as well as large and frequent changes in salinity and temperature over time. In particular, salinity progressively increases with proximity to the ocean and a surface layer of riverine fresh water sits atop denser, saltier ocean water, resulting in a vertical salinity gradient (Figure 20.23).

Salinity and temperature in these environments vary seasonally, but weather events such as storms or storm surges also cause large and sudden fluctuations (Figure 20.24). Given the shallow depth of many lagoons (often ~1 m), changes in land runoff and storm surges primarily cause sudden changes in all of the physicochemical variables in lagoon basins.

These environments often exhibit strong hydrodynamics, especially those immediately adjacent to the open ocean. Wind forcing, in particular, mixes lagoon waters and causes impressive sediment resuspension. These resuspension events increase turbidity, but also promote continuous exchange between benthic and pelagic compartments. This process provides the primary mechanism through which lagoons, which trap debris, become compartments for re-mineralizing bioavailable C and support high concentrations of inorganic nutrients.

The high concentrations of inorganic nutrients resulting from land and from high rates of remineralization of organic detritus, together with abundant light in these shallow environments, result in high rates of primary production typical of these transitional systems and, coastal lagoons in particular. In fact, high primary and secondary production rates (sometimes

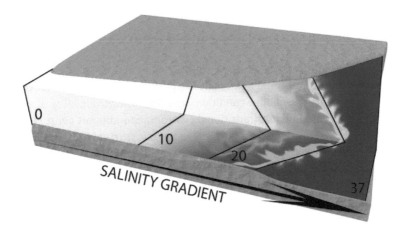

Figure 20.23 Horizontal and vertical salinity gradients (left) typical of transitional environments.

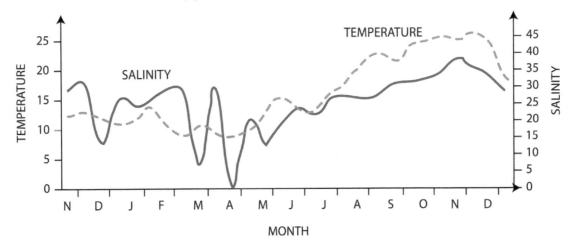

Figure 20.24 Fluctuations in temperature and salinity within a Mediterranean coastal lagoon where extreme fluctuations occur in both salinity (red line, which fluctuates annually 0–40) and temperature (green dashed line) ranging <12 °C to >25 °C.

exceeding 5–10 g C m^{-3} d^{-1}) in coastal lagoons support high biomasses of autotrophic and heterotrophic organisms that may be 15x greater than those in adjacent marine waters. Allochthonous fertilizers add to high primary productivity of lagoon ecosystems and come from land in temperate-humid regions (such as lagoons on the French coast or those of the Adriatic), whereas marine sources play a greater role in arid and semi-arid regions (lagoons of Spain's Mediterranean coasts, southern Italy, and the African coast). Total productivity (autotrophic and heterotrophic) of a Mediterranean coastal lagoon can exceed 7,000 g C m^{-2} y^{-1}. Food webs that efficiently transfer energy to apical consumers, mainly supported by a network of microbial debris, support this productivity. High productivity of coastal lagoons explains numerous species of fishes and invertebrates that use these sites for reproduction and growth, including commercial species. For example, one relatively small lagoon in Thau, France, hosts about one-fifth of ~500 species of known mollusks in the entire Mediterranean Sea.

20.2.3 Lagoon Functioning

As sites of high primary and secondary productivity, lagoons deliver abundant natural goods, and provide ecosystem services. High levels of primary productivity provide food for many wild species, including an often impressive and diverse avian fauna, but also support commercial fisheries as well as intensive or semi-intensive aquaculture.

Wide variability of environmental conditions offers opportunities for numerous microenvironments with which plant and animal biodiversity associate, often including some endemic and/or rare species. These features collectively make each lagoon ecosystem unique, thus meriting some degree of protection. In current climate change scenarios, the complex combination of increasing temperatures, increasing atmospheric CO_2 concentrations (and related changes in pH), variation in intensity and frequency of natural events such as storms will result in important consequences for coastal environments worldwide, including lagoon ecosystem structure and functioning. Increased temperatures could lead to increased total respiration rates within basins, potentially increasing frequency of hypoxia/anoxia episodes in any basins insufficiently flushed by oxygen-rich water. Changes in thermal regimes could also increase spread of "alien" species within lagoons, particularly where those species exhibit characteristics suited to warmer conditions. Many Mediterranean lagoon systems already illustrate this inevitable new reality. Expected increases in CO_2 concentrations in the next 15–20 years will likely lead to major changes in species composition of primary producer from systems largely dominated by macroalgae to systems dominated by seagrasses. Such a change would significantly alter trophic structure of these environments, with possible repercussions for resource extraction. Increases in sea level could also result in significant consequences for lagoon systems, potentially increasing salinity and reducing available mesohaline conditions and decreasing availability of nesting territories for birds.

At their locations adjacent to land-based plants, transitional environments support dense terrestrial **halophilic** ("salt-loving") vegetation (Figure 20.25), that contributes significantly to organic detritus in lagoons. This "terrigenous" component of organic detritus, in contrast to phytoplankton and macroalgae, is much more refractory for herbivores, largely because it contains structural material required by land plants but unnecessary in the ocean where water helps to support macroalgae and keep them upright. This refractory material requires "reworking," predominantly mediated by heterotrophic prokaryotes, and channeling to higher trophic levels primarily through the microbial circuit.

Transitional systems dominated by organic material of algal rather than plant origin differ greatly in flow of matter and energy. In systems characterized by a strong supply of relatively labile macroalgal detritus (Figure 20.26 left) and relatively little plant biomass, such as seagrasses (Figure 20.26 right), different pools of organic matter result in clear consequences for benthic food webs.

Figure 20.25 Plants typical of transition systems: (a) *Salicornia fruticosa*, (b) *Phragmites* spp., (c) *Ruppia maritima*. *Sources:* Giancarlo Dessì / Wikimedia Commons / CC BY-SA 3.0, Gordon Leppig & Andrea J. Pickart / Wikimedia Commons / Public domain.

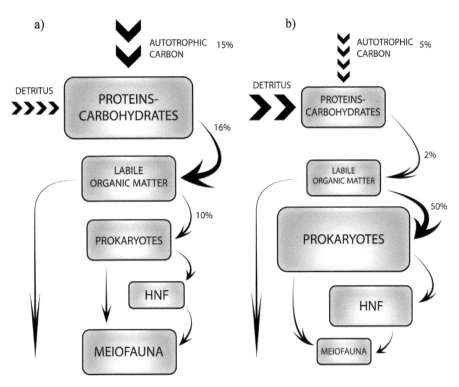

Figure 20.26 Food webs in two lagoon environments differing in origins of organic detrital matter: Left – lagoon dominated by organic material produced from algae; Right – lagoon dominated by organic material derived from seagrasses (or terrestrial plants). Substantial pools of labile organic matter (left) transfer more energy to higher benthic trophic levels, whereas detritus derived from seagrasses (right) largely supports growth of microbial biomass and limited growth in meio- and macrobenthic biomass. Legend: P-C refers to proteins and carbohydrates, Meio = meiofauna; HNF = heterotrophic nanoflagellates.

20.2.4 Models of Functional Zonation of Coastal Lagoons

Descriptive models often differentiate lagoon types based on salinity including **oligohaline** (<5), mesohaline (5–18), **polyhaline** (18–30), and **euryhaline** (>30–35) lagoons. Other classification schemes based on different approaches fall into three basic types: (1) ecotones, (2) paralic sediments, and (3) ergoclines or energy gradients.

Ecotones, which represent transition habitats between two adjacent ecosystems and lagoons, in this instance constitute a brackish water interface between two biomes, the ocean and continental waters. Thus, as a transition zone between two biotic communities, ecotones form a marginal zone between distinct communities, each of which is independently self-sustaining, largely defined by their species composition.

The strong salinity gradient generated by mixing between freshwater and seawater forms the basis for the **paralic model**. Strong physico-chemical gradients in transitional environments also vary as a function of distance from the ocean. In this case, isolation of lagoon populations compared to more typical marine ecosystems plays a dominant structuring role. Lagoons contain freshwater populations as well as a marine component. However, lagoon biotas typically exhibit intermediate adaptations to the paralic domain. In addition to species that extend into lagoons from neighboring biomes and complete most of their life cycle outside the lagoon, other species find ideal conditions for survival here. In the **confinement model**, declines in marine species penetrating into lagoons also depend on how quickly basin waters renew rather than on the salinity gradient alone. This model considers lagoon communities as a "dilute" version of marine communities from which typical lagoon communities evolved. The confinement model identifies bionomic sub-domains, each characterized by different species and organized according to degree of confinement (Figure 20.27). More specifically, researchers define the paralic domain as a strip of varying size between the ocean and land, sometimes considered an intermediate strip of circles radiating from land, that differ from each other depending on the influence of the marine domain. The paralic domain also contains different subsystems: The subdomain closest to the ocean (the close paralic, I) and the subdomain farthest from the ocean (the far paralic, VI). This subdomain also contains two different poles: the

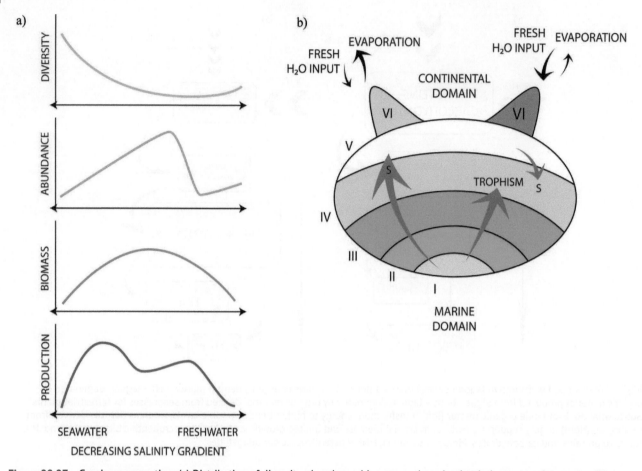

Figure 20.27 Graphs representing: (a) Distribution of diversity, abundance, biomass, and production in lagoon environments; (b) confinement is delineated along a gradient linking marine and land domains through six bionomic groups (I–VI). Decreasing wave, food supply, and increasing salinity characterize the confinement gradient. Confinement VI includes two sub-systems, one evaporitic (in which high rates of evaporation result in increased salinity – S), and the other largely freshwater (high input of fresh water off land decreases salinity).

evaporitic pole and the freshwater pole (VI). Finally, the intermediate subdomain (representing "typical" paralics II–V), spans from ocean to land and defines actual area of ecological transition.

The **ergocline model** assumes that spatial and temporal gradients in the ergocline or energy gradient define distributions of lagoon communities, and physical processes can generate ecological structures with high organic production. According to this interpretation, the bidirectional energy gradient in hydrodynamics ensures that no resource becomes limiting and system productivity therefore always remains high.

20.2.5 Lagoon Biodiversity

High variability in the physical-chemical characteristics of lagoons implies that the resident species must be able to withstand very wide and frequent environmental changes. In particular, changes in salinity and temperature typically play a key role in defining community composition. These fluctuations represent an important source of disturbance in lagoon communities, resulting in lower levels of biodiversity relative to those in the adjacent ocean (Figure 20.28). In lagoons, shallow depth, high nutrient availability, and abundant light support abundant macrophytes. Lagoon shoreline vegetation provides important habitat for many species of migratory or resident birds, and offers ideal substrate for numerous epi- and periphytic invertebrates, providing not only breeding and foraging areas, but also protection from predators. Within lagoons characterized by eutrophication, superabundant nitrophilous macroalgae may replace macrophytes and strongly compete with phytoplankton for nutrients and light, thus altering both hydrodynamics and functioning of lagoons.

In some Mediterranean lagoons, seagrasses such as *Zostera noltii*, *Cymodocea nodosa*, and *Zostera marina* also occur. These ecosystem engineers, when present, provide services to the whole lagoon ecosystem. Their progressive deterioration over the last ~25 years threatens the survival of these fragile transitional ecosystems. For example, tens of thousands of pigeon

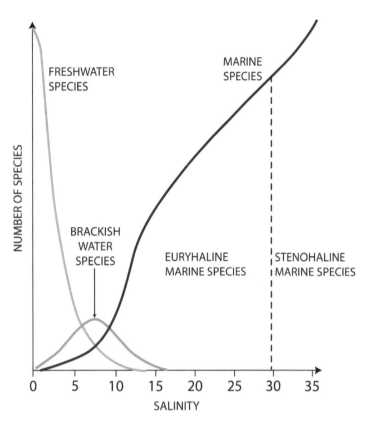

Figure 20.28 Transitional environments typically support lower biodiversity than marine or freshwater environment, as seen in changes in numbers of species in relation to salinity. Mixohaline (or brackish) species can cope well with strong environmental variation typical of lagoons.

geese (*Branta bernicla*) overwinter in the lagoon in Arcachon, France, feeding on intertidal seagrasses. In lagoons widely connected to the ocean, seagrasses such as *Zostera* spp. and *Ruppia* spp. dominate submerged vegetation in more sheltered areas, whereas *Cymodocea* spp. dominates the deepest areas. In brackish lagoons, characterized by low water renewal rates, *Ruppia* spp. and *Potamogeton* spp. dominate submerged vegetation. In both cases, these macrophytes represent an important fraction of total primary production, especially in relatively shallow lagoons where abundant light radiation penetrates all the way to sedimentary substrates. Herbivores can directly use only a fraction of the biomass produced by these macrophytes, and bacteria and fungi must break down most of this material before it becomes available for higher trophic levels.

Several examples illustrate how some marine species have developed the capacity to withstand environmental fluctuations typical of lagoons. For example, the bivalve mollusk *Cerastoderma glaucum* occurs in many Mediterranean lagoon environments; its marine counterpart, *Cerastoderma edule*, cannot tolerate thermohaline fluctuations typically observed in lagoon environments. Three species of gastropod, though all similar in size and feeding strategy, differ in tolerance to variation in salinity: *Peringia ulvae* generally occurs in marine waters, whereas *Hydrobia neglecta* and *Ecrobia ventrosa* frequent low-salinity lagoon environments, such as those along Denmark's coast. Typical Mediterranean lagoons support many zooplankton species linked to lagoon systems and rarely observed in offshore environments, such as *Paracartia latisetosa*, *Acartia margalefi*, *Acartia tonsa*, and *Calinipeda aquaedulcis*. Coastal lagoons generally support rich and diverse phytoplankton such as *Skeletonema costatum*, *Cyclotella* spp., and *Amphora* sp. Foraminifera such as *Ammonia beccarii*, *Cribrononion granosum*, and *Haynesina pauciolocula* are also usually present.

Although large environmental fluctuations that characterize lagoons result in lower species diversity relative to the adjacent ocean, the landscape-scale scenario differs greatly. In fact, sandbanks, and small and large islands support coexistence of numerous complex habitats, often increasing local biodiversity relative to less heterogeneous coastal regions. On a larger scale, high species richness characterizes coastal lagoons as a result of abundant groups of animals, plants, and microorganisms. In fact, many transitory species of fishes and invertebrates exploit lagoons for reproduction and growth, and migratory birds feed and rest in these environments. In summary, different factors control biodiversity in coastal lagoons, including presence of marine and freshwater biota, presence and maintenance of ecotones that transversely connect dryer to more humid habitats, locations influenced more by the ocean and other locations influenced more by freshwater, presence of ecosystem engineers, and complex and heterogeneous mosaic habitat structure of habitats within lagoons.

20 Intertidal Ecosystems and Lagoons

These transitional environments exchange water with the ocean and with land and, through that process, they receive relatively continuous and often massive inputs of organic and inorganic matter in both dissolved and particulate forms. This material tends to accumulate within the system and forms the starting point for lagoon production cycles. These characteristics increase vulnerability of transitional environments relative to many other marine and coastal sites, and potential eutrophication triggered by both natural and anthropogenic input of surplus inorganic nutrients. Unpredictable environmental characteristics have obvious consequences for abundances and composition of both benthic and planktonic communities. More generally, exogenous forces on these transitional ecosystems produce non-linear responses, whether natural or anthropogenic in origin. These ecosystems add significant natural and economic value.

20.3 Mangroves

"Mangrove" refers both to a type of vascular plant and the coastal forests those plants comprise, though **"mangal"** refers specifically to the biome (Figure 20.29). Mangroves includes 12 families and >70 species; this diversity of taxa and families indicate frequent, independent adaptive phenomena within several families of tropical angiosperms. Mangroves reflect **convergent evolution**, where phylogenetically distinct organisms (in this case plants of different angiosperm families) have evolved similar adaptations. Mangroves include shrubs and trees that occupy intertidal and shallow subtidal environments in tropical and subtropical areas. They dominate 75% of coasts between 25°N and 25°S spanning a total area of

Figure 20.29 Mangroves in Indonesia and the Red Sea. From top to bottom and from left to right note mangroves with elongate seeds ripe for deployment, immature plants at the bud stage, and individuals at different stages of pneumatophore root development. Images show mangroves at both low and high tides, and growing on an exposed carbonate cliff. The bottom right image shows a well-developed mangal with pneumatophore roots emerging from the sandy beach. *Source:* Roberto Danovaro (Book Author).

170,000 km² (Figure 20.30). Halophytic trees and shrubs that live in sediments rich in salt characterize these environments. Salt glands in mangrove leaves excrete salt, so these plants tolerate both fresh water and seawater and, therefore, avoid competition with other plants that cannot tolerate high salt concentrations. Mangals encompass a mosaic of different types of forest, each providing different habitats, topologies, niches, microclimates, and food resources for different animal communities. Mangroves occupy muddy and clay bottoms in the mid-tidal zone of tropical and subtropical intertidal coasts near estuarine environments.

The 24 °C isotherm typically defines latitudinal distributional limits of mangroves, which often corresponds to that of tropical corals and tropical seagrasses. Mangroves that form habitats for other species form two species clusters: the Western center includes Africa, the Atlantic coasts of South America, the Caribbean, Florida, Bermuda, Central America, and the Pacific coasts of North and South America. The eastern center from East Africa to the western Pacific Ocean contains >40 species of mangroves and forms the second species cluster. The vast mangrove forests of Southeast Asia are among the most species-rich in the world, with the Indo-Malay region as the center of diversity.

Several mangrove species typically dominate mangal forests, depending on distance from the coast (Figure 20.30).

Red (*Rhizophora mangle*), black (*Avicennia germinans*), and white (*Laguncularia racemose*), mangrove, as well as the button mangrove, *Conocarpus erectus*, can grow along the same coastline. When these species coexist, each species occurs in just one tidal exposure zone. Tidal exposure and salinity differences primarily determine this zonation (Figure 20.31).

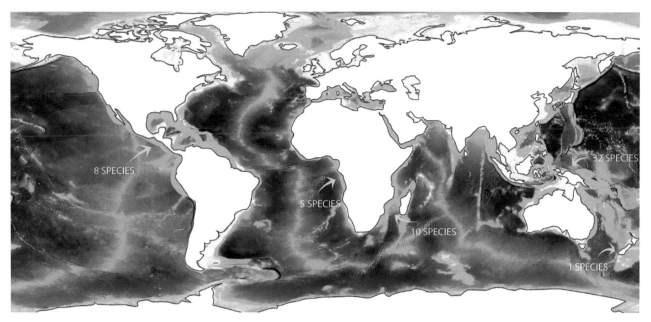

Figure 20.30 Global distribution of mangroves.

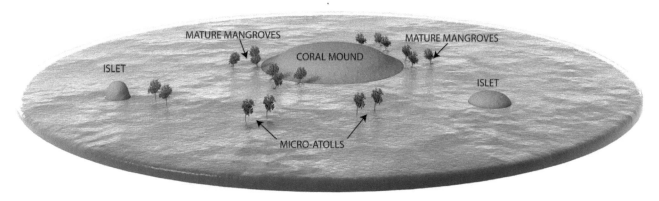

Figure 20.31 Zonation of mangroves from most marine species (seaward) to most terrestrial species (landward at center) showing structure and zonation in relation to coral reef communities.

Aerial roots of red mangrove form a dense barrier in the tidal zone, in contrast to black mangrove that occurs more landward with only periodic immersion by the tide. Its dark green leaves (6–10 cm long) produce salt crystals to eliminate excess salt. In contrast to some other mangroves, it reproduces through comparatively small seeds (4–5 cm). Smaller white mangroves occur in lower densities than other mangroves, often growing as a shrub in dryer sediments with other plants. Two salt glands in each leaf enable them to tolerate exposure to salt.

Because mangals grow in fine sediments (mud, clay, and silt), they typically occur in estuaries that deliver sediments that the mangrove's complex root system and vertical elements trap. Many species of mangrove can survive in freshwater because they tolerate wide salinity variations; this adaptation allow them to dominate transitional tropical environments, because no seagrasses can tolerate such low salinities. Numerous biological adaptations in mangroves help them thrive in this environment, including salinity tolerance highlighted above, reproductive biology, and growth form. Mangroves reproduce through seeds that grow attached to the plant. Above sea level, mangroves resemble typical terrestrial shrubs, with trunks, stems, leaves, and flowers. The specialized roots or **pneumatophores** extend above bottom and enable oxygen uptake in highly anoxic sediments. Root systems vary in form (anchors, aerial, crankshafts, and propagation roots) depending on specific habitat (Figure 20.32). Mangroves often occur landward near coral reefs and some grow rooted to existing micro atolls or re-emerged fossil coral reefs.

20.3.1 Biodiversity Associated with Mangals

A succession of populations of different species occurs along a salinity gradient moving gradually away from the coast in mangrove forests or moving between coastlines that differ in circulation, temperature, salinity, and water masses. Each species possesses adaptations to specific conditions, including strong salinity variation, high concentrations of hydrogen sulfide, and reducing conditions that arise because the mangals accumulate muddy material rich in organic matter that tends to become sub-oxic below the subsurface layers.

Mangals provide habitat for >1,300 species of animals, including 628 species of mammals, birds, reptiles, fishes, and amphibians, specifically because they provide breeding, foraging, and shelter environment. Mangrove forests provide numerous microhabitats for many resident, seasonal, and transient organisms associated with adjacent marine and terrestrial habitats. Many larger, mobile species do not restrict their movements to mangals, but instead visit seasonally or opportunistically. However, many species of resident invertebrates and some species of vertebrates totally depend on mangroves for survival and for completing important parts of their life cycle. Frequent "visitors" to mangrove communities include marine fishes and invertebrates as well as mammals, reptiles, and birds from adjacent terrestrial systems. Much like seagrasses, diverse substratum enables associations of various other vertebrates or invertebrates with the rhizomes.

Mangroves provide both hard and soft substrata, and microhabitats above and below the water line that can host a great variety of invertebrate species. Invertebrates that colonize the extensive root system and surrounding muddy habitats feed on fallen leaves, debris, plankton, and other small animals. Gastropods, barnacles, bryozoans, tunicates, mollusks, sponges, polychaetes, isopods, amphipods, shrimps, and jellyfish live both near and within the mangrove root system. Some invertebrates thrive in the mangrove canopy, including the dominant crab species, *Aratus pisoni*, which feeds mainly on red mangrove leaves. Other crabs live in the intertidal mudflat, feeding on fallen leaves and debris. Horseshoe crabs, *Limulus* sp., frequently occur in mangroves, feeding on algae and invertebrates and scavenging on carcasses and other material. Some species of vertebrates use mangal environments for years, tracking variation in food availability such as during flowering and fruit production, a period in which populations of invertebrates and small vertebrates vary in response to changes in food resources.

Red mangrove roots create a specific microhabitat for resident species (tunicates, crustaceans, mollusks, and fishes) that spend their entire life cycle attached to or moving within the root system. This root system offers an important nursery habitat for organisms such as crustaceans, mollusks, and fishes that grow there through early life history stages and complete their adult lives elsewhere. The dominant fish species in mangrove forests, members of the family Poeciliidae and *Gambusia* spp., *Heterandria formosa*, and *Poecilia latipinna* in particular, form the essential link between primary producers and the highest food web levels. Mangrove fish fauna of mangroves usually includes a mixture of representatives of marine and freshwater species. Five amphibian species feed and reproduce within mangrove habitat, dominated by tree frogs, *Hyla* spp., and marine cane toads, *Bufo marinus*. Researchers to date have reported 22 reptile taxa associated with mangroves. Moving from low tide to high tide habitats, gastropod abundances increase; during low tide gastropods aggregate in order to maintain high humidity, whereas *Metopograpsus* spp. crabs aggregate near rhizomes. Mangroves also

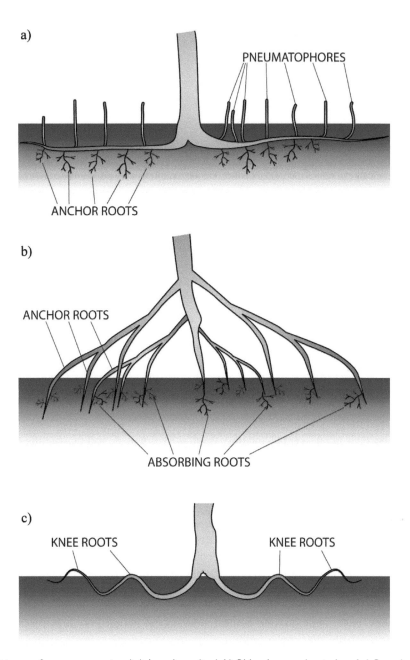

Figure 20.32 Different types of mangrove roots: a) *Avicennia* sp. (top), b) *Rhizophora* sp. (center) and c) *Bruguiera* sp. (bottom).

provide important habitat for many species of birds, such as *Vireo aliloquus*, *Coccyzus minor*, *Dendroica petechia*, and *D. discolor*, both for nesting and for foraging. Terrestrial species that use mangrove forest include high abundances of the crabs *Sesarma* spp. and *Uca* spp. In *Bruguiera* forests crabs occur "everywhere" and, in many localities, mud lobsters (*Thalassina* spp.) live in burrows. In *Rhizophora* spp. habitats, more mobile sediments support blue crab, *Metopograpsus* spp., which lives among the aerial roots, associated with abundant shrimp communities that move within the mud. Gobies characterized by eyes positioned on the back of their head occur in high abundance in channels and in higher abundances toward the ocean.

Some mangal species use both terrestrial and marine habitats. For example, modified pectoral fins on mudskippers enable locomotion on land, where they use multiple adaptations for terrestrial life such as highly vascularized lips and other surfaces that come into contact with air and water (swim bladder, stomach, intestine). Lungfish, *Periophthalmodon*

Figure 20.33 Lungfish *Periophthalmodon kalolo*. (a type of mudskipper) uses terrestrial and marine environments.

schlosseri (Figure 20.33) use their tail as a supplementary respiratory organ. Apical positioning of eyes helps them avoid predators, enabled by 360° mobility.

20.3.2 Mangal Ecosystem Functioning

Mangrove forests play multiple important ecological roles in that they: (1) Provide coastal protection against storms; (2) enhance soil formation by trapping debris; (3) filter rainwater flowing from land and remove organic matter; (4) create habitat for many species of fishes, and rich communities of epiflora and invertebrate epifauna, insects, and birds; (5) contribute significantly to primary production; (6) produce significant leaf debris that forms the basis for a complex and diverse ecosystem, including food for many consumers such as crabs, and a complex detrital chain.

Mangroves play an important role in estuarine food webs, producing large amounts of leafy debris. Together with seagrass beds, macrophytes, phytoplankton, and benthic microalgae, mangroves form the base of the food web in tropical transition systems. Globally, mangroves produce an average dry organic matter biomass of 100–200 metric tons per hectare (i.e., 10–20 kg m^{-2}), with highest primary production at the equator. Mangroves export an average of 5–10 metric tons of organic detritus ha^{-1} y^{-1}. Considering the 40% carbon content of organic matter, this export corresponds to transport of detritus of ~200–400 g C m^{-2} y^{-1}. The amount of debris transported by tidal flows in mangrove forests depends on land elevation, which in turn, determines forest flooding frequency by seawater. Each tidal incursion exports ~95% of detritus, leaving relatively little detritus material within the mangal. Invertebrates and bacteria consume some 30–80% of organic matter in the form of "waste" (dead leaves, flowers). Crabs utilize 30% of plant biomass produced by mangroves, and thus play an essential role in organic matter recycling (see Box 20.1, mangrove detritus degradation"). In general, these systems support high faunal diversity and high secondary production where mangrove leaves provide the engine, crabs mediate degradation, and a wide range of grazers and decomposers eventually use these different components.

Fungi and bacteria quickly decompose leaves that fall from mangroves. Tides aid flow of this decayed matter within estuaries, which forms a trophic resource for many marine species including economically important species of shrimp, crabs, and fishes. Algae play a primary role in mangrove food webs. Many organisms feed directly on micro- and macroalgae that thrive within mangrove communities. The network of aerial roots provides an ideal substrate for epiphytic algae, such as benthic diatoms, and cyanobacteria. Phytoplankton originating from adjacent open-ocean, freshwater, and estuarine environments can also play an important role within mangrove system. Birds and insects exploit associated fruits, but most users cannot directly digest and use tannin-rich leaves.

> **Box 20.1 Crabs and Mangrove Detritus Degradation**
>
> Caging experiments demonstrate the importance of crabs in this ecosystem (Figure 20.34), which play a fundamental role in decomposing exported organic matter. One experiment deployed packets of leaves in locations in which cages eliminated crab access, as well as unprotected packets of leaves. In the presence of crabs, rapid degradation occurred, thus demonstrating how crabs accelerate organic matter recycling by processing detritus and increasing accessibility for microorganisms. Brachyuran crabs therefore play a vital role in secondary biomass production because they consume leaves directly.
>
>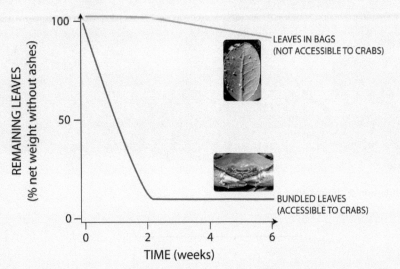
>
> **Figure 20.34** Effect of crab presence of the genus *Uca* on decomposition rates of plant debris (leaves of mangrove). Crabs are responsible for degrading >90% of plant debris. *Sources:* Peripitus / Wikimedia Commons / Public domain, Roberto Danovaro (Book author).

The remarkable production of detrital material by these ecosystems leads to nutrient release that other primary producers (seagrasses, macroalgae) in adjacent ecosystems can use, particularly coral reefs characterized by limited availability of nutrients. In fact, coral reefs frequently depend on material exported from mangroves. These strikingly different and apparently separate ecosystems can interact closely. Coral reefs form natural breakwaters that reduce wave impacts on shorelines, helping to stabilize substrate and enable mangrove root establishment and growth, which in turn, through primary production, provide surplus nutrients, detritus, and organic material essential to coral reef biota.

20.4 Salt Marshes

Salt marshes refer to coastal intertidal wetlands that tides flood and drain with salt water through numerous tidal channels. Detritus, nutrients, plankton, and fishes move in and out with the tides. Salt marshes span from temperate to Arctic latitudes, replacing mangals that occupy a similar ecological niche in the tropics (Figure 20.35). Tides, which vary from centimeters-meters, depending on location, result in zonation that parallels other ecosystems at the land-sea interface, where species composition varies with tidal elevation and grades from cordgrasses such as *Spartina alterniflora* to a mix of other cordgrasses such as *Spartina patens* as well as salt tolerant plants such as *Salicornia* spp. and *Distichus spicata* (Figure 20.36).

This saltwater inundation limits vegetation to **halophylic** (salt adapted) plants that include predominantly grasses, but also shrubs and herbs. They typically occur in association with estuaries, and particularly thrive where a sufficient sediment source allows salt marshes to develop from a young marsh to an old marsh (Figure 20.37). Salt marsh development requires fine-grained sediments, quiescent conditions without strong waves or currents, salty water to eliminate potential competitors, moderate or cool temperatures, and a reasonable tidal range to dissipate energy from water movement and accumulate sediments. Cordgrass *Spartina alterniflora*, a low marsh plant, dominates young marshes in the eastern United States. *S. alterniflora* can colonize sediments as part of a succession series, where benthic diatoms arrive first and bind together sediments with mucus. Filamentous algae arrive next, further stabilizing the shoreline so that other species such as *S. alterniflora* can then colonize. Over time, in the presence of abundant nutrients delivered by flooding tides, *S.*

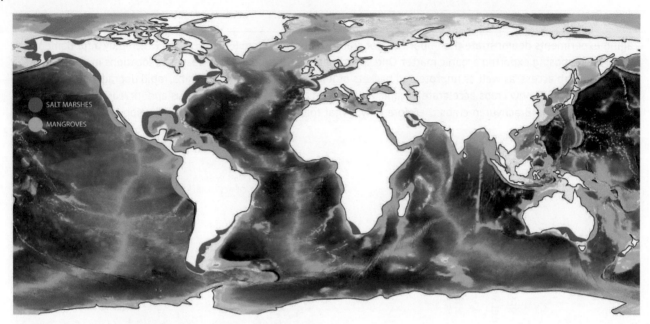

Figure 20.35 Global distribution of salt marshes.

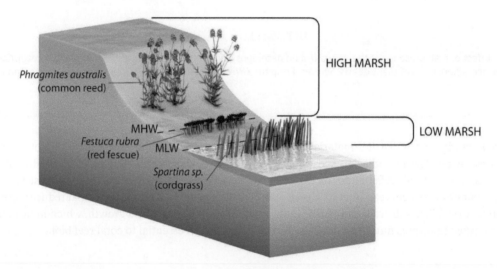

Figure 20.36 Salt marsh zonation.

alterniflora flourishes and forms thick stands that trap finer sediments building up deep mud mixed with sand and peat, creating a marshy texture of organic-rich sediment. Plant roots and rhizomes further reduce erosion and stabilize new habitats. Eventually sediment builds above the high-water mark, at which time high marsh plants colonize and outcompete *S. alterniflora* to form high marshes, which extend from the high tide mark to the highest spring tide mark. These "old marshes" receive sand and mud delivered by streams and rivers, eventually forming "new" terrestrial habitat effectively disconnected from the ocean. This process can extend the shoreline km into the ocean.

As with other intertidal systems, salt marsh biota requires adaptations to cope with their harsh and dynamic environment. Because plants that evolved from terrestrial ancestors dominate much of the flora, they must overcome osmotic pressure to take up water by generating negative hydrostatic pressure through transpiration by their thin and sometimes fleshy leaves. Structurally, in order to withstand water movement their tissues have strong lignification, and a thick epidermis. Succulent leaves and stems, and scale-like hairs limit evaporation, and plants accumulate salt in their tissues that enable normal osmosis. Salt gland cells on the undersides of leaves of other plants excrete salt from their tissue. Anoxia creates another set of challenge in that water saturates their roots and plants require oxygen in order to respire. Salt marsh plants get around this problem using **aerenchyma**, specialized tissue that delivers oxygen to submerged roots. The

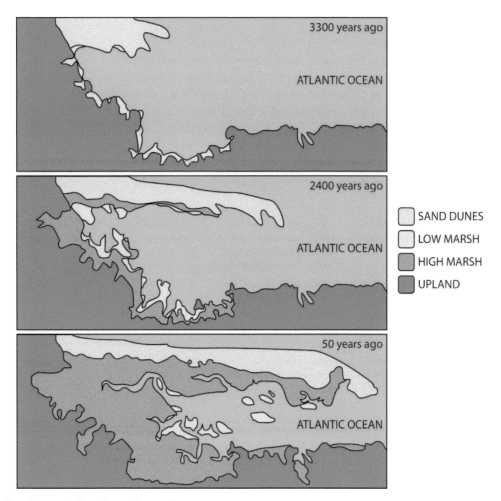

Figure 20.37 Development of a salt marsh over time.

perennial, thick roots do not penetrate deeply into anoxic sediments and stabilize the substrate with short-lived, thin, branched roots with numerous root hairs that facilitate nutrient uptake.

20.4.1 Biodiversity Associated with Salt Marshes

Salt marshes support a relatively low diversity of fauna, in part because of fluctuating environmental conditions and also because high organic content results often in oxygen-poor sediments. These species include a mix of terrestrial and marine species, including small mammals, small fishes, birds, insects, spiders, and marine invertebrates. Common marine invertebrates include amphipods, isopods, anemones, shrimps, crabs, turtles, mollusks, and gastropods. Within tidal channels, common fishes include sticklebacks, silversides, eels, and flounders. A variety of waterfowl uses salt marshes for feeding and overwintering, including ducks, herons, and oyster catchers.

20.4.2 Salt Marsh Ecosystem Functioning

Salt marshes provide a variety of ecosystem functions. Their emergent structure traps sediment and reduces water flow, thus stabilizing coastlines. Their capacity to trap sediments and take up nutrients such as nitrate and phosphate as well as toxins, thereby helps to regulate water quality in adjacent coastal waters as well as groundwater systems. Their roles in providing important habitat, nursery grounds for different fishes and invertebrates, providing food and nesting grounds for wading birds and other species, and their role as a predator refuge define some of their most critical functions. For example, in some regions of the eastern United States salt marshes provide essential food, refuge, or nursery habitat for >75% of commercial species, including shrimp, blue crab, and many finfish.

20.5 Summary

Intertidal ecosystems share several key features including strong environmental variability, and typically low diversity, but they also differ in geographic extent, vegetation, and in some key functions (Table 20.2). Because they occur at the interface between land and the ocean, many of these environments face significant and numerous pressures from human populations that tend to alter the biodiversity and functions of thee critical transition zones.

Table 20.2 Comparison between intertidal ecosystems.

Main Characteristics	Habitat / Ecosystem			
	Lagoons/ Transitional	Rocky Intertidal and Sedimentary Intertidal (Sandflats and Mudflats)	Mangals	Salt Marshes
Extent	Generally of limited dimension	Narrow band of habitat and thus limited in extent.	137,760 km^2	22,000–400,000 km^2
Distribution	Wide diffusion along the coasts of the world	Along the coasts of all continents	Tropical and subtropical environments. Generally between 30°N and 25°S	Estuarine temperate to subpolar shoreline environments
Depth	Difficult to give a global average but many lagoons are on average typically ~1 m deep.	From 0 to a few m depending on tidal extent	0–2 m	From 0 to a few m depending on tidal extent
Type of substratum	Generally soft	Both soft and hard bottoms	Soft bottoms in estuarine or coastal environments, they can grow also on ancient carbonate platforms	Generally organic rich mud
Biodiversity	Low species richness but high beta diversity between lagoons	Very low species richness	Diverse mix of terrestrial species in upper branches and marine species in roots that provide habitat for sessile organisms and shelter from predation	Relatively low
Endemism	Very few endemic forms	Few endemic forms	Not numerous	Low
Functioning	Very high primary production both from algae and from seagrasses	Relatively low primary and secondary production but important habitat for species such as migratory birds	Very high primary and secondary production, significant detritus export (mangrove leaves)	Highly productive coastal habitats that provide critical nursery grounds for a wide range of species.
Trophic networks	Simplified systems depend on detritus from primary production	Based largely on phytoplankton consumed during low tide and use of macroalgae and lichens	Systems rich in nutrients and organic detritus mainly based on detritivores and filter feeders living on roots	Systems rich in nutrients and organic detritus.
Peculiarity	Very dynamic systems subject to strong variation in environmental variables linked to shallow depths	Extreme environments because of the desiccation, salinity and temperature	Ecosystem engineers that interact strongly both with terrestrial and marine systems	Ecosystem engineers that interact strongly both with terrestrial and marine systems
State of art	Well studied given their historical importance to humans	By far the most well-studied marine habitat because of ease of access and simplicity	Less studied systems in the major part of the areas where they occur	Well studied, easily accessible systems

Questions

1) What effects does exposure have on zonation patterns in rocky intertidal environments?
2) How do biotic and abiotic variables affect barnacle distribution in the rocky intertidal?
3) What is the intermediate disturbance hypothesis?
4) Provide a definition and examples of keystone and dominant species.
5) What are the main differences between soft and hard bottom intertidal environments?
6) How does niche displacement reduce interspecific competition for resources?
7) The lagoon food web traps detritus – how?
8) Describe the paralic theory of transitional systems.
9) Which is the latitudinal range of mangrove systems?
10) Describe a mangrove system, its biodiversity, and its functioning.
11) Describe salt-marsh zonation.
12) Describe biodiversity associated with salt marshes.
13) Describe the unique features of rocky shores, salt marshes, mangroves and lagoons?

Suggested Reading

Benedetti-Cecchi, L. (2001) Variability in abundance of algae and invertebrates at different spatial scales on rocky sea shores. *Marine Ecology Progress Series*, 215, pp. 79–92.

Benedetti-Cecchi, L., Pannacciulli, F., Bulleri, F. et al. (2001) Predicting the consequences of anthropogenic disturbance: Large-scale effects of loss of canopy algae on rocky shores. *Marine Ecology Progress Series*, 214, pp. 137–150.

Bertocci, I., Maggi, E., Vaselli, S., and Benedetti-Cecchi, L. (2005) Contrasting effects of mean intensity and temporal variation of disturbance on a rocky seashore. *Ecology*, 86, pp. 2061–2067.

Blanchette, C., Denny, M., Engle, J.M. et al. (2016) Intertidal ecosystems, in Mooney, H. and Zavaleta, E. (eds.), *Ecosystems of California*. UC Press. pp. 24–25.

Brown, E.J., Vasconcelos, R.P., Wennhage, H. et al. (2018) Conflicts in the coastal zone: Human impacts on commercially important fish species utilizing coastal habitat. *ICES Journal of Marine Science*, 75, pp. 1203–1213.

Bulleri, F., Chapman, M.G., and Underwood, A.J. (2005) Intertidal assemblages on seawalls and vertical rocky shores in Sydney harbour, Australia. *Austral Ecology*, 30, pp. 655–667.

Eggleston, D.B., Dahlgren, C.P., and Johnson, E.G. (2004) Fish density, diversity, and size-structure within multiple back reef habitats of Key West National Wildlife Refuge. *Bulletin of Marine Science*, 75, pp. 175–204.

Eslami Andargoli, L., Dale, P., Knight, J., and McCallum, H. (2015) Approaching tipping points: A focussed review of indicators and relevance to managing intertidal ecosystems. *Wetlands Ecology and Management*, 23, pp. 1–12.

Fraschetti, S., Terlizzi, A., and Benedetti-Cecchi, L. (2005) Patterns of distribution of marine assemblages from rocky shores: Evidence of relevant scales of variation. *Marine Ecology Progress Series*, 296, pp. 13–29.

Konar, B., Iken, K., Pohle, G. et al. (2010) Surveying nearshore biodiversity, in McIntyre, A.D. (ed.), *Life in the World's Oceans: Diversity, Distribution and Abundance*. Wiley-Blackwell, pp. 27–41.

Lefcheck, J.S., Hughes, B.B., Johnson, A.J. et al. (2019) Are coastal habitats important nurseries? A meta-analysis. *Conservation Letters*, 12, p. e12645.

Levin, L.A., Boesch, D.F., Covich, A. et al. (2001) The function of marine critical transition zones and the importance of sediment biodiversity. *Ecosystems*, 4, pp. 430–451.

Mazzella, L., Scipione, M.B., and Buia, M.C. (1989) Spatio-temporal distribution of algal and animal communities in a *Posidonia oceanica* meadow. *Marine Ecology*, 10, pp. 107–129.

Menconi, M., Benedetti-Cecchi, L., and Cinelli, F. (1999) Spatial and temporal variability in the distribution of algae and invertebrates on rocky shores in the Northwest Mediterranean. *Journal of Experimental Marine Biology and Ecology*, 233, pp. 1–23.

Mistri, M. and Rossi, R. (2001) Taxonomic sufficiency in lagoonal ecosystems. *Journal of Marine Biological Associated of the United Kingdom*, 81, pp. 339–340.

Pérez-Ruzafa, A., Pérez-Ruzafa, I.M., Newton, A., and Marcos, C. (2019) Coastal lagoons: Environmental variability, ecosystem complexity and goods and services uniformity, in Wolanski, E., Day, J., Elliott, M., and Ramesh, R. (eds.), *Coasts and Estuaries, the Future*. Elsevier, pp. 253–276.

Toth, L.T., Aronson, R.B., Vollmer, S.V. et al. (2012) ENSO drove 2500-year collapse of Eastern Pacific coral reefs. *Science*, 337, pp. 81–84.

Viaroli, P., Bartoli, M., Bondavalli, C. et al. (1996) Macrophyte communities and their impact on benthic fluxes of oxygen, sulphide and nutrients in shallow eutrophic environments. *Hydrobiologia*, 329, pp. 105–119.

21

Subtidal Hard Substrata Ecosystems

21.1 Introduction

The "**coastal subtidal**" extends from the intertidal zone to the edge of the continental shelf at ~200 m, and spans a range of community composition and diversity, depending on: (1) temperature, (2) hydrodynamics, (3) light, (4) energetic intensity and space availability, particularly in rocky subtidal environments. This chapter considers rocky subtidal habitats, including kelp forests, coral reefs, coralligenous habitat (including rhodolith beds and maerl) and caves. These environments include some of the most productive and dynamic marine ecosystems (kelp) and some of the most diverse (coral reefs).

21.2 Subtidal Distributions

In temperate zones two major zones comprise this shallow subtidal environment: (a) the **infralittoral** zone, the shallowest subtidal zone extending to ~5 m depth and dominated by macroalgae such as kelp, and (b) the **circalittoral** zone just below it, dominated by a community of invertebrate sessile epifauna. The hard-substrate communities dominated by primary producers include photophilic algal communities, whereas those dominated by animals can include biogenic habitat-forming species such as coral and coralligenous reefs. These zones typically feature: (a) high biodiversity (driven by the presence of plants and animals and high substrata heterogeneity), (b) two-dimensionality (the vast majority of organisms grows on a primary or secondary substratum), (c) lack of protection from predators or wave energy, and (d) dominance of clonal organisms. In this environment, more than in any other, organisms themselves create three-dimensional structure and thus the seascape.

In the infralittoral zone, substratum inclination and composition strongly influence seafloor communities; macroalgae dominate relatively flat substrata, whereas animals dominate steep rocky substrata. Bottom slope plays a primary role in determining local community structure, whether dominated by macroalgae or animals. The dominance of algae on gentler slopes likely links to greater light availability. Higher sedimentation rates on gentler slopes also play an important role, as does habitat selection in invertebrate larvae that often favors more shaded environments.

21.2.1 Effect of Physical Variables and Disturbance on Benthic Communities

Physical properties vary greatly within the first meters of subtidal environments, strongly affecting biodiversity and population distributions. Major sources of disturbance include (Figure 21.1):

1) temperature and associated heat stress;
2) mechanical damage;
3) hydrodynamics;

In addition to these disturbances, light also plays a critical role in defining subtidal communities.

Marine Biology: Comparative Ecology of Planet Ocean, First Edition. Roberto Danovaro and Paul Snelgrove.
© 2024 John Wiley & Sons Ltd. Published 2024 by John Wiley & Sons Ltd.
Companion Website: www.wiley.com/go/danovaro/marinebiology

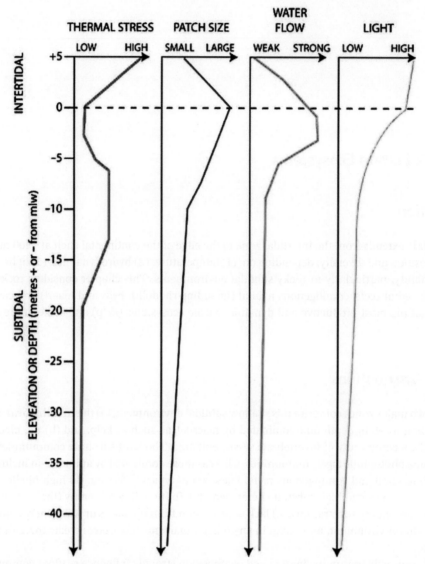

Figure 21.1 Physical variables and disturbance gradients in subtidal environments exposed to wave action. Diagrams show the relative levels of the key variables (mlw = medium low water level).

Temperature: In the shallowest subtidal habitats, strong variation in temperature represents a major stress factor for sessile organisms. Temperature variability in the subtidal zone depends on interaction between warmer surface waters and the degree to which they mix with deeper and cold waters. Depth range and intensity of the thermocline and its fluctuations determine temperature variation and, consequently, potential thermal stress. Thermoclines influence not only benthic populations, but plankton communities and nutrient availability as well.

Physical disturbance: Waves breaking in the shallow subtidal produce the strongest physical disturbance and their intensity and impact typically decrease with depth. Patches created by disturbance can vary greatly in size, from centimeters to 10s of meters. Patch dynamics follow two trajectories: Type I disturbances cause death in some individuals, leaving an empty patch surrounded by "surviving" individuals (usually members of the same community). Recovery from this type of disturbance typically depends on vegetative growth and asexual reproduction, establishment of fast-growing opportunistic species and short-lived larvae from adjacent habitats. Type II disturbance produces larger patches somewhat isolated from surrounding populations, resulting in a highly variable recovery trajectory that depends on species availability over time.

Hydrodynamics: Water movement can strongly influence organism distributions. These effects directly limit benthic organisms through exposure to strong currents, whereas secondary effects of oxygen and nutrient supply can also determine patterns. Water movement (waves, currents) typically decreases with depth, so the shallowest regions experience the greatest turbulence and drag. Bottom topography modulates currents at both small and large scales, altering near bottom flows and their effects on benthic communities. As we will discuss in the context of seamounts in Chapter 26, flow speeds increase over banks, pinnacles, and other topographic features, influencing benthic communities often dominated by filter-feeding invertebrates. Flow structures these communities by increasing flux of suspected food particles and larval supply to the seafloor, favoring benthic filter feeders in particular. Sedimentation rates vary inversely with water movements, and, consequently, sedimentation typically increases with depth in subtidal environments or where biota such as kelp forests baffle and reduce flow. At small spatial scales, differences in sedimentation can alter diversity and distribution of algal communities in subtidal rocky environments.

Light: Total light intensity decreases with depth, and irradiance levels partly define distribution limits of the major groups of algae, including macroalgae. Light promotes algal coverage, and thus associated biota. However, strong light limits distributions of some species.

21.2.2 Biotic Factors

Consumer pressure in subtidal environments often exceeds consumer pressure in the intertidal zone. Thus, biotic variables largely determine the lower distribution limits of photoautotrophic organisms and their associated epiphytes. In contrast, physical factors such as heat and desiccation stress define upper distribution limits. The influence of biological processes also varies with depth (Figure 21.2).

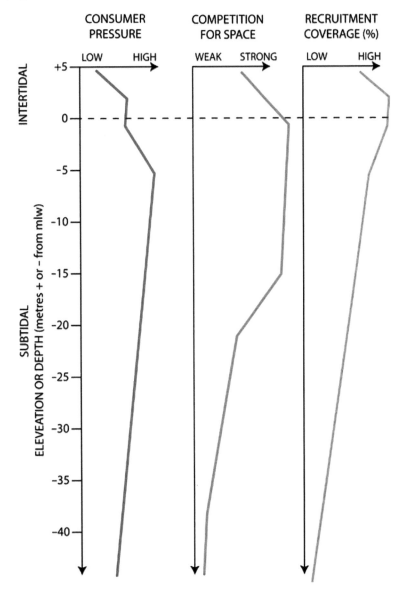

Figure 21.2 Changes in relative levels of biological processes affecting subtidal rocky communities exposed to wave action. Competition refers to that between algae and invertebrates, or between different algal taxa.

Primary consumers role: Consumers typically exert the least pressure in the shallowest depths exposed to strong wave action, which reduces feeding ability and survival of mobile consumers. For example, strong hydrodynamic forces typically exclude sea urchins from the upper few meters of subtidal environments. The ability of predators to feed in these habitats depends on how their mobility relates to when hydrodynamic stress returns. Their high mobility allows birds and fishes, for example, to feed more effectively than mobile benthic fauna in exposed habitats with strong wave action, where wave periodicity varies from seconds to minutes. Highly mobile benthic invertebrates such as crabs cannot feed effectively when wave action is too strong, but with weak flow conditions they can significantly impact communities on which they feed.

Competition: Space can limit subtidal communities, in that space on hard substratum provides a (limiting) resource that can guarantee access to light and food. Several mechanisms influence space competition, including allelopathy and overgrowth. Competition for space, though important, generally plays a lesser role than other biological processes in subtidal environments.

Recruitment: Strong recruitment often occurs in subtidal habitats, potentially leading to high percentage (up to 100%) of seafloor cover. Settling larvae tend to aggregate in the shallowest subtidal depths, because of typically lower predation rates compared to the deepest zones where organisms often form patches.

Predation and trophic cascades: Like recruitment, predation influences community structure on a small scale, but can transcend the spatial scale of a single site to link and similarly influence many sites. Trophic cascades illustrate how predators can produce effects that span large spatial scales. As noted in previous chapters, trophic cascades refer to food webs controlled by strong interactions in which consumers at a given trophic level regulate prey abundance at the trophic level below them, resulting in an alternating sequence (cascade) of predator control down to the basal level of primary producers. Otters and sea urchins in kelp forests discussed in Chapter 18 illustrate one of the best-known examples of trophic cascades.

21.3 Kelp Forests

Brown algae (order *Laminariales*) form the foundation species for kelp ecosystems, one of the most productive and dynamic marine ecosystems on Earth. These ecosystem engineers create complex three-dimensional habitat that supports large numbers of species (echinoderms, mollusks, crustaceans, fishes, and marine mammals).

Kelp forests dominate rocky substrates in cool, nutrient-rich temperate waters at medium and high latitudes. They form underwater kelp forests (genera *Macrocystis, Nereocystis*) and kelp beds (*Laminaria* spp.) at 20–40 m depth that stretch for 5–10 km outward from the coastline, depending on bottom slope (Figure 21.3). Kelp ecosystems grow from the low tide mark to depths where light limits their growth. Intensive studies over the last century, particularly from a trophic perspective, continue to provide new ideas about their importance in supporting biodiversity and ecosystem services. They consist mainly of brown algae, particularly species of the order *Laminariales* (Figure 21.3b).

Their geographical distribution (Figure 21.4) roughly complements that of coral reefs, in that kelp forests generally occur from subpolar latitudes to the 20 °C summer isotherm (the 20 °C isotherm defines the poleward limit of coral reefs); kelp forests of the cold waters of tropical Ecuador offer a notable exception, first observed in 2007. Despite the limited number

Figure 21.3 Examples of leafy kelp: (a) *Macrocystis pyrifera*; (b) a kelp forest of *Laminaria digitate*. *Sources:* Claire Fackler / NOAA / Wikimedia Commons / CC BY 2.0, Stein Fredriksen.

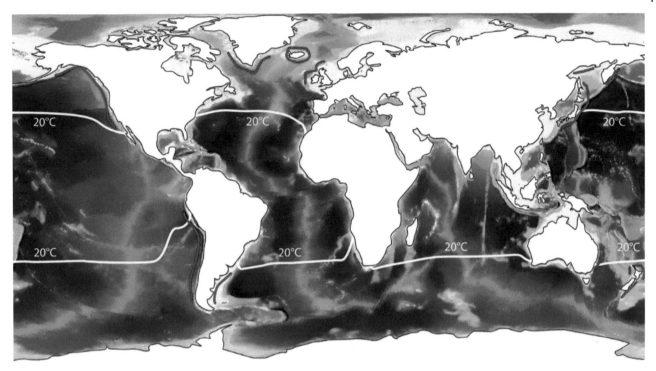

Figure 21.4 Kelp forest distributions globally.

of different kelp species, they vary in structural and functional terms, and possess adaptations that enable growth to large sizes of 10s of m in length, a fair bit taller than a 10-story building! Under ideal conditions, giant kelp (*Macrocystis* spp.) can grow vertically about 30–60 cm d^{-1}. These impressive growth rates have inspired proposals to use kelp mariculture to reduce atmospheric carbon, which they convert to living tissue. Some species, such as *Nereocystis* spp., which generally replaces *Macrocystis* spp. at higher latitudes, are annual, whereas others, such as *Eisenia* spp., can live > 20 y. *Macrocystis* spp. growth rates peak during upwelling months (typically spring and summer), with reduced or no growth during periods with reduced availability of nutrients, decreased photoperiod, and increased storm frequency.

These forests are sensitive to both climate change and human impact, particularly the effects of overfishing that can deplete predators that normally control herbivores. Left unchecked, these herbivores graze kelp excessively (**overgrazing**), gradually transforming lush kelp forests into biological deserts (**kelp barrens**). The major environmental factors that influence kelp survival include available hard substratum (usually bedrock or stable rocks), high concentrations of nutrients (especially nitrogen and phosphorus), and abundant light (typically meaning clear water). For this reason, the most productive kelp forests occur near coastal upwelling. The combination of high productivity and vulnerability of kelp result in strongly ephemeral kelp forests. Storms or rapid increases in herbivores can rapidly eliminate entire kelp beds, which may disappear completely within a year but then quickly reappear.

Based on morphology, researchers recognize three types of kelp forests:

1) **Canopy forests**: formed by kelps that grow largest in size and produce floating canopies. Primarily comprising giant kelp, *Macrocystis* spp., which can grow to lengths of 45 m, these forests dominate the west coasts of North and South America. Smaller species, including *Nereocystis luetkeana*, that belong to this group reach 10 m in length and occur from California to Alaska; its southern hemisphere counterpart, *Ecklonia maxima*, occurs in South Africa, and *Eularia fistulosa* occurs in Alaska and Asia.
2) **Stipitated forests**: Kelp fronds grow above the seafloor aided by rigid stem-like **stipes** that support the fronds. This group includes some species of *Laminaria* sp. in Europe, *Ecklonia* spp. in Australia, and *Lessonia* spp. in Chile, and some species belonging to the genera *Pterygophora*, *Eisenia*, *Pleurophycus*, and *Thalassiophyllum*. Kelps in these habitats reach a maximum length of 10 m, though averaging lengths of ~5 m.
3) **Prostrate kelp or kelp beds**: this group encompasses smaller kelps (e.g., some *Laminaria* spp.) that tend to blanket the seafloor with their fronds. Kelp beds refer to smaller kelp habitats than those that form forests.

The different types of kelp forests often associate with a turf of macrophytes and/or a seafloor of encrusting coralline algae. Each of these structural components provides habitat and food for their associated organisms. Herbivores or other forms of disturbance often remove part of the kelp covering, allowing growth of other algal types. Sometimes, portions of kelp ripped from forests by strong waves become entangled, forming "drift kelp." If waves and currents transport these kelps into environments with favorable conditions, they can settle and create new forests.

Mediterranean kelps include *Saccorhiza polyschides*, *Laminaria ochroleuca* and *Laminaria rodriguezii*. Distributions of these algae may vary from the surface to > 70 m, with maximum leaf lengths of ~3 m. The strong currents, cool waters, and rich nutrients in the Strait of Messina support large kelp forests. These forests are declining because of increased sedimentation from land and associated reduction in water clarity.

21.3.1 Biodiversity Associated with Kelp

In addition to the brown algae that structure kelp ecosystems, kelp forests support many associated species, including sea urchins, mollusks, crabs, and a diverse epibiota, in addition to fishes and marine mammals that add to the productivity and dynamic nature of these coastal habitats. As ecosystem engineers these kelps create a complex three-dimensional habitat that includes three main regions: (1) leaf surfaces, (2) water between algal blades, and (3) substratum beneath the kelp (Figure 21.5).

Figure 21.5 Schematic of a typical kelp forest showing distributions of common species and their microhabitats: (a) Canopy communities, consisting of isopods, tube worms, and bryozoans; (b) plankton community with jellyfish, fish larvae, crustaceans, and diatoms; (c) benthic community associated with kelp holdfasts include sea urchins, crustaceans, brittle stars, and anemones; (d) adjacent seafloor community with sea urchins, sea stars, algae, tunicates, and demersal fishes; (e) erect epifaunal community: coralline algae, anemones, corals, sponges, and bryozoans.

The structure and function of kelp forests differs greatly from their angiosperm and gymnosperm terrestrial counterparts, including generally higher diversity and productivity. Their shorter life expectancy and reduced need for structural support (because water provides significant support) allow a more rapid turnover time. Terrestrial forests grow to heights of 15–30 m over 20–30 y, with longevities that may span from centuries to millennia. Kelp forests, in contrast, grow to 15 m in 2–3 y with longevity of up to 25 y. The 10 major phyla associated with these systems include Chordata, Arthropoda, Annelida, Echinodermata, Bryozoa, Cnidaria, Molluska, Platyhelmintha, Porifera, and Brachiopoda.

The presence of macrophytes (whether macroalgae or seagrasses) play an important protective role for coastlines. Through their fronds, these ecosystems reduce wave energy and thus coastal erosion, trapping suspended sediments and enhancing benthic recruitment.

The canopy formed by kelps reduces light penetration, creating conditions suitable for those photosynthetic species specifically adapted to low light, thus reducing competition by creating light microhabitats. Kelp fronds provide important substrata for many photosynthetic and animal species, and some organisms feed directly on kelp. Kelp forests also provide feeding and nursery habitat for diverse benthic and pelagic organisms. The kelp holdfast and basal structure formed by coralline algae provide a settlement surface for invertebrates such as polychaetes, amphipods, decapod crustaceans, gastropods, and brittle stars. Beyond this habitat, sponges, tunicates, anemones, corals, and bryozoans dominate (Figure 21.6).

21.3.2 Trophic Networks

Kelp forests concentrate biomass and provide a significant source of food and nutrients for coastal marine ecosystems through a food chain based on macroalgal detritus. Herbivores typically consume ~10% of the biomass, leaving remaining production available to other organisms in the form of detritus. Kelp forests have enabled important research on top-down and bottom-up processes. Abiotic conditions, and particularly those required for primary producers, largely define where kelp forests occur. Thus, availability of light and nutrients play critical roles, and lead to consequent transfer of energy to higher trophic levels. Top-down processes also occur, in which predators limit the biomass of species at the lowest trophic levels; in the absence of predation, these species could proliferate because of the non-limiting nature of resources

Figure 21.6 Representatives of kelp forest invertebrate communities: (a) bryozoan, *Membranipora* sp. on *Macrocystis pyrifera*; (b) gorgonian coral *Leptogorgia chilensis*; (c) a crab living in kelp forests (North Pacific); (d) rock greenling, *Hexagrammos lagocephalus*, in kelp (North Pacific); (e) brooding anemone (*Epiactis prolifera*) growing on kelp in the North Pacific; (f) epibiont anemone beneath a kelp forest. *Sources:* Eugene van der / wikimedia Commons / Public domain, Gabriella Luongo, Gabriella Luongo.

supporting their energy requirements. Because kelp forests typically occur in turbulent waters with low sediment loading and abundant nutrients and light, kelp beds host a rich community of herbivores, including gastropods, amphipods, isopods, and limpets. Many of these herbivores feed on algae growing on rocks or leaves, with relatively little impact on mature kelp. Kelp forests also support a wide variety of mobile grazers that feed primarily on algae associated with kelp. However, some species, such as sea urchins (*Strongylocentrotus* spp.), can completely remove kelps by grazing on the holdfast, thereby detaching it from the substrate.

Sea urchins interact with kelp forests in two ways:

1) they hide from predators between cracks in rocks where abundant kelp occur; wave energy breaks off pieces of kelp that fall to the bottom and supply food to the sea urchins;
2) they become more abundant and emerge from cracks during periods of low kelp abundance to feed directly on kelp blades; in this case, sea urchins cluster together in compact patches to defend themselves from predators. This feeding can sometimes completely wipe out all the kelp in that location, producing urchin barrens.

Three theories describe sea urchin proliferation:

1) urchins become excessively abundant when some factor "represses" their predators;
2) environmental factors influence sea urchin population dynamics;
3) environmental conditions favor intensive recruitment of urchins, eventually leading to reduced kelp abundance.

Other species, such as the gastropod *Tegula* spp. graze on the entire kelp thallus, from the substrate to the surface. Several predators, including sea stars, gastropods, and crabs, live in association with kelp forests, but we lack information on the dynamics of these organisms. Fish diversity and abundance decrease when kelp disappear, and strong storms can cause mass mortality of kelp forest fishes. Seals and sea lions commonly occur in kelp forests, although sea lions only sporadically feed in these ecosystems. However, gray whales use kelp forests as a refuge from predation by killer whales and as foraging areas where they feed on invertebrates, and particularly crustaceans. One of the most iconic mammals associated with kelp forests, sea otter, *Enhydra lutris*, feeds on invertebrates associated with kelp forests. They also use kelp forests both as areas of refuge from predation by white sharks and as a nursery area for females and juveniles (Chapter 18 describes the role of sea otters as a keystone species in kelp forests).

21.3.3 Macroalgal Forests in the Mediterranean

The brown algae *Cystoseira* spp. and *Sargassum* spp. play the most important role in Mediterranean macroalgal forests. These foundation species or ecosystem engineers structure ecosystem three-dimensionally. *Sargassum* spp. differs from the other fucoids in the presence of gas vesicles at the tips of the small branches that grow between stem and leaves, forming dense habitat. *Sargassum hornschuchii* occurs in high abundance in association with coralligenous habitat of the western Mediterranean, whereas *Sargassum trichocarpum* form dense populations in the eastern Mediterranean. Some 50 species of *Cystoseira* spp. form a key component of complex structures in lower subtidal hard bottoms of the Mediterranean Sea. The lifespan of some species may exceed 40 y (e.g., *Cystoseira montagnei*).

In coralligenous habitat of the upper subtidal zone, *C. montagnei* commonly occurs, with some varieties extending into the lower subtidal. In bright, clear waters, *Cystoseira corniculata* dominates throughout the upper and lower subtidal of the Aegean, Adriatic, and Ionian Seas. These algae produce anti-herbivory compounds called **terpenes**. Their abundant, interlacing branches also create complex three-dimensional structures that provides habitat for an impressive variety of vertebrates and invertebrates, which gain food and protection from predators. Moreover, their high level of primary production helps to feed adjacent systems. However, some *Cystoseira* spp. forests (Figure 21.7) are declining as a result of over-exploitation of predatory fishes such as the Gilt-head bream, *Sparus aurata*, which leads to increased numbers of sea urchins, *Paracentrotus lividus*, that subsequently graze benthic macrophytes to excess. Removal of *Cystoseira* spp. creates "urchin barrens" (Figure 21.7) covered only by encrusting coralline algae. Recovery of *Cystoseira* spp. forests, once removed, generally occurs quite slowly because of their slow growth, a pattern evident in other macrophyte species such as *Sargassum* spp. Sometimes, simplified habitat replaces these canopy habitats, once removed, including mussel beds, algal mats, or red algae, *Gracilaria* spp. in particular.

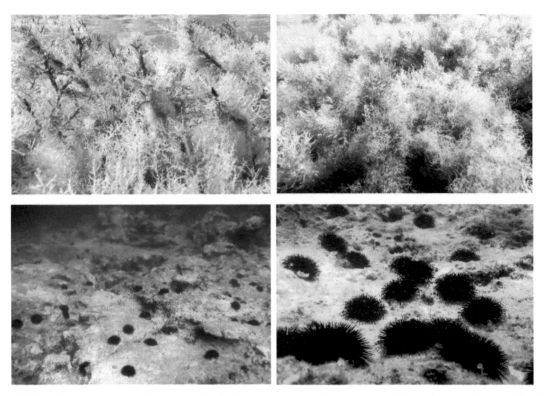

Figure 21.7 *Cystoseira* spp. forests at shallow depths (top), and overgrazing by sea urchins (*Paracentrotus lividus* and others) that leads to the formation of urchin barrens (bottom).

21.4 Coral Reefs

Coral reefs represent the most diverse and structurally complex marine ecosystem, occupying tropical waters with water temperatures >18 °C. Hard coral (Scleractinia, Hexacorallia) bioconstructors contain symbiotic algae called zooxanthellae, and build living habitat that hosts a complex food web characterized by strong competition, speciation, and biodiversity. Different forms of coral reefs include: (1) barrier reefs; (2) fringing reefs; (3) atolls; and (4) platforms. The ~5000 species of corals globally do not account for the bulk of coral reef biodiversity, in that most of the diversity comes from organisms that live within the coral. Indo-Pacific coral reefs support the greatest biodiversity in the ocean, within a region known as the **coral triangle**. Complex food webs within three-dimensional coral reefs include a rich diversity of fishes that play different important roles, including herbivores and predators of other fishes and benthic or planktonic invertebrates. In fact, reef environments support small, medium (mesopredators such as small sharks, groupers, and barracudas) and large predators (apex predators such as large sharks) that provide indicators of reef health. Coral reefs include some of the most productive ecosystems in the ocean (1,000 g C m^{-2} y^{-1}). Strong competition occurs within these systems, stimulating speciation processes that increase biodiversity; however, high vulnerability to human pressures also characterizes these systems.

Coral reefs host 32 of the known 35 marine phyla. Although they cover just 0.2% of the global ocean in area, coral reefs house perhaps more than 25% of all known marine species. Given that temperature primarily limits their distribution (and can never drop below 18–20 °C), most reefs occur between 30°N and 30°S, with elevated abundance near the Equator in three main geographical regions: (1) Indo-Pacific (from Hawaii to Africa); (2) Western Atlantic/Caribbean (from Florida to Brazil); and (3) Red Sea (Figure 21.8). This division, based on biogeography, implies the presence of different communities in each area. Researchers recognize the coral triangle between the Philippines, Indonesia, and Malaysia, as the center of global marine biodiversity, which hosts 75% of all known coral species and >3,000 species of fishes, many species of mammals (including many endemics), and most extant species of marine reptiles. The term "coral reef" sometimes refers to benthic communities dominated by corals, tropical hard substratum communities, or tropical coastal marine ecosystems. From a biological perspective, they represent calcareous structures formed by skeletons of hermatypic corals. From a geological point of view, they represent carbonate platforms consisting primarily of corals. Coral reefs first appeared hundreds of millions of years ago, and modern coral reefs evolved over the past 200–300 million y; many extant corals began to appear ~50 million y ago, although the majority date back 5,000–10,000 y. Some current reefs may be 2.5 million y old.

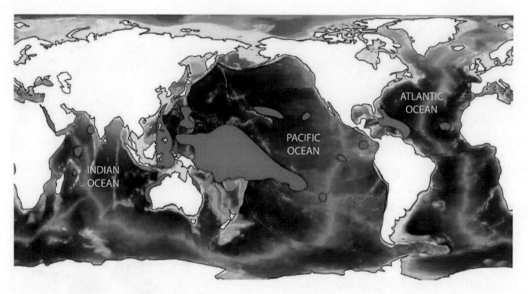

Figure 21.8 Global distribution of tropical coral reefs.

21.4.1 Zonation within Coral Reefs

Numerous species of reef-building corals comprise coral reefs, and the same species can vary greatly in morphology depending on specific location. Clear zonation characterizes coral reefs, though patterns vary depending on upwind or downwind location.

Two different breaker zones occur on atolls: Internal and external zones (Figure 21.9). Within reefs, landward of the reef crest (the **back reef**, for example, in the inside of atolls or breaker zones), temperature and salinity vary greatly, and small reefs (and microatolls) consisting of massive species and remarkably developed algal species can occur. These reefs form patches rather than a continuous coral structure. Lagoons may also develop in the inside of the back reef. These habitats include delicate and fragile species, but they rarely form the continuous unified structures seen elsewhere. If sea level changes, massive corals can form microatolls with warmer and more turbid waters than in the oceanic zone because of limited seawater exchange. The platform between the back reef and the crest experiences the strongest wave action. Hermatypic corals grow under high light conditions with less variable salinity and temperature variation than occurs in back reefs. The corals are small in size, stenohaline, and stenothermal, and occur intermixed with **coral rubble** (coral fragments) formed by wave action.

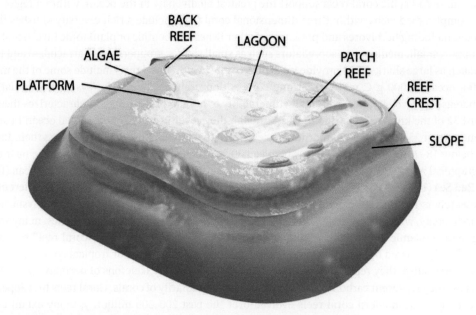

Figure 21.9 Zonation within and adjacent to a coral reef.

The **reef crest** (or edge) refers to the outer portion of reefs, with the strongest wave action. Within this zone, coralline algae dominate and corals often grow to large size. The strong wave action favors robust corals, often dominated by a single species because few species can tolerate the high energy conditions. Often, the crest develops sand channels that may extend into deep water, helping to mitigate wave impacts on the reef. Descending deeper into the seaward outer margin of the reef, the slope encompasses upper (10–15 m) and lower (15–20 m) habitats defined by depth. Greater diversity and larger forms dominate the **upper slope**. Competition for light results in higher bottom coverage, often favoring large forms including table corals. The **lower slope** favors shade tolerant (**sciaphilous**) corals, lower species diversity, and larger corals. Caves may develop along the slope or escarpment, providing habitat for some species. Finally, a region with strong currents at 20–30 m hosts gorgonians and pelagic fish, as well as massive corals, supported by strong currents that deliver suspended food particles.

21.4.2 Types of Coral Reef

Coral reefs occur in four main forms (Figure 21.10): (1) **barrier reefs**, which are compact structures parallel to the coastline, located at a specific distance from land that results in the formation of a large lagoon between the reef and landmass. The best-known example of this type of reef, the Great Barrier Reef, extends >2,000 km along the Australian coast; (2) **fringing reefs**, over which waves break, run parallel to shore, where reef flats or small lagoons separate the reef from the coast. These types of reef characterize the Caribbean Sea and Indian Ocean; (3) **atolls**, or isolated circular structures surrounded by deep water, characteristic of Polynesia and Micronesia; (4) **platforms**, or flattened structures that rise above the seafloor at shallow depths. The Maldives are perhaps the best known example of this reef type.

21.4.3 Theory of Coral Reef Formation

Charles Darwin first formulated a theory on atoll formation when he visited Bora Bora aboard the H.M.S. Beagle. He based his subsidence theory on plates tectonic. Darwin hypothesized that as plates move, volcanic cones form on the seafloor, sometimes forming islands. Coral reefs then form around these volcanoes, which gradually subside (sink) as a result of the increased mass of material (including coral), and thus begins to form an atoll. As the volcano continues to subside, the corals continue to grow on top of the previous reef at the periphery of the volcanic cone, forming a ring of coral structure around the central lagoon (Figure 21.11).

In 1915, Reginald Daly published his **"glacial control hypothesis"** linking atoll formation to changes in sea level, caused by alternation between glacial and interglacial periods. When sea level drops during glacial periods, the calcareous coral structure emerges from the ocean, exposing it to chemical erosion by aerial exposure as well as bioerosion by drillers and grazers. This process links to **karst theory**, which integrates concepts of subsidence and sea level variation. According to this theory, quaternary (the most recent period in the Cenozoic era) reefs grew on foundations of old emerged and

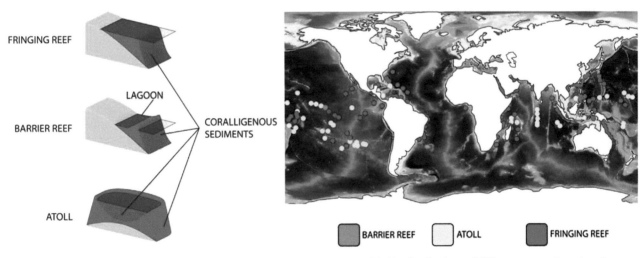

Figure 21.10 Different types of reef: fringe, barrier and atoll, as well as worldwide distributions of different types of coral reefs.

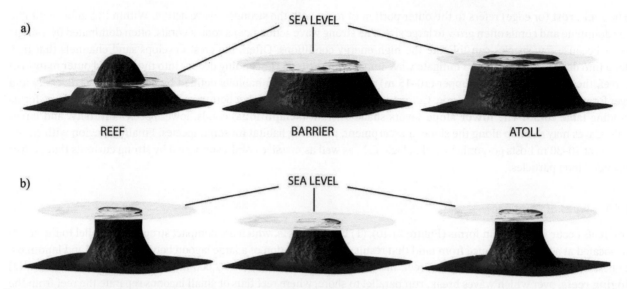

Figure 21.11 (a) Subsidence theory of reef formation, and (b) glacial control hypothesis.

karstified cliffs. Following sea level changes and exposure to air, coral erosion led to the formation of small islands and channels between them. Once sea level rose again, water flooded the islands and corals continued to grow adjacent to them.

21.4.4 Characteristics of Reef Building (Bioconstructor) Corals

Scleractinian (hexacorals) comprise the major reef building corals. Although hexacorals occur globally, in temperate seas they form small, isolated colonies, in contrast to their diversification and reef-building activities in tropical and subtropical environments. Coral reefs take on several different morphologies (Figure 21.12):

1) branching corals: staghorn or elkhorn corals, with a main branch and several secondary branches that sometimes form bush-like structures;
2) columnar corals that develop as a single column without branches;
3) globose or massive corals (such as brain coral) that form globular structures and convolutions that sometimes resemble human brains;
4) encrusting corals that form a thin layer over the bottom;
5) tabular or plate corals that form a flat, fan-shaped structure on a horizontal plane;
6) foliose corals that resemble lettuce leaves;
7) solitary mushroom corals that also resemble dorso-ventrally flattened bars of soap.

Scleractinian corals usually consist of colonies of small polyps within a calcareous matrix (Figure 21.12). Polyps, whose size can vary from 1 to 50 mm, consist of a sac-shaped gastrovascular cavity and an apical stoma (mouth) surrounded by tentacles. **Mesenteric filaments**, which represent folds of the gastrovascular cavity at the polyp base, can be everted as weapons of defense or prey capture. The **ectoderm**, or outer tissue that comes into contact with the external environment and helps to protect the coral, consists of cells that produce mucus, nerve/sensitive cells, muscle cells, and nematocysts, all of which sit atop the endoderm and "skeleton" (i.e., calcareous matrix). The ectoderm also deposits calcium that forms the calcareous skeleton (or **corallite**) around each polyp. The endoderm consists of contractile and nutritive cells. The connective tissue between various polyps forms the **coenosarc** (Figure 21.13), through which nutrient exchange occurs. Gonads also occur inside the gastric cavity. During the day, polyps avoid predators by remaining largely inside the protective cups, opening at night in order to feed on plankton.

21.4.5 Coral Reproduction

Corals use a wide variety of reproductive strategies. In some cases, they reproduce asexually without producing gametes, either through **fission** (intra-tentacular) or **budding** (extra-tentacular). The coral colony grows by increasing polyp number through budding (and therefore increasing colony size). A single polyp buds and divides, giving rise to two or more

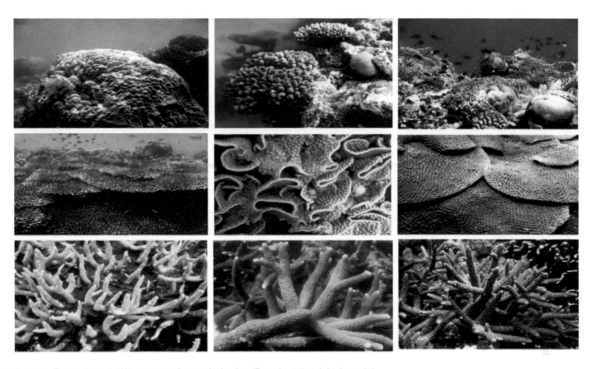

Figure 21.12　Examples of different coral morphologies (Bangka Island, Indonesia).

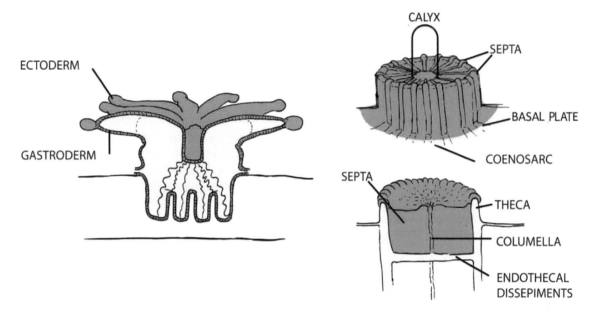

Figure 21.13　Polyp structure. Left: living external tissue, ectoderm, deposits skeleton material.

individuals, or by fragmentation associated with physical impact. A coral may also intentionally fragment through a reproductive process called **autotomy**. Individual polyps can migrate and form new colonies or release individual or groups of polyps when stressed. Asexual reproduction does not enrich genetic diversity in the species but does allow rapid increase in numbers of individuals. In contrast, sexual reproduction through spawning (Figure 21.14) enriches the gene pool and enables large-scale dispersal (up to 1,000 km). Rarely, sexual reproduction can occur internally by brooding, more commonly through release of gametes (individually or in groups) through the mouth (**broadcast spawning**) or through external development of embryos. Most hard corals are hermaphrodites, able to produce male or female gametes but some **gonochoric** (separate sexes) species release either eggs or sperm. The meroplanktonic larvae, called **planulae**, feed on

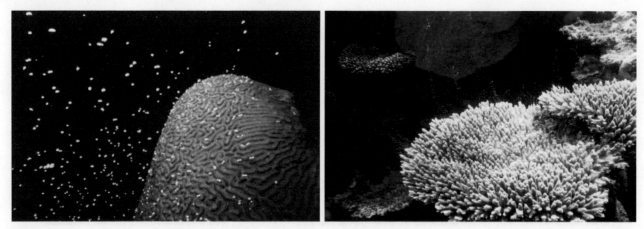

Figure 21.14 Corals during spawning (reproduction), showing gamete release from brain corals (left); and gamete release from stony corals, *Acropora* sp. (right). *Source:* NOAA / Wikimedia Commons / Public domain.

microplankton in the pelagic environment for time periods ranging from one to a few weeks. They settle once they encounter a suitable substratum, and then develop into a colony. Sexual reproduction characterizes the corals that dominates reef systems.

Zooxanthellae occur within the gastrodermis below the ectoderm. The skeletal structure (or **corallite**) deposited between polyps, performs different functions, and include the **theca** around the polyp, the inner exposed surface or **calyx**, the **coenosarc** or skeletal gap between polyps, the **septa** or vertical divisions inside the calyx, and the **columella** that forms the central part of the corallite.

21.4.6 Coral Feeding and Symbiosis with Zooxanthellae

Corals prey primarily at night by producing mucus and **nematocysts** (stinging cells) to capture plankton. They can remove plankton with an estimated removal efficiency (**clearance rate**) of 91% for diatoms and 60% for zooplankton. Corals can also feed on organic particles, and exhibit **mixotrophy**, in which they obtain part of their diet from predation or from organic particles and part from symbiotic algae called **zooxanthellae** that release carbohydrates produced through photosynthesis. Although many corals obtain only a small percentage of the energy required for growth from predation, prey availability nonetheless strongly influences their distribution, growth rate, and morphology. Dissolved organic matter provides another food source for corals, which they absorb through microvilli on their polyps. Symbiosis between polyps and zooxanthellae represents the single most important symbiosis in the ocean, and forms the basis of coral reef formation. Zooxanthellae are dinoflagellates (Chromista) that lost their flagellum over evolutionary time, thus contrasting free-living forms that depend on their flagellae for movement. The term zooxanthella derives from their yellow color (Xanth=gold) and because they live within animals (zoo- = animal). The spherical zooxanthellae vary in dimensions from 5 to 12 μm, and most belong to the genus *Symbiodinium* (Figure 21.15). They occur in densities of ~1,000,000 cm^{-2} (a healthy coral can accommodate some 10^{10} symbionts m^{-2}); sensitivity to different light wavelengths characterizes different species of zooxanthellae.

Zooxanthellae (Figure 21.16) give coral its color, which can range from orange-brown to pink, green, and blue, depending on specific pigments (usually carotenoids) within the zooxanthellae. In hard corals, planulae released from parental polyps carry zooxanthellae from the parent colony. This direct transmission of zooxanthellae, from one generation to another, means that each coral species transmits a single genetic line of zooxanthellae over time. Corals evolved for millions of years, developing hundreds of species from ancestral forms; their symbionts likely evolved in response to habitat variation, and live within polyp tissue. Zooxanthellae release organic nutrients (carbohydrates and amino acids), which polyps then take up. Zooxanthellae provide 50–95% of the nourishment for a polyp, and the photosynthetic algae pass 95% of their products to the host in the form of glycerol, peptides, amino acids, sugars, and complex carbohydrates. Studies conducted directly on coral reefs that covered different coral colonies with opaque and transparent domes demonstrated that only colonies with transparent covers survived, whereas those blocked from light expelled their symbionts and died after three months. Corals with larger polyps depend less on zooxanthellae for survival; smaller polyps cannot capture zooplankton as effectively as larger polyps, and therefore depend more on their zooxanthellae for food. The mutualistic relationship between polyp and zooxanthellae can also involve other microalgal species. Corals benefit by obtaining food and oxygen, and removing carbon dioxide (thus increasing their capacity to deposit calcium carbonate) and excreta. Zooxanthellae gain shelter as well as food derived from waste material from the polyps (ammonia, phosphates, amino acids, urea, and carbon dioxide).

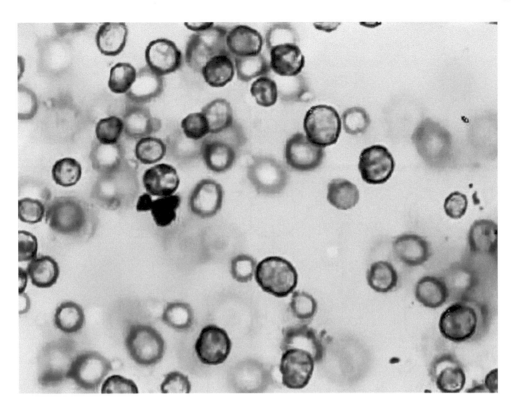

Figure 21.15　The primary tropical coral symbiont, dinoflagellate microalga, *Symbiodinium microadriaticum*.

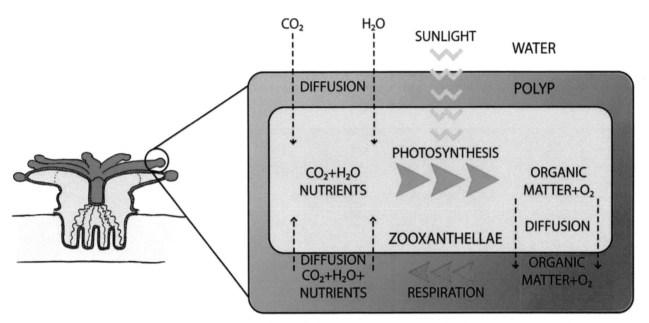

Figure 21.16　Interaction between polyps and zooxanthellae. Carbon dioxide occurs within the tissue both as dissolved gas in water and as a product of respiration. When in contact with water, carbon dioxide forms carbonic acid, a weak acid that zooxanthellae assimilate in the presence of nutrients such as ammonium excreted from polyps. Zooxanthellae then produce organic matter (carbohydrates), which they release to the polyps.

The mechanism by which zooxanthellae use polyp waste materials reflects the efficiency of this relationship. Polyps release ammonia as a metabolic waste product, and nitrogen is the only nutrient zooxanthellae cannot otherwise obtain; polyps therefore provide the only missing "ingredient" required by these symbionts. Zooxanthellae respiration releases carbon dioxide in water, where it would normally combine with water molecules to form carbonic acid (Figure 21.17). However, in performing photosynthesis, zooxanthellae remove excess carbon dioxide, reducing acidity inside the polyps and increasing the coral's ability to deposit carbonate skeleton (Figure 21.18). Indeed, the species that host zooxanthellae

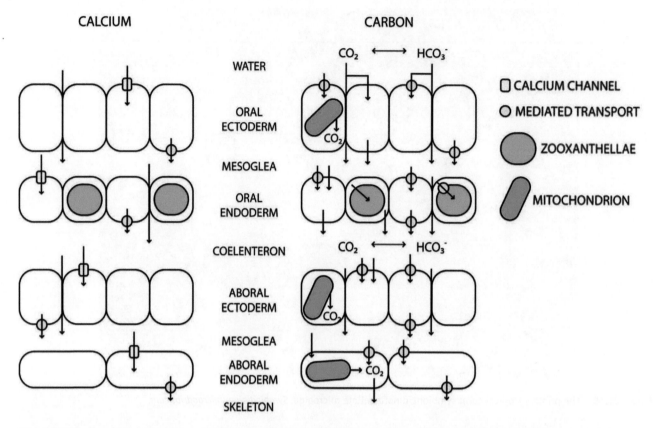

Figure 21.17 Deposition of calcium carbonate (aragonite) in scleractinian skeleton depends strongly on temperature and pH. Aragonite (a form of fibrous crystalline calcium carbonate) forms almost the entirety of coral skeletons. Coral polyps absorb calcium ions from the water and transfer them using an active pumping mechanism at the calcification site. They must keep calcium ions at very low levels to avoid compromising cell functioning. They therefore bind most of the calcium to membranes or organic molecules deposited within the coral skeleton. The process requires basic pH conditions so that removing carbon dioxide for photosynthesis allows the symbiont system to maintain these conditions. The figure illustrates involvement of metabolic pathways of calcium (a) and carbon (b) in photosynthesis. The non-calcifying fluid cytoplasm sits between the aboral ectoderm and the skeleton.

Figure 21.18 Growth over 7 months of hard coral, *Pocillopora damicornis*, in a coral breeding nursery.

are those that produce sufficient calcium carbonate to form reefs. Neither of the two mutualistic species can live without the other in this important relationship.

The basal part of the polyp of hard corals skeletons continuously secretes material in two distinct phases: (1) The basal epithelium of the polyps synthesizes an organic matrix that includes various mucopolysaccharides, proteins, glycoproteins, and phospholipids, which it secretes into the sub-epithelial space; (2) the matrix facilitates calcium carbonate nucleation ($CaCO_3$), providing structure for aragonite crystals that form the coral skeleton. Periodically, the polyp rises from its base and secretes a new basal plate atop the old one, creating a small chamber in the skeleton. Coral growth rate can vary greatly among species. Generally, massive corals grow more slowly than branched corals. Some *Acropora* spp. can grow >10 cm y^{-1}, in contrast to growth rates generally <1 cm y^{-1} in the more massive form, *Porites* spp.

21.4.7 Primary Factors Limiting Coral Growth

The delicate symbiosis between polyps and zooxanthellae also determines which factors limit coral growth. Light determines success of the symbiosis with zooxanthellae, and thus represents the main limiting factor for coral. Appropriate intensity and wavelength of light supports coral survival and growth, and explains why corals prefer transparent waters in which they can occur in depths to 50–60 m (below 40 m, only isolated formations typically occur, comprising corals adapted to lower light intensity). The intensity of light reaching corals also depends on concentrations of suspended material (i.e., phyto-, zooplankton, dissolved organic matter and inorganic matter). In contrast, excessive light causes photoinhibition and damages coral colonies.

In order to thrive and build reefs, corals require water temperatures that never fall below 18 °C. Waters with these characteristics occur in the tropics, roughly between the Tropic of Capricorn and the Tropic of Cancer. Corals live at the upper level of their temperature tolerance (between 18 and 32 °C), resulting in relatively rapid metabolisms, which has become a problem in warming oceans. When water temperatures become too high, coral health becomes compromised and the reef may experience **bleaching** (see Boxes 21.1, 21.2 and 21.3). However, at low temperatures, metabolism decreases along with calcification, and calcium carbonate becomes soluble. Thus, corals are stenotherms, species that can only live within a narrow temperature range.

Even within the same location, species may differ in their tolerances in that some species can cope with a wide range of conditions but typically only for short periods of time. Some species can withstand 4–5 °C temperature variation, in contrast to only 1–2 °C in other species. This constraint explains the absence of corals from west coasts of continents, where upwelling often brings cold water to the surface.

The balance between precipitation and evaporation, and riverine input in coastal environments, primarily determines changes in salinity. Mixing of water masses by tides and currents can also alter salinity. Osmoregulation and other physiological processes depend on salinity, and salinity therefore influences distribution patterns of biota. Corals are typically adapted to living in high salinity environments and, as stenohaline organisms, they can generally only tolerate small variation in salinity (many can survive salinities of 5 but only for short periods). This limitation explains their absence from the mouths of major rivers, other than a very few specially adapted species.

Sedimentation influences water transparency and thus light penetration. Light reduction itself limits coral distributions but particles in suspension, moved around by waves, can abrade coral colonies, and sediment deposition on colonies can damage them. Corals protect themselves through morphological adaptations, such as constantly moving their tentacles and producing mucus to clear away sediments; however, these mechanisms require energy that they might otherwise use for skeleton construction.

Hydrodynamics strongly influence coral reef zonation and reef-building species. In higher energy environments, massive and smaller colonies dominate over large branched colonies. In deeper zones with less wave energy, flattened and encrusting forms that can capture the limited available light dominate. Often, coral larvae settle in proximity to existing coral reefs. Indeed, **coral rubble**, which includes fragments of dead corals and bivalves, often provides a suitable substrate for settling larvae. In eutrophic conditions, abundant nutrients promote phytoplankton growth that reduces water clarity, inhibiting calcification and increasing bioerosion. Growing bottom algae can outcompete corals for space, and attract potentially damaging herbivores. The complexity of the combined effects of these different environmental factors complicates efforts to evaluate which variables play the most important roles.

21.4.8 Coral Reef Biodiversity

Most biodiversity associated with coral reefs comes not from the corals (about 700 species of corals in a reef), but rather from the organisms that live within habitat the corals create. The immensity of this diversity makes any full assessment almost impossible, although we do know that diversity within and between various coral reefs varies greatly. A single study that subsampled 15,000 hectares in the Philippines documented >5,000 species of mollusks alone, many of them new to science. Studies so far suggest that the ~93,000 reef species described to date likely represent some 5% of the total diversity associated with coral reefs, not including microbes. Acknowledging the difficulty in quantifying the total number of species associated with coral reefs globally, significant uncertainty surrounds projections of unknown species. Estimates range from 100,000 to several millions of species, which could represent close to 25% of overall marine biodiversity. The "coral triangle" of the Indo-Pacific Ocean hosts the highest coral reef biodiversity followed by the Caribbean region.

Soft corals (Figure 21.19a), or Alcyonarians, unlike the stony corals, do not produce an organic matrix of calcium carbonate and instead produce spicules that aid in classification and taxonomy. Many contain zooxanthellae that also contribute to their diet, although planktonic organisms provide most of their nutrition.

In addition to cnidarians, fishes occur in high abundances on fringing and barrier reefs and with various intensities of association that range from mutualism (such as clown fish) to predation (butterfly fish). Coral reef fishes contribute significantly to global fish diversity (Figures 21.20 and 21.21), even at the family level, across a gradient of decreasing diversity from west to east (Figure 21.19b). Characteristic families include labrids, scarids, pomacentrids, acanthurids, siganids, zanclids, chaetodonts, and pomacanthids.

Life cycles of most reef fishes include a pelagic larval stage that disperses offshore. Reef fishes span all trophic groups from planktivores to herbivores to predators (on other vertebrates or invertebrates). Most exhibit some form of diurnal

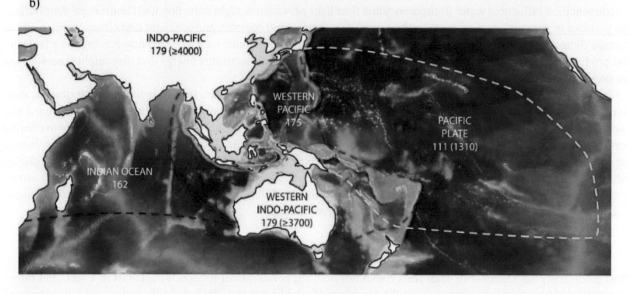

Figure 21.19 (a) Example of soft corals (Philippines); (b) Map illustrating the diversity gradient in coral reef fishes, where numbers indicate numbers of families followed by species numbers in parentheses.

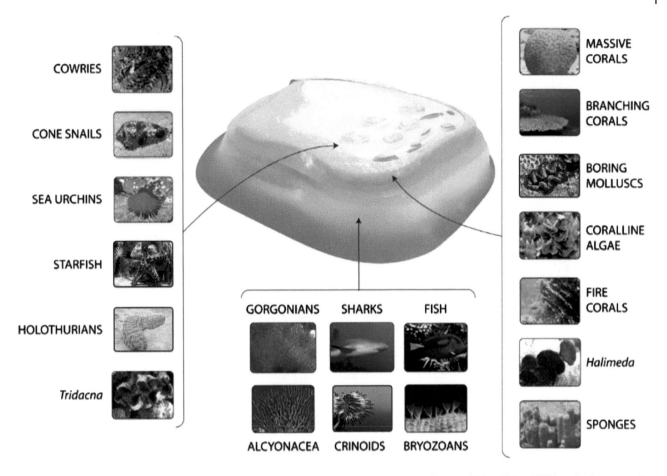

Figure 21.20 Typical groups of reef communities and their distribution on coral reefs. *Sources:* Richard Ling / Wikimedia Commons / CC BY-SA 3.0, Adrian Pingstone / Wikimedia Commons / Public Domain, Frédéric Ducarme / Wikimedia Commons / CC BY-SA 4.0, Tewy / Wikimedia Commons / CC BY 2.5, Eugene van der Pijll / Wikimedia Commons / Public domain, Frédéric Ducarme / Wikimedia Commons / CC BY-SA 3.0, Jan Derk / Wikimedia Common / Public domain, Albert Kok / Wikimedia Commons / Public domain, Nick Hobgood / Wikimedia Commons / CC BY-SA 3.0, Photo2222 / Wikimedia Commons / CC BY-SA 3.0.

or nocturnal behavior: typically, planktivorous fish aggregate within the water column in the daytime (to reduce predation) and toward the direction of the current. Their distribution depends on size and swimming ability. The largest planktivorous fish, in fact, feed primarily on open ocean prey, providing the main trophic link between reef communities and the open ocean. Zooplanktivorous fishes "spill" a large fraction of ingested food from their mouths as they feed, which other fishes can use.

The reef community changes dramatically between day and night. During a "quiet" period at dusk, most fishes hide from view because of predation risk. Planktonic organisms then rise from the bottom at night, although they face nocturnal planktivores with larger eyes and mouths than diurnal feeding species in order to help capture larger prey. Planktivores form schools during the day whereas nocturnal fishes feed individually at night. Many nocturnal planktivores consume organisms inhabiting reefs, which in turn consume smaller open ocean plankton.

During daytime, most small organisms feed on algae, taking advantage of their ability to camouflage against their surroundings. At night, the largest organisms come out under cover of darkness, including fishes such as Lutjanidae and Haemulidae. Numerous invertebrates, including commercial species such as lobster, find cover in algae and coral.

Herbivores feed mainly on algal turf, and some species "clean" surfaces of algae, thereby increasing rates of photosynthesis. Some fishes feeding on invertebrates largely specialize on choice of prey, and their diet may vary seasonally. Piscivores (fish-eaters) occur commonly in all reef systems in high abundance and diversity, and include species with different strategies (Figure 21.21): open-water species that track their prey, cryptic species that often ambush their prey, species that elude their predators through mimicry, species that stalk their prey, and species that attack prey within their dens and caves. Defense strategies against predators involve morphological adaptations (body shape and defensive structures such as thorns, aposematic (warning) or cryptic coloration, production of toxins, or behavioral adaptations (school formation, spawning aggregations, diel (day versus night) differences in activity).

Figure 21.21 Different types of reef fishes with different feeding strategies, showing planktivorous species, micropredators of polyps and corals, herbivorous species, open water species (top two rows), species of micro and macro predators typical of the reef (central two rows), and benthonektonic species (bottom two rows). *Sources:* Jason Marks / Wikimedia Commons / CC BY-SA 3.0.

Box 21.1 Coral Bleaching

"Bleaching" or "coral bleaching" involves the expulsion of symbiotic zooxanthellae from coral tissues (Figure 21.22), and results from stressors such as increases in temperature (the most frequent cause), lowering of temperature, pollution, or freshwater input (riverine or rain). Recent studies demonstrate that bleaching can be reversible. Some bleaching events coincide with phenomena such as **El Niño Southern Oscillation** (ENSO), during which ocean temperatures in some regions increase, but bleaching phenomena have also increased in frequency and intensity in recent years (Figure 21.23). Bleaching can cause drastic declines in coral cover and, in some cases, modifies the entire structure of coral communities and thus the entire ecosystem. These phenomena link to an increase in "coral diseases" and to loss of valuable habitats for fishes and other reef inhabitants.

Figure 21.22 Example of *nubbin* (apical fragments) of *Acropora* and other stony corals during bleaching phenomena.

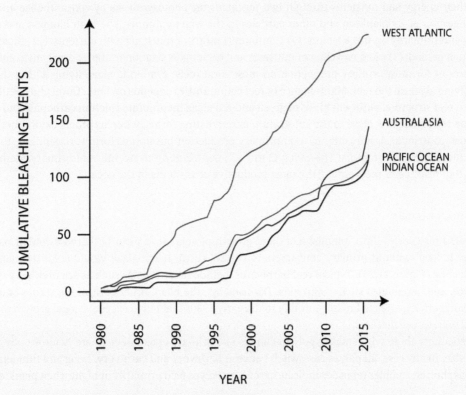

Figure 21.23 Reports of bleaching phenomena of coral reefs globally.

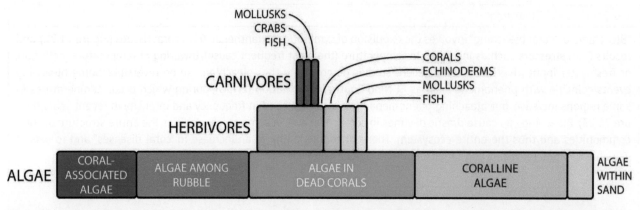

Figure 21.24 Trophic pyramid for a coral reef, illustrating various components that contribute to primary production in tropical reef systems, and their relative importance in terms of biomass and primary production. Note how the pyramid shrinks quickly at higher trophic levels.

21.4.9 Coral Reefs Functioning and Trophic Food Webs

Scleractinian corals and other bioconstructors give reefs their large, three-dimensional structure, and create the habitat for an entire community with a complex food web. Fishes play many important roles as herbivores, or predators of other fishes, benthic invertebrates, or plankton. At the base of the food web, primary producers include coral symbionts (zooxanthellae), macro-algae, phytoplankton, and seagrasses but often only in specific reef habitats (Figure 21.24). On coral reefs, scarce nutrients limit biomass and limit growth of phytoplanktonic primary producers. Encrusting algae represent a substantial fraction of biomass in coral reefs, with biomass values up to 40x greater than that of secondary consumers (Figure 21.24). Coral reefs are highly productive systems generally living within highly oligotrophic waters (noting many exceptions). Corals acquire most of their energy and nutrients through two mechanisms: photosynthesis by zooxanthellae and heterotrophy (through direct ingestion of zooplankton and other particles in the water column). The high biomass and productivity of these systems depend primarily on three factors: (1) Continuous influx of nutrients with currents; (2) efficient recycling of nutrients within polyps, and; (3) rapid recycling of nutrients and bacterial N fixation in the external environment.

Two groups account for most primary production on most coral reefs: symbiotic algae living within the tissue of the polyps, and free-living algae on the reef. Maintenance of reef communities depends on both. Coral and coralline algae form much of the actual reef structure. Different kinds of green and red algae incorporate calcium carbonate and magnesium in their tissues. Some red algae grow close to the substratum as encrusting forms, whereas others exhibit branched growth. Coralline algae also contribute significantly to total primary production but do not form turf. Similar productivity rates in coralline and filamentous algae range 160–500 g C m^{-2} y^{-1}, contributing to overall reef primary production rates of 1,000 g C m^{-2} y^{-1} that place coral reefs among the most productive ecosystems in the ocean.

21.4.10 Primary Consumers

Primary consumers (herbivores) limit establishment and growth of algae that would otherwise inhibit recruitment and survival of corals. Indeed, without primary consumers to control them, macroalgae would colonize many fringe reefs, leading to their decline (Figure 21.25). Typical reef herbivores include invertebrates such as sea urchins, gastropods, polychaetes, sea turtles, and vertebrates such as Sirenids. *Diadema* sp., the black urchin with long spines and an important grazer of Caribbean reefs, experienced mass mortality in the early 1980s that led to excessive algal growth that killed many corals and caused widespread diseases.

Corals, and particularly those with smaller polyps, can also be primary consumers where heterotrophy increases with increasing polyp size. In any case, all polyps can switch between herbivory and carnivory. Long and thin tentacles indicate herbivory, whereas shorter, stubbier tentacles indicate carnivory. Polyps feed primarily at night when plankton migrate into

Figure 21.25 Examples of primary producers on coral reefs: (a) green algae (*Caulerpa* sp.), (b) coralline algae (*Halimeda tuna*), and (c) hard corals containing microalgae; all generate primary production.

shallow, near surface waters where corals dominate. During the day, polyps contract their tentacles to limit access by predators, although some species cannot retract their tentacles. Numerous invertebrates live on reefs, such as filter feeding organisms (e.g., sponges and ascidians), a wide range of herbivorous (e.g. sea urchins) and deposit-feeding (e.g., sea cucumbers) species. Bioeroders, consisting primarily of fishes and invertebrates, play a particularly important role. Several reef fishes feed on algae, including Acanthuridae (surgeonfish) that feed on large algal thalli, Siganidae (rabbit fish) that feed on macroalgae, Scaridae that feed on algal microfilms that grow on coral, and Pomacentridae that feed on small algae. Finally, parrotfish feed on algae by breaking hard corals with their sharp teeth and ingesting their symbiotic zooxanthellae.

21.4.11 Deposit Feeders / Scavengers

Deposit feeders / scavengers fill the fundamental function of ingesting sediment and digesting the sometimes abundant organic matter it contains, thus enabling organic matter recycling and nutrient release. In tropical reef systems, holothurians and some species of polychaetes, mollusks, and crustaceans fill this role.

21.4.12 Secondary Consumers

Secondary consumers include coral predators such as the many species of obligate corallivorous butterfly fish. These fishes feed by sucking polyps from the corallite. Secondary consumers also include corals that prey on plankton during intensive nighttime feeding by polyps. Many planktivores live within the reef, where they receive protection. No fishes feed exclusively on one group of organisms, with the exception of some butterfly fish as noted above. Triggerfish use their small but robust mouth to blow bottom sand and flush invertebrates from the substrate. The mixed diet of angelfish, *Pomacantus* spp. combines small invertebrates, zooplankton, sponges, algae, and corals (Figure 21.26).

On a very different note, crown of thorns sea star, *Acanthaster plancii*, prey voraciously on corals, ingesting polyps after everting their stomach (Figure 21.27), resulting in bleaching. These sea stars increase greatly in abundance with unpredictable periodicity.

21.4.13 Tertiary Consumers

Tertiary consumers of reefs, the main carnivorous predators, typically prey on secondary consumers (Figure 21.28). Many organisms feed on corals, gorgonians, and fishes. Sparids, serranids, jackfish, barracuda (*Sphyraena* sp.), and groupers (*Epinephelus* spp.) ingest prey of different sizes. The presence of these predators, particularly barracudas, groupers, and jackfish, indicates a healthy reef ecosystem. The presence of small (mesopredators) and large (top predator) sharks also indicates high environmental quality and reef health. Sharks rarely hunt during the day. The most common species of coral reefs include gray reef shark (*Carcharhinus amblyrhynchos*), whitetip shark (*Triaenodon obesus*), and blacktip shark (*Carcharhinus blacktip*). Whitetip shark never descend below 20 m, whereas gray shark occur a little deeper.

Figure 21.26 Examples of vertebrate herbivores and micropredators on coral reefs.

21.4.14 Competition for Space in Coral Reefs

Within reefs, species compete strongly for space, which means competition for optimal conditions of light and food supply. Sessile organisms use three main strategies to dominate living space: (1) Rapid growth in free space; (2) aggression toward species occupying adjacent space; (3) release of toxic substances. When two different species of coral come into contact, one or both species extrude mesenteric filaments (digestive organs) that, in a few hours, come into contact with the contiguous species to kill its polyps (in situ digestion of tissue). Corals of the genera *Favia*, *Favites*, *Scolymia*, *Pavona*, and *Cynarina* all possess these characteristics. Other polyps attack the edge of the adjacent coral with powerful nematocysts causing necrosis of neighboring polyp tissue (e.g., *Goniopora* spp.). Others have developed tentacles with which they attack neighboring corals. The most aggressive corals include isolated fungids, whereas less aggressive species include *Porites* spp.

Figure 21.27 Crown of thorns sea stars, *Acanthaster plancii*, preying on corals.

Figure 21.28 Examples of large predators of coral reefs. The top predators (large sharks) also prey on mesopredators (barracuda, lutjanids, and jackfish). *Sources:* ArteSub / Alamy Stock Photo, Roberto Danovaro (Book author).

Box 21.2 Impact of global change on coral reefs

Coral reefs are globally recognized as the marine ecosystem most vulnerable to ocean acidification. Variation in seawater chemical characteristics that result in increased absorption of CO_2 negate the basal functions of calcium carbonate production that characterizes many organisms inhabiting reefs and form the base of coral reef structure. Ocean acidification, therefore, influences both biological and geological components. Skeletal formation in many organisms secreting one of the various forms of calcium carbonate (aragonite and calcite) varies when exposed to high CO_2 conditions. On coral reefs, calcifying corals and macroalgae represent the two most calcified groups. Acidification even affects non-calcifying organisms (Figure 21.29). Acidification causes decreased coral and coralline algal skeletal growth, a process that seems to be reversible.

(Continued)

Box 21.2 (Continued)

Figure 21.29 Models predict increasing ocean acidification, such as the model here showing expected CO_2 levels of 550 ppm by 2050 (compared to 386 ppm in 2010), will decrease the overall saturation state of the aragonite in surface waters. Time-series data confirm that pH reduction in seawater is already well underway and actually began with the industrial age. Lower images: Example of corals threatened by acidification.

Box 21.3 Pollution by Personal Care Products

Sunscreens and Coral Bleaching

Over the past 20 years, massive bleaching events have increased dramatically both in frequency and in intensity. This phenomenon is associated with positive anomalies of temperature, excess UV or alteration of availability of photosynthetic radiation, and the presence of pathogenic bacteria and pollutants. Because the sunscreens we use to protect our skin are lipophilic, their UV filters can bio-accumulate in aquatic animals, producing effects similar to those caused by other xenobiotic compounds. Recent studies demonstrated that various agents present in sunscreens can significantly increase viral production in the ocean by inducing the lytic cycle in prokaryotes with lysogenic infection. A similar phenomenon occurs in corals. With the addition of sunscreen, viral abundance in the water surrounding coral branches increases significantly, reaching values 15x higher than those in nearby control treatments. Because researchers washed corals with virus-free water prior to treatment, coral symbionts likely released the viruses. Laboratory studies that excluded other factors inducing the lytic cycle led to the conclusion that sunscreens cause coral bleaching (Figure 21.30) through the induction of the lytic cycle in symbiotic zooxanthellae with latent viral infections, which commonly occurs.

Box 21.2 (Continued)

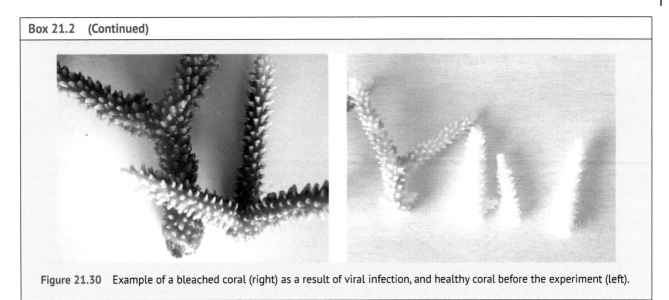

Figure 21.30 Example of a bleached coral (right) as a result of viral infection, and healthy coral before the experiment (left).

21.4.15 Interactions Between Coral Reefs and Adjacent Ecosystems

Despite their muddy substrate, mangrove forests and seagrass beds often occur near coral reefs (Figure 21.31). As discussed in Chapter 20, mangroves provide the function of physical barriers against storms, and these highly productive ecosystems export energy and organic material to adjacent ecosystems: invertebrates and bacteria consume a large percentage (30–80%) of the biomass in the form of detritus (dead leaves, flowers). The most productive mangroves occur in low salinity, humid climates

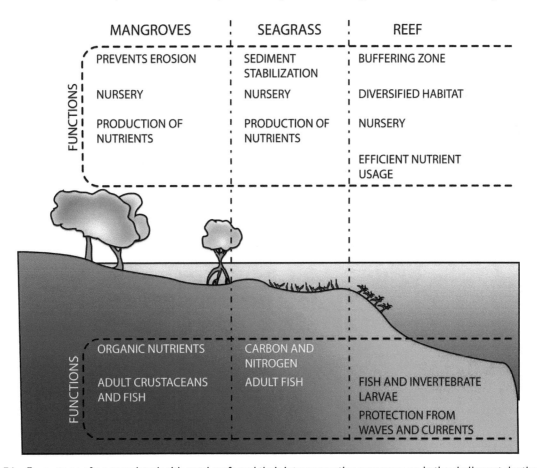

Figure 21.31 Ecosystems often associated with coral reefs and their interconnections; mangroves, in the shallowest depths adjacent to reefs, help to prevent erosion, trap sediments, release nutrients, and provide nursery habitat, as well as seagrass beds, which live slightly deeper than mangroves and also stabilize sediments. Corals occur seaward of these ecosystems in deeper water.

with freshwater inflow. These excellent nursery areas provide ideal habitat for juveniles of many species of fishes and crustaceans. Seagrass beds occur in muddy and often anoxic sediments, rich in organic matter. Roots, rhizomes, and physiological adaptations help seagrasses thrive near coral reefs where they help the reefs by: (1) stabilizing sediments, (2) protecting terrestrial environments from coastal erosion, and (3) filtering terrigenous sediments and keeping them off coral reefs. As with mangroves, these highly productive ecosystems provide an optimal nursery habitat for many species of reef fishes.

21.5 Coralligenous Habitats

Coralligenous concretions (Figure 21.32) represent the only calcareous biogenic formations in Mediterranean shallow benthic environments. Coralligenous habitat differs from rhodolith beds (see below) formed by coralline algae in more superficial littoral or sublitoral habitats, and instead develop at 20–120 m depth. The French biologist A.F. Marion proposed the term "coralligenous" in 1883 because when studying calcareous concretions in the Gulf of Marseilles he found pieces of red coral within samples. For this reason, he incorrectly called this formation "coralligenous" meaning "generator of coral".

During the Barcelona Convention, experts agreed on the following definition: "Coralligenous is a complex of biocenosis, rich in biodiversity and forming a landscape of sciaphilous and perennial plant and animal organisms with a more or less important concretion constituted by calcareous algae." In surface waters, *Mesophyllum alternans* represent the main algal bioengineer. In deeper waters, other coralline algae (*Lithophyllum frondosum*, *Lithophyllum cabiochiae*, *Neogoniolithon mamillosum*) become more important as bioengineers. Green algae, such as *Halimeda tuna* and *Flabellia petiolata*, generally cover superficial banks and may become so dense that they mask calcareous algae. At greater depth, erect algae decrease in density and coralline algae dominate photosynthetic populations.

Two key aspects characterize coralligenous habitat:

1) Calcareous algae provide most of the bioconstruction in environmental conditions marked by low light levels (but sufficient for sciaphilous algal photosynthesis), relatively low and constant temperature, uniform salinity, clean water, and weak hydrodynamics;
2) organogenic construction develops both on rocky substrates (coralligenous habitat in the lower rocky littoral), and on soft substrates comprising coastal debris, including large rhodoliths, gravels, and organogenic sands (coralligenous habitat platform). Colonies of organisms building and living in calcified structures characterize coralligenous habitat.

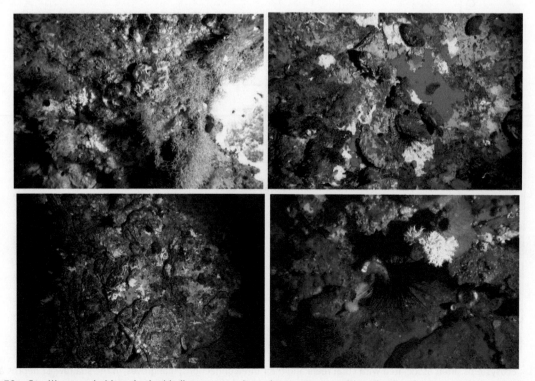

Figure 21.32 Coralligenous habitats in the Mediterranean, where they support coralline algae and numerous invertebrates such as sea stars, gorgonians, sponges, ascidians, and sea urchins.

These bioconstructions occupy the entire substratum, in some cases covering up to 80% of total surface area, creating structures that vary greatly in size and shape. In these aggregates, only the outermost layer consists of living organisms, because new individuals grow on calcareous remains of past inhabitants.

As in tropical coral reefs, three different biologic-structural components comprise coralligenous concretions: (a) Primary bioengineers that grow *in situ*, creating the characteristic rigid carbonate structures (e.g., *Corallinaceae* and *Peyssoneliaceae* coralline algae); (b) secondary bioengineers, which consolidate the colony by cementing it together; this group includes polychaetes (e.g., *Spirorbis* spp.), bryozoans (e.g., *Myriapora truncata*) and gorgonians (e.g. *Paramunicea clavata*); (c) agglomerating organisms that use sediment to fill gaps, this group includes sponges (e.g., *Geodia* spp.), bryozoans (e.g., *Beania* spp.), and soft corals (e.g., *Epizoanthus arenaceus*). The small size of coralligenous structures compared to those on tropical coral reefs means that, despite their important role in coastal system functioning, they lack resistance to wave action and cannot compete with adjacent ecosystems. However, recent studies suggest that coral play an important role in the biodiversity and functioning of the mesophotic "twilight" zone, supporting large numbers of species dominated by filter feeders: sponges, hydroids, anthozoans, bryozoans, serpulids, mollusks, and tunicates (Figure 21.33). Interstices between these species host an impressive variety of mobile polychaetes and crustaceans.

Environmental factors: Light probably represents perhaps the single most important environmental determinant of benthic organism distribution on rocky bottoms of the continental shelf, and it also plays an important role in coralligenous habitat development and growth. In fact, macroalgal bioconstructors require specific conditions that include modest but not strong light (i.e., from about 1 to 100 MJ m^{-2}y^{-1}), relatively calm waters, temperatures 10–23 °C, and salinity of ~37–38. Coralligenous communities apparently do not need high concentrations of inorganic nutrients. Diverse coralligenous benthic community includes at least four different layers: (1) an upper layer consisting of large gorgonians, sponges (including the yellow sponge *Axinella damicornis*), colonial bryozoans, sea lilies (*Antedon mediterranea*) and *Parazoanthus axinellae*; (2) an intermediate layer containing calcareous algae covered by epibionts, sponges, smaller cnidarians, tunicates (*Halocynthia papillosa* and *Microcosmus vulgaris*), polychaetes (*Filograna implexa*), and bryozoans (such as *Pentapora fascialis*); (3) a baseline layer, containing many species quite small in size; (4) a sub-baseline layer, characterized by a rich interstitial fauna, with decapod crustaceans such as *Alpheus* spp. and *Athanas* spp., and species capable of boring into hard organogenic substratum.

As with many other biogenic structures, bioeroders play a key role in coralligenous habitat. In this case sponges are the main bioeroders, in particular the family Clionaidae (e.g. *Cliona viridis*), followed by the bivalve, *Lithophaga lithophaga*, and several species of annelids. Bioeroders include three main categories: (a) **browsers** (grazers), including the sea urchins, *Sphaerechinus granularis* and *Echinus melo*; (b) **microborers** or micro-drillers, including cyanobacteria, green algae, and fungi; (c) **macroborers** or macro-drillers, including mollusks (*Lithophaga lithophaga*, *Rocellaria dubia*, *Petricola lithophaga*, *Hiatella arctica*), sipunculids (*Aspidosiphon mulleri*, *Phascolosoma granulatum*), polychaetes (*Dipolydora* spp., *Dodecaceria concharum*), and various excavator sponges (Figure 21.34).

Coralligenous algal biodiversity in the Mediterranean encompasses some 316 known species, including ecological engineers and bioeroders. The ability of algal species to build these habitats first requires mineralization of calcium carbonate crystals (in the form of Mg-calcite, as in Lithophyllaceae or aragonite as in Peyssonneliales and Udoteacee) on cell walls or within cells, a process called **calcification**.

This mineralization mechanism originated in calcareous algae from cytoplasmic vesicular structures (endoplasmic reticulum and Golgi apparatus). Furthermore, photosynthetic activity does not result in mineralization, noting that

Figure 21.33 Gorgonians produce complex structures that increase topographic complexity of the system and provide habitat for many species, such as crinoids that hold on to gorgonians to feed on organisms transported by currents (a). Many species of sharks use gorgonians as a point of attack and for spawning (b). In some cases, gorgonians can become so numerous that they form "animals" forests.

Figure 21.34 Examples of coralligenous algal reefs illustrating substrate spatial heterogeneity, high biodiversity, and predators (such as sea stars).

mineralization also occurs in parasitic and heterotrophic species. Calcification helps to balance the carbon dioxide deficit that accompanies photosynthesis, and thus maintains seawater alkalinity. Limestone production from bioconstructors varies 170–465 g of $CaCO_3$ $m^2 \cdot y^{-1}$.

As a diversity hotspot, coralligenous habitat supports some 400 known species. A large part of the fauna uses secondary biogenic hard substrate rich in micro-cavities that add structural complexity and supports specializations and symbioses. Coralligenous habitats, with a few exceptions, generally develop below 20 m and, therefore, include mostly sciaphilous species. Numerous fish species live within this mesophotic habitat, including red damselfish (*Anthias anthias*) and other planktivorous fishes, such as picarels (*Spicara smaris, S. maena*), bogues (*Boops boops*), and saddled sea bream (*Oblada melanura*). Predators include amberjack, *Seriola dumerili*, barracuda (*Sphyraena viridensis*) and sea bream (*Dentex dentex*). Characteristic wrasse species include whistle thrush (*Labrus mixtus*), canine thrush (*Lappanella fasciata*) and bottom thrush (*Acantholabrus palloni*). Many species of bream, scorpion fish, and Sciaenidae, associate with coralligenous habitat. Generally nocturnal predators include damsel fish, *Coris julis*, brown grouper, *Epinephelus marginatus*, golden grouper, *Epinephelus costae*, and rare red grouper, *Mycteroperca rubra*, in addition to conger (*Conger conger*) and moray eels (*Muraena helena*). In total, researchers have documented ~50 fish, species associated with coralligenous habitat.

Most organisms cannot graze calcium carbonate produced by photosynthetic bioconstructors, resulting in interesting trophic relationships. Other organisms use chemical defense mechanisms that make them unpalatable or even toxic. The near totality of large sessile invertebrates in coralligenous communities do not feed directly on species associated with coralligenous habitat, and instead depend on pelagic food resources. The sea urchin, *Sphaerechinus granularis*, offers a notable exception as the largest consumer of encrusting coralline algae, and different invertebrates (opisthobranchs, amphipods, and copepods) feed on the green alga, *Halimeda* spp. Examples of carnivores include many fishes, as well as prosobranch mollusks, echinoderms, mobile polychaetes, and crustaceans. Box 21.4 summarizes some examples of underwater sampling methods for these habitats.

Box 21.4 Underwater Sampling Methods for Coralligenous and Other Shallow Hard Substrate Habitat
Producing highly detailed maps (generally at scales of 1:2.000 or greater) requires integration and cross-calibration of different methods. Tools such as aerial photography or multibeam acoustics can provide samples at relatively large spatial scales of kilometers to 10s or even 100s of kilometers, but generally require ground truthing using finer-scale sampling with scuba diving, remotely operated vehicles, small submersibles, or other such approaches (Figure 21.35). Depending on the study goal, researchers may space transects evenly or space them randomly. Direct observation provides the advantage that divers can examine each biocenosis as needed and detect even the smallest formations. Finally, researchers can use a Remote Operated Vehicle (ROV), which can operate at depths much greater than divers. These systems require an underwater vehicle equipped with camera, lighting, preferably depth and bottom sensors, and ideally a temperature sensor, compass, and even sonar. An operator on a surface vessel controls the underwater unit and views the camera on a monitor while recording imagery and maneuvering. Researchers sometimes use airplanes or satellites to study the upper limits of seagrass beds, but this approach can also facilitate studies of shallow biogenic concretions.

Box 21.4 (Continued)

Figure 21.35 Scuba and snorkel underwater visual census activities, spanning from coral reefs to Mediterranean biodiversity.

Simple echo sounder surveys
Echo sounders, including the relatively inexpensive depth sounders on many boats, bounce high or low frequency sound waves off the seafloor to provide largely one-dimensional representations of seabed topography but little information on substrate or biota.

Advanced acoustic methods
Sidescan sonar, like echosounders, uses sound rather than light but can produce a reasonably accurate topographic seafloor map. By emitting sound pulses towards the seafloor at an oblique angle, sound reflects back in different ways depending on the nature of the substrate. Multibeam takes this approach a step further by emitting sound at multiple wavelengths, therefore producing considerably more information about seafloor composition.

Conventional aerial photography
On clear days with favorable sea conditions, aerial photographs provide good and rapid mapping of shallow benthic biocenosis. Under these conditions, researchers can piece together aerial photographs to produce precise maps that support effective monitoring of temporal changes in shallow biocenosis, even if photo-interpretation limits fine-scale resolution of some features.

Airborne Remote Sensing, ARS
In shallow, clear waters remote sensors and/or analog or digital cameras carried on airplanes can produce color images with spatial resolution of a few meters.

Geographic Information System, GIS
The expansion of human pressures on natural systems requires advanced spatial tools to aid conservation efforts. Biological and mathematical disciplines alone cannot assemble and compare spatial changes over time. The development of the Geographic Information System (GIS) has revolutionized conservation efforts because it allows researchers to integrate and map, within a computational framework, the types of complex data described above. GIS integrates theoretical approaches from geography and ecology into a powerful spatial "database" with statistical functions.

21.6 Rhodolith Beds (Maërl)

Unlike coralligenous habitat, rhodolith beds (also known as maërl) owe their existence to a single species or sometimes several species of coralline red algae (Figure 21.36). Rhodoliths can occur as isolated benthic nodules but with appropriate conditions they aggregate to form complex beds of many individuals between the lower intertidal and the lower limits of the photic zone. Individual rhodoliths can be monospecific or they may include multiple species, with a thallus varying from twig-like forms to spherical nodules of various shapes. The beds vary in size from 100s of m^2 to 1000s of km^2 and, because rhodoliths do not attach to the substratum, the beds occur where currents are not sufficiently strong to move rhodoliths around, but bottom flows can resuspend sediments, aided by bioturbators and weaker flow environments. Living rhodoliths typically occur atop a sedimentary layer that includes a hash of live and dead rhodolith skeletons. Their worldwide distribution spans from the poles to the tropics, with notable beds in the Gulf of California, the northeast, northwest, and southwest Atlantic, the Caribbean Sea, the Mediterranean Sea, the South and northwest Pacific, Australia, and even along Greenland and the Arctic coastline. Although little studied, these habitats rival seagrasses, kelp, and coralligenous habitat as a major macrophyte dominated habitat.

Rhodoliths grow slowly by depositing layers of calcified cells around a central core, rarely exceeding growth rates of a few $mm \cdot y^{-1}$ and even less in cold water. Temperature and light limit their distributions, noting that they are adapted to low light conditions where they can outcompete other benthic primary producers.

The variable shapes and sizes of rhodoliths create a complex three-dimensional habitat that supports a diverse community of epiphytes, epibenthos, cryptofauna, and infauna. Rhodoliths also provide critical habitat for micro and macroalgae to complete their life cycles and provide a nursery for several ecologically and economically important species. In addition to their ecological role of habitat provisioning for other species, rhodoliths produce about one third of continental shelf carbonate and therefore play a significant role in global carbon uptake and storage. Their longevity (>5500 years) means that they store carbon for a long time, and they provide an additional benefit in the information they store in their skeletons, which can be used to reconstruct oceanographic conditions over their lifetime. Some particularly long-lived rhodoliths provide information dating back more than 1200 y.

Rhodolith beds face multiple threats from humans. Climate change creates a particularly serious concern (Box 21.2) because of the highly soluble calcite formed by rhodoliths and decreases in net calcification rates with increasing ocean acidity and temperature. Effects on carbon storage and impacts on the structure and function of rhodolith beds remain largely unknown. Presumably, the combined and potentially synergistic effects of ocean acidification and global warming over the long term will alter calcification processes and significantly impact the distribution, diversity, and abundance of rhodoliths and their associated biota. Commercial harvesting of rhodoliths in the northeast Atlantic supports a variety of agricultural and horticultural products, and dredging in some areas smothers rhodolith beds and their inhabitants. Bottom trawling damages rhodoliths and homogenizes the complex habitat with strong impacts on biota. Finally, excess nutrients and sedimentation associated with sewage discharge or aquaculture facilities decreases diversity and function of rhodolith beds, favoring growth of competing macrophytes.

Figure 21.36 Rhodolith bed (maërl) habitat (left) with detail of a rodolith (right). *Sources*: ArranCOAST / Wikimedia Commons / CC BY-SA 3.0, ArranCOAST / Wikimedia Commons / CC BY-SA 3.0.

21.7 Underwater Caves

The term **marine cave** generally refers to a cavity with burrows, branches, and large chambers, totally or partially flooded (Figure 21.37) by the ocean, and sufficiently large and deep to enable human access. A progressive reduction in number of phyla, species, and biomass characterizes caves, proceeding from the entrance to the innermost portion. Horizontal gradients in light, oxygen concentration, hydrodynamics, and input of organic matter along the longitudinal axis characterize these habitats and largely drive distributions of organisms that can survive there (Figure 21.38). Within caves, marked gradients of environmental variables such as light and wave energy occur over distances of just a few m, which span distances of 10s–100s of m outside caves.

Several aspects of environmental conditions in marine caves superficially resemble deep-sea environments, including:

1) absence of light (at least inside caves);
2) limited food resources;
3) reduced hydrodynamics;
4) stable temperature over time.

The absence of light and limited food in caves, in tandem with abundant sulfur in some cases, leads to chemolithoautotrophic production resembling hydrothermal vents. Several other aspects of marine caves resemble deep-sea ecosystems,

Figure 21.37 Example of large chamber in a partially flooded cave in the Mediterranean Sea.

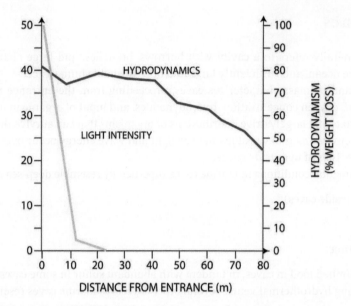

Figure 21.38 Decreased light intensity and hydrodynamics with increasing distance from cave entrance.

including formation of microbial mats and polymetallic nodules, which give peculiar blackish hues to deep rocks and walls of the innermost underwater caves.

In caves, local hydrodynamic regimes affect temperatures and stratification, in that water circulation intensity determines temperature inside caves, which depends primarily on cave topography. For example, the innermost elevated parts of a cave may remain cool relative to the entrance, which warms more quickly in the summer. Rapidity of light decrease along the longitudinal axis depends on the cave's topography, depth, and geographic location. Light affects algal distributions but does not affect sciaphilous species. Circulation inside the cave determines **turnover** or replacement time of the water, which affects organic matter sedimentation and transport, oxygen concentration, and larval supply and dispersal. Water circulation inside caves depends on topography, bathymetry, and local geography.

Generally, salinity does not vary significantly between the inside and outside of a cave (with the exception of karstic caves that receive freshwater injections), whereas the interaction between water circulation and respiration determines oxygen concentrations inside caves. In some cases, extremely low oxygen values characterize the innermost cave habitats, with oxygen concentrations similar to those in deep ocean oxygen minimum zones (1,000–2,000 m depth).

Biotic factors also influence distributions of organisms that colonize marine caves, particularly competition and larval supply.

In 1964, Jean-Marie Pérès and Jacques Picard differentiated two types of biocenosis. Sponges and anthozoans dominate biocenosis of semi-dark caves whereas serpulids, stony corals, sponges, and mobile crustaceans dominate caves in total darkness. Riedl, in his book *Biologie der Meereshöhlen* (in 1966), describes six different biotic zones (Figure 21.39):

1) shady phytal zone;
2) cave entrance zone;
3) cave anterior zone;
4) cave central zone (comparatively rich in biodiversity and abundances);
5) posterior zone (totally dark);
6) "empty quarter", characterized by the near total absence of fauna; this zone occupies a larger area in big caves, and may be absent in small caves

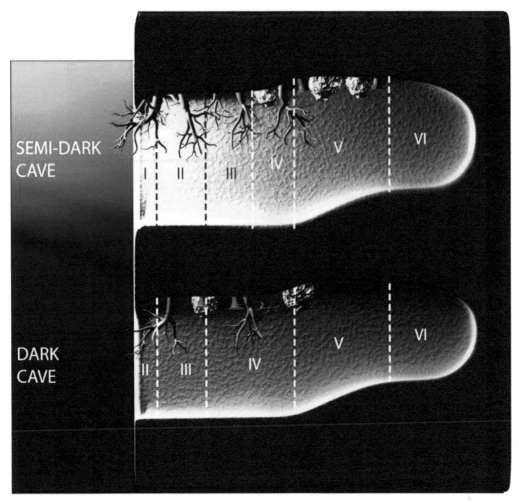

Figure 21.39 Zonation in a shallow marine cave, illustrating hard surface population coverage and biodiversity starting from the entrance of a semi-dark cave (top) and a dark cave (bottom): (I) shady phytal zone; (II) entrance zone; (III) anterior zone; (IV) central zone; (V) posterior zone; (VI) "empty quarter." This classification of marine caves uses hydroids as a biological model.

21.7.1 Cave Biodiversity

The study of biodiversity in marine caves typically requires scuba diving (see Box 21.5). Light conditions result in the effective absence of photophilous algae, favoring sciaphilous species (especially red algae) including *Peyssonnelia* spp., coralline algae, and some green algae (*Halimeda tuna* and *Palmophyllum crassum*). Gradients in plankton populations resemble those in the benthos. Mysids typically move outside caves to feed at night and refuge in caves during the day, resulting in negligible grazing inside caves. Some fishes, such as *Grammonus ater*, live in the deepest sections of caves, feeding only on benthic organisms (Figure 21.40). A wide variety of taxa comprise the benthos, spanning from protistans to representatives of many invertebrate phyla, including mysids such as *Hemimysis speluncola*, which sometimes occurs in high abundance, forming large swarms that represent an important component of cave food webs. Crustaceans and fish migrate between the inside and outside of caves, utilizing the same mechanisms and playing a similar ecological role to vertically migrating plankton, channeling significant amounts of organic matter that provide significant food resources for benthic organisms (detritivores and filter feeders), which depend on these external sources.

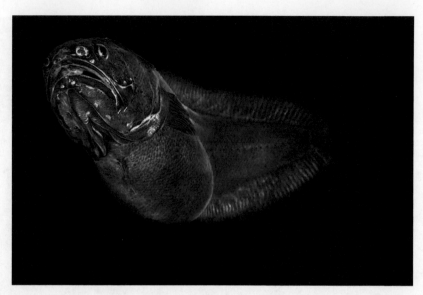

Figure 21.40 Example of fish (*Grammonus ater*) living in dark caves.

21.7.2 Adaptations in Marine Invertebrates to Life in Caves

Reduction and total disappearance of solar light limits organisms hosting autotrophic symbionts or organisms that need light to synthesize specific molecules (such as melanin). Weakened flows also reduce potential food supply for suspension feeders by reducing seawater exchange. These effects influence mobile organisms, which often migrate to and from caves. Some sponges, such as the demospongia *Petrosia ficiformis*, host large populations of symbionts that provide organic matter. The sponge *Lycopodina hypogea* captures and feeds on mysids inhabiting caves, and lacks many of the specialized cells characteristic of the phylum. The reduced hydrodynamics can lead to the following adaptations in sponges:

1) Reduced size of individuals;
2) increased diameter of exhalent channels to reduce friction;
3) cylindrical shape with a single exhalent apical chimney-like opening, through which to remove wastewater.

21.7.3 Food Webs and Functioning in Caves

The rapid decrease in light inside caves leads to a rapid decrease in photosynthetic organisms, which completely disappear a few meters from the entrance. This change greatly alters food webs (Figure 21.41). Inside the caves, reduced organic matter supply creates oligotrophic conditions similar to those in the deep sea; indeed, respiration rates within caves closely resemble those at 1,000–2,000 m depth. Even energy balance and fate of organic matter reaching the seabed parallels the processes and transformations that occur at bathyal/abyssal depths, with dark caves resembling abyssal systems far from continents, and semi-dark caves resembling bathyal systems typical of continental margins. However, chemoautotrophic production dominates caves with sulfur emissions through sulfur bacterial chemoautotrophy that supplies organic matter inside caves and supports dense populations of filter feeders. Mobile organisms (especially crustaceans and fishes) also import organic matter into caves. High abundance of mysids often occurs in oligotrophic caves and therefore represent an important food source for filter feeders and sessile predators. Within caves, organisms have adapted to the unusual trophic structure, which often lacks primary producers, the first trophic level, as well as somewhat unpredictable food supply. These characteristics typically result in low feeding specificity, a capacity to fast for long periods, and the capacity to compete well for available space (for example, preferable filtration locations), all important characteristics of cave populations.

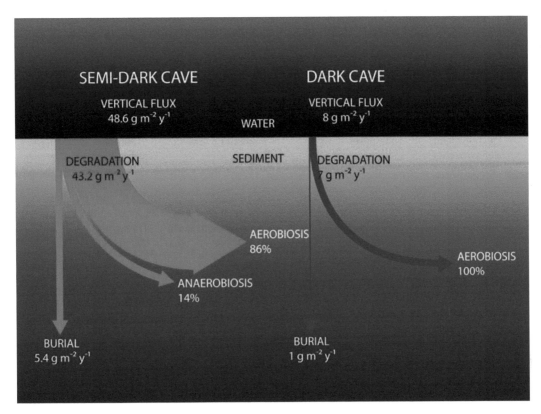

Figure 21.41 Fate of organic carbon at the water-sediment interface in semi-dark and dark caves.

Box 21.5 Methods of Study within Underwater Caves

Studies on the biology and ecology of caves generally require scuba diving. Most studies to date have focused on sessile populations living on hard substrata, followed by studies on plankton and ichthyofauna.

Benthos – The most widely used research method removes a portion of hard substratum, such as 20 × 20 cm squares. Samples are then stored in formalin or alcohol and seawater and identified in the laboratory. Researchers often use manual push cores to sample sediments on the cave floor. Photo surveys and/or visual censuses offer a non-destructive method for studying hard bottom benthos in marine caves. Laboratory analysis of photographic samples typically valuates percent cover of "conspicuous taxa" (and easily identified taxa). In situ benthic visual surveys usually require experienced operators visually evaluating the taxa present and their percent cover within a quadrat or along a transect line. Researchers evaluate supply of larvae and propagules of sessile benthic species using passive traps or artificial substrata that larvae settle into.

Mobile fauna

Researchers typically collect plankton with nets manipulated underwater. Sampling of mobile epifauna often uses destructive sampling, involving removal of the substrate. Some studies on the ecology and/or systematic of fish species, sample using small nets, often spraying anesthetics inside small caves and cavities. Researchers have developed visual census techniques to study fish populations associated with different submerged habitats, such as rocky bottoms, mixed substrates, and vegetated substrates.

Table 21.1 Comparison between subtidal hard substrate ecosystems.

Main characteristics	Habitat / Ecosystem				
	Kelp forests/ Macroalgae	Coral reefs	Coralligenous habitat	Rhodolith beds/Maërl	Submarine caves
Extent	Not still quantified but comparable to that of the mangroves and of the seagrass beds	Largest and most widespread biogenic structure in the ocean	Limited to the mesophotic zone	Shallow subtidal to limits of photic zone	Very limited, depends on geological characteristics
Distribution	All the coastal environments of elevated latitudes in cold environments (upwelling)	All oceans between 20°N and 20°S	In the Mediterranean and in many warm-temperate systems	Tropics to poles in clear waters	In all areas with hard bottoms
Depth	From 0 to 70 m of depth at elevated latitudes but up to 200 m at low latitudes	From 0 to 60 m deep	Typically between 30–200 m depth	~10 m to >250 m	From 0 to hundreds of meters deep; from few meters to tens of meters horizontally
Type of substrate	Typically on hard bottoms	Hard bottom built by corals	Generally hard bottoms	Mixture of sediment and rhodolith hash	Hard bottoms, generally carbonaceous
Biodiversity	Very high, both invertebrates and vertebrates organisms	One of the most species-rich ecosystems on Earth, may host >100,000 species	High biodiversity, >500 associated species	High	High biodiversity in external zones, low in internal parts
Endemism	Numerous endemic species	Large numbers of endemic species, potentially the most on Earth	Numerous endemic species		Rare endemism
Functioning	High production, high export of matter	High secondary production linked to symbionts and tight nutrient cycling.	Primary production limited to coralline algae, high secondary production through predators and filter feeders	High carbon storage, habitat provisioning	Photosynthetic primary production (almost) absent. Secondary production very low
Trophic networks	Trophic network based on grazing with many apical predators	Very complex, based on zooxanthellae –coral symbiosis; strong presence of herbivores (corallivores) and predators)	Very complex, linked to many sessile forms	Slow growing autotrophs, system largely depends on external food inputs	Simplified systems usually depend on organic matter produced outside of the system and on migrating plankton
Unique attributes	System depends on one or a few species functioning as ecosystem engineers	System with very high complexity, highly diversified and relatively isolated with biological interactions with no equivalent in marine ecosystems	Hard bottoms almost exclusively organogenic, interacting species	Slow growth and longevity make rhodoliths an excellent recorder of past climate	Often extreme conditions make these environments superficially similar to bathyal deep environments
State of knowledge	Historically well studied, one of the best known marine habitats	Well studied but major part of coral reef diversity remains undocumented.	Systems well studied, but limited to few geographic areas	Very limited relative to known distribution	Well studied but with mainly visual census methods, they need studies with new approaches

21.8 Summary

The subtidal hard substrate habitats described in this chapter span a wide range of diversity, functions, and characteristics (Table 21.1). Their shallow distribution often results in significant interactions with humans, which results in numerous pressures. Nonetheless, the important roles they play point to a compelling need for special consideration in conservation planning.

Questions

1) What role does space play in rocky substrates?
2) What adaptive strategies do organisms living on rocky substrates use?
3) What main physical variables influence the distribution of benthic communities in subtidal ecosystems?
4) Which algae comprise kelp forests?
5) Describe trophic networks in kelp forests.
6) Describe macroalgal forests and barrens in the Mediterranean.
7) Describe zonation on coral reefs.
8) Explain the different theories of coral reef formation.
9) How is it possible that coral reefs are highly productive in environments with limited nutrient concentration?
10) Describe the characteristics of reef building corals.
11) What adaptations have mangroves developed to live in tropical ecosystems?
12) Which symbionts characterize tropical stony corals?
13) Which factors limit coral growth?
14) How do coral reefs interact with adjacent ecosystems?
15) What is coral bleaching? What factors cause coral reef degradation?
16) What are coralligenous habitats and which bioconstructors create them?
17) What are rhodolith beds?
18) Which environmental variables control distributions of marine cave biota?
19) Describe cave biodiversity and its zonation.
20) Describe adaptations of marine invertebrates that live in caves.

Suggested Reading

Ainsworth, T.D., Fine, M., Roff, G., and Hoegh-Guldberg, O. (2008) Bacteria are not the primary cause of bleaching in the Mediterranean coral *Oculina patagonica*. *The ISME Journal*, 2, pp. 67–73.

Ballesteros, E. (2006) Mediterranean coralligenous assemblages: A synthesis of present knowledge. *Oceanography and Marine Biology: An Annual Review*, 44, pp. 123–195.

Bavestrello, G., Bianchi, C.N., Calcinai, B. et al. (2000) Biomineralogy as a structuring factor for marine epibenthic communities. *Marine Ecology Progress Series*, 193, pp. 241–249.

Benzoni, F., Galli, P., and Pichon, M. (2010) Pink spots on porites: Not always a coral disease. *Coral Reefs*, 29, p. 153.

Bourne, D.G., Garren, M., Work, T.M. et al. (2009) Microbial disease and the coral holobiont. *Trends in Microbiology*, 17, pp. 554–562.

Brzana, R. and Janas, U. (2016) Artificial hard substrate as a habitat for hard bottom benthic assemblages in the southern part of the Baltic Sea – A preliminary study. *Oceanology and Hydrobiological Studies*, 45, pp. 121–130.

Bussotti, S., Di Franco, A., Bianchi, C.N. et al. (2018) Fish mitigate marine caves oligotrophy. *Scientific Reports*, 8, p. 9193.

Carricart-Ganivet, J.P., Cabanillas-Tera, N., Cruz-Ortega, I., and Blanchon, P. (2012) Sensitivity of calcification to thermal stress varies among genera of massive reef-building corals. *PLOS One*, 7(3), p. e32859.

Castaldelli, G., Welsh, D.T., Flachi, G. et al. (2003) Decomposition dynamics of the bloom forming macroalga *Ulva rigida C agardh* determined using a 14C-carbon radio- tracer technique. *Aquatic Botany*, 75, pp. 111–122.

Cattaneo-Vietti, R., Albertelli, G., Bavestrello, G. et al. (2002) Can rock composition affect sublittoral epibenthic communities? *Marine Ecology*, 23, pp. 65–77.

Cognetti, G. and Maltagliati, F. (2000) Biodiversity and adaptive mechanisms in brackish water fauna. *Marine Pollution Bulletin*, 40, pp. 7–14.

Fabricius, K.E., Langdon, C., Uthicke, S. et al. (2011) Losers and winners in coral reefs acclimatized to elevated carbon dioxide concentrations. *Nature Climate Change*, 1, pp. 165–169.

Flemming, H.C. and Wuertz, S. (2019) Bacteria and archaea on Earth and their abundance in biofilms. *Nature Reviews Microbiology*, 17, pp. 247–260.

Garrabou, J. and Ballesteros, E. (2000) Growth of *Mesophyllum alternans* and *Lithophyllum frondosum* (Corallinales, Rhodophyta) in the northwestern Mediterranean. *European Journal of Phycology*, 35, pp. 1–10.

Georgiadis, M., Papatheodoru, G., Tzanatos, E. et al. (2009) Coralligène formations in the eastern Mediterranean Sea: Morphology, distribution, mapping and relation to fisheries in the southern Aegean Sea (Greece) based on high-resolution acoustics. *Journal of Experimental Marine Biology and Ecology*, 368, pp. 44–58.

Graham, M.H., Kinlan, B.P., Druehl, L.D. et al. (2007) Deep-water kelp refugia as potential hotspots of tropical marine diversity and productivity. *Proceedings of the National Academy of Sciences of the United States of America*, 104, pp. 16576–16580.

Grzelak, K. and Kuklinski, P. (2010) Benthic assemblages associated with rocks in a brackish environment of the southern Baltic Sea. *Journal of the Marine Biological Association of the United Kingdom*, 90, pp. 115–124.

Guidetti, P., Bianchi, C.N., Chiantore, M. et al. (2004) Living on the rocks: Substrate mineralogy and the struc- ture of subtidal rocky substrate communities in the Mediterranean Sea. *Marine Ecology Progress Series*, 274, pp. 57–68.

Heckel, P. and Jablonski, D. (1979) Reefs and other carbonate build-ups, in Fairbridge, R.W. and Jablonski, D. (eds.), *The Encyclopedia of Paleontology*. Dowden, Hutchinson & Ross, pp. 691–705.

Hoegh-Guldberg, O., Mumby, P.J., Hooten, A.J. et al. (2010) Coral reefs under rapid climate change and ocean acidification. *Science*, 318, pp. 1737–1742.

Johansen, H.W. (1981) *Coralline Algae, the First Synthesis*. CRC Press.

Kamenos, N.A., Moore, P.G., and Hall-Spencer, J.M. (2004a) Nursery-area function of maerl grounds for juvenile queen scallops *Aequipecten opercularis* and other invertebrates. *Marine Ecology Progress Series*, 274, pp. 183–189.

Kamenos, N.A., Moore, P.G., and Hall-Spencer, J.M. (2004b) Small-scale distribution of juvenile gadoids in shallow inshore waters; what role does maerl play? *ICES Journal of Marine Science*, 61, pp. 422–429.

Knowlton, N. and Jackson, J. (2013) Corals and coral reefs, in Levin, S.A. (ed.), *Encyclopedia of Biodiversity*. 2nd Edition. Academic Press, pp. 330–346.

Papenmeier, S., Darr, A., Feldens, P., and Michaelis, R. (2020) Hydroacoustic mapping of geogenic hard substrates: Challenges and review of German approaches. *Geosciences*, 10, p. 100.

Ros, J., Romero, J., Ballesteros, E., and Gili, L.M. (1985) The circalitoral hard bottom communities: The coralligenous, in Margalef, R. (ed.), *Western Mediterranean*. Pergamon Press, pp. 263–273.

Rosen, B.R. (1991) Reefs and carbonate build-ups, in Broggs, D.E.G. and Crowther, P.R. (eds.), *Paleobiology: A Synthesis*. Blackwell, pp. 341–345.

Rosenberg, E., Koren, O., Reshef, L. et al. (2007) The role of microorganisms in coral health, disease and evolution. *Nature*, 5, pp. 355–362.

Sheehan, E.V., Bridger, D., and Attrill, M.J. (2015) The ecosystem service value of living versus dead biogenic reef. *Estuarine and Coastal Shelf Science*, 154, pp. 248–254.

Stuart-Smith, R.D., Brown, C.J., Ceccarelli, D.M., and Graham, J.E. (2018) Ecosystem restructuring along the Great Barrier Reef following mass coral bleaching. *Nature*, 560, pp. 92–96.

Taylor, R. (1998) Density, biomass and productivity of animals in four subtidal rocky reef habitats: The importance of small mobile invertebrates. *Marine Ecology Progress Series*, 172, pp. 37–51.

Wahl, M. ed. (2009) *Marine Hard Bottom Communities: Patterns, Dynamics, Diversity, and Change*. 1st Edition; *Ecological Studies*, Springer.

Waldie, P.A., Blomberg, S.P., Cheney, K.L. et al. (2011) Long-term effects of the cleaner fish Labroides dimidiatus on coral reef fish communities. *PLOS One*, 6(6), p. e21201.

Wernberg, T., Krumhansl, K., Filbee-Dexter, K., and Pedersen, M.F. (2019) Status and trends for the world's kelp forests, in *World Seas: An Environmental Evaluation: Volume III: Ecological Issues and Environmental Impacts*. Second Edition. Academic Press, pp. 57–78.

Willis, T.J. (2001) Visual census methods underestimate density and diversity of cryptic reef fishes. *Journal of Fish Biology*, 59, pp. 1408–1411.

Part IIIC

Comparative Marine Ecology: Habitat Types, Their Biodiversity, and Their Functioning

…

22

Estuarine, Seagrass, and Sedimentary Habitats

In this chapter, we focus on subtidal, sediment-covered habitats of the continental shelf. Estuaries and seagrasses occupy the land-sea interface primarily below low tide, whereas sedimentary habitats extend to the edge of the continental shelf at ~200 m and beyond. A separate chapter will address habitats beyond the continental shelf, which often differ markedly in environmental conditions (see Chapter 2) and biota from those at shelf depths.

22.1 Estuaries

An **estuary** refers to a semi-enclosed coastal body of water that has a free connection with the open sea, and within which fresh water from land drainage mixes and measurably dilutes sea water. Estuaries differ from lagoons in that lagoons lack appreciable freshwater input. Estuaries occur from the tropics to the poles and vary in size from small rivers to the largest rivers in the world. Humans have often settled at the mouths of estuaries because they provide a source of freshwater and ready access to the sea. The high productivity of some estuaries provides food and jobs for people living nearby. The generally unidirectional flow of estuaries has often resulted in widespread use for sewage and industrial waste.

Estuaries originate through several different mechanisms that can vary in how recently they formed.

1) **Tectonic estuaries** form when plate tectonics uplift seafloor that partially blocks the entrance of an embayment, thereby limiting exchange with the open ocean.
2) **Fjords** form when melting ice from glacial retreat floods low lying areas of glacial valleys and sea level rises. This process produces long, narrow estuaries with steep, cliff-like sides and depths up to 1,000s of meters. Glacial moraines (ridges of sediment deposited at the front of glaciers) often form a sill at the mouth of fjords, which limits exchange with the open ocean, sometimes even to the point where the basin inside the sill becomes hypoxic or even anoxic.
3) **Bar built estuaries** form when sand spits and sand bars accumulate sediment and partially block an embayment entrance and thus restrict tidal exchange.
4) **Drowned river valley estuaries** form when rising sea level floods the lower reaches of river valleys.

Regardless of their origin or form, the same major variables regulate estuaries. Gravity obviously plays an important role in that rivers and streams flow from higher to lower elevation, thus determining flow rates as well as influencing volume of water delivered to the estuary. The ebb and flow of tides results in substantial temporal and spatial changes in temperature, salinity, food availability, and a whole range of biological processes. Finally, thermohaline circulation, driven by density differences between water masses, determines estuarine flow, and also contributes to spatial and temporal (e.g., seasonal) changes.

Scientists group estuarine types based on their circulation and mixing (Figure 22.1).

Salt-wedge estuaries occur where river outflow greatly exceeds tidal inflow, resulting in a highly stratified water column with a strong halocline that limits mixing between the salt and freshwater layers. In these estuaries, a saltwater wedge extends landward below a freshwater wedge that extends seaward. Within these systems, strong freshwater currents flow along the halocline and may generate internal waves along the halocline. These internal waves can steepen and break to mix the saltwater and freshwater as the water moves seaward. This loss of seawater results in weak currents in the

Marine Biology: Comparative Ecology of Planet Ocean, First Edition. Roberto Danovaro and Paul Snelgrove.
© 2024 John Wiley & Sons Ltd. Published 2024 by John Wiley & Sons Ltd.
Companion Website: www.wiley.com/go/danovaro/marinebiology

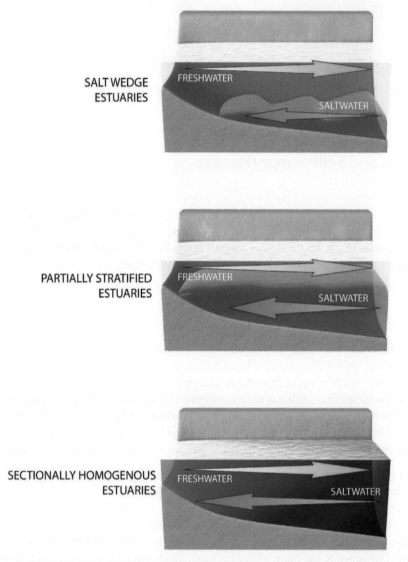

Figure 22.1 Schematics of different types of estuarine flows, illustrating differences in vertical and horizontal stratification.

saltwater bottom layer that moves too slowly to transport much sediment into the estuary from the ocean. Thus, much of the sediment in the estuary arises from river sand delivered to the landward edge of the saltwater wedge, and clays and silts delivered by the river further seaward.

Partially stratified (or partially mixed) estuaries occur where neither river inflow nor tidal mixing dominate. Within these estuaries, tidal currents promote greater mixing, weakening stratification and the halocline. The mixing of more saltwater into freshwater can generate strong bottom currents, which can transport significant volumes of sediment through the tidal inlet. Shelf sediments cover the seaward end of the estuary whereas river sediments dominate the landward end. Where weak currents occur, any suspended material sinks into the halocline, forming a turbidity maximum in the water column. In some of these estuaries, filter feeders remove much of the suspended material, concentrating it into fecal pellets. Clay particles stick together by flocculation, forming larger aggregates that sink rapidly. Mud accumulates in regions with weak currents, forming mud shoals.

Well-mixed estuaries lack an obvious halocline or stratification because tidal turbulence destroys any such structure.

Vertically homogenous estuaries and sectionally homogenous estuaries also occur. In wide estuaries, the Coriolis effect deflects river flow to one side of the estuary, and tidal inflow to the other, resulting in a **vertically homogenous estuary**. In a **sectionally homogenous estuary**, some salinity gradients may also occur across the estuary, but in both cases vertical stratification does not exist. In well-mixed estuaries, strong inflow of seawater transports abundant sediment into the estuary, resulting in dominance of marine sediments throughout.

Estuaries also take "positive" and "negative" forms (Figure 22.2), where rainfall and runoff exceed evaporation in **normal** (positive) estuaries, whereas evaporation exceeds rainfall and runoff in **hypersaline** (negative) estuaries.

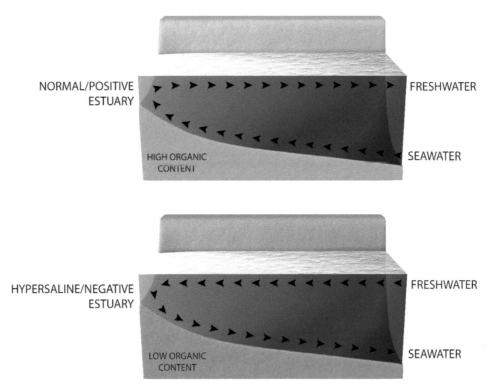

Figure 22.2 Schematic of positive and negative estuaries.

22.1.1 The Complexity of Estuarine Environments

Despite high productivity in many cases, estuaries create significant challenges for biota. In addition to potential temperature change over short periods of time, salinity can also change both in space and in time, often in just a few hours or even minutes, depending on location in the estuary (Figure 22.3).

22.1.2 Survival Strategies for Living in Estuaries

Biotic response to estuarine gradients includes a wide range of morphological, physiological, and behavioral adaptations. In some cases, these adaptations enable an expansion of niche breadth as weak competitors lose; as a result, broad distributions often characterize estuarine species with respect to sediment, resource utilization, and other aspects of living in estuarine environments. These adaptations often result from genetic divergence from open ocean conspecifics.

Physiological Adaptations in Estuaries – Within estuaries, salinity creates major physiological problems because organisms must be capable of osmotic regulation. For example, the complete absence of echinoderms, brachiopods, and nuculid bivalves from estuaries reflects their inability to cope with osmotic fluctuations. Organisms that can tolerate varying salinities do so in one of three ways (Figure 22.4). Some species, such as stickleback fish, use active transport of ions to ensure that cell ion concentrations remain constant as salinity changes (**osmotic regulators**). Other species, such as hydrozoans "go with the flow" and simply tolerate and match ambient salt ion concentrations (**osmotic conformers**). Some **partial regulators** partially regulate, in that they maintain constant ion concentrations over a range of salinities but eventually conform over part of the range of salinities they encounter. Most species cannot cope with this variation in salinity,

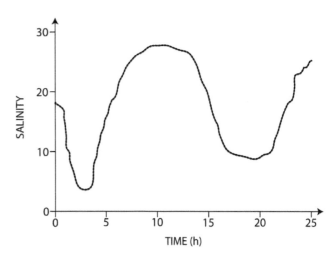

Figure 22.3 Example of potential salinity change at a given point in time in an estuary.

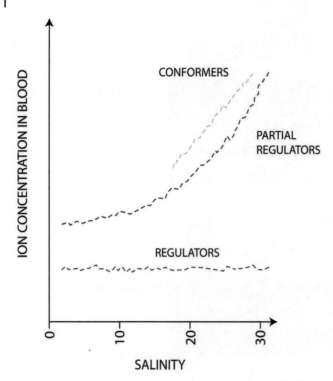

Figure 22.4 Osmotic regulation and the challenge of living in an estuary. Different species employ different strategies.

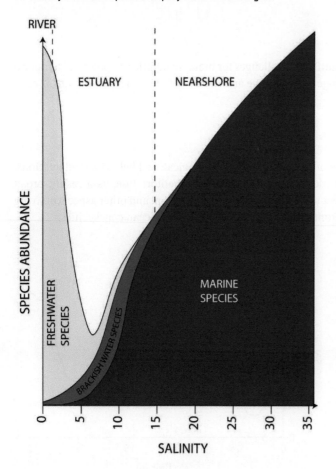

Figure 22.5 Distribution of species along an estuarine salinity gradient.

which constrains the number of species present because truly freshwater or truly marine species cannot survive. Key transition areas occur near the mouth and lower reaches of an estuary, the most downstream portion, where organisms require physiological adaptation to salinity change. A second key transition occurs at salinities of 5–8, where minimum benthic diversity occurs (Figure 22.5).

Well-oxygenated river water and general water movement in proximity to an estuary help to oxygenate waters. However, in situations such as fjords, low oxygen levels may occur where estuary morphology limits circulation. Moreover, because many estuaries are productive, and organic material may accumulate on the seafloor along with fine sediment particles, sediments may become anoxic, particularly within a few cm of the sediment-water interface. Thus, estuarine animals that live in these conditions sometimes must cope with low oxygen concentrations.

Morphological features and adaptations in estuaries – Many organisms that live in estuaries exhibit specific morphological adaptations to cope with the complex environmental conditions. Turbidity results in dominance of phytoplankton as opposed to bottom photosynthesizing species. Reduced body size, lower reproductive rate, and lower fecundity characterize many estuarine species, presumably reflecting the added physiological costs of living in an estuary. Many of these adaptations reflect genetic adaptations, which means that the genetic composition of estuarine species may differ from one location to another, resulting in distinct (and non-interchangeable) populations.

Behavioral adaptations in estuaries – Many estuarine species also use behavioral adaptations to maintain their positions within the estuary. Keeping in mind that unidirectional flow often characterizes parts of the estuary, the estuary would tend to transport organisms to the ocean. Some species therefore undertake vertical migrations during contrasting tidal phases in order to take advantage of different flow direction. Indeed, some pelagic species remain within a given location in the estuary by moving up during flood tide and down during ebb tide, an adaptation seen not only in holoplanktonic species but also in meroplankton of benthic species. Feeding adaptations also come into play. These adaptations include flexibility in feeding in time and space, omnivory, sharing of food resources among species, exploitation of food webs at different levels, ontogenetic change in diet with rapid growth, and short food webs based on detritus-algal feeders.

On a final note, mangroves and salt marshes, along with other intertidal habitats, typically occur within estuaries and display specific adaptations (discussed in Chapter 20). We consider seagrasses later in this chapter.

22.1.3 Estuarine Food Webs

The diverse sources of food within many estuaries contribute to their productivity and attract many species to estuaries for at least one phase of their life history. Organic material transported downriver sometimes settles onto the seafloor within an estuary as the estuary widens, near where it meets the ocean. Despite the refractory nature of much of this material, which renders it unavailable to most animals, microbial decomposition releases nutrients that can support phytoplankton, salt marsh plants, seagrasses, and mangroves (Figure 22.6). These primary producers, along with the partially decomposed detritus and dissolved organic matter, provide the "fuel" that supports particle consumers and predators.

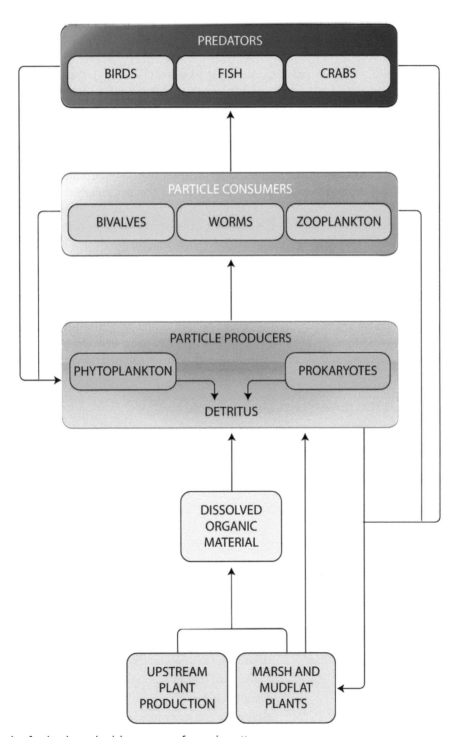

Figure 22.6　Estuarine food web emphasizing sources of organic matter.

22.1.4 Why are Estuaries Important?

The high productivity of many estuaries arises from abundant nutrients in some situations, resulting from differing levels of riverine input of dissolved and particulate nutrients (e.g., phosphorus), oceanic import of nutrient-rich deep water, and regeneration of nutrients from an active microbial community on the seafloor created by decomposition of organic material and re-injection of dissolved nutrients. The nature of this productivity varies and includes habitats ranging from seagrasses to mangroves to salt marshes to mudflats. These habitats create key nursery areas for some species, including some commercially important taxa, and often serve as "filters" for coastal runoff, in that vegetated habitats help to remove sediments and some pollutants, but also export significant organic matter and nutrients. Some estuaries also support significant fisheries.

22.1.5 Pressures on Estuaries

Many estuaries around the world face complex and often interacting pressures (Table 22.1). These pressures arise from dense human populations that often congregate around estuaries to take advantage of their productivity, as well as access to freshwater for drinking and the ocean for transport, aquaculture, and fisheries. The combination of sewage effluent and runoff of fertilizers from agricultural activities leads to expanding hypoxia around the world as excess phytoplankton sink to the seafloor and decomposers consume available oxygen. Hypoxia produces multiple effects. At low levels, benthic fauna migrate closer to the sediment water interface and become more vulnerable to predators and, as oxygen levels decrease further, no animals can survive and "dead zones" appear. Industrial activities often increase contaminants in the water and in sediments, and shipping activity results in increased invasive species. Coastal development often leads to habitat destruction, particularly damage to biological habitat at the land-sea interface such as seagrasses, salt marshes, and mangroves. Fisheries can add to this destruction, particularly when using bottom contact fishing gear such as trawls and dredges. Box 22.1 provides an example of an estuarine system.

22.2 Seagrass Beds

Seagrasses are higher plants. This attribute places them among a select few types of true plants in the marine realm, along with mangroves and some salt marsh vegetation; the phytoplankton that dominate marine primary producers are technically not true plants. Seagrasses are adapted to live in the ocean, and can form dense beds along coasts, bays, and estuaries. Water transparency (light penetration), salinity, and hydrodynamics define where they occur, along with tidal exposure

Table 22.1 Threats to estuarine environments.

Human Pressures	Stressors
Commercial fishing	Biomass removal / food web alteration
	Habitat damage (gear, ghost fishing)
	Invasive species (transported on hulls)
Oil and gas	Seismic exploration (noise, ship strikes)
	Exploratory drilling (debris, spills, noise)
Marine transportation	Invasive species (hulls, ballast)
	Sewage, bilge, other discharge
	Antifouling
Aquaculture	Invasive species (including disease)
	Habitat damage (nutrients, feces)
Land based activities	Habitat loss (wetlands, estuary)
	Nutrient and contaminant loading
	Altered hydrology
Climate change	Habitat loss (sea level rise, warming)
	Altered processes

> **Box 22.1 St. Lawrence Estuary Example**
>
> The St. Lawrence River estuary in eastern Canada illustrates the complexity of estuaries (Figure 22.7). This is the largest estuary in the world and connects the Great Lakes to the Atlantic Ocean, supporting a distinct ecosystem with
>
>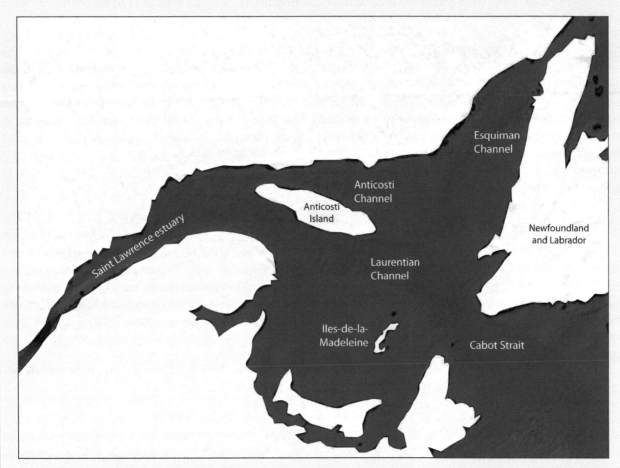
>
> **Figure 22.7** St. Lawrence estuary in Eastern Canada.
>
> high biological productivity and diversity spanning from seals and whales to phytoplankton. In addition to numerous pelagic and benthic fisheries, this estuary provides a major conduit for shipping to and from North American markets. Deep channels were cut during glacial melting 10,000 years ago, and complex circulation generates productive upwelling regions. Winter ice cover contributes to strong seasonality. Its massive drainage basin encompasses large cities as well as extensive agricultural land and industry. These activities have resulted in accumulation of contaminants that increase in concentration upstream, as well as expanding hypoxia fueled by excess nutrient loading. As in many estuaries, the hypoxic region of the St. Lawrence estuary has expanded in extent and in intensity in recent decades, eliminating habitat and compromising function of the estuarine system.

and substrate. Primary production values can reach 3,000 g m^{-2} y^{-1} supporting high abundances of organisms that live within seagrasses, including many endemic species. They span from sub-Arctic to tropical environments and, in the tropics where seagrasses flourish, they provide food for marine organisms as large as dugongs and turtles.

As **angiosperms**, or higher plants, seagrasses establish themselves in soft-sediment habitats along coastlines, bays, and estuaries. Some species, such as *Phyllospadix* spp. and *Amphibolis* spp., grow on hard substrata and *Posidonia oceanica* can settle between crevices on rocky bottoms. Seagrasses evolved from terrestrial higher plants that colonized the marine environment during the Cretaceous, about 120 million years ago (the era when *Posidonia cretacea*, the first *Posidonia* fossils appeared). Their evolutionary development culminated in the Eocene (Cenozoic Era, ~30 million years ago). These plants colonized the marine environment through gradual adaptation to brackish water and, eventually, total immersion in seawater. Cornelis Den Hartog developed the most accepted hypothesis regarding their evolutionary migration into marine environments, namely that terrestrial angiosperms evolved species capable of withstanding short immersion in saltwater within the intertidal zone. Once pollination evolved from **anemophilous**, or "wind loving," to **hydrophilic**, or "water loving," these plants were able to live completely submerged in seawater.

Comprising <0.02% of angiosperms, seagrasses include a very small portion of marine photosynthesizers. The 58 recognized species of seagrasses span 13 genera globally, more than half of which occur along the Australian coasts. In the United States, > 90% of seagrass species occur in the Gulf of Mexico. South Florida supports the most extensive seagrass community in the world. Three genera – *Halophila*, *Zostera*, and *Posidonia* – include about 55% of the species, whereas *Enhalus*, the most recently differentiated genus, contains just a single species, *Enhalus acoroides*. Among the 13 extant genera, six of them (*Amphibolis*, *Heterozostera*, *Phyllospadix*, *Posidonia*, *Zostera* and *Pseudalthenia*) occur primarily in temperate waters, whereas the other seven (*Cymodocea*, *Enhalus*, *Halodule*, *Halophila*, *Syringodium*, *Thalassia*, and *Thalassodendron*) occur mainly in tropical waters. Many seagrass beds are monospecific, especially in the temperate Northern hemisphere where a single genus tends to dominate (*Zostera* spp. in soft sediments and *Phyllospadix* spp. in rocky bottoms). However, in the Mediterranean and in tropical and subtropical environments, multi-species seagrass beds of 2–12 species occur.

As angiosperms, seagrasses share major characteristics with terrestrial plants, including differentiation of roots, stem (called **rhizomes**, which run horizontally beneath the sediment surface) and leaves (blades), but they have several characteristics allowing them to live in the ocean: they grow in a saline medium from which they draw nutrients, and require strong anchoring systems necessary to withstand wave action. All seagrasses share the following characteristics: (1) adaptations to live in salt water; (2) the capacity to live completely submerged; (3) a true anchoring system; (4) pollination using water rather than air (they complete their entire life cycle in the ocean); (5) successfully competing with other marine primary producers, particularly with algae. In order to grow in marine environments, seagrasses became a rhizomatous clonal species. The rhizome extends clones spatially, and also connects neighboring shoots. Rhizome growth rates can vary from a few centimeters per year in slow-growing, large species, up to more than 5 m y^{-1} in smaller species (larger size increases construction "cost" because of added structural support required). Seagrasses tend to form beds that provide preferred habitat for many species, which sometimes use them only as a nursery. Seagrasses absorb nutrients through their roots and shoots. The plants use nutrients as they build tissue, which they eventually lose to breakage or plant death. Microbes and other species remineralize the resulting detritus, releasing the nutrients once again. Seagrass beds play several important ecological functional roles in that they: (a) provide food and habitat for many species, including endangered species such as some marine turtles and manatees; (b) support a rich biodiversity of mollusks, gastropods, insects, fishes and other organisms and (c) yield high primary production; (d) export substantial carbon, nitrogen, and phosphorus to coastal food webs; (e) stabilize seabed sediments, store carbon, and improve water quality.

Seagrasses play a particularly important role in stabilizing the seabed, potentially dissipating 30–40% of wave energy and 60–70% of energy from ocean currents, and facilitating the establishment of coastal ecosystems such as infralittoral grasslands. This capacity to create three-dimensional underwater landscapes qualifies seagrasses as ecosystem engineers. These macrophyte beds can also extend up into the intertidal zone but develop mainly in the shallowest subtidal to depths up to 40–60 m in the clearest waters in bright sunlight. Factors that influence seagrass distribution include transparency (light intensity and penetration), salinity, exposure to air, hydrodynamics, presence of suitable substrata (seagrasses colonize environments with different sediment particle size), available nutrients, and concentrations of pollutants, noting their sensitivity to contamination by hydrocarbons and heavy metals.

Seagrasses occur along coastlines globally, except Antarctica (Figure 22.8). The worldwide distribution of the genus *Posidonia* spp. suggests that the species in the genus likely arose from a common ancestor. Their distribution in the Tethys Sea, an ancient ocean from 65.5 million to 261 million years ago, spanned from the Mediterranean to what is now the Indian Ocean and Oceania. These seagrasses gradually disappeared from the central regions, evolving through allopatric speciation to eventually give rise to a single endemic species in the Mediterranean and eight other species in Australia.

Posidonia oceanica reproduces both sexually and asexually. The asexual or vegetative mode occurs through **stolonization**, in which an entire plant consists of a horizontal rhizome, from which other rhizomes, both vertical and/or horizontal, originate and bear tufts (Figure 22.9).

One m^2 of sediment can contain up to 700 tufts of *Posidonia* sp., each consisting of 6 or more leaves with lengths that can exceed 1 m. The ribbon-like tufts are distributed like a fan, with the oldest and longest leaves external to younger leaves. The base and flap form the photosynthetic leaves, separated by a concave "ligule" from which the leaves detach during the

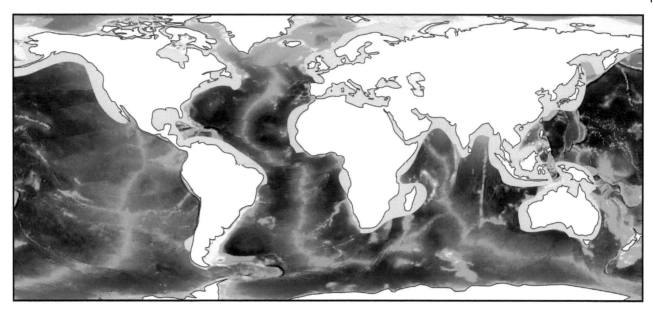

Figure 22.8 Global distribution of seagrasses shown in green.

Figure 22.9 (a) Structure of a tuft of *Posidonia oceanica*, and its components: Leaves, flower (or inflorescence) with bract, peduncle, fruit, seed, rhizome scales, rhizome, and roots. (b) flower of a marine seagrass.

autumn. The leaves grow from the bottom upwards and from the inside of each tuft outwards. The edible fruit, sometimes called "sea olive," contains a porous pericarp rich in oily material that help it float and thus increase dispersal in the environment. The pericarp opens to release the seed once it lands on the seabed. Typically, the small white flowers appear in the autumn (Figure 22.9b). *Posidonia oceanica* distributions extend 0–40 m in oligotrophic waters, which allow light penetration to greater depths. *Posidonia* sp. beds are particularly complex ecosystems that produce large amounts of organic matter and oxygen, providing excellent habitat for spawning and development of many aquatic organisms from invertebrates to fishes. A significant biota of epiphytes grows on their leaves (see section below on biodiversity). *P. oceanica* forms **mattes** (Figure 22.10), a block of compacted sediment comprised of a dense network of rhizomes, roots, and plant debris. Mattes form through the progressive burial of rhizomes under the sediment, which accumulate as a result of the presence of the seagrasses.

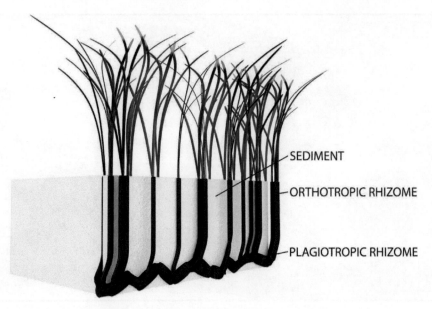

Figure 22.10 Seagrass matte with vertical and horizontal rhizomes, and sediment trapped between them.

This process forms a "terrace" (matte in French), which can grow 1 m per century. This slow, continuous process, in which rhizomes develop in the vertical direction, allows the plant to access light while sedimentation buries the basal portions. Each section of the rhizome can produce other rhizomes. Furthermore, at the level of the roots, leaves tend to greatly attenuate the amount of light up to 100x with respect to their surface. The rate of rise of the matte depends on plant growth rate and on hydrodynamics. In very sheltered areas with weak flow, the seagrass bed can grow so the leaves extend above the water, forming a natural coastline barrier together with the mattes. In exposed areas, matte erosion results in the seagrass bed regressing. If the plants die, the matte persists and algae or other seagrasses such as *Cymodocea nodosa* and *Nanozostera noltii* can colonize.

The large volume of leaves produced can accumulate on the beach, forming banquettes of leaves (Figure 22.11) that provide a unique habitat for many species of invertebrates, particularly crustaceans.

Living clones of the seagrass, *Posidonia oceanica*, may be the oldest organism on Earth. Some seagrasses of the Mediterranean that extend up to 15 km in width may be 100,000 y old!

Cymodocea nodosa (Figure 22.12) occurs widely in less transparent and oligotrophic waters than those required by *Posidonia oceanica* (Box 22.2). *C. nodosa* occurs in the Mediterranean (particularly in the Northern Adriatic Sea), along the coasts of the Canary Islands, in the Central Atlantic (off the coast of Senegal, which defines the southern limit of its distribution), and in the North Atlantic (the Sado estuary in Portugal defines its northern limit). By settling into an environment, this "pioneer species" "prepares" the substrate for other macrophytes. It can tolerate anoxia and the presence of hydrogen sulfide in the sediment. Its leaves host a rich epiphyte community almost as rich as that of *P. oceanica*. Many species of fish breed between its leaves. This widespread seagrass, second only to *P. oceanica* in the Mediterranean, forms extensive beds up to 20 m deep in locations with weak flows and fine-grained, organic-rich sediments.

Thalassia testudinum, the largest and most robust species among the macrophytes present in Florida and in the Caribbean, grows in clean clear waters, and occupies a somewhat rather narrow depth range up to 10 m depth, and occasionally to 25 m. It stabilizes different types of substrates, including corals, rocky material, sand, and very coarse-grained sediments (Figure 22.12). Dead leaves and rhizomes accumulate between the living portions of the plant.

22.2.1 Biodiversity Associated With Seagrasses

Organisms associated with these systems fall into four main groups based on the microhabitats they occupy: (a) Mobile and sessile organisms that live in the leaf layer; (b) organisms that occupy the water column above the bed and between the leaves; (c) mobile and sessile organisms living among the rhizomes; (d) fauna of the matte consisting of both sessile organisms (dominated by hydroids, bryozoans, and serpulid polychaetes) and mobile organisms (dominated by gastropods such

Figure 22.11 Seagrass bed mixed with macroalgae (top) and examples of a banquette (bench) consisting of beached and dried *Posidonia* sp. leaves (bottom).

Figure 22.12 Seagrass bed comprised of *Cymodocea* sp. and *Thalassia* sp. in a tropical environment.

> **Box 22.2 Grasslands of the Sea**
>
> Mediterranean Seagrasses
>
> Only six seagrass species occur in the Mediterranean but collectively they form a wide belt along the coast. Among them, the Mediterranean endemic species, *Posidonia oceanica* (Figure 22.13), forms one of the most productive ecosystems on the coast and, subsequently, experiences strong anthropogenic impacts. The sheer size of these seagrass beds, their high productivity, and their stability in persisting for thousands of years all contribute to their ecological importance. Its habitat characteristics, extension, and large standing biomass provide physical and trophic support for diverse associated plant and animal communities. In addition, leaf debris and scales provide unique, preferred microhabitats for many scavengers and cryptic organisms. Increasing anthropogenic impacts (human population growth, urbanization, industrialization, pollution, global climate change) place coastal ecosystems, and *P. oceanica* in particular, among the most threatened and impacted marine ecosystems.
>
>
>
> **Figure 22.13** Seagrass, *Posidonia oceanica*, forms dense meadows that can almost totally shade the seabed from light (penetration reduced by 99% between upper leaves and rhizomes).

as *Rissoa* spp., *Gibbula* spp. and *Bittium* spp., crustaceans amphipods, isopods such as *Synischia hectica* and decapods such as *Palaemon xiphias*). At the base of plants, sciaphilous species include Foraminifera such as *Miniacina miniacea*, sea urchins such as *Paracentrotus lividus*, polychaetes, and decapod crustaceans, include *Upogebia pusilla* that digs tunnels all live within the matte. High abundances of filter feeders such as the polychaete, *Sabella spallanzanii*, grazers such as sea urchins, ascidians, and carnivores also occur. Scavengers such as amphipods and isopods live below decaying leaves, gaining food and protection from predators, degrading the leaves as they feed on them. Some animals have adapted specifically to the seagrass bed environment, including homochromic and homomorphic species that blend into their environments based on color or morphology, respectively. For example, the green color of the shrimp, *Palaemon xiphias*, blends in with *Posidonia* spp. leaves. Some polychaetes live exclusively at the base of seagrass scales, obtaining nourishment from the scales and protection from predation as they dig tunnels. Interestingly, these polychaetes dig tunnels at the basal part of the living leaves, increasing leaf loss up to 50%, thus accelerating the natural process of leaf loss.

Dominant nektonic components in warm waters include pipefish (*Syngnathus typhle*), peacock fish (*Thalassoma pavo*), damselfish (*Chromis chromis*), damsel (*Coris julis*), and herbivorous fish, such as sea bream, *Sarpa salpa*. Seagrasses also support a rich hyperbenthos, including cephalopods, decapod crustaceans, signatids (sea horses), mysids, opisthobranchs and some gastropods that feed on the apical part of the epiphyte covered by leaves (Figure 22.14). The high tannin concentrations in seagrass blades mean that few organisms feed directly on them. In cooler waters, juvenile stages of important commercial fish such as Atlantic cod use seagrasses as nursery habitat. Seagrass beds also support many sessile organisms, such as epiphytic sponges on rhizomes, hydroids on rhizomes and leaves, and bryozoans and crustaceans on leaves. Many species depend upon the substrate created by seagrasses, including sea squirts (especially at the base of the bed), and a diverse community of juvenile fishes unequalled in diversity in sandy environments.

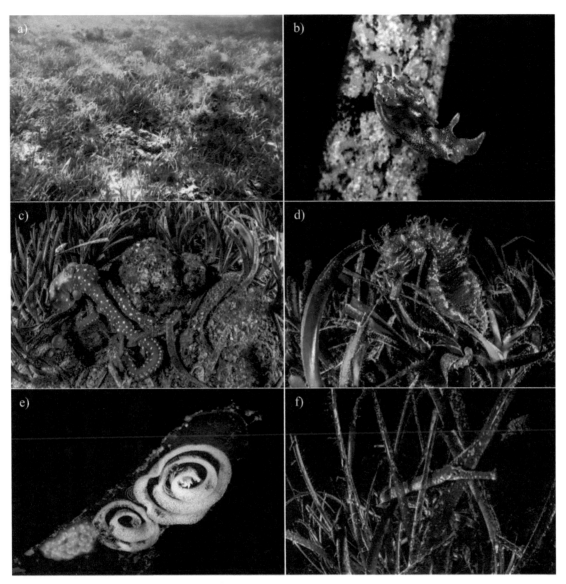

Figure 22.14 Examples of organisms inhabiting seagrass beds. (a) Blue starfish grazing; (b) encrusting algae on a leaf of *Posidonia oceanica* with a small opisthobranch on it (*Aplysia* spp); (c) during the night, predators such as *Octopus macropus*, look for prey hiding in *P. oceanica* meadow; (d) seahorse in seagrass meadow; (e) eggs on seagrass leaf; (f) pipefish hiding among seagrass leaves.

Seagrass beds offer an excellent habitat for juvenile stages of many species for two main reasons: (1) the abundance of organic matter, resulting from both primary and secondary production (epiphytes and macroinvertebrates), and (2) the sheltering effect of the leaves. Seagrasses continually produce new leaves, roots, and rhizomes, whereas the older portions of the plant "drop off" and enter the detrital food chain. For many species, the high turnover of leaf material injects significant organic input to the sediment detrital pool, while simultaneously exporting a portion of this organic matter to greater depths. Seagrass detritus also contributes to the organic matter on beaches, which organisms inhabiting these transitional environments between land and sea can then use. *Posidonia* spp. therefore provide a trophic resource to direct consumers and an indirect source to indirect consumers such as species that feed on epiphytic organisms, and secondary consumers such as crustaceans that feed on detritus or on organisms associated with it. Major grazers include sea urchins, *Paracentrotus lividus*, followed by isopods of the genus *Udotea* and by sea bream, *Sarpa salpa* (Figure 22.15). In tropical environments, where seagrasses thrive, they also provide food for large marine organisms such as dugongs and turtles.

On the leaves, hydroids, bryozoans, and serpulid polychaetes dominate, and gastropods, holothurians, and starfish graze. A diversity of crustaceans live among the leaves.

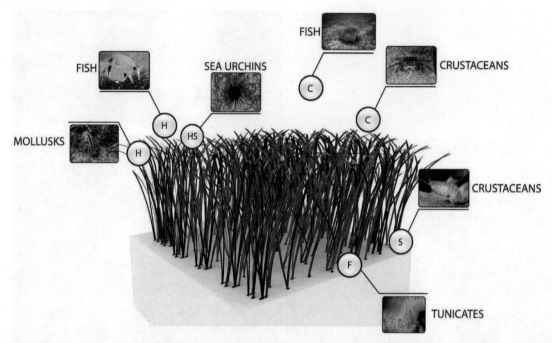

Figure 22.15 Simplified diagram of a *Posidonia oceanica* food web in the Mediterranean, including herbivores (E), herbivore/scavengers (HS), carnivores (C), scavengers (D) and filter feeders (F). *Sources:* Emmanuelbaltasar / Wikimedia Commons / CC BY-SA 4.0, FredD / Wikimedia Commons / CC BY-SA 3.0, Bernard Dupont / Wikimedia Commons / CC BY-SA 2.0, Jay Fleming/Corbis Documentary/Getty Images, Sirrob01 / Wikipedia Commons / CC0 1.0, Virginia Institute of Marine Science, Dann Blackwood / Wikimedia Commons / Public Domain.

22.2.2 Seagrass Functioning

Seagrass beds, as macrophyte forests, support a highly productive, large photosynthetic biomass (Figure 22.16). These hotspots of primary production reach values up to 10x higher than the surrounding habitat with no seagrass (see Box 22.3). Primary production values in seagrasses with tuft density exceeding 200 tufts m^{-2} vary <300–3,000 g C m^{-2} y^{-1}, a value that greatly exceeds that of the benthic microalgae that dominate sandy bottoms without macrophytes, and that achieve primary production rates typically less than 100 g C m^{-2} y^{-1}. Mediterranean net production averages approximately 400 g C m^{-2} y^{-1} and biomass averages about 180 g C m^{-2}. Sometimes the epiphytic biomass exceeds that of the leaves, adding another aspect of primary production to *P. oceanica* (or other seagrasses for that matter) that merits consideration. This added production, combined with additional epiphyte production on the rhizomes and roots and micro- and macroalgal production, can also exceed that of *P. oceanica*. Macroalgal production can be so high that, in calm waters with stable chemical and physical conditions, touching the surface of the *P. oceanica* leaves releases air bubbles (Figure 22.17). These estimates of production, when considered in tandem with estimates of seagrass global coverage, suggest a contribution of ~1.1% to global marine primary production. Because of the generally low herbivory rates in most seagrass beds, sediment storage and export to the surrounding ecosystem account for the fate of the majority of the primary production. Seagrass beds trap about 27 gT C y^{-1}, representing ~12% of the total carbon stored in marine ecosystems. Seagrasses therefore contribute significantly to the marine carbon cycle, accounting for a major fraction of CO_2 uptake by marine biota.

22.3 Sedimentary Habitats

The continental shelves cover ~ 8% of the seafloor and encompass significant marine habitats beyond just those at the land-sea interface. Sediments spanning from cobble to gravel to fine silts and clays cover much of the continental shelf, creating a wide variety of habitats defined in large part by sediment composition. These shelf environments support many of the most productive commercial fisheries globally and also play a significant role in carbon and nutrient cycling. Because continental shelves span depths from the intertidal to ~200 m, they also span a wide range of wave energies. Noting that wave energy typically dissipates by depths of 60 m at most, shelf environments also typically exhibit decreasing energy with increasing distance from shore. This pattern affects sediment composition, whereby finer sediments typically do not accumulate in the shallowest depths because bottom flows tend to resuspend them, leaving behind coarser sediments such as sand or gravel (Figure 22.18).

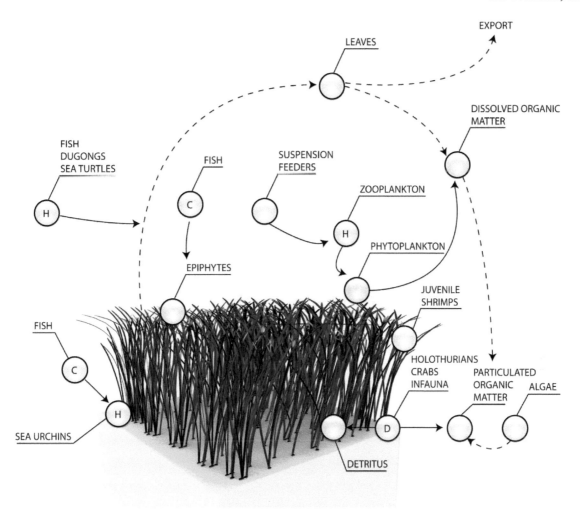

Figure 22.16 Example of a food web associated with seagrass beds in a tropical environment, including herbivores (H), carnivores (C), and deposit feeders (D).

Box 22.3 Studying Seagrass Beds

Posidonia oceanica sampling and study methods

Researchers often identify seagrass beds using markers such as buoys anchored to the bottom with weights to facilitate periodic resampling of the same location to count seagrass bundles as well as horizontal and vertical rhizomes. By marking the bundles with small holes, scientists can quantify increases in leaf length, and by sampling some bundles to estimate biomass it becomes possible to correlate length and biomass. This approach enables the study of *P. oceanica* bed biomass and production as a change in biomass over time. Scales on this plant also allow reconstruction of changes over time through lepidochronologic analysis (from the Greek, lepidos that means "scales"). **Lepidochronology** studies changes in ligules and scales at the base of the plant leaves. Because ligules change in thickness with the changing seasons, depending on temperature and plant growth rate, analyzing their thickness allows researchers to determine the age of the plant. Slow growth occurs, in fact, in late summer–autumn and during the winter, with highest growth rates in spring. This variation enables researchers to distinguish summer leaves from winter leaves, and this information can help deduce the age of the eelgrass bed. Each leaf consists of a lower base, to which the photosynthetic leaf blade attaches. The **ligule**, the point where the base attaches to the leaf blade, defines where the leaf detaches and leaves the base attached to the rhizome once it finishes its growth cycle. This scale can persist on the rhizome for a long time. This feature, together with annual variation in scale thickness allows lepidochronological dating of plants. Examination of the insertion order of the scales on the rhizome reveals variation in scale thickness from gradually decreasing values to a minimum thickness, followed by gradually increasing values to a maximum. These cyclical variations in maxima and minima correspond to a period of one year. Typically, lepidochronology normally uses horizontal rhizomes, which clearly show this periodicity.

Figure 22.17 High primary production in macrophytes yields abundant oxygen, which they release when touched.; Left, seagrass in the Red Sea (*Halophila stipulacea*). Right, a macroalgal field dominated by *Cystoseira* spp.

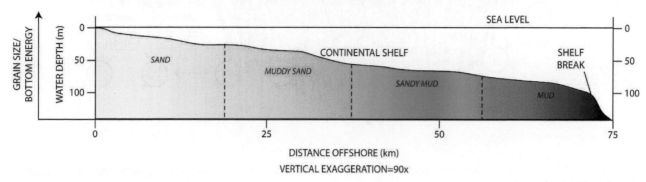

Figure 22.18 Inverse relationship between grain size and energy, and associated changes in grain size with depth. Shelf break depths typically vary from ~100–200 m.

Sediment composition at a given shelf locale strongly influences the fauna, in part because sediment composition influences biota as a habitat and food source, but also because the same variables that influence sediment composition (bottom flow, sediment sources) also influence the biota (Figure 22.19). But just as sediments influence the biota, the biota influence sediments.

Bottom flow and sediments affect biota through direct influence of bottom flow velocity on suspension feeders, food flux, transport, and settlement of larvae, and limitations on distributions by physically moving seabed fauna. Bottom flow also affects biota through velocity shear effects, which influence turbulent mixing and food supply for suspension feeders, as well as larval delivery to the seabed.

However, biota also influence flow and sediments. Animal shells and tubes can extend above the bottom into the boundary layer, thereby altering bottom roughness and particle flux to the seabed. Animals crawling across the sediment can leave behind tracks known as **"lebensspuren"** that also alter roughness. Microbes and some other organisms as well, can increase sediment "stickiness" by excreting mucus or alternatively decrease stickiness by breaking up sediment clumps as they move through it. Living organisms can also alter velocity profiles as illustrated by seagrasses whose emergent structures (the blades) extend well above the bottom and produce a **baffling effect** on flow, slowing it down and enhancing particle deposition. Finally, as noted earlier, deposit feeders repackage sediment grains as fecal pellets, often redistributing them vertically within the sediment in the process. This process can sometimes lead to **"bed armoring"** where animals deposit coarser sediments on top of finer sediments, thereby reducing the likelihood of resuspension of the finer sediment fractions beneath them.

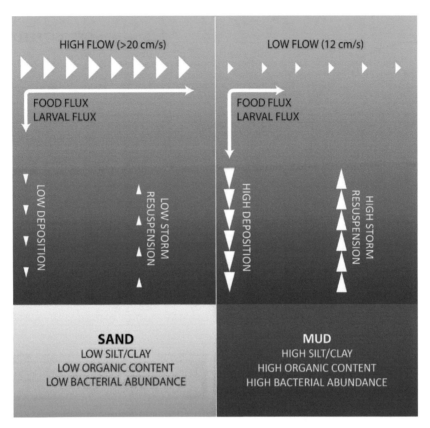

Figure 22.19 Simplified summary of complex interaction between currents, food and larval supply, and sedimentary environments.

22.3.1 Food Sources for Sedimentary Fauna

Although sedimentary environments contain a wide range of feeding types, the two primary modes, deposit feeders and suspension feeders, tend to dominate in different sedimentary habitats (Figure 22.20). To a large extend this difference reflects food supply.

Deposit feeders use a range of food sources they ingest from sediments, including sinking phytoplankton and its breakdown products (**phytodetritus**), fecal pellets produced sinking from the water above or produced within the sediment, and potentially containing undigested phytoplankton, animal tissue, and bacteria. In some shallow-water sediments, actively growing algae such as benthic diatoms provide an important food source, though the requirement for light limits this food source to relatively shallow shelf environments. Bacteria attached to sediment particles or detritus provide a critical food resource for many taxa, where bacteria may contain greater food value than the detritus itself. Detritus can also provide important nutrition including sources from land, such as leaf material, wood, and even sewage, as well as marine sources such as seagrasses, saltmarshes, and mangroves. Often, terrestrial sources produce material that is more refractory than marine sources, though marine plants such as seagrass can also be highly refractory. In locations with abundant deposit feeders, the entire surface layer of bottom sediment may pass through some animal's digestive tract several times each year! Indeed, deposit feeders in fine sediments can completely pelletize bottom sediments to produce a "fluff" layer of repackaged sediment at the sediment-water interface that goes into resuspension quite easily when disturbed. Not surprisingly, the relative importance of terrestrial sources declines with distance from land. Emerging tools such as lipid analysis and stable isotopes increasingly help researchers identify specific food sources in the marine environment. Food webs even within sediments can be quite complex (Figure 22.21a), involving transfers between detritus, microbes, meiofauna, and potentially to megafauna such as flatfish. Burrowing metazoans live near the oxygenated sediment-water interface; however, a complex community of microbes lives in hypoxic and anoxic sediments deeper within the sediment (Figure 22.21b). These microbes use different molecules as energy sources as species composition transition from those that use respiration in the aerobic layer to others that depend on methanogenesis and eventually fermentation in order to obtain energy. The efficiency of these reactions becomes less and less as availability of oxygen and oxygenated compounds decreases with depth within the sediment, meaning that living biota obtain less nutritional value.

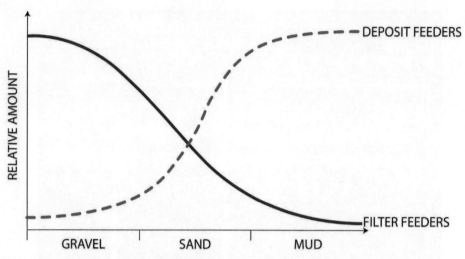

Figure 22.20 Relationship between substrate and feeding type.

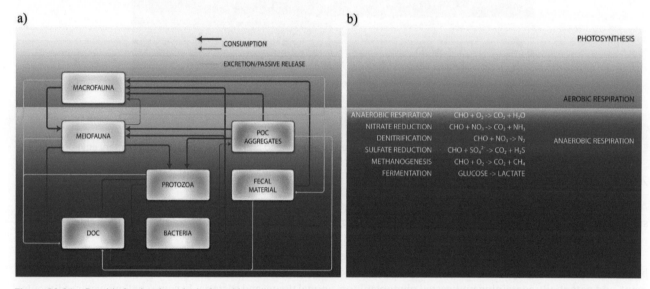

Figure 22.21 Benthic food web typical of muddy sediments, dominated by deposit feeders. (a) Protozoa provide a link between dissolved organic carbon (DOC) and particulate organic carbon (POC) and other food web components. Light brown sediment indicates the oxygenated layer and black sediment indicates the hypoxic through anoxic layer. (b) Changing metabolic pathways used by microbes as oxygen decreases with increasing sediment depth. In the uppermost layer, oxygen acts as the electron acceptor whereas nitrate and then sulfate serve that role with increasingly lower oxygen levels. Eventually, fermentation becomes the dominant process.

In contrast to deposit feeders, suspension feeders use a diversity of suspended particles that includes phytoplankton ranging in size from a few microns to hundreds of microns, bacteria suspended in seawater (< 5 μm), bacteria and other microbes living on detritus and particles in suspension, particulate matter, and dissolved organic matter (usually a minor food source). Suspension feeders include active and passive forms: active suspension feeders actively pump water across a filtering surface whereas passive suspension feeders capture particles as they flow past, using net-like structures or mucus.

Deposit feeders and suspension feeders in shelf sediments both play critical roles in marine food webs, though their relative importance varies among locations. Nonetheless, they provide a critical element of marine food webs and feedbacks to the pelagic realm (Figure 22.22).

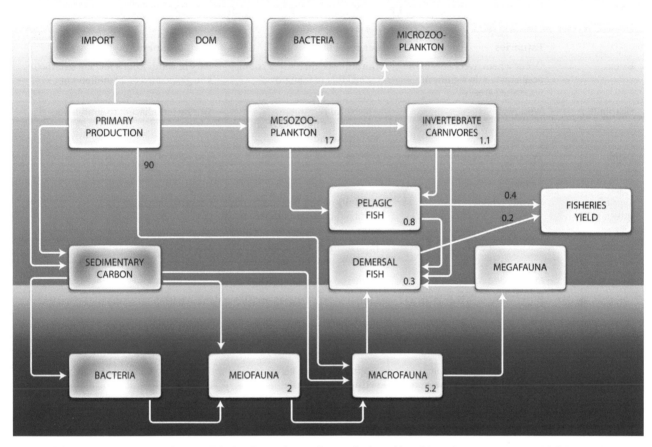

Figure 22.22 North Sea carbon flow model, showing pathways from pelagic and benthic pathways to commercial fisheries. Numbers adjacent to boxes denote production rates.

22.3.2 Sedimentary Environments and Ecosystem Functioning

Feeding by marine benthos plays an important role in marine ecosystem functioning in several keyways:

1) Sedimentary fauna play a significant role in global carbon cycling and **carbon sequestration**, referring to long-term burial as opposed to remineralization and release as carbon dioxide. Their feeding also transfers organic carbon out of sediments through food webs and organisms feeding on benthos (Figure 22.22), a process that increases availability of that energy to a broader array of taxa.
2) Microbes in sediments can metabolize some pollutants, and macrofauna in particular can influence rates of breakdown and burial depending on the degree to which they mix sediments. Macrofauna also create a potential mechanism to transfer pollutants up the food web, and thus help to mobilize them.
3) Macrofauna in particular significantly modify sediments, redistributing sediment particles vertically within the seabed and repackaging sediment grains as fecal pellets as they form a fluff layer. These activities can strongly influence sediment transport processes.
4) The complex interplay between microbes and metazoans alters sediment geochemistry, including oxygenation of sediment and release of nutrients through decomposition.
5) Suspension feeders can significantly improve water column clarity by removing phytoplankton, thereby altering phytoplankton dynamics and increasing water clarify.

Table 22.2 Comparison Among Coastal Sedimented Ecosystems

Main Characteristics	Habitat / Ecosystem		
	Estuaries	Seagrass beds	Shelf sediments
Extent	Occur where rivers flow to the ocean and drain significant watershed area	Not still well quantified but may exceed that of mangroves	Vary in width from a few km to >1000 km, global area of 27 million km^2.
Distribution	Worldwide	All coastal ecosystems (excluding Antarctica) both with mud and sand	Worldwide adjacent to continents
Depth	Typically 10s of meters but large estuaries can extend to 100s of meters.	0–70 m	0 to ~200 m
Type of substrate	Usually muddy	Typically on soft bottoms (sands or muds) but some species can grow on hard substrata	Gravel to fine mud
Biodiversity	Relatively low	High diversity of associated invertebrates and vertebrates	Moderate
Endemism	Species endemism low but genetically distinct populations occur	Numerous endemic species	Generally low
Functioning	Often highly productive and critical nursery habitat for juveniles	High production, high export of organic matter	Provide trophic support for commercial fisheries and key sites of remineralization, carbon storage, and nutrient regeneration
Trophic networks	Relatively simple trophic structure reflecting low diversity and (often) abundant nutrients	Trophic food web based on detritus production	Detritus plays a key role, with food webs based on suspension and deposit feeders
Peculiarity	Species adapted to fluctuating conditions.	Systems depend on one or a few species of seagrasses that function as *ecosystem engineers*	Highly dynamic environments
State of art	Generally well-studied, facilitated by ease of access and importance to humans.	Very well studied historically, among the most known habitats in the world	Shelf environments in developed countries well studied but major gaps in tropics and developing countries.

In summary, the coastal habitats provided by estuaries, seagrasses and shelf sediments all support critical ecosystem function and biota, but vary in their extent and distribution (Table 22.2).

Questions

1) List the characteristics of estuarine ecosystems.
2) Describe survival strategies of estuarine biota.
3) Describe an estuarine food web.
4) Why are estuaries important?
5) What are the main ecological characteristics of seagrass systems?
6) Why are seagrass meadows important?
7) Provide definitions of: rhizome, matte, and banquette. Describe their ecological role.
8) How do seagrass meadows interact with their adjacent habitats?
9) Describe the biodiversity associated with seagrasses.
10) What are the main food sources for sedimentary fauna?
11) What are the main differences between estuaries, seagrass beds, and shelf sediments?

Suggested Reading

Arnaud-Haond, S., Duarte, C.M., Diaz-Almela, E. et al. (2012) Implications of extreme life span in clonal organisms: Millenary clones in meadows of the threatened seagrass *Posidonia oceanica*. *PLOS One*, 7, p. e30454.

Bertolino, M., Calcinai, B., Capellacci, S. et al. (2012) *Posidonia oceanica* meadows as sponge spicule traps. *Italian Journal of Zoology*, 79, pp. 231–238.

Buia, M.C. and Mazzella, L. (1991) Reproductive phenology of the Mediterranean seagrasses *Posidonia oceanica (L) delile, Cymodocea nodosa (ucria) aschers*, and *Zostera noltii hornem*. *Aquatic Botany*, 40, pp. 343–362.

Calvo, S., Tomasello, A., Di Maida, G. et al. (2010) Seagrasses along the Sicilian coasts. *Chemical Ecology*, 26, pp. 249–266.

Cochran, J.K. (2014) *Estuaries. Reference Module in Earth Systems and Environmental Sciences*. Elsevier.

Den Hartog, C. and Kuo, J. (2007) Taxonomy and biogeography of seagrasses, in Larkum, A.W.D., Orth, R.J., and Duarte, C.M. (eds.), *Seagrasses: Biology, Ecology and Conservation*. Springer, pp. 1–23.

Diaz-Almela, E., Marbà, N., Álvarez, E. et al. (2006) Patterns of seagrass (Posidonia ceanica) flowering in the western Mediterranean. *Marine Biology*, 148, pp. 723–742.

Gambi, M.C., Lorenti, M., Russo, G.F. et al. (1992) Depth and seasonal distribution of some groups of the vagile fauna of the *Posidonia oceanica* leaf stratum: Structural and trophic analyses. *Marine Ecology*, 13, pp. 17–39.

Neira, C., Grosholz, E.D., Levin, L.A., and Blake, R. (2006) Mechanisms generating modification of benthos following tidal flat invasion by a *Spartina* hybrid. *Ecological Applications*, 16, pp. 1391–1404.

Norling, K., Rosenberg, R., Hulth, S. et al. (2007) Importance of functional biodiversity and species-specific traits of benthic fauna for ecosystem functions in marine sediment. *Marine Ecology Progress Series*, 332, pp. 11–23.

Pergent-Martini, C., Leoni, V., Pasqualini, V. et al. (2005) Descriptors of *Posidonia oceanica* meadows: Use and application. *Ecological Indicators*, 5, pp. 213–230.

Rife, G. (2018) Ecosystem services provided by benthic macroinvertebrate assemblages in marine coastal zones, in Hufnagel, L. (ed.), *Ecosystem Services and Global Ecology*. IntechOpen.

Short, F., Carruthers, T., Dennison, W., and Waycott, M. (2007) Global seagrass distribution and diversity: A bioregional model. *Journal of Experimental Marine Biology and Ecology*, 350, pp. 3–20.

Terlizzi, A., Anderson, M.J., Fraschetti, S., and Benedetti-Cecchi, L. (2007) Scales of spatial variation in mediterranean subtidal sessile assemblages at different depths. *Marine Ecology Progress Series*, 332, pp. 25–39.

Vacchi, M., Montefalcone, M., Bianchi, C.N. et al. (2010) The influence of coastal dynamics on the upper limit of the *Posidonia oceanica* meadow. *Marine Ecology*, 31, pp. 546–554.

23

Polar Ecosystems

Polar environments, extending from ~66° N and S to the North and South Poles, receive less energy and heat from the sun relative to lower latitudes, resulting in substantial sea ice cover for a large part of the year. The Arctic and Antarctic (or Southern) Oceans both sit above the **polar circles**, meaning within regions that experiences at least one 24-h period when the sun remains below the horizon and another period where the sun never drops below the horizon. Despite sharing these major features, the two oceans differ from each other in many important ways. In the Antarctic, the Southern Ocean surrounds the ice-covered Antarctica landmass with typical water temperatures of about −1 °C (to −1.9°C in the deeper portions). The Antarctic Circumpolar Current that flows around Antarctica without interruption by a landmass further isolates the fauna from adjacent water bodies. In the Arctic, in contrast, land masses surround the Arctic Ocean, partly covered by ice ranging in thickness from a few centimeters to 2 m, with temperatures of about −1.9 °C. The low temperatures, presence of ice, and absence of light for many months of the year require specific adaptations in the biota, contributing to high rates of endemism, particularly in the Antarctic. In contrast to other environments, polar ecosystems support not only pelagic and benthic communities, but also **sympagic** communities (i.e., associated with sea ice), that form a kind of "inverted garden" on the underside of the ice. Sympagic communities include many microalgae (including >200 diatom species), protozoans, meiofauna such as turbellarians, nematodes, and rotifers, and macrofauna, such as the endemic gammarid amphipods that thrive on the underside of ice floes.

Because of the curvature and inclination of the Earth and Earth's orbit around the sun, polar regions effectively experience just two seasons: summer and winter. During summer the sun never sets, or sets for short periods, whereas in winter it either never rises, or rises only for short periods. Geographically located at Earth's antipodes, the Arctic and Antarctic Oceans differ from each other in many aspects. Sea ice covers the vast Arctic Ocean during far northern (**boreal**) winter. Arctic waters hover close to the freezing point of seawater at a salinity of 35 and pressure of 1 atm. Starting in late boreal spring and during the summer, Arctic ice becomes thin enough for icebreakers to cross. In recent years, global warming has reduced the thickness and winter extent of Arctic ice, opening up traditionally inaccessible routes, such as the Northwest Passage, to commercial navigation without the need for icebreakers. The Southern (Antarctic) Ocean surrounds Antarctica, a continent perpetually 98% covered by continental ice with thicknesses in excess of 4 km (Figure 23.1). In fact, the mass of Antarctic continental ice pushes most of the continent below average sea level and contains the largest reservoir of fresh water on the planet (about 85% of all fresh water, including rivers and lakes). The ice cap increases the average elevation of Antarctica by 2,500 meters, about 1,500 m more than the average height of other continents. With a maximum ice thickness of ~4,000 m, its immense weight compresses the continental land mass beneath it, so that it actually sits below average sea level.

Marine Biology: Comparative Ecology of Planet Ocean, First Edition. Roberto Danovaro and Paul Snelgrove.
© 2024 John Wiley & Sons Ltd. Published 2024 by John Wiley & Sons Ltd.
Companion Website: www.wiley.com/go/danovaro/marinebiology

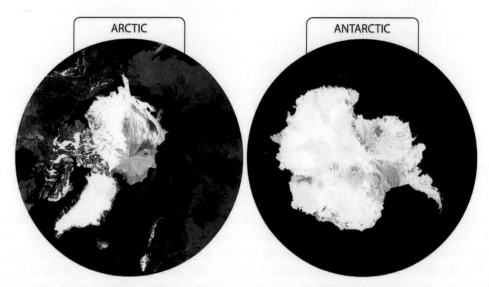

Figure 23.1 Comparison between geographic extent of Arctic and Antarctic, with maps scaled the same. *Source:* Michael Tangerlini.

Box 23.1 What are Polynyas?

Polynyas (also called Polynias) refer to persistent open waters surrounded by sea ice. This term originally referred to the navigable polar sea zone, but today indicates a thawed sea surface surrounded on one or more sides by compact sea ice (pack ice). Polynyas form primarily through two processes: 1. Warmer waters rising from depth maintain surface water above the freezing point (slowing or even preventing formation of new ice); 2. Drifting of newly formed ice from the coast caused by strong, cold, and dry **katabatic winds** (a wind that carries high-density air from a higher elevation down a slope under the force of gravity) pushes ice far from the coast (Figure 23.2). Typically, Antarctic polynyas generated by katabatic winds occur in areas of super-cold, super-salty, and super-dense water formation, which sink and flow at great depth to multiple oceans around the world.

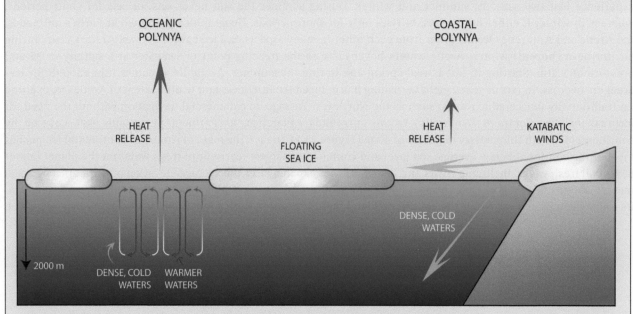

Figure 23.2 Schematic representation of a polynya and of the main factors responsible for ice not forming and thus enabling polynyas.

The Arctic Ocean averages only ~1,200 m in depth, reaching its lowest depths on the vast continental shelf along the Siberian and Scandinavian coasts, where its deepest point measures 5,450 m. In contrast, the Southern Ocean averages ~4,500 in depth, with a maximum depth of 7,235 m.

The geological histories of the two systems also differ greatly, with major implications for marine speciation. Over 170 million years ago, Antarctica was part of the supercontinent **Gondwanaland**. When Gondwanaland broke apart ~25 million years ago, one of its fragments formed the Antarctica continent we see today. In the past, Antarctica sat further to the north, in a more temperate zone, covered by forests and diverse life forms (including terrestrial marsupials now confined to Australia and South America). These characteristics sharply contrast today's cold, dry, and ice-covered landscape. The Arctic ice cap that covers the Arctic Ocean has contracted and expanded over the past million years through some 26 different glaciations. The current Arctic ice cap dates back at least five million years, with some estimates up to 15 million years.

In addition to their polar antipodal separation, large and deep oceans with strong thermal gradients have helped to isolate the Arctic from the Antarctic for millions of years. This separation has largely driven biodiversity divergence between the two polar regions of this planet, to the point that distinctive faunal elements characterize each pole with total absence of some taxa at the opposite pole. For example, polar bears live only in the Arctic, whereas penguins, leopard seals, and many other species occur only in the southern hemisphere and especially in Antarctica; polar bears and penguins meet only in comic strips, and possibly in zoos! These differences also result in very different food webs. In Antarctica, krill (euphausiids) play a pivotal role in linking primary producers to pelagic fishes, penguins, whales, and leopard (and other) seals. Rapid increase in depth with distance from the Antarctic continent leads to a reduced role for the benthic component compared to Arctic food webs, where a wide, comparatively shallow continental shelf enables an important benthic food web in parallel with the pelagic food web where copepods link to small fishes, which link to large fishes and then seabirds, whales, seals and (of course) polar bears.

23.1 Biogeography and Characteristics

The Arctic Ocean surrounds the North Pole, surrounded by the North American, Asian, and European continents (Figure 23.3). The name Arctic originates from the polar constellation Arktos (Greek for "bear"). The Arctic Ocean connects with the Pacific Ocean through the Bering Strait. Underwater mountain ranges that extend from Scotland to Greenland to Baffin Island at depths of 500–700 m define the broader boundary with the Atlantic Ocean. Numerous rivers flow into the Arctic Ocean whose total area covers approximately 14 million km^2, including the Greenland Sea, Norwegian Sea, North Sea, and Barents Sea. Its unique characteristics include: (1) Largely ice covered (mostly continuous, but some seasonal), with winds and currents driving ice movements; (2) Relative isolation from other oceans, somewhat like the Mediterranean Sea and other semi-enclosed basins. Ice areal coverage ranges 7–16 ×10^6 km^2 (in summer and winter respectively), averaging 2–3 meters in thickness.

Relatively warm and salty Atlantic surface waters enter into the Arctic Ocean near Europe and cool, sink, and intermingle with deep waters (Figure 23.4). Because low temperatures increase oxygen solubility, when waters sink in the Arctic and move toward more temperate latitudes they oxygenate deep layers of the ocean. Within this dynamic exchange, waters originating from the Gulf Stream sink in the Arctic to form **North Atlantic Deep Water** (NADW). The same process also occurs in Antarctica, forming **Antarctic Bottom Water** (AABW). These water masses form one of the main climate drivers of our planet, known as thermohaline circulation or in this specific example, the "deep ocean (or great ocean) conveyor belt" (Figure 23.5).

Cold and relatively variable temperatures characterize Arctic waters. Alternating warm and cold layers in the Arctic Ocean vary in thickness over 3–5 y periods, likely reflecting cyclic events rather than progressive changes. The large Russian Pechora and Northern Dvina Rivers deliver an average freshwater supply of ~246 km^3 into the Barents Sea.

Figure 23.3 The Arctic Ocean.

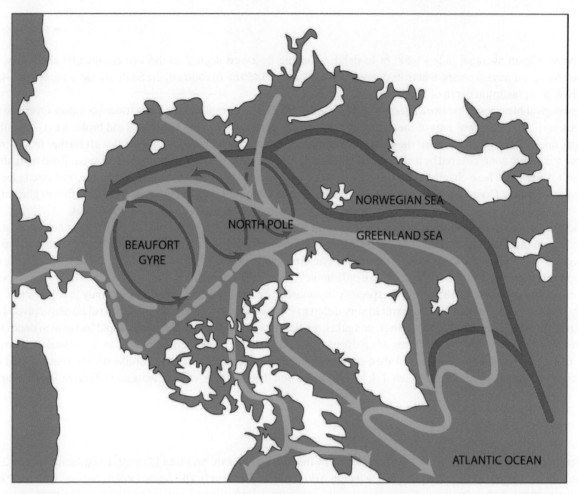

Figure 23.4 Schematic illustration of Arctic circulation. Red lines denote surface currents, and blue lines denote deep currents. A warm current originating from the Gulf Stream runs north along the west side of Greenland, resulting in less ice near Greenland compared to Canada.

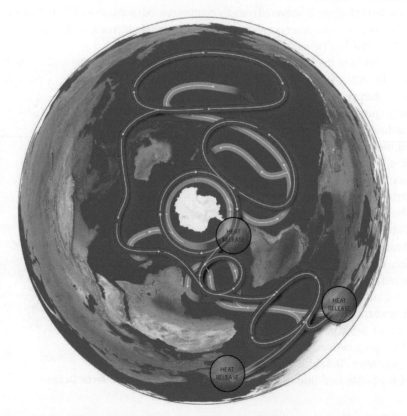

Figure 23.5 Thermohaline circulation of the global ocean. Red lines denote shallow, warm currents whereas blue lines denote deep, cold currents.

The northern extension of the mid-Atlantic Ridge divides the deep basins of the Greenland and Norwegian Seas, which connect widely with the Arctic Ocean. The shallow Barents and White Seas number among the less contaminated seafloors globally. The Barents Sea connects to the North Sea through the gap between Norway and the island of Svalbard and opens to the North into the Arctic Ocean.

The Arctic Ocean generally resembles the Antarctic continent in dimensions and shape. Three large parallel mountain ranges separate the seabed between Greenland and Siberia. The Arctic mid-ocean ridge (the Nansen-Gakkel Ridge) stretches across the northern basin of western Russia. The polar abyssal plain, an elongated depression on the flat sea floor, stretches between the Arctic Ridge and the Lomonosov Ridge, an underwater mountain range that rises to 3,084 m on average from the bottom of the abyssal plain. Between Canada and Alpha ridge, the largest sub-Arctic basin, the Canadian Basin, spans much of the Arctic Ocean averaging 3,658 m in depth.

In the Arctic Ocean, algal blooms, dominated by diatoms and the brown alga, *Phaeocystis pouchetii*, support a high zooplankton biomass with abundant copepods and euphausiid crustaceans. Arctic waters host some 240 species of fish as well as a rich benthic fauna (Table 23.1). To date, of the 4,000+ species of known multicellular invertebrates in the Arctic, > 90%

Table 23.1 Total taxa: number of registered scientific names at all taxonomic levels (sum of species, genera, families, etc.). Data from the Arctic Marine Species registry (ARMS), noting ongoing updates from the Conservation of Arctic Flora and Fauna (CAFF, the Arctic Council's biodiversity working group) and Circumpolar Biodiversity Monitoring Program (CBMP).

	Total taxa	Confirmed species
Biota	10,872	5,849
Animal Kingdom	9,443	5,137
Phyla		
Annelida	1,033	569
Arthropoda	3,051	1,918
Brachiopoda	57	11
Bryozoa	504	281
Cephalorhynca	27	13
Chaetognatha	22	10
Chordata	965	523
Cnidaria	577	285
Ctenophora	30	6
Echinodermata	322	130
Echiuroidea	10	2
Gastrotricha	35	13
Gnathostomulida	4	1
Hemichordata	5	3
Mollusca	1,213	590
Nematoda	581	308
Nemertea	136	81
Phoronida	2	1
Platyhelminthes	273	105
Porifera	433	229
Rotifera	30	14
Sipuncula	39	18
Tardigrada	17	4

(Continued)

Table 23.1 (Continued)

	Total taxa	Confirmed species
Xenacoelomorpha	52	24
Kingdom Plantae	153	49
Subkingdom Biliphyta	111	36
Subkingdom Viridiplantae	40	13
Subkingdom Protozoa	2	0
Kingdom Chromista	1269	663

live in sediments. The highly structured and complex Arctic ecosystem includes pre-apex cephalopod and fish predators that birds and mammals (particularly pinnipeds) prey upon. Recent research shows that some seastars in the benthic food web play a role as apex predators, much as polar bears do for the pelagic food web.

23.2 Biodiversity

Despite >100 years of Arctic research, until recently we lacked a single, relatively complete list of species in the Arctic Ocean and surrounding seas. One of the projects of the *Census of Marine Life*, called *Arctic Ocean Diversity* (ArcOD), examined Arctic Ocean biodiversity and created an *Arctic Register of Marine Species* (ARMS), summarized in Table 23.2. This register includes all known Arctic multicellular marine animals, but, in recent years, researchers have gradually added data on phytoplankton, macrophytes, and eukaryotic protistan biodiversity.

The low temperatures, ice presence, and absence of light for many months of the year require significant adaptation and specialization in Arctic species, resulting in numerous endemic species. For convenience, we can group biodiversity into three habitats: sympagic, pelagic, and benthic.

23.3 Biodiversity Within Sea Ice

Sea ice is a unique characteristic of polar oceans. Sea ice extent and thickness vary with season, with rapid ice forming during winter months and significant melting in summer. **Multi-year ice**, which persists through the year, enables the evolution of an endemic community associated with that unique environment. Ice formation leaves small spaces between ice crystals filled with **brine**, which refers to seawater with particularly high salt content and high concentrations of inorganic nutrients. The ice crystals and brine channels form a complex three-dimensional network of small channels with diameters ranging from less than one millimeter to a few centimeters. These channels provide habitat for a specialized sympagic community adapted to the unique conditions in this matrix. Organisms living within the ice tolerate a wide range of

Table 23.2 Numbers of currently known sympagic species.

Phyla	Group	Species richness
Diatoms		731
Other protists		296
Cnidaria	Hydromedusan jellyfish	1
Nematodes		11
Annelids	Polychaetes	4
Crustaceans	Copepods	12
	Amphipods	7
Fish	Gadidae	2 (species of cod)

environmental conditions and rapid changes in light, temperature, and salinity. These fluctuations influence the distribution of the biota within the ice, concentrating most of the biomass in the lower ice layers at the interface between ice and sea water. During periods of complete ice cover, the sympagic and planktonic communities interact strongly. The sympagic community can live both inside pores on the lower surface of the ice and inside the ice channel, or at the ice-water interface. Permeating the complex ice channels dictates a smaller size than "typical" Arctic pelagic and benthic organisms. Bacteria, unicellular algae, and small metazoans dominate ice community biomass (Table 23.2), whereas turbellarians, nematodes, rotifers, and crustaceans dominate in abundance.

Diatoms are the most important sea ice primary producers, and ice algal primary production contributes some 4–26% of total primary production annually in waters covered by seasonal ice. This contribution can exceed 50% in waters covered by multi-year ice, because ice reduces penetration of short wavelengths of light into the water column. Researchers also recently demonstrated significant photosynthetic activity a few meters beneath the ice that had previously gone undetected. Dissolved organic matter production within ice, mainly deriving from algal "waste" and "sloppy feeding" by sympagic crustaceans, supports a large prokaryote biomass within sea ice. Viruses and fungi exhibit surprisingly high diversity within these extreme habitats, which also support small flagellates, of significant but unknown total biodiversity. Protists and meiofauna, especially turbellarians, nematodes, and rotifers can reach high abundances within the ice; during the spring, larvae and juveniles of zooplankton and various benthic animals (including fish and crustaceans) migrate towards the ice and coasts to feed on ice algae, in some cases entering brine channels to feed on the rich sympagic community and escape pelagic predators.

A specialized endemic fauna, consisting mainly of large gammarid amphipods (e.g., *Thermisto* spp.), thrives on the underside of ice floes at abundances in excess of 100 indiv. m^{-2}. These amphipods represent the main prey of *Boreogadus saida* (polar cod), a species preyed upon, in turn, by seals, birds, and whales. In spring, calanoid copepods (e.g., *Calanus glacialis*) undertake daily vertical migrations from the deepest portions of the water column to feed on abundant phytoplankton in the upper water column, close to the underside of ice floes. Eubacteria and Archaea both occur within Arctic ice; the most common groups include Alphaproteobacteria, Gammaproteobacteria, and Crenarchaeota. Seasonal studies show abundance patterns similar to those in primary producers. Diatoms represent 50–75% of total protists typically found within sea ice. Until recently, diatoms were considered among the most productive and abundant microorganisms in sea ice. But in addition, researchers now recognize extraordinary biodiversity that includes hundreds of species of unicellular primary producers. Experts have identified >700 species, most of which occur throughout the entire Arctic Ocean (Figure 23.6), both within ice and in the pelagic environment. Colonies of pennate diatoms, such as *Fragilariopsis oceanica* or *Nitzschia frigida*, usually dominate, in addition to abundant colonies of centric species, such as *Eucampia groenlandica*. German scientists recently reported that the centric diatom, *Melosira arctica*, may represent the single most productive algal species in the Arctic.

Arctic sea ice also supports endemic species of nematodes, such as *Theristov melnikov*, even if we still lack information on their basic biology. All three main groups of copepod crustaceans (Calanoida, Cyclopoida, Harpacticoida) number among the most common invertebrates in the Arctic Ocean, and sea ice is no exception with high abundances of the cyclopoids, *Cyclopina gracilis* and *C. schneideri*, as well as the harpacticoids *Harpacticus* spp., *Halectinosoma* spp., and *Tisbe furcata*. These species, the primary consumers of algae, and sympagic protistans, have adapted to live on the ice surface and within the micro-channels. Adults and nauplii of different pelagic species (e.g., *Pseudocalanus* spp., and *Calanus glacialis*) also live

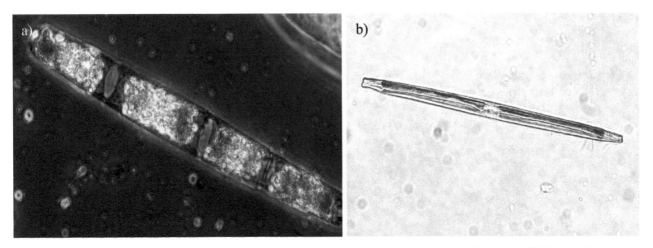

Figure 23.6 Examples of microalgal organisms associated with the sympagic (sea ice) community: (a) colonies of *Nitzschia frigida*, and (b) *Eucampia groenlandica*, two of the most abundant Arctic sympagic species. *Source:* Kristian Peters / Wikimedia Commons / CC BY-SA 3.0.

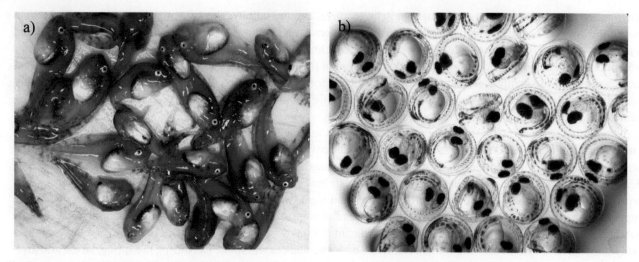

Figure 23.7 (a) Larval stages; (b) juvenile of fish sampled in polar environments. *Source:* Francesco Regoli.

on and within sea ice. Three amphipod genera *Gammarus, Onisimus*, and *Apherusa* dominate the underside of Arctic ice floes. Researchers recently discovered that many Arctic species considered exclusively benthic also living in close association with the ice. Frequently, large numbers of Arctic cod larvae associate with the ice, feeding on ice biota (Figure 23.7).

23.4 Pelagic Biodiversity

The microbial component of the water column includes prokaryotes, phytoplankton, and a complex and heterogeneous group of protistans (micro-zooplankton). Over 100 years of research on Arctic phytoplankton has yielded significant understanding of the most abundant groups (diatoms and dinoflagellates), but we know relatively little about some of the other 500–1000 phytoplankton species. In contrast, intensive studies currently underway address major knowledge gaps on bacterial and protistan diversity. Molecular techniques suggest that the Arctic Ocean supports many thousands of bacterial species.

Snow and ice cover, the low sun angle, and relatively short seasons limit phytoplankton seasonal growth in the Arctic Ocean. Generally, phytoplankton productivity begins in April and lasts until September, with a unimodal peak in biomass between late June and early July. Typically, the annual planktonic herbivore biomass peak coincides with this phytoplankton biomass peak with almost no time lag. Biological activity mainly develops in pelagic shelf waters, where seasonal ice retreat enables algal blooms to develop (Figure 23.8), mainly dominated by diatoms and the alga, *Phaeocystis pouchetii*.

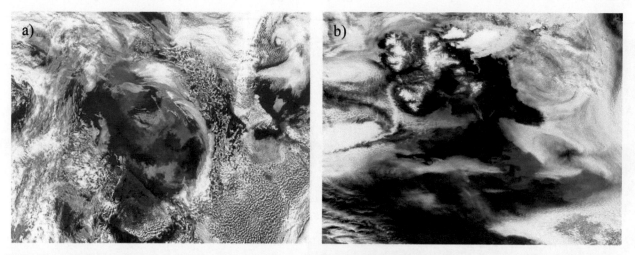

Figure 23.8 Satellite images of two phytoplankton blooms: (a) the Barents Sea and (b) off the Svalbard Islands. The green color of the bloom contrasts the deep blue of non-bloom locations and increases their visibility. *Sources:* Norman Kuring / Wikimedia Commons / Public domain, NASA / Public Domain.

For more than a century, scientists sampled animals drifting in the water column with plankton nets. Towed blindly through the water over the sides of ships, these nets selectively sample small, more robust plankton. As a result, science has sampled only a small fraction of pelagic diversity, and cannot accurately predict when, where, and how organisms regulate the flow of matter and energy through marine food webs. Comparatively more information exists on dominant mesozooplanktonic crustaceans, primarily copepods and euphausiids. Until recently, researchers considered the mesozooplankton unimportant for ecosystem functioning. However, recent studies demonstrate higher rates of ingestion, growth, and reproduction than in other crustaceans, allowing mesozooplankton to respond more rapidly to changing primary production – especially in polynyas (see Box 23.1, "Polynyas"). Off Canada's Arctic coast, ctenophores consume up to 9% of the largest copepods. Therefore, other gelatinous predators (such as jellies and siphonophores), when abundant, could presumably achieve similar ecological impact.

Because of their high abundance and ease of sampling, scientists have accumulated substantial knowledge on the taxonomic composition and life cycles of the most common Arctic copepods (Table 23.3). In contrast, comparatively poor knowledge exists on the smallest species of copepods, deep-sea species, and gelatinous zooplankton, which present greater sampling challenges. Historically, many studies focused on the genus *Calanus*, which dominates mesozooplankton biomass. In many oceans, smaller copepod species dominate numerically, but few studies have assessed their role in the Arctic Ocean. These groups probably include many undiscovered Arctic heterotrophic planktonic species. Relatively few studies have focused on appendicularians, for example, a unique group of pelagic tunicates that reach high abundances in Arctic polynyas and in the Central Arctic. Similarly, despite their abundance and importance, few studies have addressed chaetognaths, amphipods, ctenophores, and cnidarians. Researchers almost certainly underestimate the importance of grazing by ctenophores and cnidarians, both in surface and deep waters.

At least 70 species of planktonic cnidarians occur in the Arctic, including siphonophores (8 known species), hydrozoans, and ctenophores (10 known species, 5 of which were only recently discovered; Figure 23.9).

The Arctic Ocean contains >150 species of planktonic copepods, although we know little about their biology. The larger species can live 3–4 years, whereas smaller species complete one or more generations per year. The 9 species of known pelagic ostracods include *Boroecia maxima*, which dominates the group. The known Arctic planktonic fauna also includes a dozen species of amphipods (Figure 23.10), nine euphausiid (krill) species, some rare taxa, and 3 species of pteropods (Figure 23.11a).

Table 23.3 Numbers of currently known species of pelagic invertebrates in the Arctic Ocean.

Phyla	Taxonomic group	Global diversity	Arctic Species Number	Central Arctic Species Number
Cnidaria	Hydromedusan jellies	650	~50	15
	Siphonophora	190	8	7
	Scyphozoa	150	7	3
Ctenophora		80	12	12
Nemertea		97	2	2
Annelida	Polychaetes	120	6	4
Mollusca	Heteropoda	35	0	0
	Pteropoda	160	3	2
	Cephalopoda	370	8	6
Crustacea	Cladocera	8	4	0
	Ostracoda	169	9(?)	8(?)
	Copepoda	2000	156	97
	Mysidacea	700	33	13
	Amphipoda	400	10	8
	Euphausiacea	86	7	3
Chaetognatha		80	5	5
Tunicata	Larvacea	64	5	5
	Pyrosomatida	8	0	0
	Dolioida	17	0	0
	Salpa	45	0	0

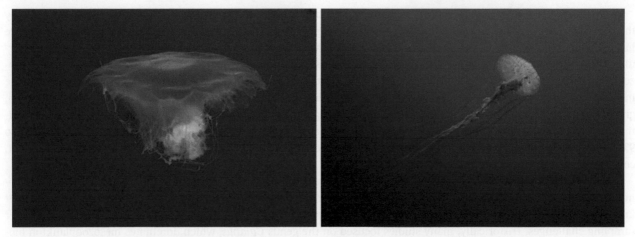

Figure 23.9 Two gelatinous Arctic megazooplankton taxa: *Cyanea capillata* (left) and *Chrysaora* sp. (right). *Sources:* Brian Gratwicke, Katrin Iken / NOAA / Wikimedia Commons / CC BY 2.0.

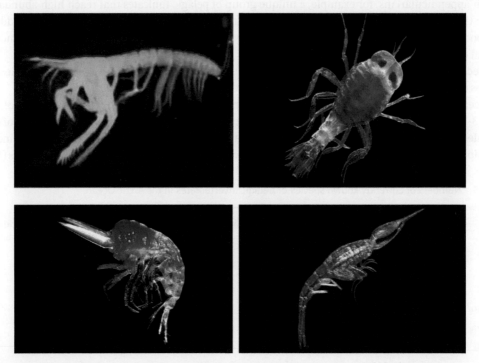

Figure 23.10 Crustaceans (isopods and amphipods) that occur in the Arctic Ocean. *Source:* Russell R Hopcroft.

The Arctic Ocean houses only a few known species of holoplankton polychaetes (Figure 23.11c), including the deepwater species, *Bathypolaris carinata*, and 5 known species of chaetognaths.

23.5 Fishes

The known 240 species of Arctic marine fishes include some anadromous and diadromous species (Figure 23.12). Two taxonomic groups comprise more than half of the species (55%): with 30% from the suborder Cottoidei (order Scorpaeniformes), and 25% from the suborder Zoarcoidei (order Perciformes). Most Arctic marine fishes are benthic or demersal, with few pelagic species such as *Benthosema glaciale*, which live at mesopelagic depths below 1,250 m during the day and ascend into epipelagic depths at night. Fisheries catch data, historical collections, and published papers form the basis for distribution maps of Arctic fishes. With relatively few photographs of Arctic species in their natural habitat, shipboard or laboratory photographs represent much of the current photographic record.

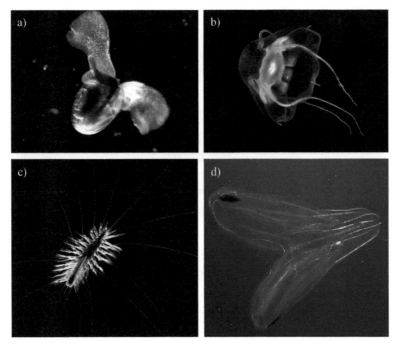

Figure 23.11 Examples of Arctic planktonic organisms: (a) pteropod; (b) jellyfish; (c) polychaete; and (d) jellyfish. *Sources:* Russell R Hopcroft, NOAA/Russ Hopcroft/Public Domain" with "Russell R Hopcroft.

23.6 Marine Mammals

Twelve species of marine mammals spend all or much of their life history in the Arctic Ocean (Figure 23.13), including four species of cetaceans (two Mysticetes – gray whales and bowhead whales, and two Odontocetes – narwhals and belugas), polar bears, walruses and six seal species. Several species of primarily terrestrial mammals that live in the Arctic interact with the coastal ocean, including Arctic fox, ermine, pine marten, polar wolf, caribou, reindeer, musk oxen, lemming, and Arctic hare. Despite their wide taxonomic range, all Arctic marine mammals are predators that live relatively long lives and provide a critical food source for Arctic Indigenous peoples.

23.7 Benthic Biodiversity

The rich Arctic benthic fauna consists of numerous species of invertebrates adapted to near-freezing water temperatures through enzymes that work at low temperatures and at low metabolic rates. Many Arctic species live 10x longer than their tropical counterparts. In shelf environments, crustaceans (especially amphipods), polychaetes, and bivalves dominate the macrofauna, many living within sediments. In contrast, echinoderms, and especially ophiuroids, dominate the epibenthic megafauna, reaching densities >200 indiv. m^{-2}. Other abundant epibenthic fauna include urchins in the Barents Sea and holothurians in the Laptev Sea. In contrast to the continental shelf, near perennial ice cover and the resulting challenges for research vessels and sampling tools limit access to deep Arctic environments. The dominant groups of organisms within Arctic deep-sea sediments closely resemble those in shelf sediments, namely polychaetes, crustaceans, and bivalves. Less frequently, these sediments also support sponges, cnidarians, tunicates, ophiuroids, and

Figure 23.12 Examples of Arctic fishes (elasmobranchs and Osteichthyes). *Source:* Kitty Mecklenburg / Wikimedia Commons / Public Domain.

Figure 23.13 Arctic marine mammals on the ice pack are dominated by pinnipeds. *Sources:* Alan Wilson / Wikimedia Commons / CC BY-SA 3.0, Blatant World / Flickr / CC BY 2.0, Merrill Gosho / NOAA / Wikimedia Commons / Public domain, Dr. Kristin Laidre / NOAA / Public Domain, Dr. Kristin Laidre / NOAA / Public Domain, Greg5030 / Wikimedia Commons / CC BY-SA 3.0, U.S. Fish and Wildlife Service / Wikimedia Commons / Public Domain, NOAA Seal Survey / Wikimedia Commons / Public Domain, Shawn Dahle/NOAA/Public Domain, Matthieu Godbout / Wikimedia Commons / CC BY-SA 3.0, NOAA Fisheries / Wikimedia Commons / Public domain.

other worm-like animals. To date, scientists have documented > 4,000 species of multicellular invertebrates in the Arctic Ocean (Table 23.4), > 90% of which live on the seabed. The Central Arctic supports ~350–400 species. Up until now, biodiversity studies in Arctic waters have primarily focused on macro- and megafauna, because of availability of simpler, standardized methodologies. Information on Arctic meiofaunal and microbial communities has recently begun to emerge.

Among the invertebrates, sponges (Figure 23.14a) thrive in the Arctic Ocean with ~160 known species, as well as cnidarians (Figure 23.14b), including Pennatulaceans, Ceriantharia, and gelatinous meroplankton benthic stages. More than 135 species of echinoderms (Figure 23.14b) live in the Arctic, and some of their distribution ranges span from the intertidal to the greatest depths of the central basin. In 1994, North American scientists published startling discoveries about the biology of this region, documenting major differences between Central Arctic species and surrounding regions, and demonstrating much more intense biological activity in both ice and water column habitat than previously recognized.

23.8 Food Webs and Functioning

The complex and highly structured Arctic food web includes significant primary production by sympagic diatoms, which provides critical trophic support for meiofauna and herbivorous zooplankton near the water-ice interface. Carnivores such as birds and mammals feed on pre-apex predators such as cephalopods and fishes (Figure 23.15). Seasonal ice melting releases significant particulate organic matter, including sympagic algae into the water column, which eventually settles to the seabed to support a rich and diverse benthic community. Some sympagic species released during melting become part of the phytoplankton community. Many benthic herbivores and carnivores spend portions of their life cycle as plankton (meroplankton), transported by ocean currents. Gelatinous zooplankton can achieve impressive concentrations during summer, when plankton blooms develop. Even the largest macro- or mega-plankton, such as ctenophores, can feed on the sympagic community once released into the water column.

Food supply and sediment composition, more than water temperature, influence functioning of the Arctic Ocean benthic compartment. On the Arctic continental shelf, the benthos receives substantial food input from the water column and, therefore, plays a particularly important role in secondary production of the entire ecosystem. High abundances of mammals that feed on the seabed reflect this remarkable secondary productivity. However, of this organic matter only a small amount reaches the deep ocean floor, resulting in much lower food availability and benthic biomass in deep Arctic benthos than on the shelf.

Table 23.4 Numbers of currently known species of benthic invertebrates in the Arctic Ocean. Note that experts regularly update these numbers as they make new discoveries.

Taxonomic group	Arctic species number
Macrophytes	191
Poriferans	162
Cnidarians	161
Nemerteans	77
Annelids	485
Bryozoans	331
Mollusks	480
Arthropods	1,317
Tunicates	57
Echinoderms	150
Sipunculids	12
Platyhelminths	116
Nematodes	296
Gnathostomulids	1
Tardigrades	4
Hemichordata	3
Brachiopods	9
Cephalorhynchs	13
Gastrotrichs	12
Echiurans	2
Phoronideans	1

23.9 Antarctica

Ice ranging in thickness of 0.5–10 m covers Antarctica for much of the year. The presence of ice helps differentiate habitats. As in the Arctic, brine channels in coastal ice can accommodate a diverse sympagic community. Researchers define Antarctica as a **high nutrient, low chlorophyll system**, which means high concentrations of nutrients, but low phytoplankton biomass because of iron limitation. Even so, the diatoms that number among the most abundant organisms in the ice, drive the high primary production (0.5 g of organic matter m^{-2} d^{-1}) that supports a high biomass of crustaceans (euphausiids, or krill, and amphipods) and benthic invertebrates such as the Antarctic bivalve, *Adamussium colbecki*, whose biomass may exceed 4–5 kg m^{-2}). The Southern Ocean supports quite high marine biodiversity, and pycnogonids, echinoderms, and polychaetes occur in particularly high abundances. In the Southern Ocean, notothenioid fishes include 174 species. Antarctica harbors impressively high levels of endemism, averaging ~50% of known total species from that region. Abrasion of ice against rocky substrates denudes the intertidal and subtidal by preventing encrusting organisms from settling at shallow depths. Sediments contain a rich and abundant biota, with numbers exceeding >130,000 indiv. m^{-2}.

Air temperatures in Antarctica average −60 °C in winter and −28 °C in summer (inland temperature), colder than the comparatively "balmy" Arctic with average winter temperatures of −40 °C in winter and 0 °C in summer. Two factors explain this difference. The continent of Antarctica receives less heat from the ocean. In comparison, relatively thin ice covers the Arctic, with a large ocean beneath it. Though hardly warm, polar seawater temperature, typically around −1 °C, provides comparative warmth to the frigid air temperatures. The ice absorbs a portion of this relative warmth, helping to moderate air temperature. Altitude also plays a significant role given the average elevation of Antarctica of ~2.3 km compared to much of the Arctic, which sits at sea level; these differences result in much colder air temperatures in Antarctica than in the Arctic.

23.9.1 Zonation, Extent, and Distribution

Antarctica, Earth's fifth largest continent in surface area, sits entirely below the 60th parallel of latitude (Figure 23.16). From satellites, Antarctica appears as a white, circular mass surrounded by the Atlantic, the Pacific, and the Indian Oceans, and whose waters constitute the only continuous oceanic environment. This body of water is somewhat inappropriately called the Southern Ocean or Antarctic Ocean, given that land masses rather than other oceans normally border true oceans. A long arm of land, the Antarctic Peninsula, extends from Antarctica toward South America, and two large bays, the Ross and Weddell Seas, occupy opposite sides of the continent. Antarctica spans a total area of ~13.8 million km^2, perpetually 98%

Figure 23.14 Polar benthic invertebrates. *Source:* Gianmarco Veruggio.

covered by ice. Antarctica and its ice cap significantly influence global climate and ocean circulation; complete melting of the Antarctic cap would raise sea level by ~70 m on average, with significant effects on Earth's geography, rotation speed, and ocean salinity. Increased coastal salinity, in tandem with warming of surface waters, could slow down or even completely stop dense, deep-water formation, thereby eliminating deep ocean circulation globally. The Antarctic continent receives sunlight for half of the year, during the austral summer from October to February, but sits in complete darkness during winter. The same phenomenon occurs in the Arctic, but with a reversed light-dark sequence. A cold circumpolar current surrounds Antarctica up to the Antarctic Convergence at ~54°S. Strong atmospheric perturbations occur at latitudes between 40° and 65°S, an area of strong wind formations known as the **roaring forties** and the **furious fifties**. The Antarctica ice mass consists of several glacial bodies and, in particular, two icecaps that cover East and West Antarctica (Figure 23.17).

About 245 million years ago, Antarctica sat in the subtropical zone, moving to its current polar location with continental drift and gradually accumulating ice. From the Cambrian onwards, Gondwana and Antarctica positions changed several

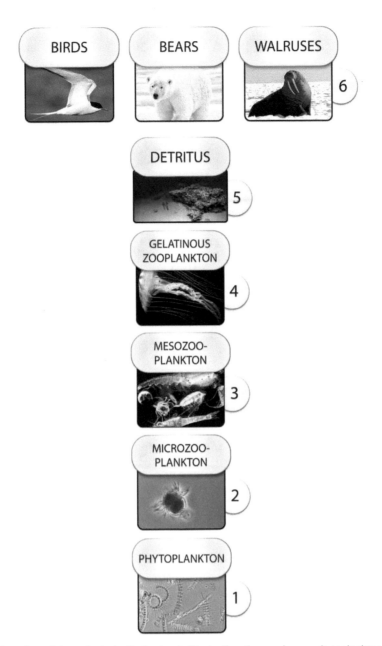

Figure 23.15 A simplified version of the pelagic Arctic food web, illustrating: 1 = producers: phytoplankton and algae within the ice that form the base of the food chain; 2 = microzooplankton feeding on phytoplankton; 3 = mesozooplankton, the larger zooplankton such as copepods and krill; 4 = gelatinous zooplankton, fish larvae, adult fishes, and whales that feed on mesozooplankton; 5 = detritus, including sympagic algal material, dead organisms, and fecal pellets that sink to the bottom; this material provides food for benthic organisms, including sea urchins, starfish, and bivalves that form an important benthic Arctic food web; 6 = top predators that include seals, walruses, and polar bears feeding on fishes. Sources: Alan Wilson / Wikimedia Commons / CC BY-SA 3.0, Joel Garlich-Miller / Wikimedia Commons / Public domain, National Oceanic and Atmospheric Administration (NOAA), Kevin Raskoff / Wikimedia Commons / Public domain, Matt Wilson/Jay Clark / Wikimedia Commons / Public Domain, NASA / Wikimedia Commons / Public domain.

times, passing through the geographic South Pole in the Silurian (430 Ma), migrating northward in the Triassic (200 Ma) and returning to the South Pole in the Cretaceous (100 Ma) when the fragmentation of the supercontinent had already started.

Antarctica is the last unexplored land mass on Earth. Its remote geographic location means less contamination by humans than the other continents, but Antarctica is not pristine. The presence of pollutants in Antarctic ice have allowed scientists to reconstruct the history of production and accumulation of contaminants of industrial or volcanic origin. Nonetheless, Antarctica's isolation makes it an excellent natural research laboratory.

Despite average annual temperature in the central plateau of Antarctica below –50 °C, warmer temperatures characterize coastal regions, especially the Antarctic Peninsula, which extends toward the North, beyond the Polar Circle. Monthly summer temperatures average ~0 °C. During the austral summer, coastal temperatures oscillate around 0 °C, peaking at 15 °C in the warmest locations. Inland environments experience lower temperature and vary from –15 °C to –35 °C.

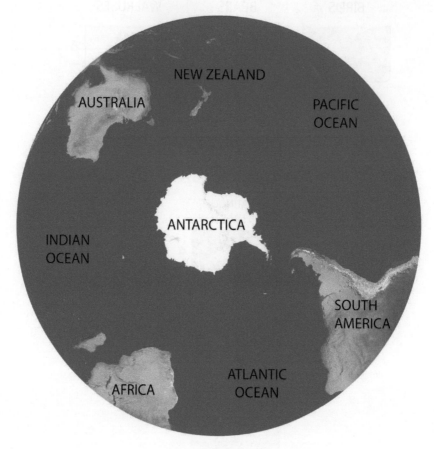

Figure 23.16 Antarctica.

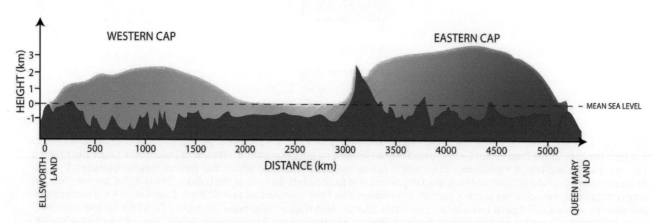

Figure 23.17 Transverse section of the two Antarctic ice caps showing ice thickness > 4 km in some locations.

In winter, temperatures range −15 °C to −30 °C along the coast, and from −40 °C to −70 °C inland. By comparison, researchers recorded the lowest temperature on Earth, −89.6 °C, in 1983 at 3,488 m altitude, at the Russian Antarctic base of Vostok. The central plateau of Antarctica, in particular, qualifies as a desert in that dry climate and scarce rainfall deliver no more than 10 mm of equivalent water annually; even the most arid areas of the Sahara receive ~20 mm of water annually! Fresh water is extremely rare on the surface, except in some coastal zones. A permanent anti-cyclonic weather system sits atop the continent, whereas the surrounding ocean experiences relatively low and variable pressure systems. The continent also experiences sudden katabatic winds that can reach speeds of 300 km h^{-1}. These winds become particularly frequent in spring, when warming coastal polynyas create updrafts that draw cold, dense air masses. These masses funnel along the slopes towards the sea at exceptional speeds.

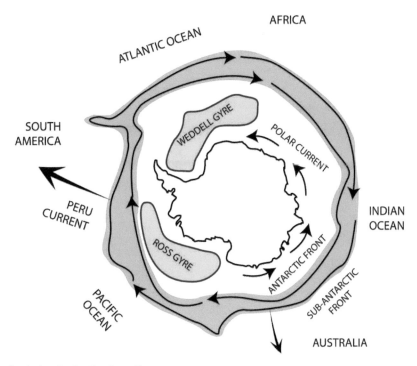

Figure 23.18 Surface circulation in the Southern Ocean.

The permanent Antarctic Circumpolar Current (ACC, Figure 23.18), the only surface current that flows around Earth uninterrupted by continents, creates a vortex that maintains physical isolation and thermohaline characteristics that differ from surrounding water masses. This physical isolation from about 60°S, plays a major role in defining the relatively distinct Antarctic biota. The ACC spans a width of ~2,000 km and moves the greatest mass of water on the planet, reaching depths 2,000–4,000 m. The prevailing winds, along with complex bottom topography characterized by large underwater mountain ranges, drive this current, generating branches that feed into all the adjacent oceans, and create upwelling along the coast.

Ice thickness varies between 0.5–10 m, with structural "anomalies", such as internal melting channels and bubbles. In contrast to salty seasonal sea ice, icebergs that detach from the numerous glaciers along the coast contain fresh water. Noting the near absence of rainfall, Antarctic Sea ice thickens with the onset of winter through progressive freezing of the water layer at the sea-ice interface. Ice begins to form close to the coast, successively expanding out to sea. Along some Antarctic coastal regions, permanent polynyas form super-dense and super-cold water in winter. When the thin layer of ice begins to form at the end of southern summer in mid-February, the thickness of this new ice rarely exceeds 10 cm, sometimes forming a layer <1 cm. The coastal ice shelf (**fast ice**), the ice that forms and remains attached to the edge of the continent for a long time, sometimes persisting until late summer, can detach and become forms part of the **pack ice**, referring to large areas of floating ice. When the coastal ice shelf detaches, it often leaves behind an **icefoot** that can potentially give rise to multi-year ice.

The production of new ice begins around mid-February with the formation of ice crystals that can form needles or small plates of ice, constituting the "oily" stage (**grease ice**), named for its resemblance to an oil slick. This stage can alternate with **frazil ice**, which resembles crushed ice slush. As freezing progresses, **pancake ice** forms, consisting of round plates up to meters in diameter that move with each other, colliding and coalescing (Figure 23.19). Freezing of seawater from the bottom thickens the ice, forming **young ice**, which becomes **gray ice** and then **gray-white** ice as the ice progressively thickens from below. The young ice (or first ice of the year) can vary widely in thickness, depending on the timing of formation. In particularly cold regions with strong winds, first-year ice can attain thickness ranging 120–180 cm. Usually, ice thicker than 130–140 cm consists of slabs (floes) produced previously that did not dissolve completely during the summer, and the freezing process recommenced during the following winter. **Floes** refer to ice islands that drift during the summer, ranging in size from 100s of meters to 10 km in diameter.

Figure 23.19 Images of ice development in Antarctica, from pack ice to polynyas.

23.9.2 Antarctic Habitats

Ice presence creates different habitats such as brine channels. These channels can extend long distances, potentially from the sediment-water interface to approximately half of the ice thickness. Near the water-ice interface during Antarctic spring, **brown ice** develops (Figure 23.20), supporting abundant life, and particularly primary producers. In general, this layer supports a highly diverse community. High microalgal abundance, dominated by diatoms, define this "inverted garden" that lives in this porous layer 2–20 cm within the ice. Sympagic communities occur in all types of ice, spanning from ice thinning by melting to thick slabs, creating these "inverted gardens".

Sea ice supports a particularly diverse and dynamic microbial community that includes phytoplankton and prokaryotes, often referred to as SIMCO (the **Sea Ice Microbial Community**). The growth of the ice microalgae enriches microbial production and supplies food for herbivorous metazoans. The brown layer microalgal community interacts with the rich SIMCO communities, which consists of prokaryotic and eukaryotic organisms that live in polar seas. The dissolved organic and inorganic nutrients in brine channels host a rich, abundant prokaryotic community.

Numerous studies in recent years have explored SIMCO bacterial diversity, focusing on evaluating the phylogeny of bacterial populations. Scientists have not fully characterized archaeal populations because SIMCO were only discovered in 2004, whereas bacteria were first discovered in Antarctic ice in 1966. Low productivity characterizes SIMCO bacteria, which are often larger than pelagic forms, and include both free-living forms and bacteria associated with algal cells within the dense ice communities. Sea ice bacteria likely include many obligate psychrophiles with optimal temperatures below 15 °C. Molecular analysis using 16S rRNA analysis shows that, despite some organisms in common between sea ice and the water layers below, many psychrophilic species characterize sea ice. SIMCO autotrophic and heterotrophic microbes live at temperatures often below freezing, requiring significant adaptations; their sensitivity to temperature changes constrains efforts to culture these bacteria in the laboratory. Ice algae also experience low periods of light availability and are adapted to grow rapidly during the few months of reduced ice. These algae cannot tolerate high light intensity, and according to recent studies they can also be active under sea ice, exploiting the available light even before peak ice melt.

During winter, reverse stratification of the water column can occur (Figure 23.21), a scenario that also occurs in alpine lakes at temperate latitudes. Colder waters occur at the surface and warmer waters at depth, even if the layers differ in temperature by only tenths of a degree. This stratification creates a particularly important barrier between water and ice; as water temperature increases, the ice typically melts from the bottom because water temperature exceeds atmospheric temperature. The melting ice places the communities trapped within the ice in contact with seawater, resulting in extremely intense algal blooms at the ice-water interface. Over just a few days, the ice-water interface can totally change in color from white-gray to green-brown. The dark layer grows over time, and the increasing temperatures slowly melt the ice. During the initial melting phase, numerous channels develop within the ice. Where ice comes into direct contact with seawater, it becomes very porous,

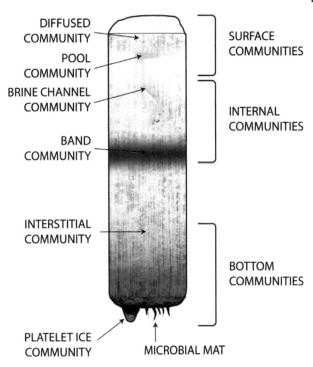

Figure 23.20 Sea ice microbial communities and brown layer in Antarctic ice.

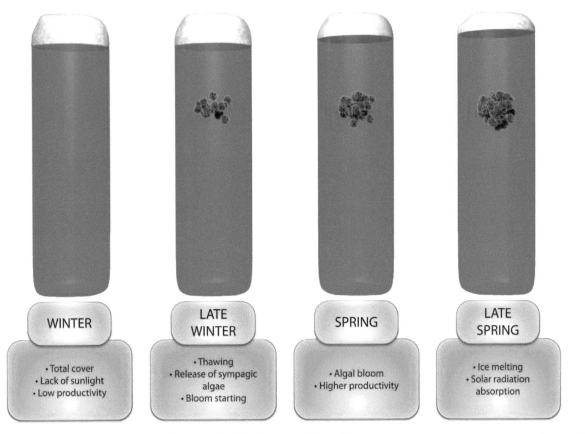

Figure 23.21 Seasonal development of primary production links to melting of ice. In winter, complete ice coverage and absence of solar radiation results in effectively no primary production. Ice melting triggers release of sympagic microalgae, resulting in strong primary production that eventually ends with re-cooling of surface waters and formation of new pack ice.

and microalgae that were initially limited to the interface penetrate inside the channels, ascending upward, until the ice dissolves to the point that it releases microalgae into the water column.

Initially, researchers considered this microalgal release as a sort of **seeding**, so that algae released from melting ice determined late spring-summer blooms. Only recently, researchers realized that a large number of the species present in sea ice cannot photosynthesize effectively in open water. Moreover, the phytoplankton community that dominates late spring and summer blooms differs greatly from the sympagic community. Benthic species dominate the sympagic algae found within brown ice, and, within a few days of release from ice, they sink to the bottom.

Pack ice creates particularly difficult and challenging conditions for research (Figure 23.22), such as struggling with compact sea ice when coring and diving through a hole in the pack ice. Of course, analysis of the microbial community associated with ice requires the use of sterile techniques.

At depths of ~3–4 m coastal algal populations grow to significant biomass of up to ~4 kg m^{-2}. This rich population of sympagic algae in the brown layer supports large populations of amphipod and euphausiid crustaceans. An estimated 98% of annual primary production occurs during the Antarctic summer, despite the short season (a total duration of 30–40 d). This short burst of production necessitates special adaptations by benthic organisms. For example, the flow of organic matter during the summer at Terra Nova Bay affects Antarctic communities associated with the bivalve, *Adamussium colbecki* (Figure 23.23). As a result of this flux, many macrobenthic species attain high biomass; sponges, for example, which characterize the coastal Antarctic seabed, can reach densities and biomass up to 4–5 kg m^{-2}.

23.9.3 Biodiversity

The Southern Ocean hosts rich biodiversity that thrives despite living in an environment dominated by glaciation and extreme environmental conditions. Some researchers believe numbers could rival tropical sediments. Antarctic ecosystems depend on a complex combination of factors that defy any simple definition. Interactions between many different combinations of physical and biological factors create a complex set of contrasting ecosystems of poorly known extent and distribution. Major physical factors that affect Antarctic ecosystems include sea ice, substrate, light, oceanic fronts, depth,

Figure 23.22 Ice coring (core sampling) to study sympagic biodiversity showing stages involved when a researcher cuts a hole in the ice (a) prior to diving and sampling (b-d).

temperature, isolation, geomorphology, seasonality, currents, and physical disturbance by iceberg. Major biological factors include primary production, biological substrate, dispersal ability, and community interactions. Much of the Southern Ocean remains unsampled, including large swaths of the intertidal zone and the deepest habitats. Known numbers of species in the Antarctic vary greatly among the various taxa: the updated taxonomic list of all Antarctic marine species created by the Register of Antarctic Marine Species (RAMS) currently includes > 8,200 species (Table 23.5).

Knowledge regarding the biodiversity of the Southern Ocean (Figure 23.24) strongly reflects sampling effort. Low diversity characterizes some Antarctic groups, such as decapods, stomatopods and bivalves, and, unlike the Arctic, the Antarctic almost completely lacks common crabs (brachyurans). However, unusual patterns of diversity characterize other groups, including a particularly rich diversity of pycnogonids, echinoderms, and polychaetes, with pycnogonids as potentially the most species-rich group. Notothenioidei fishes have also greatly diversified in the Antarctic. Despite the generally poor state of knowledge of Antarctic fish species, the Notothenioidei have received considerable attention. Of 174 known benthic or demersal species in the Southern Ocean, the Notothenioidei account for 55%, and represent ~90% of captured specimens. These fishes are endemic to the Antarctic region, acknowledging a recent observation of a Patagonian "toothfish" from deep waters off Greenland; this observation supports the idea of trans-equatorial dispersal events in cold, deep waters. Numerous scientists have studied phyla such as mollusks and crustaceans, resulting in a better (though nonetheless incomplete) state of knowledge on these groups. Current estimates of species numbers for other groups, such as nematodes and tardigrades, undoubtedly represent significant underestimates. Since 1993, scientists have doubled the number of known species in this region from 4,000 to 8,200, largely thanks to the international team working on RAMS. RAMS currently reports 8,806 described species in Antarctica, but experts project that Antarctic species could eventually number ~17,000, suggesting about 50% remain undiscovered. More than 90% of the Antarctic Ocean sits below 1,000 m, but most sampling has focused on greater depths. The stability of these environments over geological time scales likely contributed to the rich marine diversity and abundance around Antarctica, although low temperatures and slow growth potentially favored the development of gigantism in some species.

Although recent studies reduce estimates of Antarctic endemism rates that were based on morphological taxonomic analysis, endemism rates for some of the most common classes hover around 50% or more: 47% for Cyclostomata, 56% for Cheilostomata, 54% for Cephalopoda, 43% for Bivalvia, 74% for Gastropoda, 55% for Pycnogonida, and 44% for Ascidiacea. In shallow Antarctic environments spanning the intertidal, sublittoral, and circalittoral, ice abrasion of rocky substrates prevents settlement of encrusting organisms, resulting in coastlines of largely bare rocks.

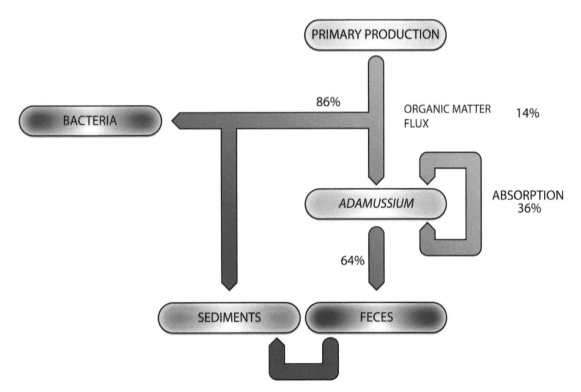

Figure 23.23 Diagram showing flow of organic matter in Antarctic coastal habitat; a large portion of water column production quickly sinks to the bottom, where it provides a major food source for benthic biota. This example shows the role of the Antarctic bivalve, *Adamussium colbecki*, a species that grows fairly large in size (shell diameter >8 cm) and density (60–80 indiv. m^{-2}).

Table 23.5 Number of taxa and species documented in the Register of Antarctic Marine Species (modified).

Taxonomic group	Species number
Kingdom Chromista	256
Kingdom Plantae	
Chlorophyta	24
Rhodophyta	70
Kingdom Protozoa	
Dinoflagellata	75
Foraminifera	179
Kingdom Fungi	
	> 51*
Kingdom Animalia	
Porifera	267
Cnidaria	459
Platyhelminthes	125
Mollusca	740
Anellida	536
Crustacea	2900
Bryozoa	316
Echinodermata	565
Tunicata	114
Other invertebrates	586
Fish	314
Other vertebrates	284
Total regional diversity	> 8700

* indicates incomplete data

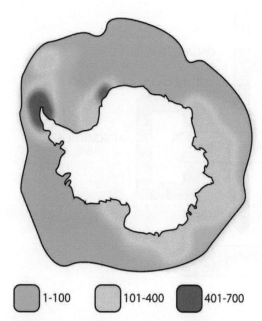

1-100   101-400   401-700

Figure 23.24 Total number of marine species recorded from the SCAR Marine Biodiversity Information Network.

At greater depth, however, biodiversity and organism abundance increase significantly. Sessile suspension feeders, the richest communities, live in the deepest parts of the continental shelf and on shallow escarpments (cliffs). Sediments can support high abundances, sometimes >130,000 individuals m^{-2}; the sponges that characterize these habitats perform similar functions (filtering, habitat provisioning) as seagrasses in temperate environments. In addition to abundant filter feeders (sponges, soft corals, polychaetes, bivalves), the benthos includes deposit feeders and grazers (polychaetes, isopods, seastars), and predators (polychaetes, crustaceans, amphipods, pycnogonids, fishes). Taxonomists have not yet formally described many of the 158 recognized Foraminifera. Most nematodes derive from generally cosmopolitan genera, but they include more than 57 species new to science. Of the >100 recognized species of ostracods, 70% remain undescribed. Within the macrofauna, the ~10,000 isopod specimens recently examined by scientists from deep-water samples represent an impressive 674+ identified species, compared to the 317 known species documented for the entire Antarctic continental shelf. In recent years, some 86% of the isopod species sampled are new to science, which span from macro- to megafauna that occur only in the Southern Ocean. Recent advances in technology have greatly increased the number of new species discoveries (Figure 23.25).

Antarctic exploration for economic gain and exploitation of its marine resources began in the eighteenth century, whereas scientific research on marine ecosystems really started only in the mid-nineteenth century. Material collected from ships, such as *Challenger*, *Belgica*, and *Discovery*, led to the first cataloging of benthos and plankton, and formed the basis for modern taxonomy for that region.

Reproductive strategies in Antarctic fishes differ from those typical of other marine habitats, with many adults producing only a few large eggs they often protect until hatching. Larvae grow quite slowly, and completion of life cycles in this environment usually takes much longer than elsewhere, often lasting three years or more. Antarctic herring (*Pleuragramma antarctica*), one of the few pelagic fishes, usually congregate in schools. Herring feed on krill and provide a critical trophic resource for penguins and seals. Seals and cetaceans that normally feed at depth prey upon another pelagic species, Antarctic cod, *Dissostichus mawsoni*, a species that may be declining because of over-fishing. The Notothenioidei (Figure 23.26) that dominate Antarctic fishes, specialize on surviving at low temperatures. During past glaciation and cooling, ancestral species likely faced tremendous selective pressure to cope with progressively colder conditions. This pressure favored adaptations to lower the freezing point of body fluids, in order to avoid physiological disruption. Their extensive ecological success relates to the presence of special antifreeze glycoproteins within their blood (AFGPs) that enable survival in ice cold waters. In addition to AFGPs, various polar and subpolar fishes have developed three other types of structurally different antifreeze proteins, which suggests these unusual proteins evolved independently at least four times. Cartilaginous fishes, though rare, also occur in the sub-Antarctic zone.

Antifreeze proteins reduce the freezing points of body fluids of fishes, with freezing points (from –0.7 to –1 °C) significantly higher than that

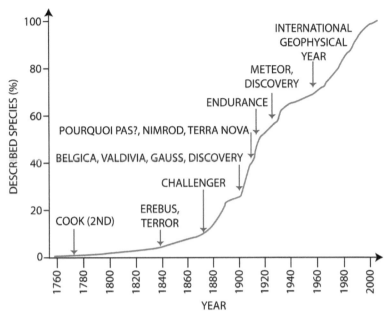

Figure 23.25 Cumulative percentage of species described over time during different Antarctic research expeditions.

Figure 23.26 Images of Notothenioidei fishes, where the AFGP genes evolved (see text). *Source:* Gianmarco Veruggio.

of seawater (−1.9 °C). These proteins inhibit development of small crystals of ice within the body. By binding to the surface of an icy crystal, the antifreeze prevents water molecules from attaching to and increasing the size of crystals.

Although many vertebrate mammals populate the continents that surround the Arctic Ocean, Antarctica largely lacks permanent terrestrial mammals. Most Antarctic animals concentrate around coastal environments, especially on Antarctic islands located at latitudes that overlap the maximum extent of winter ice; birds and small invertebrate species dominate this fauna. In contrast, the high primary production in spring–summer in Antarctic waters supports large numbers of marine mammals, including large whales and pinnipeds.

Investigating Antarctic biodiversity and ecosystems requires specific infrastructure and approaches for exploring assemblages living within marine ice and beneath the ice cover (Box 23.2).

Box 23.2 Methods to Study Antarctic Marine Biodiversity
The international nature of scientific research in Antarctica complicates any effort to calculate the number of scientists currently working on material sampled in this region. Nonetheless the number of researchers, programs, and scientific research vessels has been increasing year by year. Antarctic marine environments create many research challenges relative to other marine environments, including rapid freezing of seawater or the extreme care that scuba divers must take to avoid hypothermia or getting trapped under ice.

(Continued)

> **Box 23.2 (Continued)**
>
> During oceanographic cruises to collect sediment samples from offshore or deep-sea environments, researchers use three primary tools: grabs and buckets, box corers, and gravity or piston corers (Figure 23.27). Geophysicists also use acoustic instruments such as sub-bottom profilers and sparkers that emit sound pulses, which penetrate the seabed and bounce back to the surface where transducers capture and record their characteristics.
>
> Pack ice and icebergs add to differences between studying Antarctic biodiversity compared to that of other regions, creating hazards for research vessels and for sampling equipment. In many cases, scientists use robots and wire-guided vehicles to sample zooplankton.
>
>
>
> **Figure 23.27** Two images of Antarctic sampling using (a) an ice corer and (b) a box corer, which collect ice cores and marine sediments respectively: the ice corer can penetrate up to 2 m into the ice, whereas the box corer penetrates sediments to a depth of ~50 cm. *Source:* Antonio Dell'Anno.

23.9.4 Birds and Mammals

Among all species of penguins, only one, *Aptenodytes forsteri* (the Emperor penguin) completes its entire life cycle in Antarctica, whereas others (such as Adélie penguin – *Pygoscelis adeliae*) only breed in Antarctica, subsequently moving to lower latitudes in winter. Many species nest in colonies where individuals group in traditional locations where previous generations spawned and fledged their chicks. Grouping in colonies allows penguins to defend themselves from predators and especially from the cold. As voracious predators of fish and krill, penguins fertilize adjacent coastal marine environments through the production of guano at locations where they group together and hatch their eggs.

Many cetaceans, especially Mysticetes, breed in the northern hemisphere and migrate during austral summer to feed on krill (with timing depending on species): *Balaenoptera musculus*, blue whales, arrive first, and *Balaenoptera borealis*, sei whales, arrive last. After 3–4 months, all migrate back to warmer waters. Antarctic Odontocetes include three families: Ziphiidae, Delphinidae, and Physeteridae, especially Pinnipeds. Some species, such as leopard seals, find their ideal habitat in Antarctica (Figure 23.28).

23.9.5 Trophic Webs and Functioning

The classic Antarctic food web links ice communities with pelagic communities. This link occurs during pack ice formation in austral summer (January–February), when formation of grease ice (July–August) traps phytoplankton. The first phase of heating in summer releases microalgae trapped within the ice into the water, leading to two linkages. The first link, the **plume inoculum**, provides the inoculum for the bloom, meaning that these algae create an **ice edge bloom** (blooms at the edge of the ice). The second link, the **ice melt POM**, consists of particulate organic material (POM) previously trapped within the ice that sinks through the water column to enrich the benthic community (Figure 23.29).

Figure 23.28 Images of Antarctica mammals (a-d) and birds, including Adélie penguins and an emperor penguin (e-f). *Sources:* NOAA / Wikimedia Commons / Public domain, cianc / Wikimedia Commons / CC BY 2.0, Jason Auch / Wikipedia Commons / CC BY 2.0, Hannes Grobe/AWI / Wikipedia Commons / CC BY 3.0.

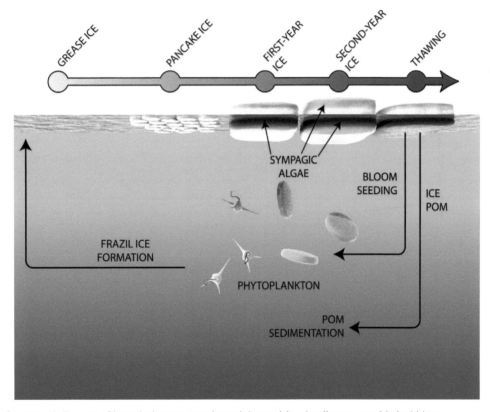

Figure 23.29 Conceptual diagram of inoculation process through ice melting, leading to a rapid algal bloom.

Recent research forced scientists to revise this scheme and add Antarctica to the High Nutrient Low Chlorophyll systems group (HLNC), characterized by large concentrations of nutrients, but low concentrations of chlorophyll in the water column. Primary production does not reach levels as high as might be expected. Antarctic waters, in fact, feature high concentrations of nitrogen and phosphorus salts but poor nitrogen bioavailability. Most ocean environments contain sufficient iron essential for algal growth, but Antarctica perennial ice cover prevents continental erosion and wind transport that provides an iron source elsewhere. Scientists therefore predicted that adding iron ("iron fertilization") to Southern Ocean surface waters could reduce atmospheric CO_2 by stimulating phytoplankton growth (and sinking). However, iron enrichment experiments increased algal production only in some cases. In addition to low iron availability, other factors such as grazing pressure may play a role.

The characteristics and vertical distributions of key ice variables such as inorganic nutrients (phosphates and nitrates) and organic components (biopolymers carbon consisting of proteins, carbohydrates, and chlorophylls) vary depending on the ice layer considered (Figure 23.30). Proteins occur in high abundance and generally increase relative to carbohydrates closer to the ice-water interface. This pattern reflects the typical distribution of chlorophyll concentrations, both in **bottom ice** (lower 10 cm of the ice column) and in **platelet ice**, where values up to 90 $\mu g\,l^{-1}$ occur in concentrations comparable to those in shallow eutrophicated lagoons during a phytoplankton bloom. The low temperatures and extremely high primary production in this ice layer result in oxygen concentrations over 100% saturation, creating particularly well-oxygenated conditions for sympagic organisms.

Heterotrophic communities, largely dominated by harpacticoid copepods such as *Stephos longipes* and *Harpacticus furcifer*, use these elevated autotrophic biomasses. High biodiversity of the autotrophic layer does not correspond to high heterotroph biodiversity. Analysis of concentrations of chlorophyll pigments through the year shows high values in November in the sea ice because of high concentrations of inorganic nutrients. Concentrations of nitrates and phosphates are two orders of magnitude higher at the water-ice interface relative to surface ice, even though such chlorophyll abundance might be expected to result in **nutrient depletion**, because of algal uptake. Researchers subsequently attributed this paradox to high organic matter degradation rates that reduced bacterial production could not counterbalance.

Surface phytoplankton produce the dominant energy flow in the Southern Ocean, followed by sinking of particulate organic matter and detritus, and then the microbial loop. The seemingly simple Antarctic food web depends on this primary production, starting with phytoplankton dominated by diatoms and a scarcity of dinoflagellates. Zooplankton comprise the next level of the Antarctic food web, characterized by few species with many individuals. Krill, the most important zooplankton

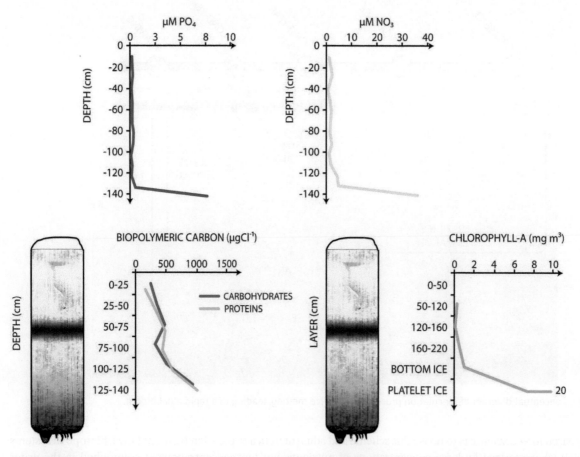

Figure 23.30 Characteristics of organic and inorganic nutrients contents within pack ice, showing vertical profiles of nutrient concentrations, biopolymer carbon content, and chlorophyll (ice core surface in contact with air, whereas ice core bottom in contact with water).

component, consists mainly of *Euphausia crystallorophias*. This species also plays a central role in balance and maintenance of the Antarctic marine ecosystem by providing the most important sources of food for almost all species at the highest trophic levels, including fishes and birds as well as large marine mammals (Figure 23.31).

Antarctic benthic filter feeders belong to invertebrate groups that occur in other oceans (sponges, cnidarians – particularly gorgonians and hydroids – bryozoans, crinoids, holothurians, tunicates, etc.) and play an essential ecological role in austral food webs. Organic matter sinking from surface waters during the summer continuously reaches the seafloor, providing the main food source for filter feeders. Filter feeders therefore represent the key mechanism to introduce organic matter into the benthic food web during austral winter. Their role parallels that of krill, channeling organic matter production to higher food web levels. Antarctica lacks terrestrial predators (no polar bears); killer whales and seals, especially leopard seals, comprise the top predators.

Historically, hunting of seals and whales was so intense that it nearly led to the extinction of some species. Regulators established the first restrictions just as this activity ceased to be profitable. Existing agreements between countries now protect most seal and whale species by regulating their removal, even though countries such as Japan continue to hunt whales under the pretext of scientific research. As a result of effective protection practices, populations of many species of Antarctic seals have increased, however, most whale populations remain low.

One of the emerging problems in Antarctic region has been the progressive reduction of the ozone layer that protects the system from UV radiation. The Antarctic ozone hole in 2020 was one the largest and deepest observed in recent years. Analyses show that the hole grew rapidly from mid-August and peaked at around 24 million km^2 in early October. It now covers approximately 23 million km^2, above average for the last decade and spreading over most of the Antarctic continent (Box 23.3).

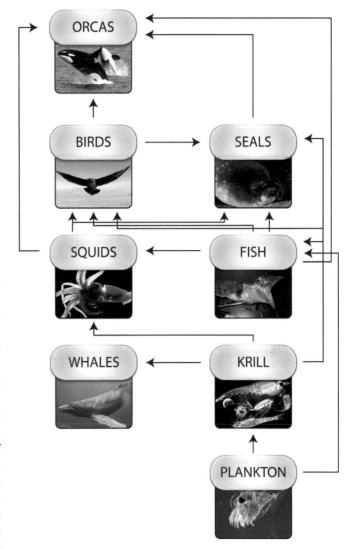

Figure 23.31 Antarctic Ocean food web. *Sources*: L. Madin / Wikimedia Commons / Public Domain., National Oceanic and Atmospheric Administration (NOAA), Wilson/Jay Clark / Wikimedia Commons / Public Domain, Mª Carmen Mingorance Rodríguez/Wikimedia Commons.

Antarctica has no government and no permanent residents, but a sparse human population occupies the various scientific stations (1,000 people in winter and > 4,000 in summer). In contrast, human populations in the Arctic exceed 4 million people, including citizens of Canada, Greenland (Denmark), Russia, Alaska (USA), Iceland, Norway, Sweden, and Finland.

Box 23.3   The Ozone Hole

In the upper atmosphere, a layer of ozone (O_3) acts as a filter that removes ultraviolet radiation particularly dangerous for life. This layer absorbs 100% of UV-C and 90% of UV-B wavelengths, the latter of which can cause serious skin diseases such as melanomas and other cancers, inhibit photosynthesis, and degenerate phytoplankton cells. Starting in the 1980s and 1990s, progressive thinning of the ozone layer caused a 40–70% reduction above the Arctic and Antarctica, respectively. The prevalence of the phenomenon at the poles, especially above Antarctica, links with the dynamics of atmospheric circulation and exchange of air masses between tropical regions, where ozone normally accumulates because of the higher intensity of solar radiation, and the poles. Thinning of the ozone layer was attributed to emission of chlorine compounds (chlorofluorocarbons) that catalyze ozone photo-oxidation from natural UV rays (Figure 23.32).

(Continued)

Box 23.3 (Continued)

These products were used for many years as inert propellants in spray cans and cosmetic products. Cycles of "natural" thickening and thinning of the ozone related to the formation and accumulation of natural "clouds" of chlorine relate to solar activity and seasonal thermal regimes, or from the emission of reactive compounds into the atmosphere from volcanic activity. UV rays strongly impact prokaryotic communities (which lack effective DNA repair systems) and phytoplankton, with important repercussions for the entire Antarctic food web.

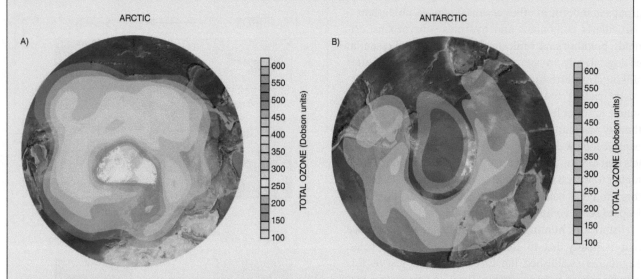

Figure 23.32 Ozone concentrations in the ozone layer above the Antarctic continent. Note that the lowest concentrations correspond with the continent.

Table 23.6 Comparison between polar ecosystems.

	Arctic	Antarctic
Extent	An expanse of ice-covered ocean rather than a continent. Ice extent spans ~13 million km^2	A continent larger in size than Oceania, perpetually covered by ice and spanning an area of over 16 million km^2
Distribution	Corresponds to Arctic Ocean above 60° N	Corresponds to Southern Ocean below 60° S
Depth	Average depth little more than 1,000 m	Average depth 4,500 m (typically between 4,000 and 5,000 m), and maximum depth over 7,200 m inside the Sandwich Southern Trench
Type of substrate	Almost exclusively soft bottoms. Presence of important communities associated with seasonal ice	Dominated by soft bottoms but with wide hard bottom habitat; key communities linked to seasonal ice
Biodiversity	Relatively high for some invertebrate groups and marine mammals but low for fishes	Very high compared with the Arctic, including numerous species of cetaceans and sea birds
Endemism	Modest levels of endemism	Elevated endemism
Functioning	Mainly based on seasonal phytoplankton production beginning with the melting of ice (despite some primary production beneath the ice)	Abundant nutrients but low iron availability limits primary production. High biomass of benthic invertebrates, plankton (krill), and cetaceans
Trophic webs	Relatively simple trophic network dominated by plankton and important role for marine mammals (pinnipeds in particular), but also major role for benthic compartment	Complex trophic networks despite direct link between phytoplankton, krill, and cetaceans, important links between water column and sediments
Peculiarities	Extreme temperature conditions. Pervasive ice has led to the development of a unique habitat and sympagic community	Extreme temperature conditions. The prevalence of ice supports unique habitat colonized by sympagic organisms
State of knowledge	Increasing in recent years but still limited and complicated by complex seascape and oceanography	Rapid advances in the last 20 y through many international collaborations

23.10 Summary

In summary, Arctic and Antarctic systems share key attributes related to their latitudinal extremes, but they also differ markedly in their basic geomorphology and physical oceanography, with profound effects on how they function as well as their biodiversity (Table 23.6)

Questions

1) What are the main differences between Arctic and Antarctic ecosystem?
2) What are the main characteristics of communities associated with polar ecosystems?
3) What do the terms "inverted garden" and sympagic communities mean?
4) What are the primary producers in polar ecosystems and how are they controlled?
5) What are the main components of the Arctic food web?
6) What are the main components of the Antarctic food web?
7) What is krill? What is its role in the food web?

Suggested Reading

Barber, D., Hop, H., Mundy, C. et al. (2015) Selected physical, biological and biogeochemical implications of a rapidly changing Arctic Marginal Ice Zone. *Progress in Oceanography*, 139, pp. 122–150.

Bargelloni, L., Marcato, S., and Patarnello, T. (1998) Antarctic fish hemoglobins: Evidence for adaptive evolution at subzero temperature. *Proceedings of the National Academy of Sciences of the United States of America*, 95, pp. 8670–8675.

Berge, J., Renaud, P., Darnis, G. et al. (2015) In the dark: A review of ecosystem processes during the Arctic polar night. *Progress In Oceanography*, 139, pp. 258–271.

Bright, M., Arndt, C., Keckeis, H., and Felbeck, H. (2003) A temperature-tolerant interstitial worm with associated epibiotic bacteria from the shallow water fumarolfes of Deception Island, Antartica. *Deep-Sea Research II*, 50, pp. 1859–1871.

Cattaneo-Vietti, R., Chiantore, M., Misic, C. et al. (1999) The role of pelagic-benthic coupling in structuring littoral benthic com- munities at Terranova Bay (Ross Sea) and in the Straits of Magellan. *Scientia Marina*, 63, pp. 113–121.

Cheng, C.H.C. (1998) Origin and mechanism of evolution of antifreeze glycoproteins in polar fishes, in di Prisco, G., Pisano, E., and Clarke, A. (eds.), *Fishes of Antarctica: A Biological Overview*. Springer, pp. 311–328.

Convey, P., Aitken, S., di Prisco, G. et al. (2012) The impacts of climate change on circumpolar biodiversity. *Biodiversity*, 13, pp. 134–143.

Dayton, P.K., Mordida, B.J., and Bacon, F. (1994) Polar marine communities. *American Zoologist*, 34, pp. 90–99.

DeLong, E.F., Ying Wu, K., Prezelin, B.B., and Jovine, R.V.M. (1994) High abundance of Archaea in Antarctic marine picoplankton. *Nature*, 371, pp. 695–697.

Fountain, A., Saba, G., Adams, B. et al. (2016) The impact of a large-scale climate event on Antarctic ecosystem processes. *BioScience*, 66, pp. 848–863.

Gambi, M.C., Lorenti, M., Russo, G.F., and Scipione, M.B. (1994) Benthic associations of the shallow hard bottoms off Terranova Bay, Ross Sea: Zonation, biomass and population structure. *Antarctic Science*, 6, pp. 449–462.

Gradinger, R., Bluhm, B.A., Hopcroft, R.R. et al. (2010) Marine life in the Arctic, in McIntyre, A.D. (ed.), *Life in the World's Oceans: Diversity, Distribution, and Abundance*. Wiley-Blackwell, pp. 183–202.

Griffiths, H.J., Barnes, D.K., and Linse, K. (2009) Towards a generalized biogeography of the Southern Ocean benthos. *Journal of Biogeography*, 36, pp. 162–177.

Guglielmo, L., Granata, A., and Greco, S. (1998) Distribution and abundance of postlarval and juvenile *Pleuragramma antarcticum* (Pisces, Nototheniidae) off Terranova Bay (Ross Sea, Antarctica*). Polar Biology*, 19, pp. 37–51.

Gutt, J., Bertler, N., Bracegirdle, T.J. et al. (2015) The Southern Ocean ecosystem under multiple climate change stresses—an integrated circumpolar assessment. *Global Change Biology*, 21, pp. 1434–1453.

Hunt, G., Drinkwater, K., Arrigo, K. et al. (2016) Advection in polar and sub-polar environments: Impacts on high latitude marine ecosystems. *Progress in Oceanography*, 149, pp. 40–81.

Ladau, J., Sharpton, T.J., Finucane, M.M. et al. (2013) Global marine bacterial diversity peaks at high latitudes in winter. *The ISME Journal*, 7, pp. 1669–1677.

Murphy, E.J., Cavanagh, R.D., Drinkwater, K.F. et al. (2016) Understanding the structure and functioning of polar pelagic ecosystems to predict the impacts of change. *Proceedings of the Royal Society B: Biological Sciences*, 283(1844), p. 20161646.

Pusceddu, A., Dell'Anno, A., Vezzulli, L. et al. (2009) Microbial loop malfunctioning in the annual sea ice at Terranova Bay (Antarctica). *Polar Biology*, 32, pp. 337–346.

Rogers, A., Johnston, N., Murphy, E., and Clarke, A. eds. (2012) *Antarctic Ecosystems: An Extreme Environment in a Changing World*. Blackwell.

Saggiomo, V., Catalano, G., Mangoni, O. et al. (2002) Primary production processes in ice-free waters of the Ross Sea (Antarctica) during the austral summer 1996. *Deep-Sea Research II*, 49, pp. 1787–1801.

Smith, W.O., Ainley, D.G., and Cattaneo-Vietti, R. (2007) Trophic interactions within the Ross Sea continental shelf ecosystem. *Philosophical Transactions of the Royal Society of London B: Biological Sciences*, 362, pp. 95–111.

Xavier, J., Cherel, Y., Allcock, A. et al. (2018) A review on the biodiversity, distribution and trophic role of cephalopods in the Arctic and Antarctic marine ecosystems under a changing ocean. *Marine Biology*, 165, pp.1–26.

24

Neritic Aquatic Ecosystems

24.1 Introduction

Scientists divide the water column of the ocean and seas into two realms, sometimes called **provinces**: the oceanic province and the neritic province (Figure 24.1). The **oceanic** or **pelagic province** encompasses the vast open ocean beyond the continental shelf break. The **neritic zone**, also known as the **coastal ocean** or **coastal waters**, includes the waters and biological communities living in the intertidal and subtidal water column adjacent to the shore, above the continental shelf. Its proximity to land contributes to relatively advanced state of knowledge on this zone, noting that the dynamics of the neritic province affect all of the other habitats described later in this chapter, and of course, the seabed below it. The neritic zone encompasses nearshore marine systems from the surface, through the upper water column and pycnocline, to the lower water column as it transitions into the benthic boundary layer that forms the interface between the water column and benthic system (we considered the seafloor and benthic boundary layer in Chapters 9 and 10). The maximum depth of the neritic zone therefore typically occurs at the continental shelf break, or ~200-m depth.

24.2 Zonation, Extent, and Distribution

From a primary production perspective, varying intensities and wavelengths of light penetrate the neritic zone, depending on water depth and clarity. The availability of light in this zone enables photosynthesis, which forms the basis of almost all production within neritic ecosystems, both by (phyto)planktonic organisms and by photosynthetic benthic organisms at shallow depths. Plankton, the prevalent biotic component of this water column habitat, varies in space and time. Although the neritic zone covers <10% of the ocean surface, the abundance of sunlight and nutrients in this zone explains why it includes some of the most productive habitats on Earth per unit of volume and the highest abundances and biomasses of marine organisms. In addition to the inshore to offshore zonation that demarcates the neritic zone, the mixed layer and pycnocline described in Chapter 16 produce vertical zonation, with generally higher productivity above the pycnocline and often a peak in biomass at the pycnocline as sinking material accumulates. The reduced photosynthesis below the pycnocline typically means lower primary and secondary production, although in close proximity to the seafloor, where organic matter often accumulates, resuspension of material in the benthic boundary layer can support a rich hyperbenthos with elevated biomass relative to the waters above it. Of course, this complexity depends on depth and season, in that winds may actually mix water all the way to the bottom, eliminating any vertical structure. Generally speaking, light and nutrient availability as well as water column stratification, determine phytoplankton composition and abundance, which in turn determines zooplankton community composition with influences up the food chain to fishes, marine mammals, and seabirds.

24.3 Biogeography and Characteristics

Several researchers, including Alan Longhurst and Ken Sherman, have divided the global ocean, including the neritic zone, into biogeographic provinces largely based on biogeochemistry and productivity. More recently, Mark Spalding and colleagues added a biodiversity element to their classification, with their 62 biogeographic provinces roughly paralleling the provinces identified by previous schemes and subdividing those provinces into 232 ecoregions, which they define as "areas of relatively homogeneous species composition, clearly distinct from adjacent systems." These classifications illustrate the diverse nature of neritic ecosystems around the world, despite their fluidity (Figure 24.2).

Biodiversity patterns also vary among neritic habitats globally. Coastal bony fish and shark diversity decrease towards with poles, with highest diversity in tropical coastal waters (though noting some temperate hotspots for functional diversity as opposed to species diversity). This pattern generally seems to hold for phytoplankton, apparently linked to total energy availability and warmer temperature (Figure 24.3). In contrast, highest diversity in pinnipeds appears to occur in polar regions.

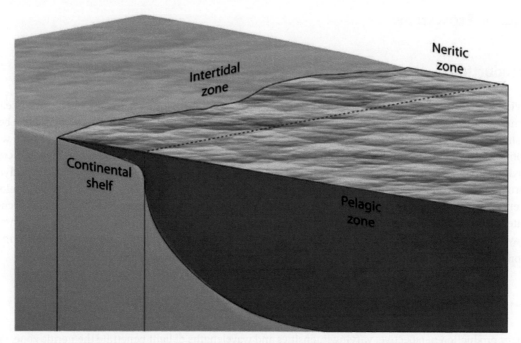

Figure 24.1 Boundaries of neritic province out to and beyond the continental shelf break to the vast oceanic zone.

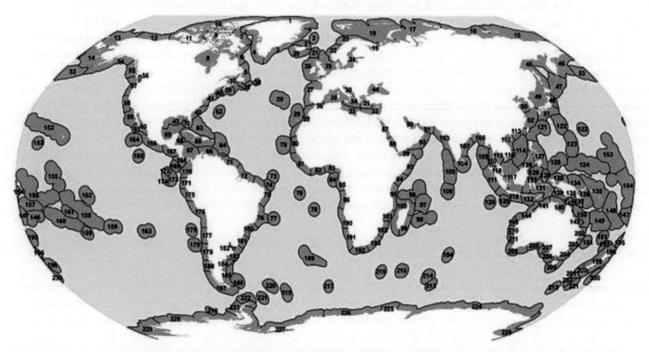

Figure 24.2 Biogeographic ecoregions, each delineated by a different number. From Spalding, M.D., Fox, H.E., Allen, G.R., Davidson, N., Ferdaña, Z.A., Finlayson, M.A.X., Halpern, B.S., Jorge, M.A., Lombana, A.L., Lourie, S.A. and Martin, K.D., 2007. Marine ecoregions of the world: a bioregionalization of coastal and shelf areas. *BioScience*, 57(7), pp.573–583.

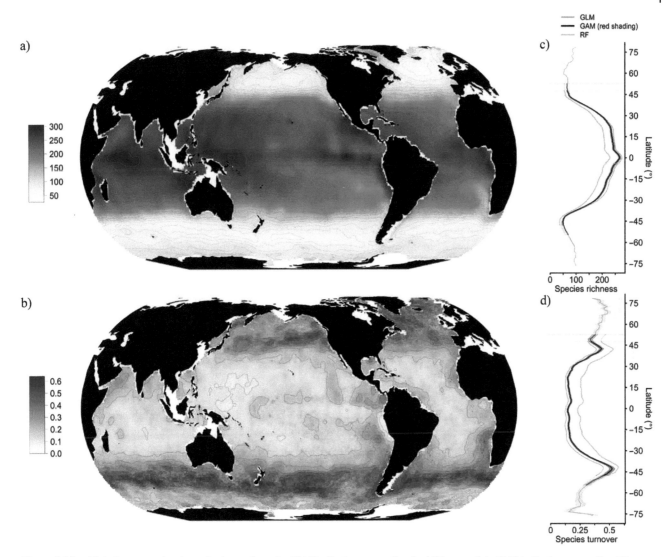

Figure 24.3 Global patterns in phytoplankton diversity. GAM indicates generalized additive models, GLM indicates generalized linear models, and RF indicates random forest models. Copyright © 2019 Damiano Righetti et al. Sci Adv 2019;5:eaau6253, some rights reserved; exclusive licensee American Association for the Advancement of Science. No claim to original US Government Works. Distributed under a Creative Commons Attribution.

24.4 Biological Characteristics

The neritic zone supports an abundance of life, and neritic waters from the tropics to the poles support a wide variety of both plankton and nekton. A third category, the **pleuston**, refers to organisms that straddle the air-water interface (e.g., the siphonophore *Physalia* and the gastropod *Janthina*, the former maintaining its position using a gas-filled float, the latter a raft of air bubbles). These organisms rarely occur exclusively in neritic waters, often spreading to oceanic waters, depending on local currents.

Neritic plankton includes phytoplankton and zooplankton that form the base and first trophic levels of marine food webs. The heterotrophic zooplankton range in size from microscopic single-celled organisms to enormous jellyfish a meter or more in diameter. The holoplankton, include the many diatoms, copepods, krill, and jellyfish that spend their entire lives drifting in the water column, often cohabiting with meroplanktonic egg and larval stages of benthic species such as sea urchins, mussels, crabs, some snails and many fishes that eventually settle to the bottom. Below we summarize some of the common biota within the neritic zone.

24.4.1 Primary Producers

The composition of the phytoplankton responsible for primary production in neritic waters varies spatially and seasonally. In highly productive temperate as well as high latitude environments, diatoms (class Bacillariophyceae) such as *Chaetoceros* spp., *Skeletonema* spp., and *Thalassiosira* spp. play a central role in spring and fall blooms and in overall primary production. They vary in size from 2 to 1,000 μm and include solitary and chain-forming species with an external shell composed of silica that results in a tendency to sink, which they offset with structures such as spines that reduce sinking rates. When they die and sink to the seafloor, they form diatomaceous oozes. Because they require relatively high nutrient concentrations, their numbers begin to decline as nutrient concentrations decrease.

The second most abundant phytoplankton group, the dinoflagellates (class Pyrrophyceae), includes both autotrophs and small heterotrophs and often bloom in concentrations up to 1–20 million cells l^{-1} immediately after the diatoms. Unlike diatoms, dinoflagellates possess flagella that enable limited movement. Importantly, this group includes representatives from genera such as *Alexandrium*, *Pyrodinium*, and *Gymnodinium* that produce lethal neurotoxin (saxitoxin) that can lead to "red tides" and paralytic shellfish poisoning, as well as fish kills. Because these dinoflagellates produce highly resilient benthic cysts that can quickly stimulate a bloom once they become active, they complicate efforts to model and forecast red tides.

A third key group, the Coccolithophorids (class Prymnesiophyceae) contain ~150 species, and they occur in both oceanic and neritic waters. Although most species favour warm waters, coccolithophorids can form intense blooms in coastal (neritic) waters at high latitudes (Figure 24.4). Their external shell consists of a series of calcareous plates known as coccoliths that can sink to the seafloor and form thick calcareous oozes; the White Cliffs of Dover formed in this way.

24.4.2 Zooplankton

Zooplankton in the neritic zone include both microzooplankton (20–200 μm), meso- and macrozooplankton. Among the microzooplankton, the colorless, strictly heterotrophic zooflagellates, 2–5 μm in diameter, can reproduce quickly and comprise 20–80% of nanoplankton numerically. Ciliates, as their name implies, possess cilia they use for locomotion and sometimes for capturing small phyto- and zooflagellates, small diatoms, and bacteria. Tintinnids, which vary from 20 to 640 μm in size, form another large and ecologically important subgroup that may consume 4–60% of coastal phytoplankton. Their external protein shells can occur in high abundances in seafloor sediments.

Within the mesozooplankton, copepods (order Calanoida) represent one of the most abundant animals on Earth, comprising 70% or more of net plankton. Thus, copepods are to the macrozooplankton what polychaetes are to the benthic

Figure 24.4 Coccolithophore bloom south of Newfoundland, Canada in July, 2002, shown as light blue in this NASA satellite image. *Source:* NASA.

sedimentary macrofauna – a dominant group in most regions. These crustaceans vary in size from mostly <6 mm but a few as large as 1 cm, and their ~1850 species provide critical links between lower and upper trophic levels by feeding on phytoplankton and representing important prey for small fishes. Their strongly seasonal cycle follows or coincides with the phytoplankton bloom, developing through 6 naupliar and 6 copepodite stages, with lesser numbers of individuals through the fall as they begin to enter winter **diapause**, a hibernation-like period during which growth or development halts and physiological activity diminishes.

Despite the numerical dominance of copepods, other larger crustaceans (macrozooplankton) also play significant roles in neritic systems. Euphausiids, or krill, are relatively large (up to 5 cm in length) macrozooplankton that commonly occur in high productivity environments worldwide, often attaining sufficiently high abundances that they form a deep scattering layer that can be seen on echosounders. One species, *Euphausia superba*, forms a major link between lower and upper trophic levels in Antarctic waters and thus helps to support the abundant marine mammals that inhabit the Southern Ocean. Various species of shrimp can also occur in high abundances, some supporting important commercial fisheries. The ~210 species of shrimp globally range from 10 to >100 mm in size, with species of omnivores and predators that include hyperbenthic species that move between the seafloor and water column as well as others that primarily occupy the plankton.

Amphipods such as *Parathemisto* spp. can sometimes dominate but typically comprise a small fraction of total zooplankton, often as carnivores that feed on copepods and other crustaceans, and sometimes even larval fishes. Other phyla also contribute to the neritic zooplankton community. Chaetognaths, or arrow worms, for example, occur throughout the global ocean and particularly in coastal waters, where their comparatively large size (<4 cm but up to 10 cm) allows them to prey on other zooplankton. Holoplanktonic gastropods known as **pteropods** can sometimes occur in high abundances in surface waters, especially in cold ocean environments where they feed on phytoplankton, detritus, and small zooplankton; they can become a key food resource for mackerel, herring, salmon and sometimes cod. Local fishermen in Newfoundland, Canada, dread pteropods, called "blackberry" based on their resemblance to berries, because cod take on a peculiar smell and texture when they ingest large numbers of pteropods that contain dimethyl sulfide. Pteropods also produce carbonate shells that sink to the seafloor to form pteropod ooze.

Gelatinous zooplankton are typically the dominating component of megazooplankton and can play a significant role in the dynamics of other zooplankton. These organisms include the Cnidaria and Ctenophora, two related but different phyla that differ in morphology but both are particularly significant carnivores on copepods and fish eggs and larvae. Cnidarians, better known as jellyfish or jellies, swim using propulsion of their bell, whereas ctenophores, known as comb jellies, swim via rows of beating cilia. These organisms occur throughout the global ocean, sometimes in high abundances, but sampling nets largely destroy them, and the normal preservatives used to fix and store animals do not work well, limiting efforts to evaluate their abundance and impact.

The phylum Tunicata form a third gelatinous group in neritic waters. Cold water planktonic tunicates known as larvaceans feature a small tadpole-like body that lives in large mucous house 5–40 mm long. They use this house to filter fine particles from the water, but quickly discard these houses when disturbed, sometimes resulting in concentrations of discarded houses in coastal waters of up to 1000 indiv. m^{-3}. These mucous houses can coat fishing nets with a slime coating fishermen refer to as "slub." Other tunicates known as salps occur in warm, temperate, and polar surface waters, reaching particularly high abundances in the Southern Ocean. They pump water through their open-ended body to create a current that brings phytoplankton/bacteria to an internal mucous net. Through this process they can transport large amounts of carbon from surface waters to the seafloor.

24.4.3 Nekton

Bony fishes (class Osteichthyes) represent some 96% of the ~30,000 species of fishes, of which 59% occur in marine and brackish waters. The Chondrichthyes, or cartilaginous fishes, include ~700 species, most of which occur only in marine waters. The neritic zone supports a wide diversity of fishes, and here we describe a few of the best-known groups. The Clupeidae, or herrings, include small, schooling fishes whose oily flesh creates a flavorful fish in species such as *Clupea harengus* a major northern commercial fishery. The Salmonidae (i.e., salmons and trouts), include many important genera that migrate between fresh and salt water, and support valuable commercial and recreational fisheries in both coastal and fresh waters. These carnivorous species are native to the northern hemisphere. The Gadidae, which includes the cods and hakes, comprise only some 70 species, but include some of the most important commercial fishes in coastal waters. Most are temperate or polar. The Scombridae includes the mackerels and tunas that school in surface waters and support important coastal and pelagic fisheries. The Scorpaenidae, the scorpion and rockfishes, all live in marine environment but vary in depth and latitude. The Pleuronectiformes, or flatfishes, associate closely with the bottom, explaining their asymmetrical flattened bodies of adult forms. All of the species are marine, and this group includes key commercial taxa from temperate environments to the Arctic.

Cetaceans, the infraorder that contains the whales, spans from blue whales, the largest animal on Earth at close to 30 m length, to 1-m long dolphins. The parvorder Mysticeti, the baleen whales or great whales, includes 11 species, all greater than 7 m in length. Many of these species move into neritic waters to feed primarily on plankton such as krill, as well as forage fish such as capelin. The series of plates, or baleen, create a sort of sieve system by which they filter massive volumes of water to remove large numbers of small prey items. Unlike the toothed whales, baleen whales lack teeth, which occur only during embryonic development. Many of these species undertake extensive seasonal migrations, often through oceanic waters, as they move between feeding grounds and calving/nursery grounds. The other major whale group, the parvorder Odontoceti, includes ~68 species of toothed whales that often feed on individual prey items of invertebrates, fishes, or even other cetaceans. Like the Mysticeti, some species occupy neritic environments at least some of the time, though others occupy oceanic environments, and a few live in rivers.

Other mammals play a relatively minor role in neritic environments in that they live primarily in estuaries with modest interaction with neritic species. The 4 species of the order Sirenia, the dugongs and manatees, are herbivores that feed on seagrasses, though some species occasionally feed on fishes and zooplankton such as jellies. Marine mammals from the order Carnivora evolved from a carnivorous ancestor and their body plan resembles terrestrial mammals with powerful and well-developed hind limbs. The Fissipedia or "split-footed" sea otters prey largely on benthic species such as urchins and clams, whereas the Pinnipedia, the seals and sea lions, often feed on fishes (though walruses also feed on benthos).

Seabirds include species that live primarily in the neritic zone as well as others that live in oceanic environments. Most feed on smaller fishes and water column invertebrates such as squid and krill, using strategies that vary from surface feeding to deep diving to plunge diving. Most of these species breed on shore and therefore move significant food material from the neritic zone to feed their offspring. Their eggs, as well as adults, have provided significant resources for land land-based species, including humans.

24.5 Ecosystem Functioning in the Neritic Zone

Particularly given the small and largely "invisible" microbes, the water column biota give the impression of a sparse community dominated by large gelatinous zooplankton and nekton (mostly vertebrates such as sharks, fish, sea turtles, cetaceans, and seals). Except when they form intense blooms that give the ocean a green (or sometimes white) color, most planktonic primary producers are invisible to human eyes, as are the small crustacean copepods that usually dominate the grazing community.

These systems function in pulses, where herbivorous zooplankton blooms follow phytoplankton blooms and, in turn, support blooms of carnivores (including fish larvae and juveniles) that feed on zooplankton and become prey for nekton. Phytoplankton blooms, typically occur in late winter – early spring when solar energy increases, and storms and riverine run-off increase nutrient availability. Zooplankton blooms typically follow closely behind or strongly overlap with phytoplankton blooms, depending on locality. Plankton abundance decreases in summer, when surface waters strongly stratify, and in some regions a second, less intense bloom may occur in the fall, when cooling of surface waters and increased winds reinject nutrients from deeper layers into the photic zone. Low production occurs in winter when light becomes more limiting and intense mixing occurs and thus limits growth of phytoplankton, although winter storms resuspend nutrients and resting stages that set the stage for the following spring.

These bloom phenomena vary in intensity with location and particularly latitude, in that reduced seasonality in day length and temperature in tropical regions result in very limited fluctuations, in contrast to strong blooms in temperate regions and brief, intense blooms as ice breakup occurs in polar regions. These differences strongly influence functioning of the contrasting systems. Many species of both zoo- and phytoplankton, especially in coastal regions, deposit resting stages in sediments during adverse conditions. Other types of blooms can occur in neritic ecosystems including rapid increases in gelatinous zooplankton, such as hydrozoans and ctenophores, as well as toxic dinoflagellate blooms (red tides) that sometimes follow spring blooms. Some types of phytoplankton produce biotoxins that accumulate in the animals that consume them, potentially causing illness and death in seabirds and mammals that occupy higher trophic levels. Other types of phytoplankton produce surfactant-like proteins that create foam on the water's surface during blooms. Seabirds exposed to the foam may lose the waterproof coating on their feathers, potentially restricting their ability to fly and resulting in death from hypothermia. Harmful algal blooms (HABs) can affect fishes and marine mammals, as well as humans, causing losses of natural resources, economic losses to coastal communities, and sometimes leading to human illnesses and deaths (Figure 24.5). Although these blooms can occur in open ocean habitats, humans feel their effects most acutely in the coastal zone where fisheries concentrate. Strong seasonal cycles in plankton characterize the neritic zone.

These tiny microbes form the basis of the productive neritic food web, supporting higher trophic levels, from zooplankton to fishes (including commercial species) to marine birds and mammals (Figure 24.6). Numerous highly migratory and

Figure 24.5 Red tides are phenomena sometimes evident in the coloration of coastal waters because of microalgae pigmented by carotenoids. *Source:* Unknown author / Wikimedia Commons / Public Domain.

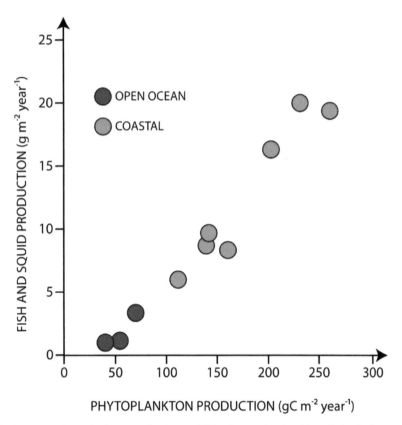

Figure 24.6 Relationship between phytoplankton production and fisheries production. Blue circles indicate oceanic habitats, green circles indicate neritic habitat.

Figure 24.7 Seabirds such as Atlantic puffin (a) dig burrows (b) in the ground along the shoreline where they nest and then fly short distances to feed on forage fish such as capelin.

schooling nekton species use neritic habitat, including many species of invertebrates, fishes, birds, and marine mammals. Some species travel and forage exclusively within this habitat, whereas others move through occasionally. Most major commercial fisheries focus on neritic species, and neritic zones include many hotspots for critical ecological processes that support thriving water column and benthic systems.

Many nearshore species use open water neritic habitat during their life history. For example, many forage fishes such as northern anchovy, herring, smelts, and sand lance feed in this open water neritic habitat. Juvenile bentho-nektonic species also occur within the water column. **Central place foragers**, animals that travel from a home base to distant foraging locations rather than simply passing through feeding locations, include breeding birds such as puffin and common murre that feed on forage fishes such as capelin and sand lance while nesting (Figure 24.7).

24.6 Fisheries Production

Fisheries experts estimate that the neritic zone provides 90% of the world's wild fisheries harvest (Figure 24.8). The combination of light, abundant oxygen (normally), suitable salinity and temperature, nutrient runoff from nearby land, regeneration from the seafloor, internal recycling (e.g., microbial loop), and upwelling from the continental shelf, support

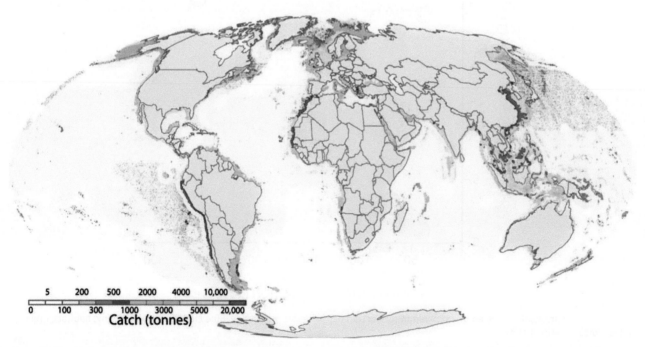

Figure 24.8 Global fisheries catch annually, showing the dominant role of neritic habitats in fisheries production. *Source* FAO.

a wide range and abundance of marine life. Processes acting at micro- to meso- scales determine the ecology of neritic habitats, and contribute to strong spatial variation in fisheries production, even within the neritic habitats that support most of the world's major marine fisheries.

The largest of the world's oceans, the Pacific, encompasses 169.2 million km^2 of ocean and some 25,000 islands, over half of the islands in the world; many of these islands depend on marine fisheries as a major source of protein and most are south of the equator. Temperate habitats support fisheries for species such as herring, salmon, sardines, snapper, swordfish, and tunas, as well as mollusks and crustaceans. Tropical fisheries target coastal fishes such as anchovy and coral reef fishes such as groupers, snappers, trevallies, parrot fish, surgeonfish, butterfly fish, and goatfish, as well as invertebrate fisheries that target lobster, squid, octopus, and cuttlefish. The Atlantic, the second largest ocean at 106.4 million km^2 supports some of the most productive fisheries worldwide, including major temperate latitude fisheries for cod, haddock, hake, herring, and mackerel, as well as tropical fisheries for flying fish, dolphin fish, and many of the same types of fishes as the tropical Pacific. Major invertebrate fisheries include lobsters, various crabs and mollusks (e.g., scallops, surf clams), and an increasingly broader diversification of target species (e.g., sea urchins, sea cucumbers). The narrow continental shelf and generally warm and nutrient depleted waters of the Indian Ocean, the third largest ocean at 73.6 million km^2, limit many fisheries to subsistence levels, although many countries remove fishes such as tuna and shrimp. Fisheries also target ponyfishes, croakers, mullets, carangids, sardines, anchovies, mackerel, sharks, prawns, lobsters, cockles, and cephalopods. The Southern Ocean, the fourth-largest ocean at 20,327,000 km^2 contains a very limited continental shelf and has supported fisheries for krill (*Euphausia superba*), toothfishes, and mackerel icefish. Historically, the Southern Ocean also supported major whaling as well as seal fisheries.

24.7 Factors Influencing Functioning of Neritic Systems

In spite of its apparent homogeneity, different oceanographic features and the peculiarities of the seafloor beneath neritic systems result in highly diverse and variable ecosystems related to the presence of specific physicochemical conditions often generated by local and regional hydrodynamism. Currents, gyres, eddies and upwelling functionally connect different components of the marine realm by transporting organisms and material among them, including deep-sea and coastal systems, depending on the shape and topographies of the shelves and coastlines.

24.7.1 Intertwining of Vertical and Horizontal Currents

The intertwining of vertical and horizontal currents creates far from homogenous conditions and **"invisible habitats"** that comprise the neritic domain. The water column realm lacks the degree of delineation seen in the benthic realm in that water column characteristics define "habitat maps" and thus tend to change and integrate more easily than the benthic realm. Currents often change seasonally, modified by winds and meteorological conditions, resulting in a highly dynamic system.

24.7.2 Physical and Chemical Factors

Numerous physical and chemical factors affect the ecology and physiology of neritic organisms. These factors include light, salinity, temperature, depth (hydrostatic pressure), physical mixing, biogeochemical processes, atmospheric exposure and influence of surface and subsurface currents, waves and swell, water mass movements, and terrestrial influences. Neritic habitats encompass many water column habitats that shift, expand, and contract over time in both predictable and stochastic patterns. Not surprisingly, many of these factors vary by location and time of year.

24.7.3 Large-Scale Currents

Large-scale currents can strongly influence the neritic zone, circulating nutrients regionally. Because its boundaries extend from the intertidal zone to the continental shelf break, the neritic domain strongly depends upon coastal morphology and the seafloor. Kelp and macroalgae provide food and habitat provisioning for some neritic species in the inner sublittoral, whereas the outer shelf, which, lacks benthic macroalgae, depends largely on phytoplankton primary production as the main food source. Neritic waters constantly move, with frequent turbulent mixing. Largely driven by physical forcing, though recent studies show that biology can also play a role, particularly in semi-enclosed embayments, in that active swimming by meso-zooplankton can contribute up to 30% of overall turbulent mixing.

As an example of ocean current complexity, the west coast of North America includes influences from the California Current System, seasonal upwelling and downwelling, El Niño/La Niña events, and changes in Pacific Decadal Oscillation that collectively move waters in this habitat over varying time and spatial scales. In the neritic zone, replacement or turnover of water occurs many times during an average human lifetime, and sometimes on annual time scales.

24.7.4 Terrestrial Inputs

Terrestrial and related coastal processes play a major role within the neritic zone. Terrestrial runoff can provide an important source of dissolved inorganic nutrients and particulate organic matter. However, excessive runoff can increase turbidity and reduce light, thus limiting photosynthesis while potentially favoring filter feeders. Increased sedimentation can also affect larval settlement, often reducing survival of early life stage in marine organisms such as corals. Together with sediment loading, runoff brings municipal and industrial pollutants that potentially include micro and macronutrient loading, heavy metals (e.g., copper, zinc) or hydrocarbons that can reduce fecundity, fertilization success, and growth of marine organisms.

Human populations widely use riverine waters for agricultural irrigation, hydroelectric power generation, and for cooling of industrial plants that add to the input of pollutants (including warming) into the neritic zone. Rivers historically often brought large sediment loads to the coastal zone, particularly during winter, replenishing beaches and helping to compensate for losses from wave erosion. In some cases, damming of rivers, mining of rivers for sand, and diverting waters for irrigation has reduced flow to the coast, reducing freshwater input as well as sediment supply. Almost invariably, society tackles this problem by building coastal defenses made up of either rocks or concrete structures. This practice has transformed long stretches of sandy shorelines into hard bottom (e.g., concrete) shores.

Terrigenous input has potentially led to different biological adaptations to cope with particulate loading. For example, a non-perforated transparent corneal membrane provides protection in myopsid squids from particles in turbid coastal waters worldwide (e.g., *Doryteuthis opalescens*). Oegopsids, a sister group to myopsids that live in clear open ocean waters, lack such a membrane.

24.7.5 Coastal Upwelling

Coastal upwelling (see Chapters 3 and 17) also plays a major role in defining neritic habitats with its alternating upwelling-relaxation events. Upwelling occurs when a steady longshore wind pushes surface waters offshore. This transport moves deep, cold, and nutrient-rich waters to the surface near the coast to replace the water pushed offshore. The nutrients brought to the photic zone enhance primary production and thus help propagate and sustain the highly productive biota of coastal waters. When upwelling winds briefly cease or reverse, the upper water layer moves back toward shore, transporting food and planktonic, larvae, juveniles, and adults. When predominant winds shift, often in fall and winter months, they tend to push surface and upper water layers shoreward and downward in a process known as **downwelling**. Downwelling helps to mix oxygen-rich waters from the upper layers downward in the water column. Importantly these upwelling and downwelling events can cause rapid changes in surface temperatures, potentially even of 10 °C, over time periods of hours.

24.7.6 Large-Scale Changes in Water Masses

Large-scale changes in water mass, typically linked to changes in temperatures and currents, often alters planktonic species composition and abundance, and indeed the survival and distribution of organisms within coastal and oceanic ecosystems. For example, El Niño or other warming events bring warm water species into nearshore environments and produce major changes over multiyear or even decadal time scales. Recently, scientists have made strides in understanding how El Niño/La Niña events and the warm and cool regimes of the Pacific Decadal Oscillation influence coastal ecosystems and their many implications for the neritic zone and its temperatures and productivity, as well as hurricanes, droughts, and productivity along coastlines. Of course, nektonic marine organisms such as adult crustaceans, mollusks, and vertebrates can swim against currents, and move at will between these water masses, but their dependence on lower trophic levels results in significant effects of water mass changes on distributions and biology.

24.7.7 River Plumes

River plumes add another water column component that affects neritic habitats. Numerous rivers discharge into coastal waters, along with groundwater that seeps freshwater into the ocean, affecting both sedimentation and coastal circulation. Riverine waters often carry also high concentrations of nutrients, but they also create gradients in salinity, enhance physical mixing, and increase turbidity. These effects are most pronounced at shallow depths, and at major river (and close to minor river) discharges, noting that meteorological events, such as hurricanes, can exacerbate their impacts. Rivers contribute significant volumes of sediments to ocean basins and provide much of the total freshwater that supplies the oceans and offsets evaporative loss. Riverine inputs lower salinities in the entire basin, but most dramatically in estuaries and deltas. Although rivers deliver significant terrigenous sediments, the waves and tidal currents within intertidal and shallow subtidal environments may limit accumulation of sediment particles by resuspending and transporting finer sediments, leaving behind coarse-grained material at the shallowest depths. Freshwater typically reaches the ocean as a distinct plume, which can strongly affect neritic waters. Large river plumes may create microhabitats of reduced salinity within the neritic zone, while potentially acting as biogeographic barriers between marine regions adjacent to the river mouth. But river plumes can stretch for hundreds of kilometers to offshore habitats (Figure 24.9), often shifting predictably over the course of each year and spreading as a function of river discharge volume and local currents.

24.7.8 Fronts

Fronts refer to interfaces between water masses that differ in density, form when rivers or upwelled waters come into contact with other water masses that differ in temperature and salinity. Oceanographic fronts in marine systems often generate high productivity (i.e., elevated primary and secondary production) that supports many species of invertebrates, fishes, seabirds, and marine mammals. These features contribute to the capacity of neritic waters to support many major commercial fisheries.

24.7.9 Neritic Food Webs

Neritic food webs typically include forage fish species such as northern anchovy, Pacific herring, and Pacific sardine that feed on phytoplankton and zooplankton, thus illustrating the close link between fisheries production and phytoplankton production. In addition to the generally greater availability of nutrients in neritic environments and thus greater primary production, the nature of food webs also contributes to high fisheries production in neritic environments. As noted in an earlier chapter, transfer efficiencies between trophic levels average ~10%. Oceanic environments typically begin with small phytoplankton and thus require greater numbers of trophic steps to move energy into trophic levels that support fisheries (Figure 24.10). In contrast, shorter food webs in upwelling regions and other neritic environments move energy much more efficiently to upper trophic levels.

Figure 24.9 Examples of river plumes, which can extend for hundreds of kilometers into the neritic zone. *Source:* Osadchiev2 / Wikimedia Commons / CC BY-SA 4.0.

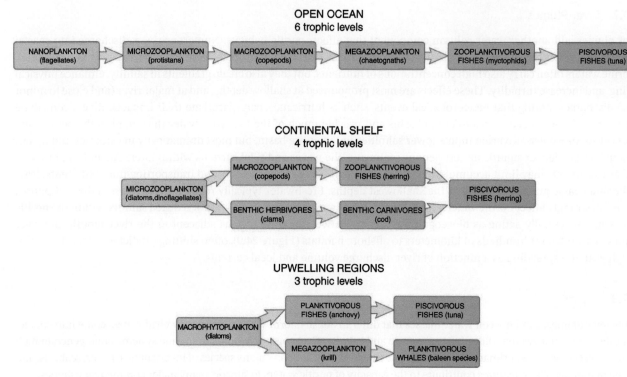

Figure 24.10 Contrast between open ocean and coastal food webs, including shelf and upwelling environments. Note differences in number of trophic levels among systems.

Human impacts on neritic ecosystems (discussed in greater detail in Chapter 28) provide insight into factors influencing functioning of neritic systems. Recent studies reported an interesting pattern in fish fauna of exploited North Atlantic shelf ecosystems. Top-down control by predators appears to dominate cold water, low diversity food webs, whereas bottom-up control dominates neritic habitats with warm water temperatures and high biological diversity. However, this pattern appears to be far from universal.

This generalization suggests that predation should structure the Ross Sea (Antarctica) neritic ecosystem, which indeed seems largely dependent on the irregular distribution of the remaining uppermost trophic levels, consisting of cetaceans, flightless seabirds, seals, and large predatory fish; humans largely extirpated these groups from the North Atlantic long ago. The development of a large polynya (persistent open water area within the pack ice) in the Ross Sea drives blooms of the colonial haptophyte *Phaeocystis antarctica*, which, in part, makes the Ross Sea the most productive stretch of water of comparable size south of the Polar Front. These intense blooms, fueled by sunlight and abundant macronutrients but constrained by availability of micronutrients (and eventually sunlight as summer fades), lasts for only a few months. The main grazers appear to be pteropods, with a presumably sparse fish fauna at the next higher trophic level given the avoidance of this area by higher-level predators. Much primary production is either re-mineralized or sinks to the benthos. Though species poor, primary production truly drives this system.

In contrast, the food web of the marginal ice zone (MIZ) ringing the polynya, and especially to the west, begins with photosynthesis by single-celled diatoms that support a much higher diversity and abundance of grazers, piscine predators, and upper-level predators. These diatoms, which experience the same nutrient and light limitations as the haptophytes, can dominate where melting pack ice stabilizes (and thus stratifies) the water column, unlike the wind-mixed central portion of the shelf where excessive mixing limits photosynthesis. Despite high production of the MIZ, an appreciable portion of phytoplankton remains ungrazed, a pattern consistent with somewhat unexpectedly low densities of grazers (especially euphausiids) and the cannibalistic nature of the major piscine predator on zooplankton, silverfish *Pleuragramma antarcticum* (the "anchovy of the Antarctic"), in the late summer. The apparent paucity of large grazers seems inconsistent with huge populations of baleen whales (minke, *Balaenoptera bonaerensis*), fish-eating killer whales (*Orcinus orca*), seals, and penguins. Indeed, the tight predator–prey relationship means that the seasonal arrival of whales forces penguins to switch their diet from krill to silverfish and dramatically increase their foraging efforts. This pattern suggests that the thousands of marine mammals and millions of penguins, along with abundant predatory Antarctic toothfish (*Dissostichus mawsoni*) deplete the grazers, with repercussions for phytoplankton biomass (i.e., a trophic cascade). Top-down processes see to force the food web in this portion of the Ross Sea, despite higher biodiversity.

24.8 Summary

In summary, neritic zones occupy far less of the global ocean than oceanic zones, but generally support much more productive fisheries and higher diversity, thanks to a greater diversity of habitats and generally greater productivity (Table 24.1)

Table 24.1 Comparison between neritic and oceanic (pelagic) ecosystems.

Main characteristics	Habitat / Ecosystem					
	Neritic	Oceanic				
		Epipelagic	Mesopelagic	Bathypelagic	Abyssopelagic	Hadopelagic
Depth	0–200 m	0–200 m	200–1000 m	1000–4000 m	4000–6000 m	6000–11,000 m
Extent	<10% of ocean, covers continental shelf	~5% of global oceanic volume	~18% of global oceanic volume	~59% of global oceanic volume	~19% of global oceanic volume	<<1% of global oceanic volume
Distribution	Continental shelf	Upper layer of open ocean	Subsurface layer beyond edge of continental shelf	Beyond edge of continental shelf	Beyond edge of continental shelf to the bottom of the ocean in most locations	Ocean trenches
Habitat characteristics	Often but not always productive	Photic zone but limited nutrients	Disphotic, permanent thermocline with major temperature change	Uniformly cold (4 °C), aphotic	Uniformly cold, aphotic	Uniformly cold, aphotic, sometimes subject to slumping
Biodiversity	Relatively low	Relatively low	Poorly known	Poorly known	Surprisingly high given low biomass	Low
Endemism	Low	Very low	Poorly known	Poorly known	Some	Relatively high
Functioning	Encompasses most commercial fisheries,	Produces primary production that supports oceanic habitats beneath. Low production rate but massive area.	Key component of biological pump, diel vertical migrations	Zooplankton diel vertical migration moves energy and carbon	Poorly known	Poorly known
Trophic webs	Often short food webs	Longer food webs but low production	Depends on organic matter export from the epipelagic zone	Except for chemosynthetic habitats depends on organic matter from epipelagic	Depend on detritus from upper layers except for limited chemoautrophy	Detritus based with limited chemoautrophy
Peculiarities	High human use and impacts		Large biomass of small fishes with potential commercial potential	Low biomass, bioluminescence common	Very low biomass, bioluminescence	Very low biomass, deepest habitats on Earth, most in Pacific Ocean

Questions

1) What is the neritic zone of the ocean?
2) What is generally the maximum depth of the neritic zone?
3) Does the neritic domain contain a specialized biota within the water column?
4) How are water column and benthic habitats connected?
5) What do we mean by "invisible habitats"?
6) What is the role of river inputs in the neritic domain?
7) How do ocean currents influence the neritic zone biota?

Suggested Reading

Ainley, D. (2007) Insights from study of the last intact neritic marine ecosystem. *Trends in Ecology and Evolution*, 22, pp. 444–445.

Cotner, J.B. and Biddanda, B.A. (2002) Small players, large role: Microbial influence on biogeochemical processes in pelagic aquatic ecosystems. *Ecosystems*, 5, pp. 105–121.

Duffy, J.E. and Stachowicz, J.J. (2006) Why biodiversity is important to oceanography: Potential roles of genetic, species, and trophic diversity in pelagic ecosystem processes. *Marine Ecology Progress Series*, 311, pp. 179–189.

Fonda Umani, S. and Alfred, A. (2003) Seasonal variations in the dynamics of microbial plankton communities: First estimates from experiments in the Gulf of Trieste, Northern Adriatic Sea. *Marine Ecology Progress Series*, 247, pp. 1–16.

Kideys, A.E., Roohi, A., Bagheri, S. et al. (2005) Impacts of invasive ctenophores on the fisheries of the Black Sea and Caspian Sea. *Oceanography*, 18, pp. 77–82.

Maugeri, T.L., Carbone, M., Fera, M.T. et al. (2004) Distribution of potentially pathogenic bacteria as free living and plankton associated in a marine coastal zone. *Journal of Applied Microbiology*, 97, pp. 354–361.

Obernosterer, I., Catala, P., Reinthaler, T. et al. (2005) Enhanced heterotrophic activity in the surface microlayer of the Mediterranean Sea. *Aquatic Microbial Ecology*, 39, pp. 293–302.

Patten, N.L., Mitchell, J.G., Middelboe, M. et al. (2008) Bacterial and viral dynamics during a mass coral spawning period on the Great Barrier Reef. *Aquatic Microbial Ecology*, 50, pp. 209–220.

Sanders, R.W., Caron, D.A., and Berninger, U.G. (1992) Relationships between bacteria and heterotrophic nanoplankton in marine and fresh waters: An inter-ecosystem comparison. *Marine Ecology Progress Series*, 86, pp. 1–14.

Smith, E.M. and Kemp, W.M. (2003) Planktonic and bacterial respiration along an estuarine gradient: Responses to carbon and nutrient enrichment. *Aquatic Microbial Ecology*, 30, pp. 251–261.

Taylor, G.T., Hein, C., and Iabichella, M. (2003) Temporal variations in viral distributions in the anoxic Cariaco Basin. *Aquatic Microbial Ecology*, 30, pp. 103–166.

Part IIID

Comparative Marine Ecology: Habitat Types, Their Biodiversity, and Their Functioning

25

Deep-Sea Ecosystems along Continental Margins

25.1 Introduction to the Deep Sea

The open ocean covers more than 65% of Earth's surface and thus dominates the surface of our planet. Far from uniform, this environment encompasses a diversity of habitats spanning depths from ~200 m to ~11,000 m. From a human perspective, these environments seem extreme in their absence of sunlight, temperatures below 4 °C, and an average hydrostatic pressure of 400 atm. The absence of photosynthetic primary production results in almost exclusively heterotrophic systems dependent on organic material supplied from shallower depths, with the notable exception of chemoautotrophic primary production that occurs mainly at hydrothermal vents or cold seeps.

The deep-sea floor hosts a wide variety of habitats and ecosystems spanning from continental slopes, cold-water corals, and canyons to mid-ocean ridges, seamounts, abyssal plains and trenches that support unique communities. In the nineteenth century scientists generally considered the deep sea as devoid of life because of the perceived extreme conditions and (incorrectly) assumed lack of oxygen. In reality, these dynamic and heterogeneous environments host life through the deepest ocean depths to 11,000 m. Scientists have proposed two main hypotheses to explain the origin of deep fauna: (1) "Recent" fauna resulting from the cooling of deep-sea waters; and (2) "archaic" fauna that include organisms that found refuge in deep waters during mass extinctions, noting that many endemic taxa of the deep sea date back to the Mesozoic or Cenozoic.

Generally, scientists associate high biodiversity levels with regions of high food/resource availability. The deep sea puzzled scientists for a long time given its high biodiversity despite extreme food limitation and dependence on production from the photic zone that decreases exponentially with increasing depth.

Scientists estimate they have mapped the bathymetry of ca 20-25% of the ocean depths and explored superficially less than 5% of the deep-sea environment, using remote tools (nets, echosounders, ROV, etc.). Sampling and direct studies of the seabed below 3,000 m have assessed in detail <0.001% of biodiversity and other habitat/ecosysrtem properties (Figure 25.1). The remoteness, inaccessibility, and complexity of deep-sea environments contribute to their status as one of the last major frontiers of exploration on our planet. The development of new technologies, the synergy of interdisciplinary skills, and greater attention to resources and ecosystem services provided by deep-sea environments create a strong impetus to accelerate research in these ecosystems. The deep sea remains the least studied habitat on Earth, particularly its biological component, offering a compelling frontier for scientific research. Deep environments, in fact, host life (Figure 25.2) at all depths, even in the Mariana Trench at ~11,000 m depth.

The physical properties of deep-sea environments below the permanent thermocline vary little, with the notable exception of hydrostatic pressure that increases by one atmosphere for every 10 m. Thus, the average hydrostatic pressure in deep oceans of ~400 atm represents a major increase from 1 atm at sea level. Furthermore, the currents that move through the deep ocean can sometimes rival the variability in direction and intensity often seen in surface waters. Water temperature in deep environments varies from −2 to >400 °C near hydrothermal vents, where spilling magma and hydrothermal venting rapidly heat nearby water. Typical temperatures in deep environments, are uniformly 4 °C or less and are largely invariant, supporting a plethora of life including diverse, metabolically active metazoans. The Mediterranean Sea and Red Sea offer a significant exception to generally cold deep-sea habitats with temperatures of ~13 °C from 600 to 4000-m depth (in the Mediterranean) and 21.5 °C at 2000 m depth (in the Red Sea).

Relatively constant salinity in the deep sea varies from 34.8 (± 0.3) below 2000 m depth, to slightly fresher (34.65) at greater depths (>4000 m).

Marine Biology: Comparative Ecology of Planet Ocean, First Edition. Roberto Danovaro and Paul Snelgrove.
© 2024 John Wiley & Sons Ltd. Published 2024 by John Wiley & Sons Ltd.
Companion Website: www.wiley.com/go/danovaro/marinebiology

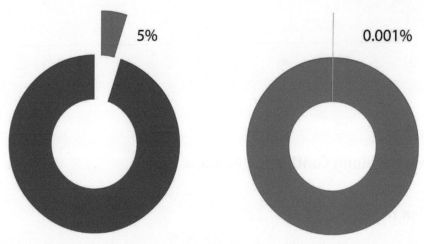

Figure 25.1 Estimate of percentage of explored seabed below 3000 m depth (5%, left) and percentage of seabed sampled and studied in detail (0.001%, right).

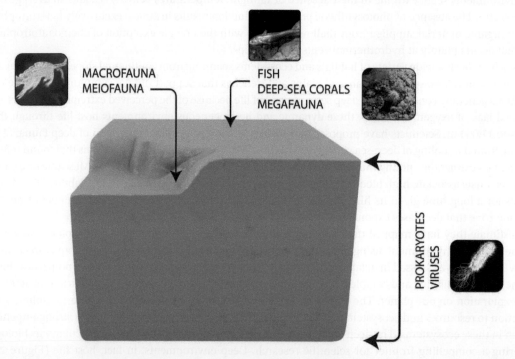

Figure 25.2 Benthic life in deep-sea ecosystems spans organisms that vary greatly in size, from viriobenthos to megafauna. Most metazoans live only in the upper few cm of the sediment, whereas Bacteria and Archaea can live to 100s of meters below the surface. *Source:* Science RF / Adobe Stock, NOAA / Wikimedia Commons / Public Domain.

In deep-sea environments, oxygen concentrations generally approach saturation, but the most extensive hypoxic and anoxic portions of the planet occur within some deep-sea environments, such as areas with an oxygen minimum zone, usually at 500–1500 m depth. These zones occur in locations with strong upwelling that promote primary production, and in closed basins such as the Black Sea where permanently anoxic conditions occur below 250 m depth.

Unlike in coastal waters, solar radiation plays no direct ecological role in the deep sea, given that all light (other than bioluminescence) disappears completely by 500–1,000 m depth. Light, however, plays a critically important indirect role by providing energy for surface phytoplankton production, a portion of which enters deep-sea ecosystems through the trophic network. In particular, an increasing number of studies on pelagic-benthic coupling in deep marine systems demonstrate the dominant role that food supply plays in determining deep-sea densities and biomass. Chemoautotrophy provides the only autochthonous primary production in the deep sea but largely within hydrothermal vents or in cold seeps, where prokaryotes (Archaea) produce the organic matter. However, the modest overall contribution of chemoautotrophic primary production supports relatively few deep-sea organisms. Therefore, the negative energy balance and effective absence of primary production in the deep sea results in an essentially heterotrophic system.

In contrast to the traditional view of the deep sea as a homogeneous and static system with invariant chemical and physical conditions, researchers have recently demonstrated major disturbance phenomena in the form of benthic storms, landslides, and slope and bottom currents that re-suspend sediments. Persistent hydrodynamic activity and intense bottom currents often characterize the **benthic nepheloid layer** (BNL, the layer at the water-sediment interface). In addition, scientists have discovered frequent deep storms; such phenomena extend 50–200 km, with energy up to 100x that of typical bottom flow, occurring with variable frequency but generally lasting less than 100 d. Vortices may move at speeds greater than 15 cm s^{-1}, re-suspending sediments, especially at bathyal depths. Furthermore, intense phytoplankton blooms in surface waters can generate periods of rapid organic flux to the deep-sea bed lasting for periods of weeks to a month or more. In short, despite some aspects of constancy, deep environments can experience extremely high variability in some physical variables over short time scales, characterized by unpredictability in food intake, variable hydrodynamics at the bottom, and strong interannual differences. Moreover, despite a seascape that may appear homogeneous, the reality is a set of heterogeneous systems composed of a mosaic of habitats of varying extent, with different particle size composition and aggregate distribution of food resources.

25.1.1 Is the Deep Sea on a Diet?

Researchers have long considered deep-sea environments as food-poor environments, given their heavy dependence on organic matter from the euphotic zone that progressively and rapidly degrades as it sinks through the water column to the bottom. Sustaining heterotrophic life forms in the deep sea critically depends on export of organic particles from the euphotic zone. Of ~48.5–54 PgC y^{-1} marine surface primary production, sediments incorporate only 0.2–0.79 PgC y^{-1} (~1%). Heterotrophs efficiently recycle organic matter so most marine primary production recycling and/or consumption takes place before reaching the deep-sea floor. This recognition supported a long-held paradigm of a food-poor environment.

Although oligotrophy apparently characterizes most deep-sea sedimentary environments, time-lapse photography has demonstrated rapid and massive phytodetritus deposition from surface waters to the sedimentary deep-sea floor at 4,000 m in 40 d. These events deliver fresh organic material with excellent nutritional value and important consequences for abundance, biomass, biodiversity, metabolism, and distribution of deep-sea species. Researchers have long debated whether these events are highly episodic and limited to short time scales, or occur more frequently, as demonstrated by reports of multiple pulses of cyanobacteria deposition to the seafloor. But what proportion of organic material delivered to the deep sea arrives through vertical versus lateral flux? Clear evidence shows that lateral advection delivers much of the organic flux on continental margins with massive and frequent down-canyon transport. Recent evidence demonstrates examples of ecological engineering on a large scale in that large coral carbonate mounds can alter local hydrodynamics to draw down relatively fresh labile surface production. In short, relatively eutrophic deep-sea habitats fundamentally challenge our view of food-poor trophic conditions and low energetics in deep-sea ecosystems.

In addition, we traditionally quantify food availability to the benthos simply by measuring bulk organic matter, without considering the importance of food quality. In systems rich in organic matter, rapid transformation of organic molecules, and particularly biopolymers, leads to complexation processes that produce high molecular weight compounds (e.g., humic and fulvic acids) that consumers cannot fully digest. For a detritus feeder, a potential 20–50% digestible fraction in deep-sea sediments compares favorably with the 5–15% digestible fraction reported for most continental shelf sediments. Therefore, the more palatable deep-sea fraction may significantly offset the low overall quantity of organic C observed in many deep-sea sediments, reducing differences from their shallow counterparts. Indeed, optimal foraging theory suggests that selective deep-sea consumers may find quality food sources more easily than their shallow-water counterparts. Results acquired in recent years indicate that several types of deep-sea ecosystems and a potentially important portion of the deep-sea biome are less food limited than previously thought, requiring significant revision of the past ecological paradigm of a "deep-sea on a perpetual diet".

25.1.2 Extreme and Harsh Conditions?

In harsh ecosystems, organisms must expend energy to compensate for physical and chemical stressors that create environmental conditions many organisms cannot tolerate. Defining "extreme ecosystems" can be complex and depends on the approach (biological, physical or chemical), and whether one uses a human frame of reference. Some researchers define extreme as: (1) an ecosystem in which one or more key environmental variable (e.g., temperature, pH, salinity, pressure, oxygen etc.; Figure 25.3) approaches the limits that most life forms can tolerate; (2) an ecosystem where one or more environmental stresses (physical or chemical) combine to limit available food/energy; (3) an ecosystem completely lacking

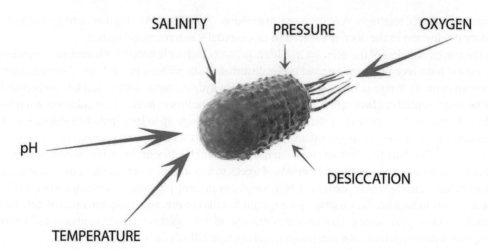

Figure 25.3 Environmental variables that can result in extreme environments. All of these extreme conditions can occur in marine ecosystems.

some major taxonomic groups; (4) an ecosystem in which organisms must spend large amounts of energy to compensate for physical and/or chemical stress. Extreme ecosystems include:

a) very cold environments (i.e., polar ecosystems, discussed in Chapter 23) with temperatures consistently below 0 °C;
b) extremely hot environments with temperatures ranging from 40 to 50 °C up to 400 °C in some hydrothermal vents;
c) acidic environments with pH values in some cases approaching 2 or basic environments with pH values up to 10;
d) extremely salty environments with salinity 20–30%, such as the brine lakes on the bottom of the Mediterranean;
e) extremely deep environments with high pressure (i.e., hadal zones discussed in Chapter 26);
f) hypoxic and anoxic environments (oxygen minimum zones and anoxic basins) or with extreme redox potential;
g) environments that contain toxic substances of natural origin (hydrogen sulfide or heavy metals).

The next few sections consider extreme conditions resulting from pH, high temperatures, high salinity and lack of oxygen, as well as high pressure. The chapters on coastal environments described desiccation and intense light radiation, low temperature conditions in Chapter 23 on polar environments.

At hydrothermal vents, superheated seawater rich in hydrogen sulfide, methane, and other elements flows from the seabed, supporting intense chemoautotrophic primary production. To date, researchers have identified ~700 species associated with these ecosystems, including many endemic species. Several species of mollusks, siboglinid worms (polychaetes), and other species possess endo- and exo-symbiont bacteria that provide a source of food for their hosts. The high densities and biomasses (up to 8 kg m^{-2} or more) have led to likening of these environments to "oases" with abundant food that depends on geothermal energy and chemosynthetic production. In shallow-water hydrothermal vents (generally characterized by temperatures <130 °C) both chemosynthetic and photosynthetic production can occur, and these habitats lack endemic fauna. Cold seeps refer to habitats where cold seawater containing methane and/or other hydrocarbons flows from the seabed. Cold seeps and hydrothermal vents share some features including: (a) high biomass and high production; (b) a complex community based on chemoautotrophy; (c) megafauna with some shared species or genera; (d) extensive mats of chemoautotrophic bacteria; (e) presence of toxic substances.

Chemoautotrophic primary production in cold seeps depends on the intensity of chemical flux. The high biomass in these ecosystems (up to 51 kg m^{-2}) depends on *Beggiatoa* spp. bacterial mats and organisms that feed mainly by filtering particles (including bacteria) from the water column. Carcasses of large cetaceans (whale carcasses) create habitats used by communities similar to those at cold seeps. These transient environments support high species diversity that includes >400 species, including worms (*Osedax* spp.) that lack a digestive tract, and depend on symbiotic bacteria that convert the fat of whale bones into sugars.

Hypoxia and anoxia occur in locations with low or no dissolved oxygen. Oxygen Minimum Zones can occur at 10–1500 m depth, and favor opportunistic species and giant bacteria that use nitrate to oxidize hydrogen sulfide. These environments typically support just a few species of foraminifera, mollusks, annelids, and nematodes. Deep anoxic hypersaline basins represent the most extreme environments in the ocean; these anoxic brines contain high concentrations of methane and hydrogen sulfide that host abundant prokaryotes (both Bacteria and Archaea) and viruses. Recently, researchers discovered three new species of Loricifera that live their complete life cycle in the absence of oxygen.

Some researchers include highly variable environments among those considered extreme, such as those characterized by: (1) sharp **thermoclines** or **chemoclines** (i.e., an abrupt change in the physical or chemical properties of water with a small change in depth); (2) large temporal fluctuations, or (3) frequent intense or catastrophic episodic events.

Studies of extreme environments help researchers understand the conditions that require special physiological functions in organisms that allow them to survive. Extreme environments provide many opportunities to deepen knowledge to understand physiological mechanisms of adaptation and survival of organisms, their ability to develop physiological responses to combined stresses, and how they develop novel mechanisms for energy acquisition. In addition, taxonomic studies of extreme environments enable exploration and understanding of important evolutionary processes such as convergence, and development of hypotheses on "primordial" life forms or how organisms might live on other planets.

In applied fields, studies of extreme environments provide valuable information on ecosystem responses to specific conditions. They also enable studies on the capacity of living organisms to cope with environmental conditions close to the limits that support life, or conditions that vary greatly. For example, researchers can use information from extreme marine environments to develop models that predict impacts of global changes such as: (a) rapid increases in ocean temperature; (b) expansion of hypoxic and anoxic zones; or (c) exposure to toxic substances such as hydrogen sulfide. These environments also hold promise for the discovery of new enzymes, proteins, and chemicals used by organisms to cope with extreme conditions, which offer potential applications in industrial processes and biotechnology.

In the next few chapters, in addition to more "benign" and relatively stable deep-sea environments, we will also consider extreme habitats "par excellence" such as hydrothermal vents (high temperatures and toxic chemicals), cold seeps (toxic methane or other hydrocarbons), hypoxic and anoxic environments (low concentrations of dissolved oxygen) and hypersaline environments. In addition to extreme environmental conditions, these systems share similarity in functioning, in that chemosynthetic primary production partly or entirely supports them.

25.1.3 Are Deep-Sea Ecosystems Depauperate?

These recent discoveries of dynamic and sometimes novel habitats completely change how we view deep-sea environments and their functioning and may help to explain how abundances of many benthic groups (especially the smallest) in deep-sea environments can sometimes exceed those in some coastal depths (Figure 25.4).

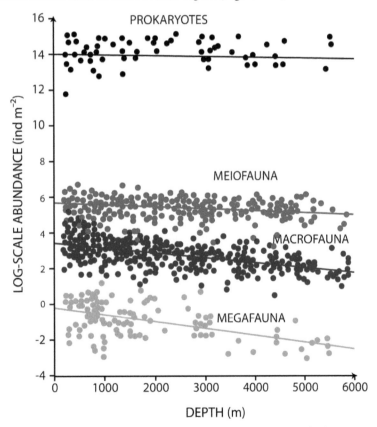

Figure 25.4 Patterns of abundance in prokaryotes, meiofauna, macrofauna, and megafauna in deep-sea sediments. The log-transformed data show densities of indiv. m^{-2} in relation to depth calculated at intervals of 200 m. Thus, megafauna and macrofauna decline with depth, whereas meiofauna show no pattern, and prokaryotes increase with depth.

Because of their massive area, deep-sea ecosystems define the main reservoir of abundance and biomass of the blue planet.

25.1.4 Metabolism and Functioning of Deep Ecosystems

Consistent with other paradigms, researchers inferred low metabolism in deep-sea organisms as a result of low temperatures and food limitation. However, new experimental evidence and meta-analysis of existing data indicates that the metabolism of abyssal organisms depends on biomass (and not temperature), and that metabolic rates of individual organisms may approximate those of coastal species living at similar temperatures. In addition, discussions neglected the assumption that dwarfism in deep-sea environments results from the scarcity of food. Despite the reduced average size of many organisms such as deep-sea gastropods and nematodes, examples of gigantism occur in many taxa (isopods, amphipods, pycnogonids, ostracods and anemones). In addition, dwarfism in nematodes in association with depocenters of organic matter at hadal depths such as the Atacama Trench at 8,000 m led to the hypothesis that dwarfism represents a specific adaptation to high pressures.

We currently lack sufficient knowledge about food webs in deep-sea environments to understand fully their functioning. Below 3,000 m prokaryotes dominate overall biomass, but researchers recently discovered that heterotrophic bacteria do not represent the totality of microbial biomass, because below 1000 m the Archaea, which are largely chemosynthetic primary producers, comprise more than 40% of prokaryotic abundance in the water column, and almost 50% of biomass and (probably) total production.

Energy limitation represents the most defining ecological characteristic of abyssal ecosystems, in that these food-limited communities depend on detrital organic matter produced in the euphotic zone 1000s of meters away (Figure 25.5). Most organic matter arrives as a slow rain of small particles, typically representing only 0.5–2% of net primary production from the euphotic zone; this supply declines with depth and varies locally depending on levels of surface primary production. Organic matter flow greatly increases when carcasses of large animals such as whales fall to the bottom.

Before organic debris mixes into the sediment, suspension feeders consume part of the food content by filtering organic matter from the water mass above the bottom. Deposit feeders, benthic scavengers of varying size, and bacteria quickly use inputs of food resulting in increased metabolic activity, growth, and reproduction. Predators consume scavengers that, in turn, become prey for larger predators, such as fishes.

The key functions of the community, including organic matter mineralization and bioturbation, rapidly decrease with decreasing particulate organic carbon (POC) flux to abyssal environments. These functions play an important role in ecosystem services provided by the abyssal environment by influencing nutrient regeneration, organic carbon deposition, and organic carbon remineralization rates on the seafloor. Thus, POC flux to the bottom apparently controls rates and patterns through which abyssal environments help modulate atmospheric CO_2 levels and levels of calcite saturation in the ocean.

Life forms that occupy the subsurface layers of sediments also play an essential role in deep-sea ecosystem functioning. For example, callianassid shrimp burrow >1 m deep into sediments, and researchers recently discovered that metabolically active prokaryotes and viruses occur up to 100s of meters below the sediment surface. They hypothesized that these organisms survive at these sediment depths by relying on a supply of matter and energy from the deep subfloor.

On a global scale, the average biomass of the two largest size classes of benthic animals (macro- and megafauna) dramatically decreases with increasing depth, and therefore with decreasing POC flux. Furthermore, the maximum size of gastropods decreases substantially between 3,000 and 5,500 m depth in the North East Atlantic. Presumably, decreasing food supply limits size, resulting in dominance in biomass by smaller groups such as bacteria and meiofauna below 3,000 m. The **metabolic theory of ecology**, which posits that the metabolic rate of organisms defines the fundamental biological rate that governs most ecological patterns, predicts that the dominance of small organisms should lead to decreased production/biomass rates at the community level, making abyssal ecosystems particularly inefficient at producing biomass compared with bathyal systems. Researchers have documented such a reduction in ecosystem efficiency in abyssal environments. Thus, extremely low energy flow characterizes some abyssal regions, but also less efficient transfer of energy to higher trophic levels.

Experiments using isotopes to trace organic matter in marine sediments have clarified some details about energy flow in abyssal ecosystems. Experiments in shallow environments (intertidal to ~140 m) that added phytodetritus (labile organic matter) reported rapid assimilation and respiration, dominated by bacteria. In contrast, carbon uptake and respiration occur ~10x slower at abyssal and bathyal depths, with macro- and meiofauna dominating initial phytodetritus uptake, rather than the bacterial component. The results are surprising considering microbial dominance of abyssal biomass and suggest that, despite dominance of prokaryotes and meiofauna in deep-sea sediments, macrobenthos respond quickly to

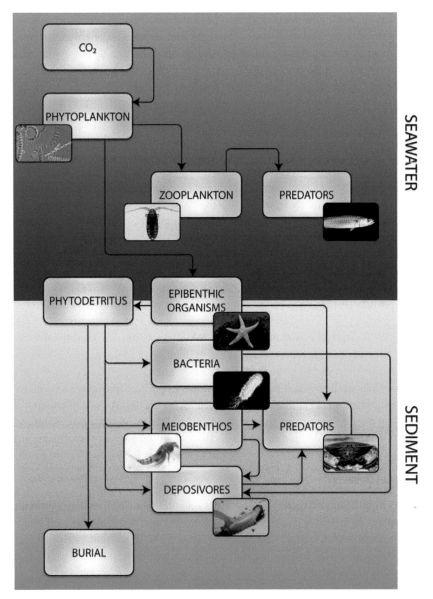

Figure 25.5 Material and energy supply to the deep sea and subsequent use of the food web by deep pelagic and benthic communities. *Sources:* NASA / Wikimedia Commons / Public domain, jkirkhart35 / Wikimedia Commons / CC BY 2.0, Science RF / Adobe Stock, H. Limen and H. MacIsaac. / Wikimedia Commons / Public Domain, davemhuntphoto / Adobe Stock.

"fresh" material, whereas microbes respond more slowly and rely on the more refractory organic components. Optimal foraging offers one possible explanation for the dominance of these components in abyssal habitats, if strong selection for higher foraging efficiency in food-poor environments led to a weaker impact of low temperatures on the functional response of macro- and meiofauna compared to bacteria. However, the type of organic material available may also explain this response. When organic material arrives in large pulses in a limited area, it benefits large organisms that quickly use it. However, when material arrives more steadily in low quantities and distributed over large areas, it favors microbial components and small animals. Alternatively, a significant portion of bacterial biomass in abyssal sediments consists of cells settling from shallow depths that may lack adaptations to survive in abyssal temperature and pressure conditions. Whatever the explanation, experiments to date show that macro- and meiofauna play important roles in the initial processing and redistribution of "fresh" food arriving in abyssal environments. The rapid removal of phytodetritus by abyssal holothurians suggests that larger organisms (megafauna) may play an important functional role in abyssal energy flow, irrespective of bacterial dominance.

Autochthonous primary production in the deep sea occurs only at hydrothermal vents and cold seeps, where chemosynthetic bacteria produce organic matter. But deep-sea ecosystems that depend on chemoautotrophic production may be much more widespread than previously thought. Cold seeps actually occur frequently along continental margins, but other systems may use chemoautotrophic production. Growing evidence points to significant contributions by "dark energy" (methane, hydrogen sulfide, and other inorganic elements) in supporting deep ecosystem functioning.

25.2 Deep-Sea Biodiversity

Early anecdotal evidence suggested an abundance of life in deep waters, as far back as 1819 when Sir John Ross recovered a basket star from a sounding line he had lowered to some 500–600 fathoms (900–1100 m) in the Strait of Messina. Around 1840 James Clark Ross and J. Hooker found living organisms in mud on the sounding lines they had lowered to 1800 m on the Antarctic continental slope. But these observations were not described in such a way that they became well known. Enter Edward Forbes, Professor of Natural History (1815–1854) at the University of Edinburgh, who convinced the British Association for the Advancement of Sciences to support a dredging expedition in the deep Aegean Sea in the Eastern Mediterranean. Low abundances in a relatively unproductive body of water, combined with a coarse net that did not effectively sample the seafloor, resulted in an absence of specimens from depths >600 m. This expedition led him to formulate his "azoic theory" which postulated that deep-sea environments cannot support life because of the limited food, high pressure, and (assumed) lack of oxygen. Charles Wyville Thompson, also from the University of Edinburgh, debunked this theory when he organized the *Challenger* expedition (1872–1876), an expedition around the world with well-defined science objectives to study a range of ocean depths, including the seabed. They found living biota everywhere they sampled, including the deep trenches. Sir John Murray, the lead zoologist on board, noted that deep-sea samples contained more species and less dominance than shallow water samples, but these important findings, reported in the 50-volume series published after the expedition, still did not extinguish the idea of a "deep-sea desert," which persisted for years. Quantitative sampling in the 1960s put the matter to rest once and for all, although this misperception continues to circulate. Scientists up to that time believed the deep sea would hide living fossils that would fill gaps in evolutionary history.

25.2.1 Oases or Biological Deserts?

Deep-sea studies originate from human curiosity about "unreachable" and frontier environments. The vast unknown inspired fantastical ideas about monsters and archaic fauna retreating into the deep, as well as theories about deep-sea environments and life that scientists have largely refuted over time. First, the abundance of oxygen (> 4 ml l^{-1}) in most deep-sea environments through to hadal depths illustrates how a scientific approach based on incorrect and untested assumptions can lead to entirely false conclusions.

In 1967, Robert Hessler and Howard Sanders reported their discovery of high abyssal biodiversity, correcting the first false paradigm of the deep-sea biology. However, other paradigms proffered by researchers also required debunking. A growing number of studies demonstrated the presence of a rich biota spanning from macro- to micro-organisms, including many taxa – some at a high taxonomic rank – not seen in shallow-water coastal systems. In general, researchers use the term biodiversity as a synonym for species richness (number of species present in a location). However, diversity can be measured at different levels, depending on whether the goal is to distinguish genetic diversity, species diversity, phyletic diversity, and functional diversity, and over spatial scales spanning from community to habitat to ecosystem. Typically in deep-sea ecosystems, despite large numbers of species, only a few individuals represent each species in a given sample. The large numbers of rare species result in unusually high ecological diversity despite relatively few individuals (Figure 25.6).

Despite the immense size of deep-sea environments and their importance for the functioning of the entire planet, relatively few studies have addressed these lesser-known biomes, with many undiscovered species. The logistical challenges in sampling, the need for specialized equipment, and the high operating costs for research vessels all contribute to the limited availability of data and surveys. But the shortage of taxonomic experts ultimately creates the largest barrier to discovery of new species.

The number of studies (and therefore available data) declines exponentially when moving from the continental shelf to the hadal trenches. Perhaps unbelievably, the biodiversity data on abyssal environments obtained up until now represents the combined analysis of just a few hundred square meters of seafloor, a truly negligible fraction of these massive ecosystems. The large size and remote locations of these environments have stymied studies on their structure and functioning.

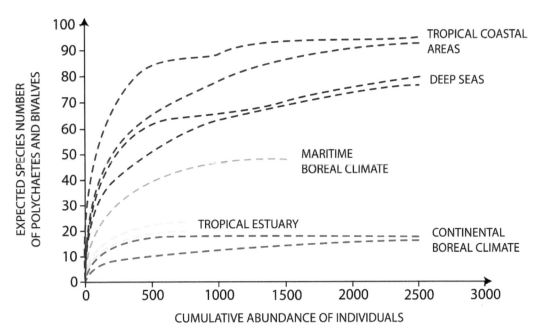

Figure 25.6 Comparison of biodiversity showing expected number of polychaetes and bivalves species with increased numbers of samples and individuals. The high diversity of deep-sea environments compares favorably with those of coral reefs and exceeds those of tropical estuarine or temperate or boreal marine ecosystems. Data based on Howard Sanders and Robert Hessler's pioneer work in the 1960s.

Recent estimates of species diversity (the number of species found in a given region) suggest that the number of species in the deep sea could rival or exceed that of the Amazonian rainforest. For example, scientists have described only ~5,000 species of nematodes (the most abundant metazoans in the deep sea), but projections estimate that 20,000–30,000 species remain undescribed. Comparisons between pelagic and benthic domains indicate higher diversity in the benthos, despite the much more extensive volume of the pelagic environment. This discrepancy between system size and number of species, and the broader representation of different phyla on the seafloor, suggest that marine life may have originated and begun to diversify on the seafloor.

Some experts believe that marine systems contain 700,000–1,000,000 undescribed animal species, many from the deep sea, and dominated by meiofauna and macrofauna. Often, >80% of the invertebrate species sampled in newly studied locations at abyssal depths (> 4,000 m) are new to science. Biodiversity on a local scale, ranging 0.1–1 m^2 in deep-sea sediments, generally spans from moderate to high. For example, a 0.25 m^2 area of seafloor might contain more than 100 macrofaunal species and 150 polychaetes that encompass 50 species. Biodiversity in deep-sea environments represents an interesting and complex topic as well as a frontier area for scientific research. Indeed, deep-sea environments not only contain large numbers of "new" species but also house the resting cysts of some important planktonic organisms; these resistant benthic forms help planktonic species survive during unfavorable environmental conditions.

High levels of diversity of some benthic taxa at bathyal depths 1,000–2,500 m, may result from the relative stability of environmental conditions and a predictable food supply from the adjacent shelf. The peak of diversity observed at ~2,000 m depth may correspond to intermediate levels of disturbance, which favor diversity (Joseph Connell's intermediate disturbance hypothesis, Chapter 20). More frequent physical disturbance in shallower depths and extremely infrequent disturbance at greater depths both dampen diversity. As described in Chapter 2 (see benthic zonation), the bathyal zone spans seabed environments from ~200 to 3,000/4,000 m depth, from the edge of the continental slope to the abyssal zone (Figure 25.7).

The bathyal zone may therefore have acted as a biodiversity reservoir for recolonization during periods of loss of biodiversity in continental shelves and abyssal environments. Numbers of species, geographical units, and faunal limits generally decrease between bathyal and abyssal zones and increasing depth. The bathyal zone accommodates some relict (ancient) species, including the most ancestral extant forms of some taxonomic groups. However, further studies on the diversity of other benthic components, such as nematodes or bacteria, do not indicate a decreasing trend in species number with increasing depth (Figure 25.8).

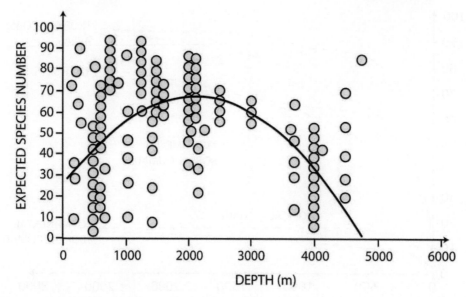

Figure 25.7 Patterns of biodiversity (expressed as expected number of species of macrofauna and meiofauna) with increasing depth. Biodiversity shows a unimodal peak at ~2,000 m depth.

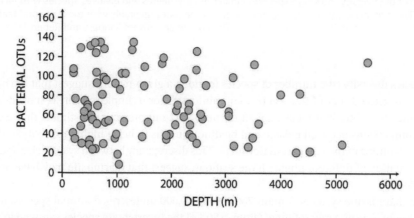

Figure 25.8 Patterns of biodiversity (such as number of operational taxonomic units of bacteria, OTUs) with increasing depth. Deep-sea benthic bacterial biodiversity does not show a clear bathymetric trend.

25.2.2 Deep Faunal Origins

Researchers have proposed several theories on the origin of deep-sea fauna, the timing of its formation, and its rate of evolution. The two most extreme hypotheses posit that: (1) current deep-sea fauna evolved relatively recently because cooling of deep-sea environments to temperatures less than 4 °C may have catastrophically affected the biota, so that only relatively eurybathic and eurythermal species survived; (2) current deep-sea fauna may conserve relatively older characteristics, and abyssal benthic organisms may represent "refugees" of archaic forms. In most cases, discrepancies in viewpoint arise from how researchers consider the relative role of older taxonomic elements.

Analysis of endemic taxa of different taxonomic rank and evolutionary age can help clarify the issue of youth or archaicity of deep-sea fauna, and eurybathic species that occur in shallow waters, likely represent the youngest deep-sea species. However, endemism within the abyssal and hadal benthic fauna spans not only species and genera, but also extends to family level and often even order or suborder (for example, Cribellosa from the Asteroidea, Psychropotina from the Holothuroidea, ascidians from the Octacnemida and Hexacrobilida). Furthemore, based on paleontological data, many endemic deep taxa date back earlier than the Mesozoic era, mostly from the Cenozoic era. The predecessors of the oldest extant Echiuroidea moved from the sublittoral to the abyssal no earlier than the Jurassic, and probably during the Cenozoic. All of the modern orders of Asteroidea appeared no earlier than the Mesozoic, and the only endemic suborder in deep-sea environments (Cribellosa), and members of the Porcellanasteridae likely evolved in the late Mesozoic. The present taxonomic composition of deep-sea mollusks suggests that none of the oldest families originated at great depths and that modern deep-sea mollusks originated from groups that moved into deeper waters during the late Mesozoic or Cenozoic. On the basis of phylogenetic analysis, deep-sea isopods (order Isopoda) began in the Miocene, originating from the fauna of cold Antarctic shelf waters. This order contains few endemic genera and no characteristically deep-sea families, and most abyssal species are closely related to other deep-sea species. Most of the available data, therefore, indicate that the present deep fauna originates from groups formed largely in the Cenozoic and no earlier than the Mesozoic.

Consistent with plate tectonics theory, trenches formed through subduction. Marginal trenches move somewhat like a downward elevator, with continuous subduction beneath the continental plate and a progressive replacement with a plate from a shallower depth. The biota that inhabit these environments sink together with the abyssal seafloor near the trench, but this extremely slow process occurs over hundreds, thousands, or even millions of years. This prolonged duration allows sufficient time for the evolution of new species, and sometimes for the evolution of higher taxonomic levels. Any biota that could not adapt over this evolutionary time persisted only at depths where they previously occurred, and did not penetrate into the trenches. Other organisms, derived from groups that can adapt to the ultra-abyssal depths, underwent adaptive changes, ultimately evolving into endemic species. Multiple trench species derive from Antarctic origin, including many isopods, holothurians, and actinians. Relatively few organisms could adapt to the extreme pressures that occur in ocean trenches, and the abiotic conditions in these features favored the evolution of endemic species and, in some cases, endemic genera.

25.2.3 Mechanisms of Generation and Maintenance of Deep-Sea Biodiversity

The presence of high animal and prokaryote diversity in deep-sea environments raises important questions about the mechanisms that generate and maintain deep-sea diversity. In the past, theories on high deep-sea biodiversity assumed high spatial homogeneity and temporal uniformity. Let's revisit the ideas introduced in Chapter 8.

Researchers have proposed four main theories to explain high biodiversity levels in the deep sea, and these theories fall into two groups: the first defined as "equilibrium theories" and the second as "disequilibrium theories".

The first group includes: (1) the *time-stability hypothesis*, which assumes that stability over time can allow species to specialize as an effect of time; (2) the second equilibrium theory is based on spatial heterogeneity, which might allow species to evolve to exploit the highly heterogeneous environment. Both theories in this group hinge on the concept of specialization determining speciation.

The "disequilibrium theories" include (1) physical disturbance and (2) biological disturbance. In both cases high biodiversity results from the lack of dominance of single/few species resulting from repeated mortality events associated with storms or landslides (physical disturbance), predation (multiple predators, including deep-sea sharks occur to depths of ~4000 m). However, these hypotheses lack definitive experimental confirmation so far.

A broad range of recent studies support the influence of spatial heterogeneity on deep-sea diversity and coupling between geosphere and biosphere. We now know that deep-sea environments can experience temporal variability and physical stress, as well as significant spatial heterogeneity. Bottom topography, hydrodynamics, and mineralogical diversity help to create complex mosaics at different spatial scales, further complicated by emission of fluids and materials from the seafloor that support very different forms of life. In addition, numerous canyons bisect continental margins, each adding unique geo physical characteristics and associated biodiversity. Large swaths of highly unstable seafloor characterize some portions of the margins, with frequent landslides that create stochastic "roulette" conditions and new opportunities for new colonization of defaunated deep-sea environments. Finally, the presence of cold seeps, canyons, seamounts, deep-water coral reefs, trenches, and depocenters of organic matter add to overall complexity, potentially adding spillover benefits akin to those of marine protected areas in coastal environments.

25.3 Deep-Sea Habitats

A wide diversity of geological structures cover a hundred million km^2 and add complexity and heterogeneity to the continental slope and abyssal plains. To illustrate some of these habitats, in this chapter we introduce seamounts, deep-water corals, abyssal plains, and trenches; the next chapter will treat these habitats in more detail, but scientists continue to discover many more habitats, including some that are new to science. Within slopes and deep basins, geological structures that include oceanic ridges, canyons, seamounts, deep-water coral reefs, hydrothermal vents, and ocean trenches support unique biological communities. The third major dimension (depth) adds 1 billion km^3 of water to pelagic ecosystems, providing habitat for animals and microorganisms to live and grow, feed, and reproduce. Since 1840 (the year Forbes officially began dredging for deep-sea life) scientists have discovered some twenty-two new deep habitats/ecosystems (Table 25.1).

Table 25.1 Year of discovery of new habitats and/or ecosystems and publication date of the study since Forbes Azoic Theory through to today.

Habitats/Deep marine ecosystems	Year
Fine sediments (400 m)	1840
Fine sediments (600 m)	1849
Fine sediments (2000 m)	1862
Submarine canyon	1863
Seamounts	1869
Sponge fields	1870
Deep open waters	1876
Fine sediments (abyssal sediments)	1876
Manganese nodules	1876
Deep corals (as distinct ecosystem)	1922
Pelagic OMZ (Oxygen Minimum Zones)	1925
Benthic OMZ (Oxygen Minimum Zones)	1928
Whale carcass (as food source)	1934
Mud volcanoes	1934
Trenches	1948
Sunken tree trunks	1952
Mid-Atlantic Ridge	1963
Forearc basin	1971
Mid-Atlantic Ridge (fast extension)	1977
Xenophyophores fields	1979
Hypersaline anoxic basins	1983
Cold seeps	1984
Mid-Atlantic Ridge (slow extension)	1986
Whale carcass (as chemosynthetic habitat)	1989
Brine (as chemosynthetic habitat)	1990
Asphalt habitat (Chapopote)	2004
Big bare region in the Pacific Ocean	2006

So far, almost all deep-sea environments remain largely unexplored. However, based on knowledge from geologists and geochemists and accurate bathymetry, we can roughly estimate proportional coverage of these habitats (Table 25.2). The abyssal plains account for >70% of the seabed, followed by continental margins (10–20%) and ocean ridge systems (~10%). Despite their limited spatial coverage, seamounts (3–4%) and trenches (2–4%) form a significant part of the seabed, whereas other habitats, such as chemosynthetic ecosystems, whale carcasses, OMZs and deep-water corals occur much more infrequently. The limited spatial extent of these habitats elevates the importance of the studies completed on these ecosystems.

25.4 Submarine Canyons

Submarine canyons are deep incisions that typically extend from the edge of the continental margin down the continental slope, bisecting many continental margins. Studies indicate that canyons differ greatly in topography, ranging from systems of connected channels to deep, wide valleys. These near ubiquitous features vary considerably in length, width, height, and morphological complexity, making each canyon somewhat unique. Canyons provide conduits for the material transported from terrestrial to deep marine environments (Figure 25.9). Their actual number remains unknown.

Two mechanisms create submarine canyons: (1) Particularly during postglacial periods with reduced sea level, rivers (and melting glaciers) carve our channels through erosion, which become flooded as sea level rises; (2) in situ erosion, such

Table 25.2 Known or estimated area encompassed by major deep-sea habitats, and proportion studied to date.

Habitat	Area (km^2)	% of oceanic surface	Studied portion
Deep pelagic waters	1,000,000,000 km^2	95% of seas and oceans	<<0.0001%
Deep seafloor	326,000,000 km^2	100%	~0.0001%
Abyssal plains	244,360,000 km^2	75%	<1%
Continental margin (150 to 3000–4000 m)	40,000,000 km^2	11%	minimal
Ridges	55,000 km long, 30,000,000 km^2	9.2%	10%
Seamounts	8,500,000 km^2	2.6%	0.25–0.28 % (250–280 seamounts sampled of ca. 100,000)
Hadal trenches	37 trenches (area not estimated)	1%	minimal
Canyons	448 canyons with a total estimated length of 25,000 km. Area unknown.	unknown	minimal
Zones with minimum oxygen (OMZ)	1,148,000 km^2	0.35%	<1%
Cold-water coral reefs	estimated 280,000 km^2	0.08%	minimal
Hydrothermal vents	~2000 vents. Area unknown	Unknown	10% (known 200 of ~2000)
Cold seeps	10,000 km^2	0.003%	2%
Whale carcasses	~35 km^2 (690,000 whale carcass with ca. 50 m^2 per carcass)	0.00001%	0.005% (~30 of an estimated 690,000)

Figure 25.9 Multibeam map of a submarine canyon system (left) and detailed view of a shallow canyon from the top (right).

as through slumping events when accumulated sediments collapse and form **turbidity currents**, a rapid, downhill flow of water and sediment akin to mudslides on land. Because of their strong topography, submarine canyons facilitate the exchange of water and matter between shelf and deep environments, channeling large amounts of sediment down the continental slope. Some canyons temporarily store sediment and carbon. Episodic events can mobilize large volumes of this sediment, transporting it to the abyss and smothering large swaths of benthic habitat, which then go anoxic for some period of time because they lack living bioturbators.

25.4.1 Canyon Biodiversity

High habitat diversity within canyons compared with adjacent continental slopes (Figure 25.10) supports diverse filter feeders such as corals, sponges, hydroids, pennatulaceans, sabellids, holothurians, and anemones, as well as other species. Canyons also host many species of decapods, micronekton crustaceans, gelatinous zooplankton (including many hydromedusae with several recently described new species known only from canyons), and fish communities (adults, juveniles, and larvae). Species diversity,

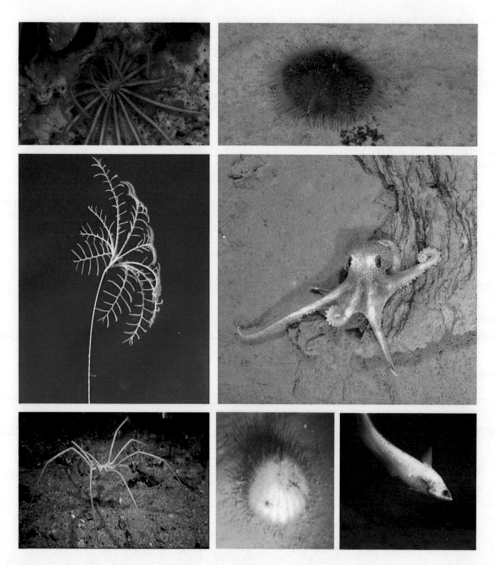

Figure 25.10 Examples of typical canyon megafauna: Crinoids, gorgonians, sea urchins, cephalopods, pycnogonids, and fishes.
Source: NOAA / Public Domain, Bret Tobalske, National Oceanic and Atmospheric Administration (NOAA).

abundance, and composition can vary from canyon to canyon and apparently relate to particle transport, topography, and hydrography of each canyon. Submarine canyons host numerous benthic species, including several species of undescribed holothurians and important commercial species such as lobsters, crabs, shrimps, hake, and plaice. The high productivity of canyons attracts many pelagic species to important feeding areas, and a large variety of whales that come to feed in some canyon environments, such as The Gully in Canada, a Marine Protected Area. Clear zonation also differentiates the head, canyon axis, and deep canyon base. For example, a more diverse fauna lives near the head of the canyon, dominated by pelagic species (copepods, amphipods, salps, other jellies) as well as cold-water corals. In addition, within canyons, endemic as well as non-endemic species occur in abundances that progressively increase with increased total organic matter flux. Benthic stages of planktonic organisms, such as resistant cysts that have settled onto the bottom, represent another important component of canyon biodiversity.

25.4.2 Canyon Functioning

Canyons typically support high faunal abundances and biomass relative to adjacent open slope environments, depending on sedimentation, organic and inorganic fluxes, and upwelling and downwelling processes. Their morphological characteristics and associated physical, chemical, biological, and geological processes differentiate submarine canyons from the adjacent slope. As a result, submarine canyons influence their entire food web, from phytoplankton to marine mammals. Canyon benthic communities differ from those in slope environments, especially canyons with high organic matter deposition. In these cases, intensified oxygen flux within sediments supports higher abundances of benthic organisms and higher rates of microbial respiration. Recent studies show that local primary production can strongly influence canyon trophic resources. Furthermore, in some canyons dense surface waters cascade through the canyon toward the bottom, directly affecting deep ecosystem functioning by supplying fresh organic matter high in nutritional value toward the bottom. Canyon characteristics strongly influence plankton dynamics because they deflect deep currents and thus transport nutrients into the photic zone, increasing primary production. Canyons can therefore act as efficient supply corridors for deep nutrients to coastal or slope areas. In addition, increased water exchange between the shelf and slope associated with canyons impacts dispersal and recruitment of fish larvae.

The discovery of a unique community of planktonic hydromedusae within canyons suggests that canyon flow and storage of organic matter from the continental shelf may play a vital role in supporting canyon endemic species. The specific composition and abundance of hydromedusae varies from canyon to canyon and seems to relate to vertical flows and the topographic and ecological characteristics of each canyon. Meroplanktonic species that seem to spend their entire life cycle close to the canyon characterize these hydromedusae.

25.5 Deep-Water Corals

As important bioconstructors, deep-water coral systems can form reefs or isolated colonies. A **deep-sea reef** refers to a rigid skeletal structure in deep water that influences the deposition of sediments in its proximity and promotes higher topographic complexity than the surrounding sediment. More specifically we often define a shallow coral reef as "a skeletal structure with hermatypic corals as the major constituent", whereas **reef mounds** refer to less rigid accumulations with organic sediment. But hexacorals, also known as "white corals", bioconstruct deep-water reefs and differ from their tropical counterparts in that they lack the zooxanthellae (symbiotic dinoflagellates) that give tropical corals their color and high productivity. Deep-water corals include members of some of the major taxa in the phylum Cnidaria. The most important species forming **arborescent** (tree-like) colonies include *Desmophyllum pertusum* (previously *Lophelia pertusa* and whose taxonomic name causes ongoing debate) and *Madrepora oculata*, the latter of which often co-occurs with the solitary species, *Desmophyllum dianthus* (Figure 25.11). Deep-water corals preferentially occur at depths of 40–7,000 m, distributed on topographic irregularities, along steep sides of mountains, canyons, and underwater volcanoes.

Cold-water corals have special environmental requirements, such as hard bottom on which to settle and grow, including exposed rock or dead coral and smaller substrata such as pebbles and worm tubes. Deep corals presumably settle at the top of topographic irregularities to benefit from greater flux of particles and potential prey. Some research points to zooplankton as the main food source for deep-water corals, along with detritus and phytodetritus. Prokaryotes add an additional food source for maintenance and growth of invertebrates associated with these structures. No detailed studies have examined the abundance and diversity of the microbial ecosystem associated with deep-water corals, but several studies on bacteria in the water in and around polyps suggests possible **bacterial gardening** (i.e., entrapment of bacteria within the coral's gastrovascular cavity, where they grow and provide a potential food source).

The scant knowledge on the distribution of deep-coral reef systems currently available stems largely from detailed studies of limited geographical areas. Studies on both the structural complexity of deep-coral ecosystems and deep-reef geographical distributions continue (Figure 25.12), noting that we still lack good estimates of the numbers of species these habitats host. Many of the taxa surveyed in these ecosystems occur in greater abundance within coral habitat rather than in the surrounding seabed. In addition, reefs host communities of animals that distinctly differ from those in the surrounding deep-sea environment; not surprisingly, suspension feeders dominate these species-rich habitats. Deep-water coral mapping, a current research priority, will provide a fundamental tool for defining and prioritizing deep-sea environments that need protection.

Our view on food supply to deep coral reef systems has evolved over the last ~20 y, from thinking that some form of seepage supported these systems to detailed understanding that local accelerated currents, tightly coupled to surface productivity, fuel them. The ecosystem functions of these reefs clearly go beyond structural habitat provisioning to include nutrient and carbon cycling.

Desmophyllum pertusum reefs support approximately a faunal diversity triple that in surrounding sediments, suggesting that deep-water corals potentially influence the density and diversity of associated species. Coral skeleton, which offers an ideal refuge for other species, often supports colonies, including other bioconstructors, such as polychaetes, and bioeroders, such as sponges. Deep-water corals may well form some of the most complex three-dimensional habitat in the deep ocean, thereby offering spatial niches for many species. Nonetheless, some groups of animals (the same coral constructors, for example) exhibit lower diversity compared to that of tropical reef systems. The European Atlantic Coral Ecosystem Study (ACES) estimated that *D. pertusum* reefs in the North Atlantic host >1,300 known species.

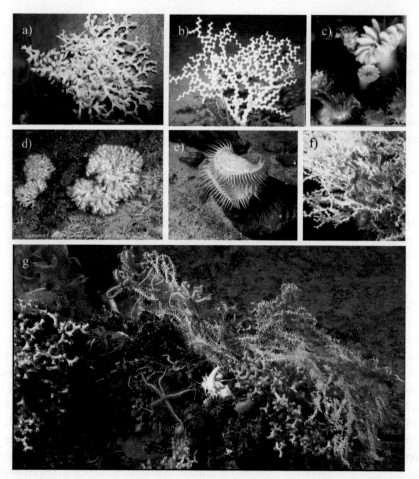

Figure 25.11 Typical cold-water corals, or deep corals: (a) *Desmophyllum pertusum*; (b) *Madrepora oculata*; (c) *Desmophyllum dianthus*; (d) *Oculina varicosa* and some examples of associated fauna: (e) fly trap anemone, possibly *Phelliactis* sp.; f-g) cnidarians, sponges, and crinoids. *Sources:* Bioluminescence 2009 Expedition / Public Domain, NURC/UNCW and NOAA/FGBNMS / Public domain, NOAA/ Monterey Bay Aquarium Research Institute / Wikimedia Commons / Public domain, Dr. Ken Sulak / Wikimedia Commons / Public domain, Dr. Les Watling / Wikimedia Commons / Public domain.

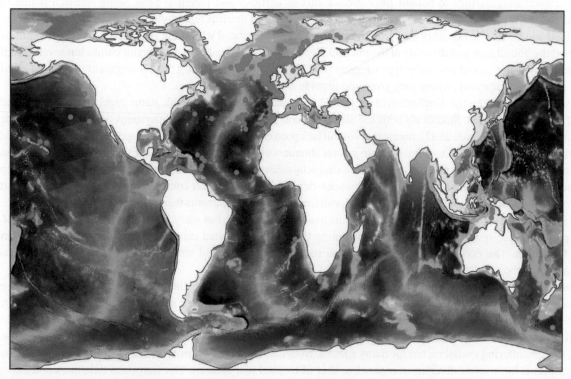

Figure 25.12 Global distribution of known deep-water corals.

Coral skeletons affected by bioerosion or mechanical damage fall to the bottom around the reef perimeter, trapping and consolidating sediments and creating new mounds on which coral can settle and grow. Settlement and development of the colony can be thought of as a cyclic process in which the associated community composition varies with reef stage of development and microhabitat availability. The sub-habitats created include living corals themselves, spaces between the branches, "**coral rubble**" (fragments of corals and bivalves), exposed dead corals, and coral fragments trapped by sediments. Studies in the Mediterranean demonstrated an extraordinary variety of species within coral rubble (Figure 25.13). Deep-water corals create a true "oasis" for a variety of specialized species. In addition, some fishes lay their eggs on deep-water corals, which provide nursery habitats. They also create a surface favored by settling larvae of some marine species. These characteristics led the International Union for the Conservation of Nature and Natural Resources (IUCN) to develop an initiative to conserve these important habitats.

25.6 Cold Seep (Hydrocarbon-Based) Ecosystems

Cold seep ecosystems occur where hydrocarbons such as methane, sometimes associated with sulfides and brines, leak from the ocean floor, as opposed to locations with geothermal emissions. These typically deep-sea ecosystems, much like hydrothermal vents, depend on chemoautotrophic production for their functioning. However, they differ from hydrothermal vents in that they lack hot fluids and because they do not occur specifically in tectonically active locations but instead occur where hydrocarbon deposits sit beneath surface sediments. Some seep fluids can reach temperatures ranging 45–55 °C. Cold

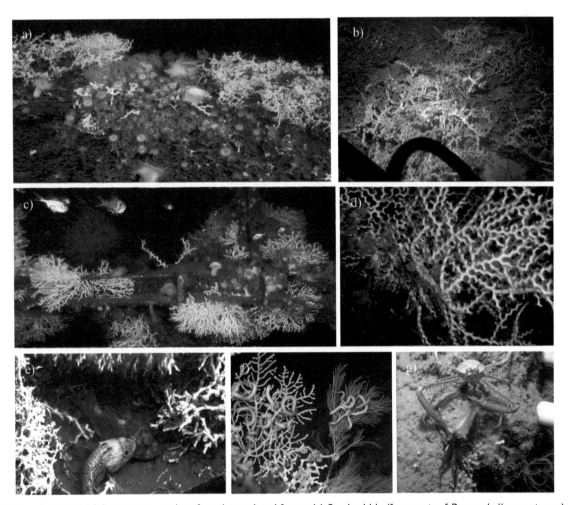

Figure 25.13 Examples of deep-water coral reefs and associated fauna: (a) Coral rubble (fragments of *Desmophyllum pertusum*); (b) isolated colonies; (c) colonies of *D. pertusum* with deep-sea lobster; (d) redfish; (e) ophiuroids attached to gorgonian coral; (f) crabs and squid. *Sources:* NOAA Photo Library / Wikimedia Commons / CC BY 2.0, Islands in the Stream 2001, NOAA/OER, NOAA-OER/BOEM, Expedition to the Deep Slope 2007 / Wikimedia Commons /Public domain, NOAA-OER/BOEMRE., Lophelia II / Wikimedia Commons / Public domain, Lophelia II Team 2009, NOAA-OER / Wikimedia Commons / Public domain.

seeps fall into three groups based on the "origin" of the emerging fluids: (1) from oils; (2) from fresh water; (3) from spills of terrestrial mantle material (Figure 25.14).

Cold seeps also take different forms: (a) Release of water from the bottom; (b) spills from salt deposits; (c) methane gas emissions; (d) mud volcanoes; (e) seeps in subduction zones; (f) asphalt volcanoes (Figure 25.15). Scientists first discovered cold seeps associated with brines and oil on the Louisiana continental slope in the Gulf of Mexico, where these habitats occur in high abundance. Following the discovery of organisms associated with a cold seep on the Florida continental margin in 1983, researchers have used submersibles such as *Alvin*, *Nautile*, *Shinkai* and *Johnson Sea Link* to discover numerous other cold seeps located along the margins of the Pacific Ocean, the Atlantic Ocean, and the Mediterranean Sea. Cold seeps occur at depths generally ranging 400–6,000 m, in a variety of geological contexts, both on **active margins**, where boundaries of actively moving plates occur, and on **passive margins**, where no plate boundaries occur. The emission of cold fluids containing chemicals and reduced substances provide energy to rich benthic communities based on a chemosynthetic food web.

Underwater photographs, video recordings, and samples obtained from submersibles provided the initial description of cold seep biology. Cold seeps support a specialized fauna that superficially resembles that at hydrothermal vents but with potentially higher biodiversity, noting limited study to date. In addition, interactions between biological and geological systems at cold seeps are more complex than many other ecosystems, including vents. Cold seep communities mostly associate with methane-rich emissions of biogenic and thermogenic origin. Most megafauna living near cold seeps resemble that associated with hydrothermal vents, with high densities and biomass of animals supported by intense chemosynthetic bacterial activity. These environments share many affinities with hydrothermal vents, most notably: (a) high biomass; (b) high production; (c) complex community structure; (d) megafauna with shared species or genera (*Calyptogena* spp. and *Vesicomya* spp. mussels, bivalves, and crustaceans); (e) extensive mats of chemoautotrophic bacteria (*Beggiatoa* spp., *Sulfobacter* spp.); (f) substances toxic to most marine organisms.

25.7 Cold Seep Biodiversity and Symbiotic Organisms

Cold seeps support relatively high overall species richness, but we almost certainly underestimate seep species diversity globally. Generally, species richness of organisms that host symbionts decreases with depth, and the known deepest cold seeps support only a few known species hosting symbionts. Soft sediments at cold seeps seem to support higher numbers of species than hard substrates. Changes in flow also result in changes in biodiversity, where species diversity may depend on recent seep history and the ephemeral nature of seep flow.

The most common species, vestimentiferans *Lamellibrachia* spp. and vesicomyid bivalves, families Vesicomyidae and Mytilidae, contain symbionts and widely dominate in terms of abundance and biomass (Figure 25.16). Some seep regions

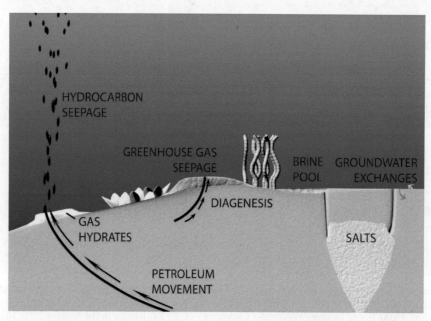

Figure 25.14 Mechanisms of hydrocarbon emission, either as oil leakage in the form of oil droplets (left), hydrate gasses that release methane and associated precipitates and rich biota (center), and as brines associated with hydrocarbon (methane) or hydrogen sulfide emissions.

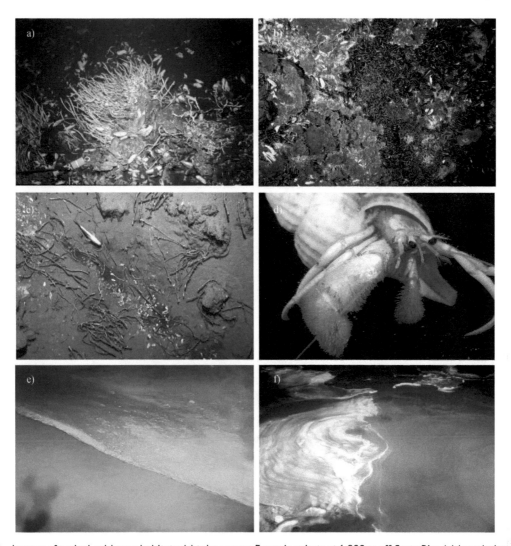

Figure 25.15 Images of typical cold seep habitats: (a) tube worms *Escarpia spicata* at 1,900 m off Costa Rica (although these tubeworms occur off the Pacific coast of Panama, communities resemble seeps in the Gulf of Mexico, Caribbean, and Atlantic Ocean); (b) bacterial mats, bivalves, and tube worms at 340 m in the Gulf of Mexico); (c) vestimentiferans and bivalves in New Zealand; (d) hermit crab sampled from a New Zealand cold seep (note dense bacterial filaments on the claws); (e) brine with hydrocarbon emissions and associated bacterial mat; (f) methane seep. *Source:* National Oceanic and Atmospheric Administration (NOAA).

support other species of bivalves and sponges, but vesicomyids (*Vesicomya* spp. and *Calyptogena* spp.) occur in almost all cold seep sites studied to date. Vesicomyid bivalves within soft sediments of seeps position the anterior part of their body in the mud. *Calyptogena* spp. shells reach a length of 20 cm, although many species in this family remain poorly known or undescribed. The bathymetric limit of these species may result from their planktotrophic larval stage and/or pressure effects on larval stages that limit larval dispersal of these deep species across topographic barriers. Vesicomyids and their chemoautotrophic symbionts may tolerate changes in flows and variation in sulfide concentration. For example, the endosymbiont chemoautotrophic bacteria hosted by *Abyssogena phaseoliformis* oxidize sulfur.

The genus *Bathymodiolus*, the often-pervasive mussels at cold seeps, has a more restricted geographical and bathymetric distribution than the genus *Vesicomya*, and occurs only in Atlantic and Western Pacific cold seeps at depths ranging 400–3,000+ m. Mussels occur in dense aggregates in seep locations with relatively high methane concentrations. Electrophoretic studies on these mussels indicate 7–8 undescribed species of *Bathymodiolus* spp. with remarkable genetic distances among species. *Calyptogena* spp. live in cold seeps that expel natural gas, composed primarily of methane, through the sediment or directly into the water. Researchers have clearly demonstrated their chemoautotrophy and ability to use methane. Bivalves from the families *Solemyidae*, *Thyasiridae* and *Lucinidae* also occur at seeps. These organisms live in burrows deep in the sediment, complicating observation and sampling efforts. *Solemyidae* occur in low abundances but more commonly in locations with low fluid emission rates, and studies have documented *Thyasiridae* and *Lucinidae* in seeps to 3,800 m depth. A reduced digestive tract characterizes all three families, further demonstrating that they must derive most of their organic carbon from endosymbiosis with chemoautotrophic bacteria. All three families, in fact, host sulfide-oxidizing bacteria.

Figure 25.16 Examples of biodiversity associated with cold hydrocarbon seeps: (a–b) *Lamellibrachia* (vestimentiferans); (c) an *Alvinocarid* sp. shrimp on a bed of mussels; (d) *Calyptogena* sp. (mollusks), partially buried in sediments; (e) crabs and starfish; (f) holothurians and bivalves on soft bottom at the periphery of a hydrocarbon seep. *Sources:* NOAA / Flickr / CC BY 2.0, Aquapix and Expedition to the Deep Slope 2007 / Wikimedia Commons / Public domain, NOAA-OE / Wikimedia Commons / Public domain, Tim Shank / National Oceanic and Atmospheric Administration (NOAA), New Zealand-American Submarine Ring of Fire 2005 Exploration /. Wikimedia Commons / Public domain, Aquapix and Expedition to the Deep Slope 2007 / Wikimedia Commons / Public domain.

Some species, such as the gastropod *Thalassonerita naticoidea* and the shrimp *Alvinocaris* cf. *muricola* are probably endemic to seeps. Siboglinids, previously known as vestimentiferan worms, occur in many seeps and include species that live in reduced sediments, others that live close to hydrothermal vents and cold seeps, and others in decaying woods, sometimes within reducing sediment of surface water and close to cold seeps. Siboglinid worms occur commonly and have been reported from some 17 of 24 cold seeps sampled, where they generally live in small groups. The genera *Sclerolinum* and *Polybrachia* rarely occur in seeps and also live at great depths, including the deepest known cold seeps (the Peru Trench, for example). The genera *Lamellibrachia* and *Escarpia* occur more widely at shallow sites. All vestimentiferans depend on their chemoautotrophic symbionts, and studies on the genera *Escarpia* and *Lamellibrachia* in cold seeps indicate that these genera depend on sulfur oxidation.

In the 1990s, researchers working in the Barbados Trench discovered an unexpected symbiosis between methane oxidizing bacteria and a new species of deep-sea carnivorous sponge (genus *Cladorizha*). This species lives in dense bush-like clusters of hundreds of individuals. Similar associations between sponges and methanotrophic bacteria were reported in *Hymedesmia* spp., a sponge encrusting vestimentiferan tubes in Gulf of Mexico hydrocarbon seeps.

Less information exists on meio- and macrofauna living near seeps. In some seeps, only those species hosting symbionts occur in high densities, whereas non-symbiotic species can occur in similar abundances at other seeps. Very limited sampling of meiofauna in seeps suggests little difference in abundance between seeps and ambient sediments. However, exceptionally high densities of meiofauna occur at cold seeps known as "mud volcanoes" (Figure 25.15). Researchers offer two explanations for this phenomenon; production from chemosynthetic free-living bacteria may enrich sediments locally, or detrital organic matter derived from groups of symbiotic organisms may provide the key food source. Among the macrofauna, suspension and deposit feeders comprise the most common invertebrates devoid of symbionts, and some of these species occur in similar densities both outside and inside seep environments.

25.7.1 Functioning of Cold Seep Ecosystems

Production in cold seeps depends on the composition and intensity of the emitted flow (Figure 25.17), which influences: (1) specific composition, density, and biomass of bivalve aggregates; (2) density and species composition of mussel beds; (3) density and biomass of clams. Flow emissions vary markedly, both within individual cold seeps and between different sites. The high biomasses reported in different seep locations (16–51 kg m^{-2}) depend on ephemeral fluid supply and associated chemosynthetic production that leads to irregular distributions. Particulate organic matter in suspension that originates from chemosynthetic activity provides the main food source. Observations of *Beggiatoa* sp. bacterial mats within groups of serpulid polychaetes suggest that these bacteria likely contribute to their diet. Furthermore, the frequent presence of sponges at seeps and documentation of numerous bacteria in their tissues tends to confirm that these species feed by filtering out free-living bacteria.

Some seep fauna include species typical of other deep-sea regions but without symbionts, presumably attracted to the enriched organic matter around seeps: some species feed on free-living bacteria in the seep fluid or on the bacterial mat, whereas carnivorous species feed on bivalves and other species scavenge. Mobile individuals dominate the many

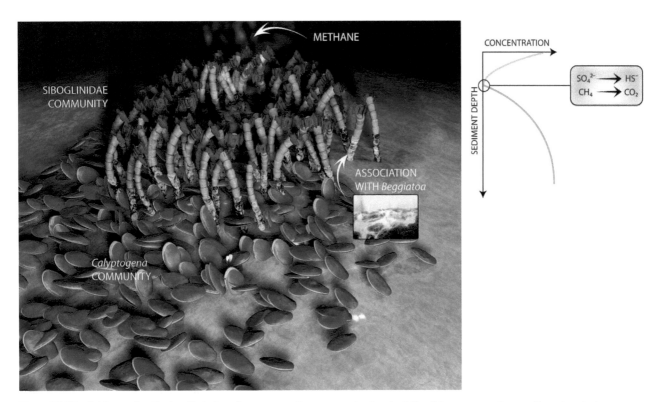

Figure 25.17 Cold seep functioning. Emission of gaseous methane supports abundant *Beggiatoa* spp. growth as well as abundant *Calyptogena* spp. Further from the seep flow, burrowers such as *Acharax* spp. occur, including locations with high concentration of hydrogen sulfide. Within sediments, anaerobic oxidation of methane converts carbon dioxide with simultaneous reduction of sulfates into sulfide ions.

scavengers at seeps, with the exception of holothurians that occasionally colonize these environments. Suspension feeders and scavengers can occur in relatively high abundances in locations colonized by a mature chemosynthetic faunal community. The diversity of species with and without symbionts, the proportion of living and dead animals, and the presence of different trophic groups define maturity of cold seep communities. A less mature community consists of individuals of small species, interspersed with symbiotic species with a sparse associated fauna. The most mature communities include different species hosting symbionts as well as numerous endemic species without symbionts that adopt different trophic strategies (filter feeders, deposit feeders, scavengers). Mature ecosystems likely occur only at seep locations where fluid flow remains relatively constant over an extended period of time. In contrast to hydrothermal vents communities where a small number of species strongly dominate, cold seeps support higher species richness.

Carnivores generally occur in low abundances at seeps, where larger forms (cephalopods, shrimp, crabs, and fish) are usually transient visitors. Some predators strongly associate with seeps. In contrast to hydrothermal vents, studies do not support large abundances of specialists, suggesting they play a lesser role in structuring cold seeps communities.

25.8 Hypoxic and Anoxic Systems (Dead Zones)

Marine vertebrates (e.g., fishes) and invertebrates require oxygen for their metabolism. Although oxygen concentrations in the ocean can sometimes fall well below saturation levels, low concentrations have little effect on respiration in many marine invertebrates. Dissolved oxygen concentration in seawater usually varies from 7 to 8 mg l^{-1} (Figure 25.18). If concentrations fall below 4 mg l^{-1}, organisms begin to react, for example by moving (in the case of mobile organisms). For many benthic invertebrates, this threshold occurs around 2.0 mg l^{-1}, but for some species this limit drops to 1.0 mg l^{-1}. **Anoxia** occurs in waters with dissolved oxygen concentrations <0.2 mg l^{-1}, which cannot support most life forms. **Hypoxia** refers to waters with dissolved oxygen content <2.0 mg l^{-1}. The term **microxic** describes waters with oxygen concentration up to 0.1 ml l^{-1}. **Disoxic** or **disaerobic** refers to oxygen concentrations 0.1–1.0 mg l^{-1}, and fully **oxic** (aerobic) refers to oxygen content >1.0 mg l^{-1}.

Hypoxic conditions cause aberrant behaviors in some species, such as abandoning their sediment burrows, with potentially serious consequences including mass mortality events. In this context, **dead zones** refer to anoxic bottom waters that lack the minimum oxygen conditions necessary to support higher organisms, whether photosynthetic or animal (Figure 25.18). Oxygen concentrations that limit survival vary, depending on the specific taxon (Table 25.3). A comparative analysis of different benthic marine organisms shows considerable variation in hypoxia thresholds among different taxa. Although a concentration of 2.0 mg l^{-1} offers a general transition threshold from oxic to hypoxic conditions, in some species a different threshold may be more appropriate. Empirical studies on sub-lethal and lethal conditions indicate that significant effects can occur at oxygen values above hypoxic conditions for half of the tested species.

Hypoxic and anoxic basins occur in the Pacific Ocean, Indian Ocean, Mediterranean Sea, and Black Sea. Dead zones have now been reported in > 400 locations that encompass >245,000 km^2 of seabed (Figure 25.19). The Gulf of Mexico dead zone extent fluctuates during the year, forming in spring, peaking in summer, and fading in autumn. Different factors lead to dead zone formation; 41% of the continental territory of the United States drains into the Mississippi River, which flows into the Gulf of Mexico; agricultural runoff (fertilizers) account for 70% of nutrient discharges that lead to hypoxia. Moreover, 12 million people live in urban areas adjacent to the river, discharging treated wastewater into the Mississippi. The Yangtze and Pearl River outflows in China also illustrate dead zones resulting from anthropogenic activity. Other locations illustrate seasonality in hypoxia, including the Central-Northern Adriatic Sea and North Sea. The most extensive hypoxic habitat, a significant portion of the Baltic Sea seabed, encompasses an area of ~70,000 km^2.

The amplitude and number of dead zones has increased exponentially over the last 50 years, mainly reflecting increased primary production and global changes. Increased primary production results in accumulation of particulate organic matter, which encourages microbial activity and consumption of oxygen in bottom waters (Figure 25.20). Hypoxia and anoxia usually occur in bottom waters, but Oxygen Minimum Zones (OMZ) also occur in intermediate depths of 500–1,500 m.

25.9 Oxygen Minimum Zones, OMZs

Oxygen Minimum Zones exhibit permanent low levels of oxygen in mesopelagic waters (oxygen concentrations <0.5 ml l^{-1} or with saturation values of approximately 7.5%; < 2µM). These zones, also known as "oxygen deficient zones" occur in the upper bathyal zone onto the continental shelf at depths of 10–1,500 m, but typically >500 m. OMZs result from oxygen consumption by living organisms, but circulation patterns affect its distribution and position within the water column. Hypoxia destroys benthic habitats; only a few mobile organisms survive in such conditions whereas others return when the hypoxic episode

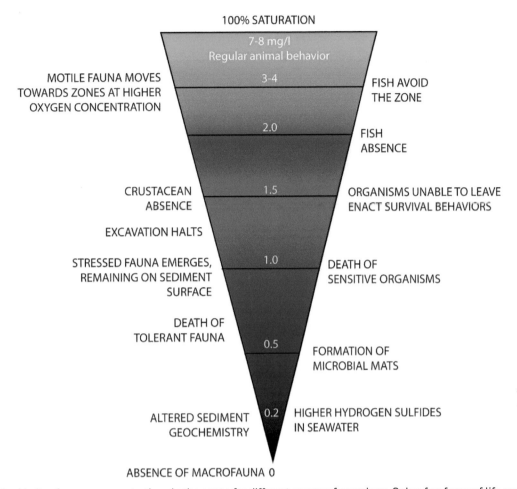

Figure 25.18 Limits of oxygen concentrations in the ocean for different groups of organisms. Only a few forms of life can survive at oxygen concentrations <2.0 mg l^{-1}, but a rich microbial community thrives under such conditions.

Table 25.3 Hypoxia thresholds for different groups of benthic organisms showing lethal oxygen concentrations for 50% of population (LC$_{50}$), sub-lethal oxygen concentrations for 50% of the population (SLC$_{50}$), and lethality time for 50% of population (SLC$_{50}$ hours). Numbers indicate Mean ± SE, standard error).

Taxa		LC 50 mg O$_2$ l^{-1}	SLC 50 mg O$_2$ l^{-1}	LT$_{50}$ h
Fishes				
	Average ± SE	1.54 ± 0.07	4.41 ± 0.39	59.9 ± 12.3
Crustaceans				
	Average ± SE	2.54 ± 0.14	3.21 ± 0.28	55.5 ± 12.4
Gastropods				
	Average ± SE	0.89 ± 0.11		
Bivalves				
	Average ± SE	1.42 ± 0.14		
Mollusks				
	Average ± SE		1.99 ± 0.16	412.9 ± 37.3
Annelids				
	Average ± SE		1.20 ± 0.25	132.2 ± 37.3
Echinoderms				
	Average ± SE		1.22 ± 0.22	201.1 ± 44.8
Cnidarians				
	Average ± SE		0.69 ± 0.11	232.5 ± 114.4
Priapulids				
	Average ± SE			1512.0 ± 684.0

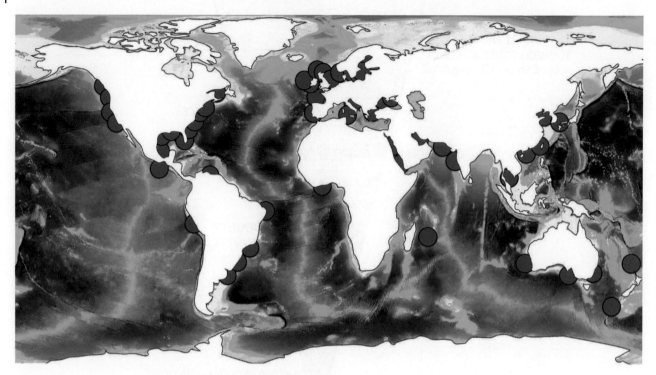

Figure 25.19 Distribution of hypoxic environments and dead zones.

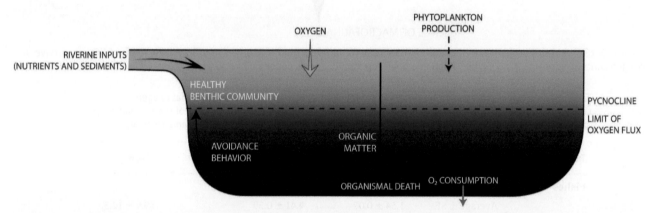

Figure 25.20 Mechanisms leading to formation of hypoxic or anoxic environments. Riverine inputs of nutrients and organic matter, together with elevated primary production and sedimentation of organic matter (senescent/dead phytoplankton, fecal pellets, organic detritus) onto the seafloor, when combined with strong water column stratification (pycnocline) and/or reduced circulation and ventilation of sediments, leads to progressive decline in oxygen at the sediment–water interface that can also extend to shallower depth layers. Oxygen depletion affects both respiration (especially microbes) and chemical oxidation of labile organic matter. These conditions lead to mortality of benthic invertebrates and demersal fishes.

ends. However, hypoxia favors opportunistic species with short life cycle, leading to reduced biodiversity. Oxygen Minimum Zones occupy large regions of the Eastern Pacific Ocean, South Atlantic Ocean, and North Indian Ocean (Figure 25.21). Despite generally similar oxygen profiles with increasing depth in different OMZs, width and depth vary depending on region.

Giant bacteria that use nitrate to oxidize hydrogen sulphide commonly occur in OMZs; formation of sulfur crystals gives the cells a white appearance. Two taxa of bacteria, *Thioploca* spp. and *Beggiatoa* spp., characterize sediments within OMZs. *Thioploca* spp. cells usually align to form long filaments that collectively span over 10,000 km^2 of continental slope in the extensive Peruvian-Chilean OMZ (Figure 25.22), at biomasses >120 g m^{-2} (Figure 25.23). Despite the extreme reduction in oxygen, protistan and metazoan communities thrive in these environments. Foraminiferal communities consist mainly of calcareous forms, but encrusting and agglutinating species can also achieve high abundances. Nematodes tolerate low oxygen concentrations better than other meiofaunal groups. Thin shells characterize gastropods (*Astyris permodesta*, for example) and bivalves (*Amygdalum anoxicolum*, for example) that can live in oxygen concentrations below 0.15 ml l^{-1}.

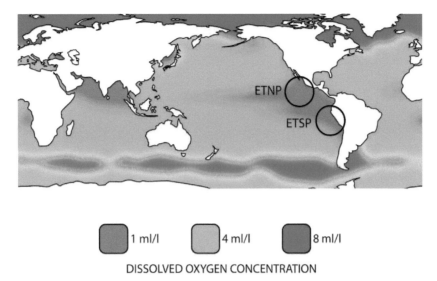

Figure 25.21 Map of average yearly dissolved oxygen concentrations at 200 m depth. Note largest regions with strong depletion of oxygen: Arabian Sea, Eastern Tropical Atlantic Ocean (corresponding to the Namibia upwelling coast), Northeastern Tropical Pacific (ETNP, corresponding to an upwelling region off California), and the South Eastern Tropical Pacific (ETSP, corresponding to the Peru-Chile upwelling). Smaller regions add to these large regions, near regional-scale upwelling locations.

Figure 25.22 Schematic three-dimensional view of the OMZ (in red) on the Peru-Chile margin. This OMZ extends thousands of miles latitudinally and longitudinally, spanning 500–1,500 m depth.

Periodic anaerobic metabolism likely leads to the production of lactic and pyruvic acid, buffered by dissolution of calcified structures. A very thin inner surface, indicative of dissolution, characterizes many bivalves living in "reducing" environments. Calcifying echinoderms commonly occur in OMZs. Amphipods that live in OMZs such as *Ampelisca* spp. have enlarged gill surfaces, a phenomenon also seen in mysids, fishes, and cephalopods. Polychaetes have developed extended tentacles that serve a similar function as gills, likely facilitating oxygen absorption. Invertebrates in OMZs often help stabilize the muddy sediment by building tubes and other structures. The mysid *Neognathophausia ingens*, one of the most common crustaceans in OMZs, illustrates numerous adaptations to low oxygen environments, including excellent ventilation supported by the large surface area occupied by the gills, minimum distance between blood flow and water, the ability to absorb >90% of oxygen from inhaled water, and the presence of respiratory pigments. Copepods of the genus *Lucicutia* have modified their life cycle to different oxygen concentrations. The crustacean *Orchomenella obtuse*, however, moves into OMZs, taking advantage of abundant food and lack of predators in hypoxic water, swimming outside to obtain additional oxygen through an "eat and run" strategy. Despite few examples of endemism in OMZs, the nematode *Glochinema bathyperuvensis* is endemic to the Peruvian margin and the polychaete *Meganerilla bactericola* and gastrotrich *Urodasys anorektoxys* that

Figure 25.23 Bacterial mat of *Thioploca* spp. covering surface sediments collected off the coasts of Peru and Chile. *Source:* Victor Ariel Gallardo.

occur in high abundances appear endemic to the Santa Barbara basin. Oxygen Minimum Zones (OMZs) offer food-rich environments with the effective absence of predators. Very frequently, "soft bodied" organisms dominate the macrofauna, particularly annelids (polychaetes and oligochaetes) that may represent more than 90% of total macrofaunal abundance.

As organic matter settles through the water column to the bottom, it experiences reduced rates of oxidation because of less efficient microbial degradation with anaerobic metabolism. Anoxia can also alter the microbial loop. Animals living in OMZs obtain energy and carbon from the detritus formed by the death of phytoplankton that sink from surface waters. Scientists initially assumed that this abundant organic matter formed the base of OMZ food webs, but recent research demonstrates that chemosynthesis by sulfur-oxidizing bacteria also plays a significant role in OMZ energy supply.

In the oxygen minimum zone adjacent to Namibia, gobies exploit this hypoxic habitat to escape predation by pelagic fish, performing daily migrations within the water column to feed on gelatinous zooplankton. These fish also consume jellies deposited on the hypoxic seabed. To withstand the extreme OMZ conditions these gobies have developed strong resistance to hydrogen sulfide toxicity and mechanisms to slow their heartbeat when moving within the OMZ.

Thioploca spp. bacteria store the nitrogen required to oxidize hydrogen sulfide. The nitrogen substitutes for the oxygen many organisms use, and the cell vacuoles function as lungs. The presence of sulfur-oxidizing bacterial endosymbionts, which fix and translocate carbon to their host, occur throughout OMZs. In the Santa Barbara Basin (off California) at least four species of foraminifera, a euglenoid flagellate, five species of ciliates, and a bivalve house endosymbiont bacteria. In contrast, four species of euglenozoid flagellates, ten species of ciliates, one nematode species and one polychaete species house ectosymbiont bacteria. For many of these taxa, researchers have not yet fully determined host-symbiont metabolic interactions, but they likely occur in most species. When other organisms digest bacteria, the energy and the carbon stored in their cells becomes available to the food web. Therefore, chemosynthesis forms the major base of the food webs that support OMZ animal communities.

25.10 Summary

In summary, the continental margins encompass a wide range of habitats varying in extent, size, diversity, and function (Table 25.4). In tandem, these habitats represent some of the dominant features on Planet Ocean and harbor a species-rich fauna that continues to yield discoveries of new species and processes.

Table 25.4 Comparison between continental margin deep marine ecosystems

Main characteristics	Habitat / Ecosystem					
	Sediment covered slopes	Submarine canyons	Deep-water corals	Cold seeps	Dead zones	Oxygen minimum zones (OMZ)
Extent	Occur along the continental margins of all continents	Limited with respect to other environments but can span 20% of margins	Negligible proportion of the seafloor	Relatively limited numbers, widely dispersed	The largest is the Black Sea but small size dead zones occur in small basins and areas	Very extensive along continental margins and progressively increase
Distribution	Occur along the continental margins of all continents	Occur along the continental margins of all continents	Occur along continental margins at middle and high latitudes	Occur along all continental margins and many areas of abyssal plains	Occur along the continental margins of all continents	Pacific, Indian, Atlantic
Depth	200–4000 m	150–4,000/5,000 m	150–6,000 m	From continental shelves down to 7,326 m	from 900-1400 m to >3,500 m	100–1,500 m
Type of substrate	Mud and sand	Hard and soft bottoms depending on location in canyon	Settle on hard substrate but often surrounded by soft bottoms	Typically soft bottoms	Typically soft bottoms	Typically soft bottoms
Biodiversity		Typically elevated, especially megafauna	Very high, may compares with that of tropical systems. Thousands of species associated with each reef	>600 described species	Very low number of species	Very low
Endemism		Probably 30% or higher	Very high, many undescribed species	Very high	Very high	Very low
Functioning		Elevated primary production and organic flux to the seafloor	Strong currents favor transport of organic matter and zooplankton that provide food resource for corals	Chemosynthetic production supplied by hydrocarbons	Chemosynthetic production	Phytodetritus and chemosynthetic production
Trophic webs		POC fluxes used by zooplankton or hyperbenthos that sustain a high number of trophic levels through to large cetaceans	Very complex, numerous fish species use corals, resulting in high concentrations of fish	Complex based on interactions between macro-organisms and endosymbionts	Largely unknown	Predominantly microbial with scavenger and predator higher components
Peculiarities		Strong currents occur in active canyons. Upwelling of deep waters and downwelling of surface water can occur	Extreme habitat complexity including microhabitats	Extremely variable in time and in space	Extremely stable in space but largely unknown	Expanding because of climatic changes and warming of surface waters
State of Knowledge		Relatively well known along several continental margins	Relatively well known along several continental margins	Still limited compared to hydrothermal vents	Still very limited	Still limited and finite

Questions

1) Why are deep-sea ecosystems considered "extreme"?
2) What is the azoic theory? Is this theory still valid today?
3) Is biodiversity related to depth?
4) What is the main food source for deep-sea environments?
5) What is the role of deep-sea benthic communities (from prokaryotes to megafauna), and how do they react to food pulses?
6) Contrast the main deep-sea habitats.
7) Are submarine canyons rich in biodiversity? Why?
8) Do symbiotic relations exist in deep-sea ecosystems? Why? Where?
9) Discuss oxygen availability in the deep sea.
10) Do hypoxic (or anoxic) conditions occur in deep-sea ecosystems? Why?
11) Are OMZs devoid of life?

Suggested Reading

Auster, P.J. (2005) Are deep-water corals important habitats for fishes? in Freiwald, A. and Roberts, J.M. (eds.), *Cold-Water Corals and Ecosystems*. Erlangen Earth Conference Series. Springer, pp. 747–760.

Bachelery, P. and Villeneuve, N. (2022) Hot spots and large igneous provinces. in Shroder, J. (ed. in Chief), *Treatise on Geomorphology Volume 5: Tectonic Geomorphology*. Academic Press, pp. 195–228.

Baco, A.R. (2007) Exploration for deep-sea corals on North Pacific seamounts and islands. *Oceanography*, 20, pp. 108–117.

Baillon, S., Hamel, J.F., Wareham, V.E., and Mercier, A. (2012) Deep cold-water corals as nurseries for fish larvae. *Frontiers in Ecology and the Environment*, 10, pp. 351–356.

Barone, G., Varrella, S., Tangherlini, M. et al. (2019) Marine fungi: Biotechnological perspectives from deep-hypersaline anoxic basins. *Diversity*, 11, p. 113.

Beuck, L., Freiwald, A., and Taviani, M. (2010) Spatiotemporal bioerosion pat- terns in deep-water scleractinians from off Santa Maria di Leuca (Apulia, Ionian Sea). *Deep-Sea Research II*, 57, pp. 458–470.

Bosley, K.L., Lavelle, J.W., Brodeur, R.D. et al. (2004) Biological and physical processes in and around Astoria submarine Canyon, Oregon, USA. *Journal of Marine Systems*, 50, pp. 21–37.

Cambon-Bonavita, M.A., Nadalig, T., Roussel, E. et al. (2009) Diversity and distribution of methane-oxidizing microbial communities associated with different faunal assemblages in a giant pockmark of the Gabon continental margin. *Deep-Sea Research II*, 56, pp. 2248–2258.

Canals, M., Puig, P., Durrieu de Madron, X. et al. (2006) Flushing submarine canyons. *Nature*, 444, pp. 354–357.

Carlier, A., Le Guilloux, E., Olu, K. et al. (2009) Trophic relationships in a deep Mediterranean cold- water coral bank (Santa Maria di Leuca). *Marine Ecology Progress Series*, 397, pp. 125–137.

Cartes, J., Company, J.B., Maynou, F. et al. (2004) The Mediterranean deep-sea ecosystems: An overview of their diversity, structure, functioning and fishing impacts. *Contribution from the World Wide fund for Nature (WWF) and the International Union for the Conservation of Nature (IUCN) to the 2004 Session of the SAC/GFCM Sub-Committee on Marine Environment and Ecosystems*, pp. 1–27.

Cerrano, C., Arillo, A., Azzini, F. et al. (2005) Gorgonian population recovery after a mass mortality event. *Aquatic Conservation*, 15, pp. 147–157.

Cerrano, C., Bavestrello, G., and Bianchi, C.N. (1999) A catastrophic mass-mortality episode of gorgonians and other organisms in the Ligurian Sea (Northwestern Mediterranean), summer 1999. *Ecology Letters*, 3, pp. 284–293.

Cerrano, C., Danovaro, R., Gambi, C. et al. (2010) Gold coral (*Savaglia savaglia*) and gorgonian forests enhance benthic biodiversity and ecosystem functioning in the mesophotic zone. *Biodiversity and Conservation*, 19, pp. 153–167.

Danovaro, R. (2010) *Methods for the Study of Deep-Sea Sediments, Their Functioning and Biodiversity*. CRC Press.

Danovaro, R., Fabiano, M., Albertelli, G., and Della Croce, N. (1995) Vertical distribution of meiobenthos in bathyal sediments of the Eastern Mediterranean Sea: Relationship with labile organic matter and bacterial biomasses. *Marine Ecology*, 16, pp. 103–116.

Danovaro, R., Fabiano, M., and Della Croce, N. (1993) Labile organic matter and microbial biomasses in deep-sea sediments (Eastern Mediterranean Sea). *Deep-Sea Research I*, 40, pp. 953–965.

Danovaro, R., Snelgrove, P.V.R., and Tyler, P. (2014) Challenging the paradigms of deep-sea ecology. *Trends in Ecology and Evolution*, 29, pp. 465–475.

De Leo, F.C., Smith, C.R., Rowden, A.A. et al. (2010) Submarine canyon hotspots of benthic biomass and productivity in the deep sea. *Proceedings of the Royal Society of London B: Biological Sciences*, 277, pp. 2783–2792.

Duineveld, G., Lavaleye, M., Berghuis, E., and de Wilde, P. (2001) Activity and composition of the benthic fauna in the Whittard Canyon and the adja- cent continental slope (NE Atlantic). *Oceanologica Acta*, 24, pp. 69–83.

Emig, C. (2004) *The Mediterranean deep-sea fauna: Historical evolution, bathymetric variations and geographical changes.* Carnets de Géologie (free download).

Fernandez-Arcaya, U., Ramirez-Llodra, E., Aguzzi, J. et al. (2017) Ecological role of submarine canyons and need for canyon conservation: A review. *Frontiers in Marine Science*, 4. doi: 10.3389/fmars.2017.00005.

Fisher, C., Roberts, H., Cordes, E., and Bernard, B. (2007) Cold seeps and associated communities of the Gulf of Mexico. *Oceanography*, 20, pp. 118–129.

Gage, J.D., Lamont, P.A., and Tyler, P.A. (1995) Deep-sea macrobenthic communities at contrasting sites off Portugal, Preliminary Results. 1 Introduction and diversity comparisons. *Internationale Revue der gesamten Hydrobiologie und Hydrographie*, 80, pp. 235–250.

Gage, J.D. and Tyler, P.A. (1991) *Deep-Sea Biology: A Natural History of Organisms at the Deep Sea Floor.* Cambridge University Press.

Garcia, K.A., Koho, H.C., de Stigter, E. et al. (2007) Distribution of meiobenthos in the Nazaré canyon and adjacent slope (western Iberian Margin) in relation to sedimentary composition. *Marine Ecology Progress Series*, 340, pp. 207–220.

Heijs, S.K., Sinninghe Damste, J.S., and Forney, L.J. (2005) Characterization of a deep-sea microbial mat from an active cold seep at the Milano mud volcano in the Eastern Mediterranean Sea. *FEMS Microbiology Ecology*, 54, pp. 47–56.

Herring, P.J. (2002) *The Biology of the Deep Ocean.* Oxford University Press.

Joye, S.B. (2020) The geology and biogeochemistry of hydrocarbon seeps. *Annual Review of Earth and Planetary Scence*, 48, pp. 205–231.

Kennicutt, M.C. (2017) Oil and gas seeps in the Gulf of Mexico, in Ward, C.H. (ed.), *Habitats and Biota of the Gulf of Mexico: Before the Deepwater Horizon Oil Spill: Volume 1: Water Quality, Sediments, Sediment Contaminants, Oil and Gas Seeps, Coastal Habitats, Offshore Plankton and Benthos, and Shellfish.* Springer, pp. 275–358.

Levin, L.A. (2002) Deep-ocean life where oxygen is scarce. *American Scientist*, 90, pp. 436–444.

Levin, L.A. (2005) Ecology of cold seep sediments: Interactions of fauna with flow, chemistry and microbes. *Oceanography and Marine Biology: An Annual Review*, 43, pp. 1–46.

Levin, L.A. and Gage, J.D. (1998) Relationships between oxygen, organic matter and the diversity of bathyal macrofauna. *Deep Sea Research Part II: Topical Studies in Oceanography*, 45, pp. 129–163.

Li, Z., Pan, D., Wei, G. et al. (2021) Deep-sea sediments associated with cold seeps are a subsurface reservoir of viral diversity. *The ISME Journal*, 15, pp. 1–13.

Lohrer, A.M., Thrush, S.F., and Gibbs, M.M. (2004) Bioturbators enhance ecosystem function through complex biogeochemical interactions. *Nature*, 431, pp. 1092–1095.

Loreau, M. (2008) Biodiversity and ecosystem functioning: The mystery of the deep sea. *Current Biology*, 18, pp. 126–127.

McClain, C.R., Nekola, J.C., Kuhnz, L., and Barry, J.B. (2011) Local-scale faunal turnover on the deep Pacific seafloor. *Marine Ecology Progress Series*, 422, pp. 193–200.

Merlino, G., Barozzi, A., Michoud, G. et al. (2018) Microbial ecology of deep-sea hypersaline anoxic basins. *FEMS Microbiology Ecology*, 94(7), p. fiy085.

Mura, M. and Cau, A. (1994) Community structure of the decapod crustaceans in the middle bathyal zone of the Sardinian channel. *Crustaceana*, 67, pp. 259–266.

Narayanaswamy, B.E., Renaud, P.E., Duineveld, G.C.A. et al. (2010) Biodiversity trends along the Western European margin. *PLOS One*, 5, p. e14295.

Olu, K., Caprais, J.C., Galéron, J. et al. (2009) Influence of seep emission on the non-symbiont-bearing fauna and vagrant species at an active giant pockmark in the Gulf of Guinea (Congo-Angola margin). *Deep-Sea Research II*, 56, pp. 2380–2393.

Ondréas, H., Olu, K., Fouquet, Y. et al. (2005) ROV study of a giant pockmark on the Gabon continental margin. *Geo-Marine Letters*, 25, pp. 281–292.

Ramirez-Llodra, E., Brandt, A., Danovaro, R. et al. (2010) Deep, diverse and definitely different: Unique attributes of the world's largest ecosystem. *Biogeosciences*, 7, pp. 2361–2485.

Rex, M.A. (1981) Community structure in the deep-sea benthos. *Annual Review of Ecology and Systematics*, 12, pp. 331–353.

Roberts, J.M., Wheeler, A.J., and Freiwald, A. (2006) Reefs of the deep: The biology and geology of cold-water coral ecosystems. *Science*, 312, pp. 543–547.

Roberts, J.M., Wheeler, A.J., Freiwald, A., and Cairns, S. (2009) *Cold-Water Corals: The Biology and Geology of Deep-Sea Coral Habitats*. Cambridge University Press.

Santelices, B. (2007) The discovery of kelp forests in deep-water habitats of tropical regions. *Proceedings of the National Academy of Sciences of the United States of America*, 104, pp. 19163–19164.

Santora, J., Zeno, R., Dorman, J., and Sydeman, W. (2018) Submarine canyons represent an essential habitat network for krill hotspots in a Large Marine Ecosystem. *Scientific Reports*, 8, p. 7579.

Schiaparelli, S. and Hopcroft, R.R. (2011) The Census of Antarctic Marine Life: Diversity and change in Southern Ocean ecosystems. *Deep-Sea Research II*, 58, pp. 1–4.

Schulze, A. (2019) *Cold Seep Ecosystems. Encyclopedia of Ocean Sciences*. 3rd Edition. Academic Press, pp. 677–683.

Sibuet, M. and Olu, K. (1998) Biogeography, biodiversity and fluid dependence of deep-sea cold-seep communities at active and passive margins. *Deep Sea Research Part II: Topical Studies in Oceanography*, 45, pp. 517–567.

Talling, R.P., Wynn, J.B., Masson, D.G. et al. (2007) Onset of submarine debris flow deposition far fromoriginal giant landslide. *Nature*, 450, pp. 541–544.

Vanreusel, A., Fonseca, G., Danovaro, R. et al. (2010) The contribution of deep-sea macrohabitat heterogeneity to global nematode diversity. *Marine Ecology*, 31, pp. 6–20.

Venrick, E.L., McGowan, J.A., Cayan, D.R., and Hayward, T.L. (1987) Climate and chlorophyll A: Long term trends in the central North Pacific Ocean. *Science*, 238, pp. 70–72.

Vetter, E.W. and Dayton, P.K. (1998) Macrofaunal communities within and adjacent to a detritus-rich submarine canyon system. *Deep-Sea Research II*, 45, pp. 25–54.

Vincx, M., Bett, B.J., Dinet, A. et al. (1994) Meiobenthos of the deep Northeast Atlantic. *Advances in Marine Biology*, 30, pp. 2–88.

Weaver, P.E., Billett, D.M., Boetius, A. et al. (2004) Hotspot ecosystem research on Europe's deep-ocean margins. *Oceanography*, 17, pp. 132–143.

WWF/IUCN (2004) *The Mediterranean Deep-Sea Ecosystems: An Overview of Their Diversity, Structure, Functioning and Anthropogenic Impacts, with a Proposal for Conservation*. WWF/IUCN (eds.), 64 p.

Yakimov, M.M., Cappello, S., Crisafi, E. et al. (2006) Phylogenetic survey of metabolically active microbial communities associated with the deep-sea coral *Lophelia pertusa* from the Apulian plateau, Central Mediterranean Sea. *Deep-Sea Research I*, 53, pp. 62–75.

26

Deep Ocean Basins

26.1 Introduction

Deep-ocean basins cover the greatest portion of the Earth's surface of all habitat types, yet scientists long considered them to be the most monotonous marine ecosystems in the world. However, we now know that deep-ocean basins include a wide variety of geographic and topographic features, including abyssal plains, hadal trenches, ocean ridges and rises, as well as submarine mountainous regions. These systems host a significant portion of global marine biodiversity, largely undiscovered.

26.2 Abyssal Plains

The abyssal plains occur at depths of 4,000–6000 m (some researchers also set the upper limit at 3,000 m), encompassing >40% of Earth's surface. Generally flat or gently sloping seabed, these habitats form from new oceanic crust that spills from the mid-oceanic ridges and spread at a rate of 20–100 mm y^{-1}. The newest seabed formed at ridges consists of rough, young bedrock but fine-grained sediments, consisting primarily of clay, silt, and the skeletons of planktonic organisms, accumulate at a rate of 23 cm every 100 y. The major characteristics of abyssal ecosystems include low biomass and abundance, high species diversity, and vast habitat extent. Despite a generally homogenous appearance, topographic and hydrodynamic complexity characterize abyssal ecosystems. Small invertebrates living on or within the sediment (Figure 26.1) dominate the metazoans.

The lack of substantive topographic barriers means that the small organisms (larvae, juveniles, and adults) that characterize abyssal plains can disperse long distances. Indeed, abyssal plains support a lower percentage of endemic species than other deep-sea habitats. However, some abyssal plain habitats support a relatively specialized community of organisms. For example, when whale or fish carcasses sink to the bottom, they attract a succession of specialized organisms that feed on the carcass for months or years. The manganese nodules that occur on some abyssal plains also support a distinct fauna.

26.3 Abyssal Biodiversity and Adaptations

We know far less about abyssal plain biota than shallower depths, and below we contrast the warm Mediterranean with the cold Atlantic to illustrate current knowledge of abyssal biodiversity patterns and adaptations. The fauna includes polychaetes such as *Fauveliopsis brevis, Tharyx marioni, Prionospio cirrifera*, with several Mediterranean endemic species (*Aricidea annae, Aricidea mediterranea, Aricidea monicae, Aricidea trilobata, Paradoneis lyra*). Decapods such as *Polycheles typhlops* and *Stereomastis sculpta* and cumaceans include numerous species endemic to abyssal environments (i.e., *Diastylis jonesi, Leptostylis bacescoi, Bathycuma brevirostre, Procampylapsis armata* in the Mediterranean). The amphipods include numerous deep-sea species, some of which live only in abyssal environments.

In abyssal environments, crustaceans dominate both the mega- and macrofauna, as illustrated in the Mediterranean. Among them, baited traps and light traps have captured the crab, *Geryon longipes* and *Chaceon mediterraneus* to depths of 3,800 m. The shrimp *Acanthephyra eximia*, the major "street sweeper" or scavenger (together with *Coryphaenoides mediterraneus*) occurs in abundance at depths up to 4,200 m. These species commonly occur in the Atlantic, but as

Marine Biology: Comparative Ecology of Planet Ocean, First Edition. Roberto Danovaro and Paul Snelgrove.
© 2024 John Wiley & Sons Ltd. Published 2024 by John Wiley & Sons Ltd.
Companion Website: www.wiley.com/go/danovaro/marinebiology

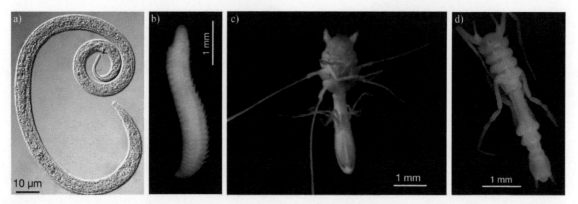

Figure 26.1 Examples of species living in the abyssal sediments of the Weddell Sea: (a) the nematode *Molgolaimus* sp.; (b) the polychaete *Ophryotrocha* sp.; (c) the isopods *Ischnomesus* sp.; and (d) *Munnopsis* sp. *Source:* Wiebke Brökeland.

eurythermic species they have also colonized the deep Mediterranean. *A. eximia* was also reported swimming close to deep hydrothermal vents on the mid-Atlantic ridge.

Until recently, researchers thought that acorn worms (Phylum Hemichordata, class: Enteropneusta) were limited to shallow water until their surprising discovery in the Kuri-Kamchatka trench at 8,100 m. Most Enteropneusta live in U-shaped burrows within sediments, from which the posterior part of the body emerges to produce accumulations of spiral fecal residues on the sediment surface. Many photographs of the abyssal seabed showed these enigmatic formations, but their true nature was understood only after the *Vema* expedition captured an image that clearly showed a large animal (1 m long) releasing spiral feces.

Some deep abyssal sediments in the Mediterranean apparently support largely monospecific cultures of the polychaete *Galathowenia fragilis*, oddly enough, one of the most abundant polychaete species in the Arctic Ocean. A large number of Mediterranean species (138) span six mollusk classes: Caudofoveata, Polyplacophora, Gastropoda, Bivalvia, Scaphopoda, and Cephalopoda. Macrofaunal abundance varies among different regions: at similar depths, abundances can range from a few individuals to >750 ind. m^{-2} in the relatively oligotrophic Mediterranean, from 1,000–2,000 ind. m^{-2} in the Atlantic, and 5,000–10,000 ind. m^{-2} in the Pacific. The highest densities of meiofauna in abyssal environments occur under surface waters with high primary production. Total meiofaunal density may vary from a few individuals to over 10^6 indiv. m^{-2}, with a biomass that can even exceed that for macrobenthos (1 g C m^{-2}). Nematodes dominate all abyssal environments numerically, increasing in importance with depth until they account for >90% of meiofauna. In almost all deep-sea systems other than high flow environments, such as seamounts and canyons, deposit feeders (selective and nonselective) dominate, with poor representation of predators. The ~180 known species of Siboglinidae tube worms occur mainly at abyssal depths <6,000 m.

Most knowledge of zooplankton comes from the epipelagic zone or the upper mesopelagic. Most quantitative information of the deep layers of the water column focuses on meso- and micro-zooplankton, with almost a complete absence of information on protozoa and bacteria. Deep-sea zooplankton biomass varies greatly among locations, noting deep zooplankton biomass in the Mediterranean Sea 7–16x lower than in the Atlantic Ocean and 3–5x lower than in the Indian Ocean. In the Mediterranean, a west to east gradient of decreasing biomass occurs below 1,050 m depth. Zooplankton abundance also decreases by a factor of 100–1,000 from epipelagic to bathypelagic zones. Deep-sea zooplankton consists largely of mesopelagic species with a smaller number of truly deep species (such as the copepod, *Lucicutia longiserrata*). This characteristic must relate to the extreme oligotrophy of the deep Mediterranean. Some species of calanoid copepods (*Calanus helgolandicus* and *Subeucalanus monachus*) provide a potential indicator of changing deep thermohaline conditions such as those that took place after 1992 as a result of a transient phenomenon. The mesozooplankton represent an important fraction of the flux of organic carbon towards the bottom (from >20–55% at 3,000 m depth).

Three years of analyses on swimmers captured in sediment traps deployed at three-month intervals, 1,200–2,250 m depth, identified copepods as the dominant zooplankton taxon in Mediterranean deep-sea pelagic environments. However, ostracods dominated the deep Ionian Sea and cnidarians dominated the Mediterranean's Ionian and Alboran Seas. Mollusks (Thecosomata), isopods, amphipods and euphausiids also occur, along with limited numbers of polychaetes and nematodes, perhaps resulting from resuspension from the benthic boundary layer.

Deep demersal fish populations in the Mediterranean apparently differ from those in the Atlantic, not only in species composition, but also in maximum size and reproductive characteristics. Studies over the last 20 years in the Mediterranean report some 43–47 fish species in the abyssal Mediterranean and 80–104 species in the abyssal Atlantic. Moreover, the

Figure 26.2 Details of the jade colored eye of *Centroscymnus coelolepis* (Portuguese dogfish) that lives at depths to almost 4,000 m.

Mediterranean shows greater similarity and more species in common between different regions (low β-diversity). The dominant families change with bathymetric bands: while macrourids in the deep Atlantic feed on small macroplankton, benthopelagic decapods in the Mediterranean use this resource and, in turn, become prey for deep-sea fishes. The most abundant species in common between the abyssal Atlantic and Mediterranean include *Coelorinchus caelorhincus*, *C. labiatus*, *Notacanthus bonaparte*, *Nezumia aequalis*, *Polyacanthonotus rissoanus*, *Phycis blennoides*, *Coryphaenoides mediterraneus*, *Mora moro*, and *Helicolenus dactylopterus*. In the Eastern Mediterranean, *C. mediterraneus* commonly occurs to 4,262 m, but the Portuguese dogfish is particularly important as top predator at bathyal/abyssal depths (*Centroscymnus coelolepis*).

Among benthic organisms, animals living >6,000 m depth do not vary much. Abyssal and ultra-abyssal (hadal) environmental conditions only really differ in substantially higher hydrostatic pressure in hadal habitats. Adaptations to high pressure typically involve physiological and biochemical specialization rather than marked morphological variations in organisms. Because of the complete absence of light at abyssal depth, body color in benthic animals has no adaptive value, resulting in widespread absence of pigmentation. Nonetheless, many species have conserved coloration even at these depths, such as green in some echiurids, yellow in *Bathycrinus* sp. crinoids, and violet for some polynoid polychaete species. All siboglinids that live at great depth appear red in color because of the presence of hemoglobin in their blood. The body of the fish *Pseudoliparis amblystomopsis* appears pink. The only available specimen of the ultra-abyssal trachomedusae, *Pectis profundicola*, is brown, whereas photographs show orange anthomedusae at abyssal depths and greater (to 8,260 m depth).

The vast majority of abyssal animals lack eyes, but some crustaceans and gastropods, even those living at great depth, retain rudimentary eyes with no functional value. Two species of isopods in the eurybathic genus *Antarcturus* maintain normal eyes, likely indicating relatively recent colonization of deep water for those species. Some fishes have retained eyes but much reduced in size or in a degenerative state compared to their shallow congenerics. The only shark species that lives in abyssal depths, the Portuguese dogfish, *Centroscymnus coelolepis*, has well-developed eyes (Figure 26.2).

Some deep-sea organisms display a "rickets" like phenomenon, expressed as a thinning and reduction in calcification rate in skeletal structures. This phenomenon links to a calcium carbonate deficiency because of undersaturation of calcium carbonate in seawater at great depth and in associated sediments. This undersaturation creates problems for benthic organisms living at these depths that requite calcium carbonate for skeletal development. The **carbonate compensation depth**, which occurs at 4,000–5,000 m, refers to the depth below which dissolution rates of calcite and aragonite exceed deposition rates. These organisms face a constant challenge in maintaining their activity and renewing skeletal structures to offset dissolution.

26.4 Abyssal Gigantism and Dwarfism

Some benthic invertebrates show increasing body size with increasing habitat depth. This phenomenon of **gigantism**, or unusually large size, has been reported in several orders of crustaceans (Figure 26.3). Isopods of the suborder Asellota illustrate the most striking examples of size increase with depth and large individuals among ultra-abyssal species, including *Alicella gigantea*, the largest known abyssal amphipod species. In 1984, researchers described the largest known species of

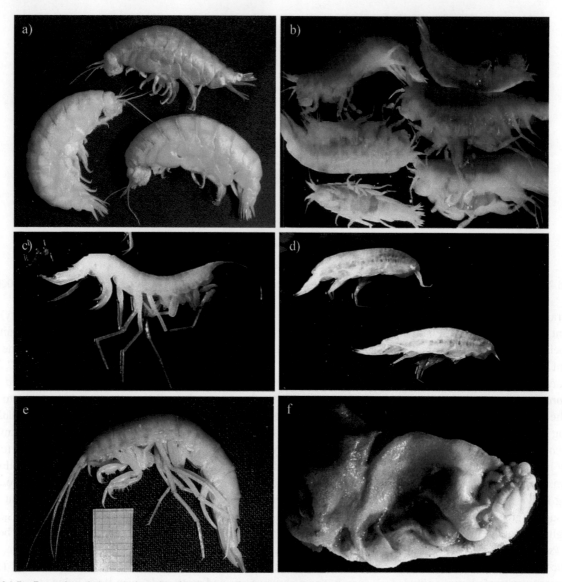

Figure 26.3 Examples of abyssal gigantism in crustaceans: (a) amphipod, *Eurythenes gryllus*; (b) amphipods of the genus *Hirondellea*; (c–e) deep-sea amphipods from Atacama trench; (f) holothurian, *Elpidia atakama*.

tanaidacean, *Gigantapseudes adactylus* (75 mm in length). The largest member of the order Amphipoda reaches a length of 282 mm, a size that greatly exceeds many ultra-abyssal amphipods. Still, only scattered examples of gigantism at depth exist within other taxonomic groups.

Comparisons between body sizes of various siboglinids indicate larger size in the hadal genera *Spirobrachia*, *Heptabrachia* and *Diplobrachia* than their shallow-water congeners. Among the *Elpidia* spp. holothurians, the size of many abyssal species exceeds those that live at shallower depths. However, in addition to gigantism (Figure 26.4), examples of dwarfism also exist. For some fishes, for example, abyssal species never grow to the sizes seen in bathyal or coastal species. Below 4,000 m depth giant isopods and amphipods represent the most impressive predators.

Each species (and sometimes even related species and genera) contain a roughly constant number of cells, but increased body size during ontogeny reflects increased cell size. These data align with the assumption that increased body size in deep-sea isopods results from prolongation of their life cycle. Similar assumptions were made for giant lysianassid amphipods. In some cases, increased size occurs in closely related species of crustaceans in shallow depths of polar-regions, reflecting relatively low temperatures. In other cases, large size correlates with great depth and does not depend on temperature. Instead, hydrostatic pressure may cause this phenomenon, but more likely, both factors play a role. In any case, the biological mechanism that causes gigantism remains unclear because it occurs so selectively within just a subset of some taxonomic groups.

Figure 26.4 Examples of deep-sea gigantism: (a) abyssal holothurian (unidentified); (b) pycnogonid; (c) Scorpaenidae; (d) abyssal fish; abyssal holothurians; (e) *Amperima rosea*; and (f) *Molpadia musculus*. *Sources:* NOAA / Public Domain, Scott C. France / Wikimedia Commons / CC BY-SA 2.0, NOAA / Public Domain, NOAA / Wikipedia Commons / CC BY 3.0, NOAA / Public Domain, NOAA / Craig Smith / Public Domain, Smithsonian Institution / Trang Nguyen / Wikimedia Commons / CC0 1.0.

26.5 Functioning of Abyssal Systems

Studies using underwater camera systems documented periodic, massive deposition of phytodetritus produced in the photic zone onto the seabed. Aggregation of material into mats of organic matter accelerates deposition and material can sink to 5,000 m depth in just a few days (Figure 26.5). This mechanism deposits large amounts of material with high nutritional quality on the abyssal seabed, providing a vital source of food for abyssal communities.

The activity of deep-sea deposit feeders increases considerably during these deposition events, but even if the visible material disappears completely in a few weeks, some accumulates in the sediment to create a long-term food reserve.

26.5.1 Seamounts

Seamounts (Figure 26.6) refer to topographic structures that rise above the seabed but do not penetrate the ocean surface. Generally, scientists identify these topographic features as geological structures that rise from the bottom to a minimum height of 1,000 m, though biologists include structures <1,000 m in height, because ecologically, a 1,000 m cutoff has no basis. Seamounts often occur in a series of semi-linear features corresponding to mid-ocean ridges, but they can also occur as individual structures. Globally, seamounts are numerous, with an estimated 30,000–100,000 seamounts in the Pacific alone, spanning ~6% of the ocean floor. The Census of Marine Life estimated that 63,000 seamounts >1,000 m in elevation exist globally, the majority of which remain undiscovered, unsurveyed, and unstudied.

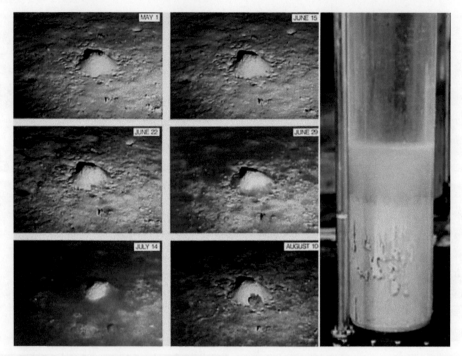

Figure 26.5 Deposition of microalgal and cyanobacterial phytodetritus at ~5,000 m depth, seen as a green blanket of "fluff" several centimeters thick on the sediment surface. A few days after the deposition event, holothurians and other deposit feeders have consumed most of the visible organic matter.

Figure 26.6 Distribution of the estimated 63,000 seamounts globally based on data collected and verified by regional datasets, or estimates based on satellite data, or information from registered vessels.

Seamounts usually occur as isolated or small clusters of cone-shaped structures, often of volcanic origin, and usually dominated by hard substrate. These structures sometimes extend above the ocean surface as oceanic islands. For example, the islands of Hawaii, the Azores, and Bermuda were once underwater seamounts. Seamounts represent important and interesting habitats for several reasons. First, they offer an excellent case study for understanding marine biodiversity patterns. Seamounts vary greatly in their associated biodiversity, sometimes supporting endemism and centers of speciation, and acting as a "stepping stone" for dispersal of some species, including some coastal taxa. Some seamounts support high production and important commercial species, but the fragility of these habitats demands careful management of human use that must be handled with care and with better scientific information.

Seamounts typically arise from tectonic plate movement or through intra-plate volcanic activity. **Hotspots** refer to areas of the seabed in isolated regions of plates where breaks in the crust allow magma to rise from the mantle to create an underwater volcano. Seamounts formed by hotspots are easily identifiable through their geochemical composition, whose rubidium, strontium, cesium, thorium, and potassium content reflect their lower mantle origin. Intriguingly, these hotspots remain stationary, whereas the oceanic plates above them actually move. Seamounts and islands form on plates above hotspots, as magma released through the break in the crust accumulates. If a volcano grows to the point that it breaks the ocean surface it becomes an oceanic island or an atoll. The Hawaiian Islands in the Pacific Ocean and the Azores and associated seamounts in the Atlantic Ocean provide excellent examples of seamount hotspots. The movement of a plate over a hotspot produces a chain of volcanoes with the newest island over the hotspot and the oldest furthest away. With time, these islands tend to sink below the surface and become seamounts such as the oldest Hawaiian island, which is now a seamount.

Not all seamounts originate from hotspots. Seamounts also form near the boundaries of tectonic plates, whether subduction or spreading. At mid-ocean ridges, magma rises to fill gaps between spreading plates, whereas near subduction zones, buoyant magma may arise as one plate subducts beneath the other, creating volcanoes and seamounts. Seamounts rise up to several kilometers from the bottom and, depending on ocean depth, take on different shapes and steepness (Figure 26.7). Semi-enclosed seas such as the Mediterranean also contain seamounts. The numerous Mediterranean seamounts occur in groups in the western basin in contrast to the dispersed distribution in the eastern basin, which contains some of the largest seamounts, such as Marsili and Palinuro in the Tyrrhenian Sea (Figure 26.8).

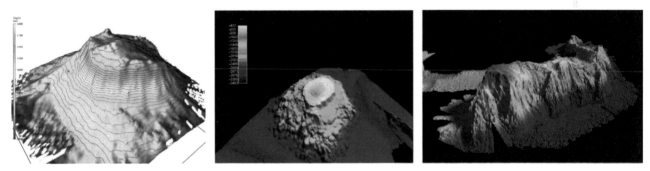

Figure 26.7 Multibeam images illustrating seamount morphologies.

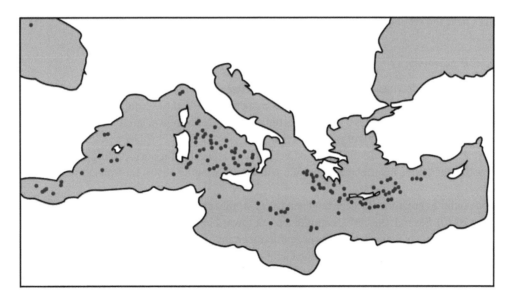

Figure 26.8 Mediterranean seamounts showing 169 known structures, of which only 50 have been surveyed and mapped to date.

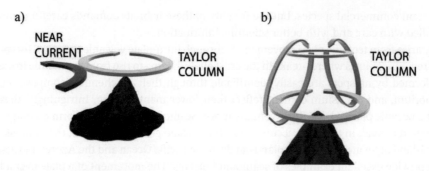

Figure 26.9 Potential effects of seamounts on water masses, and associated currents: (a) rotary clockwise motion and (b) subsequent downwelling associated with a Taylor column.

Seamounts vary in shape from elliptical to circular or conical and sometimes elongate, varying from sharp to flat-topped summits. Their comparatively steep sides often do not accumulate sediment, resulting in a predominantly rocky substrate. The ocean includes three types of submerged mountains: (1) **seamounts**, the isolated peaks rising 1,000 m or more above the seafloor; (2) **knolls,** the isolated peaks rising <1 km in elevation; (3) **pinnacles,** the small mountain ranges that rise above the seabed.

Seamounts create significant topographic obstacles that interact with ocean currents, potentially forming internal waves, eddies, and rotary flows around the sides and summits. On a scale of 10 km seamounts may cause: (a) the formation of vortices that become trapped on the top of the mountain or scatter on the flanks of the seamount; (b) waves that become trapped around the seamount; (c) amplification, distortion, and reflection of internal waves. These circulation features can upwell cold, nutrient-rich waters to near the ocean surface. Particularly when interacting with other currents, this upwelling can also dramatically increase turbulence around the seamount (Figure 26.9). For example, one study reported levels of mixing in a 100 km^2 area around Cobb Seamount in the Pacific similar to 100,000 km^2 in the open ocean. Under certain conditions associated with complex seamount-current interactions, a "trapping" vortex known as a "Taylor column" may form (Figure 26.9). This phenomenon involves an **anti-cyclonic** (clockwise movement) vortex developing at the top in tandem with downwelling at the center of the vortex. Researchers proposed Tayler columns as a mechanism to retain water masses, nutrients, and larvae of benthic and pelagic species, and thus enhance retention and recruitment of many marine populations.

The **seamount effect** refers to increased concentrations of nutrients and particulate organic material around seamounts that can support high abundances of benthos, zooplankton, and fishes. Furthermore, increases in internal waves and turbulence around the seamount can enhance food supply to benthic filter feeders. This mechanism can increase food availability for many organisms, including fish larvae and zooplankton. Seamount currents may force zooplankton that perform vertical migrations to the top of the seamount during periods when they would normally descend, creating a potential food source for predators near the seamount summit.

26.5.2 Seamount Biodiversity

Despite the initial discovery of deep seamounts during the *Challenger* expedition from 1872 to 1876, researchers only really began to appreciate the diversity and exceptional species associated with these unique environments after 2000. The biology and life history of all benthic species in seamounts remains largely unknown, but the fauna includes some long-lived species with lifespans on the order of 100s of years.

Several seamount studies report potentially high biodiversity with respect to the surrounding open ocean and large numbers of endemic species of benthic invertebrates on some seamounts, ranging 5–50% for isolated seamounts or seamounts chains.

Researchers debate whether seamounts focus biodiversity, or simply provide additional habitat within abyssal surroundings, or both (Figure 26.10). Several aspects of seamounts point toward high biodiversity and high endemism. These attributes include the isolated nature of many seamounts or seamount chains, a limited larval dispersal strategy adopted by many seamount species, and periodic formation of retention currents. However, some seamounts, especially those organized in chains or located near escarpments, largely offer a steppingstone for abyssal organisms during dispersal at great depth, over large spatial scales, and through numerous bio-geographical barriers. The richness and diversity of fish and invertebrate communities associated with seamounts attract other forms of marine life, such as seabirds and cetaceans (Figure 26.11).

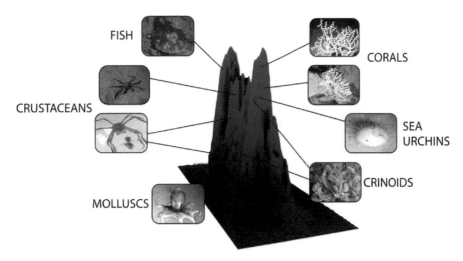

Figure 26.10 Typical fauna associated with a seamount. *Sources:* NURC/UNCW and NOAA/FGBNMS / Wikimedia Commons / Public domain, National Oceanic and Atmospheric Administration (NOAA).

Figure 26.11 Seamounts are hotspots of invertebrate and vertebrate biodiversity such as: (a) *Chimaera* sp. on Manning Seamount; (b) deep-sea lobster (*Munidopsis* sp.); c) Crinoids and pennatulaceans on Balanus Seamont; (d) basket star; (e) the squid *Gonatus onyx*; and (f) the Picasso sponge (*Staurocalyptus* sp.) on Davidson Seamount ad depths >1,000 m depth. *Sources:* unknown author / Wikimedia Commons / Public domain, NOAA/Monterey Bay Aquarium Research Institute / Wikimedia Commons / Public domain, Dr. Les Watling / Wikimedia Commons / Public domain, NOAA/Monterey Bay Aquarium Research Institute / Wikimedia Commons / Public domain.

For example, the albatross, *Phoebastria nigripes*, occurs in abundances 14x higher above Cobb Seamount relative to surrounding waters. Numerous other seabirds, such as *Oceanodroma furcata* and *O. leucorhoa*, occur in greater numbers near the seamount. Migratory birds and marine mammals also use seamounts as feeding areas grounds during their annual migrations. Using hydrophones, researchers identified Lo'ihi seamount (Hawaii), and Cordell Bank Seamount as feeding grounds for whales and marine mammals.

One study reported >850 species in the Tasman Seamounts (South-eastern Australia), including many species associated with deep-water corals. Recent studies of a seamount off New Caledonia identified >2,000 species, half of them new to science. The Vema Seamount (Southern Atlantic) supports ~28% endemic species. Two endemic fish species and four endemic crustaceans were reported in the Great Meteor Seamount (in the North Atlantic). Horizon Guyot in the Pacific contains 36% endemic invertebrate species. These examples illustrate the potential for new species discoveries in seamounts globally.

The rocky bottom and steady currents along the sides and base of seamounts create ideal conditions for benthic filter feeders such as crinoids, ascidians, holothurians, sponges, and deep-water corals such as *Desmophyllum pertusum* (see below). Biogenic species such as corals increase the complexity of seamount environments by adding numerous microhabitats, including empty spaces between branches, living and the dead coral surfaces, and cavities within coral skeletons and within sediments. Deep-water corals support a high diversity of animals reminiscent of tropical reefs. These sedentary (or sessile) species typically require hard substratum for adhesion and strong currents to supply food, remove waste, and aid larval dispersal. Currents thus play a key role in defining distributions of suspension feeders, which occur in particularly high abundances on the tops and sides of seamounts.

On small scales, higher densities of animals occur on peaks and terraces; these hard substrates support many different types of organisms, including lobsters, crabs and other crustaceans, echinoderms (sea stars, holothurians), polychaetes, and mollusks. However, the bases, terraces, and flattened tops of some seamounts (guyots) sediment can accumulate and host diverse life forms, including segmented and unsegmented worms, crustaceans, gastropods, bivalves, and ascidians. Cnidarians, pennatulaceans (sea pens), sponges, barnacles, cerianthid sea anemones, crinoids, and other echinoderms often occur on sediment surfaces. Moreover, xenophyophores can occur within sediments around the seamount; these large single-celled protozoans build complex baseball-sized structures using sediment particles. Many of these organisms are suspension feeders and thus prefer areas exposed to strong currents. Endemism also occurs in meiobenthos; one study of community composition in sediments at the base of two Mediterranean Sea seamounts identified 23% of nematode species as endemic.

Retention of zooplankton over seamounts plays a critical role in increased secondary production. Noting generation times for most zooplankton on the order of weeks, for these secondary producers to benefit from increased primary production requires that they remain above the seamount for at least several weeks. In the absence of retention or lower residence time of zooplankton above seamounts, such as when closed vortices move away them from the seamount, secondary production would move toward the base of this structure and away.

Communities of fish on different seamounts have developed morphological, ecological, and physiological adaptations to colonize an environment characterized by strong currents and high flux of organic matter, a markedly different pattern from most deep-sea environments. Some fishes have evolved powerful swim performance with a hydrodynamic body and high metabolic rates, resulting in a greater demand for energy.

Slow growth and high longevity characterize seamount fishes, which can occur in high abundances, especially on seamounts near continents, such as Bowie and Cobb Seamounts near British Columbia, Canada. Since 1960, commercial fishing on seamounts has increased greatly, particularly by Japanese and Russian trawlers. These trawlers have captured as estimated 133,400 metric tons of fishes on seamounts in just over 10 years. Pelagic species such as *Pentaceros wheeleri* comprise the majority of this catch (90%). In addition to fishes, commercial fisheries on seamounts also target invertebrates such as crabs and lobsters. The high diversity, the discovery of species new to science, the presence of high endemism, the functioning of seamounts as a habitat for spawning and mating, and the extreme longevity of some organisms have motivated legal protection of some seamounts, particularly from overfishing. Researchers have proposed several hypotheses to explain the high concentrations of fishes around seamounts, including high primary production associated with upwelling and trapping of diel migrating plankton by seamount vortices. Mesopelagic organisms that migrate during the day contribute significantly to the diets of fish stocks as prey, in an otherwise food-poor environment (Figure 26.12).

To reproduce successfully, species that live on seamounts must: (1) settle and colonize their primary habitat or (2) colonize new seamounts. Most deep-sea seamount taxa reproduce by producing larvae. These larvae can remain in the plankton for weeks or months and may colonize the same seamount, or move between seamounts, swept by currents generated by the interaction between seamounts and tides, or by large-scale flows. Larvae could then settle on seamounts distant from their point of origin. Researchers hypothesized that Taylor columns can accumulate organisms by increasing residency time of the associated water

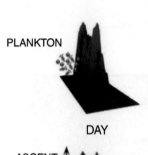
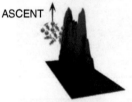

Figure 26.12 Plankton migration on a seamount provides a key food resource for seamount fish.

mass. Fluid dynamics suggests that these flows may retain fluids and particles, including larvae, for time scales of weeks, a period sufficiently long to allow some larvae to re-colonize their seamount of origin. In other seamounts, larval development time exceeds water mass residence time and adjacent habitats apparently resupply seamounts. Direct evidence points to larval retention in benthic fishes, but populations of many species differ genetically among individual seamounts, indicating limited scales of dispersal in some species. Understanding community structure requires knowledge about larval dispersal: sessile (benthic) larvae are unlikely to be swept away from a given seamount, in contrast, to a high probability that currents will transport free-swimming larvae away. To maintain a resident community, seamounts regularly supply new recruits, whether locally retained or delivered from adjacent habitats. Studies of a species of fish in the family Sternoptychidae on some seamounts shows insufficient reproducing adults to maintain the population, suggesting allochthonous recruitment. A study on dispersal modes of various larval forms of benthic species on other seamounts reported a very short planktonic stage or lack of planktonic stage in many species. This pattern suggests the presence of a recirculating current (with a cycle of about 17 d), because organisms with a longer planktonic stage would lack a mechanism to re-colonize the seamount.

26.6 Deep-Sea Hydrothermal Vents

In 1977, scientists working off the Galapagos Islands discovered a water mass with unexpectedly high temperatures at ~2,500 m depth, in an area where new seabed had recently formed. When geologists first visited the site with the submersible ALVIN, they were completely unprepared for what they discovered: an extraordinarily rich biological community dominated by giant bivalves and enormous tube worms, consisting of animals mostly new to science. Since that time, scientists have discovered similar hydrothermal vent communities in other deep locations, all in tectonically active areas (Figure 26.13). In these locations, movement of tectonic plates can create cracks in the seabed where magma superheats seawater deep in the crust; that superheated water subsequently spews out of the seabed. As it encounters ambient seawater at a temperature <4 °C, minerals precipitate, forming pinnacles of mineral-rich basalts; the chimneys emit a gray or

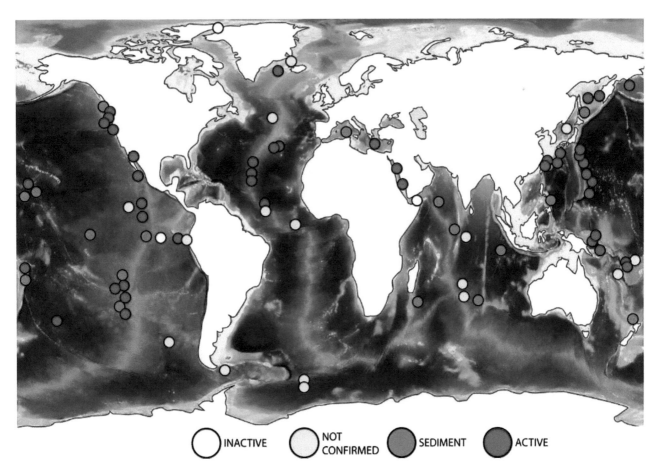

Figure 26.13 Distribution of known deep-sea hydrothermal vents.

white smoke at rates of 1–5 m s^{-1}. This smoke is produced when water and the magma come into contact. These habitats occur in highest abundance along the oceanic ridges that globally extend 60,000 km in length with intermittent volcanoes ("seafloor spreading centers"). They may also occur near isolated seamounts formed from hotspots. Mid-ocean ridges typically occur at 2,000–5,000 m, meaning completely darkness, high pressure, and ambient waters temperatures of only 2 °C.

Geologists estimate that in ten million years the entire volume of the global ocean could circulate through hydrothermal vents, heating and altering ocean chemistry in the process. Vent emissions form a sort of underwater geyser that emits hot water rich in dissolved gases and isotopes of helium, methane, and hydrogen. Hydrothermal vent "fields" consist of multiple venting locations, or **vent fields**, of 10s–100s of km^2, whereas individual vents can extend for 100s of m^2 to several hectares.

In these environments, salinity can exceed that of typical seawater by approximately one-third to two times. Hydrothermal circulation takes place on mid-ocean ridges, where cold, dense water percolates into the fractured oceanic crust near ridge crests. Geothermal energy from the magma chamber that supplies the ridge heats the vent fluid to temperatures >400 °C. This heating, in combination with interactions with the surrounding crust chemically modifies the fluid, strongly acidifying it to pH 2–3, eliminating all dissolved oxygen and accumulating high concentrations of reduced gases such as methane and hydrogen sulfide. The union of hot basalt and seawater produces hydrothermal fluid, typically in large volumes (Figure 26.14). Rapid mixing with cold, ambient seawater produces a weaker vertical than horizontal temperature gradient, and researchers have recorded temperatures of ~10 °C at 2–3 m above the vent opening.

Black chimneys (also known as pinnacles, chimneys, smokestacks, or fumaroles) that can rise hundreds of feet from the seabed eject the hot fluid, which generally appears as a dark-colored smoke (Figure 26.15). Hydrothermal fluids emitted

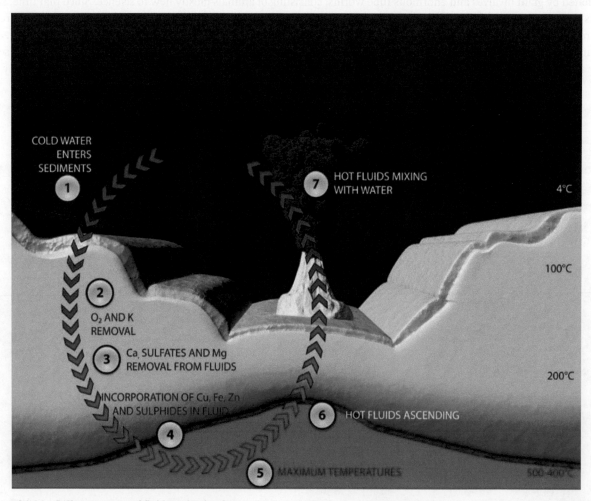

Figure 26.14 Different stages of fluid production from a deep-sea hydrothermal vent: 1–2. Cold seawater enters seafloor sediment, which removes oxygen and potassium; 3. Calcium, magnesium, and sulfates are removed from the fluid; 4. The fluid leaches copper, iron, zinc, and sulfides from the crust; 5. The fluid reaches maximum temperature; 6. Hot fluids ascend and flow from the seafloor; 7. Fluid and smoke mixes with cold seawater rich in oxygen, causing precipitation.

Figure 26.15 A hydrothermal vent chimney emitting fluid and smoke (left) and an ROV measuring and sampling (right).
Sources: NOAA / Public Domain, National Oceanic and Atmospheric Administration (NOAA).

from the chimneys emerge in the surrounding cold water, forming vortices filled with hydrothermal precipitates. The vent openings vary in size, but are typically 10 cm in diameter or less.

The vent fluids contain high concentrations of metals, particularly iron, manganese, copper and zinc, but also precious metals such as silver, gold and platinum, and highly toxic metals such as cadmium, mercury, and arsenic. As the vent fluids mix with cold deep waters, small flakes of iron sulfides rapidly precipitate, producing black smoke. This vent fluid provides the source of chemical energy necessary for the development of lush biota associated with hydrothermal vents. But this same venting means that animals living near vent openings may experience considerable variation in temperature and low oxygen concentrations (or even anoxia). These organisms may also experience a rain of inorganic precipitates, thereby exposing them to high concentrations of dissolved heavy metals and precipitates.

Immediately after the discovery of deep hydrothermal vents (also known as hot vents, because of the hot venting), researchers hypothesized that life originated in these types of environments, given the similarity between ocean conditions long ago, including high temperatures and low pH, and those at hydrothermal vents. They proposed that organic matter first appeared as water at high pressure rapidly changed in temperature from 350 to 2 °C. This organic matter began with the synthesis of peptides and other polymers that subsequently led to the evolution of living organisms. However, temperatures of 350 °C tend to hydrolyze and destroy the main organic compounds in living organisms (peptides, RNA, and DNA), pointing to a need for a more complex mechanism that perhaps began with simple protein formation at vents but development elsewhere.

The undersea ridges that host hydrothermal systems extend onto the terrestrial crust, but submarine volcanism represents >75% of total volcanic activity on Earth.

26.6.1 Biodiversity Associated with Deep-Sea Hydrothermal Vents

The discovery of hydrothermal vents and their unique fauna represents one of the most extraordinary scientific discoveries of the twentieth century. Endemic species dominate vent fauna, and ~95% of the associated animals, which currently number ~700 known species, were unknown to science prior to the discovery of these ecosystems. Hydrothermal vents support a community with high biomass but low species diversity compared to most other deep-sea environments, for major size groups, from meio- to macro- and megafauna. Some taxa rarely or never occur at vents, including cnidarians (except anemones), some echinoderm groups, sponges, xenophyophores, brachiopods, bryozoans, and fishes (with a few exceptions, including some elasmobranches that lay eggs near vents). Bivalves and vestimentiferans (siboglinids), characterized by the complete absence of a digestive tract, dominate in terms of visibility because of their large size.

In 1985, taxonomists proposed that the Vestimentifera, the large tube worms that characterize many hydrothermal vent environments, merited designation as a new phylum of animals, but subsequent molecular studies demonstrated that vestimentiferans belong to the phylum Annelida.

Other organisms associated with vents include crabs, gastropods, several unique polychaetes, benthic siphonophores, and Enteropneusta worms. In the Galapagos Rift where vents were first discovered, researchers have described extensive populations of mollusks, particularly *Bathymodiolus thermophilus* mussels, followed in abundance by Brachyura crabs

(*Bythograea thermydron*), as well as associated species of gastropods, limpets, polychaetes, and a zoarcid fish. Other vent communities include anemones, sponges, echinoderms, benthic foraminifera, and copepods. The biomass of giant bivalves that occur at some vents, such as *Turneroconcha magnifica* and *Bathymodiolus thermophilus* may exceed 10 kg m^{-2} (Figure 26.16). The symbiotic bacteria that constitute 75% of the mass of the giant bivalve gills convert energy from sulfates emitted at vents to produce organic matter, and release exudates that mussels can filter out and use as a food source.

Among the most spectacular animals, the siboglinid (vestimentiferan) worm, *Riftia pachyptila* (Figure 26.17) may be the single most emblematic vent animal because of its unusual size, abundance, and complete dependence on endosymbiosis for survival. *R. pachyptila* also provides habitat for anemones, shrimps (especially the genus *Alvinocaris* sp.), limpets (especially the genus *Neomphalus*), and mussels.

Polychaetes of the family Alvinellidae (Figure 26.18) live in vents, even at high temperatures close to where fluid flows from the outer chimney. These polychaetes build impressive biogenic structures comprised of tubes of an organic matrix, in which they live. They do not host intracellular bacteria, as observed in vestimentiferans, but rather **exo-symbiotic** bacteria that live outside the organism in close association with the dorsal epithelium. Difficulties associated with maintaining specimens in suitable condition complicate physiological investigations, but studies to date show that the alvinellid worm, *Alvinella pompeiana* can tolerate temperatures >105 °C. Researchers hypothesize that they can survive at such high temperatures by: (a) creating a physical barrier (their tube) against the hot fluid, and (b) internal convective cooling mechanisms that continuously move fluid along the length of their body, dispersing heat.

Figure 26.16 Bivalve communities, such as *Calyptogena* sp. occur in particularly high abundance in some vent habitats but generally not in the immediate vicinity of vent emissions. *Source:* NOAA / Public Domain.

Figure 26.17 Example of a siboglinid community (vestimentiferans). The external vivid red plumes of *Riftia pachyptila* result from hemoglobin in their blood. They retreat into their white tubes when water temperatures rise too much. Different species of invertebrates and vertebrates live in association with this sibloglinid worm population. *Source:* NOAA / Wikimedia Commons / Public domain.

Figure 26.18 Alvinellid polychaetes at a hydrothermal vent. The pink tufts indicate *Alvinella pompejana*. *Source:* NOAA / Wikimedia Commons / Public domain.

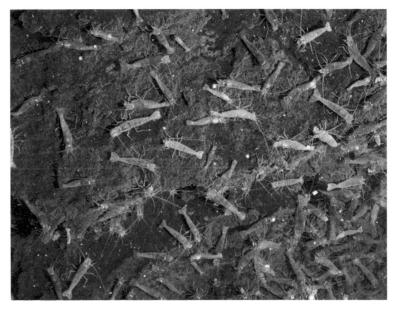

Figure 26.19 Aggregation of *Alvinocaris* sp. crustaceans around a hydrothermal vent. These crustaceans can swarm in dense concentrations at different parts of the vent. *Source:* Submarine Ring of Fire 2006 Exploration / Wikimedia Commons / Public domain.

Among crustaceans, alvinocarids occur in high abundances at vents, especially the genus *Alvinocaris* (Figure 26.19). In fact, their high abundance suggests a species highly adapted to environmental conditions at vents. *Rimicaris exoculata* illustrates adaptation of a crustacean sensory system to hydrothermal vents; this shrimp lives along the walls of vent chimneys in massive, dense aggregates of up to 3,000 indiv. l^{-1}, feeding on the symbiotic bacteria that live in the gill, on the outer shell, and on ingested particles. The visual system of *R. exoculata* has evolved into a large plate on its dorsal surface, in which photoreceptors probably cannot identify a distinct image, but can sense dim light. This sensitivity probably reflects an adaptation to respond to "light" from the openings of vent chimneys in the form of black body radiation that allows shrimp to orient themselves to chimney walls.

The "gastropod with the scaly foot" recently discovered in the central Indian ridge offers another example of adaptation. These mollusks host **thiotrophic bacteria**, which use sulfide compounds for energy, contained in a large esophageal gland covered by an operculum modified with hundreds of aligned scales covered by a thin layer of sulfur. Spaghetti worms, *Saxipendium coronatum* (class Enteropneusta) also occur in high abundance at some vents. These bacterivorous or filter-feeding organisms attach to rocks at the periphery of vent fields.

Vent macrofauna include different species such as *Thermanemertes valens* (phylum Nemertea), one of the first colonizers of vents, which lays on the rocks or on the microbial mats, anchored to the substratum by its tail while the head moves freely in the water. As with all Nemertea, these carnivores and scavengers feed on a variety of species, and on vents these include ciliates, larvae, and bacterial mats. The polychaete *Branchiplicatus cupreus* lives in *Riftia* sp. tubes and in

Figure 26.20 Some species that contribute to unique deep-sea hydrothermal vent diversity: (a) galatheid crabs and shrimp that feed on filamentous bacteria on the shells of the bivalves; (b) an anemone near the crater of the West Rota volcano; (c) sea urchins; and (d) *Bythites hollisi*, the only known fish species that lives exclusively at hydrothermal vents. *Sources:* National Oceanic and Atmospheric Administration (NOAA), NOAA Photo Library / Wikimedia Commons / CC BY 2.0, NOAA Photo Library / Flickr / CC BY 2.0, National Oceanic and Atmospheric Administration (NOAA).

Calyptogena sp., whereas the polychaete *Branchipolynoe pettibonae* lives in the mantle of vent bivalves (Figure 26.20). Among the anemones, species such as *Actinostola callosa*, reach a length of 1.5 m. Several endemic species of fishes (Figure 26.20) and crustaceans (Figure 26.20) also occur at vents. Among the meiofauna, 80–90% of free-living nematodes consist of 4–5 species living in the habitat closest to the vent. Nematode species form part of the **thiobios**, the fauna inhabiting environments with hydrogen sulfide and characterized by a high length-width ratio that enables greater cuticular uptake of oxygen and dissolved organic matter. Nematodes lack endemic vent forms, at least at the genus level.

As dynamic ecosystems, hydrothermal vents undergo significant and rapid change. Geological processes cause these dynamic changes, resulting in creation of new vents as new flows begin and "death" of other vents as flow ceases. Vent communities typically persist for a few years to decades at a precise location, as specific venting patterns change over time. Geophysical and geochemical evidence suggest short "bursts" of hydrothermal activity for periods of tens of years. On a finer time scale (daily), temperatures in vent communities can fluctuate rapidly and sometimes unpredictably, reflecting changes in near bottom flows (e.g., tides) and sub-bottom flows (vent fluid geophysics), suggesting that animals can experience significant environmental change. Variation in growth ring microstructure on bivalve shells that resemble seasonal growth bands in shallow-water bivalves attest to this change, as do short-term changes in vent biota.

26.6.2 Ecosystem Functioning at Hydrothermal Vents

Exceptionally high densities and biomass of specialized animals make hydrothermal vent systems real deep-sea "oases" rich in life. Biomasses can reach ~8.5 kg m^{-2} (wet weight) in "cool" vent habitats compared to 2–4 kg m^{-2} (wet weight) in the hottest vent habitats (200–360 °C), and 0.1 g m^{-2} (wet weight) in non-vent deep-sea habitats. The discovery of this impressive benthic biomass in deep-sea waters, far from photosynthetic production in surface waters, immediately raised the question of how animals could obtain the energy required to attain such production. Many studies since have documented a food web driven entirely by geochemical energy and therefore not dependent on solar energy. Researchers proposed two possible nutrient sources to support these rich communities: (a) transport of organic matter from surrounding waters through hydrothermal circulation, and (b) organic matter produced by bacterial chemosynthesis. Specific tests excluded the first hypothesis by showing that heat oxidizes and quickly transforms organic particles within the fluid before yielding sufficient organic matter to support a broad vent community. This observation, in combination with isotopic and other studies, clearly supported the second hypothesis, that chemoautotrophic bacterial production forms the base of the food web that supports most hydrothermal vent biota. Hydrothermal vents also support significantly elevated production in the surrounding seabed, where biomass can reach 10–50 kg m^{-2}, with physiological rates that often differ from related shallow-water species (See Box 26.1).

To date, researchers have isolated >250 forms of free-living bacteria from deep-sea hydrothermal vents. These forms include methane, hydrogen, iron, and manganese oxidizing bacteria, and vast abundances of sulfur bacteria that use material within the vent plume spilling from the chimneys. The hydrothermal plume contains the highest bacterial concentrations, including thermophilic and hyper-thermophilic bacteria that can withstand temperatures >100 °C (methanogenic bacteria have been isolated at 110 °C). Vent bacterial production rates of ~50 g C m^{-2} d^{-1} translate to almost 20 kg C m^2 y^{-1}.

Chemosynthesis indirectly relies on photosynthesis in that chemoautotrophic bacteria use oxygen dissolved in seawater to oxidize reduced organic compounds. They then use energy from this reaction to synthesize complex organic molecules using carbon dioxide (in the form of bicarbonate ions) as a carbon source. The biomass in these ecosystems compares favorably with the most productive marine ecosystems, but in this case driven by chemoautotrophic production by microbial communities that use hydrogen sulfide and other compounds emitted from vents as a source of energy they convert into microbial biomass, and eventually transform into organic matter and energy that form the base of a complex food web (Figure 26.21).

Chemosynthetic bacteria form the base of the food web and generate significant biomass that higher trophic levels can use. Primary consumers include animals that can filter bacteria from the water, others that graze bacterial films from rocks, and some that form symbiotic associations with bacteria. Animals such as limpets can graze on hard surfaces whereas suspension feeders can filter bacteria suspended in the water. As a result, clams, mussels, tubeworms, and other species reach truly impressive biomass and density.

At least 10 phyla include species that host endosymbionts, particularly siboglinid annelids (vestimentiferans) that lack an intestine and depend entirely on their endosymbionts for energy. However, some mollusks and flatworms also fully depend on their intracellular symbionts. Species from other phyla, such as protistans, sponges, annelids, arthropods, echinoderms, nematodes, and priapulids, can have both extracellular and intracellular symbionts.

Many predators feed on organisms with symbionts or filter feeders in systems with high abundances of *Riftia pachyptila*, including anemones, shrimp (which may swarm at densities up to 300 indiv. m^{-2}), crabs, and octopus. Recent manipulation experiments showed that vent predators can modify the structure and functioning of benthic communities. The high

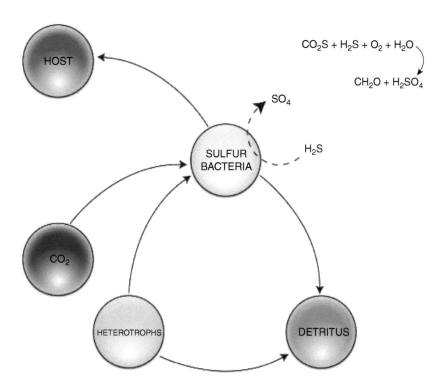

Figure 26.21 Chemosynthesis in hydrothermal vents. Sulfur bacteria use hydrogen sulfide as an energy source to build organic molecules and grow. Carbon dioxide dissolved in water provides the carbon source. Heterotrophic organisms feed directly on bacteria that live in symbiosis with larger organisms and provide organic material, or they die and become part of the particulate organic material (detritus), which heterotrophs, in turn, may use as food.

concentrations of H_2S at both vents and seeps beg the question of how organisms manage to avoid H_2S poisoning. In some organisms (bivalves and vestimentiferans at seeps, see example below) different parts of the body separate sulfide and oxygen and transfer it to the bacteria. In other organisms (such as vent mussels and vestimentiferans, for example), they absorb both molecules through respiratory surfaces. Despite the high toxicity of sulfides for most animals, mussels in these environments convert H_2S to thiosulfate prior to "transporting" it to symbionts. Many other groups of organisms bind H_2S to specific proteins for transport, thus reducing toxicity. However, not all endosymbionts use reduced sulfur compounds; some use reduced carbon compounds, and methane in particular.

Two species that inhabit vents, *Turneroconcha magnifica* and *Riftia pachyptila*, have developed detoxification systems that allow them to live in hydrothermal environments despite potentially lethal concentrations of sulfides. *T. magnifica* produces a protein that binds the sulfides ("**sulfide-binding protein**"), protecting hemoglobin in the erythrocytes and the cytochrome-c oxidase system in cells from toxic effects of sulfides. Some species use a sulfide oxidizing enzyme system in the outer layer of muscle tissue.

Vestimentiferans localize symbiotic bacteria exclusively in their trophosome, a highly vascularized organ situated between two coelomic cavities containing a hemoglobin-rich fluid. *Riftia pachyptila* controls toxicity of sulfide molecules not only by binding them and preventing their accumulation in large amounts in hemoglobin and cells, but also through oxidizing the molecules. In this species, oxidation of sulfides protects superficial tissues of animals from sulfur poisoning. In *Riftia pachyptila*, hemoglobin has a high affinity for oxygen, allowing it to capture oxygen at low concentrations, including hypoxia, accumulate, and use it during short periods of anoxia. However, their hemoglobin also transports sulfur with a similar affinity as for oxygen, allowing them to simultaneously transport both molecules inside the trophosome without reacting within the blood (Figure 26.22). The high affinity of these hemoglobin cells for sulfur coupled with its high abundances within the vascular liquid allow bound sulfur to reach levels one or two orders higher than ambient sulfide concentration, thereby keeping free sulfur concentrations in their blood at a much lower level than in the ambient environment.

Strong spatial gradients characterize vents emissions, whose intensity decreases exponentially moving from the chimneys to the periphery. For this reason, the communities that depend on the emissions also vary along this gradient (Figure 26.23). Closer to the vent flow, species that host autotrophic bacteria such as vestimentiferans (or mussels or bivalves) colonize the

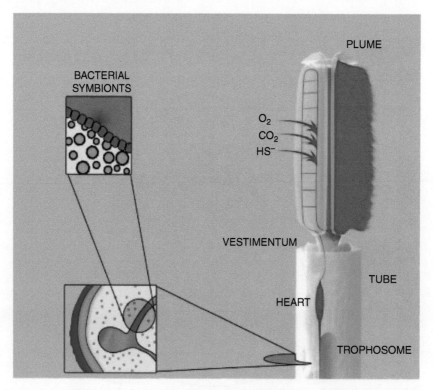

Figure 26.22 Example of the symbiotic association in the vestimentiferan, *Riftia pachyptila*. Right side shows the whole organism with the richly vascularized apical plume and collar (vestimentum) connected to the rest of the animal (trophosome) filled with bacteria and lacking a digestive tract. A protective tube that covers the animal attaches to the bottom with an opisthosoma. The left image shows the vascularized plume, where the animal exchanges oxygen, sulfur, and carbon dioxide with the ambient environment to supply the symbiotic bacteria.

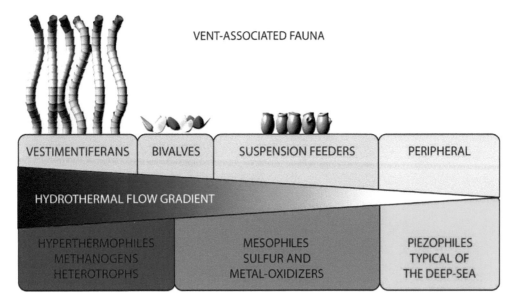

Figure 26.23 Example of zonation along a gradient of geothermal emissions from a deep-sea hydrothermal vent. Siboglinids (vestimentiferans) occur adjacent to vent emission, followed by a zone of bivalves with chemoautotrophic symbiotic bacteria that transitions into a zone dominated by filter feeders (suspension feeders). At the periphery, communities gradually become indistinguishable from the ambient deep-sea soft bottom community. Lower part of image shows parallel zonation in microbial components, with domination by methanogenic and hyper-thermophilic prokaryotes (in addition to heterotrophic prokaryotes) closest to the hot vent emissions, transitioning to mesophilic prokaryotes (40–50 °C) that oxidize sulfur and metals. Finally, towards the periphery, prokaryotes adapted to low temperatures and high pressures typical of deep-sea environments, dominate.

vent. Vestimentiferan abundances decrease when concentrations of sulfides decrease, whereas bivalves with mixotrophic abilities increase. Finally, farther from the chimney, the dependence on bacteria declines and organisms that feed by filtration dominate. If a large number of heterotrophic organisms dominates these communities, the vent is probably dying.

The life history strategies of vent organisms favor large individuals to maximize fertility. In addition, the short lifetime of individual vents requires that species can colonize at substantial distance (>100 km) to reach other vent locations suitable for the development of new communities. Most species associated with vents have lecithotrophic larvae with poor dispersal ability. Vent species include opportunistic forms (i.e., r strategists) with planktotrophic or teleplanic larvae (mussels, decapod crustaceans) that have great dispersal potential. Lecithotrophic larvae increase dispersal success by using intermediate steps (stepwise dispersal) (see Box 26.2 on large whale carcasses). This hypothesis suggests that when a given vent begins to die, meroplanktonic lecithotrophic larvae of resident organisms disperse elsewhere, perhaps using decomposing carcasses as waypoints on their journey. This scenario offers a plausible explanation for the presence of vents many kilometers from the nearest similar habitat. Researchers calculate that ~500 gray whale carcasses arrive on the seafloor each year within an area of 8×10^5 km^2 (=1 indiv. 1,600 km^{-2}). They calculated the distance between carcasses at the same stage of succession at 5–40 km (varying depending on stage of decomposition).

Box 26.1 Coastal Hydrothermal Vents

In coastal vents, unlike deep hydrothermal vents, abundant sunlight means that micro-phytobenthic and micro-phytoplankton components may play an important role. In immediate proximity to vents, at shallow depth, sediment temperature reaches ~130 °C and water can boil, noting that high pressure in deep-sea hydrothermal vents keeps water from actually boiling, even at 300–400 °C. Researchers have described shallow hydrothermal vents from the western Pacific, Indonesia (Figure 26.24), along the California coast, Iceland, Azores and Mediterranean Sea. Biogeochemical processes in shallow hydrothermal vents differ considerably from those in deep-sea environments, not only because of the dual contributions of photosynthetic and chemosynthetic primary production but also because coastal vents essentially emit water, heat, and gas comprised primarily by CO_2, but also CH_4, H_2S, and H_2. Fluids influence an area of

(Continued)

Box 26.1 (Continued)

Figure 26.24 Coastal vent just a few m deep in an equatorial environment (North Sulawesi, Indonesia). Left, hot fluid up to 90 °C flows from a vent opening 30–40 cm in diameter. Also note a small pomacentrid fish passing quickly through the boiling flow, likely to shed itself of parasites. Right, sampling vent sediments with test tubes.

a few m^2, but trace effects may extend up to 1 km away. The carbon dioxide originates from geological processes, from acid degradation of carbonate sediments accelerated by fluid acidity and high temperatures. Most of this gas diffuses into the atmosphere, but its upward movement through the water column generates local microcirculation patterns with important implications for migrating organisms. H_2S commonly occurs in volcanic emissions and the emitted sulfur forms precipitates in flocs creating actual sulfurous brine lakes. Waves remix the precipitated matter formed by stratified plumes in the water column that form concentric communities of animals around the vent opening.

Deep-sea vent communities consist primarily of animals with symbionts that form well-defined spatial patterns, which differentiate them from shallow vents that can accommodate bioconstructors such as coralline algae and scleractinians, forming communities that largely resemble those in surrounding habitats. In warm coastal vents, high concentrations of iron, sulfur, and zinc offer a great variety of habitats and micro-niches for microbes, which can form three different types of bacterial mats that can grow to thicknesses of 30 cm: (1) diatom mats, (2) mixtures of algae and prokaryotes, and (3) prokaryote mats. Diatom mats grow in the absence of hydrogen sulfide emissions, dominated by the genera *Melosira* and *Thalassiosira*. Mixed mats that include diatoms associated with microorganisms form at temperatures between 55 and 60 °C, able to reduce or oxidize sulfur compounds. Prokaryote mats can be white or various colors, comprised of iron-sulfur-oxidizing bacteria and photosynthetic prokaryotes including the genera *Thiobacillus*, *Thiomicrospira*, and *Thiosphaera* or filamentous sulfur bacteria such as *Thiothrix* spp. and *Beggiatoa* spp. Prokaryotic abundances in sediments decline with increasing distance from hydrothermal vent openings. Archaea may represent some 40% of total prokaryote abundance within hydrothermally active sediments. Bacterial sulfate reducers (BSR) in **calderas** (collapsed volcanoes) may represent ~90% of total bacterial abundance, declining in percentage with increasing distance from the vent.

In contrast to obligate microbial communities at deep-sea hydrothermal vents that can only live near vent openings, obligate species rarely occur in coastal vents. Here, blooms of autotrophic microplankton dominated by diatoms and phytoflagellates can occur and photosynthetic rates exceed those in non-vent areas. Zooplankton abundance and biomass is higher near vents compared to adjacent coastal environments, and communities differ in species composition. The species that dominate coastal vents resemble those found in unstable environments, including taxa such as capitellid polychaetes, abundant scavengers, and a non-dominant bacterivorous fauna. Scavengers benefit from the rich food supply provided by organisms killed or weakened by the extreme environmental conditions. Nematodes generally dominate coastal hydrothermal vent environments, whereas copepods can dominate where coarse grained sediments occur. Size and abundance of organisms generally decrease close to coastal vents, accompanied by increased abundances and larger individuals of a single species of nematode that hosts symbiotic bacteria.

26.7 Whale Carcasses

The great whales represent the largest forms of life on Earth with a biomass on the order of tens of metric tons; individuals of eight different species of cetacean reach a biomass ranging 30–170 metric tons. A whale carcass that sinks to the bottom therefore represents a massive food input to the seafloor, enabling the development of dense animal communities. In 1989, the first discovery of a chemoautotrophic community on a whale carcass catalyzed a series of studies of these novel ecosystems. Although their initial discovery suggested that these habitats resembled hydrothermal vent fauna, further studies revealed the unique nature of whale carcass communities, and high species diversity and endemism. They also share many species with chemoautotrophic ecosystems, such as cold seeps and, to some degree, with hydrothermal vents. This means that whale carcasses could facilitate the establishment and spread of species that live in chemoautotrophic systems, supporting the **stepping stone theory** (see Box 26.3). The arrival of a whale carcass on the seafloor transforms the surrounding environment, altering spatial patterns, composition, and diversity, and greatly increasing the amount of available organic matter. A series of successional stages of different colonizers occurs with each carcass deposition. The amount of organic carbon supplied to the system by a carcass of 40 metric tons roughly equals the amount of organic carbon normally delivered to one hectare of abyssal seabed over 100–200 y. Sediments below the carcass (~50 m^2) quickly becomes anoxic. The succession of species that colonizes whale carcasses (Figure 26.25) helps in understanding how deep-sea environments respond to episodic enrichment whether of natural or anthropogenic origin.

Whale carcasses passes through four main successional stages characterized by dramatic shifts in dominant species and trophic structure: (1) mobile scavenger stage; (2) enrichment stage with opportunistic species; (3) sulfophilic stage; 4) reef stage. During the first stage (0.5–1.5 mo after carcass arrival) scavenging organisms such as hagfish (mainly *Eptatretus deani* and *Myxine circifrons*), amphipods and crabs, and especially sharks (*Somnionus pacificus*) up to 3.5 m long attack the carcass soft tissues. Over time, smaller scavengers arrive to feed on the carcass (sharks first, followed by amphipods and, eventually, copepods).

During the second stage (4–24 months depending on carcass size) scavengers remove >90% of the whale soft tissue at a rate of 40–60 kg d^{-1}, and only a few megafaunal scavengers remain. Dense communities of heterotrophic macrobenthos colonize sediments and the organic-rich bones. Errant polychaetes (*Vigtorniella* spp., Dorvilleidae), small white gastropods, and bivalves colonize sediments 1–3 m from the carcass at densities of 20,000–45,000 indiv m^{-2} (the highest densities ever reported for macrofauna below 1,000 m). "Fresh" bones of great whales may contain 60% lipids (~5–8% of total body mass), supporting anaerobic microbial decomposition. Sulfur reduction leads to hydrogen sulfide leakage from the bones, which some species can tolerate and use for chemoautotrophic production.

During the third stage (sulfophilic stage, which lasts 5–50 y) a diverse community of both heterotrophic and chemoautotrophic bacteria grows on and within the bones. Large populations (>10,000 indiv.) of mussels, *Idas washingtonius*, with chemoautotrophic endosymbionts, isopods (*Ilyarachna profunda*), galatheid crabs, dorvilleid worms, limpets, and gastropods develop. vesicomyid and lucinid bivalve occur with in sediments adjacent to the bones and, occasionally, vestimentiferans as well.

The fourth, or reef, stage describes the presence of only bones on the seabed. Chemoautotrophic production stops because the biota have completely consumed all of the organic matter' leaving suspension feeders as the dominant biota. The bones

Figure 26.25 Recently arrived gray whale carcass at the seafloor with hagfish as early colonizers (left). Deep-sea sharks also arrive early to attack the carcass (center). Hagfish also occur at the last successional stage when mostly just bones remain (right).
Source: Craig R. Smith, N. King, Oceanlab, University of Aberdeen, UK, Craig Smith and Mike DeGruy.

 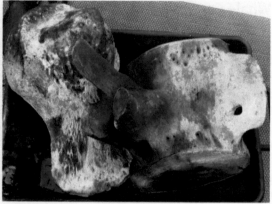

Figure 26.26 A 17-m long carcass in the Mediterranean Sea: whale vertebrae 8 months after sinking (left); vertebrae sampled for faunal analysis 14 months after sinking (right).

provide a hard substratum in bathyal and abyssal environments on which species such as *Idas washingtonius, Ilyarachna profunda, Pyropelta craigsmithi, Pyropelta corymba, P. musaica*) can settle and grow. Many of the abundant species associated with whale skeletons rarely or never occur in other habitats.

Ecologists have proposed three models to explain successional sequences: (a) facilitation, (b) tolerance, and (c) inhibition. In the **facilitation model**, the first species to arrive at the carcass modify the habitat to facilitate establishment of subsequent species. The **tolerance model** implies that strong competitors follow and eventually outcompete the initial colonizing species. The **inhibition model** suggests that, over time, species that prevent other species from using the resource become established and eventually dominate. Opportunistic or "r" strategists characterize early stages, whereas specialists or "K" strategists arrive later, exploiting scant resources remaining from previous stages.

In November 1987, during a mission on the submersible Alvin to study the seabed near Santa Catalina, California, Craig Smith, a marine biologist at the University of Hawaii, discovered a blue whale skeleton 21 m long at 1,240 m depth. This discovery initiated the study of **whale fall** or **whale carcass ecosystems**. Since this first discovery, scientists have found other carcasses and deliberately sunk others (obtained from stranding events) for scientific purposes, to allow researchers to examine succession in these ecosystems.

These studies show that large cetacean carcasses (Figure 26.26) can also promote microbial colonization and taxonomic diversity of benthic microbial communities (dominated by Archaea) of adjacent marine habitats.

26.7.1 Whale Carcass Biodiversity

Whale carcasses seem to number among the most diverse deep-sea ecosystems per unit area (Figure 26.27). The low species richness of the scavenger and opportunist stages transitions into the high diversity sulfophilic stage, with more than 190 macrofaunal species per carcass. To date, researchers have documented ~407 total species associated with a whale carcass, in contrast to 469 and 230 species, respectively for the best-studied hydrothermal vents and oil seeps. Of the different phyla in whale carcasses, annelids (polychaetes) dominate, accounting for 47–60% of total species. Abundant communities of vesicomyid bivalves, mussels, and gastropods supported by sulfur oxidizing endosymbionts apparently use the sulfate derived from anaerobic decomposition of lipids concentrated in the bones.

26.7.2 Functioning of Whale Carcass Systems

Chemosynthetic processes support primary production in whale carcasses. The ability to remove and use lipids within the bones, the exploitation of the sulfur by chemoautotrophic symbiotic and free-living bacteria, and the presence of extensive microbial mats associated with organic enrichment support these ecosystems, extending well beyond simple use of carcass tissues by scavengers. Energy sources and trophic structure change depending on the stage of the carcass, from: (1) soft tissue and lipid in the carcass bones; (2) hetero- and chemoautotrophic free-living bacteria; (3) chemoautotrophic endosymbiont bacteria; (4) predation on primary consumers; (5) debris deposited in sediments near the carcass. This variety of resources supports high species diversity that includes highly specialized species of predators, deposit feeders, bacterivores, filter feeders, organisms with chemoautotrophic endosymbionts, and organisms specialized in utilizing lipid from whale bones (Box 26.2 and 26.3).

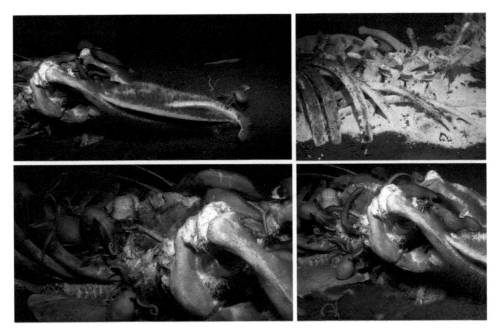

Figure 26.27 Example of fauna associated with whale carcasses: a group of octopuses feed on the bones of a whale (top left); later, they are joined by fish and other deep-sea animals (bottom left and bottom right); the bones of the whale become a fertile ground for the growth of deep-sea microorganisms (top right). *Source:* NOAA / Public Domain, NOAA / Craig Smith / Public Domain.

Box 26.2 Whale Carcasses: Bone-Eater Worms

Bone-eater worms (polychaetes) lack a mouth and digestive system and use special appendages full of bacteria to feed on whale carcasses. Researchers infer that these worms use a feeding mode that differs from any other known strategy, and their discovery highlights the extraordinary diversity of deep-sea life. The worms (Figure 26.28), discovered in the Pacific Ocean on whale skeletons at nearly 3,000 m depth, have developed a symbiotic relation with rod-shaped bacteria (order Oceanospirillales) that can convert oils and fats contained in whale bones into sugars that the worms use as a food source. *Osedax* sp., a new genus, means "eaters of bones" in Latin. These worms have the ability to feed on whale skeletons, showing that a dead whale can provide a significant energy source for nutrient-poor regions of the ocean. This symbiosis illustrates a particular adaptation to an "extreme" and anomalous environment that links whales that live largely in the upper ocean and biodiversity of deep-sea benthos. For this reason, the overharvesting and (in some cases) extinction of large cetaceans may have important repercussions for deep-sea biota.

Figure 26.28 *Osedax sp.* (bone-eater worms) colonizing the bones of a whale carcass. *Source:* Robert C / Wikimedia Commons / CC BY 2.0.

> **Box 26.3 Whale Carcasses As "Stepping Stones"?**
>
> The stepping stone ecology theory suggests that whale carcasses allow hydrothermal vent species to persist. Vent fields often occur far from one other, often spanning distances >1,000 km. This sparse distribution raises the question of how larvae of vent species manage to "move" between vents; whale carcasses therefore offer potential stepping-stones between vent fields where they can settle, mature, and reproduce, increasing their ecological importance globally. The reduction of large cetacean populations by commercial whaling has potentially impacted hydrothermal vent fauna, perhaps increasing the likelihood of extinction of organisms that depend on whale carcasses for dispersal.

26.8 Affinities Between Vent and Seep Communities

The idea that skeletons of whales provide important stepping stones for vent and seep faunal dispersal has generated some controversy. A review of chemosynthetic communities reported that vents and whale carcasses share 11 known species of macro/megafauna, with many also occurring in soft sediments near vent sites; whale carcasses and cold seeps share 20 known species (Box 26.4). Therefore, all three habitat types share only a small percentage (2–10%) of species and each habitat distinctly differs from the other. Presumably future sampling of carcasses, vents, and seeps will increase the known species shared among these habitats. The limpet genus *Pyropelta*, for example, includes two species that specialize on whale carcasses but also occur at vents. Two genera of gastropods known from vents also occur in whale carcass habitats.

> **Box 26.4 Cold-Seep Biodiversity Associated with Sunken Wood**
>
> Branches and trunks from trees along with wood from other sources can sink to the seafloor, even to considerable depth. Rivers transport much of this wood to the ocean including locations far from the coast, and along the way they become saturated with water and sink to the seabed. Wood hosts a relatively depauperate fauna that includes gastropods, bivalve and Polyplacophora mollusks, decapod crustaceans, polychaetes, and echinoderms; Figure 26.29) with close taxonomic affinities to hydrothermal vent, cold seep, and whale carcass faunas. Animals use sunken wood as a source of food or shelter. The gradual decomposition of the organic portion releases small amounts of methane, which some colonizers can use.
>
>
>
> **Figure 26.29** Colonization of sunken wood by mollusks such as *Idas washingtonia* and *Cocculina craigsmithi*. *Source:* Craig R. Smith.

26.9 Anoxic Basins

Anoxic environments areas can also occur naturally. Examples include the deep Black Sea (Figure 26.30), the Cariaco Basin in Venezuela, and Saanich Inlet off Canada's Vancouver Island. One of the most extensive anoxic zones occurs in the deep basin of the Black Sea (>150 m depth). The Black Sea has a **negative water balance**, meaning less input than output: the 200 km^3 y^{-1} of incoming water from the Mediterranean Sea does not balance the annual output of ~320 km^3. The limited

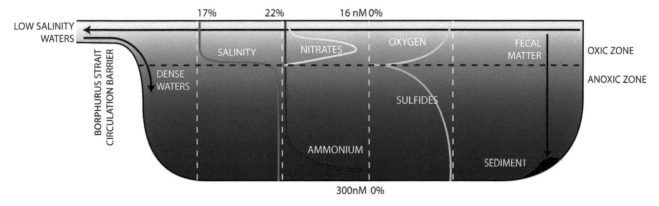

Figure 26.30 Cross section of the Black Sea basin illustrates water exchange between the Black and Mediterranean Seas. Dense waters enter the Black Sea through the Bosphorus Strait (left) toward the bottom, whereas low salinity and low density water exits at the surface. Oxygen concentrations within the water column decline drastically with depth. In contrast to most systems with readily available ammonium ion in surface layers with relatively constant organic nitrogen concentrations, anammox prokaryotes rapidly oxidize ammonium ions to nitrate.

exchange of water between the Mediterranean and Black Seas through the Bosphorus Strait results in mixing only above 150 m depth. Significant contribution of freshwater (and thus nutrients) from the Danube, Don and other smaller rivers contribute to the development of anoxic conditions. Oxygenated waters occur above the pycnocline but below that, a transition zone called the SubOxic Layer (SOL) sits above the 90% of Black Sea waters that experience permanent anoxic conditions.

Water inputs from the Mediterranean Sea and freshwater inputs from the rivers create lateral flow resulting in a dynamic rather than stationary anoxic layer. Until the 1980s, researchers considered the deep Black Sea to be devoid of life other than microbes, given its anaerobic nature. A "rain of corpses" of planktonic organisms characterizes the sediments in the anaerobic zone. The Black Sea oxic zone supports prolific benthic communities of polychaetes, crustaceans, echinoderms, and mollusks, as well as bacteria and Archaea. These communities have high abundances but few species. Benthic animals form three zones: (1) shallow depths to 120–150 m, where macrofauna, meiofauna and microbenthos proliferate; (2) between 120 and 150 and 250–300 m, with only meiofauna and microbenthos; (3) >250–300 m, where only the microbial component survives. However, researchers working in the northwestern region of the Black Sea at a depth of 645 m recently discovered nematodes (*Desmoscolex* sp., *Tricoma* sp., *Cobbionema* sp.) and tanaids (about 670 individuals m^{-2}). Further studies to 2,250 m depth documented the presence of typical continental slope species, including some unknowns. Moreover, they confirmed the presence of **anammox prokaryotes**, microorganisms capable of oxidizing ammonium under anoxic conditions and liberating gaseous N_2 into the atmosphere. This process uses O_2 bound to other molecules.

26.9.1 Hypersaline Anoxic Systems

Deep hypersaline anoxic basins (DHABs) represent deep extreme environments characterized by a complete lack of oxygen, extremely high salinity (with values up to 30%, "brine" concentrations may be 10x higher than seawater). Often, high concentrations of methane and hydrogen sulfide occur in these anoxic basins, now known from the Eastern Mediterranean Sea, the Gulf of Mexico, and the Red Sea (Figure 26.31a). The six anoxic basins surveyed to date in the Mediterranean Sea (Bannock, Urania, Discovery, Atalante, Tyro and La Medee; Figure 26.31b) occur at depths of 3,200 and 3,600 m.

The origin of Mediterranean anoxic basins dates back to interactions between tectonic processes, fluid expulsion, and dissolution of Messinian evaporitic rocks. These events took place 5.9–5.3 million years ago, probably forming brines through dissolution of **evaporites**, or water-soluble minerals produced through concentration and crystallization of an aqueous solution. These evaporates accumulated and became entrapped in seabed depressions.

Brines create a stable and somewhat isolated habitat because of seabed topography, the high density of brine, and its unusual chemistry. These factors prevent mixing with seawater or salt diffusion. Thus, the sharp halocline keeps the well-oxygenated water near the DHABs from mixing with brine water. The combination of salt concentrations close to saturation, high hydrostatic pressure, absence of light, anoxia, and the sharp halocline place these basins among the most extreme habitats on Earth. Despite similar ionic composition in the brines of L'Atalante, Bannock and Urania Basins, the salinity in Urania Basin is much lower than the others.

High concentrations of magnesium (Mg^{2+}) and low concentrations of sodium (Na^+) in Discovery Basin brine make it unique, and perhaps the single most extreme marine environment. Recent studies demonstrate metabolically active bacteria and archaea populating these basins. Bacteria dominate the Discovery, L'Atalante and Bannock Basins, whereas Archaea dominate the Urania Basin. Bacterial diversity always exceeds that of Archaea, with dominant bacterial groups including

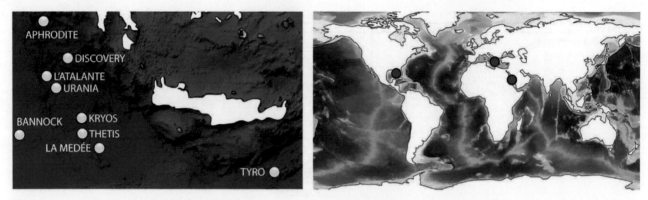

Figure 26.31 Global locations with permanently anoxic deep basins (left), and distribution of the main deep hypersaline anoxic basins discovered to date in the Mediterranean (right).

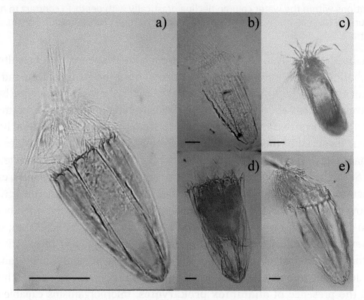

Figure 26.32 Light microscope images of the Loricifera discovered in the deep hypersaline anoxic L'Atalante basin: (a) an undescribed *Spinoloricus* sp. stained with Rose Bengal showing the presence of an oocyte; (b) another specimen (not stained) of *Spinoloricus* sp.; (c) an undescribed species of *Pliciloricus* sp.; (d) an undescribed species of *Spinoloricus* sp, stained with Rose Bengal showing an oocyte; (e) molting with exuvia of an undescribed species of *Spinoloricus* sp. Scale bars indicate 50 microns.

Gamma-, Delta- and Epsilon- Proteobacteria, Sphingobacteria and Halobacteria. Many of the dominant Archaea sequences (akin to "species") belong to new taxonomic groups absent from seawater adjacent to the basins. These differences likely relate to specific geochemical conditions within the different basins, and their physical separation from oxygenated deep waters. This isolation may have led to the development of specific microbial communities within each anoxic basin. Analysis of prokaryotic diversity within the brine column of the Bannock, L'Atalante, and Urania Basins documented the existence of many prokaryotic phylogenetic taxa that collectively form a unique and highly stratified community. In addition to prokaryotes, these basins host high abundances of viruses that represent the equivalent of major predators of this microbially dominated system. However, recent research in L'Atalante Basin demonstrated that some metazoans can live in anoxic conditions (Figure 26.32). Three newly discovered species within the phylum Loricifera (*Spinoloricus* nov sp., *Rugiloricus* nov sp. and *Pliciloricus* nov sp.) live exclusively in the sediments of this basin. Different techniques demonstrated metabolic activity and adaptations in these organisms specific to the extreme conditions in the basin, such as the absence of mitochondria and a large number of organelles similar to the hydrogenosomes associated with prokaryotic endosymbionts.

26.10 Ocean Trenches

Subduction of oceanic crust results in destruction of seabed and creation of the ocean's deepest habitats, **ocean trenches**, which typically extend from 6,000 to 10,000 m depth (Figure 26.33). The best-known examples of trenches, the Mariana and Philippines Trenches, extend to >10,500 m depth. The deepest trench in the South Pacific Ocean, Atacama Trench,

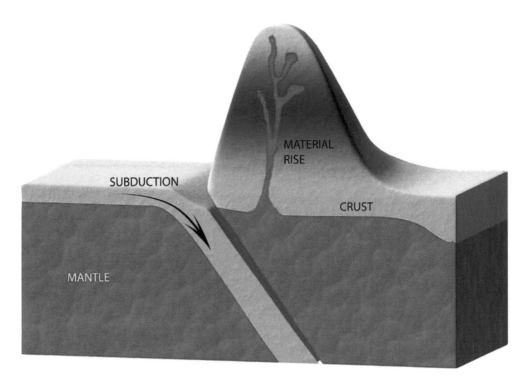

Figure 26.33 Process by which oceanic trenches form.

extends to ~8,000 m. Depending on their distance from the continents, trench characteristics differ: a relatively depauperate fauna characterizes trenches far from continents, in contrast to a richer and more productive fauna close to continents. Although many people imagine trenches as biological deserts, these environments can trap organic matter exported from the continental shelf and thus support surprisingly high densities and biomass of organisms. Trenches located below productive surface waters may support abundances and biomass up to three orders of magnitude greater than those in oligotrophic areas of similar depth. Ocean trench biota generally exhibit reduced size (dwarfism) relative to similar taxa at shallower depths. Unfortunately, we lack sufficient information to define general rules on how food availability limits meiobenthic size and abundance at hadal depths. In any case, analysis of comparatively rich environments such as the Atacama Trench identify pressure rather than food limitation as the most important driver of trench community composition. Mediterranean trenches do not reach hadal depths but exceed 3,000 m depth in the Eastern basin near Crete, where recent studies indicate highly abundant and active microbial communities.

The study of life in the deep ocean trenches essentially began <5 decades ago. Before 1948, researchers largely discounted the possibility that life could exist at such depths. The development of suitable tools that extended traditional sampling methods for deep organisms, and adoption of new study methodologies have led to great progress in the last 25 years. In 1950, the Danish *Galathea* expedition undertook the first biological study of five deep trenches. American and Russian expeditions in the last decade have led to significant developments in the study of trench biota, leading to descriptions of numerous new species and genera and, in some cases, new families. This work has increased the number of known species of multicellular animals from depths >6,000 m to >700.

The term "ocean trench" does not specifically refer to a range of depths, but rather to geomorphological formations and can therefore include trenches shallower than 6 km. However, "ultra-abyssal" or "hadal" zones refer specifically to depths >6,000 m.

A relatively small total area exceeds 6,000 m compared to the total area occupied by the abyss (3–6 km in depth): in fact, trenches represent about 2% of total ocean area in contrast to the 76% covered by abyssal habitat. Currently, of the 37 known deep-sea trenches globally, most (28) occur in the Pacific Ocean (which hosts the 9 deepest trenches), and the remainder in the Atlantic and Indian Ocean (5 and 4 respectively). Hydrostatic pressure differentiates ultra-abyssal trenches from shallow trenches, reaching levels of 600–1,100 atmospheres at depths of 6,000–11,000 m. For organisms inhabiting these environments, which collectively account for almost 99% of total ocean water mass, this pressure exceeds normal conditions for most ocean life. In addition to pressure effects, which can limit the qualitative diversity of ultra-abyssal populations, other environmental conditions may favor those organisms able to tolerate extremely high pressures. Other than pressure, environmental conditions in deep-sea trenches generally resemble those at abyssal depths. Nonetheless, additional characteristics and some unique properties exclude many animals from trenches (especially at the greatest depths), whereas other aspects favor some organisms relative to other ocean environments.

26 Deep Ocean Basins

Temperature remains exceptionally stable at depths >6 km. Available data show that temperatures at depths 6–11 km vary among different trenches from −0.27 to +4.49 °C. Excluding the trenches, with the highest (Banda and Cayman) and lowest (the sub-Antarctic South Sandwich trench) values, the temperature range for all other trenches differs by only 2 °C. The temperature within a single trench varies much less, and from 6–11 km depth, variation within a trench never exceeds 0.9 °C. At depths >6 km temperature usually increases with increasing depth as a direct result of increased pressure known as an **adiabatic temperature increase**. Water temperature below 6 km slightly exceeds that of abyssal depths for a given oceanic region. Salinity at depths of 6–10 km does not differ from typical ocean values, with most trenches at 34.7 ± 0.2. Dissolved oxygen content varies greatly among different trenches and with season and depth range considered, even within a given trench (4.0–6.9 mg l^{-1} or ~55–70% saturation). The lowest recorded oxygen concentration recorded (2.0–2.4 mg l^{-1}, ~30% saturation) was in Banda Trench, but, even this saturation level appears sufficient to support an abundant and diverse fauna.

26.10.1 Hadal Biodiversity

Overall, hadal faunal diversity is at least 3–4x lower than that of abyssal fauna. Researchers have documented 37 classes of animals spanning ~210 families and 720 different morphotypes that represent different species, but many of which have not yet been described (Figure 26.34). These 720 taxa include 660 benthic and 58 pelagic species (Hydromedusa, Calanoida,

Figure 26.34 Examples of abyssal and hadal megafauna and hyperbenthos including; (a-b) decapod crustaceans; (c-e) macrourids; (f-g) holothurians; (h) Portoguese shark.

Ostracoda, Gammarida, for example). At ultra-abyssal depths (in almost all trenches), holothurians (sea cucumbers) play a dominant role, usually followed by bivalves. Other important groups include polychaetes, sea anemones, isopods, and amphipods, with minor roles for other animal groups. More obvious differences exist among trench faunas at the species level and, to a lesser extent, at the genus level. The entire benthic fauna at depths >6 km exhibits 56% endemism, compared to ~41% endemism in the pelagic fauna. Endemism for metazoans overall usually increases with depth. The transition zone between abyssal and hadal depths usually occurs at 6,000–7,000 m, characterized by minimal levels of endemism compared with greater depths. Endemism at 6,000–6,500 m averages ~26%. The highest percentage of endemic species (86–100%) occurs in the deepest trenches (Tonga, Mariana, and Philippines). For all other trenches endemic percentages vary from 37–81%. Researchers have documented just two classes of sponges, Hyalospongiae and Demospongiae at depths >6,000 m, represented almost exclusively by the order Cornacuspongidae. At greater depths (8,950–9,020 m) several new small sponges remain unidentified. The spread of sponges within trenches apparently relates not to depth but rather to the presence of substrate micro-relief that provides an ecological niche suitable for the existence of these organisms.

Researchers have documented three cases of high abundances of sponges at >6,000 m, dominated by *Hyalonema apertum*. Several studies of Pacific Ocean trenches documented hydromedusa polyps, and researchers have found scyphopolyps produced by sponges of the genus *Stephanoscyphus* (class Scyphozoa) in many trenches to 10,000 m depth. Within the class Anthozoa, soft corals, gorgonians, and pennatulaceans also occur in trenches, with representatives of the genera *Khophobelemnon* and *Umbellula* (Pennatula) from depths >6,000 m. Within deep-sea trenches, as well as in abyssal environments, diverse polychaetes are among the most common and abundant groups of benthic invertebrates. On average, among the macrofauna only holothurians and bivalves occur in greater number and biomass, and polychaetes nonetheless dominate species number. Researchers have documented polychaetes in the Philippines, Mariana, and Tonga Trenches at depths to 10,730 m.

Many different taxa have been reported at depths >6000 m including Copepoda, Cirripeda and Ostracoda, but also Mysidacea, Cumacea, Tanaidacea, Isopoda, Amphipoda and Decapoda. These taxa include many new species of both pelagic (calanoid) and benthic (harpacticoid) copepods. Benthic ostracods have been sampled as deep as 7,950–8,100 m in the Puerto Rico Trench and benthic hadal mysids, and cumaceans to ~8,000 m. Tanaids occur in high diversity at depths >6,000 m, with 53 documented species spanning 15 genera and 9 families. Isopods are particularly characteristic of hadal depths. With 122 known species, the numbers of isopod species below 6,000 m exceeds that of any other crustacean group or any other class of multicellular animals. Impressive abundances of two genera, *Eurycope* and *Storthyngura*, occur even at depths >9,000 m. Some 63% of isopods at hadal depths are endemic. Amphipods also characterize trenches to depths of 10,500 m, with 20 known pelagic species (suborders Hyperiidea and Gammaridea) and 36 known benthic species (11% endemic). Below 10,000 m the only known amphipod species, *Hirondellea gigas*, occurs in high abundances. Few decapods occur at depths >5,000 m.

Gastropod mollusks occur in all trenches to depths >10,150 m, with 68% endemism. Scaphopod gastropod mollusks occur at depths >6,000 m, noting the greatest documented depth for this class is 7,650 m. High biodiversity characterizes bivalve mollusks in deep-sea trenches but, unlike the gastropods, they often occur in high abundances below 6000 m, second only to holothurians, with 47 known species (especially the family Ledellidae with 19 species) and 68% endemism. Only a few cephalopods live within trenches at depths greater than 6,000 m. Several sipunculids have been reported at depths to 7,000 m (*Nephasoma schuttei*). Crinoids have been reported at depths < 9,735 m with 8 endemic species (no species in trenches occur at depths <6,000 m). Many asteroids have been reported from 15 trenches in three oceans to depths of 10,000 m (family Porcellanasteridae and Freyellidae) with 40% endemism. Most studies of deep-sea trenches showed ~43% endemism in ophiuroids, one of the most characteristic groups of abyssal depths >6,000 m. Some scientists refer to deep-sea trenches as "the kingdom of holothurians," because they dominate ultra-abyssal depths (followed by bivalves) and represent 25–50% of all animals caught by trawling. Some trawls retrieve several thousand holothurians, the largest benthic animals at abyssal and hadal depths. Trench holothurians follow crustaceans and polychaetes in taxonomic diversity, with 69% endemism. The order Elasipoda forms the most important benthic biocenosis in trenches, playing a vital role in energy transport and biomass production in the ocean's greatest depths. Echinoids occur only at <7,000 m depth with two endemic species and one endemic subspecies (38% of species). Only 7 of these species occur at greater depths, with 22 species endemic to the hadal zone (76%). Siboglinid worms have been reported at 9,735 m depth (*Heptabranchia subtilis*, a species endemic to this depth). A benthic fish, *Holcomycteronus profundissimus*, was caught for the first time at hadal depths during the *Princesse Alice* expedition in 1901 and identified as a new species and a new genus. Among the metazoans, no documented examples exist at hadal depths for Calcispongiae, ctenophores, Kinorhyncha, Merostomata Kamptozoi, Phoronoida, pterobranchs, and benthic decapods (suborder Reptantia), but we cannot say whether these gaps reflect true absence, undersampling of trenches, or difficulty in sampling these groups.

Table 26.1 Comparison among deep ocean basin ecosystems.

Main characteristics	Habitat / Ecosystem						
	Abyssal plains	Seamounts	Deep hydrothermal vents	Whale carcasses	Anoxic zones (Dead zones)	Deep hypersaline anoxic basins	Hadal trenches
Depth	3,000/4,000–6,000 m	Base at >1,000– with elevation >1,000 m	200–5,000 m	10–3,000 m (to date)	150–>2,500 m	2,500–3,700 m	6,000–>11,000 m
Extent	40% of Earth's surface and the most widespread of all biomes	Still not precisely known but < 5% of the seafloor	Relatively limited	Very limited, especially in recent decades because of the whaling	Very reduced and generally localized	Extremely localized	Occupy 1–4% of ocean area
Distribution	Occur in all seas and oceans of the world	Occur in all seas and oceans particularly along oceanic ridges	All oceans	All oceans	Pacific, Atlantic, Black Sea	Pacific, Atlantic, Mediterranean, Red Sea	Occur along subduction zones
Type of substrate	Almost exclusively soft bottoms	Dominated by hard bottoms but sediments may accumulate on flat topography	Typically hard bottoms	Typically soft bottoms	Typically soft bottoms	Typically soft bottoms	Almost exclusively soft bottoms, but presence of hard bottoms possible
Biodiversity	High equitability and high numbers of species but low abundances of most species	Very high especially for megafauna, often associated with deep-water corals	Almost 700 described species	~400 described species	Very low, almost exclusively microbial	Very low, almost exclusively microbial	Numbers of species relatively low given the extreme environmental conditions
Endemism	Not well quantified	>50% at many locations	Very high	High and similar to that of cold seeps	Very low	Very few species- potentially three Loricifera species complete their entire life cycle in anoxic conditions	Not higher than 10–30%
Functioning	Low supply of organic matter to the bottom, particularly in locations far from continents	Often high primary production with Tyler column	Chemosynthetic production supplied by geothermic fluids	Chemosynthetic production based on carcasses and sulfhydryl acid	Phytodetritus and chemosynthetic production	Chemosynthetic production probably low	Some rich in organic matter, particularly those near upwelling, poor in locations far from continents

Trophic webs	Lesser role for megafauna and macrofauna. Greater relevance for the smaller size classes (meiofauna, protistans, and prokaryotes)	Based mainly on sessile filter feeders, with presence of fish and apical predators	Complex, based on interactions between macro-organisms and endosymbionts	Very complex with elevated numbers of scavengers, predators, and the microbial food web	Only microbial with virus-prokaryotic-protozoans loop	Only microbial with virus-prokaryotic-protozoans loop	Unknown but dependent on accumulation of detritus from surface waters
Peculiarities	Wide extent, with some heterogeneity, can favor parapatric speciation	Strong isolation between different seamounts, important for connectivity studies	Temperatures up to >400 °C	Most organic rich systems in the world	Total absence of oxygen and variability in physical-chemical conditions	Extraordinary salinities (20–30%) and variable pH	Extreme pressures and low temperatures, systems may host psychrophilic relict fauna
State of knowledge	Very limited	Some seamounts well known but most unsampled	Very high for some locations, others likely undiscovered	Limited	Limited but improving	Extremely limited	Extremely limited

26.11 Summary

Abyssal plains are widespread but other basin habitats are much more patchy spatially (Table 26.1). All of these habitats are characterized by high endemism and characteristic biota, and poor overall state of knowledge. Like other deep-sea environments, they offer tremendous opportunity for discovery of new species and novel processes.

Questions

1) What are the main food sources for deep-sea organisms?
2) What mechanisms support life on the abyssal plains?
3) How do cold-seep habitats form?
4) What are the main differences between shallow hydrothermal vents and those that characterize the adjacent continental slope?
5) Describe the main ecological features of seamounts.
6) What is the seamount effect?
7) Describe hydrothermal vent zonation.
8) What are the similarities and differences between anoxic systems and hypersaline basins?
9) Describe the general characteristics of hadal environment (those environments below 6,000 m depth).

Suggested Reading

Aguzzi, J., Fanelli, E., Ciuffardi, T. et al. (2018) Faunal activity rhythms influencing early community succession of an implanted whale carcass offshore Sagami Bay, Japan. *Scientific Reports*, 8, p. 11163.

Baco, A.R. and Smith, C.R. (2003) High species richness in deep-sea chemoautotrophic whale skeleton communities. *Marine Ecology Progress Series*, 260, pp. 109–114.

Bailey, D.M., King, N.J., and Priede, I.G. (2007) Cameras and carcasses: Historical and current methods for using artificial food falls to study deep-water animals. *Marine Ecology Progress Series*, 350, pp. 179–191.

Beliaev, G.M. and Brueggeman, P.L. (1989) *Deep Sea Ocean Trenches and Their Fauna*. Nauka Publishing House.

Bennet, B.A., Smith, C.A., Glaser, B., and Maybaum, H.C. (1994) Faunal community structure of chemoautotrophic assemblage on whale bones in the deep northeast Pacific Ocean. *Marine Ecology Progress Series*, 108, pp. 205–223.

Bergquist, D.C., Ward, T., Cordes, E.E. et al. (2003) Community structure of vestimentiferan-generated habitat islands from Gulf of Mexico cold seeps. *Journal of Experimental Marine Biology and Ecology*, 289, pp. 197–222.

Billett, D.S.M., Bett, B.J., Jacobs, C.L. et al. (2006) Mass deposition of jellyfish in the deep Arabian Sea. *Limnology and Oceanography*, 51, pp. 2077–2083.

Blankenship-Williams, L.E. and Levin, L.A. (2009) Living deep: A synopsis of hadal trench ecology. *Marine Technology Society Journal*, 43, pp. 137–143.

Bo, M., Bertolino, M., Borghini, M. et al. (2011) Characteristics of the mesophotic megabenthic assemblages of the Vercelli seamount (North Tyrrhenian Sea. *PLOS One*, 6, p. e16357.

Brandt, A., Gooday, A.J., Branda, S.N. et al. (2007) First insights into the biodiversity and biogeography of the Southern Ocean deep sea. *Nature*, 447, pp. 307–311.

Brazelton, W.J., Ludwig, K.A., Sogin, M.L. et al. (2010) Archaea and bacteria with surprising microdiversity show shifts in dominance over 1,000-year time scales in hydrothermal chimneys. *Proceedings of the National Academy of Sciences of the United States of America*, 107, pp. 1612–1617.

Bruun, A.F. (1956) The abyssal fauna: Its ecology, distribution and origin. *Nature*, 177, pp. 1105–1108.

Clark, M.R., Rowden, A.A., Schlacher, T. et al. (2010) The ecology of seamounts: Structure, function, and human impacts. *Annual Review of Marine Science*, 2, pp. 253–278.

Clark, M.R. and Tittensor, D.P. (2010) An index to assess the risk to stony corals from bottom trawling on seamounts. *Marine Ecology*, 31, pp. 200–211.

Clarke, M.R., Schlacher, T.A., Rowden, A.A. et al. (2012) Science priorities for seamounts: Research links to conservation and management. *PLOS One*, 7, p. e29232.

Cocito, S., Bedulli, D., and Sgorbini, S. (2002) Distribution patterns of the sublit- toral epibenthic assemblages on a rocky shoal in the Ligurian Sea (northwestern Mediterranean). *Scientia Marina*, 66, pp. 175–181.

Colangelo, M.A., Bertasi, R., Dall'Olio, P., and Ceccherelli, V.H. (2001) Meiofaunal biodiversity on hydrothermal seepage off Panarea (Aeolian Island, Tyrrhenian Sea). *Mediterranean Ecosystems: Structures and Processes*, 46, pp. 353–358.

Connelly, D.P., Copley, J.T., Murton, B.J. et al. (2012) Hydrothermal vent fields and chemosynthetic biota on the world's deepest seafloor spreading centre. *Nature Communications*, 3, p. 620.

Craig, C.H. and Sandwell, D.T. (1988) Global distribution of seamounts from Seasat profiles. *Journal of Geophysical Research*, 93, pp. 10408–10420.

Dando, P.R., Aliani, S., Arab, H. et al. (2000) Hydrothermal studies in the Aegean Sea. *Physics and Chemistry of the Earth, Part B: Hydrology. Oceans and Atmosphere*, 25, pp. 1–8.

de Forges, B.R., Koslow, J.A., and Poore, G.C.B. (2000) Diversity and endemism of the benthic seamount fauna inthe southwest Pacific. *Nature*, 405, pp. 944–947.

De Leo, G.A. and Levin, S. (1997) The multifaceted aspects of ecosystem integrity. *Conservation Ecology*, 1, pp. XIII–XIV.

Debenham, N.J., Lambshead, P.J.D., Ferrero, T.J., and Smith, C.R. (2004) The impact of whale falls on nematode abundance in the deep sea. *Deep-Sea Research I*, 51, pp. 701–706.

Dell'Anno, A., Mei, M.L., and Danovaro, R. (1999) Pelagic-benthic coupling of nucleic acids in an abyssal location of the Northeastern Atlantic Ocean. *Applied Environmental Microbiology*, 65, pp. 4451–4457.

Deming, J., Reysenbach, A.L., Macko, S.A., and Smith, C.R. (1997) The microbial diversity at a whale fall on the seafloor: Bone-colonizing mats and animal-associated symbionts. *Microscopic Research Techniques*, 37, pp. 162–170.

Desbruyéres, D., Chevaldonné, P., Alayse, A.M. et al. (1998) Biology and ecology of the "Pompeii worm" (Al- vinella pompejana Desbruyéres and Laubier), a normal dweller of an extreme deep-sea environment: A synthesis of current knowledge and recent developments. *Deep-Sea Research II*, 45, pp. 383–422.

Ebbe, B., Billett, D.S., Brandt, A. et al. (2010) Diversity of abyssal marine life, in McIntyre, A.D. (ed.), *Life in the World's Oceans: Diversity, Distribution, and Abundance*. Wiley-Blackwell Publishing, pp. 139–160.

Epp, D. and Smoot, N.C. (1989) Distribution of seamounts in the North Atlantic. *Nature*, 337, pp. 254–257.

Fisher, C.R., Takay, K., and Le Bris, N. (2007) Hydrothermal vent ecosystems. *Oceanography*, 20, pp. 14–23.

Forges, R., Koslow, J., and Poore, G. (2000) Diversity and endemism of the benthic seamount fauna in the southwest Pacific. *Nature*, 405, pp. 944–947.

Fortunato, C.S., Larson, B., Butterfield, D.A., and Huber, J.A. (2018) Spatially distinct, temporally stable microbial populations mediate biogeochemical cycling at and below the seafloor in hydrothermal vent fluids. *Environmental. Microbiology*, 20, pp. 769–784.

Fricke, H., Giere, O., Stetter, K. et al. (1989) Hydrothermal vent communities at the shallow subpolar Mid-Atlantic Ridge. *Marine Biology*, 102, pp. 425–429.

Fujiwara, Y., Kawato, M., Yamamoto, T. et al. (2007) Three-year investigations into sperm whale-fall ecosystems in Japan. *Marine Ecology*, 28, pp. 219–232.

Gambi, C., Vanreusel, A., and Danovaro, R. (2003) Biodiversity of nematode assemblages from deep-sea sediments of the Atacama Slope and Trench (South Pacific Ocean). *Deep-Sea Research I*, 50, pp. 103–117.

Geange, S.W., Connell, A.M., Lester, P.J. et al. (2012) Fish distributions along depth gradients of a sea mountain range conform to the mid-domain effect. *Ecography*, 35, pp. 557–565.

Gebruk, A.V., Galkin, S.V., Vereshchaka, A.L. et al. (1997) Ecology and biogeography of the hydrothermal fauna of the Mid-Atlantic Ridge. *Advances in Marine Biology*, 132, pp. 93–144.

Genin, A., Dayton, P.K., Lonsdale, P.F., and Spiess, F.N. (1986) Corals on seamounts provide evidence of current acceleration over deep-sea topography. *Nature*, 322, pp. 59–61.

Goffredi, S.K., Paull, C.K., Fulton-Bennett, K. et al. (2004) Unusual benthic fauna associated with a whale fall in Monterey Canyon, California. *Deep- Sea Resseach*, 51, pp. 1295–1306.

Haymon, R.M., Baker, E.T., Resing, J.A. et al. (2007) Hunting for hydrothermal vents. *Oceanography*, 20, pp. 100–107.

Hays, G.C. (2003) A review of the adaptive significance and ecosystem consequences of zooplankton diel vertical migrations. *Hydrobiologia*, 503, pp. 163–170.

Heptner, M.V. and Ivanenko, V.N. (2002) Copepoda (Crustacea) of hydrothermal ecosystems of the World Ocean. *Arthropoda Selecta II*, 2, pp. 117–134.

Hessler, R. and Lonsdale, P. (1991) Biogeography of Mariana Trough hydrothermal vent communities. *Deep-Sea Research Part A. Oceanographic Research Papers*, 38, pp. 185–199.

Jamieson, A.J., Fujii, T., Mayor, D.J. et al. (2009) Hadal trenches: The ecology of the deepest places on Earth. *Trends in Ecology and Evolution*, 25, pp. 190–197.

Kelly, M. and Rowden, A. (2001) Rock sponges on seamounts: Relicts from the past. *NIWA Biodiversity Update*, 4, p. 3.

Kelly, N., Metaxas, A., and Butterfield, D. (2007) Spatial and temporal patterns of colonization by deep-sea hydrothermal vent invertebrates on the Juan de Fuca Ridge, NE Pacific. *Aquatic Biology*, 1, pp. 1–16.

Koslow, J.A. (1997) Seamounts and the ecology of deep-sea fisheries. *American Scientist*, 85, pp. 168–176.

Koslow, J.A., Gowlett-Holmes, K., Lowry, J.K. et al. (2001) The seamount benthic macrofauna off southern Tasmania: Community structure and impacts of trawling. *Marine Ecology Progress Series*, 213, pp. 111–125.

Little, C.T. (2010) Life at the bottom: The prolific afterlife of whales. *Scientific American*, 302, pp. 78–82.

Lundsten, L., Barry, J.P., Cailliet, G.M. et al. (2009) Benthic invertebrate communities on three seamounts off southern and central California. *Marine Ecology Progress Series*, 374, pp. 23–32.

Maugeri, T.L., Lentini, V., Gugliandolo, C. et al. (2009) Bacterial and archaeal populations at two shallow hydrothermal vents off Panarea Island (Eolian Islands, Italy). *Extremophiles*, 13, pp. 199–212.

Mitchell, N.C. (2022) Seamounts, in Shroder, J.F. (ed.), *Geomorphology*. 2nd Edition. Academic Press, pp. 901–918.

Nakagawa, S. and Takai, K. (2008) Deep-sea vent chemoautotrophs: Diversity, biochemistry and ecological significance. *FEMS Microbiology Ecology*, 65, pp. 1–14.

Nakamura, K. and Takai, K. (2014) Theoretical constraints of physical and chemical properties of hydrothermal fluids on variations in chemolithotrophic microbial communities in seafloor hydrothermal systems. *Progress in Earth and Planetary Sciences*, 1, p. 5.

Nussbaumer, A.D., Fisher, C.R., and Bright, M. (2006) Horizontal endosymbiont transmission in hydrothermal vent tubeworms. *Nature*, 441, pp. 345–348.

O'Hara, T.D., Rowden, A.A., and Williams, A. (2008) Cold-water coral habitats on seamounts: Do they have a specialist fauna? *Diversity and Distributions*, 14, pp. 925–934.

Oliver, S.P., Hussey, N.E., Turner, J.R., and Beckett, A.J. (2011) Oceanic sharks clean at coastal seamounts. *PLOS One*, 6, p. e14755.

Pitcher, T., Hart, P.J.B., Morato, T. et al. eds. (2007) *Seamounts: Ecology, Fisheries & Conservation*. Blackwell.

Priede, I.G., Froese, R., Bailey, D.M. et al. (2006) The absence of sharks from abyssal regions of the world's oceans. *Proceedings of the Royal Society of London B: Biological Sciences*, 273, pp. 1435–1441.

Pusceddu, A., Gambi, C., Zeppilli, D. et al. (2007) Seamounts effects on food sources, meiofaunal abundance and Nema- tode diversity in neighbouring deep-sea sediments. *Deep-Sea Research II*, 56, pp. 755–762.

Ramirez- Llodra, E., Shank, T.M., and German, C.R. (2007) Biodiversity and biogeography of hydrothermal vent species thirty years of discovery and investigations. *Oceanography*, 20, pp. 30–41.

Richer de Forges, B., Richter, B., Koslow, J.A., and Poore, G.C.B. (2000) Diversity and endemism of the benthic seamount fauna in the southwest Pacific. *Nature*, 405, pp. 944–947.

Roden, G.I. (1987) Effect of seamounts and seamount chains on ocean circulation and thermohaline structure, in Keating, B.H., Fryer, P., Batiza, R., and Boehlert, G.W. (eds.), *Seamounts, Islands and Atolls Geophysical Monography*. American Geophysical Union. 43, pp. 335–354.

Rogers, A., Billett, D., Berger, W. et al. (2003) Life at the edge: Achieving prediction from environmental variability and biological variety, in Wefer, G., Billett, D.S.M., and Hebbeln, D. et al. (eds.), *Ocean Margin Systems*. Springer-Verlag, pp. 387–404.

Rogers, A.D. (1994) The biology of seamounts. *Advances in Marine Biology*, 30, pp. 305–350.

Rowden, A.A., O'Shea, S., and Clark, M.R. (2002) Benthic biodiversity of seamounts on the northwest Chatham Rise. *Marine Biodiversity and Biosecurity Report*, 2, pp. 1–21.

Schlacher, T.A., Rowden, A.A., Dower, J.F., and Consalvey, M. (2010) Seamount science scales undersea mountains: New research and outlook. *Marine Ecology*, 31, pp. 1–13.

Schreier, J.E. and Lutz, R.A. (2019) Hydrothermal vent biota, in Cochran, J.K., Bokuniewicz, H.J., and Yager, P.L. (eds.), *Encyclopedia of Ocean Sciences*. 3rd Edition. Academic Press, pp. 308–319.

Smith, C.R. and Baco, A.R. (2003) Ecology of whale falls at the deep-sea floor. *Oceanography and Marine Biology: An Annual Review*, 41, pp. 311–354.

Smith, C.R., Baco, A.R., and Glover, A.G. (2002) Faunal succession on replicate deep-sea whale falls: Time scales ans vent- seep affinities. *Cahiers de Biologie Marine*, 43, pp. 293–297.

Smith, C.R., De Leo, F.C., Bernardino, A.F. et al. (2008) Abyssal food limitation, ecosystem structure and climate change. *Trends in Ecology and Evolution*, 23, pp. 518–528.

Smith, C.R., Glover, A.G., and Treude, T. (2015) Whale-fall ecosystems: Recent insights into ecology, paleoecology, and evolution. *Annual Review of Marine Science*, 7, pp. 571–596.

Smith, C.R., Kukert, H., Wheatcroft, R.A. et al. (1989) Vent fauna on whale remains. *Nature*, 34, pp. 127–128.

Smith, W.O. and Jordan, T.H. (1988) Seamount statistics in the Pacific Ocean. *Journal of Geophysical Research*, 93, pp. 2899–2919.

Southward, E.C., Tunnicliffe, V., Black, M.B. et al. (1996) Ocean-ridge segmentation and vent tube worms (Vestimentifera) in the NE Pacific, in MacLeod, C.J., Tyler, P.A., and Walker, C.L. (eds.), *Tectonic, Magmatic, Hydrothermal and Biological Segmentation of Mid-Ocean Ridges*. Geological Society Special Publication 118, pp. 211–224.

Thresher, R.E., Adkins, J., Fallon, S.J. et al. (2011) Extraordinarily high biomass benthic community on Southern Ocean seamounts. *Scientific Reports*, 1, p. 119.

Tunnicliffe, V. (1988) Biogeography and evolution of hydrothermal-vent fauna in the eastern Pacific Ocean. *Proceedings of the Royal Society B: Biological Sciences*, 223, pp. 347–366.

Tunnicliffe, V. (1991) The biology of hydrothermal vents: Ecology and evolution. *Oceanography and Marine Biology: An Annual Review*, 29, pp. 319–407.

Tunnicliffe, V. and Fowler, M.R. (1996) Influence of sea floor spreading on the global hydrothermal vent fauna. *Nature*, 379, pp. 531–533.

Van Dover, C.L., German, C.R., Speer, K.G. et al. (2002) Evolution and biogeography of deep-sea vent and seep invertebrates. *Science*, 295, pp. 1253–1257.

Van Gaever, S., Galéron, J., Sibuet, M., and Vanreusel, A. (2009) Deep-sea habitat heterogeneity influence on meiofaunal communities in the Gulf of Guinea. *Deep-Sea Research II*, 56, pp. 2259–2269.

Van Gaever, S., Moodley, L., de Beer, D., and Vanreusel, A. (2006) Meiobenthos at the Arctic Håkon Mosby Mud Volcano, with a parental-caring nematode thriving in sulphide-rich sediments. *Marine Ecology Progress Series*, 321, pp. 143–155.

Vanreusel, A., De Groote, A., Gollner, S., and Bright, M. (2010) Ecology and biogeography of free-living nematodes associated with chemosynthetic environments in the deep sea: A review. *PLOS One*, 5, p. e12449.

Vinogradova, N.G. (1997) Zoogeography of the abyssal and hadal Zones. The biogeography of the oceans. *Advances in Marine Biology*, 32, pp. 325–387.

Weston, J.N. and Jamieson, A.J. (2022) Exponential growth of hadal science: Perspectives and future directions identified using topic modelling. *ICES Journal of Marine Science*, 79, pp. 1048–1062.

Wilson, C.D. and Boehlert, G.W. (2004) Interaction of ocean currents and resident micronekton at a seamount in the central North Pacific. *Journal of Marine Systems*, 50, pp. 39–60.

27

Oceanic Ecosystems

27.1 Introduction

The global ocean encompasses vast gradients from the ocean's surface to depths of almost 12,000 m, from the coast to the High Seas and from the tropics to the poles. Many organisms that live in the oceanic realm live out their entire existence without ever encountering the shore, the seafloor, or the ocean's surface. They spend their entire lives surrounded by water on all sides and never experience any other environment. Below the first 1,000 m depth in oceanic environments, organisms never even see sunlight. In contrast to benthic environments, their habitat consists of continually moving water that influences every aspect of pelagic life, water properties, and physical dynamics. Evolutionary processes in pelagic systems therefore arise as ecological adjustments to changes in water properties and circulatory regimes than paleo-biogeographic influence such as continental drift and glaciation that can strongly affect benthic systems. Seawater properties and ocean circulation provide the frame of reference in pelagic systems, and primarily define pelagic bioregions. The enormous volume occupied by the open ocean represents >95% of the habitable space on Earth. These huge ecosystems host a large variety of organisms, with wide variation in marine biodiversity and communities with depth in the water column. The major structuring variables in oceanic habitats include depth, which co-varies with temperature, salinity, pressure, and light penetration. These variables result in a layering of open-ocean pelagic ecosystems into **bathomes**, or biomes delineated by water depth.

27.2 Factors Influencing the Life and Distribution of Pelagic Organisms

In terms of spatial distribution, defining bathomes relies largely on biological elements and how relevant processes link to the environment at specific scales. Whereas physical considerations delineate boundaries, the precise, ecologically relevant boundaries require analyses of biological distributions and associations. In contrast, researchers use physical and geological variables as corroborating rather than primary factors in defining benthic provincial and bionomic regions. Depth structuring in pelagic and benthic systems also differs: the less dynamic depth structure in benthic systems sharply contrasts the three-dimensionality and temporal variability in the pelagic realm at a variety of scales, which add considerable complexity in delineating spatial bioregions. Highly depth-structured water column properties and ocean circulation regimes, along with their variation in time, require that pelagic bioregion definitions and analyses consider these factors.

Seafloor processes and benthic communities exert relatively little influence on water column habitats >50 m above bottom. Light, chemical, and physical clines create invisible barriers to marine species across depths, such as the photic/disphotic transition that delimits photosynthetic production, or seasonal or permanent pycnoclines (whether primarily thermocline or halocline) that can influence export of primary production and dampen biological exchange among specific classes of organisms. These barriers limit bathymetric distribution ranges of pelagic species, and even contribute to speciation.

27.2.1 Light, Darkness, and Nutrients

Light within the upper 200 m (epipelagic zone) of the oceanic pelagic realm provides the sole energy source for the massive water column beneath. The disappearance of light with depth, and generally low concentration of nutrients in oceanic

Figure 27.1 Examples of pigmented hadal species, (a) the fish *Pseudoliparis amblystompsis* and (b) the cnidarian, *Crossota norvegica*. *Sources*: A. J. Jamieson, Oceanlab, University of Aberdeen, UK, NOAA / Kevin Raskoff / Wikimedia Commons / CC BY 2.0.

waters, create a highly food-limited environment. The complete absence of light below ~1,000 m also means little adaptive value for body coloring in pelagic animals and many organisms lack pigmentation. However, many species have preserved coloring over evolutionary time even at hadal depths, as illustrated by the pink body of the hadal fish, *Pseudoliparis amblystomopsis* (Figure 27.1), the brown color of the only known specimen of the hadal cnidarian, *Pectis profundicola* (Figure 27.1), and the orange color of an anthomedusa photographed at 8,260 m.

The overwhelming majority of hadal animals lack eyes, although some Crustacea retain rudiments of eyes with no functional role and likely indicate comparatively recent arrival of representative of this group at great depth. Known fishes from hadal depths also retain eyes, but they are much smaller in size or degenerated relative to shallower representatives of the same genera.

27.2.2 Pressure

Pressure can profoundly affect the biology of pelagic organisms at abyssal and hadal depths. All abyssal and hadal animals totally lack gas-filled cavities such as sinuses or swim (air) bladders, and fluids fill all body cavities, resulting in balanced internal and external pressure with no need to expend energy. This lack of air space, in tandem with the largely incompressible nature of fluids, spares animals from the crushing external pressure. However, the hadal fish, *Holcomycteronus profundissimus*, represents one known exception in that it retains a small swim bladder and has never been observed at depths <5000 m.

27.2.3 Shallow-Deep Connectivity

Tight interconnections characterize some shallow and euphotic habitats and deep-sea systems. The life cycles of many pelagic (and benthic) species, particularly macro and megafaunal components, include a planktonic larvae stage dispersed by currents within oceanic waters. Downslope and upslope currents can transport larvae, propagules, and juveniles across depths and ecosystems, resulting in frequent exchange between shallow and deep waters. Several commercially important species that typically occupy the seafloor move seasonally or periodically into pelagic habitats and shallow-water ecosystems, helping replenish overexploited local populations. For example, in the precious red coral, *Corallium rubrum*, a widely distributed species that occupies depths >30–1,000 m depth, deep-sea populations supply larvae to shallow-water populations. Additionally, the red shrimp, *Aristeus antennatus*, an important fisheries species, generally recruits at depths >1,000 m.

27.2.4 Vertical Migrations

Given the fluid habitat, all oceanic species move, but many display degrees of active mobility that allow them to use various vertical zones of the ocean during different life stages or during daily excursions (vertical migrations). The three-dimensional structure and connectivity of deep-sea pelagic ecosystems also has relevance for refuge habitats, as illustrated by different groups of plankton (either meso- macro- or megaplankton, ranging 0.2 mm^{-2} m in size or greater). Most zooplankton species undertake daily vertical migration, which, depending on their size, can cover distances up to 2 km or more as they move to the surface at night in order to feed and then return to deeper layers to avoid visual predation during daylight hours. This migration contributes to the **biological pump**, the ocean's biologically driven sequestration of carbon to the ocean interior and seafloor sediments. Nekton also undertake vertical migrations, including many fish species, giant squid, sea turtles, and most marine mammals. Some cetacean such as sperm whale easily dive to ~2500 m depth to forage, and sea elephants dive to >1700 m depth in search of prey (Figure 27.2). Several charismatic species, from sea lions to penguins, hunt large plankton in the deep sea to depths of >2,000 m.

27.2.5 Feeding and Recruitment in the Deep

In Antarctica, krill find refuge from predation at abyssal depths and several species of plankton including copepods, euphausiids, decapods, and fishes (e.g., the Antarctic silverfish, *Pleuragramma antartica*) recruit at depths >500 – 1,000 m.

27.2.6 Body Size

Pelagic fauna lack the increase in body size with depth and gigantism seen in some hadal benthos, and the biological basis for "gigantism" in a small subset of a few taxonomic animal groups remains unclear. Some deep-sea animals exhibit a phenomenon known as deep-sea **ratcheting**, which refers to a thinning and reduced degree of calcification of the skeletal formations. This phenomenon results from a deficiency of calcium carbonate below the critical (compensation) depth (usually 4,000–5,000 m), where the water is undersaturated with calcium carbonate, resulting in negligible content in sediments.

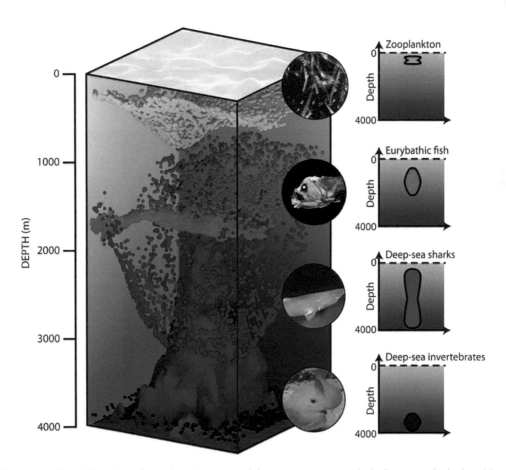

Figure 27.2 Representation of the three-dimensional structure of deep-sea ecosystems, including not only the benthic component (seafloor) but also the pelagic (water column) component and its diversity of various pelagic depth ranges across life cycles and behavioral stages. A comprehensive marine biodiversity conservation strategy must consider all of these components. *Source:* NOAA / Wikimedia Commons / Public Domain, Unknown author / Wikimedia Commons / CC BY-SA 3.0.

27.2.7 Biodiversity

In pelagic habitats, marine species richness does not necessarily decline with depth, indeed species number in some taxa appears to increase to ~400 m depth and then remain consistent to at least mid-slope depths. Even though we know that the pelagic deep sea supports a rich biodiversity that parallels the unique deep-sea habitats beneath (e.g., seamounts, submarine canyons, hydrothermal vents, cold seeps etc. Figure 27.3), undersampling of the deep pelagic biota limits our ability to describe such patterns with confidence. These benthic habitats sometimes link to marine populations and endemic species in the overlying water column.

Since 1840, scientists have discovered nearly 30 new types of deep-sea ecosystems and a wide range of newly identified marine species at depths ranging from 200 m to 11,000 m deep. A series of recent discoveries, enabled by new technologies, have increased our knowledge of habitats and ecosystems unique to the deep sea. These newly discovered deep-sea ecosystems share the common attribute that they host unknown biodiversity and novel species. Recent estimates based on species accumulation curves suggest that at least two thirds of overall deep-sea biodiversity remain undiscovered. Undersampled deep-sea habitats include seamounts, canyons, ridges, cold seeps, hydrothermal vents, manganese nodule fields, deep-water coral areas, and other unique habitats and the water column fauna above them. But even the deep pelagic fauna above lower continental slopes and abyssal plains remains poorly sampled. Seamounts and canyons are emblematic examples of the complexity and difficulty in exploring deep-sea biodiversity. For example, seamounts serve as important hotspot of pelagic species, and many also sustain endemic species. Although we lack an exact count, experts estimate possibly 30,000–100,000 seamounts globally, most of which are completely unexplored, particularly the pelagic fauna overlying them. Studies in New Zealand suggest levels of endemism of seamount habitats close to 50% that, if generalizable to seamount systems globally, would clearly represent an important contribution to marine biodiversity worldwide. Of the 9,951 seamounts mapped globally, only 6% fall partly or fully within marine protected areas, leaving an overwhelming proportion of seamount biodiversity unprotected and well below the Aichi protection target of 10% for marine ecosystems. Similar issues regarding knowledge gaps and lack of protection apply to submarine canyons, which are known hotspots of production and heterogeneous habitat, key feeding areas for large cetaceans, and target locations for intensive commercial trawling (e.g., red shrimp).

Biodiversity in these little-known marine ecosystems faces numerous threats from human activities ranging from fishing removal to sediment plumes from deep-sea mining. Fisheries depletion of some pelagic fisheries resources has already

Figure 27.3 Some components of the bathyal and abyssal fauna, including: *Beryx decadactylus* (Alfonsino), *Psychrolutes marcidus* (blob fish), *Architeuthis sanctipauli* (giant squid), *Hoplostethus atlanticus* (orange roughy), *Harriotta raleighana* (narrownose chimera), *Grimpoteuthis Robson* (Dumbo octopus). This work is made available under the terms of the Creative Commons Attribution-NonCommercial 4.0 Unported License, https://creativecommons.org/licenses/by-nc-sa/4.0. *Source:* Dr. Ken Sulak / NOAA / Wikimedia Commons / Public domain, Cristian M / Flickr / Public domain, Citron / Wikimedia Commons / CC BY-SA 3.0, NOAA / Wikimedia Commons / CC BY 2.0, NOAA / Wikimedia Commons / Public Domain.

occurred in recent decades, especially for large fishes. Fishing efforts now target greater depths, increasing by 50–100 m per decade, following depletion of shallow resources, with considerable interest in potential exploitation of midwater fishes, a massive biomass of low value small fishes that live in oceanic pelagic depths from ~200–900 m. Many fisheries worldwide now operate well beyond the edge of the continental shelf. Understanding pelagic ecosystems and ensuring their sustainability will require consideration of the 3-D properties of pelagic ecoregions and habitats, including biodiversity features within different vertical zones on the entire water column as well as differentiating those features that cross multiple water depths from those limited to unique pelagic habitats.

27.3 Classification of Pelagic Regions

In addition to vertical zonation of pelagic systems, biogeographers classify the pelagic realm based on horizontal extent and characteristics. Among the best known, of these, Alan Longhurst's marine biogeography uses biogeochemical differences among broad swaths of ocean in defining a two-dimensional map of pelagic habitats (Figure 27.4). At the broadest scale, ocean basins differ in characteristics and, to some degree, biota (Table 27.1). However, except for tightly enclosed or semi-enclosed ocean basins, broad mixing and transitional zones blur the boundaries at the open ends of the basin. In the absence of continental land masses, hydrodynamic and atmospheric features set the intrinsic scales of variability, along with external forcing from radiative input (sunlight). These features contribute to, and interact with, three-dimensional (re)circulatory systems. Continental masses disrupt and create circulatory systems arising from both hydrodynamic flow around the continents, as well as intrinsic "closed" circulatory responses and boundary currents (for example, North Atlantic circulation and the Gulf Stream) – all of which interact in complex ways with each other.

Ocean basins (Level 1) – Ocean basins define the broadest scale of pelagic structuring, in a global context, characterized by the presence of a unique set of three-dimensional ecological systems that reflect water properties and circulatory systems influenced, in turn, by continental landmasses and of course the ocean basins themselves. Analyses of water properties and circulatory dynamics at global and basin scales define the boundaries at this level.

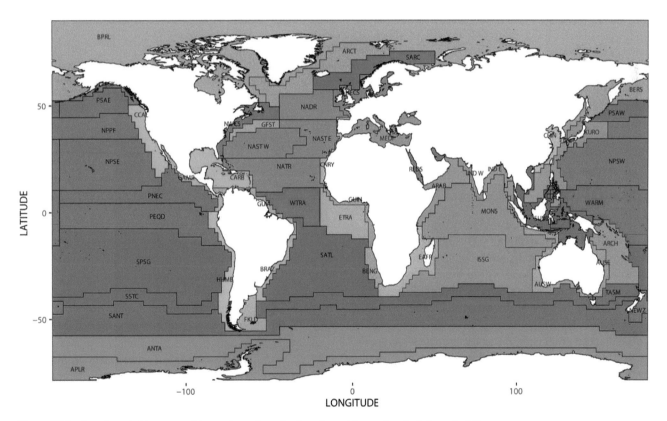

Figure 27.4 Longhurst's biogeographic provinces based primarily on biogeochemical characteristics.

Table 27.1 Classification, definitions, and examples of pelagic regions.

Level	Name	Definition	Examples	Approach and definitions
0	Oceans	In a global content, partitioning at this level recognizes the distinction between fauna (collective ecosystems) of the Indian, Pacific, and Southern Oceans. Regionalization at this level needs to be at global scales (since the distinctions are global in context). For this project, a descriptive narrative is used to distinguish the differences and transitions between the Indian, Pacific and Southern oceans.	Pacific, Atlantic, Indian, and Southern oceans.	Literature
1a	Oceanic zones	Winds, solar forcing, and geostrophy dynamically combine to drive a series of largely circumferential and latitudinally oriented water masses within oceans, each of which can be characterized by its water properties, circulation, and assemblages of biota. Transitional zones between some water masses are generally characterized by higher plankton production, which influences trophic structure and interactions. For this project the classification based on physical properties are guided and corroborated by the distribution of phytoplankton and pelagic fish on the continental shelf.	The dominant water masses surrounding Australia are the South-West Pacific Central Water, Indonesian Throughflow Water (Australasian Mediterranean Water) and Indian Central Water	Large-scale zones that extend across ocean basins. Datasets that are global (or at least basin-wide) should be used.
1b	Oceanic Substructure, Water Masses	Within the Oceanic zones, substructure is characterized by largely latitudinal bands of water masses extending through the water column. These bands represent segments of circulatory systems that may span ocean basins	Example: Antarctic Intermediate Water, Subtropical Lower Water, Tropical Surface Water.	Full suite of variables
2	Seas: Circulation Regimes	Within latitudinal bands, different ocean circulations and air/sea moisture exchanges result in different retention, mixing and transport of water properties and biological organisms. Such regions respond differently to seasonal and interannual climate variations. Consequently, temporal changes may occur in the location, extent and strength of circulation regimes and their biota. Transitional zones characterize the temporal changes and adjustments between the different circulation regimes – and again may be characterized by higher productivity	Circulation within the Mediterranean basin with inflow of surface Atlantic waters and outflow of deep Mediterranean waters.	Nested analysis of temperature, salinity, and oxygen.
3	Fields of Features	Within Circulation Regimes, structure can be characterized by regions of differing energetics; for example, mixing due to eddy activity, frontal oscillations, and boundary currents. Transition regions represent changes in the energetics at the boundary of the field and seasonal movement, and variations in strength	Regional-scale fields of eddies. Seasonal movements of the eddy fields	Sea-surface temperature data, geostrophic currents.
4a	Features	Description of structure and dynamics of individual features.	Individual eddies	
4b	Feature Structure	Internal structure and dynamics of individual features	The semi-permanent eddies	

Oceanic zones (Level 1a) – Within oceans, latitudinal processes largely determine regional substructure. At these scales, winds, solar forcing, air/sea moisture exchange, and geostrophy combine dynamically to drive a series of latitudinal circulatory processes that result in bands of water-mass structures referred to as **oceanic zones**. "Thermocline" water masses dominate these bands near the surface, having formed on the surface before subduction transported them deeper into the ocean interior. Water properties, circulation, and assemblages of biota vary with depth and span scales similar to ocean basins. Higher plankton production often occurs in transitional regions where water properties differentiate bands; this enhanced production influences trophic structure and interactions.

Water masses (Level 1b) – Bands of water types comprising a series of latitudinal core water masses and transitional water masses and fronts create substructure, including with depth, in oceanic zones. These water masses may, for example, represent segments of basin-scale circulatory systems (northern segment, central core, southern segment, etc.).

Seas: Circulation Regimes (Level 2) – Within latitudinal bands at Level 1b water masses described above, different ocean circulation patterns result in different retention, mixing, and transport of water properties and biological organisms. The resulting regions at this level respond differently to seasonal and interannual variation. Consequently, temporal changes may occur in location, extent, and strength of these regions and their biota. High productivity often characterizes the transitional zones between the different circulation regimes.

Fields of Features (Level 3)

Within circulation regimes, regions differing in energetics (eddy activity, frontal oscillations and boundary currents) create substructures demarcated by transition regions where changes in energetics at the boundary of the field. In this classification, variability in surface temperature defines fields of features.

Features (Level 4)

Within Circulation Regimes structures at smaller spatial scale can further influence biotic distributions.

27.4 Functional Classification of Pelagic Systems

In ecological terms, pelagic classifications can represent **Cells of Ecosystem Functioning** (CEFs, *sensu* Ferdinando Boero). CEFs represent a holistic approach based on the concept on intraspecific connectivity, linking metapopulations inhabiting different locations that contain similar habitat types. Scientists quantify connectivity in terms of current patterns, comparisons of species lists (Beta diversity), propagule presence in the water column, and genetic connections among populations of representative species. Some ecologists take connectivity a step further by considering metacommunities, arguing that entire communities may be considered in a connectivity framework. To fully account for ecosystem functioning, however, intraspecific connectivity of individuals of a species represents just one of the three essential connectivity components, with interspecific fluxes (food webs) and extra-specific fluxes (biogeochemical cycles) providing the other two key elements. Splitting marine space into coherent units, where ecosystem functions take place, forms the basis for the concept of Cells of Ecosystem Functioning. Because we now have a fairly good understanding of current patterns and their general dynamics, they provide a good starting point to hypothesize where the boundaries of CEFs might occur.

We use the Mediterranean Sea to illustrate the complexity of pelagic oceanic realms (Figure 27.5), in that it operates somewhat like a miniature ocean in illustrating how different water masses interact. The formation of deep water in the Mediterranean depends on generating a water mass (through excess evaporation) with higher salinity than the water masses in the adjacent Atlantic Ocean. This difference triggers the surface Gibraltar Current that flows from the Atlantic through the Strait of Gibraltar into the Eastern Mediterranean, with deeper return flow of the Intermediate Levantine Current back in the Atlantic Ocean. This flow exchange renews the Mediterranean but, over just the upper 500 m depth, constrained by a bottom sill that extends to within 280 m of the surface. The much deeper average depth of the Mediterranean basin and the lack of water renewal to the bottom would lead to anoxic crises in the deep sea, but the cold engines of the Gulf of Lyons, and the Northern Adriatic and Aegean Seas produce dense water that renews deeper layers.

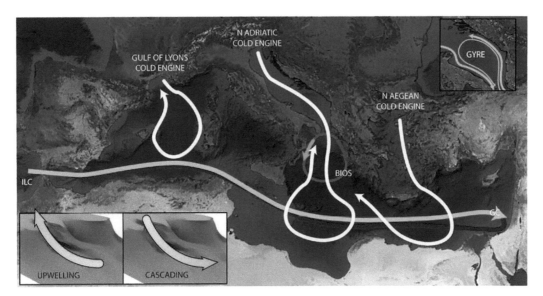

Figure 27.5 The main horizontal and vertical currents of the Mediterranean Basin, showing surface currents in blue and deeper currents in pink: GC, Gibraltar current; ILC, intermediate Levantine current; BIOS, Bimodal Oscillating System. Lower inset: canyon-generated cascading and upwelling currents. Right inset: gyre generated by coastal morphology. After Danovaro R., Boero F., 2019. Italian seas. In: Sheppard C. (Ed). World Seas. An Environmental Evaluation (second ed.). Academic Press, London. pp. 283–306.

The cold engines form cascading currents that flow through marine canyons, bringing oxygen-rich coastal waters to the deep and leading to offshore upwelling of nutrients to the surface. The persistence of life in the Mediterranean deep sea depends on water mass exchanges between the deep sea and the coast in terms of oxygen injection into the deep, and nutrient and propagule injection into coastal systems that promote production. The Mediterranean Sea contains two sub-basins that differ in ecosystem processes, generated by patterns in ocean currents. The cold engines apparently exert little effect on the Tyrrhenian, reducing the intensity of water renewal and contributing to biodiversity richness in its deepest regions. The current patterns of the Tyrrhenian Sea, in fact, differ from those elsewhere in the basin because they depend on surface and deep currents that enter from the northern coast of Sicily (Figure 27.5).

The Tyrrhenian Sea, in turn, influences the Ligurian Sea and the Western Mediterranean with its outflowing currents. Despite these connections, however, the Tyrrhenian Sea represents a compact and somewhat distinct entity determined by an internal connectivity that defines original environmental features and, possibly, ecosystem functions. The horizontal unidirectional currents and gyres of the Tyrrhenian Sea link to vertical currents driven by canyons as inshore upwelling and offshore downwelling.

Jellies (cnidarians) provide a crude tracer of current patterns in this system. *Pelagia noctiluca*, for instance, an abundant scyphozoan in the Tyrrhenian Sea, depends on individuals transported from the deep sea that reach the coast through canyon-generated upwelling, where horizontal currents then distribute them along the coasts. Tyrrhenian Sea circulation depends on an incoming surface current branch of the Gibraltar Current, Modified Atlantic Water, and Levantine Intermediate Water from the Eastern Mediterranean that flows deeper, parallel to MAW, and enters the Ligurian Sea. A branch flows southward along the Sardinian Coast and exits the Tyrrhenian basin, together with Tyrrhenian Deep Water (TDW) (Figure 27.6). The interaction of these currents with the coastline and seafloor forms a series of cyclonic and

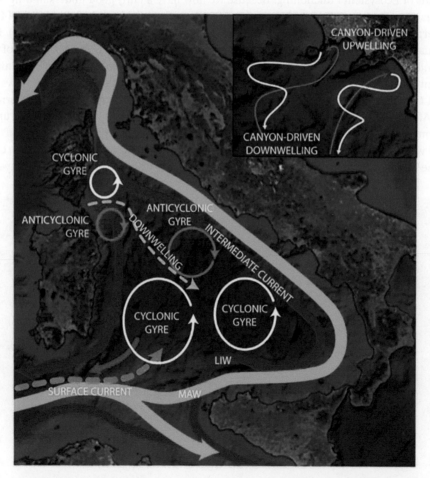

Figure 27.6 Main currents of the Tyrrhenian Sea west of Italy. LIV: Levantine Intermediate Water; MAW: Modified Atlantic Water; TDW: Tyrrhenian Deep Water; WMDW: Western Mediterranean Deep Water. Inset: Canyon-driven upwelling and downwelling. From various sources, graphics: F. Boero. After Danovaro R., Boero F., 2019. Italian seas. In: Sheppard C. (Ed). World Seas. An Environmental Evaluation (second ed.). Academic Press, London. pp. 283–306. 146.

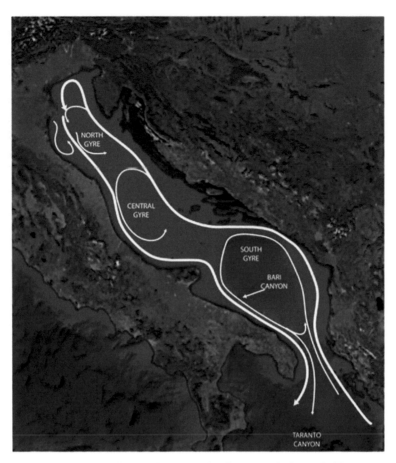

Figure 27.7 Main currents of the three regions of the Adriatic Sea. After Danovaro R., Boero F., 2019. Italian seas. In: Sheppard C. (Ed). World Seas. An Environmental Evaluation (second ed.). Academic Press, London. pp. 283–306. *Source:* Michael Tangerlini.

anticyclonic gyres. Large numbers of submarine canyons inundate the rim of the Tyrrhenian Sea, driving deep water toward the coast; upwelling currents bring deep-sea nutrients into coastal waters.

Vertical currents link the deep sea with coastal systems, triggering phytoplankton blooms at the end of winter by transporting nutrient-rich deep water to the coasts. Horizontal currents disperse the spring bloom throughout the basin where gyres concentrate and thus enhance Tyrrhenian Sea production and contribute to the formation of a series of functioning hot spots. A cold-water engine in the much shallower northern Adriatic Sea (Figure 27.7), in tandem with surface currents, generates three main gyres that connect the opposite shores of the basin, dividing the basin into three regions.

These oceanographic processes define water masses characterized by higher internal connectivity of ecological processes than with neighboring water masses. This connectivity, however, varies among groups of organisms. High mobility connects populations of some species, in contrast to greater separation of populations of other, less vagile species. Interspecific fluxes also vary among species. White corals in the Southern Adriatic, for instance, draw food from the cascading current generated by the Northern Adriatic cold engine, resulting in a trophic connection, even if populations lack similar intraspecific connectivity with that region of the basin.

27.5 Vertical Zonation in Pelagic Ecosystems

Investigations of biodiversity pattern across large spatial scales, especially in the context of 3D systems, typically begins with habitat and biodiversity mapping. Most maps present the world in two dimensions, typically viewed from above and often ignoring the third dimension of depth. In recent years, registration of land ownership has begun to account more for the third dimension, with the aim of assigning ownership and rights to the third dimension, such as the space above buildings that might block views from adjacent structures.

The seafloor often forms the basis for mapping marine habitats and marine spatial planning, as in recent global mapping of ocean geomorphology. Including the three-dimensional nature of inshore waters adds significant challenge, and even more so in offshore environments given the large volume of the third dimension of the water column and its largely unknown dynamics. These constraints challenge any marine spatial planning efforts such as designing marine protected areas for pelagic habitats. Scientists have sometimes attempted to incorporate time-series data and oceanographic processes in the past, but rarely in the deep sea.

Defining three-dimensional rather than two-dimensional marine ecoregions, the standard approach in terrestrial ecology, could significantly advance understanding of marine ecosystems. A definition of three-dimensional marine ecoregions should incorporate processes and connectivity (including larval dispersal) of species and habitats (e.g., related to upwelling, currents, and gyres) as well as biodiversity features based on species distribution ranges in horizontal and vertical dimensions, with some consideration of time. Characterization of marine habitats and species distribution in three dimensions requires understanding major biological and ecological differences among different depth zones. Scientists divide pelagic ecosystems into five different main zones (Figure 27.8):

1) **The epipelagic zone** (or upper open ocean) encompasses the part of the oceanic realm with sufficient sunlight for algae to convert carbon dioxide into organic matter. This zone accounts for 2–3% of ocean volume and spans from the ocean surface to ~200 m. This layer includes the portion of the water column with reduced light and some photosynthesis, but not enough to produce significant primary production biomass. The epipelagic zone hosts a wide range of iconic animals, such as whales and dolphins, billfishes, tunas, jellyfishes, sharks, and many other groups. Because of the large area of the epipelagic zone, its algae account for much of the global ocean original primary production and create ~40% of atmospheric oxygen (through photosynthesis). Organisms that live in the epipelagic zone may come into contact with the ocean surface.

2) **The mesopelagic zone** (or middle open ocean, the "twilight zone") stretches from the bottom of the epipelagic zone (~200 m) to the depth at which no sunlight exists, approximately 1,000 m depth. The upper limit of the mesopelagic zone coincides with the maximum depth of seasonal variability in temperature, the seasonal thermocline, and penetration of sunlight sufficient to support photosynthesis. The mesopelagic zone occupies ~10% of global ocean volume, more than double

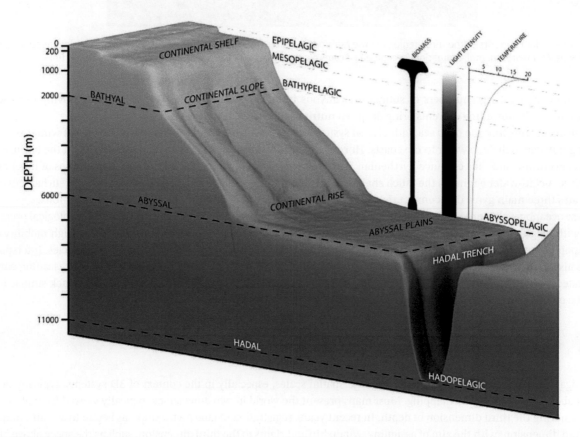

Figure 27.8 Vertical zonation in pelagic waters showing changes in biomass per unit volume, availability of solar radiation, and typical temperatures.

the volume of the epipelagic. From 200–1,000 m depth, light can penetrate and influence **nictemeral cycles** (day/night cycles, sometimes referred to as **circadian**) of many planktonic (and benthic) species, but without any possible significant photosynthetic activity signal. The most abundant vertebrates on Earth, small bristlemouth fishes, live within this zone, and scientists calculate that 90% of total fish biomass in the global ocean lives in the mesopelagic zone. Many species of fishes and invertebrates that live here migrate into shallower, epipelagic depths to feed, but only under the cover of night.

3) The next deepest zone, the **bathypelagic zone** (or lower open ocean) begins at the bottom of the mesopelagic zone (~1,000 m) and extends to 3,000–4,000 m spanning a much wider depth range than the mesopelagic with a volume >10x times greater than the epipelagic zone. The bathypelagic zone comprises nearly 40% of total ocean volume and, together with the abyssopelagic zone, represents the largest ecosystem on Earth. A complete lack of sunlight defines the upper bound of this zone, and bathypelagic organisms therefore live in complete darkness, 24 h d^{-1}. The organisms themselves can interrupt this darkness, however, by producing light or **bioluminescence** that may attract prey or potential mates. Some species have completely lost any capacity for sight.

4) **The abyssopelagic zone** stretches from 3,000 to 4,000 m to 6,000 m depth and extends to the abyssal plain seafloor. This zone represents ~30% of ocean volume and supports a large number of specialized species.

5) **The hadopelagic zone** only exists in specific locations around the world and includes the waters overlying the hadal trenches. These waters range 6000 – >11,000 m depth and occupy 1–2% of total ocean volume. Where deep, wide trenches occur in the otherwise flat seafloor, the water that fills them comprises the hadopelagic zone, which encompasses the deepest parts of the ocean.

27.6 Biodiversity of Pelagic Systems

27.6.1 Epipelagic Biodiversity

The upper 200 m of the ocean has been the focus of most scientific sampling and is therefore the best-known pelagic oceanic habitat, providing environments for many large vertebrates, including marine mammals and reptiles that must access the ocean surface to breathe (Figure 27.9). Compared to deeper layers, this zone supports much more ocean life because of sunlight that penetrates the surface and supports smaller-sized phytoplankton, particularly picoplankton that thrive in low nutrient conditions (Figure 27.10). Small size confers a greater surface area to volume ratio, which facilitates nutrient uptake at low concentrations. Small size also slows sinking rates, helping photosynthesizers remain in the photic zone, and reduces "self-shading," where individual cells shade one another. These photosynthetic groups include *Synechococcus* spp. (Class Cyanobacteria, the "blue-green" bacteria), a small photosynthetic prokaryote (~1 um), first discovered to be widespread in 1979, even though we now know they often outnumber eukaryotic phytoplankton. The even smaller **prochlorophytes**, again prokaryotic, were not discovered until the late 1980s, even though they may be the most abundant photosynthetic microbes on Earth. The larger net plankton include photosynthetic eukaryotes with some of the same major groups as shelf environments, with a lesser role for diatoms and a proportionally greater role for dinoflagellates and coccolithophores, the latter of which fall within the nanoplankton size range.

Figure 27.9 Examples of pelagic fauna inhabiting the epipelagic zone, including: (A) A turtle surfaces to breathe; (B) Pelagic sardine school.

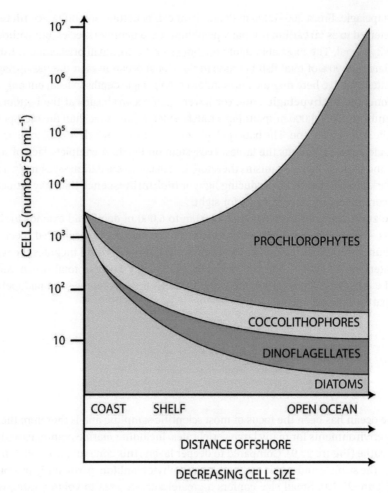

Figure 27.10 Changing phytoplankton composition moving from inshore to offshore.

Zooplankton include a wide variety of smaller forms that play a lesser role in coastal environments. Protists include members of the Acantharia, Radiolaria, and Foraminifera that occur almost entirely in the open ocean. These single-celled organisms have elaborate skeletons and use pseudopodia to feed on bacteria, phytoplankton, and small zooplankton, and, in turn, become prey for larger zooplankton such as copepods.

Nekton include diverse fishes spanning a wide range of size classes, including ocean sharks and many bony fishes such as billfish, sauries, tunas, sunfish, dolphin fish and myctophids (lanternfish) that often dominate epipelagic waters numerically. Oceanic fisheries tend to focus on high value fishes such as tuna and swordfish. Marine mammals such as sperm whales and blue whales, as well as some species of dolphins, also spend much of their life in oceanic waters; marine turtles also spend significant time in oceanic waters. The epipelagic zone is also the most important spawning area for most oceanic species, which release their gametes in seawater.

27.6.2 Mesopelagic Biodiversity

The mesopelagic zone spans from the photic zone above to complete darkness as depth increases, with bioluminescent animals as the only light source deep in this zone. The limited food supply at these depths forces some animals to migrate into the photic zone at night to feed. Many mesopelagic animals can consume individuals larger than they are, utilizing large, sharp teeth with expandable jaws and stomachs. For example, dragonfish and viperfish both have an enormous mouth, sharp pointy teeth, large eyes, and large stomachs. Some swim up at night to hunt. Both species hunt in the deep sea and in surface waters with their bioluminescent lures. Bigscale fish, *Poromitra crassiceps*, with their characteristic oversized scales and bony plated head, live at depths of 150–1,000 m. As in some other fishes in this zone, the bigscales rely on black pigments to make themselves effectively invisible to would-be predators. Their scales also easily detach if predators attempt to feed on them, enabling escape. Other fishes, such as anglerfish, use specially developed lures to catch prey and save energy they might otherwise have to expend chasing prey in a nutrient poor subzone. Hatchetfish, though frightening in appearance, measure just a few cm in length, with a huge jaw and long teeth. The photophores underneath their body and below their eyes may help this fish hide behind its own silhouette, leaving them mostly invisible to predators swimming below. Adult snipe eels look almost like a flat tapeworm with a small flat head at one end. Snipe eel jaws curve outward, helping them trap shrimp with their sharp teeth. Siphonophores, animals related to jellies that stun their prey with a special tentacle, also live in the mesopelagic zone. Other common bioluminescent organisms, ctenophores, have adapted the iridescent cilia they use for locomotion to confuse and scare predators. Ctenophores come in many forms, and typically use long tentacles to capture crustaceans and fish. Firefly squid, a more complicated bioluminescent organism, has one photophore on its head, another around its eyes, one on the body surface, and the remaining photophores on their tentacle tips. Possible uses for the photophores include predator deterrence, communication, lures for prey, and light to enable hunting in the dark.

27.6.3 Bathypelagic Biodiversity

Animals living in the bathypelagic zone rely on detritus for food or by preying on other animals in this zone. At this depth and pressure, fishes, mollusks, crustaceans, and jellies dominate the animal biota. Sperm whales occasionally hunt at these depths to prey on giant squid. Black and red dominate animal color here, and any bioluminescence usually appears blue. Vampire squid, the most common mollusk at bathypelagic depths, can turn itself inside-out to use its spiky tentacles to deter predators or capture prey as they throw their tentacles over the unfortunate prey like a net from above. Snake dragonfish resemble viperfish or dragonfish of the mesopelagic zone and also migrate upward to hunt at night but must travel further given their deep daytime habitat. Anglerfish in the bathypelagic zone are easily recognized for their enormous mouth and lure on its head that resembles a flashlight. In addition to long teeth on their jaws, the throats of anglerfish feature additional teeth. Over evolutionary time, the males of this species have been reduced to a tiny parasitic fish that attaches to conspecific females near their genitals. Both sexes occur anywhere from the surface to the deepest waters. The transparent body of amphipods in the bathypelagic zone helps them hide, a strategy shared by shrimp, another common bathypelagic crustacean.

27.6.4 Abyssopelagic Biodiversity

Relatively few organisms have evolved adaptations to survive in the abyssopelagic zone, from 4,000 m to the abyssal seafloor with near-freezing temperatures and extreme pressures. Animals capable of living at these depths include some species of deep-sea squid, and octopus that swim above the seafloor. The transparent deep-sea squid also uses photophores to lure prey and deter predators, illustrating multiple adaptation to the aphotic, food-poor environment. Echinoderms in the abyssopelagic such as swimming sea cucumbers (sea pigs), use wing-like structures to swim as they "fly" through the water. Other largely transparent crustaceans swim through resuspended sediments above the seafloor and have evolved without eyes that would serve no purpose at these depths.

27.6.5 Hadopelagic Biodiversity

The water that fills deep-ocean trenches comprises the hadopelagic zone, defining the deepest pelagic habitat on Earth. All trenches together occupy an area about the size of Australia with an average maximum depth >10,000 m. Sampling campaigns in the 1950s produced much of our knowledge of hadal biology (Danish Galathea and the Soviet Vitjaz expeditions), although two separate recent self-funded expeditions (director James Cameron and explorer Victor Vescovo) have built on this initial sampling. Far from an environment devoid of life as originally perceived, the hadal zone hosts a substantial diversity and abundance of fauna with a high degree of endemism. The hadal zone cannot simply be considered a continuation of the deep-sea environment because depths > 6,000 m split trench habitats into disjunct, isolated clusters. Furthermore, compelling evidence suggests that the characteristic V-shaped topography of trenches funnels food resources downward (as particulate organic carbon; POC), which accumulates along trench axes. The combination of extreme hydrostatic pressure, accumulation of food along trench axes and geographical isolation has likely resulted in high rates of endemism in these habitats, and an extraordinarily high abundance of some taxa and total exclusion of other faunal groups. Oceanographic and biological interactions can create and maintain "biological hotspots" in the ocean, ultimately controlling regional biodiversity, biogeography, and the evolution of deep-sea fauna. However, these same processes may also limit productivity, endemism, and diversification of populations among habitats. Hydrostatic pressure and limited food supply primarily structure the biotic assemblages (including species diversity and endemism) of hadal waters. There are no known specific adaptations for living at hadal depths, beyond the capacity to cope with extreme hydrostatic pressure. Externally, the animals living below 6 km show no obvious differences from those living in the lower abyssal zone, suggesting physiological and biochemical adaptations rather than pronounced morphological changes.

Questions

1) What are the main factors influencing life in pelagic ecosystems?
2) What are the classification levels of pelagic regions?
3) How does the Mediterranean Sea act like a miniature ocean?
4) Why would defining a three-dimensional rather than two-dimensional marine ecoregion advance marine ecosystem understanding?
5) Describe vertical zonation in pelagic ecosystems.
6) How does biodiversity change within the water column?

Suggested Reading

Belgrano, A., Batten, S.D., and Reid, P.C. (2013) Pelagic ecosystems. Levin, S.A., (ed.), in *Encyclopedia of Biodiversity*. 2nd Edition. Academic Press, pp. 683–691.

Boero, F., De Leo, F., Fraschetti, S., and Ingrosso, G. (2019) The Cells of Ecosystem Functioning: Towards a holistic vision of marine space. *Advances in Marine Biology*, 82, pp. 129–153.

Briscoe, D., Hobday, A., Carlisle, A. et al. (2016) Ecological bridges and barriers in pelagic ecosystems. *Deep-Sea Research Part II: Topical Studies in Oceanography*, 140, pp. 182–192.

Fonda Umani, S., Malisana, E., Focaracci, F. et al. (2010) Disentangling the effect of viruses and nanoflagellates on prokaryotes in bathypelagic waters of the Mediterranean Sea. *Marine Ecology Progress Series*, 418, pp. 73–85.

Genin, A. (2004) Biophysical coupling in the formation of zooplankton and fish aggregations over abrupt topographies. *Journal of Marine Systems*, 50, pp. 3–20.

Holland, K.N., Kleiber, P., and Kajiura, S.M. (1999) Different residence times of yellowfin tuna, *Thunnus albacares*, and bigeye tuna, *T. obesus*, found in mixed aggregations over a seamount. *Fishery Bulletin*, 97, pp. 392–395.

Karner, M.B., DeLong, E.F., and Karl, D.M. (2001) Archaeal dominance in the mesopelagic zone of the Pacific Ocean. *Nature*, 409, pp. 507–510.

Pusch, C., Beckmann, A., Mora Porteiro, F., and von Westernhagen, H. (2004) The influence of seamounts on mesopelagic fish communities. *Archive of Fishery and Marine Research*, 51, pp. 165–186.

Skliris, N. and Djenidi, S. (2006) Plankton dynamics controlled by hydrodynamic processes near a submarine canyon off NW Corsican coast: A numerical modelling study. *Continental Shelf Research*, 26, pp. 1336–1358.

UNEP (2006) *Ecosystems and Biodiversity in Deep Waters and High Seas*. UNEP Regional Seas Report and Studies No 178. IUCN REPORT.

Part IV

Human Impacts and Solutions for Planet Ocean: Applied Marine Biology

In some ways, humans are among the most influential organisms with respect to the ocean, but sadly that influence results from *Human Impacts*, the focus of this section. The English biologist Thomas Henry Huxley infamously stated in 1883 that "*Probably all the great sea fisheries are inexhaustible; that is to say, that nothing we do seriously affects the numbers of fish. Any attempt to regulate these fisheries seems consequently, from the nature of the case, to be useless.*" Similarly, for those of us who were undergraduates decades ago, if someone had told us that humans would measurably alter the alkalinity (pH) of the global ocean, we would have thought it impossible. Yet advancements in technology and growth of human populations over the last century have extended our footprint to even the deepest ocean trenches, where scientists have now documented the presence of plastics and synthetic contaminants in marine organisms. Thirty years ago, most marine ecologists would have identified fishing impacts as the single greatest threat to ocean life, and that threat certainly has not subsided. Indeed, local- to regional-scale impacts of fishing today show only modest improvements in recent years at best, and even then in depressingly few cases. But we now recognize that the global ocean has absorbed >90% of the heat that humans have added to the global system, and ~30% of the additional carbon dioxide. Climate change has caused significant changes to ocean life on a global scale, and thus represents the greatest and most pervasive threat to our planet because it leaves no part of the ocean untouched.

These chapters summarize the different ways that human activities have altered ocean environments, and the ensuing complex changes, but we can't solve problems until we identify them. Every play needs a protagonist, and human impacts certainly fill that niche! And we will soon move from problems to solutions!

We promised solutions, and *Marine Biodiversity Conservation* offers just that! When we talk of ocean management, we really mean ocean use management, because we will probably never manage ocean ecosystems the way we do some terrestrial systems because their large size, dynamic nature, interconnectedness, and complexity make management near impossible. But we can manage ocean use and how humans interact with ocean life, and try to minimize our impacts. In reality, humanity cannot afford not to use the tremendous variety of resources the ocean offers. The protein, minerals, and even genetic resources we extract from ocean environments contribute significantly to human well-being and quality of life.

But as we will learn from *Human Impacts*, our activities have significantly altered, and mostly negatively, a wide range of ocean ecosystems. Although these changes may lead us to give up in despair, we can actually do better as a species to reduce our impacts on ocean environments and actually improve ocean health and sustainability.

Scientists and ocean use managers have increasingly developed tools to help in marine conservation. The rapid expansion of marine protected areas over the last decade, along with other management interventions, shows great promise in improving ocean sustainability. More recently, the concept of restoration has gained momentum, initially beginning with efforts to restore coastal wetlands but more recently expanding to develop strategies to expedite recovery of a wide variety of damaged marine environments.

The three chapters that comprise this section consider how to improve our relationship with the ocean. We draw the curtain to mark the end of our "play" by thinking about how marine biology research can contribute to a better future for Planet Ocean. The health of this planet hinges upon the health of the ocean, and as the next generation you will inherit a planet with significant problems but also significant opportunity to achieve the holy grail of sustainable use. Countries around the world are rallying around the concept of "sustainable blue economy," recognizing both the untapped potential to gain further benefits from the ocean but also the need to reduce, rather than increase, our footprint. We can do better. We must. And we can all make a difference.

28

Human Impacts on Marine Ecosystems

Humanity has drawn many benefits from the global ocean, but those benefits come with a cost that has increased in tandem with human population size and technology. The nature of damaging human activities varies in time and space, and at levels ranging from subcellular to ecosystem. **Pollution** refers to the direct or indirect addition of substances or energy into the marine environment (including estuaries) by humans that causes harmful effects on living resources, human health, hinders marine activities including fishing, and alters quality of sea water and its benefits. Different forms of pollution increasingly impact marine ecosystems, resulting in significant decreases in populations and health of numerous species, habitat degradation, and species loss. In this chapter we summarize the main threats to marine ecosystems and their sources. To fully understand the present-day status of marine species, populations, and habitats, we must consider historical data and the pristine conditions prior to impacts of human activities.

28.1 Historical Data

Historical data provide a benchmark to compare past and present ocean populations, and to understand the effects of loss on natural populations. These data are valuable not only to historians, but also to ecologists and environmentalists because they provide important information about factors that influence marine populations. For this reason, comparing the ocean past to what lives in the ocean today provides a means to test theories about effects of anthropogenic activities on marine populations. Humans have long hunted in inland waters and the ocean to harvest all kinds of organisms, and the effect of loss of animal populations has attracted much interest, especially in the last century. Ecological studies can evaluate current condition of harvested populations or those indirectly affected by fishing (e.g., habitat destruction), but such studies cannot look back in time. Keep in mind that modern fisheries science really began only after World War II or later, with relatively few quantitative assessments before then. Fisheries ecologist Daniel Pauly, coined the phrase "shifting baseline syndrome," referring to our tendency to define baselines of "normal" based on our own lifetime experiences, which rarely encompass the onset of human-induced change. Marine ecologist Jeremy Jackson built on this concept in looking at historical records for the Caribbean and criticizing ecologists for assuming that we can infer the historical state of an ecosystem from the first scientific descriptions; scientific assessments often begin long after the onset of significant human impacts. Jackson used trade statistics to demonstrate the removal of many thousands of turtles annually from the Caribbean (see below). This information, in tandem with the wide use of "Tortuga" in place names (Spanish for tortoise), clearly suggests that modern analyses of Caribbean coral reefs that began in the 1950s (or later) generally ignore the loss of what was once a major player in these ecosystems. Historical approaches can help in understanding ocean change from the past to the present, as shown more and more in recent studies. Historians have identified three chronological periods for which historical documentation exists.

1) Contemporary: This category encompasses specific documentation on fishing collected since ~1900, when statistics really began, through present day. These documents vary in nature, resolution, and completeness over time and space, but enable scientists to estimate the condition of stocks of fish and other marine animals. However, data collection methods often require validation, often varying in quality from one country to the next.

Marine Biology: Comparative Ecology of Planet Ocean, First Edition. Roberto Danovaro and Paul Snelgrove.
© 2024 John Wiley & Sons Ltd. Published 2024 by John Wiley & Sons Ltd.
Companion Website: www.wiley.com/go/danovaro/marinebiology

2) Archives: This category refers to information derived from port and customs archives in Europe, America, and colonial states between 1850 and 1900, prior to the development of statistical approaches, which can be compared and synthesized with data published subsequently.
3) Historical documents: Prior to 1850, scientific data were scarce and more difficult to interpret. Monasteries and other religious libraries contain data on fishes available in local markets. Historical paintings and even restaurant menus provide additional insights into the species readily available to fishers in times past. For a few species such as anadromous fishes, data series exist from 1500 to the present, but with few such examples. Historians have therefore collected information to formulate indices of fishing pressure that, when combined with recent data, can enable long-term analysis of marine ecosystems changes.

In addition to historical archives, researchers use palaeoecological data derived from geological records, acknowledging the important need to confirm the validity of these series with contemporary data sources. Palaeoecological evidence, such as the preservation of fish scales in anoxic sediments or isotopic and species analysis of mollusk shells and bones in **middens** (ancient garbage dumps), can provide insights into the ecology of oceans past. Unfortunately, the absence of historical or palaeoecological data for many exploited populations limits efforts to reconstruct oceans past. Even where such data exist, or can be derived from historical and ecological archives, analytical constraints often add considerable uncertainty in reconstructions.

The simplest hypothesis on the effects of loss in aquatic ecosystems posits that abrupt environmental changes cause significant shifts in exploited and ecologically related species. For example, historical data may reveal temporal variation in distribution or abundance of a species, which can contribute to long-term time series, though often requiring cautious interpretation. For example, data suggesting strong effects of sudden intrusion of salty water on Baltic Sea fishes can facilitate testing of ecologically significant hypotheses of major environmental change, whether gradual or cyclic, in the absence of other data sources.

Most past studies focused on single species capture data without explicitly considering interactions with other species. For example, different sources of information (including qualitative observations, anecdotes from fishermen, reports of impressive catches over the past 500 years not seen in recent times Indigenous knowledge) suggest reduced population size at high trophic levels following large-scale fishing pressure. More recent quantitative stock assessments of the development of fisheries confirm these reports of reduced biomass. Reduced biomass at high trophic levels suggests reduced primary production in the global ocean over time, or that energy flow through different trophic levels has changed over time. These findings demonstrate the utility of combining historical, palaeoecological, Indigenous, and contemporary data in evaluating change in marine food webs, and that some ocean change may arise from alterations at the base of the food web.

28.1.1 Marine Animal Populations in Human History

Ancient Romans ate fish, which they considered a prized food. Because of difficulties in transporting and storing fish in the Mediterranean climate prior to the invention of refrigeration, fresh fish appeared at dinner tables only near the coast.

Thus, only the rich and wealthy minority of the population could afford to eat fresh fish regularly. To address the unpredictable supply to markets, commercial fishmongers and Roman elite hobbyists alike also reared fish in artificial ponds or tanks (Figure 28.1). Captured marine fishes could be kept alive in tanks or were preserved in salt or vinegar. Salted fish supplied domestic markets, but their fresh counterparts were expensive delights reserved for the tables of wealthy families.

Fishes play a key role in "De re coquinaria" of Apicius – the only cookbook preserved from antiquity. The high fuel costs and the absence of tools for cooking limited consumption of hot meals to mobile "kitchens" on street corners. The most commonly used product of fish, a sauce called **garum** made from fermented fish, was used to flavor stews, soups, and many other dishes. Locals produced this sauce by fermenting entire fish (including gills and viscera) in large tubs over extended periods: they collected the resulting liquid in amphorae for preservation and transport. Although they used different types of fishes, most garum was prepared from anchovies and mackerel. The types of fish mixed in the tubs reflected the composition of local catches and point to important geographic differences in garum products. Because garum production required abundant fish, processing systems were concentrated along fish migration routes and in straits. Scientists have discovered large groups of Roman cisterns for salting fish on the Atlantic coasts of Portugal and Morocco, in the Black Sea, along the Strait of Gibraltar, in the Gulf of Lyon, and along the Strait of Sicily. Unlike fresh or even salted fish, a large portion of the Roman Empire population consumed garum. By comparing the amount of olive oil in jars to jars of garum, researchers estimate that the fish sauce was as pervasive in Roman kitchens as ketchup in kitchens in the United States today!

Dugongs, turtles, fishes, and large sharks were the first large vertebrates to disappear from coastal ecosystems because of indiscriminate human hunting more than a century ago (Box 28.1). Low fecundity, late maturation, and long generation-times have resulted in extinctions in some species and hindered or prevented the recovery of others. For example, published studies and skeleton collections confirm recent steep declines in common dolphin (*Delphinus delphis*) from the Mediterranean,

Figure 28.1 Evidence of life and historical fishing; (a) mosaic dating to the first century BC (found in Pompeii) showing some of the main species of marine animals caught from Italian waters and served on the tables of ancient Romans; (b) Pompeii, House of Menander, mosaic with swimmers and fishermen. (c) Torre Astura (RM), remains of an ancient Roman villa of the Republican-Imperial age (first century BC to first century AD) and its adjacent fish market. *Sources:* cayojuliocaesar / Adobe Stock, Arianne King / Public Domain.

where it was abundant in many areas from Roman times up to the 1960s. Today, common dolphins remain relatively abundant only in the Alboran Sea adjacent to the Atlantic, with relict groups in the Tyrrhenian and eastern Ionian Seas.

For millennia, coastal populations exploited whales near the shore using a wide variety of tools, such as spears, harpoons, nets, and poisons. Around 1800, the United States dominated whaling with sailing ships and the use of harpoons, both in the North and in the South Atlantic and extending to the Pacific and to the Indian Ocean. As early as 1850 whaling began to decline, as serial depletion of whale populations began. In the nineteenth century, US whalers captured an estimated 400,000 **great whales** (the largest whales, including 12 baleen species as well as sperm whales), of which 2/3 were sperm whales. By the twentieth century, the availability of motorized vessels and development of efficient hunting weapons such as exploding harpoons mounted on decks of ships attracted almost all coastal nations into whaling. Different regions adopted similar methods and efficiency, and the expanded whaling industry reduced many species to the brink of extinction, including bowhead whales (*Balaena mysticetus*), gray whales (*Eschrichtius robustus*), and right whales (*Eubalena* spp.). They also targeted many species of whales in the Antarctic Ocean, from the largest animal on Earth, the blue whale (*Balaenoptera musculus*), to the smallest whales. In recent years, some of the largest species of whales have shown modest, though slow increases in number, largely resulting from moratoria on endangered whales and a shift in the few remaining whaling nations to the smallest species.

For hundreds of years, green turtles provided food to indigenous peoples, settlers, explorers, pirates, and slaves in the Caribbean. Hawksbill turtles were prized for their shells, which were carved and turned into hair combs and fashionable

---

Box 28.1 Differences Between Terrestrial and Marine Extinctions

Species extinctions in the sea

Despite some similarities in the biological basis of extinction risk in marine and terrestrial environments, extinction in the ocean may be more problematic than on land. First, marine environments lack any equivalent to pollinators. This means that marine sessile species (i.e., those with reproductive challenges like plants on land) must rely on diffusion or transport of sperm and eggs, or clones. This difference increases the probability of an **Allee effect** (or density dependent mortality, where a disproportionately low recruitment rates occur at low population levels). The Allee effect could be important for marine sessile species for fertilization, so individuals must typically live close to each other. In short, low population abundances may seriously hinder the potential for sessile species populations to recover. For example, reduced abundances of the white abalone (*Haliotis sorenseni*) on the west coast of North America could lead to extinction; despite a 15-y ban on fishing, this species has not recovered, a pattern repeated in other species of abalone. High mobility likely reduces the importance of the Allee effect in marine fishes.

A second issue with extinction in marine environments relates to larval behavior of some species. Some invertebrate species are **gregarious** settlers, meaning that their larvae attach near conspecifics, limiting re-colonization and dispersion into new areas or currently unoccupied habitats. When these species create habitat for other species (such as reef-building corals), their colonization or re-establishment in a system may play a critical role in re-establishing associated species.

jewelry in the eighteenth and nineteenth centuries. In the Cayman Islands alone, the capture of 13,000 green turtles annually provided food for Jamaicans slaves. Both hawksbill and green turtles lay their eggs on Caribbean beaches, increasing their vulnerability. Today, green turtles and hawksbill turtles remain endangered globally.

Hard corals, oyster beds, seagrasses, and macroalgal forests (kelp forests) create important underwater landscapes in coastal habitats. Their wide distribution and three-dimensional complexity help to stabilize the physical environment and provide habitat for thousands of species. Many of these ecosystems are disappearing because of direct mechanical destruction (by dredging and trawling), or because of removal of top predators and cascading effects. The growth of dominant species of coral and oysters and associated accumulation of sediment and fragments of skeletons produce large coral reefs and oyster reefs. In fact, the skeletons and shells of these species remain intact long after they die, creating habitat for a wide range of other species. In contrast, macrophytes do not produce enduring structures, and three-dimensionality within their habitats quickly disappears when they die. Habitat loss of these ecosystem engineers results in decreased recruitment of associated species and, in many cases, leads to a decline in diversity and abundance.

Many Caribbean coral reefs remained stable until the 1980s, when they collapsed and decreased by >50%. Declines in species of elkhorn coral (*Acropora* spp.) and brain coral (*Montastrea* spp.) were particularly severe and resulted from a combination of: (1) unidentified pathogens, (2) bleaching (whitening) associated with abnormal changes in water temperature, and (3) massive growth of macroalgae (at the expense of corals) associated with the disappearance of their fish and invertebrate grazers as a result of overfishing and disease respectively.

Seagrass beds provide food resources and habitat, playing a particularly vital role as nursery areas for many fishes and invertebrates. They also stabilize coastlines by buffering waves and thus increase water transparency by trapping particles. The most common species in the Caribbean are *Thalassia testudinum*, *Syringodium filiforme*, and *Halodule wrightii*, whereas *Spartina alterniflora* dominates temperate latitudes. During the 1980s, seagrasses along the coasts of Florida and Europe experienced a mass mortality event resulting from an infectious disease, exacerbated by pollution and massive exploitation of turtles and dugongs, which graze on these plants and stimulate their growth. The many "Tortuga Bays" in the Caribbean attest to extraordinary abundance of green turtles (*Chelonia mydas*) when Columbus arrived, noting that the local name for seagrass (turtle grass) illustrates their significant effect. Turtles periodically return to feed on the same area of the seagrass bed, promoting continuous seagrass growth in those patches, and merging with adjacent patches to produce continuous beds. Turtle grazing efficiently "mowed" the seagrass beds, reducing the flow of debris and nitrogen to adjacent bare sediments by about 20x by metabolizing seagrass cellulose through microbial fermentation in their intestine and dispersing feces and urine away from the seagrass bed, thereby exporting nutrient and organic matter "subsidies" to these adjacent habitats.

In Africa, fishermen exploited stocks of the gastropod *Haliotis tuberculata* for tens of thousands of years before commercial fishing beginning in the 1950s led to rapid decline. Illegal fishing of this mollusk continues, exacerbated by the loss of lobster populations that consume small sea urchins, which prey upon *H. tuberculata* recruits uncontrolled and thus further limiting recovery.

Large vertebrates in the Caribbean and other tropical regions have declined significantly. Caribbean monk seal numbers among the 18 documented marine global extinctions. Many local extirpations and precipitous declines in numbers have also occurred. Big fishes, such as groupers and sharks, monk seals, manatees, crocodiles, and sea turtles have been overfished since the eighteenth century (Box 28.2).

Box 28.2 History of Marine Animal Populations

Knowing the past is the key to understanding the present and planning the future. The project The History of Marine Animal Populations (HMAP; www.hmapcoml.org) was a project of the International Census of Marine Life, *a* global research initiative to understand the diversity, distribution, and abundance of life, past, present, and future in the global ocean. HMAP's primary objective was to study marine history, specifically the interaction of humans with the sea. The project described fishing activities (tools, fishing effort, fishing areas), qualitatively and quantitatively described marine populations over time (presence/absence, frequency, size, distribution, habitat, etc.), evaluated changes using a wide range of indicators (population, community, ecosystem), and developed models to assess the role of fisheries and other human impacts (e.g., nutrient loading, overfishing) in order to hypothesize ocean change under human exploitation prior to modern scientific data. The project reported major losses in ocean productivity, and particularly the loss of large individuals (Figure 28.2).

Box 28.2 (Continued)

Figure 28.2 Large halibut caught near Cape Cod, USA. As a result of overfishing, such massive fishes are rarely caught today in this region. HMAP studied changes in interactions between humans and marine biota over the course of history. *Source:* Artemas Ward / Wikimedia Commons / Public Domain.

28.2 Biodiversity Loss

Conservationists have debated marine extinction in the ocean, in part because of fewer documented marine extinctions compared to land or fresh water, which has led to widespread belief that marine extinctions have been, and remain, few in number (Box 28.3). Marine extinctions probably occur more often than we realize because of the many unknown species that make detection of loss much more difficult than in better-known environments. Scientists also debate the widespread and persistent perception of greater resilience in marine species because they are at least assumed to be highly fecund, widespread in distribution, fast growing and, therefore, presumably able to withstand high levels of exploitation and recover quickly when at low abundances. Growing scientific evidence indicates attributes in many marine species similar to those that increase vulnerability of freshwater and terrestrial species. Long life cycles and late sexual maturity in some marine species place them at a distinct disadvantage when subjected to exploitation. We also now recognize much more limited distributions in many marine species given increasing recognition of the heterogeneous nature of the marine environment and/or the limited dispersal ability of many juveniles and adults. In addition, examples such as the collapse of northern cod in Atlantic Canada clearly demonstrate that high fecundity does not guarantee high rates of reproduction or effective insurance against overfishing.

Understanding the extent of recent extinctions can help in understanding how modern extinction rates compare to those over geological time, to evaluate whether we should consider current trends as "normal" or a major cause for concern. A broader view of extinction processes can help to identify those species, taxonomic groups, and regions/habitats most at risk to extinction. Extinction is, however, often a non-random process in which life history characteristics and rarity, body size, and sensitivity to environmental stressors such as pollution play an important role in determining risk.

The effects of biodiversity loss on an ecosystem also depend on the sequence of species loss. Marine extinctions are likely ongoing at an enormous scale, often with local losses that cumulatively represent a significant problem. Increasing pressure from human activities places marine mammals, birds, and turtles, as well as fishes and marine invertebrates under pressure. In many cases, humans have already driven some species to extinction and place others on the brink. In the Mediterranean Sea, human activities in the last century have reduced shark populations by >90% for some species, and up to 99% for others (Figure 28.3). Human impacts seem to affect the largest species more severely than smaller species (Figure 28.4), and, in the case of sharks, large taxa are now so scarce that they no longer play a significant role in ecosystem functions.

28 Human Impacts on Marine Ecosystems

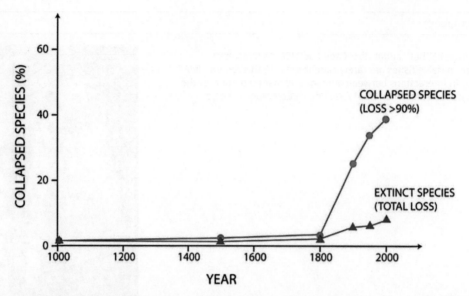

Figure 28.3 Loss of biodiversity (in terms of species) in marine ecosystems. The lines indicate collapsed species (red circles, > 90% decline) and extirpated species (blue triangles, 100% decline) over the past 100 years. The 12 regions considered were within the waters of Europe, North America, and Australia.

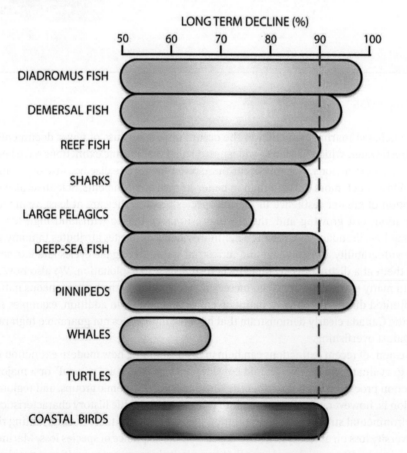

Figure 28.4 List of large, endangered taxa that declined drastically in abundance because of human activity.

The threat of marine biodiversity loss results directly from the influence of human activities and increases in human population growth along coasts. Humans have profoundly altered the ocean environment, transformed coasts and changed global biogeochemical cycles, often by fishing a wide range of species and transferring others from one marine location to another, where the often cause major change. Some of the greatest losses have occurred in coastal and wetland habitats that create a critically important interface between land and sea; for example, human activity has transformed or destroyed about 50% of mangroves globally. Humans extract an average equivalent of 8% of ocean primary production, but as much of 25% of production in upwelling and 35% in temperate continental shelf systems. Catches generally focus on top predators, whose removal often greatly alters marine ecosystems. In the North Sea, experts estimate local extirpation of about 50 species (including gray whales) in historical and prehistoric times. Direct human exploitation caused ~25 of these 50 extirpations (mainly involving large species), whereas habitat loss caused ~17 of the losses and pollution another 3.

Box 28.3   Marine Species Extinct or Close to Extinction

Potential extinctions caused by habitat destruction
- *Syngnathus affinis*, currently, among the rarest fish species in the Gulf of Mexico, may already be extinct. In the 1970s it commonly occurred in surface waters above seagrass beds in the United States and Mexico, but its habitat is rapidly disappearing.
- *Phyllaplysia smaragda*, a once common molluscan nudibranch described in the early 1970s from seagrass beds on the Atlantic coast of Florida, has not been seen since 1982.
- *Cerithideopsis fuscata*, a mollusk in San Diego Bay (United States) has not been seen since 1935 and is almost certainly extinct. The loss and modification of muddy seabed and coastal marshes likely played a major role in its disappearance.
- *Lottia edmitchelli*, a common mollusk in Pleistocene deposits from Southern California, was last collected live before 1860; increasing sedimentation and destruction of coastal habitats seems to have caused population declines.
- *Totoaba macdonaldi*, a large species of fish in the Gulf of California (USA), was reduced to very few individuals through over-exploitation and habitat alteration. Described in 1809 as very abundant, this species would migrate annually to lay their eggs near the mouth of the Colorado River, where it was intensely exploited. Dikes built in the Colorado River reduced freshwater flows and juvenile survival, resulting in population collapse.
- *Lottia alveus*, a limpet, occupied eelgrass beds on the Northeastern coast of North America, but went extinct in 1930 when disease eliminated large swaths of its *Zostera* spp. habitat.

Potential extinctions caused by episodic events
- *Millepora boschmai* (fire coral) reduced sharply in coverage following an intense El Niño event between 1982 and 1983 that caused a long period of warming seawater on eastern Pacific reefs. This recently described coral was initially considered extinct, but five surviving colonies were subsequently discovered. Its survival remains precarious. Many other species of corals have suffered massive contractions as a result of bleaching and related phenomena such as El Niño.
- *Azurina eupalama*, the Galapagos damselfish, once quite common in the Galapagos and Cocos Islands, has now disappeared from the Galapagos. Its status in the Cocos Islands remains uncertain.

Potential extinctions caused by the introduction of other species
- *Brachionichthys hirsutus*, the spotted handfish, is endemic to estuarine environments in South Australia and very limited in distribution. Predation on its benthic eggs by the starfish, *Asterias amurensis*, introduced from Japan has brought this species to the brink of extinction.

Potential extinctions caused by overfishing
- *Haliotis sorenseni*, white abalone, occurred at densities of >10,000 indiv ha^{-1} between Northern Mexico and Southern California. Recent estimates place its population size at a few tens of individuals, and its global extinction is imminent.
- *Dipturus laevis*, the barndoor skate, was once common in continental shelf surface waters, has virtually disappeared from the northern Atlantic. Remaining individuals live at depths >1,000 m.
- *Pterapogon kauderni*, the cardinal fish, occupies a very narrow distributional range in shallow lagoons within a small area of Indonesia. The absence of a planktonic dispersal phase may explain its limited distribution; this unusual characteristic for marine fishes occurs more frequently in freshwater fishes. First described in 1933, aquarists "rediscovered" this species in 1994 and enthusiastically began capturing them from the wild for the aquarium trade. Although now abundant in captivity, overharvesting has pushed this species to the brink of extinction in the wild.
- *Latimeria chalumnae*, the coelacanth, was discovered as a "living fossil" in 1938 after it had been assumed extinct for 66 million years. This extremely rare fish was known only in the Comoro Islands, until other, probably genetically distinct, populations were discovered in northern Sulawesi. Its specialized habitat (caves on rocky slopes 100–300 m depth), long life cycle with slow growth, and low recruitment rate explains its rarity. Populations in the Great Comoros decreased by more than 30% between 1991 and 1994 (from ~550–650 to 340–440 individuals), likely because of bycatch in an expanding fishery.

28.3 The Main Threats to Marine Life and Ecosystems

The predominant threats to marine biodiversity include: (1) contamination; (2) habitat degradation, fragmentation, and destruction, (3) overfishing, (4) introduction of nonnative or invasive species, and (5) climate change. The following sections briefly expand on the nature of these threats.

28.3.1 Contamination

Contamination, one of the main source of pollution (Box 28.4), arises from the introduction of different types of pollutants that span from chemical groups (hydrocarbons, **heavy metals,** referring to inorganic substances such as copper that become toxic when concentrated, nutrients that lead to eutrophication, and synthetic chemicals now >70,000 in number) to **pathogens** (referring to disease agents such as viruses and bacteria) as well as sound (ship engines and echosounders, seismic exploration) and light pollution. Pollutants occur in particularly high abundances at the seafloor, at pycnoclines, and at the air-sea interface. Pollution impacts may cause problems at different levels, from individual cells (metabolic impairment, poisoning) to whole organisms (physiological change, output, disease susceptibility, increased larval mortality) to populations (reduced reproductive efficiency, changes in size structure) and communities (changes in species composition, altered food webs). Pollution, whether chronic or acute, may result from addition of excess nutrients and associated eutrophication, pathogens (e.g., fecal coliforms, *E. coli*), heavy metals, or a variety of toxic chemicals that can affect the reproductive biology of many marine species. Collectively, these impacts can profoundly alter biogeochemical cycles and other processes fundamental to overall healthy functioning of ecosystems.

28.3.2 Habitat Degradation, Fragmentation, and Destruction

Degradation, fragmentation, and destruction of marine habitats result from profound changes often driven by economics, such as coastal development for housing and tourism. These activities can destroy, erode, and fragment natural environments such as seagrass beds, salt marshes, mangroves, coral reefs, and deep coral reefs. For example, anchors and trawling activities fragment meadows of seagrass, *Posidonia oceanica*, promoting erosion and deterioration of meadows. A wide range of factors based either on land, such as increased sedimentation and sewage input, have already seriously damaged or destroyed some 60% of coral reefs. Even deep-water (or cold-water) corals, which occur at depths of 300–1,000 m and create habitat for other species akin to tropical coral reefs, face increasing damage and fragmentation through trawling activities, which penetrate to ever increasing depths.

28.3.3 Overfishing

The single greatest effects of human activities on ocean life link to fisheries activities, which create impacts through habitat destruction (caused by fishing gear), population decreases in target species through overfishing, alteration of food webs, bycatch, and ghost fishing. Although ship's logs demonstrate significant declines in fisheries in locations such as the North Sea and New England that date back hundreds of years, commercial (market) fisheries boomed after World War II driven by national needs to rebuild economies. Moreover, significant advances in ship technology from navy research increased

---

Box 28.4   Impacts on Deep-Sea Ecosystems

Despite the remoteness of deep-sea environments from pollution sources, numerous studies demonstrate the presence of contaminants in the deep sea. Extraction of hydrocarbons and minerals, dumping of organic and inorganic waste, including radioactive and toxic compounds, and deep-sea trawling illustrate the ever-expanding footprint of anthropogenic activities. A growing body of research demonstrates the presence and, sometimes, the bioaccumulation of toxins in deep-sea organisms (such as lead accumulation in Portuguese dogfish – *Centroscymnus coelolepis*). Trawling increasingly alters continental margin habitats to depths of 1,500 m and beyond, including major damage to deep water corals; improved technologies and fishing pressure in shallow depths has depleted fish stocks and pushed fishing effort into greater depths. Recent studies suggest that fishing impacts have reduced abundances of some fishes by 5075%, with unknown effects on biodiversity and function.

capacity to fish at greater depths, greater distances, sweeping greater volumes of water or seabed, and using improved capacity to locate fish. For some species, recreational fisheries also play a role in that recreational fishers can outnumber commercial fishers, and thus carry significant political clout, particularly given the lucrative spinoff industries such as outfitters and recreational boat builders. The wide perception of marine biota and ocean environments as common (rather than private) property and a right rather than a privilege complicates management in terms of how to divide the resource and who should take responsibility for ensuring sustainability.

Fisheries encompass diverse strategies and needs, resulting in corresponding diverse impacts. Pelagic fisheries, whether in neritic or oceanic realms, typically focus on schooling smaller-bodied fishes or squids in coastal waters, or large migratory species such as tunas in oceanic waters. The highly migratory nature of some larger species such as bluefin tuna, a fishery where a single fish can fetch US$1 million, creates particularly complex management issues as they move through international waters and jurisdictional waters of different nations. In contrast, fisheries for species that reside just above (shrimp, and **groundfish** such as cod and flatfish) or in/on (scallops, lobster, clams) the seabed typically use bottom contact gear, some of which damages seafloor habitat and produces significant bycatch. For example, mobile gear such as bottom trawls and dredges dragged across the bottom often collect far more bycatch than the target species.

As noted above, in describing impacts on individual species, fishing (e.g., bottom trawling) can cause and aggravate effects of habitat degradation, often impacting multiple species by removing large numbers of individuals and biomass from ecosystems. Fishing gear also results in damage to organisms and significant bycatch in groups such as mollusks and crustaceans, thereby altering biodiversity.

Exploitation of marine resources increasingly shows clear signs of both direct and indirect impacts. Direct impact arises from existing and future exploitation of resources in marine systems, such as fishing, and extraction of oil and gas, natural deposits of metals, and bio-prospecting. Additional effects include direct seabed use such as pipelines, cable routing, carbon sequestration, and through pollution, such as contamination by waste disposal, dumping, maritime activities, and accidental oil spills during transport or mining activities (Figure 28.5).

Figure 28.5 Effects of excessive fishing activity on fish abundance and size. The images above show fishing activities from 1963 to 2001, during which catches declined globally and average fish size decreased, sometimes dramatically. In practice, fishing generally selectively removes larger fishes over smaller ones. The bottom images compare catches from a fishing contest in 1958 in Key West, Florida (USA), dominated by groupers (Epinephelinae) to the 2007 winners in which fish size decreased dramatically and species composition changed completely. *Source:* Wil-Art Studio/Florida Keys History Center-Monroe County Public Library, Loren McClenachan.

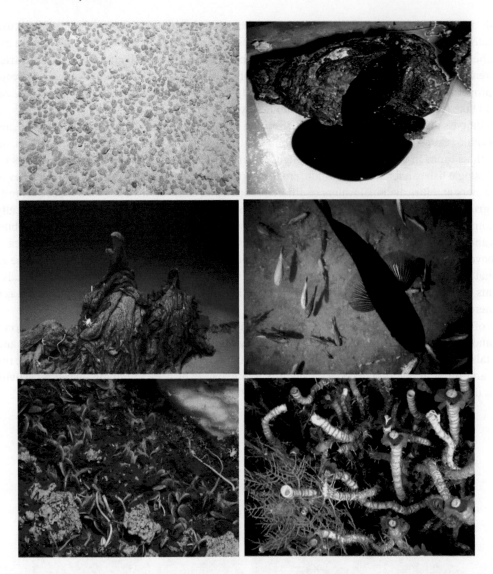

Figure 28.6 Some potential untapped marine resources, include huge reserves of polymetallic nodules, solid or semi-solid hydrocarbons, floating cages for fish farming, and untapped diversity in deep marine systems including species new to science with potential bioprospecting utility, and wind energy. *Source:* Abramax / Wikimedia Commons / CC BY-SA 3.0, Bárbara Ferreira / MARUM - Center for Marine Environmental Sciences / CC BY-NC-SA 3.0, NIWA, Ian MacDonald, Texas A&M University, Corpus Christi/Census of Marine Life.

28.3.4 Extraction of Abiotic Resources

The extraction of abiotic resources represents one of the main threats to marine life. Exploitation of hydrocarbons can cause widespread release of these contaminants, often leading to chronic pollution. In addition, accidents such as the Deep-water Horizon spill in the Gulf of Mexico or the sinking of several oil tankers are responsible for some of the most important environmental disasters in the last century.

The exploitation of minerals and geo-resources from the seabed represents another potentially important source of impact. The collection of these raw materials poses serious risks of destruction of vulnerable and unique habitats in shallow and abyssal seafloors (Figure 28.6).

28.3.5 Non-indigenous or Alien Species

Invasive species seriously threaten the biodiversity of native species, in that alien species introductions can lead to increased competition and predation. Mechanisms of introduction vary with location, but major vectors include transfer of adults and larvae in ship ballast water, individuals attached to hulls of ships or boats, through movement of brood stock and associated parasites and disease in aquaculture, inadvertent introduction by scientists, and deliberate release of invasive species in new environments. Some introductions date back centuries. The common periwinkle, *Littorina littorea*, which occurs widely through the intertidal of North America, likely arrived on the hull of a ship in the 1800s or possibly earlier on Viking ships. Globalization has increased the movement of goods around the world, and therefore the movement of ships and ballast water.

San Francisco Bay offers one of the best-known examples of major biodiversity change associated with the hundreds of invasive species that make the estuary one of the most invaded marine environments in the world. The Asian clam, *Potamocorbula amurensis*, arrived in the 1980s and exploded in number during an unusual hydrographic event that resulted in them taking a foothold. As voracious suspension feeders, they now largely strip the bay of phytoplankton, with obvious ramifications for other species. Experts attribute the arrival of another invasive species, the green crab, *Carcinus maenas*, to lobster shipments, or possibly in ballast water; this species has caused major declines in native crabs and clams. On the east coast, green crab has altered eelgrass communities, with additional impacts on American lobster and other native species.

The arrival of invasive species on ship hulls and in ballast water often occurs over time scales of decades. Similarly, major alterations of shoreline have rapidly accelerated invasions, such as the Suez Canal, known as the **Lessiepsian migration** (see also Chapter 8), from the Red Sea to the Mediterranean (Figure 28.7).

Marine biota changes frequently, even within time periods of a few years. Rare species can become common or vice versa. Natural, anthropogenic, and climatic variables primarily drive the disappearance of once common species. The interglacial phase that currently characterizes the Mediterranean means that Atlantic thermophilic species, which prefer warm waters, comprise most new arrivals.

Many non-indigenous species (NIS) enter a new system through human activities, as numerous species are intentionally or accidentally introduced through fouling on ships, ballast water, transfer of organisms for aquaculture, trade of live bait, marine food packaging with live algae, and through the aquarium trade. Migrant species penetrate discontinuously, often with occasional sightings and few direct observations on their establishment, suggesting **pseudo-populations**, referring to non-native populations, able to settle and grow, but not to reproduce. Occasional "alien" presence requires the arrival of new larvae from the source population. Most non-native or non-indigenous species are thermophilic. The list of animals and exotic plants that have newly invaded oceans and seas continues to grow. Increasing water temperatures that create corridors connecting regions partially explain variation in distribution of some species. Non-native species arrivals have displaced some native species, sometimes leading to cascading effects on food webs.

28.3.6 Global Climate Change

Anthropogenic activities are progressively increasing atmospheric concentrations of CO_2 and fluxes of other greenhouse gases (methane, CH_4, and nitrous oxide, N_2O), which trigger global climate change and the consequent warming, oxygen depletion, and acidification of the global ocean, altered precipitation regimes, as well as increased ice melting. Changes in physico-chemical conditions also induce shifts (generally, a decrease) in global primary production and carbon export to the ocean interior.

Climate change now produces the most pervasive effect of human activities on the ocean globally, with effects on almost every ocean environment and multiple direct and indirect impacts (Figure 28.8).

Increased CO_2 associated with fossil fuel emissions in particular, as well as other less abundant greenhouse gases such as methane, have created a greenhouse effect. Earth's atmosphere retains more radiative energy than in the past because of increased concentrations of these gases, much like the glass on a greenhouse lets radiative energy in but retains much of that heat. Keep in mind that the ocean contains 50x more CO_2 than the atmosphere and perhaps 16x more carbon than the terrestrial biosphere. The heat capacity of the upper 3.2 m of the ocean roughly equals that of the entire atmosphere, meaning that the ocean has also absorbed some 90% of greenhouse warming since 1971 and surface waters have warmed more than 1 °C globally since the late 1880s. Although 1 °C may not seem like much, this warming has caused multiple changes in ocean life. Warming ocean temperatures and increased dissolved carbon dioxide, changes in intensity or direction of ocean currents, ocean acidification, increased ice melt and sea level rise, changes in vertical stratification, alteration of biogeochemical and metabolic pathways, and changes in food webs and trophic links all illustrate indirect effects that result from direct effects on environmental conditions (Box 28.5).

An increasing number of studies indicate rapid changes in physicochemical conditions in the deep ocean and predict negative effects of global change in terms of all of these variables (Box 28.6). Models predict that temperatures at all depths, including the abyss, could increase by >1 °C over the coming decades. Marine habitats under areas of deep-water formation may experience reductions in oxygen concentration in the water column by as much as 0.03 ml l^{-1} by 2100. Furthermore, bathyal depths (200–3,000 m) will show the most significant reduction in pH values in all oceans by the year 2100 (0.29 – 0.37 pH units), accompanied by a decline of 3.7% in the North-East Pacific and Southern Oceans. Yet, the most noticeable predicted change will be the reduction of organic matter supply, especially in the Indian Ocean (with decreases of 40–55% by the end of the century).

All of these changes influence the biodiversity and functioning of marine ecosystems. Marine organisms are key actors in cycling key elements driving ecosystem processes. Ongoing changes profoundly influence these organisms, but to a

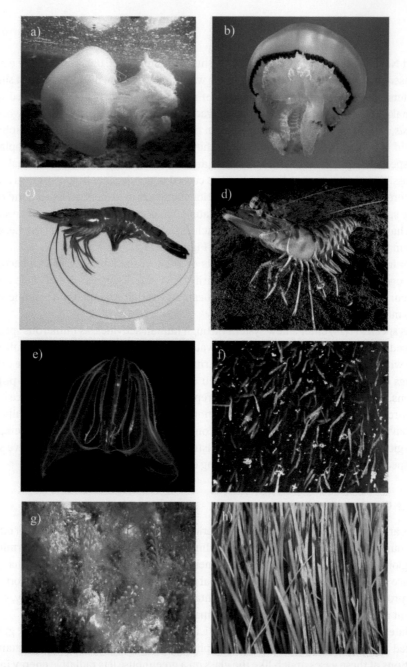

Figure 28.7 Examples of alien species and alien species that now compete with Mediterranean species: (a) the jellyfish, *Rhopilema nomadica* has replaced the indigenous species *Rhizostoma pulmo* (b) in some areas of the Levantine Sea; (c) the shrimp, *Penaeus japonicus* is progressively replacing *Penaeus kerathurus* (d); (e) the comb jellyfish *Mnemiopsis lleydi* has invaded the Black Sea, creating significant problems for fishermen because they prey on larvae and juveniles of small pelagic anchovies (f); (g) the invasive alga *Caulerpa sp.* competes, alters habitat, and reduces survival of other macrophytes (e.g., *Posidonia oceanica* (h) in Mediterranean coastal habitats. *Sources:* Roberto Danovaro (Book author.), GFDL&CC / Wikipedia Commons / CC BY-SA 3.0.

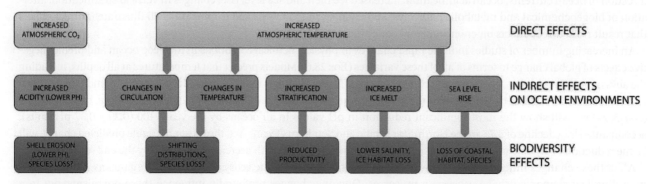

Figure 28.8 Direct and indirect effects of climate change on marine biodiversity.

> **Box 28.5 Marine Regions Most Impacted by Global Change**
>
> The environmental settings of the Mediterranean Sea and the Arctic Ocean (with an average depth of ~1,450 and 2,000 m depth, respectively, vs 3,750 m of the global oceans) suggest they will react faster to global changes than the larger ocean basins. The presence of small gyres (eddies) that characterize semi-enclosed basins such as the Mediterranean basin have implications for upwelling of deeper waters and effects on primary productivity. This upwelling consequently affects flux of organic matter settling to the deep seafloor. The trajectories of deep and bottom currents are largely unknown, but studies have documented strong currents up to 1 m s^{-1} in submarine canyons, in relation to climate driven episodic events. Rapid vertical transport of surface waters to great depth occurs as a result of dense water convection when evaporation and cooling increase surface water density. These phenomena, known as cascading, occur periodically over short time periods (weeks). Limited average depth of the Mediterranean basin results in relatively rapid (50–80 y) deep-water turnover compared with wider oceanic regions, but this advantage is largely compromised by vulnerability to climate change and the much higher rates of deep-water warming, which have accelerated in the last decades.

> **Box 28.6 Climate Change in the Abyss**
>
> Until just a few years ago, researchers considered deep-sea environments impervious to significant changes related to climate other than over extended time periods (decades to hundreds of years) associated with estimated turnover times for deep water masses. Studies demonstrate a direct link between **orbital forcing**, the effect on climate of slow changes in the tilt of the Earth's axis and shape of the Earth's orbit around the sun, as well as structure and productivity of the surface of the ocean and biodiversity in deep environments over long time scales (10^3–10^4 years). Further studies have linked long-term changes in deep-sea benthic communities at 4,100 m to climate phenomena such as El Niño – La Niña, likely driven by changes in food availability. More recent studies demonstrated that episodic events on a regional scale can significantly impact ecosystem functioning and deep-sea biodiversity. Given new recognition of more widespread and frequent episodic events in deep-sea ecosystems than previously realized, including delivery of large amounts of organic and inorganic material, this topic will likely become a priority for future research.

different extent for different latitudes and biogeographic regions, with stronger impact on marine ecosystems at high latitudes. Anticipated decreases in primary production at tropical and mid-latitudes will alter the quantity and quality of food supply to the seafloor, with downstream consequences on organic matter cycling and supply of ammonia and nitrate needed for sustaining the metabolism of all organisms.

Numerous lines of evidence suggest that warming of surface waters has already caused biogeographic shifts in distributions of a wide range of organisms from microbes to vertebrates. Warming creates a particularly difficult challenge for polar species, which cannot simply migrate further from the equator because there is nowhere further to go. Deep-sea species that tolerate extremely narrow temperature ranges cannot migrate to colder, deeper water if such habitat does not exist. For many species, habitat requirements other than water temperature may preclude simply moving in order to address temperature issues. For example, shifting their distributions poleward may not help species that occupy offshore banks surrounded by deeper water because the shallow banks may not exist poleward. Many species also depend on ocean currents to deliver eggs and larvae to suitable nursery habitat, and spatial shifts, either in species or in ocean currents, may alter this relationship.

Ocean acidification illustrates some effects of global change related to carbon dioxide emissions. Carbon dioxide concentrations have increased from 280 ppm (parts per million) in 1750, to >400 ppm today. The global ocean has absorbed about a third of this excess carbon dioxide, but at a significant cost; the average pH of the global ocean has decreased ~0.1 units from pre-industrial levels. Although this number may seem small, the fact that emissions from human activities have altered 1.35 million km^3 of seawater should give us all pause because this change has significant biological consequences. Moreover, if current carbon dioxide emission trends continue, the pH of the surface ocean will decrease 0.14–0.35 units by 2100. Ocean acidification decreases the ability of the ocean to absorb additional atmospheric carbon dioxide, implying that future CO_2 emissions will likely accelerate global warming. Increased ocean acidity reduces formation and accelerates breakdown of carbonate shells formed as calcium magnesium (coralline algae), aragonite (corals and pteropods) and calcite (some species of phytoplankton such as coccolithophores and foraminifera), with serious implications for all those species.

Changes in ocean oxygen may arise from temperature increases, in the stronger thermocline, in basin-scale multidecadal variability, in slowdown of oceanic overturning, and a potential increase in biological consumption. Although scientists have documented periodic hypoxic events in many different regions of the world, the lack of evidence of hypoxic abyssal conditions likely reflects very low inputs of organic material to the seafloor. Although increasing temperatures might increase the potential spread of oxygen minimum zones (OMZs), decreased primary productivity might balance such a risk. Direct effects from depletion of O_2 levels and rising water temperatures may impact embryonic survival rates of vulnerable oviparous (egg-laying) elasmobranchs.

Climate change can affect nutrient concentrations and rates of primary production, fluctuations in the depth of the mixed layer, and the duration of vertical stratification. In estuaries, for example, changes in intensity and seasonality of circulation patterns may influence larval dispersal mechanisms. Warmer temperatures and intensification of stratification can influence both the abundance and the production of phytoplankton. Because phytoplankton remove carbon dioxide from the atmosphere and transfer carbon to higher trophic levels, any change in timing, abundance, or species composition of primary producers could potentially affect the rest of the food web. In offshore ecosystems, recruitment dynamics affect population dynamics of many marine species, adding new individuals to the population each year. Cold-water species often synchronize recruitment with seasonal phytoplankton production cycles. If warming alters the timing of reproduction of these species, poor recruitment may occur when larval timing does not coincide with availability of fundamental resources such as phytoplankton or small zooplankton (the **match-mismatch hypothesis**).

Direct effects of human activities that operate over different temporal and spatial scales (Figure 28.9), such as habitat destruction, over-exploitation, and pollution, can also interact with indirect effects associated with global climate change and alteration of ocean biogeochemical cycles, leading to changes in species and even extirpations. Evidence at regional scales from a wide range of ecosystems such as estuaries and coral reefs indicates rapid decreases in abundances of key species, biodiversity, and entire functional groups.

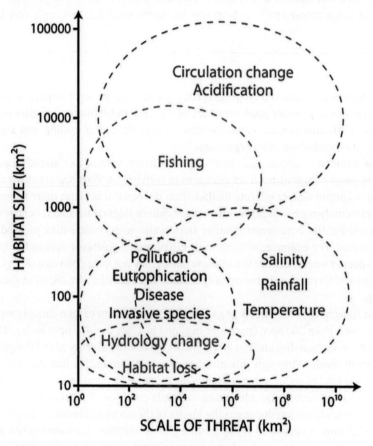

Figure 28.9 Scales of different threats that act in tandem with climate change.

28.4 Synergistic Impacts on Marine Ecosystems

Over the past five decades, human activities have left their mark on all marine systems. Several studies demonstrate that even deep-sea ecosystems, long considered pristine, now show clear signs of alteration and the presence of anthropogenic pressures as illustrated by the documentation of microplastics in the Mariana Trench at >10,000 m depth.

Available data also indicate that different anthropogenic stressors (Figure 28.9) do not act independently, rather they have cumulative and synergistic effects on the marine biota. Among the multiple-stressors impacting marine ecosystems, exploitation of biotic and abiotic resources represents one of the most significant threats (Figure 28.10). When taking cumulative / synergistic effects into account (Figure 28.11), the overall threat of all these impact sources on marine biodiversity increases considerably.

The impact of global climate change and cumulative effects associated with other current anthropogenic stressors adds particular concern (Figure 28.11). Despite our inability to predict type and magnitude of biological responses to ocean biogeochemistry change related to global change, existing knowledge suggests a major selective pressure upon species, their patterns of body size, abundance, distribution, and species richness, as well as ecosystem functioning. Cumulative effects will likely magnify biological and ecological responses, especially if they interact with other stressors, necessitating multiple physiological adaptations to survive. Coral reefs, in which massive bleaching and growth reduction have

Figure 28.10 Some of the many stressors that can compromise marine ecosystem functioning, with dramatic consequences on goods and services for humans. These include: Maritime traffic, oil and gas extraction, increased catastrophic events along the coast, trawling effects on coral reefs and other ecosystems, loss of nets that accumulate on and above the seabed and continue to "ghost fish," effect of climate change on marine ecosystems, release of waste at sea, and coral bleaching related to climate change. *Sources:* Unknown Source / Wikipedia Commons / Public Domain, Jim Evans / Wikimedia Commons / CC BY-SA 4.0, The National Guard/Wikipedia, Cliff / Flickr / CC BY 2.0, NASA / Jeff Schmaltz / Wikimedia Commons / Public Domain.

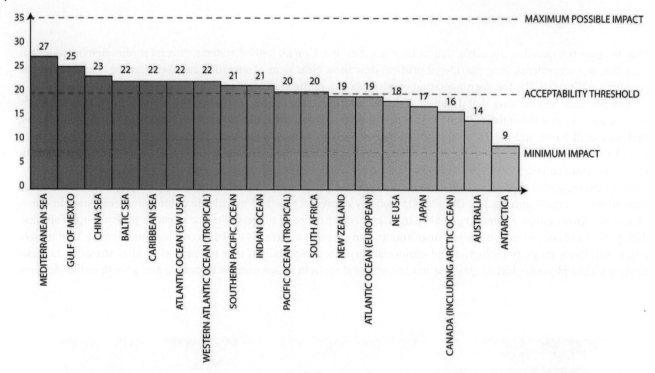

Figure 28.11 Cumulative impacts of habitat destruction, overfishing, alien species, climate change (temperature), hypoxia, acidification, and pollution (each pressure is ranked from 1 (low) to 5 (highest).

been linked to relatively modest contemporary warming and acidification, provide an excellent example of this scenario. Even deep-sea ecosystems, for which the magnitude of biogeochemical shifts will be smaller (Figure 28.12), may undergo substantial biological responses, mainly because of the stability of the deep ocean, and thus faunas adapted to narrower ranges of environmental variation than those in shallow marine habitats. Current predictions suggest that coastal habitats, such as seagrasses, mangroves, rocky reefs, and shallow soft bottoms will experience the greatest impacts. In term of species or taxa, pinnipeds, squids and other cephalopods, cetaceans, and euphausiids appear most vulnerable to cumulative impacts associated with global change.

These changes will bring many negative consequences, including:

1) changes in distribution, structure, and production of planktonic as well as benthic coastal and deep-sea communities;
2) increases in variability of environmental conditions and processes that affect marine organisms;
3) reduced connections between deep and coastal environment that alter biogeochemical cycles;
4) limited connectivity amongst marine populations and habitats;
5) altered primary production and fishery resource production;
6) altered life histories and reproductive cycles of vertebrate and invertebrate species, increasing vulnerabilities and extinction rates;
7) altered marine food webs as a result of decoupling of timing or total production and consumption;
8) decreased capacity to tolerate changes related to global climate change;
9) limited resilience to recover the pre-impact conditions;
10) decreased capacity to limit the establishment of non-indigenous species;
11) increased epidemiological phenomena such as rapid increases or decreases in disease vectors at the expense of communities;
12) increased vulnerability to mass mortality events.

Despite these daunting and seemingly hopeless challenges, we can do better as a species in working toward sustainable ocean use. The next two chapters describe some of the solutions and strategies that can halt and even reverse these declines.

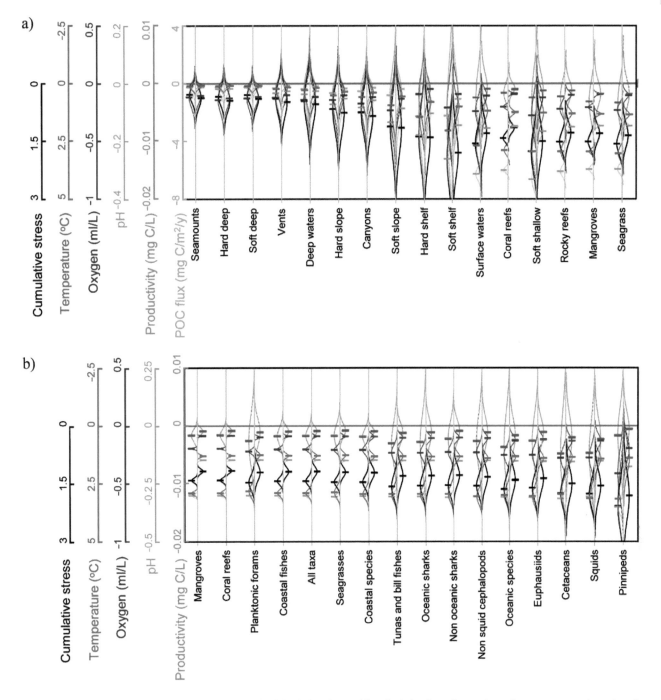

Figure 28.12 Mean (horizontal dashes) and standard deviation (curved lines) of absolute change in each parameter projected to the year 2100 for different marine habitats (Plot A), as well as biodiversity hotspots for individual taxa (Plot B). https://doi.org/10.1371/journal.pbio.1001682.g005. *Source:* Camilo Mora et al 2013 / PLOS / CC BY 4.0.

Questions

1) How can historical data help us understand current impacts on humans on marine life?
2) Which marine taxa have gone extinct?
3) What are the main differences between marine and terrestrial extinctions?
4) What is the primary threat to marine biodiversity?

5) Describe the impacts of habitat destruction.
6) Describe the impacts of overfishing.
7) Describe the impacts of pollution.
8) How do non-indigenous species impact marine biodiversity?
9) How does global climate change alter marine biodiversity?

Suggested Reading

Baillie, J.E.M., Hilton-Taylor, C., and Stuart, S.N. eds. (2004) *IUCN Red List of Threatened Species A Global Species Assessment*. IUCN, xxiv + 191 pp.

Barry, J.P., Buck, K.R., Lovera, C.F. et al. (2004) Effects of direct ocean CO injection on deep-sea meiofauna. *Journal of Oceanography*, 60, pp. 759–766.

Baum, J.K. and Worm, B. (2009) Cascading top-down effects of changing oceanic predator abundances. *Journal of Animal Ecology*, 78, pp. 699–714.

Beaugrand, G. (2014) *Marine Biodiversity, Climatic Variability and Global Change*. Routledge.

Berkes, F., Hughes, T.P., Steneck, R.S. et al. (2006) Globalization, roving bandits, and marine resources. *Science*, 311, pp. 1557–1558.

Boero, F. (2012) Animals and global change. *Italian Journal of Zoology*, 79(1). p. 1.

Brierley, A.S. and Kingsford, M.J. (2009) Impacts of climate change on marine organisms and ecosystems. *Current Biology*, 19, pp. R602–R614.

Bryan-Brown, D.N., Connolly, R.M., Richards, D.R. et al. (2020) Global trends in mangrove forest fragmentation. *Scientific Reports*, 10, pp. 1–8.

Carlton, J.T. (1985) Transoceanic and interoceanic dispersal of coastal marine organisms: The biology of ballast water. *Oceanography and Marine Biology: An Annual Review*, 23, pp. 313–371.

Carlton, J.T. and Geller, J.B. (2000) Ecological roulette: The global transport of nonindigenous marine organisms. *Science*, 261, pp. 78–82.

Casagrandi, R. and Gatto, M. (2002) Habitat destruction, environmental catastrophes, and metapopulation extinction. *Theoretical Population Biology*, 61, pp. 127–140.

Cheung, W., Watson, R., and Pauly, D.I. (2013) Signature of ocean warming in global fisheries catch. *Nature*, 497, pp. 365–368.

Cury, P., Shannon, L., and Shin, Y.J. (2003) The functioning of marine ecosystems: A fisheries perspective, in Sinclair, M. and Valdimarsson, G. (eds.). Responsible fisheries in the marine ecosystem. FAO/CAB International, pp. 103–123.

Danovaro, R., Bongiorni, L., Corinaldesi, C. et al. (2008) Sunscreens cause coral bleaching by promoting viral infections. *Environmental Health Perspectives*, 116, pp. 441–447.

Danovaro, R., Corinaldesi, C., Dell'Anno, A., and Snelgrove, P.V.R. (2017) The deep-sea under global change. *Current Biology*, 27, pp. R461–R465.

Danovaro, R., Fonda Umani, S., and Pusceddu, A. (2009) Climate change and the potential spreading of marine mucilage and microbial pathogens in the Mediterranean Sea. *PLoS One*, 4, p. e7006.

Danovaro, R., Gambi, C., Dell'Anno, A. et al. (2008) Exponential decline of deep-sea ecosystem functioning linked to benthic biodiversity loss. *Current Biology*, 18, pp. 1–8.

Danovaro, R., Fanelli, E., Aguzzi, J., Billett, D., Carugati, L., Corinaldesi, C., Gjerde, K., Jamieson, A.J., Kark, S., McClain, C., Levin, L., Levin, N., Ramirez-Llodra, E., Ruhl, H., Smith, CR., Laurenz, T., Van Dover, L.C., Moriaki, Y. (2020) Ecological variables for developing a global deep-ocean monitoring and conservation strategy. *Nature Ecology & Evolution*, 4(2), pp 181-192.

Dayton, P.K. (1998) Reversal of the burden of proof in fisheries management. *Science*, 279, pp. 821–822.

Diaz, R.J. and Rosenberg, R. (2008) Spreading dead zones and consequences for marine ecosystems. *Science*, 321, pp. 926–929.

Doney, S.C., Kleypas, J.A., Fabry, V.J., and Feely, R.A. (2008) Ocean acidification: The other CO2 problem. *Annual Review of Marine Science*, 1, pp. 169–192.

Doney, S.C., Ruckelshaus, M., Duffy, J.E. et al. (2012) Climate change impacts on marine ecosystems. *Annual Review of Marine Science*, 4, pp. 11–37.

Duffy, J.E. (2003) Biodiversity loss, trophic skew, and ecosystem functioning. *Ecology Letters*, 6, pp. 680–687.

Duffy, J.E. (2009) Why biodiversity is important to the functioning of real-world ecosystems? *Frontiers in Ecology and the Environment*, 7, pp. 437–444.

Fortibuoni, T., Libralato, S., Raicevich, S. et al. (2010) Coding early naturalists' accounts into long-term fish community changes in the Adriatic Sea (1800–2000). *PLOS One*, 5(11), p. e15502.

Galil, B.S. (2007) Seeing red: Alien species along the Mediterranean coast of Israel. *Aquatic Invasions*, 2, pp. 281–312.

Gedan, K.B., Silliman, B.R., and Bertness, M.D. (2009) Centuries of human-driven change in salt marsh ecosystems. *Annual Review of Marine Science*, 1, pp. 117–141.

Gross, K. and Cardinale, B.J. (2005) The functional consequences of random versus ordered species extinctions. *Ecology Letters*, 8, pp. 409–418.

Guidetti, P., Terlizzi, A., Fraschetti, S., and Boero, F. (2003) Changes in mediter- ranean rocky-reef fish assemblages exposed to sewage pollution. *Marine Ecology Progress Series*, 253, pp. 269–278.

Hall-Spencer, J.M., Rodolfo-Metalpa, R., Martin, S. et al (2008) Volcanic carbon dioxide vents show ecosystem effects of ocean acidification. *Nature*, 454, pp. 96–99.

Halpern, B.S., Walbridge, S., Selkoe, K.O. et al. (2008) A global map of human impact on marine ecosystems. *Science*, 319, pp. 948–952.

Heneghan, R.F., Galbraith, E., Blanchard, J.L. et al. (2021) Disentangling diverse responses to climate change among global marine ecosystem models. *Progress in Oceanography*, 198, p. 102659.

Hoegh-Guldberg, O. (1999) Climate change, coral bleaching and the future of the world's coral reefs. *Marine and Freshwater Research*, 50, pp. 839–866.

Hutchings, J.A. (2000) Collapse and recovery of marine fishes. *Nature*, 406, pp. 882–885.

IPBES (2019) *Global assessment report on biodiversity and ecosystem services of the Intergovernmental Science-Policy Platform on Biodiversity and Ecosystem Services*, Brondizio, E.S., Settele, J., Díaz, S., and Ngo, H.T. (eds.). IPBES Ssecretariat.

Jackson, J.B.C. (2001) What was natural in the coastal oceans? *Proceedings of the National Academy of Sciences of the United States of America*, 98, pp. 5411–5418.

Jackson, J.B.C. and Johnson, K.G. (2001) Measuring past biodiversity. *Science*, 293, pp. 2401–2404.

Jackson, J.B.C., Kirby, M.X., Berger, W.H. et al. (2001) Historical overfishing and the recent collapse of coastal ecosystems. *Science*, 293, pp. 629–638.

Koslow, J.A., Boehlert, G.W., Gordon, J.D.M. et al. (2000) The impact of fishing on continental slope and deep-sea ecosystems. *ICES Journal of Marine Science*, 57, pp. 548–557.

Kroeker, K.J., Micheli, F., Gambi, M.C., and Martz, T.R. (2011) Divergent ecosystem responses within a benthic marine community to ocean acidification. *Proceedings of the National Academy of Sciences of the United States of America*, 108, pp. 14515–14520.

Lasram, F.B.R., Guilhaumon, F., Albouy, C. et al. (2010) The Mediterranean Sea as a 'cul-de-sac' for endemic fishes facing climate change. *Global Change Biology*, 16, pp. 3233–3245.

Levin, L.A., Alfaro-Lucas, J.M., Colaço, A. et al. (2023) Deep-sea impacts of climate interventions. *Science*, 379, pp. 978–981.

MacLeod, M., Arp, H.P., Tekman, M.B., and Jahnke, A. (2021) The global threat from plastic pollution. *Science*, 373, pp. 61–65.

Mazzola, A., Mirto, S., and Danovaro, R. (1999) Initial fish-farm impact on meiofaunal assemblages in coastal sediments of the western Mediterranean. *Marine Pollution Bulletin*, 38, pp. 1126–1133.

Micheli, F., Benedetti-Cecchi, L., Gambaccini, S. et al. (2005) Cascading human impacts, marine protected areas, and the structure of Mediterranean reef assemblages. *Ecological Monographs*, 75, pp. 81–102.

Mirto, S., La Rosa, T., Danovaro, R., and Mazzola, A. (2000) Microbial and meiofaunal response to intensive mussel-farm biodeposition in coastal sediments of the western Mediterranean. *Marine Pollution Bulletin*, 40, pp. 244–252.

Murawski, S., Methot, R., and Tromble, G. (2007) Biodiversity loss in the ocean: How bad is it? *Science*, 316, pp. 1281–1284.

Myers, N., Mittermeier, R.A., Mittermeier, C.G. et al. (2000) Biodiversity hotspots for conservation priorities. *Nature*, 403, pp. 853–858.

Myers, R.A. and Ottensmeyer, C.A. (2005) Extinction risk in marine species, in Norse, E.A. and Crowder, L.B. (eds.), *Marine Conservation Biology: The Science of Maintaining the Sea's Biodiversity*. Island Press, pp. 58–79.

Myers, R.A. and Worm, B. (2003) Rapid worldwide depletion of predatory fish communities. *Nature*, 423, pp. 280–283.

Myers, R.A. and Worm, B. (2005) Extinction, survival or recovery of large predatory fishes. *Philosophical Transactions of the Royal Society B: Biological Sciences*, 360, pp. 13–20.

Occhipinti-Ambrogi, A. (2007) Global change and marine communities: Alien species and climate change. *Marine Pollution Bulletin*, 55, pp. 342–352.

Oguz, T. (2005) Long-term impacts of anthropogenic forcing on the Black Sea ecosystem. *Oceanography*, 18, pp. 112–121.

Parravicini, V., Thrush, S.F., Chiantore, M. et al. (2010) The legacy of past disturbance: Chronic angling impairs long-term recovery of marine epibenthic communities from acute date-mussel harvesting. *Biological Conservation*, 143, pp. 2435–2440.

Pauly, D., Christensen, V., Dalsgaard, J. et al. (1998) Fishing down marine food webs. *Science*, 279, pp. 860–863.

Pearson, T.H. and Rosenberg, R. (1978) Macrobenthic succession in relation to organic enrichment and pollution of the marine environment. *Oceanography and Marine Biology: An Annual Review*, 16, pp. 229–311.

Porzio, L., Buia, M.C., and Hall-Spencer, J.M. (2011) Effects of ocean acidification on macroalgal communities. *Journal of Experimental Marine Biology and Ecology*, 400, pp. 278–287.

Probert, P.K., McKnight, D.G., and Grove, S.L. (1997) Benthic invertebrate bycatch from a deep-water trawl fishery, Chatham Rise, New Zealand. *Aquatic Conservation*, 7, pp. 27–40.

Rabalais, N.N., Turner, R.E., and Wiseman, W.J. (2002) Gulf of Mexico hypoxia, A.K.A. "The Dead Zone". *Annual Review of Ecology and Systematics*, 33, pp. 235–263.

Raffaelli, D. (2004) How extinction patterns affect ecosystems. *Science*, 306, pp. 1141–1442.

Roberts, C.M. and Hawkins, J.P. (1999) Extinction risk in the sea. *Trends in Ecology and Evolution*, 14, pp. 241–246.

Sala, E., Kizilkaya, Z., Yildirim, D., and Ballesteros, E. (2011) Alien marine fishes deplete algal biomass in the Eastern Mediterranean. *PLoS One*, 6(2), p. e17356.

Sala, E. and Knowlton, N. (2006) Global marine biodiversity trends. *Annual Review of Environment and Resources*, 31, pp. 93–122.

Sarà, G., Sarà, A., and Milanese, M. (2011) The Mediterranean intertidal habitat as a natural laboratory to study climate change drivers of geographic patterns in marine biodiversity. *Chemical Ecology*, 27, pp. 91–93.

Smith, W.K. and Solow, A.R. (2012) Missing and presumed lost: Extinction in the ocean and its inference. *ICES Journal of Marine Science*, 69, pp. 89–94.

Solan, M., Cardinale, B.J., Downing, A.L. et al. (2004) Extinction and ecosystem function in the marine benthos. *Science*, 306, pp. 1177–1180.

Starkey, D.J., Holm, P., and Barnard, M. eds. (2012) *Oceans Past: Management Insights from the History of Marine Animal Populations*. Routledge.

Sweetman, A.K., Thurber, A.R., Smith, C.R. et al. (2017) Major impacts of climate change on deep-sea benthic ecosystems *Elementa. Science of the Anthropocene*, 5(4), pp. 23.

Theriault, J.A. (2005) Marine mammals and active sonar – A review of the potential for negative impact from the use of active sonar and emerging mitigation techniques. *Sea Technology*, 46, pp. 23–29.

Thompson, R.C., Olsen, Y., Mitchell, R.P. et al. (2004) Lost at sea: Where is all the plastic? *Science*, 304, p. 838.

Turschwell, M.P., Connolly, R.M., Dunic, J.C. et al. (2021) Anthropogenic pressures and life history predict trajectories of seagrass meadow extent at a global scale. *Proceedings of the National Academy of Sciences of the United States of America*, 118(45), p. e2110802118.

Tyack, P.L., Zimmer, W.M.X., Moretti, D. et al. (2011) Beaked whales respond to simulated and actual navy sonar. *PLOS One*, 6, p. e17009.

Utne-Palm, A.C., Salvanes, A.G.V., Currie, B. et al. (2010) Trophic structure and community stability in an overfished ecosystem. *Science*, 329, pp. 333–336.

Watling, L. and Norse, E.A. (1998) Disturbance of the seabed by mobile fishing gear: A comparison to forest clearcutting. *Conservation Biology*, 12, pp. 1180–1197.

Williams, N. (2008) Mediterranean threats. *Current Biology*, 18, pp. 3–4.

Worm, B., Barbier, E.B., Beaumont, N. et al. (2006) Impacts of biodiversity loss on ocean ecosystem services. *Science*, 314, pp. 787–790.

Worm, B. and Lotze, H.K. (2016) Marine biodiversity and climate change, in Letcher, T.M. (ed.), *Climate Change*. 2nd Edition. Elsevier, pp. 195–212.

29

Marine Biodiversity Conservation

29.1 Introduction

As discussed in the previous chapter "Human Impacts on Marine Ecosystems," marine biodiversity faces multiple threats, both as individual and as cumulative impacts. Different ocean stakeholders, from fishers to the United Nations, consider the global ocean a common for all of humanity, not a privately owned resource, raising questions regarding not only how its resource should be shared, but also who is be responsible for minimizing environmental impacts and ensuring sustainability.

In the 1960s, in an essay that otherwise raised some questionable ideas, Garrett Hardin used the analogy of shared cattle grazing commons to introduce the concept of the **Tragedy of the Commons**, referring to the dilemma arising when multiple individuals, acting independently and rationally in their own self-interest, ultimately deplete a shared limited resource, even when that depletion clearly benefits nobody in the long term. Because no individual owns "ocean space," managing ocean use falls to national and regional governments, and to international organizations in the case of the high seas, which do not fall under the jurisdiction of any one nation. The issues and examples provided in the previous chapter illustrate that our sustainability efforts have generally failed, and that we must strive to do better. Acknowledging the temptation to throw up our hands in despair, science can actually help to make a difference. Many ocean environments still thrive, with spectacular diversity and productivity that we can still hope to conserve. This chapter describes how we could achieve that goal.

29.2 Conservation Objectives

Marine biodiversity conservation efforts typically respond to specific drivers of change such as fishing or oil development and impacts associated with these activities, such as species extirpations, alteration of food webs, and loss of ecosystem functions such as habitat provisioning, productivity, and nutrient cycling. Irrespective of the drivers of change and the strategy developed to mitigate those changes, the most effective marine conservation strategies work toward specific and measurable objectives. Part of the challenge links to contrasting needs. Scientists and conservationists often identify lofty but vague objectives. For example, we can strive to "sustain biodiversity" because we think it is important, but how do we define biodiversity? Do we mean all biota from viruses to whales? Do we include species that occasionally stray into a region? Keep in mind that no study has ever quantified diversity from viruses to megafauna in even a single cubic meter of water or 1 m^2 of seafloor, let alone a geographic region. At the same time, managers of ocean use require that researchers establish specific objectives and targets that scientists can assess, such as sustaining a given species at a particular abundance level, known as a **reference point**. Some types of spatial closure strategies such as marine protected areas (MPAs) often seek to conserve species of interest and to enhance abundances of fishes that increase in number within the protected area, but then spill over into adjacent regions and thus enhance regional fisheries production. Irrespective of the objectives, ocean stakeholders such as fishers want to know what specific benefits a closure provides in exchange for them giving up access to resources within a given location.

Historically, conservation efforts and objectives have tended to focus on single species, including endangered, vulnerable, and charismatic, highly visible species such as marine mammals and seabirds, and (sometimes) commercial species such as cod. However, decades of research show that species play diverse roles in marine ecosystem functioning, and conservation efforts must therefore consider a wide range of species. Even protecting a single species can prove daunting. In some

Marine Biology: Comparative Ecology of Planet Ocean, First Edition. Roberto Danovaro and Paul Snelgrove.
© 2024 John Wiley & Sons Ltd. Published 2024 by John Wiley & Sons Ltd.
Companion Website: www.wiley.com/go/danovaro/marinebiology

instances, conservation efforts focus on protecting spawning aggregations, noting that for economic reasons, commercial fishing often targets aggregations of fish. Many species of fishes, mobile invertebrates such as crabs, seabirds, and marine mammals often aggregate at known locations in order to reproduce Capturing them in such aggregations can be much more cost effective than chasing them around after they disperse. Allowing animals to reproduce increases the probability of adding new individuals to a population. Nonetheless, at the same time, we can rarely protect all populations of a given species, so which populations should we prioritize? In some cases, we know that some populations, known as **source populations**, contribute more to subsequent generations than **sink populations**, which produce few, if any, successful offspring.

Imagine a coastline occupied by a given species that produces planktonic eggs and larvae (meroplankton) that predominant currents transport north to south (Figure 29.1), given limited or no ability of these stages to swim. In this scenario, eggs and larvae released by spawners (sink populations) at the southern limit of the species are swept offshore where they die, whereas those spawners toward the north (source populations) release propagules that can settle to the south and survive before currents would otherwise transport them offshore. From a fisheries management perspective, fishing the southern populations will have little effect on overall recruitment, and overfishing these populations will not affect overall sustainability of the species. In contrast, fishing the northern populations runs the risk of depleting source populations and driving the species to extinction.

Benthic marine organisms that produce meroplanktonic larvae that spend their initial life stages suspended in the water column illustrate how ocean currents supply larvae from one source area to a receiving area. These benthic organisms include most macrofaunal and megafaunal organisms. Conversely, currents do not transport meiofauna, which lack

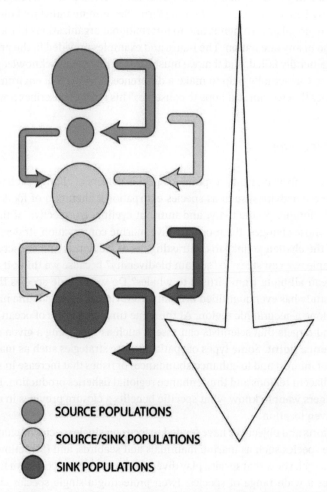

Figure 29.1 Source and sink population considerations in conservation planning. The white arrow indicates the predominant current whereas the circles indicate different populations of a given species. Green populations (sources) provide propagules to downstream populations, whereas currents sweep propagules from red populations (sinks) offshore where they experience high mortality.

meroplanktonic larvae, unless bottom flows resuspend and transport these tiny organisms, which are similar in size to meroplanktonic larvae.

Unfortunately, we generally lack detailed information on life cycles and life histories for most species and must therefore draw inferences based on knowledge of ocean currents and reproductive biology. Recent tools such as population genetics and otolith/shell microchemistry, which uses DNA and trace elements in fish ear bones or larval mollusk shells, respectively, to trace population origins. This approach helps in identifying which subsets of populations to protect. Of course, protecting only the spawners in a population may not be sufficient to protect a species. For example, if the larvae or juveniles of a given species experience high mortality then these other life stages may need targeted protection.

Although many conservation strategies focus on single species, the large numbers of species in the global ocean and lack of information currently available for most of them suggest that a species-by-species conservation strategy cannot succeed in attaining objectives related to sustaining ocean health. Species comprise just one element of biodiversity, and genetic diversity, ecosystem diversity, and nature's benefits to people all contribute to the array of biodiversity that merits consideration (Figure 29.2). We know, for example, that habitats such as seagrasses and mangroves serve as nurseries for the juvenile stages of multiple species, and other habitats such as coral reefs provide critical habitat for a rich diversity of associated species (Figure 29.3). Thus, protecting their habitat offers a means of helping to protect such species, but it also offers a means of protection that does not require complete knowledge of that species other than where it lives. Protecting

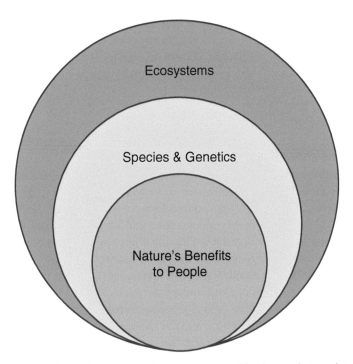

Figure 29.2 Facets of marine biodiversity that influence marine conservation objectives and strategies.

Figure 29.3 Habitat forming species such as coral (left) and mangroves (right) provide an alternative approach to conservation that focus on habitat rather than individual species.

a chosen area or environment also offers a degree of protection for all the species within that habitat, potentially including undescribed species completely unknown to science.

The first stage in any conservation framework involves mapping. Detailed maps of species distributions, biodiversity patterns, and ecosystem functions at multiple spatial scales in the global ocean greatly facilitate conservation planning, but we rarely possess such information. However, because habitat can serve at least as a crude proxy for each of these attributes, habitat maps provide a surrogate for ecosystem attributes that conservationists can use to prioritize efforts. The use of emerging tools such as multibeam seabed mapping, as well as satellite and other remote sensing imagery, provides increasingly detailed habitat maps in comparatively shorter time periods than would be possible with video transects, diver surveys, or other traditional mapping techniques. The increasing availability of bathymetric data at fine spatial resolution and improvements in modelling the spatial and temporal dynamics of marine species now provide better data for conservation applications.

Some conservation efforts therefore focus on ecosystem-level priorities, affording protection for key attributes such as biodiversity, productivity, or other ecosystem functions such as nutrient cycling, carbon sequestration, or trophic support. Protecting geographic locations that support high levels of biodiversity provides good "bang for the buck" in that one location provides benefits for many species. Globally, as discussed earlier, tropical environments often support more species per unit area than their temperate or polar counterparts. Thus, on a global scale, tropical habitats such as coral reefs merit particular conservation attention, but coastal environments fall within the territorial waters of individual nations, and those nations make their own decisions on conservation priorities based on science capacity, economic interests, and other factors. On a regional scale, some habitats support unique rather than numerous species, sometimes including high levels of endemism. Many of the unique forms that live at hydrothermal vents, for example, do not occur in any other habitat, making them an obvious candidate for conservation efforts, particularly given the patchy and ephemeral nature of vent environments. More recently, some conservation efforts have begun to focus on function, protecting locations of high productivity, such as coastal wetlands.

29.3 The Third Dimension of Marine Conservation

Despite the strong bias in marine biodiversity knowledge toward nearshore and shallow marine habitats, marine species richness does not necessarily decline with depth – far from it. For example, researchers have documented a unimodal pattern of benthic biodiversity with peaks at 1500–2500 m depths for wide portions of the oceans. We now recognize highly complex and rich pelagic and deep-sea biodiversity with unique deep-sea habitats (seamounts, submarine canyons, hydrothermal vents, cold seeps, etc.) that sustain marine populations and endemic species. Comprehensive conservation planning must consider biodiversity features of interest within different parts of the water column, including deep waters as well as the seabed. Shallow and euphotic habitats sometimes tightly connect to deep-sea systems. The life cycles of many pelagic and benthic species, particularly macro- and megafaunal components, include meroplanktonic larval dispersal by currents in offshore and deep-sea areas. Downslope and upslope currents can transport larvae, propagules, and juveniles across depths and ecosystems, enabling continuous exchange between shallow and deep waters. The deep seafloor hosts some commercially important species, a subset of which move to shallow-water ecosystems seasonally or periodically, replenishing overexploited local populations. These linkages complicate conservation planning.

Refuge habitats increase the importance of three-dimensional structure and connectivity of deep-sea ecosystems. Daily vertical migrations occur in many zooplankton taxa, depending on their size, swimming distances of 2 km or more during the night to the surface for feeding and then returning to the deep to escape predation. Several charismatic species, from sperm whales to sea lions and even penguins sometimes hunt large plankton in the deep sea, some species to depths of >2,000 m (Figure 29.4). Marine conservation should consider all of these elements from the entire water column down to the deep seafloor, thereby also helping to conserve biodiversity at shallow depths. A definition of three-dimensional marine ecoregions should incorporate processes and connectivity of species and habitats (e.g., related to upwelling, currents, and gyres) in addition to representation of biodiversity features using species distribution ranges. Connectivity within ecological systems (such as larval dispersal), can be incorporated into 3D systematic conservation planning.

29.4 Conservation Strategies

Historically, fisheries represented the single greatest threat to marine conservation, and managers and conservationists have therefore developed multiple management tools to promote sustainable fisheries. To varying degrees, these different tools have some application to other pressures as well, and all tools come with pros and cons. The decision on which

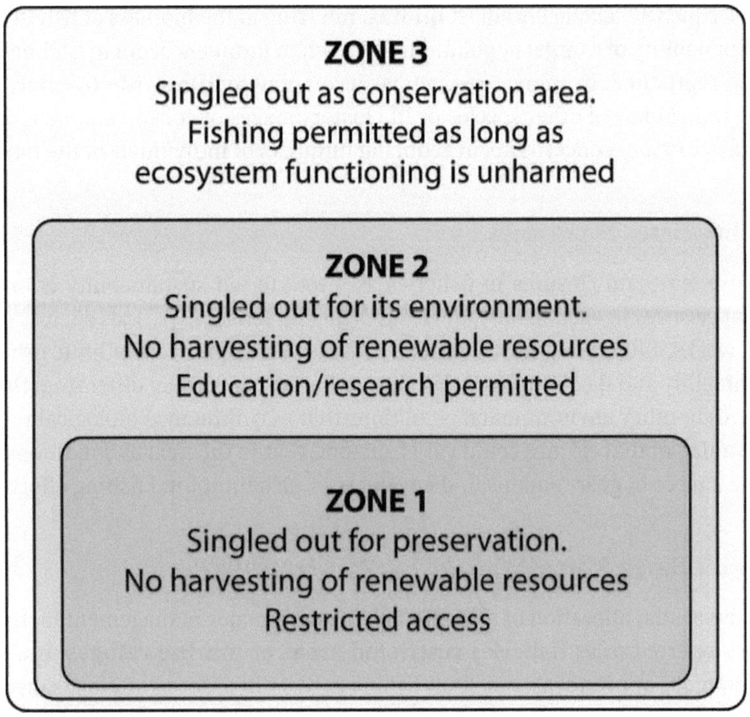

```
┌─────────────────────────────────────────┐
│              ZONE 3                     │
│   Singled out as conservation area.     │
│      Fishing permitted as long as       │
│   ecosystem functioning is unharmed     │
│   ┌─────────────────────────────────┐   │
│   │           ZONE 2                │   │
│   │  Singled out for its environment.│   │
│   │ No harvesting of renewable resources│ │
│   │   Education/research permitted  │   │
│   │  ┌───────────────────────────┐  │   │
│   │  │         ZONE 1            │  │   │
│   │  │ Singled out for preservation.│ │   │
│   │  │ No harvesting of renewable │  │   │
│   │  │       resources           │  │   │
│   │  │    Restricted access      │  │   │
│   │  └───────────────────────────┘  │   │
│   └─────────────────────────────────┘   │
└─────────────────────────────────────────┘
```

Figure 29.4 Use of zones within an MPA in order to meet contrasting stakeholder needs and objectives.

management tool to apply runs into the reality that many people and nations depend on ocean fisheries as a primary source of protein; for this reason, where and when conservation strategies play out depends on trade-offs between conservation objectives, economics (including short-term and long-term considerations), and social issues (employment, food supply). Although we may tend to think of Marine Protected Areas as the primary conservation tool available for marine environments, ocean use managers actually use a variety of approaches that vary depending on how, when, and where fishers prosecute a fishery.

29.4.1 Access to Fisheries: Who and How

The **entry** or opening of a new fishery offers the best opportunity for effective management and sustainability because managers can limit the number of fishing licenses and gears used for that fishery. Fishers therefore gain rather than lose something, and do not incur extra costs of moving or altering their existing fishing activity. In some cases, governments buy out fishing licenses in an effort to reduce fishing pressure and provide an incentive to those willing to leave a fishery. Unfortunately, these programs can be expensive, and require willingness of some portion of fishers to accept buyouts.

Gear limitations, by restricting gear types, amounts, and fishing locations where fishers can use them, offer another mechanism to advance conservation objectives; managers can advance conservation objectives without necessarily closing a fishery. For example, the most destructive fishing gear, bottom trawls, have become increasingly larger and more destructive over time. Large rollers on rock hopper gear, for example, allow the gear to move over rocks without tearing the net, enabling fishers to access increasingly rugged bottom habitat that once provided a natural refuge from fishing pressure. Limiting such gear in some locations, while allowing less destructive types of gear, may allow a fishery to remain open. Other gear modifications include hook size and shape for hook and line fisheries, mesh size and shape, and grids or openings that enable escape by non-target individuals, whether juveniles of the same species or individuals of another species not specifically targeted by the fishery. Managers may also limit the amount of gear used, such as the number of crab pots a given fisher may deploy at one time.

Vessel size limits also influence how a fishery impacts an environment. Consider, for example, the number of fish a 60- or 70-m long factory freezer trawler might be able to remove from the water and freeze for later sale, in comparison with a small open coastal skiff used by an inshore fisher that must return to shore within hours of catching fish. Small boats cannot carry as heavy or large gear as a big vessel or cover the same distances.

Fishing quotas – Most frequently, managers adjust **quotas**, referring to the biomass of fish that a fishery can remove at a given time, to ensure sustainability of a target population. Thus, when numbers begin to decline, lower quotas leave more individuals in the water to reproduce. In some cases, quotas also apply to **allowable bycatch**, meaning that if a given fishery captures too many individuals of other species or life history stages of conservation concern, managers may close the fishery even in the absence of any concerns about reducing numbers of individuals of the target species.

29.4.2 When to Fish: Time-Based Approaches

Fisheries managers also use temporal closures in fisheries as a tool to aid sustainability efforts. Most commonly, this approach involves seasonal closures in each fishery, or closures during sensitive periods such as spawning. Some fisheries may open for only a few weeks, illustrating how temporal closures can significantly limit fishing effort. These sorts of approaches also offer adaptability and flexibility, in that a closure in one region may differ from that in another region with different water temperatures or other environmental conditions that may influence biological processes such as timing of spawning. Moreover, the timing of that closure could vary from one year to the next as conditions change. Shortened **soak times** (the amount of time a piece of gear remains in the water) can also limit total fishing effort.

29.4.3 Where to Fish: Area-Based Tools

Fisheries managers often use spatial allocation of access to fisheries as a major management tool. Many countries use **fisheries closures**, sometimes referred to as **fisheries restricted areas** or **marine refuges** as a conservation tool. These closures, which can be temporary or even seasonal, limit fishing activity in order to attain a conservation objective such as rebuilding a stock or avoiding damage to habitats. Because such closures may be temporary and often focused on a single, clear objective, they are legislatively simpler and easier to sell to the public than permanent closures.

29.5 Marine Protected Areas

Many people appropriately think of **marine protected areas** (MPAs) as an essential marine conservation tool, and MPA designation includes a complex process to increase the likelihood it will help to achieve useful conservation objectives. However, the precise definition of MPAs varies among different stakeholders and governments. The *International Union for the Conservation of Nature* (IUCN) defines MPAs as natural areas that receive protective management according to pre-defined management objectives. The *World Wildlife Fund for Nature* defines MPAs as "An area designated and effectively managed to protect marine ecosystems, processes, habitats, and species, which can contribute to the restoration and replenishment of resources for social, economic, and cultural enrichment." These definitions demonstrate that the level of protection a given MPA receives varies greatly, from some cases with effectively no protection to "gold standard" MPAs that ban all types of extractive activities within their boundaries. Some countries use a zoning system within their MPAs (Figure 29.4), allowing different types of activity within each of the zones (thus resulting in varying levels of protection).

Lawmakers create MPAs for many different reasons, including enhancing commercial fisheries and biodiversity conservation. Depending on the specific objectives associated with a given MPA, optimal size and shape may vary, noting that the challenge of convincing stakeholders to agree to designation of an MPA increases with increasing size of the MPA and the degree of protection proposed. Ecologists in the 1970s and 1980s also debated the pros and cons of conservation at different spatial scales, whether single large or several small (SLOSS) protected areas yield better conservation results. The current consensus is that no "one solution fits all" and that the size, number, and shape of an MPA strategy depends entirely on the specific situation and the conservation objectives. This consideration also points to widespread interest in developing **networks of MPAs**, where the network collectively provides enhanced benefits relative to the individual MPAs. For example, population connectivity among different locations illustrates how different locations can influence one another, along with the potential benefits associated with protecting those locations.

Designation of an MPA represents just one step in a complex series of actions (Figure 29.5). The process begins with the scoping and the identification of a need, but past MPA experience demonstrates the need for early engagement and buy-in of diverse stakeholders in the MPA process.

Even successful designation of an MPA does not guarantee conservation success. Some stakeholders may not accept the legitimacy of the MPA and, noting the high cost of enforcement, may choose to ignore MPA regulations and continue to

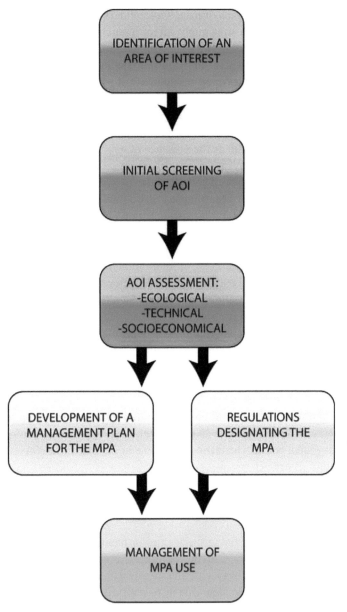

Figure 29.5 Steps typically considered in establishing a marine protected area.

apply pressures. Furthermore, if managers cannot demonstrate tangible benefits of the MPA, support from stakeholders may wane. These needs point to the importance of setting measurable and attainable objectives and associated targets, whether these relate specifically to conservation goals or to associated research objectives. Such targets become meaningless in the absence of monitoring to determine whether the MPA is meeting the needs identified during the designation process.

29.5.1 Criteria for Prioritizing Marine Areas to Protect

Choosing the location and extent of marine protected areas requires criteria quite different from those used for terrestrial protected areas. Although in many regions around the world, local populations depend largely on ecosystem services and resources provided by natural terrestrial areas, such dependence tends to be even greater for marine areas, especially in developing island nations. Some forms of fishing can occur in large areas without threatening the conservation objectives of an MPA because they do not involve habitat modification. This situation increases the feasibility of balancing conservation and the needs of local stakeholders and rights holders (e.g. Indigenous groups). Managers must weigh events outside the MPA that might affect it, such as pollution from land or adjacent waters. Following key principles, over the past few years many countries have applied a rigorous set of criteria for site selection.

The criteria for identifying MPA size take into account biogeographic and/or ecological factors. Biogeographic criteria include the presence of rare biogeographic qualities or representativity of a biogeographic "type" or types and the presence of unique or unusual geological features.

Ecological criteria might include ecological processes or life-support systems (e.g., as a source of larvae for downstream areas), the integrity, or the degree to which the area, either alone or in association with other protected areas, encompasses a complete ecosystem, a variety of habitats, as well as the presence of habitats for rare or endangered species. Ecological criteria also include: (a) presence of nursery or juvenile areas; (b) presence of feeding, breeding, or rest areas; (c) existence of rare or unique habitat for any species; (d) degree of genetic diversity within species; or (e) "naturalness," the extent to which the area has been protected from, or has not been subject to, human-induced change.

Economic criteria also require consideration, such as existing or potential economic contributions associated with protection (e.g., protection of an area for recreation, subsistence, use by traditional inhabitants, appreciation by tourists and others, or as a refuge nursery area or source of economically important species). Regarding **social importance**, additional criteria include potential existence value to local, national, or international communities because of its heritage, historical, cultural, traditional, aesthetic, educational, or recreational qualities. **Scientific importance** adds another consideration, such as research and monitoring value of national or international significance (for example, the Southern Ocean and Antarctic region). In terms of **operational convenience**, major considerations include degree of insulation from external destructive influences, social and political acceptability, degree of community support, accessibility for education, tourism, and recreation, as well as compatibility with existing uses, particularly by local people.

Ecologically and Biologically Sensitive Areas (EBSAs) refer to ocean locations of special importance for their ecological and biological characteristics, such as essential feeding grounds or protective habitat for juvenile stages or for reproduction. EBSAs encompass a wide range of ecosystem types around the world, including highly productive or species-rich sites, unique species and communities, and locations. The *United Nations Convention on Biological Diversity* (CBD) established an objective and methodical framework in 2008 to recognize these special environments for their contribution to the healthy functioning of the global ocean. The technical process in designating EBSAs requires that a given location meet at least one of seven scientific criteria:

1) Uniqueness or rarity – does the location support species that are rare or absent from other locations?
2) Special importance for life history stages of species – does the location support vulnerable life history stages of a species of particular interest, such as nesting sites?
3) Importance for threatened, endangered or declining species and/or habitats – does the location support species or habitats that have declined in abundance or coverage to the point that extinction or local extirpation could occur?
4) Vulnerability, fragility, sensitivity, or slow recovery – are the species within the location particularly sensitive and easily damaged (e.g., coral) or do their life histories (e.g., slow growth, few offspring, late maturation) contribute to slow recovery?
5) Biological productivity – does the location support high productivity, whether primary or secondary?
6) Biological diversity – does the location support regionally high species numbers (i.e., a biodiversity "hotspot")?
7) Naturalness – is the location highly pristine and thus in a highly "natural" state?

Note, however, that EBSA designation does not offer any specific protection measures as such, but instead flags them as important locations that merit attention. In some cases, this designation may provide a steppingstone toward protective measures.

The economist Robert Costanza and his team in the 1990s estimated the combined economic value of ocean ecosystem services at ~US$29 trillion/year, representing about 60% of the total "value" of all nature on Earth. They revised their ocean estimate upward to $50 trillion in 2011, though greater increases in valuation of terrestrial ecosystems dropped the global ocean contribution to Earth's total to ~40%. This **economic valuation**, while somewhat vague and based on many assumptions, creates an economic case for promoting conservation. Thus, although conservation-minded people recognize an ethical duty to avoid extinctions and compromising ecosystems, even the most profit-oriented developer could potentially see some value in maintaining nature. Unfortunately, the short-sighted aspect of many economic drivers (and election cycles) sacrifices long-term losses in species and habitats in favor of short-term economic gain. Nonetheless, this analysis has added important facets to conservation discussions, and raised awareness regarding ecosystem functioning and associated services to humanity.

The economic valuation approach described above benefits from knowledge of the ecological roles or functions that different groups of species play in ocean ecosystems. Indeed, marine ecologists increasingly emphasize the importance of conservation strategies that focus on objectives related to sustaining ecological functions critical to ocean health such as habitat provisioning, production, food web support, and nutrient cycling. This approach shifts emphasis away from individual species per se, and more toward the specific functions they contribute to ecosystem processes. It offers the advantage of moving from a species-by-species strategy toward one focused at a higher level within the ecosystem. Unfortunately, we often lack detailed information on spatial and temporal patterns in different functions, and the significant sampling and analysis required to fill such information gaps limits the likelihood that we can fill those gaps quickly. Ecologists have therefore proposed a **biological traits approach**, which considers, as part of protected area planning, morphological, biochemical, physiological, structural, phenological (timing of key developmental events), behavioral, and ecological traits of an organism such as feeding mode and habitat-forming capacity (Figure 29.6). This approach offers the advantage of focusing on, not only how organisms respond to their environment, but also the effect of those organisms on ecosystem processes. Importantly, such a strategy does not necessarily require species-level taxonomy of the individuals involved and allows scientists to predict functional impacts based on key traits of individuals. It also feeds well into some of the ecosystem-level approaches described below that ocean use managers increasingly embrace. Despite few examples of application of such approaches in protected area designations to date, this strategy may increase in popularity in the near future because of its reduced cost and time investment.

29.5.2 What Have We Learned from Existing Marine Protected Areas?

MPA science has grown almost exponentially in the last decade and experience to date has taught some important lessons. First, although regulating fishing and other activities can help to address some conservation challenges, MPAs offer unique benefits that other management tools cannot provide, such as habitat protection and protection of unknown diversity (Box 29.1). We also know that different objectives require different types of MPA strategies, depending on a variety of variables such as life histories of focal species, extent of habitat, and many others.

In terms of planning and management of MPAs, one lesson learned is that small MPAs usually yield small benefits, and although smaller MPAs may often be more convenient in terms of avoiding conflicts with stakeholders, MPAs of convenience also typically fail in meeting important conservation objectives (Box 29.2). Managers should also not consider MPAs in isolation; the best protection in the world will not help if an upstream activity such as release of pollutants compromises

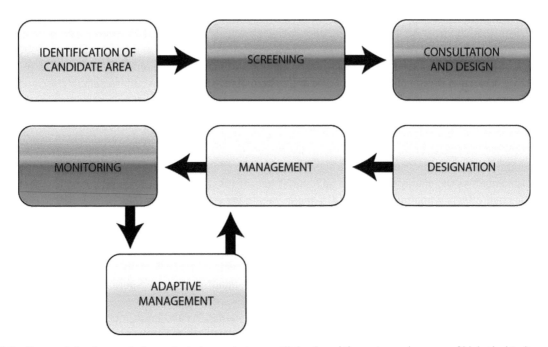

Figure 29.6 Stages of development of a protected area strategy outlining in red those steps where use of biological traits could aid in achieving important conservation objectives. Modified from Miatta M., Bates, A.E. & Snelgrove, P.V.R. (2021) Incorporating biological traits into marine protected area strategies. *Annual Review of Marine Sciences* 13, 421-433. Source: Adapted from Miatta, et., (2020) Annual Review of Marine Sciences 13, doi.org/10.1146/annurev-marine-032320-094121.

Box 29.1 Benefits of Marine Protected Areas

Scientific research has provided increasing and robust evidence of multiple significant benefits from the establishment of marine protected areas.

- Almost all MPAs contribute to the maintenance or resilience of both biological diversity and abundance.
- Marine protected areas can help to ensure continuity and future options for those benefits by protecting the health of marine ecosystems: for example, the ecosystem services of coral reefs include shoreline protection, sediment production, and sediment retention.
- MPAs benefit fisheries thanks to "no-take" reserves, which play an important role in arresting and possibly reversing the global and local decline in fish populations and productivity.
- Support for stock management. Traditionally MPAs and reserves (including specific fisheries management measures such as closures and catch restrictions) have benefited fisheries through stock enhancement and management.
- Protection of habitat benefits key life cycle stages including spawning, juvenile settlement, nursery grounds, and major feeding grounds.
- Strategically located protected areas provide sites for settlement and early growth of juveniles that, when mature, spill over into adjacent fished areas.
- Support for fishery stability: Studies of marine protected areas with core "no-take" reserves established in overfished coastal and island habitats show significant improvements in fish catch, leading to sustainable catch levels.
- Coastal and marine ecosystems contribute to beach and shoreline stability, assimilate and process waste, and contribute to the quality of life of coastal people.
- Well-managed marine protected areas with core "no-take" reserves often attract major tourism. In Australia, the Great Barrier Reef attracts ~1.8 million tourist visits, with the industry valued at over $A1 billion per year, compared to estimates of $A359 million for the annual worth of Great Barrier Reef fisheries.

Box 29.2 Key Lessons Learned for the Establishment and Management of Marine Protected Areas

- In today's marine environment we cannot divorce the questions of resource use and conservation, because different users seek marine natural resources and their living space for many different purposes;
- The tendency in some areas to oppose the recognition of fishery reserves as MPAs appears counterproductive, inhibiting cooperation between fishers and environmentalists in creating and managing MPAs;
- A long history of conflict and lack of cooperation between environmental and fisheries management agencies exists in almost all areas of the world.
- This lack of joint action inhibits progress in establishment and management of MPAs, and stymies progress on sustainable ocean use. Design of individual MPAs and system plans should serve both sustainable use and environmental protection objectives, and relevant agencies should work together in planning and management;
- Successful MPA processes must engage local people from the earliest possible stage in order for any MPA to succeed. This involvement should extend to local stakeholders receiving clearly identifiable benefits from the MPA;
- Socio-economic considerations usually determine MPA success or failure. In addition to biophysical factors, managers should address these considerations from the outset in identifying sites for MPAs, and in selecting and managing them;
- "Great" is sometimes the enemy of good. An MPA that is not ideal in the ecological sense, but which meets the primary objective, offers an improvement over striving in vain to create the "perfect MPA";
- Incomplete information rarely justifies postponing action on establishment of MPAs. Sufficient information usually exists to justify an MPA ecologically and to set reasonable boundaries;
- Design and management of MPAs must be both top-down and bottom-up;
- An MPA must have clearly defined objectives against which to evaluate its performance, and a monitoring program to assess management effectiveness. Management should be adaptive, meaning periodic review and revision as dictated by monitoring results;
- Much of the global debate on the merits of small, highly protected MPAs and large, multiple use MPAs arises from the misconception that it must be one or the other. In fact, nearly all large, multiple use MPAs encapsulate highly protected zones, which can function in the same way as individual highly protected MPAs. Conversely, a small, highly protected MPA in a larger area subject to integrated management can be as effective as a large, multiple use MPA;
- Because of the highly connected nature of the ocean, which efficiently transmits substances and forcing factors, an MPA will rarely succeed unless embedded in, or so large that it constitutes, an integrated ecosystem management regime.

conservation objectives within the MPA. Likewise, ignoring connectivity can also result in an ineffective MPA, whether referring to population connectivity or connectivity of food resources supplied by adjacent locations.

The need for monitoring MPAs, some of which may occur in remote locations, with cost-effective and reliable indicators of status is important, but challenging; however, advances in technology such as satellite sensors and autonomous underwater vehicles continue to create new opportunities for monitoring that may help to address cost and access issues. Monitoring also points to the need for patience, in that attaining many conservation objectives takes time. Even a positive response to the creation of an MPA may prove difficult to detect, given the inherent "noise" created by natural variability in marine ecosystems. Furthermore, as knowledge accumulates, reassessment of the MPA objectives and monitoring approach may become necessary. Finally, although MPAs do not offer a "silver bullet" to all the woes faced by the global ocean, they do represent an important tool in a diverse toolbox of management actions that can advance conservation goals.

29.6 Cumulative Impacts and Biodiversity Conservation

The previous section focused primarily on fisheries impacts because, historically, much of ocean use management has focused heavily on fisheries management and conservation. However, other activities cause significant impacts on ocean life as discussed in Chapter 28, and many conservation efforts today recognize diverse threats to ocean ecosystems. Globally, researchers have considered fishing impacts as the single greatest concern, but climate change has now become at least as great a concern because its effects generally extend throughout the global ocean. But these two issues illustrate that human impacts rarely act in isolation, and we must therefore consider the cumulative effects of multiple activities (Figure 29.7). Different activities result in different impacts, depending on their intensity (how strong) and frequency (how often). In some cases, when these impacts act in tandem, one effectively swamps the others, resulting in **masking**. In other cases, the effects simply accumulate in a linear **additive** fashion so that the cumulative result equals the sum of the individual effects. Often, impacts produce a **synergistic** effect, with substantially stronger combined effects compared to their mere sum. Finally, and infrequently, the impacts of different activities "cancel each other out" producing less severe **compensatory** impacts in combination than the individual activities. We currently know much less about cumulative impacts of different activities than about their individual effects, a topic that scientists have recently begun to address.

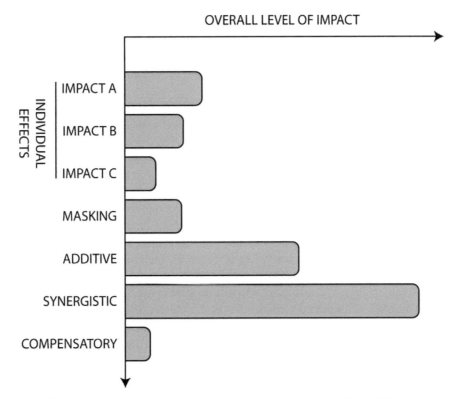

Figure 29.7 Comparison of individual effects of human impacts and their potential interaction, which may sometimes exacerbate, and other times reduce, overall impacts.

29.7 Conservation Frameworks

Conservationists and ocean use managers increasingly embrace a framework for environmental decision-making known as the **precautionary principle**, which contains four core components:

1) take preventive action in the face of uncertainty;
2) shift the burden of proof to the proponents of an activity;
3) explore alternatives to actions that could prove harmful;
4) increase public engagement in decision making.

Additional considerations include:

1) Carefully controlling access to new fisheries from the onset;
2) Cap fishing capacity and catch rates at conservative levels that do not risk overfishing, including reference points (e.g., minimum sustainable biomass) for fished stocks that cannot be exceeded;
3) Respond to any decline in stock size immediately by developing and implementing a recovery plan;
4) Establish protected areas to provide refugia and expedite habitat restoration;
5) Develop fishing gear and techniques that minimize bycatch and habitat damage;
6) Establish data collection and reporting mechanisms that provide reliable and accurate information;
7) Reduce perverse subsidies such as government programs that promote economically unsustainable fisheries;
8) Collaborate with all stakeholders when developing management policies.

Ecosystem-Based Management (EBM), an approach increasingly adopted by ocean use managers around the world, recognizes the full array of interactions within an ecosystem that includes humans, moving beyond considering single issues, species, or ecosystem services in isolation. The 1992 *UN Convention on Biological Diversity* refers to the "ecosystem approach" and defines it as "Ecosystem and natural habitats management [...] to meet human requirements to use natural resources, whilst maintaining the biological richness and ecological processes necessary to sustain the composition, structure and function of the habitats or ecosystems concerned. Important within this process is the setting of explicit goals and practices, regularly updated in the light of the results of monitoring and research activities."

Ecosystem-Based Fisheries Management (EBFM) narrows the concept, defined by the US *National Research Council* as "an approach that takes major ecosystem components and services – both structural and functional – into account in managing fisheries [...]. It values habitat, embraces a multispecies perspective, and is committed to understanding ecosystem processes [...].Its goal is to rebuild and sustain populations, species, biological communities and marine ecosystems at high levels of productivity and biological diversity so as not to jeopardize a wide range of goods and services from marine ecosystems while providing food, revenues and recreation for humans." This approach focuses explicitly on fisheries and focuses management responsibility on the users.

In 2002 the *Food and Agriculture Organization* (FAO) adopted the even broader term **Ecosystem Approach to Fisheries** (EAF) in response to concerns about some EBFM terminology, parallels of EAF with the "Precautionary Approach" to fisheries, and because the broader scope of EAF could include development, food safety, and other issues more appropriate to the *Food and Agriculture Organization* mandate. In short, EAF provides a fisheries framework that links more closely to human well-being.

In recent years, some managers have adopted **Marine Spatial Planning**, referring to a holistic process that brings together diverse ocean stakeholders – including those involved in energy, industry, government, conservation, and recreation – to coordinate informed decisions that support sustainable marine resource use. This process, which applies ideas initially developed for land use planning, seeks to resolve the needs of all users to optimize how different activities occur in different environments by linking plans, policies, and regulations. Such efforts require setting clear objectives, initial assessment of spatial use, implementation, monitoring of resources, analysis of change, and review of the efficacy of spatial use. This process works across sectors to place activities within a geographic context when making decisions on resource use, development, conservation, and managing activities in ocean space. Marine Spatial Planning strives to be: (a) multi-objective under the umbrella of increased sustainability, and inclusive of ecological, social, economic, and governance objectives; (b) spatially focused to manage use within a specific ocean region, ideally at an ecosystem level and thus sufficiently large to incorporate critical ecosystem processes; and (c) integrated to bring together interrelationships and interdependence of each component within the defined management area.

Restoration offers one other conservation strategy that merits mention here; although Chapter 30 addresses restoration it represents another management action that can contribute to conservation objectives.

29.8 Legal Instruments

Conservation efforts, through legislation of one form or another, potentially operate at a wide range of spatial scales and legal jurisdictions. Some of the earliest and best-known conservation legislation has focused on individual species. For example, the *Endangered Species Act* in the United States and the *Species at Risk Act* in Canada both work to recover species whose numbers have become critically low relative to their historical norm. Other legislation focuses specifically on fisheries and thus may encompass not only individual species but also their habitats. For example, Canada's Fisheries Act and the *Magnuson-Stevens Fishery Conservation and Management Act* in the United States focus not only on the target species of fisheries, but also on the habitats critical to their success. Many marine species span international boundaries. For example, models suggest that populations of some deep-water corals off Nova Scotia, Canada, depend on larvae supplied by populations in the northeast United States for new recruits. This means that conservation of corals in this part of Canada may well depend on conservation efforts in the United States. Similarly, Pacific salmon that spend years in the ocean move between Canadian and United States jurisdictions, so fishing quotas in one country have direct ramifications for the other country. For this reason, many countries have endorsed regional specific or general agreements and Conventions such as the *Convention for the Conservation of Anadromous Stocks in the North Pacific Ocean*, or the *Convention for the Protection of the Marine Environment* (Oslo-Paris Convention or OSPAR) for the Northeast Atlantic. Many Regional Fisheries Management Organizations, including the *North Atlantic Fisheries Organization* (NAFO) focus particularly on trying to manage sustainably fish stocks that straddle boundaries. The *UN Convention on the Law of the Sea* (UNCLOS) offers some protection for global fisheries resources from territorial seas to international waters. In the last few years, the United Nations entered and completed negotiations for a new treaty focused on Biodiversity Beyond National Jurisdiction, which addresses sustainability of the massive biodiversity (genetic, species, ecosystem) that lives in ocean waters outside the jurisdiction of any one country (Box 29.3). These negotiations include discussions on complex issues such as marine genetic resources, capacity building, and shared benefits for all of humanity.

Box 29.3 A High Seas Conservation Strategy

In keeping with an integrative approach to conversation, we describe a potential strategy by which to optimize protected areas in the deep sea, arguably one of the most complicated marine applications given the massive depths and major knowledge gaps involved (Figure 29.8). First, identify a swath of ocean of interest, in this case on a continental slope shown as light blue and gray, and divide it into a grid (e.g. 300 × 300 m) of columns (dark blue in lower left) to create 100 planning units subdivided into three depth zones (upper left). Then examine the vertical distributions of species to yield f, priority location for protection.

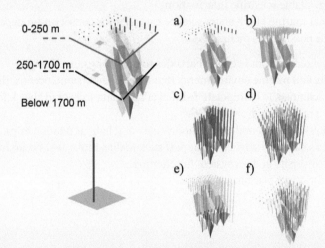

Figure 29.8 Example distribution of species enabling a protection strategy. Reported are: (A) common shallow species; (B) common medium-depth species; (C) common deep species and calculated desirable conservation characteristics such as (D) species richness and (E) conservation value; (F) prioritization preferences (high values in red-purple).

Table 29.1 Knowledge gaps that create challenges in ocean protection planning and potential strategies to deal with such gaps.

Challenge	Solution
Many unknown species and unknown distributions	Proxies for diversity, novelty, abundance, function
Life histories, including dispersal potential often not well known	Look to species with similar reproductive strategies, addressing associated risk
Remoteness & size of some habitats	Habitat mapping prioritization, "connectedness" estimates
Diverse, sometimes allochthonous, food supply	Predictive models, adaptive management
Evaluating impacts of warming oceans over time scales of decades	Temperature envelopes, sensitivity analysis
Slow replacement in some species	Prioritize large, no-take areas
Complex habitat requirements of some species	Maps, species turnover estimates
Incorporating functions and services	Biological traits analysis, function maps

29.9 Science Challenges and Solutions – Moving Science to Policy?

Our incomplete knowledge of ocean environments creates the temptation to throw up our hands in despair as we lament the challenges of trying to create the perfect protected area and to know all the answers about the biodiversity associated with it. Given an ambitious target of 30% protection of each marine habitat by 2030 set by the *International Union for the Conservation of Nature* and championed by the *Global Ocean Alliance*, governments face international pressure to move conservation protection forward. Some scientists worry that our incomplete knowledge will compromise conservation efforts, and they argue that such targets rush the process. Nonetheless, others feel that ocean life cannot wait, given increasing pressures, including climate change. Indeed, some MPAs must be better than none, and if we can develop approaches that allow adaptive management as new knowledge becomes available, then we can make a difference on sustainable oceans, even within the next decade. Below we provide a few examples of how to deal with imperfect knowledge (Table 29.1).

29.10 How Science Can Contribute

Many countries now embrace the concept of science-based decision making, which does not assume perfect knowledge but bases decisions upon the best available scientific information.

The specific contributions that marine biodiversity science can offer for conservation depend on the specific objectives and application, however, some possibilities include:

1) Baseline maps, biodiversity databases, and genetic barcodes to monitor future change.
2) Baseline data on biodiversity and marine environments that help to understand oceans, their ecosystems & processes.
3) Predictive tools that allow scientists to extrapolate from available data to extend that knowledge to new situations to infer biodiversity patterns in marine environments.
4) Analytical and sampling tools to predict and assess biodiversity and habitat relationships.
5) Decision-making frameworks based on science to support sustainable, integrated ocean management.
6) New findings on marine biodiversity and ecosystem functioning.
7) Direct advice & data input for ocean planning.

Questions

1) What is a species-by species conservation strategy?
2) Can a species-by-species conservation strategy succeed in sustaining ocean health? Why?
3) What are the advantages in protecting a specific geographic location?
4) What are "Ecologically and Biologically Sensitive Areas (EBSAs)"?
5) Can economic valuation of a marine ecosystem help to promote conservation?

6) What is a biological traits approach?
7) How can we make fisheries sustainable?
8) What are the advantages in creating MPAs and what are the limits of MPAs?
9) What are the typical steps for the establishment of MPAs?
10) What does "precautionary principle" mean? What are the main management approaches?
11) Can science help in balancing the urgency of ocean conservation with economics and political pressure?

Suggested Reading

Abdulla, A., Gomei, M., Hyrenbach, D. et al. (2009) Challenges facing a network of representative marine protected areas in the Mediterranean: Prioritizing the protection of underrepresented habitats. *ICES Journal of Marine Science*, 66, pp. 22–28.

Arrieta, J.M., Arnaud-Haond, S., and Duarte, C.M. (2010) What lies underneath: Conserving the oceans' genetic resources. *Proceedings of the National Academy of Sciences of the United States of America*, 107, pp. 18318–18324.

Beaumont, N.J., Austen, M.C., Atkins, J.P. et al. (2007) Identification, definition and quantification of goods and services provided by marine biodiversity: Implications for the ecosystem approach. *Marine Pollution Bulletin*, 54, pp. 253–265.

Carpenter, S.R., Mooney, H.A., Agard, J. et al. (2009) Science for managing ecosystem services: Beyond the Millennium Ecosystem Assessment. *Proceedings of the National Academy of Sciences of the United States of America*, 106, pp. 1305–1312.

Cartes, J.E., Maynou, F., Sardà, F. et al. (2004) The Mediterranean deep-sea ecosystems: An over-view of their diversity, structure, functioning and anthropogenic impacts, in *The Mediterranean Deep-sea Ecosystems: An Overview of Their Diversity, Structure, Functioning and Anthropogenic Impacts, with a Proposal for Conservation*. IUCN, Málaga and WWF, Rome, p. 9–38 (free download).

Claudet, J., Osenberg, C.W., Benedetti-Cecchi, L. et al. (2008) Marine reserves: Size and age do matter. *Ecology Letters*, 11, pp. 481–489.

Cognetti, G. and Maltagliati, F. (2008) Perspectives on the ecological assessment of transitional waters. *Marine Pollution Bulletin*, 56, pp. 607–608.

Danovaro, R., Aguzzi, J., Fanelli, E. et al. (2018) An ecosystem-based deep-ocean strategy. *Science*, 355, pp. 452–454.

Danovaro, R. and Pusceddu, A. (2007) Ecomanagement of biodiversity and ecosystem functioning in the Mediterranean Sea: Concerns and strategies. *Chemistry and Ecology*, 23, pp. 347–360.

De Leo, G.A., Paris, G., Gatto, M. et al. (1998) Spotlight needed on Italian policy (multiple letters). *Nature*, 391, p. 12.

ESF (2007) Response to the European Commission's Green Papers: (i) Towards a future Maritime Policy for the Union: A European vision for the oceans and seas, (ii) The European Research Area: New Perspectives. *ESF Marine Board Position Paper*, 11, pp. 1–48.

Game, E.T., Grantham, H.S., Hobday, A.J. et al. (2009) Pelagic protected areas: The missing dimension in ocean conservation. *Trends in Ecology and Evolution*, 24, pp. 360–369.

Gatto, M. and De Leo, G.A. (2000) Pricing biodiversity and ecosystem services: The never-ending story. *Bioscience*, 50, pp. 347–355.

Giakoumi, S., Mazor, T., Fraschetti, S. et al. (2012) Advancing marine conservation planning in the Mediterranean Sea. *Reviews in Fish Biology and Fisheries*, 22, pp. 943–949.

Gipperth, L. and Elmgren, R. (2005) Adaptive coastal planning and the European Union's water framework directive: A Swedish perspective. *Ambio*, 34, pp. 157–162.

Grigg, R.W. (1984) Resource management of precious corals: A review and application to shallow water reef-building corals. *Marine Ecology*, 5, pp. 57–74.

Gubbay, S. (2003) Protecting the Natural Resources of the High Seas. *Scientific Background Paper, WWF/IUCN High Seas Marine Protected Areas Project*.

Jouffray, J.B., Blasiak, R., Norstroem, A.V. et al. (2020) The Blue Acceleration: The trajectory of human expansion into the ocean. *One Earth*, 2, pp. 43–54.

Kroodsma, D.A., Mayorga, J., Hochberg, T. et al. (2018) Tracking the global footprint of fisheries. *Science*, 359, pp. 904–908.

Lotze, H.K. (2021) Marine biodiversity conservation. *Current Biology*, 31, pp. R1190–R1195.

Micheli, F., Saenz-Arroyo, A., Greenley, A. et al. (2012) Evidence that marine reserves enhance resilience to climatic impacts. *PLOS One*, 7, p. e40832.

Myers, R.A., Baum, J.K., Shepherd, T.D. et al. (2007) Cascading effects of the loss of apex predatory sharks from a coastal ocean. *Science*, 315, pp. 1846–1850.

Norse, E.A. ed. (1993) *Global Marine Biological Diversity: A Strategy for Building Conservation into Decision Making*, Vol. 2. Island Press.

Norse, E.A. and Crowder, L.B. (2005) *Marine Conservation Biology: The Science of Maintaining the Sea's Biodiversity*. Island Press.

Palumbi, S.R., Sandifer, P.A., Allan, J.D. et al. (2009) Managing for ocean biodiversity to sustain marine ecosystem services. *Frontiers in Ecology and the Environment*, 7, pp. 204–211.

Sala, E., Mayorga, J., Bradley, D. et al. (2021) Protecting the global ocean for biodiversity, food and climate. *Nature*, 592, pp. 397–402.

Townsend, M., Thrush, S.F., and Carbines, M.J. (2011) Simplifying the complex: An 'Ecosystem Principles Approach' to goods and services management in marine coastal ecosystems. *Marine Ecology Progress Series*, 434, pp. 291–301.

Van den Hove, S. and Moreau, V. (2007) *Deep-Sea Biodiversity and Ecosystems: A Scoping Report on Their Socio-economy, Management and Governance*. UNEP-Regional Seas Programme/UNEP-WCMC, p. 88.

White, C., Gaines, S., Kendall, B.E. et al. (2008) Marine reserve effects on fishery profit. *Ecology Letters*, 11, pp. 370–379.

30

Restoring Marine Habitats

30.1 A Decade For Ecosystem Restoration

As discussed in Chapter 28, direct and indirect human pressures on marine ecosystems have increased significantly over the last five decades with further increases expected in the coming years. These increases will accelerate further loss of marine habitats and their biodiversity, impairing ecosystem functioning. The most deleterious human activities include pollution, overexploitation of resources, introduction of invasive species, and habitat destruction and/or fragmentation. Today, we recognize the dominant effect of global climate change and its rapid transformation of marine ecosystems, exacerbating other deleterious anthropogenic impacts (Figure 30.1). Collectively, these impacts have already caused declines in abundances of large marine species (primarily megafauna), declines in local biodiversity, and reduced capacity of the ocean to provide ecosystem goods and services.

These activities are degrading natural capital and marine habitats worldwide. Some coastal habitats such as saltmarshes, mangroves along with seagrass, meadows, coral reefs, and kelp forests urgently need protection, noting that some areas have seen declines in spatial extent by 25–80% or more. Coral reefs are particularly threatened with projected further declines of 70–90% with an ocean temperature increase of 1.5 °C, and larger losses at 2 °C. Studies documented a 35% decline in both marine/coastal and inland natural wetland habitats between 1970 and 2015. The unprecedented declines in marine vegetated ecosystems through anthropogenic impacts has compromised associated essential ecosystem services such as coastal protection and carbon sequestration.

Despite the development of policies to protect ecosystems from further degradation (UN Agenda 2030 that aims to protect 30% of the marine environment by the year 2030), many marine ecosystems have become degraded to the point that effective recovery cannot happen without human intervention, and thus require active restoration actions.

The UN has committed to halting and reversing the decline in health and productivity of the global ocean and its ecosystems and to protecting and restoring its resilience and ecological integrity. The UN General Assembly of the Conference of the Parties to the Convention on Biological Diversity designated the decade 2021–2030 as the United Nations Decade of Ocean Sciences for Sustainable Development and the United Nations Decade for Ecosystem Restoration. These designations aim to support and scale up efforts to prevent, halt, and reverse the degradation of ecosystems worldwide and raise awareness of the importance of successful ecosystem restoration. Ecosystem Restoration will also contribute to the Paris Agreement adopted under the United Nations Framework Convention on Climate Change, and to achieving the Aichi Biodiversity Targets and the post-2020 global biodiversity framework. These restoration initiatives encourage all countries to:

1) mobilize resources, capacity-building, and scientific research, cooperation, and momentum for ecosystem restoration at global, regional, national, and local levels;
2) mainstream ecosystem restoration into policies and plans to address current national development priorities and challenges resulting from the degradation of marine and terrestrial ecosystems, biodiversity loss, and climate change vulnerability, thereby creating opportunities for ecosystems to increase their adaptive capacity and enhance opportunities to maintain and improve livelihoods for all;
3) build on and reinforce existing restoration initiatives in order to scale up good practices;
4) promote the sharing of experiences and good practices in ecosystem conservation and restoration.

Marine Biology: Comparative Ecology of Planet Ocean, First Edition. Roberto Danovaro and Paul Snelgrove.
© 2024 John Wiley & Sons Ltd. Published 2024 by John Wiley & Sons Ltd.
Companion Website: www.wiley.com/go/danovaro/marinebiology

Figure 30.1 Multiple impacts that result in degradation or destruction of marine habitats: (a) trawl marks on the deep-sea floor; (b) discharge of waters from aquaculture or desalination plants; (c) introduction of non-indigenous species (*Caulerpa* sp.); (d) marine litter (plastic and other litter along the beach strand line); (e) a trawling net full of fish indicating the overfishing of several commercial species; (f) Ghost fishing and trammel net on hard bottoms. *Sources:* Courtesy of Chris Smith, Courtesy of Thanos Dailianis, Courtesy of Donat Petricioli, Courtesy of EPILEXIS/HCMR.

30.2 Defining Ecological Restoration

The *International Standards for the Practice of Ecological Restoration* clearly distinguishes between "ecological restoration" and other forms of ecosystem repair (Box 30.1). Although **restoration** refers to the process of "assisting the recovery of ecosystems that have been degraded, damaged, or destroyed," the **"recovery"** of an ecosystem refers to the achievement of a target environment similar to an appropriate "native" pre-impact model or reference ecosystem, in terms of its specific compositional, structural, and functional ecosystem attributes (Figure 30.2). Restoration includes an action or multiple actions that jumpstart recovery and place a degraded ecosystem on a trajectory for recovery, regardless of the period required to achieve the recovery outcome. **Passive restoration** relies on the spontaneous resilience of an ecosystem to recover simply by removing the environmental stressor(s) that led to degradation, in contrast to **active restoration** that relies on active interventions to the ecosystem, whether by adding or removing structures, species, or stressors. Ecosystem restoration principally aims to establish a self-supporting habitat similar to the "original" habitat prior to the impact. The selection of the desired reference system guides ecological restoration and should be historically inspired and grounded in social processes that include multiparty stakeholders and restoration scientists and practitioners. Application of a group of restoration actions at a site can help in achieving restoration goals, including remediation, reparation, and rehabilitation.

Rehabilitation strives simply to replace structural or functional characteristics damaged by an impact, and to enhance the social, economic, and ecological value of the "new" ecosystem. This strategy represents a less ambitious goal than restoration in not necessarily achieving full recovery of processes and returning to "pre-disturbance" conditions. To define the targets of restoration requires identifying "reference ecosystems," referring to "undamaged sites." Alternatively, scientists define reference ecosystems based on multiple sources of information on diverse ecological and biological variables (e.g., biodiversity, life cycles, functional variables, and food webs) supported by abiotic measurements. In any case, restoration

Figure 30.2 Schematic representation of restoration targets: Reversing degraded habitats into three-dimensional, highly diverse habitats often dominated by habitat-forming species. Images illustrate: restoration of seagrass meadows (top); restoration of algal forests and assemblages on hard bottoms (middle); restoration of deep-sea habitats (bottom).

planning requires clear definitions of ecosystem baselines. Full recovery requires several steps, each encompassing a range or family of restorative activities that span a continuum of different restorative practices: reduced societal impacts, remediation, rehabilitation, and ecological restoration.

Ecological restoration experts have abandoned the idea of full recovery in a short time, recognizing the dynamic nature of ecosystems, and that restoration may not follow or complete the ecological trajectory necessary to reestablish the original community over time, especially considering the need for ecosystems to adapt and evolve in response to climate change. That said, the ultimate vision in ecological restoration is to get biodiversity back. Ecological restoration, when implemented effectively and sustainably, helps to protect biodiversity, improve human health and wellbeing, increase food and water security, deliver goods, services, and economic prosperity, and supports climate change mitigation, resilience, and adaptation. This solutions-based approach engages communities, scientists, policymakers, and land managers to repair ecological damage and rebuild a healthier relationship between people and nature. When combined with conservation and sustainable use, ecological restoration provides the necessary link to move local, regional, and global environmental conditions from a state of ongoing degradation to one of net positive improvement.

Box 30.1 Glossary of Marine Ecosystem Restoration, Presented in Alphabetical Order

- **Assisted regeneration:** An approach to restoration that focuses on actively triggering any natural regeneration capacity of biota remaining on a site or nearby, as distinct from reintroducing biota to the site or leaving a site to regenerate. Although managers typically apply this approach to sites of low to intermediate degradation, even some extremely degraded sites have proven capable of assisted regeneration given appropriate treatment and sufficient time. Interventions can include removal of pest organisms, reapplying ecological disturbance regimes, and installation of resources to prompt colonization.

(Continued)

> **Box 30.1 (Continued)**
>
> - **Baseline condition**: The condition of a restoration site immediately prior to the initiation of ecological restoration activities.
> - **Damage (to ecosystem):** An acute and obvious deleterious impact on an ecosystem.
> - **Degradation (of an ecosystem)**: A level of deleterious human impact to ecosystems that results in the loss of biodiversity and simplification or disruption in their composition, structure, and functioning, generally leading to a reduction in the flow of ecosystem services.
> - **Destruction (of an ecosystem)**: When degradation or damage substantially compromises all macroscopic life, and commonly ruins the physical environment of an ecosystem.
> - **Disturbance regime:** The pattern, frequency, timing, or occurrence of disturbance events that characterize an ecosystem over a given time period.
> - **Ecological restoration:** The process of assisting the recovery of a degraded, damaged, or destroyed ecosystem. Although some researchers use the terms ecosystem restoration and ecological restoration interchangeably, ecological restoration always addresses biodiversity conservation and ecological integrity, whereas some approaches to ecosystem restoration may focus solely on ecosystem services.
> - **Ecological restoration project:** Any organized effort to achieve substantial recovery of a native ecosystem, from the planning stage through implementation and monitoring. As a long-term program, such a project may require multiple agreements or funding cycles.
> - **Ecosystem resilience:** The degree, manner, and pace of recovery of ecosystem properties after natural or human disturbance. In plant and animal communities, this property strongly depends on individual species adaptations to disturbances or stresses experienced during their evolution.
> - **Full recovery:** The state whereby all ecosystem attributes closely resemble those of the reference (model) ecosystem. An ecosystem exhibiting self-organization that leads to the full resolution and maturity of ecosystem attributes precedes full recovery. At the point of self-organization, the restoration phase could be considered complete and management shifts to a maintenance phase
> - **Indicators (of recovery)**: Characteristics of an ecosystem that scientists can use for measuring progress toward restoration goals or objectives at a particular site (e.g., measures of presence/absence and quality of biotic or abiotic components of the ecosystem).
> - **Natural regeneration:** Germination, birth, or other recruitment of biota including plants, animals, and microbiota, that does not involve human intervention.
> - **Natural (or spontaneous) regeneration approach:** Ecological restoration that relies only on increases in individuals following removal of degradation drivers (unassisted regeneration).
> - **Recovery**: The process by which an ecosystem regains its composition, structure, and function relative to the levels identified for the reference ecosystem. Restoration activities usually assist the recovery that can be partial or full.
> - **Rehabilitation**: Management actions that aim to reinstate a level of ecosystem functioning on degraded sites, where the goal is renewed and ongoing provision of ecosystem services rather than the biodiversity and integrity of a designated native reference ecosystem.
> - **Restoration activities:** Any action, intervention, or treatment intended to promote the recovery of an ecosystem or component of an ecosystem, such as substrate amendments, control of invasive species, habitat conditioning, species reintroductions, and population enhancements.
> - **Restorative activities:** Activities (including ecological restoration) that reduce degradation or improve conditions for the partial or full recovery of ecosystems.
> - **Restorative continuum:** A spectrum of activities that directly or indirectly supports or attains at least some recovery of lost or impaired ecosystem attributes.
> - **Translocation:** The intentional transporting (by humans) of organisms to a different location.

Historically, research on restoration best practices and methods has mainly focused on terrestrial, rather than on marine ecosystems. Although some of the basic principles developed for terrestrial systems translate to applications in marine environments, the knowledge base regarding factors that enhance or constrain restoration success remains quite limited for marine applications. For example, we now recognize that the complex, expensive, and lengthy process involved in restoration of a degraded ecosystem requires considerable time, resources, expertise, and knowledge. Moreover, access to marine ecosystems is evidently more difficult than for terrestrial ecosystems, and in some cases extremely difficult, as in offshore environments. Importantly, underwater restoration actions in the marine environment will require different approaches depending on the specific habitat and restoration activities.

Although large-scale actions are common practices in terrestrial ecosystems, such efforts remain a challenge in marine ecosystems. Governments should promote ecological restoration in marine ecosystems and identify "green and blue industries" as promising targets to stimulate interest and carry out specific actions. The scientific community and different ocean stakeholders, including local communities, funding organizations, governmental bodies, citizens, and volunteers can rally around a common vision that strongly engages interest and involvement in restoration projects.

Three major logistical constraints hinder marine ecosystem restoration: (1) The long time scales (sometimes spanning several decades or even centuries) required to achieve restoration targets; (2) the substantial funding and high-technology equipment needed, particularly in the deep sea (e.g., submersibles or ROVs); and (3) the difficulty in scaling up any restoration intervention to sufficiently large spatial scales to achieve significant impact. The upfront costs of marine ecosystem restoration create a particularly challenging impediment.

30.3 A Global Plan for Marine Ecosystem Restoration

The *2050 vision of the Biodiversity Strategy* (https://www.cbd.int/doc/c/0b54/1750/607267ea9109b52b750314a0/cop-14-09-en.pdf) promotes the protection of natural capital and the restoration of marine habitats and their biodiversity for their essential contribution to human wellbeing and economic prosperity. Healthy seas offer multiple benefits and ecosystem services, including the maintenance of a healthy environment, food provisioning, pollutant abatement, and products needed for our wellbeing, as well as cultural, educational, and recreational benefits. However, despite marine restoration actions in many areas of the world, their success rate varies greatly. For instance, whilst >65% of tropical coral reef and salt marsh restoration projects successfully achieved their goals, seagrass restoration efforts identify a lower success rate pointing to the need for new approaches to enhance the initial establishment success of foundation species and ensure the long-term persistence of restored habitats. Variation in restoration success stems from different sources, including the inherent biology and ecology of species, their interactions, and how, where, and when restoration efforts occur. This variation leads to uncertainty in terms of conservation outcomes and economics, and the need to develop robust methodologies to restore habitats effectively and deliver the full range of potential conservation and socioeconomic benefits. We sorely lack careful studies on the cost-benefit analysis and other aspects of the economics of restoration for marine environments; this gap contributed to the decision by the United Nations to declare 2021–2030 as the *Decade on Ecosystem Restoration* (https://www.decadeonrestoration.org).

Maximizing the impact and efficiency of restoration efforts requires increased coordination of national and international action and policy. *Agenda 2030 and Legally Binding Instruments for Restoration* offer an opportunity to develop national and regional restoration goals and targets in alignment with other processes relating to the sustainable **blue economy**, referring to the "sustainable use of ocean resources for economic growth, improved livelihoods, and jobs while preserving the health of ocean ecosystems."

Successful restoration requires long-term, sustained financing. Given that some marine ecosystems require decades to achieve full recovery, funding commitments must match these timeframes. The design, implementation, and long-term monitoring of sustainable and effective restoration efforts urgently require innovative funding and cross-sectoral collaborations.

30.4 Restoring Fragile Marine Habitats

Researchers have recently begun to explore the potential for restoration in both shallow soft and hard bottoms (including mesophotic habitats) and in the deep sea. Not surprisingly, restoration actions focus largely on seafloor environments, noting that the constant motion of the water that defines pelagic habitats does not easily lend itself to specific restoration beyond termination of anthropogenic pressures.

The most fragile and vulnerable coastal and deep-sea habitats include some of the habitats described in earlier chapters: seagrass meadows (*Zostera marina*, *Z. noltii*, *Cymodocea nodosa*, *Posidonia oceanica*), algal and kelp forests, coralligenous outcrops (including *Corallium rubrum*, *Paramuricea clavata*, *Eunicella singularis*, *E. cavolini*), cold-water corals (*Callogorgia*

Figure 30.3 Successful pilot projects that used restoration actions to reverse degraded habitats include seagrass meadows (upper panels); coralligenous outcrops (middle panels) and deep-sea habitats (lower panels).

Figure 30.4 Field work to advance restoration in shallow water habitats: left panel, SCUBA divers suspending gorgonians fragments to create new colonies; right panel, SCUBA divers taking measures to dislocated the transplanted seagrasses. *Sources:* Carlo Cerrano, NOAA / Public Domain.

verticillata, Paracalyptrophora josephinae, Viminella flagellum, Lophelia pertusa), canyons, seamounts, and fjords. Many diverse habitat types could benefit from habitat restoration (Figure 30.3).

Restoration includes different practices such as species translocation/transplanting, seedling and grazer removal, the use of artificial (and biodegradable) substrates, or the re-introduction of native species that can support the recovery of specific endangered species and habitats. Each pilot action must consider "restoration success" in order to identify the criteria for selecting target species and habitats likely to promote successful restoration activities (Figure 30.4). Experience acquired to date provides evidence that extreme/episodic events (e.g., storms, heat waves) can compromise restoration activities and survival of target species in marine shallow habitats, irrespective of the methodologies used for marine ecological restoration. Experience also suggests that successful marine ecological restoration requires adequate financing, relevant policy decisions, social awareness, and engagement with the private sector.

Recent studies demonstrate that promoting positive interactions between individuals of the same species can increase restoration success, highlighting the importance of facilitative interactions in restoring ecosystem-engineering species. Facilitative interactions among ecosystem engineers may also promote resilience and recovery, but <3% of restoration projects have integrated interspecific interactions.

The success of restoration actions at large scales varies across ecosystems, and restoration success largely depends on the status of anthropogenic pressures. Knowledge developed in pilot actions can help in identifying the environmental conditions necessary to optimize restoration success, as well as the best practices (protocols) to apply at larger scales to expedite recovery of degraded habitats and endangered species in coastal soft and hard bottoms ecosystems.

30.5 Restoration of Coral Reefs

The complex biological and ecological procedure and the task of emulating natural phenomena of reef biology create significant challenges in restoring any type of degraded reef habitat. Nonetheless, coral reef restoration has become an increasingly important tool in supporting conservation measures. This type of strategy requires defining the ecological bases and most effective protocols under different environmental constraints appropriate to specific local species assemblages.

During the last decades, coral reef restoration/rehabilitation efforts have focused on three main approaches: (1) structural restoration; (2) physical restoration and (3) biological restoration.

Structural restoration involves construction of habitat structures, in this case artificial reefs, sinking of wrecks, or relocation of rocks/dead coral heads. Structural restoration aims to increase the amount of reef structure and habitat available for corals themselves, as well as to create substrate on which other reef organisms can grow. Environments in which disturbance has led to reef loss typically require structural restoration.

Environmental restoration involves addressing the environmental conditions under which corals grow to improve their health, growth rates, or reproductive ability. Some of the techniques used include mid-water coral nurseries placed in locations with superior water quality (such as in the open ocean) but maintaining ambient light conditions for which corals and their zooxanthellae are adapted. High survival rates and faster coral growth than similar colonies on natural reefs characterize corals growing on mid-water nurseries because of the decreased stress caused by sedimentation, eutrophication, predation, or pollution on the reef itself.

Biological restoration usually involves increasing the amount of living coral on a reef in locations with available structure. Scientists generally achieve this objective by collecting and rehabilitating naturally broken coral fragments (nubbins), propagating coral colonies, or through direct transplantation of coral fragments or colonies. These techniques align with the concept that newly established coral colonies in denuded reefs should originate from an external habitat as similar as possible to the recipient site. However, techniques used so far for removal of coral materials, their transportation, and re-attachment (e.g., without an acclimatization period) could compromise donor sites and transplanted material. As such, varying degrees of success in past coral restoration experiments come as no surprise.

Recently, scientists proposed the "coral gardening" tenet, a strategy resembling terrestrial forest plantations, as an alternative approach to direct transplantation of corals (Figure 30.5).

This strategy follows a two-step restoration protocol. The first stage involves growing large numbers of farmed nubbins and small coral fragments within specially designed underwater coral nurseries (Figure 30.6). Deployment of these nurseries in sheltered zones nurtures small coral fragments to grow to sizes suitable for transplantation. In the second phase, divers transplant different sizes and species combinations of the nursery-grown coral colonies to degraded reef sites. This approach draws from theories and vast experience accrued over many years of terrestrial silviculture. Compared to ex situ efforts, comparatively cheap in situ coral farming can potentially produce large quantities of coral fragments (thousands of individuals), an important pre-requisite for broad-scale reef restoration.

Furthermore, the establishment of in situ nurseries in habitats similar to recipient sites and in sheltered locations may provide cultured corals that have received an appropriate (indeed essential) acclimation period for increasing post-transplantation survivorship and growth. So far, researchers have conducted in situ coral "farming" experiments in the Red Sea, East Africa, Southeast Asia, and the Pacific. Varying results from these experiments further emphasize the need to develop protocols specifically adapted to locations differing in environmental conditions and coral assemblages.

Figure 30.5 Example of coral gardening. First (left), scientists collect nubbins of corals from donor populations and locate them in gardens for rearing. After 7 months of growth (right) healthy corals are ready for transplanting.

Figure 30.6 Example of an underwater coral nursery. *Source:* NOAA / Public Domain.

30.6 Restoration of Seagrass Meadows

The choice of restoration techniques and time scales for recovery can vary significantly amongst different seagrass species. For example, *Posidonia oceanica*, a slow-growing species, requires longer time scales for recovery than *Cymodocea nodosa* and *Zostera marina*, which exhibit faster clonal growth. Therefore, restoration efforts should align with such timeframes. However, the extreme vulnerability of seagrass meadows to anthropogenic pressures points to the importance of reducing or removing critical pressures, such as eutrophication (which limits light availability and growth) and habitat destruction, and re-establishing appropriate sediment conditions. Supportive policies and management should aim to reduce such pressures in combination with restoration.

Seagrass restoration generally uses two main approaches: (1) transplanting a portion of healthy seagrass into a degraded location. This approach raises the possibility of transplanting single rhizomes or an entire portion of the habitat (typically ~1 m², and including sediment, root mass, and above sediment plants); (2) by implanting *plantulae* (juvenile plants deriving from cultured seagrass seeds) into the degraded habitat. Selection of the receiving system should consider existing

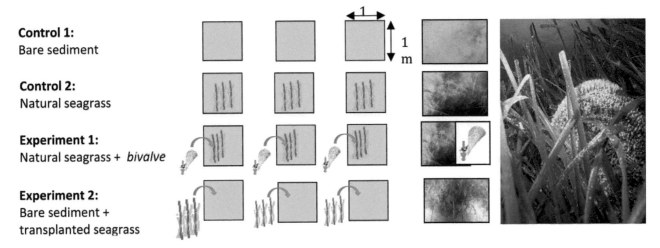

Figure 30.7 Example of an experimental approach used to test positive effects of co-existing bivalves (e.g., mussels or *Pinna nobilis*) and seagrass (e.g., *Zostera noltii* or *Posidonia oceanica*) in replicated plots.

pressures as well as historical conditions. For example, restoration efforts should prioritize locations that have hosted seagrasses historically.

Recent studies demonstrate that positive interactions between epifaunal bivalves and plants dominate marine habitats (Figure 30.7). Plants primarily promote bivalve survival and abundance by providing substrate and shelter, whereas bivalves promote plant growth and survival by stabilizing and fertilizing sediments and reducing water turbidity through filtration. The prevalence of positive interactions increases with water temperature in subtidal habitats. If done properly, co-restoration could increase initial survival, persistence, and resilience of foundation species, while promoting recovery of associated biodiversity and ecosystem services

A long-term process requires long-term monitoring, and seagrass meadows undergoing restoration require long-term management to monitor overall ecosystem response. Additional actions might include introduction of healthy populations of associated species, especially top predators, which can control algal (over)growth through trophic cascades. Researchers have repeatedly identified large-scale planting as an important method for increasing success of seagrass restoration. Greater spatial distribution allows greater opportunity for transplanting donor populations, which increases the probability of recovery success. Therefore, experts strongly encourage active restoration of seagrass meadows at large spatial scales.

30.7 Restoration of Macroalgal Forests

30.7.1 Restoration of Kelp Forests

Kelp-forest restoration requires knowledge of interactions with local predators (such as sea urchins), prevalence of turf algae, as well as an understanding of local and global conditions. Removal of all existing anthropogenic pressures (e.g., from urbanization, eutrophication, and increasing sediment loads) should precede restoration because many coastal regions have experienced loss of such forests or transformation into simpler and less productive communities such as barrens and algal turfs as a result of these pressures, which can hinder restoration efforts. Kelp forest restoration might entail transplants from adjacent sites or promotion through assisted restoration where kelp typically show rapid recovery rates following removal of sea urchins and eutrophication (Figure 30.8). Policies and management practices to support ecosystem recovery should aim to reduce such pressures in combination with restoration efforts. Implementation of restoration actions for kelp forests at large spatial scales in suitable areas typically improves the success of restoration efforts.

30.7.2 Restoration of *Cystoseira* spp. Forests

Successful planning of restoration of brown algal forests hinges upon understanding their growth and health. Significant loss of such habitat has occurred in rocky shore habitats of the Mediterranean Sea. Active restoration should not be carried out in environmental conditions where natural donor populations are in a critical state; active restoration (planting), and restoration should instead rely on recruitment enhancement and growth of propagules (Figure 30.9). In this case we should

Figure 30.8 Bottom conditions before and after a restoration intervention for kelp forests. (*Frontiers in Marine Science* 07-00074-g001).

Figure 30.9 Examples of recruitment enhancement techniques in which earthenware substrata collect propagules of *Cystoseira* spp., where the tiny propagules grow until they reach a visible size of a few cm prior to transfer to the receiving site, where they can grow further and colonize new surfaces.

expect prolonged recovery for restoration, possibly decades. An optimal restoration strategy should combine two approaches: recruitment enhancement techniques, for instance through the cultivation of propagules on appropriate substrates and transplanting them into receiving habitats, and sea urchin eradication in order to control their impact on transplanted algae. These approaches appear to be the most effective for *Cystoseira* spp. forestation in degraded shallow water barrens. The maintenance and efficacy of restoration efforts also depends on reducing human pressures, such as eutrophication, chemical pollution, coastal development, and sedimentation. Supportive policies and management interventions should aim to reduce such pressures in combination with restoration efforts.

30.8 Restoring Ecosystem Engineers: The Case of Coralligenous Outcrops

A combination of nutrient enrichment, invasive species, increased sedimentation, and mechanical impacts, mainly from fishing activities as well as climate change, currently threaten coralligenous assemblages. Managers should prioritize reduction of these pressures prior to starting any restoration actions. Experimental restoration of coralligenous habitats at the local scale has begun in pilot areas characterized by highly fragmented habitats along NW Mediterranean coasts following mass-mortality events. Active restoration on hard bottoms generally focuses on recovery of structural and/or habitat-forming species that can provide habitat for associated species. Restoration actions can involve gorgonians and red corals, or sponges, which transplant relatively simply with good success rates, resulting in high survival rates and requiring relatively low initial efforts (Figure 30.10). However, in the case of transplanting of slow growing, long-lived coralligenous habitat species with limited recruitment, such as sponges (*Petrosia fisciformis*, *Spongia lamella*, *S. officinalis*) and octocorals (*Paramuricea clavata*, *Corallium rubrum*), restoration can take decades. Some other taxa, such as bryozoans (*Pentapora fascialis*), grow at much faster rates and can restore structural complexity within five to ten years.

Depth represents the main limitation for restoration of hard bottom habitats. Work requiring SCUBA divers, long diving times, and considerable expertise can occur at depths to 30–40 m depth, but deeper environments create considerable safety challenges and require other approaches (e.g., vessel-supported deployments).

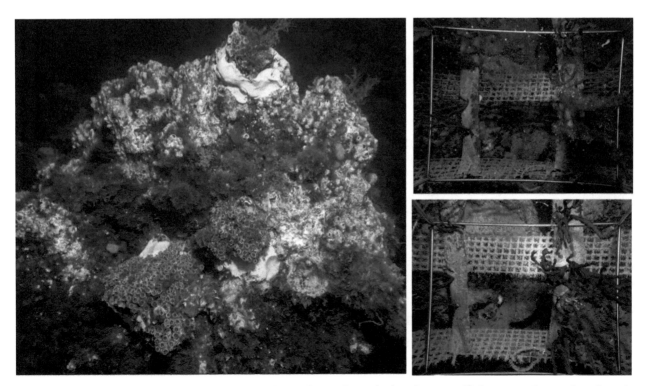

Figure 30.10 Transplantation of sessile invertebrates: Researchers collect colonies of sponges, *Cladocora caespitosa*, red corals, and gorgonians from donor sites and transplant them onto defaunated surfaces using epoxy putty (left) or incubate them prior to transplanting (right). *Source:* Courtesy of Carlo Cerrano.

30.9 Restoration of Deep-Sea Habitats

Ecological restoration of deep-sea habitats is particularly challenging because it requires advanced underwater technologies, complex logistics, and significant economic investments compared to shallow-water environments. Future availability of more accessible, cost-effective underwater technology (such as Autonomous Underwater Vehicles or cabled observatories) could enable broader application and upscaling of restoration of vulnerable and critical deep-sea habitats, such as deep-water corals and gorgonian forests, and possibly even hydrothermal vents, and methane seeps. Restoration of deep-sea ecosystems will surely challenge future decision makers evaluating the costs and benefits of maintaining ecosystem functioning and provisioning of goods against the economic benefits of exploitation. Developing new approaches and methodologies to support and accelerate natural recovery of deep-sea habitats affected by trawling, oil exploitation, and mining, remains a major scientific research challenge.

Promoting successful restoration requires removal of existing stresses and pressures. Cold-water coral habitats are highly sensitive to a range of human activities, including mineral and hydrocarbon exploration and extraction, and commercial bottom fisheries. Many experts consider the latter threat as the single greatest pressure, often resulting in the removal of entire communities. The slow recovery potential of cold-water corals points to the need to prioritize reduction of such pressures prior to and during long-term restoration actions.

The costs required for restoration along with major uncertainties of outcomes raise major concerns, in that restoration of deep-sea ecosystems could easily cost 3–4 orders of magnitude more than for shallow-water ecosystems. Pilot actions carried out in the framework of several international projects provide additional evidence in support of the feasibility of deep-sea restoration using transplant techniques (following rearing and acclimatization; Figure 30.11) or artificial habitats that enhance larval recruitment of some deep-sea species. A combination of passive and active restoration will optimize restoration success and reduce potential costs. However, this strategy requires time, and full recovery of damaged deep-sea ecosystems within a few years rather than decades or centuries requires active interventions. Moreover, given the life history traits of corals and other deep-sea organisms with great longevity, slow growth, and late reproduction, documentation of fully restored habitats will require long-term monitoring.

Combining active and passive restoration offers a potential solution for these complex habitats. The remoteness of deep-sea habitats dictates that restoration actions depend on expensive technologies (e.g., large ships and remotely operated vehicles, ROVs). The considerable cost of deep-sea habitat restoration actions in comparison with shallow water habitats may severely limit the capacity and feasibility for large-scale restoration actions, likely necessitating a combination of restoration approaches with assisted regeneration at small scales and natural regeneration (through fisheries closures, marine protected areas) at large scales. Given the slow growth rate of deep-sea organisms, the recovery of damaged deep-sea habitats will require restoration measures, punctuating the need to consider carefully the wisdom of ongoing and future extraction activities in deep-sea environments.

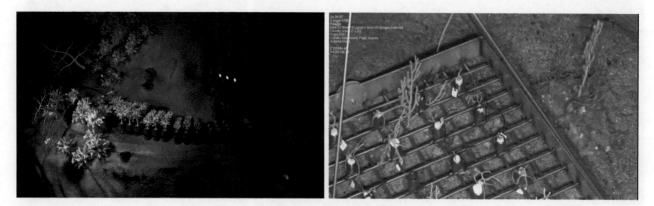

Figure 30.11 Coral gardening for subsequent transplanting of sessile invertebrates in the deep sea.

30.10 Perspectives of Marine Ecosystem Restoration

Marine ecosystem restoration requires upscaling, and efforts should develop in parallel with terrestrial and freshwater restoration efforts to deliver transformative change. In the context of climate change and urban expansion, management must prioritize actions in order to ensure effective and resilient restoration efforts, as well as delivery of maximum biodiversity benefits.

As part of a mitigation hierarchy, practitioners should pair restoration actions with supportive and robust management practices that reduce anthropogenic pressures, noting a vital need to address the root cause of ecosystem degradation and loss. Restoration policies and activities must therefore target pressure reduction, both ocean and land based, as society undertakes effective and sustainable restorative actions.

From increased social awareness within civil society to greater engagement with the private sector, effective restoration must embrace transdisciplinary collaborations. Indeed, global success of the UN Decade on Ecosystem Restoration and regional and national success of the *EU Biodiversity Strategy 2030* will hinge on cross sectoral and inter-disciplinary collaboration.

Questions

1) Which major habitats are currently decreasing in spatial extent?
2) Define "ecological restoration" and explain its aims.
3) Describe the most common restoration practices.
4) Differentiate between passive and active restoration.
5) What factors can compromise restoration actions?
6) Describe three main restoration approaches that apply to coral reef restoration.
7) Describe the two restoration approaches often applied to seagrass restoration.
8) What characteristics of the "receiving system" should researchers prioritize for transplantation?
9) Why does co-restoration promote the recovery of associated biodiversity and ecosystem services?
10) Which major groups of organisms must restoration of hard bottom habitats consider? And why is their recovery important?
11) What are the major limitations of hard bottom habitat restoration?
12) What major deep-sea habitats require restoration efforts and why?

Suggested Reading

Aronson, J., Clewell, A., Moreno Mateos, D. et al. (2016) Ecological restoration and ecological engineering: Complementary or indivisible? *Ecological Engineering*, 91, pp. 392–395.

Barbier, E.B., Hacker, S.D., Kennedy, C. et al. (2011) The value of estuarine and coastal ecosystem services. *Ecological Monographs*, 81, pp. 169–193.

Bayraktarov, E., Banaszak, A.T., Montoya Maya, P. et al. (2020) Coral reef restoration efforts in Latin American countries and territories. *PLOS One*, 15(8), p. e0228477.

Bayraktarov, E., Saunders, M.I., Abdullah, S. et al. (2016) The cost and feasibility of marine coastal restoration. *Ecological Applications*, 26, pp. 1055–1074.

Boström-Einarsson, L., Babcock, R.C., Bayraktarov, E. et al. (2020) Coral restoration – A systematic review of current methods, successes, failures and future directions. *PLOS One*, 15(1), p. e0226631.

Chen, W., Wallhead, P., Hynes, S. et al. (2022) Ecosystem service benefits and costs of deep-sea ecosystem restoration. *Journal of Environmental Management*, 303, p. 114127.

Da Ros, Z., Dell'Anno, A., Morato, T. et al. (2019) The deep sea: The new frontier for ecological restoration. *Marine Policy*, 108, p. 103642.

Danovaro, R., Aronson, J., Cimino, R. et al. (2021) Marine ecosystem restoration in a changing ocean. *Restoration Ecology*, 29, p. e13432.

Derksen-Hooijberg, M., Angelini, C., Lamers, L.P.M. et al. (2018) Mutualistic interactions amplify saltmarsh restoration success. *Journal of Applied Ecology*, 55, pp. 405–414.

Duarte, C.M., Agusti, S., Barbier, E. et al. (2020) Rebuilding marine life. *Nature*, 580, pp. 39–51.

Fabbrizzi, E., Giakoumi, S., De Leo, F. et al. (2023) The challenge of setting restoration targets for macroalgal forests under climate changes. *Journal of Environmental Management*, 326, p. 116834.

Fava, F., Bavestrello, G., Valisano, L., and Cerrano, C. (2010) Survival, growth and regeneration in explants of four temperate gorgonian species in the Mediterranean Sea. *Italian Journal of Zoology*, 77, pp. 44–52.

Fraschetti, S., McOwen, C., Papa, L. et al. (2021) *Where* is more important than *how* in coastal and marine ecosystems restoration. *Frontiers in Marine Science*, 8, p. 626843.

Gann, G.D., McDonald, T., Walder, B. et al. (2019) International principles and standards for the practice of ecological restoration. 2nd Edition. *Restoration Ecology*, 27, pp. S1–S46.

Gerovasileiou, V., Smith, C.J., Sevastou, K. et al. (2019) Habitat mapping in the European Seas - is it fit for purpose in the marine restoration agenda. *Marine Policy*, 106, p. 103521.

McAfee, D., McLeod, I.M., Alleway, H.K. et al. (2022) Turning a lost reef ecosystem into a national restoration program. *Conservation Biology*, 2022, p. e13958.

McAfee, D., Reis-Santos, P., Jones, A.R. et al. (2022) Multi-habitat seascape restoration: Optimising marine restoration for coastal repair and social benefit. *Frontiers in Marine Science*, 9, p. 910467.

McDonald, T., Jonson, J., and Dixon, K.W. (2016) National standards for the practice of ecological restoration in Australia. *Restoration Ecology*, 24, pp. S6–S32.

Moreno-Mateos, D., Barbier, E.B., Jones, P.C. et al. (2017) Anthropogenic ecosystem disturbance and the recovery debt. *Nature Communications*, 8, p. 14163.

O'Connor, E., Hynes, S., and Chen, W. (2020) Investigating societal attitudes towards marine ecosystem restoration. *Restoration Ecology*, 29(S2), p. e13239.

Palumbi, S.R., Mcleod, K.I., and Grünbaum, D. (2008) Ecosystems in action: lessons from marine ecology about recovery, resistance, and reversibility. *BioScience*, 58, pp. 33–42.

Silliman, B.R., Schrack, E., He, Q. et al. (2015) Facilitation shifts paradigms and can amplify coastal restoration efforts. *Proceedings of the National Academy of Sciences of the United States of America*, 112, pp. 14295–14300.

Török, P. and Helm, A. (2017) Ecological theory provides strong support for habitat restoration. *Biological Conservation*, 206, pp. 85–91.

van der Heide, T., Temmink, R.J.M., Fivash, G.S. et al. (2021) Coastal restoration success via emergent trait-mimicry is context dependent. *Biological Conservation*, 264, p. 109373.

van der Heide, T., van Nes, E.H., Geerling, G.W. et al. (2007) Positive feedbacks in seagrass ecosystems: Implications for success in conservation and restoration. *Ecosystems*, 10, pp. 1311–1322.

Van Dover, C.L., Aronson, J., Pendleton, L. et al. (2014) Ecological restoration in the deep sea: Desiderata. *Marine Policy*, 44, pp. 98–106.

Van Tatenhove, J.P.M., Ramírez-Monsalve, P., Carballo-Cárdenas, E. et al. (2021) The governance of marine restoration: Insights from three cases in two European seas. *Restoration Ecology*, 29(S2), p. e13288.

Zahawi, R.A., Reid, J.L., and Hollet, K.D. (2014) Hidden costs of passive restoration. *Restoration Ecology*, 22, pp. 284–287.

Zhang, Y.S., Cioffi, W.R., Cope, R. et al. (2018) A global synthesis reveals gaps in coastal habitat restoration research. *Sustainability*, 10, p. 1040.

31

How Far We Have Come: Past, Present, and Future Research on the Marine Biology of Planet Ocean

31.1 Introduction

Historically, marine biologists sampled the ocean either by working adjacent to the shoreline or by lowering various nets and grabs blindly over the side of vessels ranging from small boats to large research ships. Although boats and ships remain the central platform from which researchers sample the ocean, new platforms now enable much greater presence in ocean environments, both in space and in time. Part of this evolution stems from widespread recognition of some scientific questions that require sampling over longer time periods and over greater spatial scales than a single ship could hope to achieve. This limitation also links to the high cost of operating large ships, which can easily run to $50,000–80,000 d^{-1}, not including the salaries of the scientists on board. This high cost means that the vast majority of research cruises span periods of weeks at most, thus typically covering distances of hundreds rather than thousands of km. Finally, some types of sampling and experimentation require in situ rather than remote presence, a need that imposes severe limits on what ship-based sampling alone can accomplish. Fortunately, advances in technology and other types of sampling platforms enable greater scientific opportunities. Below, we consider alternative sampling platforms as well as new optical and acoustic sensors, revolutionary genetic tools, digital imaging and automated analyses, major advances in communications and data transmission options, and better and more powerful computers and models. In short, as we identify increasingly complex questions in marine biological research, we also develop technological solutions.

In order to understand where marine biological research will head, we should consider how we got to where we are and how that research identified knowledge gaps and opportunities. The study of marine organisms has ancient origins, from the Stone Age through to Aristotle, the "father of marine biology" who began to classify sea anemones of the Mediterranean. However, the major research advances in marine biology really came with geographic exploration in the fifteenth and sixteenth centuries, and the steady development of new tools and technologies for accessing ocean environments, both in terms of seagoing vessels and in terms of sampling organisms and measuring environmental variables. Dedicated oceanographic research ships, the development of scuba diving, and the birth of submersibles and subsequently autonomous and remote operating vehicles, have catalyzed rapid acceleration in biological marine research over the last 50 years. Over the past 10–15 years, the revolution in molecular biology has entered marine research, enabling the unambiguous identification of microscopic and other poorly known organisms, identifying sibling species, or determining source populations. The most recent frontiers in research focus on exploration of extreme marine habitats, including the deep sea, and in evaluating the smallest and least known groups such as microbes. The major scientific challenges for marine biology in the near future include understanding connections and interactions among different ecosystems (including land), complete knowledge of life cycles, life histories, functions performed by individual species, and how these variables interact. Globally, research efforts vary greatly with far more researchers in developed countries and, not surprisingly, more research within their territorial waters.

The UN promoted the *Decade of Ocean Science for Sustainable Development*, 2021-2030, in which the big challenges identified by the UN include the conservation of marine habitats to reach the target 30% of the ocean surface by 2030 (UN Agenda 2030). The UN targets also include "zero pollution" for marine waters. In parallel, the United Nations promoted the *Decade on Ecosystem Restoration* 2021-2030 for expanding active interventions and policies to promote and fund actions to aid and accelerate the recovery of compromised ecosystems of all types, including the restoration of damaged marine ecosystems.

Marine Biology: Comparative Ecology of Planet Ocean, First Edition. Roberto Danovaro and Paul Snelgrove.
© 2024 John Wiley & Sons Ltd. Published 2024 by John Wiley & Sons Ltd.
Companion Website: www.wiley.com/go/danovaro/marinebiology

31.2 The Birth of Marine Biology

Knowledge and interest in marine organisms dates back to the Stone Age; evidence such as cave drawings from 18,000 years ago suggests that humans moved towards the ocean and exploited its resources. Along the coasts of Mexico and California, large clusters of fossil shells indicate remains of meals comprised of marine organisms, and archaeological sites include many relics such as tools and instruments (harpoons and hooks made of stones or bones) for fishing. The tombs of Egyptian pharaohs contain detailed drawings of animals and fishes that suggest some ability to distinguish good prey from bad prey (e.g., toxic puffer fish). In the fourth century BC, Aristotle started the first taxonomic analysis of marine organisms, with some inaccuracies such as confusing polyps of cnidarians with flowers of "marine vegetables".

With time, human interaction with the ocean evolved from merely using living resources along the shoreline to the development of ships and navigational aids that enabled explorers to venture farther afield. Explorers dating back to the Polynesians, some 40,000 to 60,000 years, and Phoenicians, some 8000 years ago, through the end of the Middle Ages, when Francis Bacon, John Cabot, Christopher Columbus, Vasco da Gama, Ferdinand Magellan, Marco Polo, and Amerigo Vespucci searched for new lands and routes. They soon recognized, sometimes costing them their lives, the need to understand ocean currents and bathymetry, eventually extending to deeper analysis of ocean physical-chemical properties such as salinity and temperature, and its dimensions and bounds compared to the continents. This period also saw the birth of the first scientific foundations, such as the Royal Society of London in 1645. These societies showed particular interest in the marine environment, financing the first research at sea. Robert Hooke (1635–1702), curator of the Royal Society, developed tools to measure bottom depth and ocean temperature, whereas Robert Boyle (1627–1691) undertook research to understand variation in temperature and salinity in the marine environment. Guillaume Rondelet can be considered the founder of ichthyology, the study of fishes, thanks to his work "*De Piscibus Marinis*" in 1554, whereas the Swedish naturalist Linnaeus (1707–1778) introduced **binomial nomenclature**, the two-word naming system he used to classify thousands of species, which gave rise to the modern scientific classification of living organisms. In 1725, Luigi Ferdinando Marsili studied the relationship between the salinities of the Black Sea and Mediterranean Sea, which he published in the book *Histoire physique de la mer*. James Cook (1728–1779) was the first European to cross the Antarctic Polar Circle. Otto Müller (1730–1784) described the benthic fauna of the coasts of Denmark and Norway in the book *Zoologiae Danicae Prodromus*. Antonio van Leeuwenhoek (1632–1723) provided the first descriptions of plankton, Johan Ernst Gunnerus (1767), Bishop of Nideros (Trondheim), first described marine copepods. Francois Peron and David Porter produced the first highly detailed cartography of the Pacific Ocean. Charles Darwin (1809–1882) organized one of the most fascinating early oceanographic expeditions aboard the ship *Beagle* (1831–1836), visiting many of the Galapagos Islands as well as coastal Australia.

Charles Wyville Thomson (1830–1882) organized the *Challenger* expedition, the first major interdisciplinary oceanographic expedition to sample the deep ocean systematically, including the use of a dredge to sample. The Englishman Edward Forbes (1815–1854), a professor of natural history at the University of Edinburgh noted earlier for his "azoic theory," was the first researcher to study the distribution of organisms, both animal and photosynthetic, in relation to increasing depth. The German zoologist Karl Möbius (1877), studying oyster banks, first defined the term biocenosis. In 1896, the Danish fisheries biologist C.G.J. Petersen developed a seabed sampler – a **grab**, referring to an instrument lowered over the side of a boat or ship to sample surface sediment, and began the first quantitative studies of benthos. Vaughan Thompson (1828) used the first plankton nets, which Charles Darwin also used on the *Beagle*. Victor Hensen, in 1887, defined plankton as free-floating organisms in the sea, and also developed quantitative methods to study plankton and its chemical composition.

The deep ocean has long been a source of fascination and curiosity for explorers. The French used steamships such as the *Travailleur* to study these remote systems in 1880 (Bay of Biscay), 1881 (Mediterranean Sea and Atlantic Ocean), and 1882 (Canary Islands), and more modern propeller ships such as the *Talisman* in 1883 (Canary Islands, the Azores and Cape Verde). These oceanographic cruises inspired the Museum of Natural History in Paris to lead to a major expedition in 1884, creating an exhibit based on the samples collected and tools used for exploration of the seabed.

Several Scandinavian expeditions sampled the North Sea (1876), Barents Sea (1877), and then explored the Arctic from 1893 to 1896. Alexander Agassiz (1835–1910) studied fish migrations and gave his name to a trawl that scientists still use today for sampling nekton. The Prince of Monaco, Albert I (1848–1922), a pioneer in marine environmental studies, established the Oceanographic Museum in Monaco that still graces the shoreline at Monte Carlo, and the Oceanographic Institute in Paris. The early oceanographic cruises marked the onset of modern ocean science and inspired the building of the first research laboratories to support sample collection and maintenance of living organisms for behavioral and

Figure 31.1 A picture of the research vessel Hirondelle. *Source*: National Oceanic and Atmospheric Administration (NOAA).

physiological studies. These first research laboratories, dedicated to visiting scientists undertaking studies of ocean life, included Concarneau (1859, on the French Atlantic coast), Sevastopol (1871, on the Black Sea), Roscoff (1872, off the Brittany coast), the Anton Dohrn Zoological Station in Naples, Italy (1872), and the Marine Biological Laboratory in Woods Hole, Massachusetts (1888).

The Mediterranean Sea has one of the longest histories of marine biodiversity studies, though with less focus on deeper environments than some other regions of the planet. The *Washington* expedition (1881–1883) collected data on the benthos and hyperbenthos in the deep Tyrrhenian Sea (Mediterranean) with trawling nets reaching 3,115 m depth. Prince Albert I of Monaco supported numerous oceanographic expeditions in the deep Mediterranean (and beyond) on board *Hirondelle* (1874) (Figure 31.1), the *Princess Alice*, the *Alice II* (1891–1897), and *Hirondelle II*. The ship *Pola* (1890–1893) undertook the first extensive exploration of deep-sea fauna in the Mediterranean Levant basin. Danish researchers also conducted important oceanographic cruises (*Thor*, 1908 and *Dana*, 1928–29), which greatly expanded knowledge of the fishes living at depths beyond 1,000 m in the Mediterranean. After these oceanographic expeditions, the oceanographic cruise *Polymède* on board the vessel *Charcot Jean* undertook major deep-sea fish collections in the Western basin and the German expedition *Meteor* sampled the Eastern basin. Since the 1980s, numerous dedicated research projects have deepened knowledge of deep-sea life in the Mediterranean Sea.

Increased commercial fishing gave rise to new societal challenges that required deeper knowledge of marine resources and highlighted the need for international scientific collaboration. In 1902, the first international institution was founded for this purpose, the *International Council for the Exploration of the Sea* based in Copenhagen, Denmark. At the same time, in Monaco, the *Commission Internationale pour l' Exploration scientifique de la Méditerranée* was born. One fundamental task of these institutions, which continue to operate today, was to standardize methods and definitions in the various fields of marine sciences. The end of the Second World War saw greater coordination of research among international institutions, together with the introduction of improvements in scientific tools. During the *International Geophysical Year*, IGY (July 1957–December 1958), 40 nations deployed research groups on 60 ships, with the support of hundreds of marine laboratories spread along the coast, and accomplished important scientific research on the ocean. At that time, Jacques Cousteau organized cruises aboard the *Calypso* (1951–1973), which showed the general public amazing imagery of the underwater world, stimulating interest in the ocean and increasing human threats (Figure 31.2). Indeed, Cousteau also developed observation chambers akin to mini submersibles (a diving saucer) with underwater cameras through which he documented the marine environment by producing popular movies such as *Le monde du Silence* (1956), in collaboration with Louis Malle (Bouée laboratoire), and *Le Monde sans soleil* (1964).

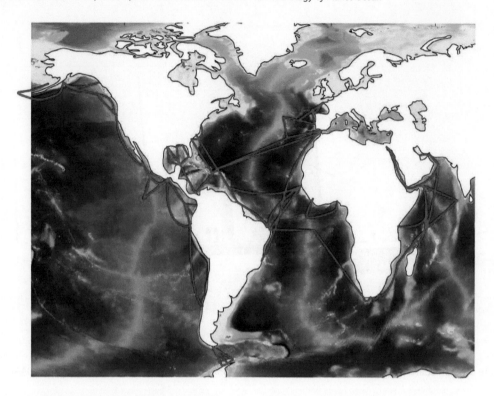

Figure 31.2 The *Calypso* routes between 1951 and 1973.

31.3 The History of Ocean Exploration

Marine biology and biological oceanography, which focus on ocean life and how life interacts with its environment, respectively, are relatively new research fields grounded in a long history of exploration started in 1768 and progressing beyond the end of World War II 1945:

1768–1771: *Endeavour expedition* on board the vessel *Banks* – Capitan James Cook

1831–1836: *Beagle expedition* with C. Darwin (see above)

1865–1868: *Magenta expedition* with Filippo de Filippi and Enrico Hillyer Giglioli

1868: *Lightning* and *Porcupine* expeditions with Wyville Thomson

1872–1876: *Challenger expedition* – Wyville Thomson

1872–1908 Several surveys and nautical charts in the Mediterranean Sea and in coastal waters of Somalia and Eritrea with the Royal Navy.

1874–1876: *Gazzelle*

1875–1876: *Discovery*

1877–1880: *Blake and Albatross* with Alessandro Agassiz expeditions in the tropical Atlantic and Pacific Ocean)

1878–1879: *Vega*

Travailleur with Milne Edwards and Richard

1881–1883 Aboard the R.N. *Washington*. Giglioli demonstrates the existence of a deep fauna and the oxygenation of the deep Mediterranean waters, refuting Carpenter's theory.

1882–85: *Vettor Pisani*, commanded by Palumbo, curated by Chierchia (circumnavigated the globe)

1885: *Princesse Alice, Hirondelle I, Hirondelle II* with Prince Albert I of Monaco (Mediterranean and Atlantic Ocean), investigated whale feeding and surface currents of the Atlantic Ocean.

1890–1898: *Pola* (Mediterranean and Red Seas)

1893–1896: *Fram* led by Fridtjof Nansen (Artic Ocean)

1897–1899: *Belgica* (Antartica)

1898–1899: *Valdivia* con Chun

1899–1900: *Stella Polare* con il Duca degli Abruzzi (polar seas)

1903–1905: the *R.V. Liguria* circumnavigated the globe.

1905–1910: *Thor* with Johannes Schmidt (Atlantic and Mediterranean). Studied eel migration

1910: *Michael Sars*

1910 The mathematician Vito Volterra formulates the famous mathematical model (Lotka-Volterra) regarding the biological equilibrium between various species of marine fish predators and prey.

1920–36: *Dana* with Johannes Schmidt. Eel migration

1922 Activities of the *Hydrographic Institute*, interrupted by World War I, resumed.

1923–1924 The R.N. *Magnaghi* conducts research on currents, weather conditions, and marine fauna in the Aegean Sea and the Red Sea.

1924–1929: *Meteor* expedition

1924–1951: *Discovery* expedition. Investigated the biology of baleen whales;

1928 With the R.N. *City of Milan*, the Navy participated in the Arctic expedition of the Italian airship, undertaking research on magnetic fields, hydrography, gravimetry, and oceanography.

1930 The Hydrographic Institute conducted a survey on Libyan waters.

1931–1935 The Royal Navy undertakes gravimetric research with the submarines *Vettor Pisani* and *De Geneys*, as well as triangulation seafloor soundings (Royal Ships *Magnaghi* and *Ostia*).

1935–1939 New charts (30 in 1938 alone) are published, and hydrographic surveys accelerate, especially in the Gulf of Aden and off the coast of Somalia and Albania (R.N. *Cherso*, *Magnaghi*, *Cariddi*, and *Berta*).

1942–1946 World War II interrupts these activities, which resume in the spring of 1947.

After World War II, the development of new technologies (bathyscaphe, sampling gears, scuba diving, underwater video cameras) enabled new exploration that produced documentaries. This was also the case with the *Galathea* expedition, which explored abyssal fauna, and *Calypso*, led by Jacques Cousteau

31.4 Present and Future of Marine Biology

Understanding the ocean past can help in predicting and preparing for the future ocean. In this regard, analysis of time-series data (repeated measurements over time) at a specific site for extended periods of time (decadal time scales) becomes increasingly important because it helps in developing ideas regarding past changes and predicting future scenarios by applying sophisticated mathematical models. These benefits point to value in developing large databases and analyses that span across multiple data sets (**meta-analysis**) to predict global climate change effects on marine ecosystems. Forecasting models particularly benefit from long-term studies in pilot areas. In addition to the need to assess temporal change, marine ecosystem studies must consider multiple spatial scales from the smallest (mm to cm) to the largest (100s to 1,000s of km) in order to evaluate the spatial extent of changes and of how variables acting at different spatial scales may influence the distribution of biodiversity, abundance, and biomass of organisms and their functions. These response variables all represent fundamental aspects of the goods and services that marine environments can provide for human wellbeing. Marine research will require increasingly technologically advanced equipment that can evaluate aspects of marine ecosystems with increasing precision, spatial and temporal extent, efficiency, and rapidity (Figure 31.3). We also now have many more options beyond ships and basic grabs and nets to achieve these objectives.

31.4.1 Sampling Platforms

Boats and ships continue to play a major role in ocean science research and specifically marine biology, keeping in mind that small boats can access only a tiny fraction of ocean environments, and sampling the deepest environments requires large ships available to only a small subset of scientists. (Remember, a **boat** can fit on a ship, but a **ship** cannot fit on a boat!)

Noting the "snapshot view" provided by research cruises that typically last for just days to weeks at best, marine scientists in the twentieth century began to develop and deploy oceanographic moorings that would provide longer periods of data collection. **Moorings** refer to lines of oceanographic instruments anchored to the seafloor but with flotation to keep the line vertical and collect measurements from instruments such as thermometers at specific locations and heights above bottom. These moorings have played a particularly important role in physical oceanography, however, success depended on recovering the mooring and downloading the data stored on the instrument, as well as working with limited battery life.

Figure 31.3 Marine research requires combined use of different sampling tools as well new technologies to study remote systems, including those characterized by extreme conditions. The use of advanced sampling technologies requires research vessels equipped for the study of marine biology, but capable of biological, physical, chemical, and geological measurements. Researchers can trigger Niskin bottles on a sampling rosette at specific depths to collect water for nutrient analyses. Microbiological analysis requires sterile sample bottles capable of maintaining constant in situ conditions. Typically, a multi-parameter CTD probe (conductivity, temperature, and depth) fitted on a rosette sampler, which records changes in salinity (indirectly, by measuring conductivity), temperature, and depth (indirectly by measuring pressure). For small benthos (micro, meio-, and macrofauna) scientists use a box corer or multicorer, which collects undisturbed sediment samples. For experimental and sustained observations, scientists lower a benthic lander equipped with multi-parameter probes (such as oxygen, nutrients, as well as cameras), onto the seabed for an extended period in order to collect frequent or continuous observations in situ. A wide array of nets still play a vital role in sampling different groups of plankton. The use of submersibles, although quite limited in availability, enables scientists to observe and explore the deep sea first hand.

Scientists have lost many moorings to storms, instrument/battery failure, and even the inability to get back to a remote location. With modern communications, however, a satellite link can receive data transmitted from depth to a surface float and relay it to researchers around the world. Moorings offer excellent temporal resolution but extremely localized spatial coverage and battery power limits deployment duration.

Within the last 20 years, increasing numbers of researchers have developed **cabled observatories**, which run a fiber optic cable from a node on the seafloor onto a receiving station on land, transmitting data and simultaneously providing power to instruments. The **node** provides the interface to connect different instruments to the cable, and thus provide continuous real-time data to researchers on shore through the internet, without the need to replace batteries. By deploying multiple nodes at a location, cabled observatories provide some spatial coverage but these structures tend to be limited in spatial scope and mobility (Box 31.1).

Nowadays, some researchers augment fixed moorings and sensors with moving platforms such as passive, drifting floats and ocean gliders. Researchers internationally have joined together to deploy thousands of **Argo floats**, small drifting platforms that measures temperature, salinity, and depth as they move passively with ocean currents. Researchers program the floats to change their buoyancy and move up and down, utilizing very little power as they profile the water column and then surface to get a location from satellites and transmit their data around the world. **Ocean gliders**, behave like underwater airplanes as they cover pre-programmed survey routes, using a similar buoyancy strategy while they profile the water column, but with the capacity to navigate actively rather than drifting passively. These platforms have characterized water column physical environments but some versions now include biogeochemical and biological sensors such as fluorometers to estimate chlorophyll abundance.

Submersibles, referring to a manned "ship" capable of submerging and operating under water, came into use by scientists a little more than 50 years ago. The initial exploration carried out by Cousteau in 1959, followed in 1960 by the *Trieste* bathyscaphe dive by Jacques Piccard and Don Walsh to 10,987 m in the Mariana Trench (Mindanao), and then *Alvin* missions since 1964 opened new frontiers in research on the deep ocean. In recent years, the use of these underwater operating systems has increased exponentially. However, the enormous purchase and maintenance costs severely limits use of these platforms, despite their particular importance in studying deep and complex environments, such as hydrothermal vents, cold seeps, canyons, seamounts, deep-water corals, and ocean trenches.

Many research submersibles and **Remotely Operated Vehicles** (ROVs) today are equipped with high-definition camera and video systems to observe marine biota and their habitats. Articulated mechanical arms on some well-equipped vehicles enable collections of different types of samples, as well as deployment and recovery of in situ experiments. These tools allow researchers to discover and explore novel, sometimes small and rare marine systems including extreme environments characterized by high temperatures, high concentrations of sulfides, or reduced oxygen. Such ecosystems and their specialized biota offer unique opportunities to develop new technologies and bioproducts with potential commercial applications.

ROVs first, and autonomous underwater vehicles (AUVs) today, increasingly push the limits of innovation, not only collecting high quality images but also using an expanding array of sensors to characterize the environment, and movable arms to collect specimens. These vehicles differ greatly from submersibles in that scientists participate remotely through a fiber optic cable contained within a protective tether (ROV), or alternatively researchers preprogram a survey route (AUVs). Larger AUVs can now accommodate instruments such as sidescan sonar or multibeam echosounder, and can collect water samples for genetic analysis or other applications such as nutrient analysis. Increasing miniaturization of sensors may soon enable swarms of tiny AUVs to characterize ocean environments and their biota at scales never before imagined (Figure 31.4).

Remote sensing via satellites revolutionized ocean sciences when they provided the first synoptic images of ocean surface temperature, ice, and wind beginning in the late 1970s. Other satellites subsequently added sensors to measure chlorophyll as well as ocean features such as altimetry (which measures surface height of the ocean). Because chlorophyll provides a measure of ocean surface productivity, marine biologists continue to use this information widely. But what satellites gain in synoptic perspective, they lose in depth and versatility; satellites "see" only the upper few meters of the ocean and generally do not work well in the presence of clouds and fog, which block their optical sensors. In addition to satellites, researchers may use airplanes equipped with special cameras to image coastal environments and provide spatial information on coral reefs, seagrasses, kelps, and other habitats at the land-sea interface.

The global positioning system (GPS), that uses triangulation from satellites to tell our cell phones or car navigational systems where we are, does not work underwater. This limitation greatly constrains underwater navigation and positioning, so that submersibles, for example, must depend either on transponders deployed in their work region at known locations, complex calculations of their position relative to their support ship (if present), or surfacing in order to obtain a satellite fix. This GPS constraint also has ramifications for a relatively new technology called **satellite tags**, which refers to miniaturized tags that researchers can attach to animals to track their movements. For marine

Figure 31.4 (a) Ocean Networks Canada Ocean Observatory; (b) an instrument platform deployment in Saanich Inlet, British Columbia; (c) hydrothermal vents at one of the nodes; (d) sediment covered seafloor at one of the nodes. *Source:* Flickr/oceannetworks, oceannetworks.

Box 31.1 Ocean Networks Canada: The Canadian Ocean Observatory

Canada supports the most complex cabled observatory ever developed for ocean science (Figure 31.5). Located off Vancouver Island in the Northeast Pacific, "Ocean Networks Canada" (ONC) sends fiber optic cables to nodes at nine locations spanning 50 km of continental shelf and 800 km of deep sea from the upper continental slope to the abyssal plain. Each node supports a range of ocean sensors, including cameras at multiple sites, as well as a wide range of environmental sensors. Researchers from around the world can log onto the ONC web site and observe the seafloor in real time and collect data in real time from thousands of meters below the ocean surface.

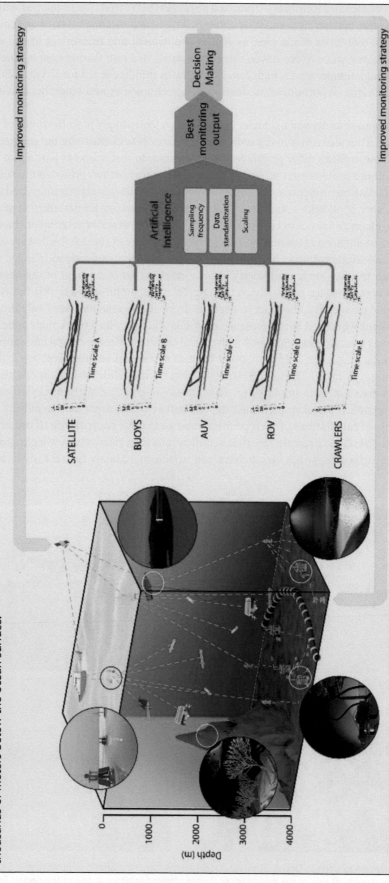

Figure 31.5 Example of networks of mobile, semi-mobile, and fixed monitoring stations (satellite, buoys, AUV, ROV, cabled, and stand-free observatory as well as crawlers) up to cyber processing layers that, in tandem with application of Artificial Intelligence, will allow scientists to acquire data from all different technologies and monitor the oceans.

mammals and turtles, which must surface in order to breathe, that short surface interval provides an opportunity for tags to obtain a position from passing satellites, and thus track movements with relatively high precision. For animals such as fishes that do not surface, scientists had to come up with another solution and therefore developed **popup tags**, which refers to tags that release from the animal after some period of time and pop up to the surface where they download their data to passing satellites. But how do these tags provide tracking information? The clever solution lays in clocks, depth sensors, and light sensors incorporated into these tags. Based on precise time of day when the sun sets at different locations, and knowledge of light attenuation with depth, researchers can back calculate the position the animal occupied at a given time. But this sensor technology represents just the tip of the iceberg in that researchers can now add other sensors to these tags that measure aspects of the animal's biology such as heart rate or body temperature, as well as how its environment changes by measuring variables such as salinity and temperature. In essence, this research uses living animals in place of ships at a tiny fraction of the cost. Some of these animals also pass through remote environments such as the coastline of Antarctica, or beneath ice, where traditional sampling may simply not be possible.

31.4.2 Implementation of Technologies Enabling Biological Observations At Sea

The challenge of monitoring the open ocean, large spatial scales, and the deep sea, particularly in international waters, relates to its vast area and 3-dimensional features and complexity. Capturing ocean heterogeneity on the multiple, relevant scales as required for the efficient and sustainable development of numerous human activities requires well-designed, highly replicated, and long-lasting monitoring approaches, focusing on ecologically relevant locations.

The future of environmental monitoring should primarily focus on organisms and their ecosystems, and then aim to include the physical and chemical environment as key/explanatory variables. Such a strategy must begin with statements of goals and objectives, and defining key variables and indicators. It must also consider appropriate spatiotemporal frequency of biological data sampling to document the ecological heterogeneity of the seabed and water column to enable assessment of biological status at depth.

Scientists can implement a global ocean monitoring strategy by expanding existing efforts or establishing new networks of multi-parameter platforms (i.e., including time-series imagery with associated environmental data). High-definition (HD) camera installation is pivotal in sustaining the scientific, management, and societal value of such a network. Combining video (current HD or new low-light) and acoustic imaging, as well as hydrophones at the seabed and in the water column could efficiently characterize high-complexity ecosystem components such as megafauna and bioturbation processes (i.e., biologically driven mixing of the sediments) of direct relevance to services for human health (e.g., fisheries and carbon burial). High-frequency (minutes) time-lapse imaging over a period of years can quantify biodiversity status and key species presence, if conducted at the appropriate spatial scales envisaged by proposed monitoring networks of stations (several km^2). This image acquisition represents one key component of the proposed biological monitoring revolution in tandem with other biological sensors on mobile robotic platforms moving in between nodes (AUVs, crawlers, etc.), thereby facilitating community monitoring along with concomitant oceanographic and geological data collection.

Pelagic gliders, bottom crawlers, instrumented buoys, and moorings should complement benthic networks of spatially distributed platforms to spearhead a comprehensive bentho-pelagic monitoring program in representative habitats within international seabeds and overlying water columns (Figure 31.6). By coupling imaging with active detection assets (e.g., tracking animals carrying acoustic tags and environmental data loggers through space and time), scientists could achieve a synergy between passive and active sampling methods, introducing "animal-mediated intelligence" (i.e., inferences on animal decisions to enter or leave areas according to a perceived eco-field). Acoustic and satellite technologies already monitor large-scale movements of key pelagic megafauna. Some megafaunal organisms spend significant portions of their time in the deep ocean and could become a priority for tagging studies. In short, deep-sea monitoring technology is advancing, but many "sensory" gaps remain.

In addition, researchers can now apply molecular analyses, including barcoding and metagenomic investigations, based on high throughput sequencing, at large spatial scales, expanding our ability to identify (new) species, (new) genes, and their putative functions. A remote, high frequency, durable, and strategic biological and environmental monitoring strategy using innovative technology could deliver essential data on ecosystem components from microbes to large fauna in order to inform policy decisions. The implementation of Artificial Intelligence in imaging (animal recognition and tracking) and data treatment (data banking and applied multivariate statistics) will facilitate operational autonomy in monitoring and delivering real-time integrated information allowing rapid decision-making, and hence a new technological function by those networks: ecosystem surveillance.

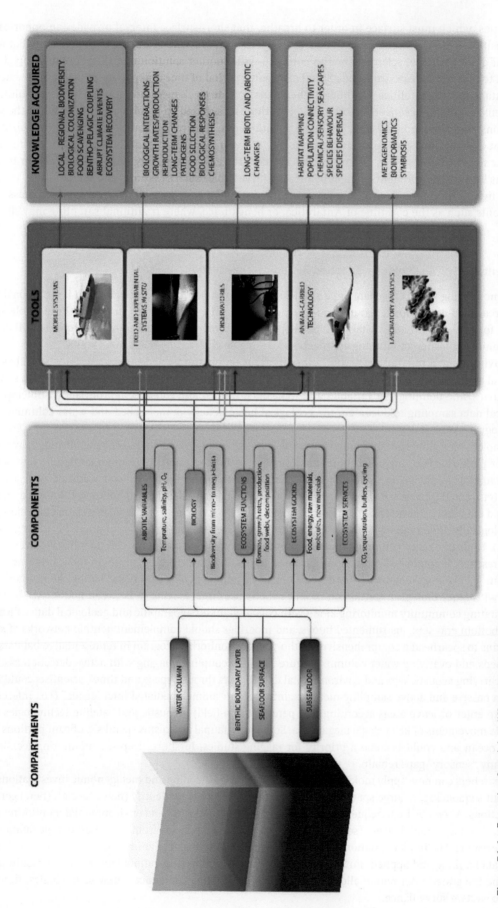

Figure 31.6 Example of applying different technologies for monitoring the 3D aspects of marine systems, from the water column to the sediment water interface and to surface and subsurface sediments.

31.4.3 Sensors

Although some ocean sensors have been around for decades and provide reliable and stable measurements, marine biological sensors have lagged far behind those used by other marine science disciplines. Most sensors use optics (light) or acoustics (sound), though salinity, for example, is inferred indirectly from conductivity measurements. Even those sensors that use wet chemistry generally use some sort of light measurement to document chemical concentration or presence. Although we tend to think of water as transparent, even the clearest water absorbs light, limiting the range of applications for optical sensors.

Optical sensors, which use natural or artificially generated light to document some aspect of the biotic or abiotic environment, have dominated some types of marine sensors for decades. Other than fluorometers, which provide reliable estimates of primary producer abundance and some information on composition, cameras have provided the most useful information on living organisms from microbes to megafauna, other than actual collection of physical specimens. Imaging technologies have advanced tremendously in the last decade or so (think of what your cell phone can do!) through digital imaging, which produces cheap, high-resolution images that support much more effective studies of ocean life than ever before. Researchers now use video (and still) cameras in advanced applications such as in situ flow cytometers, video plankton recorders, benthic landers, and ocean observatories. The development of automated analyses that use tools such as machine learning are now reducing the processing time required for imagery, and the many hours that students would spend looking at video and still images can sometimes now be duplicated in minutes. Although this type of analysis is still in development, it shows great promise, particularly as computer computational capacity quickly accelerates. Researchers are also now developing instruments that can complete in situ molecular analysis. By preprogramming these instruments or connecting them to a fiber optic cable or other communication link, scientists can collect samples over time that detect the presence of specific types of organisms based on their DNA. At present, this approach focuses on individual species, such as toxic dinoflagellates, but may soon be able to collect and analyze environmental DNA (eDNA) samples, or bits of DNA that biota have sloughed off, as described below.

As discussed in an earlier chapter, biodiversity links closely to ecosystem functioning, and studies in deep-sea environments and elsewhere indicate potentially dramatic negative consequences of biodiversity loss for ecosystem functioning (Box 32.2). This emerging concern punctuates the need to apply sophisticated methods to evaluate the biodiversity of even the smallest components, such as viruses and prokaryotes, noting few such studies exist to date, despite their fundamental ecological role in food webs, both benthic and pelagic. The ocean contains a massive reservoir of microbes and potential new discoveries, including some with biotechnology potential. New molecular approaches that include barcoding, fluorescence in situ hybridization (FISH), automated ribosomal inter-genic spacer analysis (ARISA), and massive sequencing of 16S rRNA genes enable the study of even the smallest components of biodiversity. Molecular biology allows microbial specialists to do what fish ecologists could do 100 years ago, namely describe the organisms in a particular environment. But these molecular tools also allow researchers to detect the presence of a given gene and expression of that gene. For example, these tools can now differentiate toxic versus non-toxic strains of the species of dinoflagellates that cause paralytic shellfish poisoning. Molecular techniques also enhance analysis of multicellular organisms by enabling rapid recognition of biodiversity (including cryptic species) that traditional microscopic approaches cannot distinguish. New eDNA approaches indicate the presence of organisms based on bits of DNA that organisms have shed.

Sensor technology is taking off, and scientists are now developing chemical "sniffers" that can detect oil spills or other contaminants in the water, even at low concentrations. Thus, AUVs can undertake searches to map the dispersal of an oil spill or locate chemicals that may indicate the presence of hydrothermal vents. Micro-profilers on benthic landers can measure oxygen concentrations or nutrient efflux from sediments. Our capacity to characterize the ocean environment has improved dramatically, even within the last decade, and holds great promise in understanding how the ocean works and how it changes.

Acoustics, or sound waves, also play a major role in ocean sensing. Unlike light, sound waves can penetrate through water and even solid objects such as the seabed. Broadly speaking, these sensors may use passive or active acoustics. **Passive acoustics** refers to "listening" to sounds in nature such as whale songs, grunts made by fishes, or clicks made by some crustaceans, which researchers use to infer behavior and even abundance of different animals. Researchers often deploy **hydrophones**, effectively underwater microphones, on moorings or on observatory cables to collect sound data. **Active acoustics** refers to applications that generate sound to obtain information on the natural environment. Geologists have long used active acoustics to study seabed composition, and physical oceanographers have used sound to infer water temperature. In recent decades, biologists have used active acoustics to estimate abundances of live animals. In essence, the research vessel sends out sound waves that bounce off objects in the water such as fishes, and based on the intensity and number of returns they receive they can sometimes infer numbers of fishes, for example. Although this approach cannot differentiate well among species, it can cover large areas of ocean in a short time period and involves non-destructive sampling.

> **Box 32.2 The Census of Marine Life**
>
> Contemporary marine biology began when quantitative and experimental approaches expanded on qualitative natural history approaches, with the explicit formulation and testing of hypotheses. From 2001 to 2010, the *Census of Marine Life* (CoML) project undertook the largest census of ocean life ever undertaken. Aiming to investigate the past, present, and future of marine life, CoML led a 10-year international effort to assess diversity, distribution, and abundance of marine life. To illustrate the magnitude of this project, it included 2,700 scientists from more than 80 countries (including the authors of this book), 540 research cruises, US$650 million in research funding, more than 2,600 scientific publications, over 6,000 potential new species, and more than 30 million data records related to marine organisms. This first census of marine life produced the most complete inventory of life ever compiled, providing a fundamental basis for future ocean research. Predicting, measuring, and understanding global changes in the marine environment requires such knowledge, as well as increased availability and awareness of such data, as does development of effective conservation strategies for marine resources. The Census considered all global oceanic life forms, from microbes to whales, from the surface to the bottom of the seas, from pole to pole, and involved the most influential marine biologists in the world.
>
> During these 10 y of research, scientists discovered and surveyed new species, new habitats, and previously unknown interactions/connections in the global ocean. The most surprising findings included the documentation of extensive mats of microbes that comprise the greatest living biomass on Earth, the discovery of the "Jurassic shrimp" *Neoglyphea neocaledonica*, thought to be extinct for 50 million y, and the presence of multicellular animals (three species of the animal phylum Loricifera) capable of living in the complete absence of oxygen on the seafloor, where scientists previously thought only microbes could survive. This project applied increasingly sophisticated technological systems to align the pieces of the puzzle represented by the global ocean. The Census laid a solid basis for guiding future collaborations between institutions and countries, with the shared purpose of increased understanding of ocean life for the sustainable use of goods and services that the ocean provides. New technologies represent the future for ocean exploration, both in terms of tools that enhance field observation scope, frequency and utility for laboratory and computer analysis of what those samples can tell us.

31.5 Application of Marine Technologies

In recent years, thanks to sophisticated tools such as ROVs and biogeochemical sensors, scientists have documented the growing accumulation of waste on the seabed and in the water column and its diverse and changing composition (plastic, glass, metals, bottom trawls, fishing gear, numerous chemicals). Some contaminants arrive from land runoff or from the atmosphere, and some come directly from maritime traffic. We know little about the effects of many of these contaminants, particularly for vulnerable ecosystems such as polar and deep-sea fauna, with growing concern about the high levels of contamination caused by inorganic and persistent organic contaminants on marine biota. Some contaminants directly impact major biological functions (such as enzymatic activity), whereas others stimulate hormone production, alter biological cycles, and limit the reproductive potential of species. Scientists have developed toxicology analyses to assess threats and impacts of current anthropogenic activities (including contaminant accumulation), covering different trophic levels from scavengers to top predators such as deep-sea sharks, *Centroscymnus coelolepis*. Greater awareness of impacts of contaminants in even the deepest marine environments, including bioaccumulation and biomagnification of contaminants such as heavy metals and persistent organic compounds, point to a need for more research, particularly their effects on ecosystem goods and services.

These many examples raise significant concerns regarding the many processes that occur in marine ecosystems that contribute fundamentally to global biogeochemical cycles supporting life on Earth. In addition, marine ecosystems from the intertidal to the deep sea provide direct goods and services of growing economic importance that governments and businesses refer to as the "blue economy." The need to integrate knowledge of marine ecosystems derived from natural sciences with those from economics and social science research will help to identify the social and economic implications of human activities. This integrated effort will catalyze fundamentally important socio-economic analysis of impacts of human activity, mapping of the main anthropogenic activities in marine systems, and an evaluation of ecosystem goods and services offered by the ocean (including deep environments).

The future of marine research includes expansion of **bioprospecting**, the collection, and screening of biological material for commercial purposes. Though terrestrial bioprospecting has become widespread the search for marine bioproducts has just begun and thus offers a new frontier. Scientists have used marine organisms for compounds such as pharmaceuticals, molecular probes, enzymes, cosmetics, food supplements, and agrochemicals. Indeed, microbes, algae, and marine invertebrates have yielded ~15,000 natural products, demonstrating the rich source of new metabolites that marine organisms offer, whether structurally or biologically active. Drugs derived from marine organisms could potentially encompass a variety of applications such as antibiotics, anti-cancer drugs, antioxidants, antifungals, anti-HIV, anti-tuberculosis, and anti-malarial compounds. Microorganisms, corals, sponges, and tunicates represent the primary target taxonomic groups for research on new drugs from the ocean because they offer the greatest potential in terms of bioactive compounds.

31.6 Marine Biology Research in the Next Decade

We conclude with some thoughts on what the next generation of researchers, including some of you who are using this book, can do in order to advance understanding of the global ocean and ensure its long-term sustainability. We believe research priorities for the next decade will include:

- Understanding interactions among diversity and ecosystem processes, structure, and function for increasing predictive capacity in understanding the consequences of shifts in biodiversity on ecosystem services on which humans depend;
- Providing assessment of marine ecosystem functions at different spatial scales for improving understanding of marine ecosystems;
- Understanding ecosystem response to multiple impacts and climate change;
- Assessing habitat and biodiversity loss, along with ecosystem shifts resulting from habitat loss and species invasions;
- Restoring degraded marine habitats for recovering their biodiversity loss and their capacity to deliver ecosystem services.
- Enhancing marine conservation and assessing its potential for improving ecosystem services and offsetting the effects of global change;
- Understanding links between ocean health and human health (*OneHealth* approach);
- Understanding impacts of alien (non-indigenous) species on ecosystems, their processes, and biodiversity;
- Understanding impacts of human activities as well as climate change on the deep ocean;
- Evaluating contributions of microbial-macrobial interactions and the "holobiont" on marine organism and ecosystem functions;
- Assessing impacts of emerging pollutants (e.g., personal care products, plastics and additives, pharmaceuticals), artificial light at night, noise, and toxin effects on marine species and habitats;
- Censusing global marine biodiversity by integrating traditional taxonomy with molecular, phylogenetic, and massive sequencing techniques to bridge the main gaps in basic knowledge of genomes, functions, ecology, and evolution of marine species;
- Developing new tools to identify species by integrating new technologies, such as DNA barcoding in order to assess biodiversity rapidly and unambiguously;
- Identifying and monitoring Essential Ocean Variables (EOVs), the most critical marine biology variables that researchers internationally agree should be standardized and collected routinely in order to create intercomparable data sets that enable broad-scale analyses and comparisons;
- Improving knowledge of life cycles, life histories, movements, and connections (connectivity) of different populations of marine species (e.g., habitat use, aggregation areas), and environmental influences on animal behavior and their dispersion;
- Identifying new marine bio-resources, new biomaterials, and developing research that uses marine organisms as models for developing new materials, processes, and solutions;
- Developing extensive and rapid exploration technologies to support better documentation of large portions of poorly known marine territory, as well as ocean observatories for continuous observation on ocean change, including the deep sea;
- Quantifying the economic and non-monetary value of all marine habitats, from the intertidal to the deep sea;
- Involving citizens and Indigenous groups in data collection and promoting scientific utilization of their information on marine ecosystems and biodiversity (citizen science);
- Engaging scientists from developing nations and improved knowledge of poorly-known ocean regions where science capacity currently limits exploration.

Ocean science, including the study of marine biology, has entered a major phase of innovation and discovery. Opportunities for young scientists to advance understanding and sustainable use of ocean ecosystems have never been so great. The ocean will continue to offer a source of wonder, discovery, and opportunity. We close this book by encouraging those with an interest in ocean life to become involved and make a difference. Life on Planet Ocean depends on ocean life, and we humans can do far better as a species at living in harmony with the global ocean. Our future as a species may well depend on it.

Questions

1) List the main stages in the development of marine biology research.
2) What were the major technological innovations in the history of marine biology?
3) What are the main topics / areas of research for future marine biology?
4) What are Essential Ocean Variables?
5) How can new technologies support marine monitoring?
6) What are the main priorities to consider for the future of marine biology?

Suggested Reading

Boxall, S. (2007) Oceanographers are talented—Eventually. *Oceanography*, 20, pp. 168–169.

Campbell, L., Gaonkar, C.C., and Henrichs, D.W. (2022) Integrating imaging and molecular approaches to assess phytoplankton diversity, in Clementson, L., Eriksen, R., and Willis, A. (eds.), *Advances in Phytoplankton Ecology*. Elsevier, pp. 159–190.

Changeux, T., Blazy, C., and Ruitton, S. (2020) The use of citizen science for marine biodiversity surveys: From species identification to ecologically relevant observations. *Hydrobiologia*, 847, pp. 27–43.

Costello, M.J., Bouchet, P., Boxshall, G. et al. (2013) Global coordination and standardisation in marine biodiversity through the World Register of Marine Species (WoRMS) and related databases. *PLOS One*, 8(1), p. e51629.

Costello, M.J., Coll, M., Danovaro, R. et al. (2010) A census of marine biodiversity knowledge, resources, and future challenges. *PLOS One*, 5(8), p. e12110.

Danovaro, R. (2010) *Methods for the Study of Deep-Sea Sediments, Their Functioning and Biodiversity*. CRC press.

Danovaro, R., Aguzzi, J., Fanelli, E. et al. (2017) An ecosystem-based deep-ocean strategy. *Science*, 355, pp. 452–454.

Danovaro, R., Aguzzi, J., Fanelli, E. et al. (2020) Ecological indicators for an integrated global deep-ocean strategy. *Nature Ecology and Evolution*, 4, pp. 81–192.

Davis, C.S., Gallager, S.M., Berman, M.S. et al. (1992) The video plankton recorder (VPR): Design and initial results. *Archiv für Hydrobiologie*, 36, pp. 67–81.

Duffy, J.E., Amaral-Zettler, L.A., Fautin, D.G. et al. (2013) Envisioning a marine biodiversity observation network. *Bioscience*, 63, pp. 350–361.

ESF (2001) Marine biotechnology. A European strategy for marine biotechnology. *ESF Marine Board Position Paper*, 4, pp. 1–29.

Harden-Davies, H., Amon, D.J., Chung, T.R. et al. (2022) How can a new UN ocean treaty change the course of capacity building? *Aquatic Conservation: Marine and Freshwater Ecosystems*, 32, pp. 907–912.

Harden-Davies, H. and Snelgrove, P.V.R. (2020) Science collaboration for capacity building: Advancing technology transfer through an agreement for biodiversity beyond national jurisdiction. *Frontiers in Marine Science*, 7, p. 40.

Howell, K.L., Hilário, A., and Allcock, L.A. (2020) Blueprint for a decade to study deep-sea life. *Nature Ecology and Evolution*, 5, pp. 265–267.

Kendall, Jr., J.J., Ahlfeld, T.E., Boland, G.E. et al. (2007) Ocean exploration discovery and offshore stewardship. *Oceanography*, 20, pp. 20–29.

Liang, J., Feng, J.C., Zhang, S. et al. (2021) Role of deep-sea equipment in promoting the forefront of studies on life in extreme environments. *iScience*, 24(11), p. 103299.

Miloslavich, P., Bax, N.J., Simmons, S.E. et al. (2018) Essential ocean variables for global sustained observations of biodiversity and ecosystem changes. *Global Change Biology*, 24, pp. 2416–2433.

Olson, R.J. and Sosik, H.M. (2007) A submersible imaging-in-flow instrument to analyze nano-and microplankton: Imaging FlowCytobot. *Limnology and Oceanography: Methods*, 5, pp. 195–203.

Pawlowski, J., Bruce, K., Panksep, K. et al. (2022) Environmental DNA metabarcoding for benthic monitoring: A review of sediment sampling and DNA extraction methods. *Science of the Total Environment*, 818, 151783.

Rogers, A.D., Appeltans, W., Assis, J.A. et al. (2022) Discovering marine biodiversity in the 21st century. *Advances in Marine Biology*, 93, pp. 23–115.

Scholin, C.A., Birch, J., Jensen, S. et al. (2017) The quest to develop ecogenomic sensors: A 25-year history of the Environmental Sample Processor (ESP) as a case study. *Oceanography*, 30, pp. 100–113.

Snelgrove, P.V.R. (2016) An ocean of discovery: Biodiversity beyond the Census of Marine Life. *Planta Medica*, 82, pp. 790–799.

Watson-Wright, W. and Snelgrove, P. (2021) Technological advancements to improve ocean understanding, in Hotaling, L. and Spinrad, R. (eds.), *Preparing a Workforce for the New Blue Economy*. Elsevier, pp. 17–32.

*Glossary

A

ABC method Abundance Biomass Comparison, an application of the K-dominance curves that compares abundance dominance curves and biomass dominance curves.
abiotic non-biological. Refers to the physical, chemical and geological characteristics of the environment.
abyssal zone the deep bottom of the ocean basins that extends from 4,000–6,000 m.
abyssopelagic zone zone extending from 3,000–4,000 m to 6,000 m depth at the abyssal plain seafloor.
abyssoplankton plankton living at depths between 4,000 and 6,000 m.
accidental species species that only occur in a biocoenosis by accident.
acclimatization the process by which an organism adjusts to changes in its environment to survive.
accompanying species species that typically dominate a biocoenosis numerically.
acoustic release a device that disconnects a mooring wire from the anchor weight when triggered remotely by an acoustic signal through an electronic mechanism.
acoustic sensors sensors that emit sound waves to measure ocean variables.
actinotrophs larval brittle stars.
active canyon canyons subject to strong turbidity currents and mass transport.
active margins continental margins on which plate boundaries occur.
active restoration restoration that relies on active interventions to the ecosystem, whether by adding or removing structures, species, or stressors.
active suspension feeders suspension feeders that actively create feeding currents that bring particles to them.
additive effects where multiple effects simply accumulate in a linear additive fashion so that the cumulative result equals the sum of the individual effects.
adelphophagy where nourishment comes from other embryos present in the oviduct.
adenosine triphosphate an organic compound that provides energy to drive and support many processes in living cells.
adiabatic temperature increase increases in temperature with increasing depth as a direct result of increased pressure on the water.
aerobic respiration a form of respiration in organisms that require oxygen to break down or oxidize organic substances.
Allee effect (or **density-dependent mortality**) where a disproportionately low recruitment rates occur at low population levels.
alleles the different forms of a given gene.
allochthonous or non-local production.
allopatric (or **geographic)** speciation that occurs when geographic isolation caused by physical barriers prevent contact between two populations initially belonging to the same species.
allowable bycatch the amount of bycatch managers will allow before closing a fishery.
alpha diversity the diversity of a community or a sample.
ambit the portion of the geographic region in which an organism occurs and directly interacts with the ecosystem.
ambush predators less-mobile predators that wait for prey to come close to them to attack.

Marine Biology: Comparative Ecology of Planet Ocean, First Edition. Roberto Danovaro and Paul Snelgrove.
© 2024 John Wiley & Sons Ltd. Published 2024 by John Wiley & Sons Ltd.
Companion Website: www.wiley.com/go/danovaro/marinebiology

amensalism an association between two different species in which one species inhibits or excludes the other from access to available resources, while the other remains unaffected.

ammocoetes larval stage of lampreys.

ammoniotelic excretion the excretion of nitrogen as ammonium ions, as in copepods.

amphidromous species species that move between fresh water and seawater but not for reproduction, such as mullet.

ampoules of Lorenzini (Fig. 4.29), specialized electro-reception organs in the heads of sharks.

anabolic metabolism the synthesis of proteins, nucleic acids, and many other organic molecules from the environment.

anadromous species species that live in marine waters and return to rivers to spawn, such as salmon.

anaerobic respiration a form of respiration in some microbes that releases energy by splitting organic molecules into simpler forms, thus requiring no free oxygen.

anagenesis (or **phyletic evolution**) occurs when an ancestral population undergoes a gradual change over time with splitting.

anammox a process in which specialized bacteria convert nitrite and ammonium into N_2.

anammox prokaryotes microorganisms capable of oxidizing ammonium under anoxic conditions and liberating gaseous N_2 into the atmosphere.

anemophilous "wind loving" species.

angiosperm flowering plants that protect their seeds in fruits.

anion a negatively charged ion.

anoxia absence of oxygen.

anoxic waters with dissolved oxygen concentrations lower than 0.2 ml L^{-1}, which cannot support most life forms.

anoxic hypersaline basins the most extreme ocean environments. Oxygen-free brines with varying high concentrations of methane and hydrogen sulfide.

Antarctic Bottom Water (ABW) a large, dense water mass formed from the cooling and sinking of waters around Antarctica.

anti-cyclonic clockwise movement.

anti-freeze glycoproteins specialized proteins in the blood that reduce the freezing temperature.

antibiosis interaction in which one organism damages or kills another using a chemical secretion, whereas the other remains unaffected.

aphotic zone the deep, perpetually dark portion of the water column where no light penetrates and no photosynthetic activity occurs, beginning at 500/600–1000 m depth and extending to the bottom of the trenches.

aphytal zone the deeper 92–94% of the seabed that extends roughly from the edge of the continental margin to the major ocean trenches, and where insufficient light penetrates to support bottom living photosynthetic organisms.

apoptosis programmed cell death.

aposematic warning coloration that (typically falsely) warns predators that already experienced negative encounters with organisms similar in appearance.

arborescent tree-like, as in some corals.

Archaea unicellular microorganisms without a nucleus. Archaea make up a significant fraction of prokaryotes in most marine environments. Many of these are chemosynthetic and adapted to living in extreme environmental conditions.

Archimedes principle that an object displaces its own volume in water receives an uplift equivalent to the mass displaced.

Argo floats small drifting platforms that measure ocean variables such as temperature, salinity, and depth as they move passively with ocean currents.

asexual reproduction reproduction achieved through mechanisms such as budding and fragmentation that give rise to new individuals without genetic recombination.

aspect ratio height to diameter ratio.

assimilation the biological uptake of a substance.

atolls isolated circular structures surrounded by deep water.

auricolaria larval holothuroids.

autoecology the individual ecologies and life cycle of two taxa.

autotroph an organism that obtains energy not from organic matter, but through photosynthesis or by oxidation of inorganic materials.

azoic theory Edward Forbes, theory that deep-sea environments cannot support life because of the lack of food, high pressure, and (assumed) lack of oxygen.

B

bacterial gardening entrapment of bacteria within the gastrovascular cavity of corals, where they grow and provide a potential food source.

baffling effect where a structure slows down water flow.

baleen the series of plates that form a sort of sieve system by which baleen whales filter massive volumes of water.

banquette a benchlike shoreline feature consisting of beached and dried *Posidonia* spp. leaves.

barotolerant (or piezotolerant) organisms that tolerate a wide range of pressures.

barrier reefs compact structures parallel to the coastline, located at a specific distance from the mainland that result in the formation of a large lagoon between the reef and the mainland.

Batesian mimicry instances where an organism imitates a defense mechanism present in another species but lacks the specific mechanism.

bathomes biomes delineated by water depth.

bathyal zone the upper portion of the continental slope that extends at an ~4° grade to ~3,000 m depth.

bathymetric zone the depth zone that divide the seabed and overlying water column into different layers.

bathypelagic zone (or lower open ocean) zone beginning at the bottom of the mesopelagic zone (ca 1000 m) and extending to 3,000–4,000 m.

bathyplankton plankton that live in the truly deep bathypelagic layer, totally devoid of light (1,000–3,000/4,000 m).

bed armoring where animals deposit coarser sediments on top of finer sediments, thereby reducing the likelihood of resuspension of the finer sediment fractions.

bedload transport movement of sediment particles by rolling along bottom.

benthic pertaining to association with the seafloor.

benthic domain the totality of the seabed.

benthic lander a platform equipped with multi-parameter probes that, when lowered onto the seabed for an extended period, collects frequent or continuous in situ measurements.

benthic nepheloid layer seafloor region characterized by persistent hydrodynamic activity, resuspended sediments, and intense bottom currents.

bentho-ichthyophagous taxa that feed on demersal fish.

bentho-pelagic coupling where material released from the sediments (nutrients, sediments, detritus, larvae) flows from the benthic to the pelagic compartment.

benthonekton (or nektobenthos) species such as flatfish that actively swim close to the bottom, without settling onto the seabed for extensive periods.

benthopelagic organisms that move between the water column and sea seafloor.

benthos organisms that inhabit the seabed or live in direct relationship with it.

beta diversity the degree of change (or turnover) in species composition between communities within a habitat.

binomial nomenclature the two-word naming system Linneaus used to classify thousands of species, which gave rise to the modern scientific classification of living organisms.

bio-constructions macroscopic structures produced by many plant and animal species as they build, maintain, and expand carbonate skeletons over time.

bioconstructors species of plans and animals that produce, maintain and sometimes expand macroscopic carbonate bio-constructions.

bio-prospecting the exploration of biodiversity for new resources of social and commercial value.

biobuilders organisms capable of constructing substrates or biogenic constructions mainly through the accumulation of calcium carbonate.

biocoenosis a group of interdependent organisms of relatively consistent composition, number of species, and abundance that occur in specific environmental conditions, and persists in a geographic location.

bioconstructions macroscopic structures originating from plant and animal species that produce carbonate skeletons.

biodiversity every living system at levels of organization spanning from molecules to ecosystems.

biodiversity hotspots locations of exceptional biotic richness.

bioeroders organisms that erode biological structures such as boring polychaetes in corals.

biogenic resulting from biological activity.

biogenous sediments sediments formed from the hard parts of living organisms, such as silica diatom frustules, sands derived from coral calcium carbonate skeletons, and calcium carbonate tests from foraminiferans.

biogeography the study of organisms in relation to geographic space.
biological disorder (or predator cropping) hypothesis the hypothesis that predators control or crop back the number of organisms present in a location and thus promote diversification.
biological habitat provisioning the process by which living organisms that create new habitat increase habitat spatial heterogeneity, thereby increasing biodiversity.
biological reference point a metric of population or stock status from a biological perspective.
biological traits approach a strategy in protected area planning that considers, morphological, biochemical, physiological, structural, phenological (timing of key developmental events), behavioral, and ecological traits of an organism.
bioluminescence defined as biologically produced light, to communicate with conspecifics lure prey, or deter predators.
bio-mineralogy the science that investigates the interactions between organisms (typically sessile organisms) and the mineralogy of the substrate.
biopolymeric fraction the bioavailable fraction formed by organic matter produced both by autotrophic and heterotrophic processes from biopolymers (proteins or nucleic acids) and other compounds (such as non-structural and relatively simple carbohydrates or lipids).
biosonar (or echolocation) an auditory imaging system used by various species to navigate and to estimate the location of prey in environments where visual cues are ineffective.
biosphere the total habitable portion of Earth.
biotope the geographic area occupied by a biocenosis.
bioturbation process by which living biota, and larger sedimentary organisms, move sediment grains.
bipinnaria and **brachiolaria** larval asteroids (sea stars).
bipolar species (or genera) those species (or genera) present at both poles but absent in between.
bleaching the loss of color in symbiont dinoflagellates and the host corals.
bloom rapid and short-term explosion in the number of marine organisms due to increased reproduction in one or more species in response to favorable environmental conditions.
blue economy the sustainable use of ocean resources for economic growth, improved livelihoods, and jobs while preserving the health of ocean ecosystem.
boreal northern.
bottleneck hypothesis the hypothesis that meiobenthic predators can control the composition and structure of the adult macrobenthic community feeding by on macrobenthic larvae and juveniles (i.e., temporary meiofauna).
bottom currents currents that move the portion of the water column immediately above the seabed.
bottom ice the lower 10 cm of the sea ice column.
bottom-up control processes determined by resources via lower trophic levels.
boundary layer a region where velocity increases with increasing distance from a surface.
brine seawater with particularly high salt content and high concentrations of inorganic nutrients.
brown ice ice that forms near the water-ice interface during the Antarctic spring, supporting abundant life, and particularly primary producers.
bubble-netting a gulping system in whales that produces a circle of ascending air bubbles that create a sort of barrier to prey.
burst swimming speeds the maximum swimming speeds that organisms can maintain for short periods.
bycatch the incidental removal of non-target species by a fishery.
byssus threads a series of "tethers" secreted by bivalves such as blue mussels to affix them to hard substrates (rocks, conspecific shells).

C
cabled observatories an observation system that uses a fiber optic cable to transmit data and simultaneously power instruments on the seafloor from a receiving station on land.
calcification the mineralization of calcium carbonate crystals on cell walls or within cells.
caldera depression in a collapsed volcano.
cannibalism feeding on individuals of the same species.
capsid the protein coat of viruses.
capsomeres identical repeating units that comprise the capsid and surround the viral genome.

carbonate compensation depth the depth below which the dissolution rate of calcite and aragonite exceeds the deposition rate, which typically occurs between 4,000 and 5,000 m.
carrying capacity the maximum value of population growth in a location, often described by a sigmoidal curve.
catadromous species species that spawn in the ocean but develop and typically mature in freshwater, such as eels.
cation a positively charged ion.
cavitation the formation of gas bubbles in liquids that move when pressure drops below that of water vapor while the surrounding temperature remains constant.
central place foragers animals that travel from a home base to distant foraging locations.
character displacement changes in morphological characteristics between competing species that allow them to coexist.
characteristic species species that live only in a specific kind of biocoenosis.
chemoautotrophic microbes that utilize chemicals such as hydrogen sulfide or methane as an energy source to create organic carbon.
chemoautotrophs organisms that obtain energy from chemical reactions, particularly by oxidizing reduced and/or inorganic compounds.
chemocline a chemical gradient.
chemolithoautotrophy where organisms derive energy from reduced inorganic compounds.
chemolithotrophy the process by which organisms obtain energy from inorganic compounds.
chemorganotrophy the process by which organisms obtain energy from organic compounds.
chemosynthesis process of producing organic material by microbes from inorganic compounds such as hydrogen sulfide without requiring the presence of light as an energy source.
chemotaxis the directed movement of organisms in response to concentration gradients of extracellular signals.
choanocytes specialized flagellated cells in sponges.
ciliate feeders organisms that feed primarily on ciliates.
ciphonauta larval bryozoans.
circadian cycles rhythmic behaviors within a period of 24 hours.
circalittoral zone zone encompassing depths below the low tide mark with sufficient light to support photosynthesis.
circumboreal species species that span the cold-temperate oceans and seas of the northern hemisphere.
cladogenesis when a population splits into two or more sub-populations that represent new species.
climate change a change in climate that is directly or indirectly attributable to human activities, which alters the composition of the planetary atmosphere and which adds to the natural climate variability observed over similar time intervals.
coastal subtidal the region extending from the intertidal zone to the edge of the continental shelf at ~200 m.
cohort a group of individuals belonging to the same generation.
cold seeps ecosystems where hydrocarbons such as methane, sometimes associated with sulfides and brines, leak from the ocean floor and support chemosynthetic communities.
colloblasts specialized sticky cells on the tentacles of ctenophores that they use to capture prey.
colonial species species that reproduce clonally and form expanding "colonies".
commensal benefitting from but not harming a host.
commensalism a close and sustained symbiotic relationship between two organisms or populations or species living together where one benefits but with no negative effect on the other, or one benefits and the other suffers.
community ecological unit composed of all interacting species/populations within an area.
commuting repeated round-trip movements by an organism, often daily.
compensation depth the depth at which where productivity and respiration rates for a given phytoplankton cell balance when averaged over 24 hours.
compensatory effects where impacts of different activities "cancel each other out" to produce a less severe effect than the individual activities.
competition the simultaneous need of two or more organisms (or species) to obtain an essential resource in limited supply.
conductivity the degree to which a medium such as seawater conducts electricity.
confinement model a model of coastal lagoon zonation that considers the lagoon community as a "dilute" version of the marine community and identifies bionomic sub-domains, each characterized by different guide species and organized according to the degree of confinement.

contaminants every element/molecule/compound or product that, if released into the environment, can cause harmful effects on organisms or habitats.
continental rise the region at the base of the continental slope between 3,000 and 4,000 m and then slowly grades into the abyssal zone.
continental shelf a region of the seafloor that gently slopes (1°) from the subtidal (or sublittoral) zone to the edge of the shelf at 150–200 m depth.
continental slope the zone extending from the continental shelf break and terminating at the continental rise, covering ~6% of the global seafloor.
convergent evolution where phylogenetically distinct organisms evolved similar adaptations.
coral rubble fragments of corals, often including bivalves.
coralligenous assemblages calcareous formations produced by the accumulation of encrusting algae growing in dim light.
Coriolis effect the deflection of moving objects not firmly connected to the Earth's surface as they travel long distances around Earth.
cosmogenous sediments sediments that form from tiny sediment grains from outer space delivered as space debris breaks up entering the atmosphere.
counter-shading body coloring that reduces visibility of prey to their predators against surface lighting.
coupling biological linkages between the two realms such as larvae, food, and nutrients.
covalent bond the interatomic linkage that results from the sharing of an electron pair between two atoms.
craniate animals possessing a skull.
critical depth depth at which total amount of productivity (or oxygen produced) integrated for phytoplankton over that depth range balances total respiration (or oxygen consumed).
critical erosion velocity the lowest velocity at which sediment grains of a given size (diameter) will move.
critical habitat geographical areas that contain essential characteristics for one or more endangered species.
cruising speed (or **sustained swimming speed**) speeds that organisms can sustain for extended periods.
cryptic (or **sibling**) **species** two or more reproductively incompatible species that are morphologically indistinguishable from one another.
cryptic mimicry where animals attempt to avoid being seen, by adapting coloration and/or forms that resemble the substrate on which they live.
cryptobioturbation particularly small and subtle bioturbation.
CTD electronic probe used for oceanographic measurements for the *in situ* measurement of temperature, pressure and conductivity. It can also be equipped with other sensors.
cultural services the non-material benefits people obtain from nature, such as artistic inspiration.
current any persistent movement of water masses, induced by various causes, capable of mobilizing or transporting large volumes of water and sometimes sediments.
cutaneous vasculature the blood vessels that supply oxygen to the skin.
cyphonautes larvae produced by bryozoans.
cyprid larval stage of barnacles that develops from naupliar stages.
cysts the dormant stage of a microorganism or invertebrate that helps the organism survive in unfavorable environmental conditions.
cytoplasm the gelatinous liquid inside cells.

D

dark energy methane, hydrogen sulfide, and other inorganic elements that provide a source of energy for deep-sea ecosystems.
dead zones anoxic bottom waters that lack the minimum oxygen conditions necessary to support higher organisms.
decomposer organism that transforms dead organic matter into energy forms.
deep ocean conveyor belt (or **great ocean conveyor belt**) the large system of global circulation driven by thermohaline circulation that connects the major ocean basins.
deep scattering layer (DSL) (or **reflective deep layer**) depth layer of the ocean with high zooplankton concentrations sufficient to scatter sound waves.
definitive (or final) host a host for the sexually mature stages of a parasite.

delta a particular mouth shape of a river near which the river deposits sedimentary material originating from continents, resulting in multiple branches of flow.
delta diversity the degree of dissimilarity (turnover) of species diversity between two habitats.
demersal near-bottom living, swimming across and sometimes resting on the seafloor, but dependent on the seafloor habitat for feeding, shelter, or reproduction.
denaturing alteration or breakdown, as in protein denaturing, which alters metabolic pathways.
denitrification the reduction of nitrate by microbes through a series of intermediate gaseous nitrogen oxide products to ultimately produce molecular nitrogen.
density the mass per unit volume of a substance.
density-dependent processes that regulate population growth rates dependent on the density of the population.
density-dependent mortality (or **allee effect**) where a disproportionately low recruitment rates occur at low population levels.
deoxyribonucleic acid the double stranded, helical molecule that carries genetic information for the development and functioning of most organisms.
deposit feeding obtaining nutrition by ingesting sediment particles and stripping off associated food.
dermis the inner "skin" layer of metazoans.
desiccation drying out.
determinate growth growth that stops once a genetically pre-determined structure has completely formed or growth stage has been reached.
detritus particles of debris comprising nonliving organic matter, in the form of feces or cells or tissues from bodies of marine organisms.
diadromous species of fish that migrate from freshwater to the sea and vice-versa.
diagenesis the physical and chemical processes that affect sedimentary materials after deposition and before metamorphism, and collective organic matter transformations.
diapause a biological development disruption "programmed" by the genome that suspends the life cycle of an organism, typically in advance of unfavorable environmental conditions.
dicopatric model speciation associated when a geographic barrier that individuals cannot overcome initially divides a large ancestral population with a wide distribution range into two similar sized subpopulations.
diel cycles cycles that occur over a 24-h period.
diel migration the synchronized movement of organisms (zooplankton and fish) up and down in the water column over a daily (24-h) cycle but irrespective or day or night.
dimensionless ratio a ratio that lacks units because all unit cancel out.
diplanula a type of asteroid larvae.
direct development where adults release juveniles rather than eggs or larvae.
disaerobic (or **disoxic**) waters with oxygen concentrations between 0.1 ml l^{-1} and 1.0 ml · l^{-1}.
disoxic (or **disaerobic**) waters with oxygen concentrations between 0.1 ml l^{-1} and 1.0 ml · l^{-1}.
dispersal departure from preferred breeding habitat, with distances spanning from a few centimeters to basin scales.
disphotic zone the layer of the water column with light penetration in insufficient amounts to support photosynthetic activity that balances respiration needs typically to depths of 200–500/600 m.
distal outermost position.
distribution centers regions with high organic matter content on the seabed.
diurnal related to daytime, as in diurnal vertical migration.
diversity expresses species richness described according to different measures.
DOM acronym for dissolved organic matter.
dominance the quantitative importance of a species within a community.
downwelling the wind-induced vertical currents that cause surface waters to sink to depth.
drag frictional resistance as a fluid moves past a fixed object, also often called frictional drag or viscous drag.
drop off region a zone of coral reefs at 20–30 m that hosts gorgonians and pelagic fish.

E
ebb tide low tide.
echinopluteus larval stage of echinoids (sea urchins).

echolocation (or **biosonar**) an auditory imaging system used by various species to navigate and to estimate the location of prey in environments where visual cues are ineffective.

Ecologically and Biologically Sensitive Areas (EBSAs) ocean locations of special importance for their ecological and biological characteristics.

ecosystem a community or group of living organisms that live in and interact with each other in a specific environment.

ecosystem approach to fisheries a fisheries management framework that links closely to human well-being.

ecosystem engineers a type of keystone species that directly or indirectly modulates the availability of resources to other species, often through creation or modification of habitat.

ecosystem functions the collective life activities of plants, animals, and microbes and the effects of these activities

ecosystem services the multiple benefits provided by ecosystems to humans.

ecosystem-based fisheries management (EBFM) a fisheries management framework that takes major ecosystem components and services-both structural and functional–into account.

ecosystem-based management a fisheries management framework that recognizes the full array of interactions within an ecosystem that includes humans, moving beyond considering single issues, species, or ecosystem services in isolation.

ecotone a margin or transition zone that mixes characteristics of the adjacent zones.

ectocrines (or **tele-mediators**) a broad class of organic compounds synthesized and released into seawater by marine plants and animals that influence the behavior or biological characteristics of conspecifics.

ectoderm outer tissue of corals that in contact with the external environment and helps to protect them.

ectosymbiont organisms living on the surface of another organism host.

ectotherms "cold-blooded" species for which their surrounding environment determines body temperature.

Ekman spiral a rotating column of water that forms when water moves at an angle to the wind direction due to the Coriolis Effect, resulting in increasing deflection of water with increasing depth.

El Niño Southern Oscillation (or **ENSO**) the cyclical alternation of extreme weather events in the south-central Pacific Ocean.

endemic a species whose geographical distribution is spatially limited to a given region.

endobenthos (or infauna) macrofaunal organisms living within the substrate.

endolithon an organism that lives buried in rock, usually calcareous, often causing flaking.

endopelon (or **endopsammon**) communities that live within the sediment.

endoplasmic reticulum the transportation system in eukaryotic cells responsible for protein folding and other important functions.

endopsammon (or **endopelon**) communities that live within the sediment.

endosymbiont organisms living within the body or cells of another organism.

ENSO (or **El Niño Southern Oscillation**) the cyclical alternation of extreme weather events in the south-central Pacific Ocean.

environmental DNA (eDNA) bits of DNA that biota have sloughed off into the surrounding environment.

ephyra the larvae produced through strobilation (transverse splitting) by schyphozoans.

epibenthos organisms that live upon the substrate and extend above the bottom into the water column to obtain oxygen and food.

epibionts organisms that specifically live attached to other organisms such as bivalve shells.

epidermis the outer "skin" layer.

epifauna organisms that live atop sediment (or hard substratum) surfaces.

epilithic organisms individuals that grow on rock.

epilithon communities that live on hard substrates.

epineuston organisms that live above the sea surface film.

epipelagic zone (or **upper open ocean**) the surface zone of the oceanic realm with sufficient sunlight for algae to convert carbon dioxide into organic matter, typically extending to ca 200 m.

epipelon cyanobacteria and eukaryotic algae that live on or in sediments.

epiphytic organisms that grow on marine macroalgae.

epiphyton communities that live on other plants.

epiplankton organisms that live in the zone from the surface to a depth of 100–200 m.

epipsammon generally referred to microalgae attached to grains of sand or on surface sediments.
epistrate feeders organisms that graze on primary producers, whether by scraping microbial films from hard surfaces or detaching small fragments from species such as kelp.
epizoic organisms animals that commonly live on other animals such as bryozoans, crustaceans, and mollusks.
epizoon communities that live on animals.
epsilon diversity changes in broadly regional diversity, such as the change in species diversity between marine and terrestrial ecosystems.
ergocline a model that proposes that the distribution of communities in a lagoon environment can be interpreted as a function of an ergocline, i.e., an interface system characterized by spatial and/or temporal gradients, within which physical processes can generate ecological structures with high organic production.
erythrocytes red blood cells.
essential ocean variables (EOVs) important ocean variables that researchers internationally agree should be standardized and collected routinely to create intercomparable data sets that enable broad-scale analyses and comparisons.
estuary terminal part of a river influenced by the periodic flow of the sea.
ethology the study of behavior.
eukaryotes single or multi-celled organisms whose cells contain a nucleus and other membrane-bound organelles, including all animals, plants, fungi, and protists, as well as most algae.
eulittoral zone the zone extending from the lower limit of the spring tides to the upper limit of complete immersion.
eunekton the "true" nekton that spend their entire lives below the ocean surface.
eurybathic species that span wide depth ranges can withstand these conditions.
euryhaline species that can tolerate a wide range of salinities.
eurythermal organisms organisms that tolerate wide variations in temperature.
eutrophic systems with excess nutrients, often leading to hypoxia.
eutrophication when a water body becomes overly enriched with nutrients, leading to abundant growth of photosynthetic organisms.
evaporates water-soluble minerals produced through concentration and crystallization of an aqueous solution.
evaporite basin a basin in which loss of water through evaporation exceeds freshwater input.
evenness how individuals are apportioned among the different species in a sample or location.
exopinacocytes specialized cells in sponges.
exoskeleton an exterior skeleton such as in crab or lobster.
exosymbiotic a symbiont that lives on the exterior of the body of its host.
exploitative competition a form of competition where one competitor reduces or consumes the resource.
extant currently living species, as opposed to those that are extinct.
extra-specific species organic matter processes processes in organic matter cycling that do not involve direct relationships between species but instead depend on the relationship between species and organic detrital matter.
exudates released dissolved material.

F
facies assemblage dominated by one or a few species that characterize it and which reflect the key environmental characteristics.
facilitation model where the first species to arrive at a habitat modify it in such a way that it facilitates establishment of subsequent species.
facilitative (or **positive**) **interaction** interactions between two species that benefit at least one of the participants and does no harm either of them.
facultative a potential but not required relationship.
facultative anaerobic microbes microbes that can survive with or without oxygen.
faneric mimicry mimicry that uses color to provide information to the observer.
fast ice (or **land fast ice**) the progressive freezing of the water layer at the sea-ice interface to form a coastal ice shelf.
fecal pellets rounded particles, mostly 10–500 μm in diameter and excreted by organisms.
fetch the distance over which the wind blows.
filter feeders (or **suspension feeders**) organisms that use filtering structures to capture food particles from the water surrounding them.

final (or **definitive**) **host** a host for the sexually mature stages of a parasite.
fishing down the food web the process by which fishing selectively removes larger species and leads to increases in smaller species.
floes ice islands ranging in size between hundreds of meters up to 10 km in diameter that drift on the ocean during the summer.
flood high tide.
fluvial referring to riverine inflow.
food chain the transfer of energy in an ecosystem as a linear process with linear ramifications.
food web a detailed interconnecting diagram that shows the overall feeding relationships between organisms in a particular environment.
founder effect where the frequency of genes in the new population differs from the parent population because it was founded by a small subset of individuals from the larger source population.
free-stream velocity the velocity of the fluid above the boundary layer that experiences no frictional drag from the bottom.
fringing reefs reefs over which waves break, running parallel and adjacent to the shore, where reef flats or small lagoons separate the reef from the coast.
fronts interfaces between water masses that differ in density.
fundamental niche the entire set of conditions under which an organism can survive and reproduce.
furious fifties area of strong winds formed from atmospheric perturbations at latitudes between ~50 and 65° S.
fusiform tapered at both ends.

G

gamma diversity the total number of species in a habitat or landscape.
garum in ancient Rome, a sauce made from fermented fish used to flavor stews, soups, and many other dishes.
Gelbstoff ("yellow substance") optically measurable dissolved organic matter in water.
gene flow the spread of genes through migration and subsequent reproduction between individuals belonging to different populations.
genetic bottleneck low intraspecific genetic diversity that increases the risk that a species cannot cope with atypical environmental conditions.
geographic (or **allopatric**) **speciation** speciation that occurs when geographic isolation caused by physical barriers prevents contact between two populations initially belonging to the same species.
geopolymeric fraction refers to inorganic polymers.
gigantism the phenomenon in which some organisms increase in body size with increasing depth.
gills specialized, highly vascularized structures in some types of animals used for gas exchange.
glacial control hypothesis a hypothesis that links coral atoll formation to changes in sea level.
global positioning system (GPS) the satellite triangulation system that provides accurate geographic position.
Golgi apparatus an organelle that packages proteins into vesicles.
Gondwanaland the ancient supercontinent that incorporated present-day South America, Africa, Arabia, Madagascar, India, Australia, and Antarctica.
gonochoric capable of producing sperm or eggs but not both.
grab an instrument lowered over the side of a boat or ship to sample surface sediment.
grain size the size of sediment particles.
gram-negative bacteria bacteria that do not retain the crystal violet stain used in the Gram staining method of differentiating bacteria.
gram-positive bacteria bacteria that retain the crystal violet stain used in the Gram staining method of differentiating bacteria.
gray ice (and then **gray-white** ice) ice formed as young ice progressively thickens from below.
grease ice "oily" ice formed when ice crystals develop into needles or small plates of ice.
great ocean conveyor belt (or **deep ocean conveyor belt**) the large system of global circulation driven by thermohaline circulation that connects the major ocean basins.

great whales the largest whales, including 12 baleen species as well as sperm whales.
gregarious settlement larvae settling near conspecifics.
gregarious species species whose propagules settle in proximity to other individuals of the same species.
gross secondary production the energy channeled into heterotrophic organisms.
groundfish fishes associated with bottom habitat, such as cod and flatfishes.
growth anomaly a wide range of injuries and anomalies in organisms such as corals, which arise from both tumors and altered growth in response to an external stimulus.
guild species that exploit a similar resource in a similar way, even if they do not occupy a similar ecological niche.
guyots seamounts whose summit has been flattened over time by eruption and/or wave erosion to create table-like structures.

H

habitat the place colonized by a specific set of more vegetal and/or animal species.
habitat heterogeneity hypothesis a hypothesis that attributes diversification to spatial heterogeneity rather than stability.
hadal zone the portion of the seabed below 6,000 m depth.
hadopelagic zone zone that only exists in specific locations around the world and includes the waters overlying the hadal trenches from 6000 to >11,000 m depth.
hadoplankton the plankton in the ocean trenches that extend from 6,000–11,000 m.
halocline refers to changes in salinity across water masses, as often shown in salinity-depth profiles.
halophilic salt loving.
harem a social group consisting of a single male and numerous females, typically linked to reproduction.
heat-shock proteins (HSP) specialized proteins that protect enzymes by forming enzyme complexes that protect the enzyme.
heavy metals metals such as copper that become toxic when concentrated.
hematocrit the percentage of blood volume occupied by the cellular component.
hemoglobin the blood protein responsible for oxygen transport.
herbivores animals that feed on plant tissues.
hermaphroditic capable of producing gametes of both sexes.
hermatypic reef building corals.
heterotherms (or **poikilotherms**) animals in which body temperature tracks that of their environment.
high nutrient, low chlorophyll system surface water with high concentrations of nutrients, but low phytoplankton biomass because of some other limiting nutrient, typically iron.
holobenthos organisms that live their entire life cycle in the benthic compartment.
holoplankton organisms that live their entire lives in suspension in the water column.
home range the larger habitat occupied by an organism that provides the requirements for breeding, feeding, or survival.
homeoosmotic ions ions similar to those in the surrounding environment.
homeosmotic organisms (or **osmotic regulators**) species that maintain a constant internal salt concentration.
homeotherms animals that maintain a relatively constant body temperature that may differ from their environment.
homochromia a cryptic response where individuals of a species mimick the colors of its substrate by minimizing any silhouette effect.
homomorphic species species that blend into their environments based on morphology.
hotspot area of exceptional biotic richness compared to the surrounding areas.
hybridization when two populations mix genetically.
hydrodynamics referring to the study of fluid motion.
hydrogen bond a bond between polar molecules in which hydrogen binds to a larger atom, such as oxygen or nitrogen, through electromagnetic attraction.
hydrogenous sediments sediments produced from chemical reactions in seawater, which represent a relatively minor sediment source.
hydrometer an instrument that determines the density of a liquid of known temperature.

hydrophilic "water loving".
hydrophones underwater microphones used to collect sound data.
hydropneumatic skeleton a skeleton based on fluid under pressure, as in nematodes.
hydrothermal vents the cracks in the Earth's crust near spreading and subduction zones along the edges of ocean plates where superheated seawater rich in dissolved minerals and chemicals spews out.
hyperbenthos organisms that live just above the seafloor.
hyperthermophilic organisms that thrive in extremely hot environments.
hypertonic maintaining higher concentrations of solutes compared to surrounding seawater.
hypervolume the multidimensional space that determines population survival, where each dimension represents an environmental factor.
hypostracum the middle shell layer of bivalve mollusks comprising alternating organic lamellae and calcareous layers.
hypotonic maintaining lower concentrations of solutes compared to surrounding seawater.
hypoxia waters with dissolved oxygen content lower than 2.0 ml l^{-1}.
hypoxic environments with low levels of oxygen, often defined as concentrations less than 2.0 ml·L^{-1}.

I

ice edge bloom blooms at the edge of sea ice.
ice melt POM particulate organic material (POM) previously trapped within the ice that sinks through the water column to enrich the benthic community.
ice-foot land fast ice that remains attached to the shore and potentially gives rise to multi-year ice.
ichthyology the study of fishes.
ichthyophagous taxa that feed on fish.
ichthyoplankton fish eggs and larvae present in plankton.
inbreeding limited mixing with other populations within a species for reproduction.
indirect development where embryos of a given species develop into a larval stage but subsequently metamorphoses directly into an effective adult stage.
infauna organisms that live below the sediment surface.
infralittoral (or **intertidal** or **mesolittoral** or **mediolittoral**) **zone** the stretch of coast at the land-sea interface between the highest and lowest tide levels.
infraplankton that live permanently at depths between 500 and 600 m.
inhibition model succession model in which, over time, some species prevent the establishment of other species that might otherwise utilize the resource and eventually dominate.
inquilinism where one organism shares part of the living space of another, and neither "occupant" benefits or suffers.
integumentary system the set of organs forming the outermost layer of an animal's body.
interference competition the negative interaction of organisms or species as they seek an essential but non-limiting resource.
intermediate disturbance hypothesis the hypothesis that episodic natural phenomena can lower biodiversity, but high stability leads to intense competitive extinction, whereas intermediate disturbance allows a higher diversification.
intermediate host a host in which developing and immature stages of a parasite live.
internal currents currents that move the portion of the water column below 200 m depth to the near bottom.
interstitial referring to the spaces between sediment grains.
intertidal (or **infralittoral** or **mesolittoral** or **mediolittoral**) **zone** the stretch of coast at the land-sea interface between the highest and lowest tide levels.
invisible habitat the unseen habitats that comprise the neritic domain.
iponeuston organisms that live below the sea surface film.
isophasic strategies strategy in which mobility allows species adapted to either aquatic or land habitats to enter the intertidal zone during favorable periods.
isospatial strategies strategy in which animals do not use mobility to avoid unfavorable conditions, and instead remain in the intertidal irrespective of tidal phase and manage to resist unfavorable conditions.

K

K-selected species species characterized by slow-growing, large, late maturing individuals.

karst theory a theory that integrates concepts of subsidence and sea level variation.
karstic formed from carbonate rock subsequently exposed to water dissolution.
katabatic wind a wind that carries high-density air from a higher elevation down a slope under the force of gravity.
keystone species a species that exerts a significant and disproportionate impact on its community or ecosystem relative to its abundance.
killing the winner a model that postulates that viral infection controls microbial community diversity whereas non-specific grazing controls microbial abundance.
knolls isolated peaks on the seafloor rising less than 1 km in elevation.

L

La Niña phenomenon opposite to El Niño with a strong drop in surface water temperature.
labile easily digested or clavated by enzymatic activities.
lacrimal gland tear gland.
lagoon a fully or largely enclosed coastal water body that connects with the ocean but also receives continental waters.
laminae distinct layers, as in layers of sediment.
land fast ice (or **fast ice**) the progressive freezing of the water layer at the sea-ice interface to form a coastal ice shelf.
land-sea interface where land masses meet the ocean.
Langmuir cells the orbital surface circulation cells created by wind.
larvaceans cold-water planktonic tunicates.
larvae stage of the life cycle stage that occurs in many marine species and that differs morphologically, physiologically, and ecologically from adult forms.
lateral line systems the tactile sense organs in fishes (and some other vertebrates) that detect movements and pressure changes in the surrounding water.
lebensspurren tracks created by animals crawling on the sediment surface.
lecithotrophic larvae typically pelagic larvae that depend on internal food reserves and do not feed as they develop.
lepidochronology the study of changes in ligules and scales at the base of plant leaves.
lessepsian migration the entry of alien species from the Red Sea through the Suez Canal.
Liebig's Law the law of the minimum, which states that the scarcest resources rather than total resource availability limit growth.
life cycle the developmental stages of organisms, plankton can be divided into holoplankton and meroplankton.
light scattering the degree to which a liquid changes light direction as it reflects off suspended particles.
ligule a thin outgrowth at the junction of leaf and leafstalk of many grasses and sedges.
limivorous deposit feeding mode obtaining nutrition by ingesting sediment particles and stripping off associated food in subsurface sediments.
log layer an intermediate layer that makes up about 15% of the total boundary layer thickness but in which velocity varies with the logarithm of height above bottom.
log-deficit layer the uppermost layer of the boundary later.
lower open ocean (or **bathypelagic zone**) zone beginning at the bottom of the mesopelagic zone (ca 1000 m) and extending to 3000–4000 m.
luciferase enzyme involved in bioluminescence processes in combined action with luciferin.
luciferin class of light-emitting heterocyclic compounds present in various organisms in which bioluminescence is observed. Works in combination with luciferase.
luminous light producing.
lysis the transformation of a cell into cellular debris composed of dissolved molecules.
lysogenic (or **temperate**) **viruses** viruses that infect the cell and integrate their DNA into the host genome, replicating their genome until some additional factor (e.g., UV, high temperatures, pollutants) induce the lytic cycle.
lytic viruses viruses that infect a cell, replicate, and release once the infected cell dies.

M

macroalgae multicellular algae (such as rockweed or kelp) that are visible to the naked eye.
macrobenthos (or **macrofauna**) organisms retained on a 0.5 or 1 mm sieve but too small to identify in bottom photographs.
macrofauna (or **macrobenthos**) organisms retained on a 0.5 or 1 mm sieve but too small to identify in bottom photographs.

macronutrients nutrients required in relatively high concentrations by photosynthetic organisms, specifically nitrate, phosphate, and silicon.
macrophages organisms that feed on individual particles, including masticators and shredders that feed on individual particles, whether living organisms or food parcels.
macrozooplankton relatively large plankton, visible to the naked eye.
mantle the intermediate layer of Earth, between the outer crust and inner core.
Margalef's Index an index that compares the cumulative rate between the number of species and the number of individuals in a community.
marine snow the rain of organic material that falls from surface waters into the deep ocean.
marine spatial planning (MSP) a holistic process that brings together diverse ocean stakeholders to coordinate informed decisions that support sustainable marine resource use.
maritime activity or location associated with the ocean.
masking when one impact effectively masks another.
match-mismatch hypothesis the theory that poor recruitment occurs when larval timing does not coincide with availability of fundamental resources such as phytoplankton or small zooplankton.
matte a block of compacted sediment comprising a dense network of rhizomes, roots, and seagrass plant debris.
mean spring tides the highest of monthly tides that coincides with the alignment of the sun and moon along the same axis with respect to Earth.
mediolittoral (or **infralittoral** or **mesolittoral** or **intertidal**) **zone** the stretch of coast at the land-sea interface between the highest and lowest tide levels.
megafauna organisms visible in bottom photographs.
megalopa/zoea larva of decapod crustaceans.
meiofauna animals (normally metazoans) between 20 and 30 μm and 0.5–1 mm in size.
meiofauna animals <44 μm in size.
melon structure in the head of dolphins that acts as a lens for locating prey, focusing acoustic waves in a narrow beam projected forward.
merobenthos the planktonic larvae of benthic organisms, in contrast to meroplankton.
meroplankton (or **temporary plankton**) organisms that spend part of their life cycle in the plankton.
mesenteric filaments folds of the gastrovascular cavity at the base of the polyp of corals, which they can evert as weapons of defense or prey capture.
mesoglea the jelly layer of tissue in jellyfish.
mesohaline bodies of water with salinity 5–18.
mesolittoral (or **infralittoral** or **intertidal** or **mediolittoral**) **zone** the stretch of coast at the land-sea interface between the highest and lowest tide levels.
mesopelagic zone (or **middle open ocean**) stretches from the bottom of the epipelagic zone (ca 200 m) to the depth at which no sunlight exists, approximately 1000-m depth.
mesophiles organisms that live in moderate temperatures.
mesophotic zone the layer of the water column with insufficient light penetration to support significant photosynthetic activity, and typically spanning from 50 to 150/200-m depth.
mesoplankton organisms that live in the comparatively oligotrophic mesopelagic zone (200–1,000 m).
mesopredators mid-level predators.
mesopsammon organisms of the meiofauna that live between the crevices of the sediments.
mesozooplankton heterotrophic organisms varying in size between 0.2 and 20 mm.
meta-analysis analysis that spans multiple data sets.
metabarcoding genetic analysis that considers a single gene marker for all organisms in a sample.
metabolic rate the energy necessary to carry out vital life functions.
metabolic theory of ecology a theory that states that the metabolic rate of organisms defines the fundamental biological rate that governs most ecological patterns.
metagenomic analysis genetic analysis that considers all the genes of all organisms in a sample.
metazoans multicellular organisms.
methanogens methane producing microbes.

Michaelis-Menten equation an equation that describes the enzymatic dynamics of degradation of a substrate.
microbial gardening the process by which organisms that promote microbial growth in sediments by secreting and depositing mucus on sediment grains that bacterial populations rapidly colonize.
microbial loop a trophic pathway through which dissolved organic matter (DOM) re-enters the food web through its transformation into bacterial biomass.
micronutrients nutrients required in low concentrations by photosynthetic organisms, such as iron, magnesium, and zinc.
microorganisms forms of life invisible to the naked human eye and visible only through the use of a microscope.
micropatches small-scale patches.
microphages organisms that feed on suspended particles or on small organisms.
microplankton autotrophic and heterotrophic organisms ranging from 20–200 μm in size.
microtides very small tidal excursions, often seen in semi-enclosed seas such as the Mediterranean or Baltic.
microvores organisms that feed on bacteria.
microxic waters with oxygen concentration up to $0.1 \text{ ml} \cdot l^{-1}$.
mid-ocean ridge an underwater "mountain chain".
middens ancient garbage dumps.
middle open ocean (or **mesopelagic zone**) stretches from the bottom of the epipelagic zone (ca 200 m) to the depth at which no sunlight penetrates, approximately 1000-m depth.
migration movements between clearly separated and well-defined locations.
mimicry morphological and chromatic adaptation that allows the various species to increase the chances of survival in a given environment.
mixohaline referring to brackish water.
mixotrophic organisms that are both primary producers and consumers of organic matter, depending on environmental and ecological conditions.
mollusks a phylum of (mostly) shelled invertebrate animals with a soft, unsegmented body that includes snails, mussels, and octopuses.
monomers molecules comprising a single subunit.
mooring a wire cable held vertical by buoyant floats, with a heavy weight (such as a train wheel) to anchor it to the seabed.
mosaic stages of heterogeneity and disturbance that includes different successional stages of K species and r species.
mucus-trap feeding a similar technique to microbial gardening where organisms secrete and deposit mucus to trap food particles.
Müller's larvae larvae of turbellarians.
Mullerian mimicry situations where both the original model and the mimic possess similar deterrents.
multi-year ice ice that persists through the year into the following winter.
mutualism a close and sustained symbiotic relationship between two organisms or populations or species living together, where both species benefit.
μ_{max} the maximum growth rate of a population or species.

N

nanobenthos includes all organisms between 2.0 and 20 μm in size, both autotrophic and heterotrophic.
nanoflagellates microbes with flagella within the size range 2–20 μm.
nanoplankton organisms between 2 and 20 μm in size.
natatorial adapted for swimming.
natural capital the world's stocks of natural assets which include geology, soil, air, water and all living things.
nauplii early developmental stages copepods that precede the **copepodite** developmental stages.
neap tides the lowest of monthly tides that occur during the first and the last quarter phases of the Moon when the Moon and Sun are aligned at 90 degrees with respect to each other and Earth.
nearest neighbor distance (NNR) the minimum distance between two locations simultaneously occupied by organisms.
nectochaeta pelagic larval stage of polychaetes.
negative water balance where the water output (e.g., outflow, evaporation) exceed the input (inflow, rainfall) in a basin.

nektobenthos (or **benthonekton**) species such as flatfish that actively swim close to the bottom, without settling onto the seabed for extensive periods.
nekton actively swimming organisms capable of making significant headway against currents.
nematocysts stinging cells of cnidarians.
nematodes cylindrical animals with bilateral symmetry covered by a more or less thick cuticle.
neoendemic species where a new, relatively localized species evolved relatively recently as a result of divergent adaptation of existing species to differing environmental conditions.
neritic province the relatively shallow portion of the ocean that extends from the land-sea interface to the edge of the continental shelf at 200 m, and where light often penetrates all the way to the seafloor.
net photosynthesis situations where photosynthesis exceeds respiration in primary producers, resulting in net energy gain for the ecosystem.
net secondary production the energy channeled into heterotrophic organism biomass, taking into account losses to respiration and excretion.
neuston refers to organisms adapted to living in contact with the sea surface film (produced by surface water tension).
new moon a moon phase where the dark side of the moon faces Earth, when the sun and the moon align on the same axis but with the moon between the Earth and sun.
new production primary production resulting from the utilization of oxidized forms of nitrogen as well as fixation of atmospheric nitrogen (N_2).
nictemeral cycle day/night cycles, sometimes referred to as circadian.
nitrification a microbial process that sequentially oxidizes reduced nitrogen compounds (primarily ammonia) to nitrite and nitrate.
nitrogen fixation the conversion of N_2, which most organisms cannot use, into ammonia, which many organisms can, by specialized bacteria.
nitrophilous nitrogen loving.
nocturnal related to nighttime, as in nocturnal migration.
normoxic (or **oxic**) environments with "normal" levels of oxygen necessary to support most organisms.
North Atlantic Deep Water (NADW) a large, dense water mass that originates largely from Gulf Stream waters that cool and sink in the Arctic.
North Atlantic Oscillation (NAO) a North Atlantic weather phenomenon in which atmospheric pressure at sea level varies over time periods of decades.
nutrients refers to inorganic compounds or ions used primarily in the nutrition of primary producers.

O

obligate a required relationship.
obligate aerobes organisms that can survive only in the presence of oxygen.
obligate anaerobic microbes microbes that survive only in the absence of oxygen.
obligate psychrophiles bacterial taxa that require cold conditions and die at temperatures above 4 °C.
ocean gliders underwater "airplanes" that glide over pre-programmed survey routes as they profile the water column, but with the capacity to actively navigate rather than drifting passively.
oceanic (or **pelagic**) **province** the water masses beyond the continental shelf.
oceanic trenches narrow, long topographic depressions of the seafloor that form the deepest parts of the ocean floor, extending to more than 11,000 m depth.
oceanic zones a series of latitudinal circulatory processes that result in bands of water-mass structures.
oligohaline bodies of water with salinity <5.
oligotrophic referring to low production regimes.
oligotrophic systems poor in nutrients.
omega diversity global-scale changes in species diversity.
omnivores animals that feed on a mixed diet of plants and animals.
ontogenetic developmental stages of a species.
oophagia where nourishment comes from other eggs produced by the mother.
opportunists a species that can quickly exploit new resources as they arise.
optical sensors sensors that use natural or artificially generated light to image some aspect of the biotic or abiotic environment.

optimal foraging theory a behavioral ecology model that helps predict how an animal behaves when searching for food.

orbital forcing the effect on climate of slow changes in the tilt of the Earth's axis and shape of the Earth's orbit around the sun.

organic content the amount of organic matter present, typically expressed as carbon to nitrogen ratios, total organic carbon, or biochemical components of organic matter.

osmolytes small organic molecules produced by cells.

osmoregulation physiological activity of an organism aimed at keeping internal salts and body fluids balanced within a narrow range.

osmotic balance challenges where ion concentrations inside cells of biota may differ greatly from the surrounding seawater.

osmotic conformers species in which internal salt concentrations effectively track their environment.

osmotic regulators (or **homeosmotic organisms**) species that maintain a constant internal salt concentration.

osmotrophs organisms that incorporate organic molecules obtained from organic particles through chemical breakdown, phytoplankton, or heterotrophic bacteria.

ostracum the middle shell layer of bivalve mollusks comprising an organic substance, conchiolin, and calcium carbonate.

otoliths ear bones of fish used for conducting sound.

overgrazing herbivores grazing algae excessively.

oviparous producing young by hatching eggs after the parent has laid them.

ovoviviparous producing young by hatching eggs within the body of the parent.

oxic (or **aerobic**) waters with oxygen concentrations greater than $1.0 \text{ ml} \cdot \text{l}^{-1}$.

oxic (or **normoxic**) environments with "normal" levels of oxygen necessary to support most organisms.

oxygen minimum zone regions of offshore ocean, generally between 500 and 1,500 m, where high oxygen consumption results in an O_2 concentration minimum.

P

pack ice ice formed by freezing of seawater but not attached to land.

paleoendemic species formerly widespread species now restricted to a smaller geographic region.

pancake ice round plates up to meters in diameter that develop from grease ice.

panoceanic species species that occur in all oceans.

paralic model a model of coastal lagoon zonation driven by the strong salinity gradient generated by mixing between freshwater and seawater.

paradigm of temporal stability the physical stability of marine environments allows for extreme adaptability and specialization that minimizes competitive interactions between species.

parapatric speciation occurs in widely distributed species in which reproductive isolation occurs between members of adjacent populations in the absence of geographic barriers.

parasitism a close and sustained symbiotic relationship between two organisms or populations or species living together where one benefits and the other suffers.

paratenic host a host in which larval forms of a parasite cannot complete their development.

parthenogenesis reproduction mechanism where unfertilized eggs develop into a new individual.

partial regulators animals that maintain constant ion concentrations over a range of salinities but conform over part of the range of salinities they encounter.

particulate organic matter (POM) detritus in the form of particles.

parvorder a specific taxonomic category above superfamily and below infraorder.

passive acoustics "listening" to sounds in nature such as whale songs to infer behavior and even abundance of different animals.

passive canyon canyons subject to largely vertical flow, much like the open ocean.

passive margins continental margins on which no plate boundaries occur.

passive restoration restoration that relies on the spontaneous resilience of an ecosystem to recover simply by removing the environmental stressor(s) that led to degradation.

passive suspension feeders suspension feeders that depend on currents to bring particles into contact with feeding structures.

patch mosaic theory the theory that a mosaic of stages of heterogeneity and disturbance that includes different successional stages of K species and r species will create a higher overall diversity of the mosaic created by these microenvironments than that present in a more uniform area.
pathogens disease agents such as viruses and bacteria such as fecal coliforms.
pediveligers late stages of mollusk larvae.
pelagic (or oceanic) province the water masses beyond the continental shelf.
pelagic domain the totality of the water volume.
pelagic-benthic coupling (or pelago-benthic coupling) the process by which energy and matter transfers between the water column to the seafloor.
pelago-benthic coupling (or pelagic-benthic coupling) the process by which energy and matter transfers between the water column to the seafloor.
pericapsid the exterior glycoprotein-lipid coating of viruses.
pericarp a porous part of the edible fruit of seagrasses rich in oily substances that help it float.
periostracum the outer shell layer of bivalve mollusks, comprising a substance similar to chitin.
peripatric speciation (or semi-geographic) speciation that occurs in widely distributed species where adjacent populations become reproductively isolated in the absence of geographical barriers.
periphytic organisms that live on the surface of submerged plants and other underwater objects.
permanent thermocline a persistent feature characterized by a rapid change in temperature over a short depth range that is unaffected by diurnal (day night) or seasonal changes.
permeability the rate at which water can percolate through sediment.
permeases specialized transport enzymes in the membranes of bacterial cells selectively transport substances through the membrane.
phagocytize to engulf and then digest prey fragments.
phagotrophs a heterotrophic organism that feeds by ingesting organisms or organic particles.
phenotypic plasticity morphological variation in a species, resulting from "interaction" with the environment.
phenotypic plasticity the capacity of an organism to change in response to stimuli or inputs from the environment.
philopatry the tendency of some species of migratory animals to return to a particular place to feed or reproduce.
phoresis a dispersal mechanism in which transport of a (generally small) organism occurs by adhering to the body of a larger animal.
photic (or euphotic) zone the highly illuminated portion of the water column where intensive photosynthesis can take place and where photosynthetic activity exceeds respiration activity, typically extending from the surface to a depth of 20–50 m depending on water clarity.
photoautotrophs photosynthetic primary producers, also sometimes imprecisely called autotrophs.
photophilic light loving.
photophores light organs.
photosynthetic efficiency the efficiency of light utilization.
photosynthetically active radiation (PAR) the wavelengths of light used in photosynthesis.
phototrophy the process by which light triggers electron flow in photosynthetic organisms.
phyletic evolution (or anagenesis) occurs when an ancestral population undergoes a gradual change over time with splitting.
phyletic gradualism the hypothesis that evolution proceeds as a slow and gradual process.
phyllosoma the flattened planktonic larvae produced by lobsters and other crustacean.
phylogeny the evolutionary diversification of a given taxonomic group.
phytal zone the ~6–8% portion of the seabed where sufficient light reaches the bottom to support photosynthetic seafloor flora.
phytodetritus decaying phytoplankton and its breakdown products.
phytoplankton bloom often rapid increases in phytoplankton populations.
PI curve a photosynthesis / light intensity curve.
picobenthos all living prokaryotic and eukaryotic organisms ranging 0.2–2.0 μm.
picoplankton plankton that includes prokaryotic and eukaryotic organisms ranging 0.2–2.0 μm.
Pielou's index a measure of the distribution of abundances among species.
piezotolerant (or barotolerant) organisms that tolerate a wide range of pressures.

pilidium larval nemertines.
pinnacles the small mountain ranges that rise above the seabed.
piscivores fish predators.
Planet Ocean an alternative name for Planet Earth attributed to author Arthur C. Clark.
planktology (or **planktonology**) the study of plankton.
plankton paradox many species of phytoplankton occupy the same volume of water using the same resources.
planktonekton small, fast organisms, and large organisms with relatively ineffective locomotor apparatus.
planktonology (or **planktology**) the study of plankton.
planktotrophic larvae typically pelagic larvae that feed as they develop.
planula larva of cnidarians.
platforms in geology refers to flattened structures that rise above the seafloor at shallow depths, whereas in oceanography refers to the physical setting that house instruments, spanning from ships to moorings to satellites, even including the seabed.
pleuston the floating plankton that live at the air-sea interface.
pneumatophores lateral roots that extend out of the surface of the water and facilitate the exchange of oxygen and carbon dioxide for the roots submerged in anoxic sediment.
poikilotherms (or **heterotherms**) animals in which body temperature tracks that of their environment.
polar circles lines encompassing regions that experiences at least one 24-hour period when the sun remains below the horizon and another 24-hour period where the sun never drops below the horizon.
pollution the direct or indirect addition of substances or energy into the marine environment (including estuaries) by humans that causes harmful effects on living resources, human health, hinders marine activities including fishing, or alters quality of sea water and its benefits.
polyandry where a female mates with multiple males.
polyandry where the female of some species may mate with several males in succession, or even with two at the same time.
polymers molecules formed from many similar subunits bonded together.
polygynandry multiple matings for both sexes.
polygynandry multiple males mating with multiple females.
polygyny where a male mates with multiple females.
polygyny male mating with multiple females.
polyhaline bodies of water with salinity 18–30.
polynyas persistent open waters surrounded by sea ice in polar regions.
POM acronym for Particulate Organic Matter.
poorly sorted sediments containing a mixture of grain sizes.
population a group of individuals belonging to a given species that live in an environment and interact with one another more than with individuals from other population.
popup tags tags attached to large animals that release after some time and pop up to the ocean surface where they download their data to passing satellites.
porosity the amount of pore space available between sediment grains.
positive (or **facilitative**) **interaction** interactions between two species that benefit at least one of the participants and does no harm either of them.
pre-zygotic gametes prior to egg fertilization.
precautionary approach an environmental decision-making framework increasing used by conservationists and ocean use managers to increase sustainability.
predator cropping hypothesis the hypothesis that predators control or crop back the number of organisms present in a location and thus promote diversification.
predators animals, mostly mobile, that attack, kill, and consume individual and generally mobile prey.
preferential species species that live almost exclusively or preferentially in a specific (or similar) kind of biocoenosis.
primary producers photosynthetic organisms.
primary production a series of processes through which photosynthetic organisms synthesize organic matter from inorganic carbon, typically expressed as the total amount of organic matter (or organic carbon) per unit area or volume expressed as units of mass (grams, moles) of carbon fixed by primary producers per unit area (e.g., m^{-2}), or unit volume (e.g., m^{-3}).

primordial sea (or **primordial soup**) the primitive environment from which life originated, and the evolutionary process that led to various forms of life.

primordial soup (or **primordial sea**) the primitive environment from which life originated, and the evolutionary process that led to various forms of life.

principle of competitive exclusion the idea that two species cannot occupy the same niche, because the strongest one will persist, and the weakest will disappear with time.

prokaryotes single-celled organisms without a nuclear membrane.

propagules the unit of dispersal, whether a spore, egg, larva, or other life history stage.

protandric organisms that begin as males and then become females.

protein enrichment an increase in organic matter nutritional value and quality through the transformation of detrital matter.

proteobacteria a highly diverse phylum of bacteria that include both pathogenic and free forms and include nitrogen-fixing and photosynthesizing bacteria.

proterogynic organisms that begin as females and metamorphose into males.

protists eukaryotic organisms that are not animals, plants, or fungi.

protozoa heterotrophic organisms consisting of a single eukaryotic cell.

provisioning services the products directly obtained from ecosystems that benefit ecosystems such as creation of biotic habitat structures for other species.

psammon fauna able to move within interstitial spaces in sand.

pseudo-populations non-native populations, able to settle and grow, but not to reproduce.

psychrotrophic bacteria "cold loving" bacterial taxa, usually restricted to permanently cold habitats.

pterigopods copulatory organ in sharks consisting of modified pelvic fins.

pteropods holoplanktonic gastropods.

punctuated equilibrium the hypothesis that periods of rapid and important genetic and phenotypic changes alternate with long periods of stability and genetic invariance.

pycnocline an abrupt change in the density profile, which often coincides with a thermocline or halocline.

Q

quiescence a developmental delay in immediate response to environmental stressors.

quotas the biomass of fish that a fishery can remove at a given time, to ensure sustainability of a target population.

R

r-selected species species characterized by fast growth, rapid maturation, small individuals.

radula tongue-like, toothed feeding structure in gastropods.

rafting a mechanism of passive transport whereby small organisms settle on or adhere to floating objects.

Rapoport's rule an ecogeographical "rule" stating that latitudinal ranges of species are generally smaller at lower latitudes than at higher latitudes.

ratcheting a thinning and reduced degree of calcification of the skeletal formations in deep-sea organisms.

realized niche the set of conditions actually used by given organisms, accounting for other interactions such as predation and competition.

recovery the achievement of a target environment similar to an appropriate local native model or reference ecosystem, in terms of its specific compositional, structural, and functional ecosystem attributes.

Red Queen Hypothesis the idea that species (e.g., hosts) must constantly adapt, evolve, and proliferate in order to survive against ever-evolving opposing species (e.g., parasites).

redox potential electric potential of a platinum electrode immersed in a solution containing oxidized and reduced states of a substance relative to that of a normal hydrogen electrode.

redox potential discontinuity depth (RPD) the depth of oxygen penetration into sediments.

reef crest (or **reef edge**) the outer portion of the reef with the strongest wave action.

reef edge (or **reef crest**) the outer portion of the reef with the strongest wave action.

reef mounds the large calcium carbonate skeleton left behind as a reef grows and corals below die off.

reef upper slope reef zone that occurs at 10–15 m characterized by high diversity and large corals.

reflective deep layer (DSL) (or **deep scattering layer**) depth layer of the ocean with high zooplankton concentrations sufficient to scatter sound waves.

refraction the bending of light as it passes through media of different densities, such as air and water.
refractory material a material that is resistant to decomposition.
regulating services benefits provided by ecosystem processes that moderate natural phenomena, such as coastline protection against storms.
rehabilitation the replacement of ecological structural or functional characteristics damaged by an impact, typically enhancing the social, economic, and ecological value of the new ecosystem.
relict sediment continental shelf sediment deposited during the recent geologic past that is not in hydrodynamic equilibrium with the present-day environment.
remineralization the decomposition of organic matter and release of their minerals and nutrients.
resilience the ability of a system to return to its previous state once a disturbance ceases.
resistance the ability of a system to withstand disturbances without altering its community structure, biodiversity, and functioning.
resource any aspect of the environment used to satisfy an organism's need.
restoration the process of assisting the recovery of ecosystems that have been degraded, damaged, or destroyed.
resuspend (or resuspension) the process of moving particles off the bottom so they are suspended in the water.
rete mirabilis the highly vascularized network of red muscles in tuna.
Reynolds number (Re) the ratio between inertial forces and frictional forces.
rhizome a horizontal stem that runs horizontally beneath the sediment surface.
ribonucleic acid (RNA) an important biological macromolecule that converts the genetic information of DNA into proteins in most living organism.
roaring forties area of strong winds formed from atmospheric perturbations at latitudes between ~ 40° and 50°S.
Rouse parameter the ratio of the tendency of particles to sink to the seafloor to the tendency for flow turbulence to keep them in suspension.

S

sacculinization the transformation of an organism's body into a bag-like structure by a parasite.
salinity the salt content of a water.
salt marshes coastal intertidal wetlands that the tides flood and drain with salt water through numerous tidal channels.
saltpond a basin in which exchange with the ocean no longer occurs or occurs minimally.
scattering reflection of light due to particulate material suspended in the water column.
scavengers organisms that feed on microscopic organic particles consisting of debris, bacteria, protists, and small planktonic organisms or large particles or carcasses of dead organisms.
school a group of fish that swim in synchrony at the same speed and direction with consistent distances to their nearest neighbor.
sciaphilous shade tolerant organisms such as some corals.
sea state the degree of wave motion on the ocean at a given time.
seabed the bottom of the ocean.
seabed hotspots areas of the seabed in isolated regions of plates where breaks in the crust allow magma to rise from the mantle to create an underwater volcano.
seamount effect increased concentrations of nutrients and particulate organic material around seamounts that can support high abundances of benthos, zooplankton, and fishes.
seamounts underwater promontories/mountains that rise 1000 m or more above the seafloor.
searching predators predators that look for less mobile prey.
seasonal movements occupation of different geographical locations during different times of the year, typically involving movements over varying distances.
seasonal thermocline a seasonal feature characterized by a rapid change in temperature over a short depth range that is unaffected by diurnal (day night) changes.
Secchi disk a white circular disc 30 cm in diameter used to estimate water transparency by lowering it on a rope into the water until it disappears from sight.
sectionally homogenous estuaries where salinity gradients may occur along the length of an estuary but with no vertical stratification.

sediment mixing depth the depth to which organisms mix sediments.
semi-diurnal occurring twice daily, as in tides at some locations.
semi-geographic (or **peripatric speciation**) speciation that occurs in widely distributed species where adjacent populations become reproductively isolated in the absence of geographical barriers.
semi-saturation constant the concentration at which a species reaches half of its maximum rate of growth.
sequestration removal of organic substances in a system.
sessile or **sedentary** non-mobile organisms, often attached to the seafloor or other structures.
seston Collectively, particles suspended in water including living forms, and organic aggregates.
sexual dimorphism species in which physical morphologies other than reproductive organs differ markedly between sexes.
sexual reproduction reproduction achieved by producing two gametes that fuse into a zygote.
Shannon-Weaver (or **Shannon-Wiener**) **Index** a diversity metric that examines the characteristics of a community by proportionately weighting all species present in a sample.
Shannon-Wiener (or **Shannon-Weaver**) **Index** a diversity metric that examines the characteristics of a community by proportionately weighting all species present in a sample.
shear change in fluid velocity with increasing distance from the bottom.
shear stress is a force per unit area τ.
shear velocity a measure of momentum transfer within the benthic boundary layer.
shelf break the edge of the continental shelf at 150–200 m depth.
shoal a group of fish that orient randomly in a group with variable distances to their nearest neighbor.
sibling (or **cryptic**) **species** two or more reproductively incompatible species that are morphologically indistinguishable from one another.
sill oceanic ridge that partly separate basins and sometimes limits the exchange of water between basins.
Simpson's Index a diversity metric based on the probability that two individuals randomly chosen from a habitat (in a sample) belong to the same species.
sink population population that produce few successful offspring and contributes little to subsequent generations.
sloppy feeding process by which grazers do not consume all phytoplankton they graze on, typically releasing substantial particulate and dissolved organic material in the water.
soak time the amount of time a piece of gear remains in the water.
somatic referring to the body, as in somatic growth.
sorting the degree to which a given sediment sample contains a mixture of different grain size categories.
source population population that contributes significant offspring to subsequent generations.
speciation the evolutionary process that leads to the emergence of new species.
species abundance the distribution of abundance of species.
species richness the number of species present in a sample, community, or taxonomic group.
specific gravity the ratio of the density of a substance to the density of some substance (such as pure water).
speleology the study of caves and other karst structures.
stability-time hypothesis the hypothesis that physical stability of marine environments (e.g., deep or tropical) allowed species to adapt and specialize, minimizing competitive interactions among species and promoting diversity.
standard deviation the average variability between repeated measurements.
station keeping movements within a home range.
stealth predators predators that seek highly mobile prey.
stenoecium referring to an organism that has a limited tolerance with respect to the variation of any environmental factor.
stenohaline species highly sensitivity to salinity changes in organisms.
stenothermal organisms organisms that cannot tolerate variation in temperature.
stepping-stone theory the idea that whale carcasses provide stepping-stones habitat that allow highly patchy hydrothermal vent species to persist.
stipe the relatively rigid, stem-like structure that supports kelp blades.
stolonization where an entire plant consists of a horizontal rhizome, from which other rhizomes, both vertical or and horizontal, originate.
stratify to form layers.
stress alteration of the physiological conditions and/or health conditions of an organism in response to environmental stimuli which do not necessarily result in its death.

strobilation the process of segmentation and detachment of a portion of the polyp that becomes the body of the planktonic medusa.
stromatolites laminar fossil structures produced by marine cyanobacterial activity.
structural restoration the construction of habitat structures to enhance restoration.
sublittoral (or **subtidal**) zone to the edge of the shelf at 150–200 m depth.
submarine canyons deep incisions that bisect many continental margins, typically extending from the edge of the continental margin down the continental slope.
submersible a manned "ship" capable of submerging and operating under water.
subsidence geological sinking over time.
subsidence theory Darwin's hypothesis that as plates move, volcanic cones form on the seafloor, sometimes forming islands.
substratum (or **substrate**) the material that makes up the seabed.
subtidal (or **sublittoral**) zone to the edge of the shelf at 150–200 m depth.
succession a series of progressive changes in the composition of an ecological community over time.
supply-side ecology the transfer of larvae from a source area mediated by currents to restock a receiving area down current.
support services ecosystem services necessary to produce all other ecosystem services, such as nutrient cycling.
supralittoral the zone at the land-sea interface affected by the spray of waves and encompassing the transition between a fully terrestrial environment and a fully marine environment.
surface currents currents that move the portion of the water column from the surface to 200 m depth.
surface tension the tension of the surface film of a liquid caused by the attraction of the particles in the surface layer.
surface/volume ratio (S/V) the ratio of total surface area to body volume.
surficial related to the surface, as in surficial sediments.
suspended load transport the movement of particles in suspension.
suspension feeders (or **filter feeders**) organisms that use filtering structures to capture food particles from the water surrounding them.
sustained swimming speed (or **cruising speed**) speeds that organisms can sustain for extended periods.
swell long-period waves that occur at locations away from where wind created the waves.
swim bladder an elastic structure filled with air or other gases, which essentially works as a hydrostatic organ.
symbiont an organism that lives or cooperates with an individual of a different species, irrespective of which species benefits.
symbiosis a constant and intimate relationship between different species.
symbiotic relationships where both species that participate in a relationship gain benefits.
sympagic algae algae that live attached to the underside of the ice.
sympagic communities communities associated with sea ice.
sympatric speciation speciation that occurs when two groups of individuals from the same species become reproductively isolated but without geographic barrier.
synecological ecological studies that consider the structure, development, and distribution of communities and treat their interactions and processes as "functional units".
synergistic effects where combined effects produce a substantially stronger overall effect than their mere sum.

T

taxon any systematic group from species to phylum.
taxonomy the study of the identify of organisms.
tectonics the geological process involved in forming and destroying continents and new seabed.
tele-mediators (or **ectocrines**) a broad class of organic compounds synthesized and released into seawater by marine plants and animals that influence the behavior or biological characteristics of conspecifics.
teleplanic larvae planktotrophic larvae that may remain planktonic for many months.
temperate (or **lysogenic**) **viruses** viruses that infect the cell and integrate their DNA into the host genome, replicating their genome until some additional factor (e.g., UV, high temperatures, pollutants) induce the lytic cycle.
temporary plankton (or **meroplankton**) organisms that spend part of their life cycle in the plankton.
terpenes anti-herbivory compounds produced by brown algae.
terrigenous related to or originating from terrestrial sources.
Tethys an ancient ocean from 65.5 million to 261 million years ago.

thallus A vegetative body in plants and fungi that is not differentiated into roots, stems, or leaves.
thecae the protective shells of diatoms.
thermal capacity the large amount of heat required to be added to a material such as water to raise its temperature by a fixed unit, such as 1 °C.
thermocline refers to changes in temperature across water masses, as often shown in temperature-depth profiles.
thermohaline circulation large-scale ocean circulation driven by differences in temperature and salinity, and thus density, which drives much of the circulation at depths greater than 1,000 m.
thermotolerant able to tolerate high temperatures.
thiobios the fauna inhabiting environments with hydrogen sulfide, typically characterized by a high length-width ratio that allows greater cuticular uptake of oxygen and dissolved organic matter.
thiotrophic bacteria bacteria that utilize sulfide compounds for energy.
tidal excursion the depth range spanned by tides at a given location.
tides periodic phenomen a rising sea level that result from the combined effects of the gravitational pull of the moon and sun and centrifugal forces.
tolerance model succession model that implies that strong competitors follow initial colonizing species and eventually outcompete them.
top-down processes determined by upper trophic levels such as by predators.
Tragedy of the Commons the dilemma arising when multiple individuals, acting independently and rationally consulting their own self-interest, ultimately deplete a shared limited resource, even when clearly not in anyone's long-term interest.
transitional aquatic environments typically coastal environments characterized by a more or less pronounced mixture of inland waters and seawater.
tripton the collective nonliving fraction of suspended material.
trochophores larval polychaetes.
trophic cascade changes in one trophic level that produces cascading effects on the abundance, biomass, or productivity of a community, population, or trophic level through multiple links in the food web, such as when the addition or removal of a top predator triggers indirect changes in predator and prey populations through the food web.
trophic efficiency the amount of energy transferred between trophic levels.
trophic groups referring to different species with similar dietary strategies that target similar food sources from a given trophic level.
trophic levels the position occupied in a food web by different organisms.
trophic value the quality of organic matter as a food source for consumers.
trophosome a specialized organ in hydrothermal vent tube worms that houses a rich bacterial flora of chemoautotrophic symbionts.
turbidity currents rapid, downhill flow of water and sediment akin to mudslides on land.
turnover time (biological) the time required by an organism to duplicate itself.
turnover time (physical) the amount of time required for complete exchange of water in a basin.
twilight (or mesophotic) zone the bathymetric band between the photic zone and 200 m depth.
type 1 error a statistical interpretation that results in the researcher concluding that a treatment effect exists where in reality it does not.
type 2 error a statistical interpretation that results in the researcher concluding that no differences exist between the experimental treatment and controls when they actually differ.

U

unicellular organisms consisting of a single cell.
unidirectionally moving in just one direction.
upper open ocean (or **epipelagic zone**) the surface zone of the oceanic realm with sufficient sunlight for algae to covert carbon dioxide into organic matter, typically extending to ca 200 m.
upwelling the wind-induced vertical currents that transport colder, nutrient-rich water from deep waters closer to the bottom up to the surface.
urchin barrens biological deserts formed in regions of normally high kelp abundance caused by overgrazing by sea urchins.

V

vagile mobile.
veligers early-stage mollusk larvae.
vent fields multiple hydrothermal vents with in a given location.
ventilation gas exchange, such as through gills.
vertically homogenous estuaries estuaries in which the Coriolis effect deflects river flow to one side of the estuary, and tidal inflow to the other, resulting in a vertically homogenous estuary.
vesicles sacs or bladders of some macroalgae such as *Fucus* spp.
vestigial degenerate, rudimentary, or atrophied over evolutionary time.
vicariance the geographical separation of a population, resulting in a pair of closely related species.
viral short circuit (or viral shunt) a process within the microbial loop where viruses, through infection and lysis of prokaryotic and phytoplankton cells, alter rates of transfer of living organic matter in the pool of dissolved and particulate organic matter.
viral shunt (or viral short circuit) a process within the microbial loop where viruses, through infection and lysis of prokaryotic and phytoplankton cells, alter rates of transfer of living organic matter in the pool of dissolved and particulate organic matter.
viriosphere the totality of environments in which viruses occur, spanning every environment on Earth.
viruses biological entities, sub-microscopic in size, that cannot live or reproduce outside a host cell, because they lack metabolic activity and biosynthetic functions.
viscosity the "thickness" of a fluid or gas, or its resistance to a change in shape.
viscous sublayer a thin layer of the boundary layer immediately above the sediment-water interface over smooth bottoms.
viviparous giving birth to live young that have developed inside the body of the parent.
vocalize some marine cetaceans utilize high speed of sound transmission in water to communicate among them.
volcanogenous sediments sediments that arise from material ejected from volcanoes.

W

water column the water that fills the ocean basins.
water content the amount of water contained within sediments, which links closely to sediment porosity of sediment.
well sorted sediments relatively uniform in size, for example containing just one grain size class.
well-mixed estuaries estuaries that lack an obvious halocline or stratification because tidal turbulence destroys any such structure.
winds the air masses that move from areas of high pressure to areas of low pressure.

X

xeronekton animals that must maintain a connection with the ocean surface (e.g., marine mammals, sea turtles) to obtain oxygen.

Y

young ice initial ice formed by the freezing of seawater from the bottom of new surface ice.

Z

zonation identification of zones associated with specific populations that characterize them.
zoobenthos animal component of benthos.
zooplankton animal component of plankton.
zooxanthella symbiotic algae that occur within the tissues of hard corals.
zygote fertilized egg stage.

Index

Note: *Italic* page numbers refer to *figure* and **Bold** page numbers reference to **tables**.

a
abiotic factors   170
abiotic resources, extraction of   592
abundance   149
   of meiofauna   151
   of picobenthos   147–148
abyssal   173
   biodiversity and adaptations   531–533
   ecosystems   506
   environments   506, 532
   gigantism   533–534, *534*
   megafauna   *558*
   organisms   506
   plains   19–20, 531
   systems, functioning of   535–541
   zone   18
Abyssogena phaseoliformis   519
abyssopelagic biodiversity   579
abyssopelagic zone   29, 577
abyssoplankton   182
Acanthastaer plancii   355, 413, *415*
Acanthephyra eximia   531, 532
accidental species   172
acclimate   58
acclimatization   55
accompanying species   171–172
Achanthaster sp.   356
acoustic and satellite technologies   641
acoustic imaging   641
acoustic release   307
Acropora cervicornis   356
Acropora palmata   356
Actinostola callosa   546
actinotroch larvae   266
actinotrophs   184
active acoustics   643
active canyon   306
active detection assets   641
active margins   518
active prokaryotes   506

active restoration   619
active suspension feeders   96
active transport   65
Adamsia palliate   89, 320
Adamussium colbecki   170, 269, 467, 474
adaptation   55
adelphophagy   224
adenosine triphosphate (ATP)   105, 280
adiabatic temperature increase   558
adjacent ecosystems, interactions between coral reefs
     and   417–418
adoplankton   182
advanced acoustic methods   421
adverse environmental conditions   270
aerenchyma   386
aerial exposure, adaptation to   98
aerobic   60
agnatha   224
Aichi Biodiversity Targets   619
Airborne Remote Sensing (ARS)   421
Akatopora tincta   349
albatrosses   229
algal adaptations   321
algal photosynthesis   69
Alicella gigantea   533
alien species   592–593, *594*
Alitta virens   345
allee effect   585
alleles   250
Alligator mississippiensis   241
allochthonous   141
   fertilizers   376
allopatric or geographic speciation   110–111
allowable bycatch   608
alpha diversity   113
Alvinella pompeiana   99, *99*, 544
alvinellid polychaetes   *545*
Alvinocaris   544
Alvinocaris cf. *muricola*   520
Alvinocaris sp. crustaceans   *545*

Marine Biology: Comparative Ecology of Planet Ocean, First Edition. Roberto Danovaro and Paul Snelgrove.
© 2024 John Wiley & Sons Ltd. Published 2024 by John Wiley & Sons Ltd.
Companion Website: www.wiley.com/go/danovaro/marinebiology

Amblyrhynchus cristatus 240
ambush predators 96
amensalism 165, 321
amino acids 105
ammocoetes 224
Amperima rosea 301
Amphicteis gunneri 68
amphidromous species 64, 246
amplification 302
ampoules of Lorenzini 74, 223
anabolic metabolism 280
anadromous species 64
anaerobic phototrophic bacteria 190
anaerobic respiration 60
anagenesis *108*
anammox 284
   prokaryotes 555
"ancestral" diet 95
anemones 320
anemophilous 440
angiosperms 439
anglerfish 578, 579
Anguilla anguilla 354
Anguillicoloides crassus 354
angularity 24
animal cytoplasm density 79
"animal-mediated intelligence" 641
anion 31
Anisakis simplex 354
anoxia 297, 504
anoxic 158, 522
anoxic basins 554–556
anoxic conditions 63, *63*
anoxic environments 554
Antarctica 467–481
   biodiversity 474–477
   birds and mammals 478
   habitats 472–474
   mammals *479*
   trophic webs and functioning 478–481
   zonation, extent, and distribution 467–472
Antarctic benthic filter 481
Antarctic biodiversity and ecosystems 477
Antarctic Bottom Water (AABW) 46, 457
Antarctic Circumpolar Current (ACC) 455, 471
Antarctic coastal habitat *475*
Antarctic communities 474
Antarctic ecosystems 474
Antarctic endemism 475
Antarctic fishes 476
Antarctic groups 475
Antarctic herring 476
Antarctic marine biodiversity 477–478
Antarctic marine environments 477
Antarctic Ocean food web *481*
Antarctic Odontocetes 478
Antarctic polynyas 455
Antarctic research expeditions 477
Antarctic sampling *478*

Anthopleura xanthogrammica 318
anthropogenic activities 593
anthropogenic stressors 597
antibiosis 165
anti-cyclonic (clockwise movement) vortex 538
anti-cyclonic weather system 470
Anti-Freeze Glycoproteins (AFGLPs) 59
antifreeze glycoproteins (AFGPs) 476, *477*
antifreeze proteins 476
anti-predatory strategies 346
aphotic zone 29
aphytal zone 142
apoptosis 357
aposematic warning 349
appendicularia 195
Aptenodytes forsteri 478
aquatic and terrestrial habitats **118–119**
aquatic environments 106
aquatic viruses 145
arborescent (tree-like) colonies 515
archaeocytes 95
archimedes principle 79
Architeuthis dux 227
Arctic deep-sea sediments 465
Arctic ecosystem 460
Arctic fishes *465*
Arctic food web 245
Arctic marine mammals *466*
Arctic megazooplankton taxa *464*
Arctic multicellular marine animals 460
Arctic Ocean 457, 459, **463**
Arctic Ocean Diversity (ArcOD) 460
Arctic planktonic organisms *465*
Arctic Register of Marine Species (ARMS) 460
Arctic sea ice 461
area-based tools 608
Argo floats 638
Aristeus antennatus 568
arthropods 91
Artificial Intelligence 641
asexual reproduction 403
assimilate 42
assisted regeneration 621
Asterias amurensis 589
Asterionella formosa 353
Astomonema southwardorum 321
Astropecten spp. 167
Atlantic gray whales 246
atolls 401
Aurantimonas coralicida 356
Aurelia aurita 84, *288*
Aureococcus anophagefferens 186
auricolaria 184
autochthonous primary production 508
"auto-ecological" approach 157
auto-ecology 119
autonomous underwater vehicles (AUVs) 17, 639, 643
autotomy 403
autotrophic microorganism 192

autotrophic planktonic organisms **185**
autotrophs 42
"azoic theory" 634
Azurina eupalama 589

b

bacterial diversity 555
bacterial gardening 515
bacterial sulfate reducers (BSR) 550
bacteriocytes 95
bacteriophages 157, 186
bacterioplankton 189, 197
bacterivorous microzooplankton 327
baffling effect 448
Balaena mysticetus 103
Balaenoptera borealis 478
Balaenoptera musculus 478
balistids 88
bar built estuaries 433
barotolerant (or piezotolerant) organisms 68
barrier reefs 401
baseline condition 622
basin-scale multidecadal variability 596
batesian mimicry 349
bathomes 567
bathyal region 17, 173
bathymetric patterns in marine biodiversity 128
bathymetric zones 3
Bathymodiolus spp. 519
Bathymodiolus thermophilus 543, 544
bathypelagic biodiversity 579
bathypelagic zone 29, 577
bathyplankton 182
Bathypolaris carinata 464
bed armoring 448
bedload transport 23
Beggiatoa spp. 278, 504
behavioral adaptations in estuaries 435
behavioral mechanisms 257
"bell-shaped" pattern 128
benthic anthozoans 346
benthic biodiversity 155, 465–466
benthic coastal organisms 134
benthic communities 164, **170**, 307, 391–393, *392*, 547, 567
benthic cysts *272*
benthic domain 13
Benthic ecology of viruses 157
benthic environments 164
benthic food web *450*
benthic invertebrates 9, **467**, 533
benthic larvae 269
benthic marine organisms 604
benthic nepheloid layer (BNL) 503
benthic organisms *142*, **143**, 165, **166**, 170, 250
 zonation of 172–177
benthic ostracods 559
benthic-pelagic coupling 270–273
benthic predation on merobenthos 272
benthic primary producers 69

benthic prokaryotes *148*
 compartment 308
 diversity of 147
 ecology of 157
benthic psammon 26
benthic size groupings *143*
benthic vertical zonation *175*
benthic viruses
 abundance of 145
 diversity of 146
bentho-ichthyophagous species 98
bentho-nekton 218
bentho-nektonic food webs *245*
bentho-pelagic coupling 310
bentho-pelagic food web *325*
benthopelagic organisms 136
"benthopelagic plankton" 166
benthos 155, 427
 comparison between hard and soft bottom 165–167
 inhabiting soft bottoms 167–169
 from microbes to megafauna 159–164
 organization of benthic assemblages 171–172
 in space and time 170–171
 trophic groups 164–165
 zonation of benthic organisms 172–177
Benthosema glaciale 464
beta diversity 115
binomial nomenclature 634
biocenosis 142
biochemical adaptations 56
biochemical composition of organisms 7
biocoenosis 171
bio-constructions 92
bioconstructors 92–94, *94*
biodiversity 6, 103, 313–315, 407–411, 460, 474–477, *510*, 570–571, *588*, 587, 643
 associated with kelp 396–397
 associated with mangals 382–384
 associated with salt marshes 387
 of benthos
  benthic biota 141–143
  benthos and plankton 141
  classification of benthos based on size 144–155
 definition of 113
 development 106, *106*
 graphic representation of *120*
 hotspots 125–126
 latitudinal gradient of 126–128, *127*
 measures of 119–121
 patterns 486
  broad-scale biodiversity patterns 125
  marine biogeography 128, 130–136
  processes controlling distribution of marine 125–128
  theories on evolution and maintenance of 136–138
 of pelagic systems 577–580
 of plankton 179–198
 within sea ice 460–462
biodiversity loss 587–589
Biodiversity Strategy 623

bioeroders 92, 419
biogenic material 18
biogenic species 540
biogenous sediments 25
biogeographical regions 130–131
biogeographic ecoregions *486*
biogeographic regions, species distributions within 131–132
biogeography 128
    of mediterranean 132, 134–136
biological deserts 508–510
biological-disorder hypothesis 136
biological diversification 103
biological habitat provisioning 93
"biological hotspots" 579
biological particles 181
biological restoration 625
biological traits approach 611
bioluminescence strategies 55, 71, 72, 577
"bioluminescent" sexual dimorphism 71
bioluminescent signals 70, 71, 72
bioluminescent structures 72
biomineralogy 26
biopolymeric fraction 295
bioprospecting 644
biosphere 1, 3
biotic factors 171, 393–394
biotic variables 354
biotope 171
bioturbation 26, 96, 168, *169*
bipinnaria 184
bipolar species 130
birds and mammals 478
bivalve communities *544*
black band disease (BBD) 356
"black-box" approach 158
black chimneys 542
black disease 357
bladderless redfish 85
bleaching 358, 407
"blue economy" 644
bone-eater worms 553, *553*
bony fishes 68, 489
Boreogadus saida 461
boring or endolytic organisms 26
Boroecia maxima 463
bottleneck hypothesis 161
bottom currents 45
bottom ice 480
bottom topography 392
bottom-up control 158
    of trophic food webs 329–331
boundary layer 21
box fish *88*
brachiolaria 184
Brachionichthys hirsutus 589
Branchiplicatus cupreus 545
Branchipolynoe pettibonae 546
brine 460
broadband signals 75

broadcast spawning 403
broad-scale biodiversity patterns 125
"broundo" 93
brown algae 394
brown ice 472
browsers 419
bubblenetting 246
budding 402
burst swimming or cruising swimming 86
byssus threads 95

c

cabled observatories 638
Calanus finmarchicus 201, 210, 212, 327
calcareous algae 173, 418
calcification 419
calcium carbonate *406*
calderas 550
Caligus orientalis 353
calyx 404
Calliactis parasitica 320
Callorhinus ursinus 347
Calonectris diomedea 241
Calypso routes 635, *636*
Calyptogena spp. 519
Canadian Ocean observatory 640
cannibalism 238
canopy communities *396*
canopy forests 395
canyon benthic communities 514
canyon biodiversity 513–514
canyon functioning 514–515
canyon megafauna *514*
canyons 20
capsid 144
capsomeres 144
carbonate compensation depth 533
carbon dioxide concentrations 595
carbon emission fluxes 304
carbon fluxes *303*
carbon sequestration 451
Carcharhinus falciformis 255
Carcharhinus leucas 346
Carcharodon carcharias 235, 346
Carcinus maenas 593
Caretta caretta 239, 346
Caribbean coral reefs 586
carnivora 226
carnivores 164, 242, 466
carnivorous sponges 95
carnivorous zooplankton 324
carotenoids 301
Carretta carretta 239
carrying capacity 267
cartilaginous fishes 217
Cassis cornuta 97
catadromous fishes 250
catadromous species 64
cation 31

caves
    biodiversity   425
    food webs and functioning in   426–427
    marine invertebrates to life in   426
cavitation   75
cell growth and photosynthesis   283
cell lysis   293
Cells of Ecosystem Functioning (CEFs)   573
cellular level adaptations   83
celtic biota   135
Census of Marine Life   460
central place foragers   492
Centroscymnus coelolepis   235, 533
cephalopods   **227**, 227, 236–239, *238*
Cerastoderma edule   379
Cerastoderma glaucum   379
Cerithideopsis fuscata   589
cetacea   226
cetaceans   244, *245*, 490
Cetorhinus maximus   83, *84*, 235, *235*
Cetotedium sp.   224
Chaceon mediterraneus   531, 532–533
chaetognaths   489
characteristic species   171
Chelidonichthys lastoviza   348
Chelonia mydas   239, 346
chemoautotrophic ecosystems   551
chemoautotrophic microbes   98
chemoautotrophic microorganisms   278
chemoautotrophic organisms   277, 278
chemoautotrophic primary production   504
chemoautotrophic production   508, 551
chemoautotrophic prokaryotes   **322**
chemoautotrophic symbionts   520
chemoautotrophic symbiotic bacteria   322
chemoautotrophs   147, **279**
chemoautotrophy   502
chemoclines   63, 505
chemolithoautotrophic metabolism   278
chemolithoautotrophy   278
chemolithotrophic metabolism   280
chemolithotrophy   277
chemoorganotrophic organisms   280
chemorganotrophy   277
chemosynthesis   547
    in hydrothermal vents   *547*
chemosynthetic bacteria   547
chemosynthetic communities   554
chemosynthetic ecosystems   512
chemosynthetic primary production in ocean   278–280
chemosynthetic processes   552
chemosynthetic production   504
Chilean abalone model   336
choanocytes   96
chondrichthyes   222, 235, 489
chromista   187
Chthamalus stellatus   366, *367*
Cidaris cidaris   168
ciliate feeders   161

ciliates (microzooplankton)   327
circadian   577
circalittoral zone   93, 173, 177, 391
circulation regimes   572–573
circumboreal species   135
Cladocora caespitosa   629
cladogenesis   108, *108*
classical ecology   323
clearance rate   404
climate change
    in abyss   595
    on marine biodiversity   594
clupeidae   489
C method   287
C. montagnei   398
cnidaria   266
cnidarian life cycles   *263*
cnidarians   489
coastal communities   211
coastal environments   61
coastal hydrothermal vents   549–550
coastal lagoons   *374*
    ecology of   375
    functional zonation of   377–378
coastal ocean   485
coastal sedimented ecosystems   **452**
coastal subtidal   391
coastal upwelling   494
coastal waters   485
coccolithophores   192, 193, 194, *488*
coccolithophorids   488
Cocculina craigsmithi   554
coelacanths   220
coenosarc   402, 404
cohort analysis   288–290
cold hydrocarbon seeps   520
cold-seep biodiversity   518–522
    associated with sunken wood   554
cold seep (hydrocarbon-based) ecosystems   517–518
cold seep ecosystems, functioning of   521–522
cold seep functioning   *521*
cold seeps   504
cold-water coral habitats   630
cold-water corals   515, *516*
cold-water species   596
colloblasts   96
colonial species   92
columella   404
commensalism   171, 319, 320
commercial fishing   635
community ecology perspective   370
commuting   247
comparative marine ecology   275
compensation depth   281
compensatory   613
competition   171, 394
complementarity model   317–318
complex biotic interactions   323–332
    bottom-up control of trophic food webs   329–331

detrital trophic network   326–327
microbial loop   327–328
mixed wasp-waist control   332
top-down control on trophic food webs   331–332
trophic networks   323–326
trophic networks based on dissolved organic matter   327
viral shunt   328–329
comprehensive bentho-pelagic monitoring program   641
comprehensive conservation planning   606
Concholepas   336
Concholepas concholepas   336
Conductivity, Temperature, Depth (CTD) probes   34
conductivity of seawater   34
confinement model   377
connectivity   4
consumption mechanisms   257
contamination   590
contemporary marine biology   644
continental margin deep marine ecosystems   **527**
continental margin environments   306
continental rise   18
continental shelf   13, 18
continental slope   18
conventional aerial photography   421
Convention on Biological Diversity (CBD)   317
convergent evolution   380
copepodite developmental stages   266
copepods   194, *195*
coral bleaching   411
coral diseases   355–358
coral feeding and symbiosis with zooxanthellae   404–407
coral gardening   625, *626*, *630*
coral growth, primary factors limiting   407
coralligenous algal biodiversity   419
coralligenous algal reefs   *420*
coralligenous concretions   418
coralligenous habitats   93, 418, *418*, 418–421
coralligenous outcrops, case of   629
Corallinaceae   177
Corallina elongata   177
Corallina mediterranea   92
coralline algae   177
  reefs   176, 177
corallite   402, 404
Corallium rubrum   568
coral morphologies   403
coral reefs   371, 385, 399–418, *400*, *408*, *411*, *413*, *415*, 596
  and adjacent ecosystems, interactions between   417–418
  biodiversity   407–411
  competition for space in   414
  fishes   408
  formation, theory of   401–402
  functioning and trophic food webs   412
  global change on   415
  restoration of   625–626
  trophic cascades in   *258*
  types of   401
  zonation within   400–401
coral reproduction   402–404

coral rubble   400, 400, 517
coral skeleton   515, 517
coral triangle   399
corers   152
Coriolis effect   45
Coris julis   337, *338*, 351, *351*
Corophium volutator   345, 351
Coryphella archaria   346
cosmogenous sediments   25
counter-shading   71, 347
crabs and mangrove detritus degradation   385
craniate   224
Crassostrea virginica   162
critical depth   281
critical erosion velocity   23
Crocodilus porosus   241, *241*
cruising   86
"cruising speeds"   248
crustacean cysts   273
crustaceans   194, *464*
"cryptic"   347
cryptic mimicry   348
cryptic species   113
crypto-bioturbation   168
ctenophores   96, 579
cumulative growth method   289
current   44
cutaneous vasculature   59
cuttlefish   238
cyanobacteria   147
cyanobacterial phytodetritus   *536*
cyanophyceae   150
cycliophora   11
Cyclothone spp.   83
Cymodocea nodosa   442, 626
cyphonautes   184, 266
Cypselurus heterurus   88, *89*
Cystoseira   7
Cystoseira corniculata   398
Cystoseira sp.   333
Cystoseira spp.   398, *399*
  forests restoration   627–629

d

Dactylopterus volitans   86
Dardanus arrosor   320
Dardanus spp.   320
"dark energy"   508
dark zone   306
daytime vertical migration   203
D. calidus   322
dead zones   522, *524*
Decade on Ecosystem Restoration   623, 633
Decade on Ocean Science for Sustainable Development   633
decapods   531
deep abyssal sediments   532
deep-coral ecosystems   515
deep-coral reef systems   515
deep demersal fish populations   532

deep ecosystems, metabolism and functioning of   506–508
deep environments   503
deep faunal origins   510–511
deep hydrothermal vents   543
deep hypersaline anoxic basins (DHAB)   555
deep hypersaline anoxic systems   62
deep marine systems   502
deep ocean basins
　abyssal biodiversity and adaptations   531–533
　abyssal gigantism and dwarfism   533–534
　abyssal plains   531
　affinities between vent and seep communities   554
　anoxic basins   554–556
　deep-sea hydrothermal vents   541–550
　ecosystems   **560–561**
　functioning of abyssal systems   535–541
　ocean trenches   556–559
　whale carcasses   551–554
deep ocean conveyor belt   47, *47*
deep pelagic and benthic communities   *507*
Deep Scattering Layer (DSL)   202
deep-sea biodiversity   508–511
deep-sea ecosystems   501–503, *502*, 503, 506, 508, *569*, 590, 597, 630
　cold seep biodiversity and symbiotic organisms   518–522
　cold seep (hydrocarbon-based) ecosystems   517–518
　deep-sea biodiversity   508–511
　deep-sea habitats   511–512
　deep-water corals   515–517
　depauperate   505–506
　extreme and harsh conditions   503–505
　hypoxic and anoxic systems (dead zones)   522
　metabolism and functioning of deep ecosystems   506–508
　oxygen minimum zones (OMZs)   522, 524–526
　submarine canyons   512–515
deep-sea environments   56, 131, 501, 502, 506, 508, 509
deep-sea floor hosts   501
deep-sea gigantism   *535*
deep-sea habitats   511–512, **513**, 630
deep-sea hydrothermal vents   *541*, 541–550, *542*, *549*, *550*
　biodiversity associated with   543–546
　diversity   *546*
deep-sea isopods   534
deep-sea monitoring technology   641
"deep-sea on a perpetual diet"   503
deep-sea ratcheting   *569*
deep-sea reef   515
deep-sea restoration   630
deep-sea species   595
deep-sea vent communities   550
deep-sea zooplankton biomass   532
deep-water corals   515–517, *516*, 540
　mapping   515
　reefs   *517*, 517
deep-water formation   *47*
definitive or final hosts   352
Delphinapterus leucas   247
delta   372
delta diversity   115

demersal benthos   167
demersal (near-bottom living) fishes   88
"demersal zooplankton"   166
denaturing   56
Denaturing Gradient Gel Electrophoresis (DGGE)   146
Dendropoma petraeum   93
denitrification   284
dense ice communities   472
density   3, 35–38
density dependent   346
depauperate   505–506
deposit feeders   96, 327, 413
dermis   91
Dermochelys   239
Dermochelys coriacea   254
desiccation   49, 173
Desmophyllum dianthus   515
Desmophyllum pertusum   515, 540
detrital food web   324
detrital organic matter   297
detrital transport   302
detrital trophic network   326–327
detritus   95, 293, 294, 295, 298, 449
detritus feeders   327
diadromous species   250
diagenesis   173
　of organic matter   298
diapause   270, 489
diatoms   461
dicopatric model   111
diel cycles   166
digital imaging   643
dimethyl-sulfoniopropionate (DMSP)   327
dinoflagellate algae   74
dinoflagellates   149, 488
diplanula larvae   266
Dipturus laevis   589
direct development   108
direct life cycle   352
direct or indirect development   263
disaerobic   522
"disequilibrium theories"   511
disoxic   522
dispersal   247
dissolved gases   43–44
dissolved organic carbon (DOC)   290, *450*
dissolved organic matter (DOM)   *10*, 158, 295, 295, 299–300, 326–328, *328*
　production   461
　trophic networks based on   327
dissolved (free) oxygen in sea   44
Dissostichus mawsoni   476
distribution of microbenthos   149
disturbance regime   622
divalent   84
dogfish (*Squalus* spp.)   258
dominant hydrodynamic processes   *374*
dominant nektonic components   444
downwelling   48, 494

drag 21, 95
drivers and passengers model 318
drowned river valley estuaries 433
Durinskia dybowskii 321
dwarfism 506, 533–534

e

Earth's mantle 20
"eat and run" strategy 525
ebb 49, *49*
echinopluteus larvae 184, 266
echolocation 75, 217
Ecklonia maxima 395
ecological criteria 610
ecological engineering 503
Ecologically and Biologically Sensitive Areas (EBSAs) 610
ecological niche 188
ecological or functional redundancy 318
ecological restoration 620–623
"ecological volume" 6
economic criteria 610
economic valuation approach 610
"ecosystem approach" 614
Ecosystem Approach to Fisheries (EAF) 614
Ecosystem-Based Fisheries Management (EBFM) 614
Ecosystem-Based Management (EBM) 614
ecosystem engineers 333
  restoring 629
ecosystem functioning 313–315, 451
  at hydrothermal vents 546–550
ecosystem functioning I
  chemosynthetic primary production in ocean 278–280
  photosynthetic primary production 280–286
  primary production 277–278
  respiration 290–291, **291**
  secondary production 287–290
ecosystem functioning II
  dissolved organic matter (DOM) 299–300
  extra-specific processes 293
  organic matter and detritus in ocean 293–299
  organic matter export to seabed 307–311
  pelagic-benthic coupling 300–306
ecosystem productivity 335
ecosystem resilience 622
ecosystem restoration 619
ecosystem services 313
ecotones 172, 377
Ecrobia ventrosa 372, 379
ectocrine adaptations 74
ectocrines 74
ectoderm 402
eelgrass communities 593
Ekman spiral 45
Ekman transport 46, *46*
elasmobranchii 222
elasmobranchs 223–224, 230
electrical conductivity adaptations 74
electromagnetic energy 41
Eledone cirrhosa 353

elephant seals 244
El Nino, case of 201
El Niño Southern Oscillation (ENSO) 203, 411
elongated fishes *87*
Embiotoca lateralis 349
Emerita talpoida 167
Emiliana huxleyi 188
Endangered Species Act 615
endemic species 135
endemism 540
endobenthos 166
Endolithon communities 151
endolytic species 26
endopelon communiy 151
endopsammon communiy 151
energy balance 141
energy pyramids 325
energy theories 127
energy trophic pyramid *326*
Enhydra lutris 242, *243*, 398
environmental characteristics 4
environmental decision-making 614
environmental factors 419
environmental fluctuations 9
environmental heterogeneity 170
environmental light conditions 69
environmental restoration 625
environmental variables *504*
enzymatic degradation 299–300
epibenthos 166, 167
epibionts 167
epidermis 91
epifauna 15
epifaunal organisms 108
epifluorescence microscopy 189
epilithic organisms 173
epilithon communities 151
Epinephelus marginatus 354, 420
epineuston 181
epipelagic biodiversity 577–578
epipelagic zone 576
epipelon communities 150
epiphytic species 167
epiphyton communities 151
epiplankton 181
epipsammon communities 151
epistrate feeders 96
epizoic or endozoic organisms 321
epizoon 151
epsilon diversity 115
"equilibrium theories" 511
Equitability Index 122
erect epifaunal community *396*
ergocline model 378
errant polychaetes 551
erythrocytes 59
Escarpia 520
Eschrichtius robustus 88, 103
estuaries 29, 433–434

complexity of estuarine environments 435
  food webs 437
  physiological adaptations in 433
  pressures on 438
  survival strategies for living in 435–436
estuarine environments 285, 435, **438**
estuarine flows 434
estuarine salinity gradient *436*
EU Biodiversity Strategy 2030 631
Eucampia groenlandica 461
Eudorella truncatula 68
eukaryotes 147
  symbiosis between invertebrates and 321
eulittoral zone area 362
Eumetopias jubatus 347
eunekton 217, 217
Euphausia crystallorophias 481
Euphausia superba 489
euphausiids 207
euphotic zone 29, 43
Euprymna scolopes 323
European Atlantic Coral Ecosystem Study (ACES) 515
eurybathic species 68
Eurycope 559
euryhaline species 64, 377
eurythermal adaptation 99
eurythermal and energy theories 127
eurythermal organisms 56
eurythermal species 131
eutrophication 297, *297*
eutrophic systems 145
evaporite basin 32
evaporites 555
evenness 119
evolutionary history and antiquity 7
evolution of life, fundamental events of **107**
exopinacocytes 171
exoskeletons 153
exo-symbiotic bacteria 544
exploitative competition 171, 349
extra-specific processes 293
extra specific species organic matter processes 293
extreme environments 504–505
extreme habitats "par excellence" 505
"extreme" marine environments 18

f
facilitation model 170, 318, 552
facultative anaerobic microbes 25
facultative predators 161
facultative relationships 315
facultative symbionts 321
faneric mimicry 349
fast acceleration 86
fast ice 471
fauna *553*
fecal pellets 302
feces 303
feeding

  mode and mobility 165
  and nutrition 95–98
  and recruitment in deep 569
  structures 165
femtobenthos 144–147
fetch 48
fields of features 573
filamentous algae 385
filter feeders 96
fisheries closures 608
fisheries management perspective 604
fisheries production 492–493
fisheries restricted areas 608
fishes 218, 220–224, 464–465
  classification and biodiversity of **220**
  and formation of fish shoals 233–234
"fishing down the food web" 255
fishing quotas 608
fish migration patterns 250
fission 402
fissipedia 226, 490
fjords 433
floes 471
flood 49, *49*
flow cytometry 189
Food and Agriculture Organization (FAO) 614
food availability 290
food chain 323
food limitation 308
food supply and sediment composition 467
food webs 437, 447
  and functioning 466–467
  and functioning in caves 426–427
  role of top predators in 255–259
foraging movements 247
foraminifera 151
foraminiferal communities 524
Forbes Azoic Theory **512**
founder effect 111
fractal dimension 7
fractal geometry 170
Fragilariopsis oceanica 461
fragile marine habitats 623–625
frazil ice 471
free-floating organisms 634
free-stream velocity 23
frictional drag 21
fringing reefs 401
fronts 495
fundamental niche 188, 315
Fundulus heteroclitus 345
furious fifties 468

g
Galathowenia fragilis 532
Galeocerdo cuvier 333, 346
gamete fertilization and fusion 263
gamma diversity 115
garum 584

gastropod mollusks   93, 559
gastropods   361
"gastropod with the scaly foot"   545
gear limitations   607
gelatinous zooplankton   196, *196*, 467, 489, 490
gelbstoff   327
gene flow   108
genetic and morphologic diversity   113
genetic bottleneck   262
geochemical composition   537
Geographic Information System (GIS)   421
geographic speciation   111
geopolymeric fraction   295
geothermal energy   504, 542
Geryon longipes   531
giant bacteria   524
giant squid   *227*
giant viruses   145, 186
Gigantapseudes adactylus   534
gigantism   533–534
gill circulation in fish   *61*
gills   60
"glacial control hypothesis"   401
glacial marine sediments   25
Glaucus atlanticus   83
global carbon cycling   451
global climate change   593–596
global fisheries   *492*
   resources   614
global ocean   14
   seabed   *18*
   seafloor   *19*
Global Ocean Alliance   616
global positioning system (GPS)   639
global warming   57
Glochinema bathyperuvensis   525
Glycera carcharias   345
gobies   63
Gondwanaland   457
gonochoric species   263, 403
grab   634
grain size   24, 167
Grammonus ater   425, *426*
gram-negative bacteria   190
grasslands of the sea   444
gray ice   471
gray whales   246
gray-white ice   471
grazers   96
grease ice   471
great whales   585
"green and blue industries"   623
green turtles   249
gregarious organisms   153
gregarious settlement   171, 263
gregarious settlers   585
gregarious species   92
gross secondary production   287
groundfish   591

growth anomaly (GA)   357
guild   323
gulls   229
guyots   20
Gymnura micrura   255

h

habitat degradation, fragmentation, and destruction   590
habitat destruction, potential extinctions caused by   589
habitat forming species   *605*
habitat heterogeneity hypothesis   136
habitat maps   493
hadal   173
hadal biodiversity   558–559
hadal megafauna   *558*
hadal zone   18
hadopelagic biodiversity   579–580
hadopelagic zone   29, 577
hadoplankton   182
Haematopus moquini   318
Haliotis sorenseni   589
Haliotis tuberculata   586
halocline   33
Halofolliculina corallasia   355, 357
halophilic vegetation   376
halophyli   385
hard substrata, characteristics of   26
hard-substrate communities   391
harmful algal blooms (HABs)   490
harpacticoid copepods   480
hatchetfish   579
hawksbill turtles   585
heat capacity   31
heat-shock proteins (HSP)   59
heat waves   58
heavy metals   590
Hediste diversicolor   66, 351
Heliocidaris erythrogramma   110, *110*
Heliocidaris tuberculata   109, *110*
hematocrit   59
Hemimysis speluncola   425
hemoglobin reduction   59
Henricia sanguinolenta   68
herbivores   164, 409
hermaphroditic species   263
hermaphroditism   352
hermatypic corals   515
hermatypic (reef building) corals   356
Heterandria formosa   382
Heterosigma akashiwo   188
heterotherms   55
heterotrophic bacteria   327, 506
heterotrophic communities   480
heterotrophic hypothesis   104
heterotrophic macrobenthos, dense communities of   551
heterotrophic planktonic organisms   **185**
heterotrophic zooplankton range   487
heterotrophs   147
hexacorals   515

Hexactinella   154
high-definition (HD) camera installation   639
high-energy environments   94
high habitat diversity   513
high nutrient   467
high nutrient low chlorophyll (HNLC) water   331, 479
high seas conservation strategy   615
high surface tension   31
Hirondellea gigas   559
History of Marine Animal Populations (HMAP)   586–587
"hit and run" strategy   361
Holcomycteronus profundissimus   559, 568
holobenthic species   264
holoplankton   182, 487
holoplanktonic gastropods   489
holoplanktonic species   264
holoplanktonic zooplankton   **183**
holothurians   301
homeo-osmotic ions   84
homeosmotic organisms   65
homeotherms   55
home range   247
homochromia   347
homomorphia   347
hotspots   537
human-centric perspective   3
"hyaline bodies"   321
Hyalonema apertum   559
hybridize   111
Hydrobates pelagicus   241
Hydrobia neglecta   379
hydrocarbon emission   *518*
hydrodynamic efficiency   86
hydrodynamics   25, 392–393, 407
hydrogen bonds   29, 31
hydrogenous sediments   25
hydrometers   34, 38
hydrophilic   440
hydrophones   643
hydro-pneumatic skeleton   91
hydrothermal circulation   542, 546
hydrothermal environments   548
hydrothermal fluids   98, 542
hydrothermal vents   55, 98, 504, 546, 606
  chemosynthesis in   *547*
  chimney   *543*
  communities   525
  ecosystem functioning at   546–549
hydrothermal vents "fields"   542
Hymedesmia spp.   521
hyperbenthos   166, *558*
hypersaline   434
hypersaline anoxic systems   555–556
hyperthermophiles   99
hyper-thermophilic enzymes   99
hyper-thermophilic prokaryotes   98
hypertonic marine organisms   65
hypostracum   91
hypothesis testing   343

hypotonic blood   66
hypoxia   297, 504
hypoxia thresholds   **523**
hypoxic   62
  and anoxic systems (dead zones)   522
  environments   *524*
Hypsoblennius gentilis   350
Hypsoblennius jenkinsi   349

i

ice   49
ice algae   473
ice coring   *474*
ice development in Antarctica   *472*
ice edge bloom   478
ice fishes   59, *59*
icefoot   471
ice melt POM   478
ichthyophagous species   98
Idas washingtonia   554
Idas washingtonius   551
idiosyncratic model   318
imaging technologies   643
inactivation   295
inbreeding   207
Index of Trophic Diversity (IDT)   121
indices of relative species abundance   119
indices of species richness   119
indirect life cycle   352
Indonesian coral reefs   97
infauna   15, 166
infralittoral zone   93, 163, 173, 391
infraplankton   182
inhabit intertidal depths   *175*
inhibition model   552
initial melting phase   473
inoculation process   *479*
inorganic nutrients and trace elements   42–43
inquilinism   320
in situ coral "farming" experiments   625
integument   60
integumentary system   222
interference competition   171, 349
Intergovernmental Panel on Biodiversity and Ecosystem Services   313
intermediate communities   211
intermediate disturbance hypothesis   136, 366–369
internal currents   45
International Geophysical Year (IGY)   635
International Standards for the Practice of Ecological Restoration   620
International Union for the Conservation of Nature (IUCN)   218, 517, 608
interspecific interactions II   343
  competition   349–351
  diseases of marine organisms   354–358
  methods to escape predation   348–349
  parasitism   351–354
  predation   343–348

interstitial spaces   158
intertidal ecosystems
  comparison between   **388**
  and lagoons
    comparison between soft and hard bottom intertidal environments   370–371
    mangroves   380–385
    niche displacement to reduce competition   371–372
    rocky intertidal habitats   359–361
    salt marshes   385–387
    survival strategies for rocky intertidal environments   361
    transitional environments between land and ocean   372–380
intertidal environments   315, 361
intertidal habitats   366–369
intertidal invertebrates, behavioral strategies of   361
intertidal organisms   *363*
intertidal predators   *367*
intertidal sedimentary environments   359
intertidal zonation   360, 362
intertidal zones   163, 173, *176*
inverse model   318
invertebrates
  biodiversity   *539*
  and eukaryotes, symbiosis between   321
  and prokaryotes, symbiosis between   321
  symbiosis between invertebrates and   321
  symbiosis between vertebrates and   321
  and unicellular eukaryotes, symbiosis between   321
"inverted gardens"   455, 472
invisible habitats   493
iponeuston   181
isophasic adaptation   361
isophasic strategies   361
isospatial strategy   361
iteroparous species   267

j

jawless fishes   224
Jaxea nocturna   169
jellies (cnidarians)   574
Juvenile bento-nektonic species   492

k

Karlodinium veneficum   353
karst theory   401
katabatic winds   456
kelp barrens   395
kelp forests   394–399
  distributions   395
  invertebrate communities   *397*
  restoration   627
kelp holdfast   397, 398
keystone species   170, 333, 369, *369*
Khophobelemnon   559
"killing the winner" model   188
knolls   538
"known diversity"   190
K-strategy species   **269**

l

labile organic matter   298
lacrimal (tear) gland   66
lagoon biodiversity   378–380
"lagoon channels" converge   374
lagoons   372
  functioning   376
  shoreline vegetation   378
  system   *374*
Lamellibrachia spp.   518, 520
laminae   168
Laminariales   394
Laminaria ochroleuca   396
Laminaria rodriguezii   396
land and ocean, comparison between   **4–6**
land-based species   3
land–sea interface   3, 4
Langmuir cells   87
large-scale changes in water masses   494
large-scale currents   493–494
larva   261
larval classification   **269**
larval cyprid   266
larval ecology   264–268
lateral line systems   71
lateral transport   306
Laticauda colubrine   240
Latimeria chalumnae   111, 111, 589
latitudinal gradients of biodiversity   126–128, *127*
leafy kelp   *394*
lebensspuren   448
lecithotrophic larvae   184, 267, 549
lepidochronology studies   447
Lernaeocera branchialis   353
Lessepsian migration   125, 593
Leuresthes tenuis   251
Liebig's Law   288
life aquatic
  comparison between sea and land   3–6
  fractal complexity of marine and terrestrial ecosystems   6–11
life cycles   182, 261
  and reproduction   261–265
life history strategies   267–269
life in seas and oceans   53–54
light   40–42
  darkness, and nutrients   567–568
  intensity   280, 393
ligule   447
Limecola balthica   351
limivorous animals   96
linear additive fashion   613
linear barrier islands   374
Lithophaga lithophaga   419
Lithophyllum byssoides   94, 173
Lithophyllum sp. platform   *176*
Littorina littorea   592
local and seasonal movements   247
log-deficit layer   23
log layer   23

Lolliguncula brevis 227
long-based fins 88
longitudinal gradients in tropical biodiversity 128
Loricifera 64
Lottia alveus 589
Lottia edmitchelli 589
low chlorophyll system 467
lower slope 401
low oxygen concentrations 60–64
loyalty 248
Lucicutia 525
Lucinidae 519
luminescence 71
luminous 321
lungfishes 220
Lycopodina hypogea 426
lysogenic (or temperate) viruses 144
lytic viruses 144

m

Mackeret cells 44
Macoma balthica 162, 172
macroalgae 4
macroalgal diversity 126
macroalgal forests
   in mediterranean 398
   restoration 627–629
macro and mega-zooplankton 195–196
macrobenthos 151
macroborers 419
Macrocystis 395
macrofauna *152*, 162, 451
macrofaunal abundance 532
   and diversity 152–153
macrofaunal sampling 153
macronutrients 73
macrophages 95, 164
macrophytes 142
macroplankton 179, 207
macro-scale plankton distribution, physical control of 201
macrourids (grenadiers) *86*
macrozooplankton 179, *196*, 488
Madrepora oculata 515
maerl 93
mammal migration 88
maneuvering 86
mangal 380
mangal ecosystem functioning 384–385
mangroves 380–385
Manta bisostris 83
Margalef's Index 122
marginal ice zone (MIZ) 496
marine amniotes 66
marine and terrestrial biodiversity 9
marine animals 74–76
   Arctic multicellular 460
   populations in human history 584–585
marine benthic communities 270
marine benthic environment *174*
marine benthic organisms 91

marine biodiversity *104*, 125, *605*
   bathymetric patterns in 128
   climate change on *594*
   loss 588
   mechanisms of marine speciation 107–112
   origin and evolution of marine life 103–107
   processes controlling distribution of 125–128
   quantifying marine organism biodiversity 113–122
   science 616
   studies 635
marine biodiversity conservation 603
   conservation frameworks 614
   conservation objectives 603–606
   conservation strategies 606–608
   cumulative impacts and biodiversity conservation 613
   legal instruments 615
   marine protected areas (MPAs) 608–613
   science-based decision making 616
   science challenges and solutions 616
   third dimension of 606
marine biogeography 125, 128, 130–136
marine biology 3, 634–636
   and ecology 3
   research in next decade 645–646
marine biota 593
marine cave 423
marine conservation 606
marine ecosystems 330, *330*, 570, 644
   functioning 315, 317, *597*
   human impacts on
      biodiversity loss 587–589
      historical data 583–587
      main threats to marine life and ecosystems 590–596
      synergistic impacts on 597–599
   restoration 621–622, 631
      global plan of 623
      perspectives of 631
marine environmental studies 634
marine environments 3, 6, 9, *33*, *89*, 95, 136, 238, 286, 298, 300, 313, 315, *348*, 635
marine fisheries 493
marine food chain *10*
marine food webs 326
marine habitats 593
   decade for ecosystem restoration 619
   defining ecological restoration 620–623
   global plan of marine ecosystem restoration 623
   perspectives of marine ecosystem restoration 631
   restoration
      of coral reefs 625–626
      of deep-sea habitats 630
      ecosystem engineers 629
      fragile marine habitats 623–625
      of macroalgal forests 627–629
      of seagrass meadows 626–627
marine invertebrates 266, **322**, 426
   to life in caves 426
marine life
   census of 644
   origin and evolution of 103–107

marine macroalgae   7, 162
marine mammals   **76**, 224–226, **226**, 242–247, **249**, 346, 465, 490, 578
marine organisms   6, 7, 69, 74, 95, *268*, 593
　biodiversity   113–122
　diseases of   354–358
marine organisms I
　adaptations to light   69–72
　adaptations to low oxygen concentrations   60–64
　adaptations to nutrients   73
　adaptations to produce sound and communicate in water   74–76
　adaptations to temperature   55–60
　adaptation to pressure   68
　adaptation to salinity   64–67
　ectocrine adaptations   74
　electrical conductivity adaptations   74
marine organisms II   79–89
marine organisms III
　adaptations to life on the seafloor   91
　adaptation to aerial exposure   98
　adaptation to extreme temperatures and potentially toxic chemicals   98–99
　adaptation to waves and energy   94–95
　feeding and nutrition   95–98
　support and protection structures   91–94
marine parasites   **353**
marine pelagic fish   **231**
marine picoplankton abundances   190
marine protected areas (MPAs)   603, 606, *607*, 608–613, 616
　benefits of   612
　establishment and management of   612
marine refuges   608
marine reptile ecology   239–241
marine resources   591
marine sediments   **25**, 158
marine snow   302
Marine Spatial Planning   614
marine speciation   107–112
marine species   **116–117**, 589
marine technologies, application of   644–645
marine tele-mediators   74
marine vegetables   634
marine vertebrates   522
marine viruses   188
masking   613
match-mismatch hypothesis   596
mattes   *441*
mature ecosystems   522
M. balthica communities   172
mechanical fragmentation   300
mediterranean seagrasses   444
mediterranean seamounts   *537*
megabenthos   153–155
megafauna   346
megafaunal organisms   *154*, 641
megafaunal sampling   153
megalopa stage   266
megalopa-zoea   184

Meganerilla bactericola   525
meiobenthic grazing   161
meiobenthos (meiofauna)   151–152
meiofauna   11
　abundance of   152
　ecology of   161
meiofaunal biodiversity   *129*
meiofaunal sampling   152
Melanocetus johnsonii   72
Melanopsis impressa   109
melon   75
Melosira   550
Melosira arctica   461
merobenthic species   264
meroplankton   182
meroplanktonic larvae   171, *185*
meroplanktonic lecithotrophic larvae   549
meroplanktonic species   264
mesenteric filaments   402
Mesodinium rubrum   321
mesoglea   91
mesolittoral zone   173, 359
Mesonychoteuthis hamiltoni   227
mesopelagic biodiversity   578–579
mesopelagic zone   29, 576–577
mesophiles   99
mesophotic zone   29, 41
mesoplankton   181
mesopredators   259
mesozooplankton   194–195, 207, 210, 327, 488, 494
mesozooplankton filtration   211
meta-analysis   637
metabarcoding   155
metabolic rate   289
metabolic theory   56
metabolic theory of ecology   56, 506
metagenomic analysis   155
metagenomic studies   145
metazoans   7, 132
methanogenesis   191
Metopograpsus spp.   382, 383
Metrarabdotos   109
Metridium senile   154
microalgal phytodetritus   *536*
microbenthos   149–151
microbial biogeography   192
microbial communities   157, 188, 547
microbial gardening   161
microbial loop   158, 327–328
microborers   419
micronekton   166, 179, 196, 207
micronutrients   73
microorganisms   147
micropatches   158
microphages   95, 164
microphytobenthos   150, 286
microphytoplankton   *193*
microplankton   193–194
microtides   372

microvores 161
microxic 522
microzooplankton 194
middens 584
mid-ocean ridges 13, *15*
migration 248
Millepora boschmai 589
Miniacina miniacea 444
mixed wasp-waist control 332
mixing depth 169
mixotrophic species 148, 181
mixotrophy 404
mobile fauna 427
mobile organisms 98
mobile species 167
Mola mola 354
molecular biology 643
monoculture 333
Monodon monoceros 247
monomers 104
Montipora sp. *350*
moorings 307, 637
Morbillivirus 354
morphological features and adaptations in estuaries 436
mosaic 136
mucus 404
mucus-trap feeding 161
"mud volcanoes" 521
mullerian mimicry 349
Müller's larvae 266
multi-parameter platforms 641
multi-year ice 460
mutualism 319
mutualistic 171
"mutualistic" interactions 315
Mya arenaria 95
Mycteroperca rubra 420
mysticetes 226, 478
mysticeti 248, 490
Mytilus californianus 333, *334*, 366
Mytilus galloprovincialis *58*, 68

n

nanobenthos 148–149
 diversity of 149
 ecology of 158
nanoplankton 192
narrowband signals 75
natural capital 313
natural regeneration 622
natural (or spontaneous) regeneration approach 622
nauplii 266
Nautilus spp. 85, 236
neap tides 49
Nearest Neighbor Distance (NND) 233
nearshore species 492
necrosis and detachment in gorgonians *135*
nectochaeta 266
negative water balance 134, 554

nekto-benthos 218
nekton 84, 489–490, 578
 cephalopods 227
 ecology of cephalopods 236–239
 ecology of chondrichthyes 235
 ecology of seabirds 241–247
 fishes 218, 220–224
 fishes and formation of fish shoals 233–234
 great migrations of 247–255
 main organisms and characteristics of 218
 marine mammals 224–226
 marine reptile ecology 239–241
 patterns of biodiversity in 229–232
 reptiles 228, **229**
 role of top predators in food webs 255–259
 seabirds 228–229
 sharks at risk of extinction from indiscriminate hunting 236
 species contributing to 217
nektonic animals 218
nektonic organisms *219*
nematocysts 319, 404
nematodes 354, 525
nematode trophic groups *121*
Neobenedenia girellae 353
neoendemic species 135
Neognathophausia ingens 525
Neogoniolithon brassica-florida 93, 94
Neopetrolisthes sp. 320
Nephthys incisa 345
Nereis diversicolor 266
neritic and oceanic (pelagic) ecosystems, comparison between **497**
neritic aquatic ecosystems
 biogeography and characteristics 485–487
 biological characteristics 487–490
 ecosystem functioning in the neritic zone 490–492
 factors influencing functioning of neritic systems 493–497
 fisheries production 492–493
 zonation, extent, and distribution 485
neritic ecosystems 496
neritic food webs 495–497
neritic plankton 490
neritic province 13
 distribution of *16*
neritic systems, factors influencing functioning of 493–497
neritic zones 29, 485, 497
 ecosystem functioning in 490–492
neritic zooplankton community 489
net photosynthesis 43
net secondary production 287
networks of MPAs 608
neuston 181
new moon 49
new production 284
niche displacement to reduce competition 371–372
nicotinamide adenine diphosphate (NADH) 278
nictemeral cycles 577

nitrate   42
nitrogen fixation   190, 284
Nitzschia frigida   461
noble   44
Noctiluca   84
Noctiluca miliaris   71, *84*
nocturnal hatching   239
node   638
non-indigenous or alien species   592–593
non-indigenous species (NIS)   593
non-photosynthetic organisms   93
normal (positive) estuaries   434
North Atlantic Deep Water (NADW)   46, 457
North Atlantic Oscillation (NAO)   135
northern (boreal) winter   455
North Sea carbon flow model   *451*
Notomastus latericeus   68
Notothenioidei   476
Notothenioidei fishes   475
Notothenioid fishes   467
novel optodes   44
nubbin   *411*
Nucella lapillus   366
nudibranchs   349
null hypothesis   343
nutrient depletion   480
nutrient-rich systems   327
Nuttallia nuttallii   351

O

oases   508–510
obligate   315
obligate aerobes   25
obligate anaerobic microbes   25
obligate symbionts   321
occasional "alien" presence   593
ocean, comparison between land and   **4–6**
ocean acidification   *416*, 593
ocean basins   13–15, 571
ocean bottom   15–20
ocean currents   44–50
ocean depth   17
ocean ecosystems   3
ocean exploration   636–637
ocean floor topography   17
ocean geomorphology   575
ocean gliders   638
ocean health   611
ocean heterogeneity   641
oceanic depths and land elevations   *14*
oceanic ecosystems
    biodiversity of pelagic systems   577–580
    classification of pelagic regions   571–573
    factor influencing life and distribution of pelagic organisms   567–571
    functional classification of pelagic systems   573–575
    vertical zonation in pelagic ecosystems   575–577
oceanic fisheries   578
oceanic province   14, 485

oceanic trenches   18, 20
oceanic zones   29, 572
ocean management   581
Ocean Networks Canada   640
oceanodromous species   250
ocean physical-chemical properties   634
ocean protection planning   **616**
ocean science   646
ocean surface temperature   *35*
ocean trenches   17, 556–559
Octopus vulgaris   320
Oculina patagonica   355
Odobenus rosmarus   354
odontocetes   226
odontoceti   246–247
oegopsids   494
Olavius algarvensis   323
oligohaline   377
oligophotic (food poor) mesopelagic zone   181
oligotrophics   136, 284
    environments   327
    systems   145, 329
oligotrophy   503
omega diversity   115
omnivores   164
"one solution fits all"   608
ontogenetic stages of species   261
Onychocella alula   349
oophagia   224
Open Ocean Ecosystems   275
open water communities   211
operational convenience   610
opportunists   161
optical sensors   643
optimal foraging   507
optimal foraging theory   179, 213, 503
optimal restoration strategy   629
orbital forcing   595
Orchomenella obtuse   525
organic aggregates in   302–306
organic content   25
organic debris   506
    in environment   297
organic detritus   293, 376
organic matter
    classification of   **294**
    concentration   306
    and detritus in ocean   293–299
    distribution   *296*
    export to seabed   307–311
    flow   307–308
    in sea   294
    sedimentation   300–301
    transport mechanism   306
organic polymers   103
organisms   *61*
organogenic construction   48
"ornamentation"   84
Osedax sp.   *553*

osmolytes   65
osmoregulation   407
osmotic challenges   55
osmotic conformers   65, 435
osmotic regulation   *436*
osmotic regulators   65, 435
osmotrophs   331
osteichthyes   220–222
ostracum   91
overfishing   590–591
overgrazing   395
oviparous   223
ovoviviparous   223
oxic   158, 522
oxidize organic substances   60
oxygen   43
oxygen concentrations   24, 502
oxygen deficient zones   522
oxygen minimum zones (OMZs)   43, 62, 128, 504, 522, 522, 524–526
ozone hole   481–482

p

Pacific Decadal Oscillation   494
pack ice   49, 471
Pagurus arrosor   89
Pagurus prideaux   320
Palaemonetes pugio   345
Palaemon xiphias   444
palaeoecological evidence   584
paleoendemic species   135
pancake ice   471
panoceanic species   135
Paracentrotus lividus   444, 444
"parachute effect"   80
"paradox of the plankton"   211
Paralichthys olivaceus   353
paralic model   377
parasitic   171
parasitism   319, 321, 351–354
paratenic hosts   352
Parathemisto spp.   489
parthenogenetic reproduction mechanisms   352
partially stratified (or partially mixed) estuaries   434
partial regulators   435
particulate organic carbon (POC)   *450*, 506
particulate organic matter (POM)   95, 293, 295, 302, 326, *328*
passive acoustics   643
passive margins   518
passive restoration   620
passive suspension feeders   96
patch mosaic theory   136
pathogens   590
Pearson-Rosenberg model   *297*
Pectis profundicola   533, 568
pediveligers   184
Pelagia noctiluca   574
pelagic and benthic domains   **30**
pelagic Arctic food web   *469*

pelagic-benthic coupling   300–306
pelagic biodiversity   462–464
pelagic dolphins   248
pelagic domain   13
pelagic ecosystems, vertical zonation in   575–577
pelagic fauna   569, *577*
pelagic fish abundance   229
pelagic fisheries   591
pelagic food web   *244*
pelagic invertebrates in the Arctic Ocean   **463**
pelagic lecithotrophic larvae   108
pelagic organisms, factor influencing life and distribution of   567–571
pelagic planktotrophic larvae   108
pelagic province   485
pelagic regions
  classification of   571–573, **572**
pelagic systems
  biodiversity of   577–580
  functional classification of   573–575
pelagic zone   29
pelicans   229
pelitic   26
Peltodoris atromaculata   348
penguins   229
Pentaceros wheeleri   540
Peprilus triacanthus   346
Pérès and Picard Model   172, 172
pericapsid   144
Peringia ulvae   *372*, 379
periodic anaerobic metabolism   525
Periophthalmodon kalolo   *384*
Periophthalmodon schlosseri   383–384
periostracum   91
peripatric model   111
permanent thermocline   35
permeability   24
permeases   158
persistent hydrodynamic activity   503
personal care products   416–417
Perumytilus   336
petrels   229
Petrosia ficiformis   426
Peysonelliaceae   177
Phaeocystis antarctica   496
Phaeocystis pouchetii   459, 462
phagotrophs   331
phenotypic plasticity   108, 267
Philometra lateolabracis   354
philopatry   254
phi (φ) scale   20
Phocidae group   244
Phocine (Phocid) Distemper Virus (PDV)   354
Phoebastria nigripes   539
Pholas dactylus   177
phoresis   89
Phormidium valderianum   357
phosphate   42
photic zone   43

photoautotrophic organisms   277
photoautotrophs   147
Photobacterium genus   321
Photoblepharon   71
photoelectric sensors   43
photophilic algal communities   391
photophores   70, 71
photosynthesis   69, 278
photosynthetically active radiation (PAR)   43, 69, 280
photosynthetic autotrophic organisms   323
photosynthetic efficiency   286, 286
photosynthetic primary production   280–286
photosynthetic production   504
phototrophy   277
phycodnaviridae   146
phyletic evolution or anagenesis   108
phyletic gradualism   108
Phyllaplysia smaragda   589
Phylum Loricifera   64
Physester macrocephalus   83
physical disorder   136
physical disturbance   392
physical variables and disturbance on benthic communities   391–393
physicochemical conditions   44
physiological adaptations in estuaries   435
physiological stress   57
phytal systems   172–173
phytal zone   142
phytobenthos   163, 286, 323
phytodetrital contributions   301
phytodetritus   449, 506
phytoplankton   43, 73, 79, 179, 194, *210*, 297, 303, 324, 326, 329, 384, 486, 487, 490
   blooms   283, *462*
   communities   9, 370, 467
   composition   *578*
   diversity   *487*
   photosynthesis   302
   productivity   462
   seasonality   209
   species   211
phytoplanktonic organisms   280
picobenthos   147–148
picophytoplankton   *189*
picoplankton   189–192
PI curve   280
Pielou's Index   122
pigmented hadal species   *568*
pilidium   184
pilidium larvae   266
pink-line syndrome (PLS)   357
pinnacles   538
Pinna nobilis   153
"pioneer species"   442
Pisaster ochraceus   318, 333, *334*, 366
Pisaster sp.   366
piscivores   337
Planet Earth   3

planet ocean   3
   human impacts and solutions for   581
   marine biology of   633
      application of marine technologies   644–645
      birth of marine biology   634–636
      history of ocean exploration   636–637
      marine biology research in next decade   645–646
      sampling platforms   637–641
      sensors   643–644
      technologies enabling biological observations at sea   641, 642
planktology   181
plankton   79, 179, 485
   abundance   490
   abundance comparisons among different planktonic components   197–198
   classification based on life cycles   182–184
   community   *396*
   distribution   201–208
   ecology of   208–211
   phytoplankton species   211
   planktonic classification based on water column distributions   181–182
   planktonic organisms   180–181
   size classes   184, 186–196
   zooplankton nutritional mode   211–213
planktonekton   218
planktonic copepods   *81*
planktonic crustacean species   83
planktonic foraminiferans   306
planktonic hydromedusae   515
planktonic organisms   80, 86, *180*, 270–271, 409, 514
planktonic phyllosoma   80
planktonic prokaryotes   190
plankton migration   *540*
plankton vertical migrations   *205*
planktotrophic larvae   184, 267, 269
planulae   266, 404
platelet ice   480
plate tectonics theory   511
platforms   401
Platyhelminthes   95, 354
Pleuragramma antarcticum   496
pleuronectiformes   489
pleuston   98, 181, 487
plume inoculum   478
pneumatophores   382
Pocillopora damicornis   355, 357, *406*
Poecilia latipinna   382
poikilotherms   55, 56
polar benthic invertebrates   *468*
polar circles   455
polar ecosystems   455–457
   Antarctica   467–481
   benthic biodiversity   465–466
   biodiversity   460
   biodiversity within sea ice   460–462
   biogeography and characteristics   457–460
   comparison between   **482**

fishes 464–465
   food webs and functioning 466–467
      marine mammals 465
      pelagic biodiversity 462–464
polar environments 455
Polarographic Oxygen Sensors 44
polarographic principle 44
pollution 583
polyandry 246
Polybrachia 520
Polycarpa sp. 97
polychaetes 525
polygynandry 246
polygyny 246
polyhaline 377
Polymerase Chain Reaction (PCR) analysis 146
polymers 104
polynyas 456
polyunsaturated fatty acids (PUFAs) 297
Pomacentrus amboinensis 69
population 171
popup tags 641
porcellanasteridae 510
Porcellanella triloba 320
Porites lutea 357
porosity 24
Porpita porpita 83
Portuguese dogfish 68
Posidonia oceanica 93, *441*, 445, 444, *444*, *444*, 445, *446*, 447, 590, 626
Posidonia spp. 444
positive (or "facilitative") interactions 315–319
Potamocorbula amurensis 593
potamodromous species 250
potential extinctions
   caused by episodic events 589
   caused by introduction of other species 589
   caused by overfishing 589
practical salinity units (PSU) 32
"prebiotic soup" 105
"Precautionary Approach" 614
precautionary principle 614
predation 171, 343–348
   and trophic cascades 394
"predator avoidance" 207
predator cropping 136
predator-prey relationships 346, 348
predators 96
preferential species 171
pressure 568
pre-zygotic mechanisms 111
primary consumers 394, 412–413
primary producers 4
primary production 277–278
primitive groups 60
primordial sea 104
primordial soup 104
principle of competitive exclusion 211
Prochlorococcus spp. 69, 190

prochlorophytes 577
production/biomass method 289
prokaryote mats 550
prokaryotes 147, *155*, 327, 515
   symbiosis between invertebrates and 321
   symbiosis between unicellular eukaryotes and 321
   symbiosis between vertebrates and 321
prokaryotic abundances 550
prokaryotic mortality 329
prolific benthic communities 555
propagules 88
prostrate kelp or kelp beds 395
protandric 263
"protein enrichment" 327
proteobacteria 190, 280
proterogenic 263
protozoa 187
protozoa osmoregulate 66
provinces 485
Prunum coniforme 109
Pseudoliparis amblystomopsis 533, 568
Pseudopolydora paucibranchiata 351
pseudo-populations 593
psychrotrophic 59
Pterapogon kauderni 589
pterigopods 223
pteropods 489
Puffinus puffinus 241
punctuated equilibrium 108
pycnocline 35
pycnometers 38
Pycnopodia helianthoides 318
pyranometer 43
Pyropelta genus 554

q
quiescence 270
quotas 608

r
rafting 89
Rapid Tissue Necrosis (RTN) 357
Rapoport's rule 127
realized niches 188, 315
recalcitrant 295
"recovery" of ecosystem 620
recruitment 394
recruitment enhancement techniques *628*
redox potential (Eh) 24
redox potential discontinuity depth 152
Red Queen hypothesis 352
reef building (bioconstructor) corals 402
reef communities 409, 415
reef crest 400, 401
reef fishes *410*
reef mounds 515
reefs host communities of animals 515
"reference ecosystems" 620
reference point 603

refraction   41
refractory organic components   507
regenerated production   284, *284*
Regional Fisheries Management Organizations   615
Register of Antarctic Marine Species (RAMS)   475
rehabilitation   620, 622
relative predictability   254
relict sediment   25
remineralization   42
remineralizes   277
Remotely Operated Vehicles (ROVs)   420, 639, 644
remote sensing   639
reproduction movements   248
reptiles   228, **229**
    osmoregulate   66
resilience   314
resistance   270–273, 314
resource environmental compartment   164
respiration   290–291, **291**
respiratory pigments   61, 62, *63*
restoration   620, 624
    activities   622
    of coral reefs   626–626
    of deep-sea habitats   630
    ecosystem engineers   629
    fragile marine habitats   623–625
    of macroalgal forests   627–629
    of seagrass meadows   626–627
restorative activities   622
restorative continuum   622
restoring ecosystem engineers   629
resuspend particles   20
rete mirabilis   88
reverse stratification   473
reversing thermometer   36–37
R. exoculata   545
Reynolds Number   87
Rhincodon typus   83
Rhinoptera bonasus   259
rhizomes   440
Rhizophora mangle   353
Rhizophora spp.   383
Rhizophydium planktonicum   353
rhodolith beds   422
Richelia intracellularis   190
rich epiphyte community   442
Riedl Model   173
Riftia pachyptila   62, 322, 544, 547, 548, *548*
river plumes   495, *495*
rivet or functional redundancy hypothesis   318
Rizocephalo sacculina   352, *352*
roaring forties   468
rocky intertidal communities   *368*
rocky intertidal consumers   364, 366
rocky intertidal environments   361–370
rocky intertidal habitats   359–361
rocky intertidal primary producers   363–364
rocky intertidal trophic food webs   370
rocky intertidal zonation   *360*, 361–370
Rouse Parameter   23

S

Sabellaria   177
Sabella spallanzanii   444
Saccorhiza polyschides   396
sacculinization   353
salinity   3, 31–34
    depth profile   *34*
    in sea, measures of   34
salmonidae   489
salt marshes   385–387, *386*, *387*
    biodiversity associated with   387
    ecosystem functioning   387
    zonation   *386*
salt-wedge estuaries   433
sampling platforms   637–641
Sardina pilchardus   252, *253*
Sargassum hornschuchii   398
Sargassum spp.   398
Sargassum trichocarpum   398
Sarpa salpa   444
satellite tags   639
"sawtooth" trend model   273
Saxipendium coronatum   545
scaphopod gastropod mollusks   559
scavengers   96, 164
Schizochytrium spp.   353
school   233
sciaphilous   173, 401
science-based decision making   616
scientific importance   610
scleractinian corals   402, 412
Sclerolinum   520
scorpaenidae   489
Scutellastra cochlear   349
seabed   3
    boundary layers and their characteristics   20–23
    hard substrata, characteristics of   26
    mapping   22
    ocean basins   13–15
    ocean bottom   15–20
    sediment movement   23–25
    sediments, characteristics of   20
    soft sediments, characteristics of   26
    topography   15
seabirds   228–229, 241–247, 490, *492*
seafloor processes   567
seafloor substratum   3
seagrass   440
    detritus   445
    global distribution of   *441*
    matte   442
    meadows restoration   626–627
    restoration   626
seagrass beds   438–442, *443*, 445–447, 586
    biodiversity associated with   442–446
    seagrass functioning   446
sea ice   472
sea ice bacteria   472
sea ice microbial communities (SIMCO)   472, *473*
seals   244, 245

seamount biodiversity 538–541
seamount effect 538
seamount morphologies 537
seamounts 20, 535–538
"sea olive" 441
sea otter-urchin-kelp trophic interaction 333
searching predators 96
seas 572–573
seasonal ice melting 466
seasonal stabilization 285
seasonal thermocline 35
sea state 40
sea turtles 251
seawater, properties and characteristics of 29
   density 35–38
   dissolved gases 43–44
   hydrogen bonds 31
   inorganic nutrients and trace elements 42–43
   light 40–42
   oxygen 43
   pressure 38–39
   salinity 31–34
   sound 39
   temperature 34–35
   viscosity 38
Sebastes carnatus 350
Sebastes chrysomelas 350, 350
Secchi disk 43
secondary consumers 413
secondary production 277, 287–290
second equilibrium theory 511
sectionally homogenous estuary 434
sedentary 88
sedentary species 167
sedimentary environments 451
sedimentary fauna 451
   food sources for 449–450
sedimentary habitats 446–449
   food sources for sedimentary fauna 449–450
   sedimentary environments and ecosystem functioning 451
sedimentation 171, 407
   rate 301
sediment composition 448
sediment movement 23–25
sediment thickness 16, *16*
sediment trap technologies 306
seeding 474
seeds 88
seep communities, affinities between vent and 554
seep fauna 521
Selachii 91
selective deep-sea consumers 503
semelparous species 267
Semibalanus balanoides 366, *367*
semi-enclosed seas 537
semi-geographic or parapatric speciation 111
sensors 643–644
Sepia 85
Sepia officinalis 84

septa 404
Seriola dumerili 353
Seriola quinqueradiata 353
serpulid polychaetes 92
Serranus scriba 348
Serratia marcescens 356
sessile 88
sessile fauna 166
sessile invertebrates, transplantation of *629*
sessile species 264
seston 295
sexual dimorphism 71
shallow-deep connectivity 568
shallowest subtidal habitats 392
shallow formations 177
shallow hard substrate habitat 420–421
shallow-water biodiversity *126*
shallow-water hydrothermal vents 504
Shannon index 122
Shannon-Wiener Index 121
sharks
   abundance of 236
   attacks 236
   predation 259
shear 20
   stress 21
   velocity 23
shelf break 18
"shifting baseline syndrome" 583
shoal 233–234
siboglinids 520, *549*
   community 544
Siboglinum caulleryi 68
Sicyases sanguineus 98
"side scan sonar" 17
"signature whistles" 247
sills 20
simple echo sounder surveys 421
Simpson's Index 122
single large or several small (SLOSS) 608
sink populations 604
siphonophores 579
sirenians 226, *226*
sirenids 247
size range 7
Skeletonema costatum 281
skeleton-eroding band (SEB) 355
skeleton erosion band (SEB) 357
sloppy feeding 179, 212, 293
small-scale plankton distribution 202–208
soak times 608
social importance 610
sodium chloride 31
soft and hard bottom intertidal environments 370–371
"soft bodied" organisms 526
soft corals 408, *408*
soft sediments 26, 167
solar radiation 40, *42*, 44, 502
Somnius pacificus 258
source populations 604

Spartina alterniflora 385, 586
Sparus aurata 398
spatial heterogeneity *137*
Spartina alterniflora 385
speciation 107–108
species abundance 119
species-by-species conservation strategy 605
species distributions within biogeographic regions 131–132
species diversity, composite measures of 121–122
species extinctions in sea 585
species richness 119
specific gravity 79
spermaceti 83
sperm whales 75
Sphaerechinus granularis 420
Sphyrna lewini 223
Spirobranchus sp. polychaetes *97*
Spisula solidissima 168
spring tides 49
Squilla mantis 349
S. subtruncata 172
stability-time hypothesis 136
stabilization 300
standing crop 324
station keeping 247
statistical errors 343
stealth predators 96
stenohaline organisms 64
stenothermal organisms 56
Stephanoscyphus 559
stepping stone theory 551
Sternoptychidae 541
Stichodactyla spp. 320
stipes 395
stipitated forests 395
St. Lawrence estuary example 439
stolonization 440
Storthyngura 559
stratify 35
strobilation 266
Strongylocentrotus purpuratus 317
structural restoration 625
Stylopoma 109
submarine canyons *513*, 512–515
submersibles 638
SubOxic Layer (SOL) 555
subsidence 373
substratum 6
subtidal distributions 391–394
subtidal hard substrate ecosystems
    comparison between **428**
    coralligenous habitats 418–421
    coral reefs 399–418
    kelp forests 394–399
    rhodolith beds (maërl) 422
    subtidal distributions 391–394
    underwater caves 423–427
subtidal zone 18, 173, 176
succession 172

sunfish *221*
sunscreens and coral bleaching 416–417
supply-side ecology *270*, 270
supralittoral zone 162, 172–173, 361
surface circulation in the Southern Ocean *471*
surface currents 45
surface pelagic predators 290
surface phytoplankton 480
surface/volume ratio 80
surficial (surface) sediments 18
suspended load transport 23
suspension feeders 44, 45, 451
sustainable blue economy 581, 623
"sustain biodiversity" 603
sustained swimming 86
swell 48
swim bladder 85
Symbiodinium microadriaticum 405
Symbiodinium sp. 358
symbiosis 171, 319–323
    between invertebrates and invertebrates 321
symbiotic organisms 518–522
symbiotic relationships 89
sympagic algae 49
sympagic communities 455, 472
sympagic species **460**
sympatric speciation 111
Symsagittifera roscoffensis 321
Synechococcus 69, 190
synergistic effect 613
Syngnathus affinis 589
Synischia hectica 444

t

T4 bacteriophages 144
Tagging of Pacific Predators (TOPP) project 251
taxonomy 119
tectonics 103
    estuaries 433
Teleostei 91
teleosts 229
    fishes 318
teleplanic larvae 269
temperature 392
temporary (transient) movement 86
terpenes 398
terrestrial and freshwater restoration 531
terrestrial and marine biogeography 130
terrestrial and marine extinctions 585
terrestrial and ocean ecosystems 278
terrestrial ecosystems 325
terrestrial environments 4, 6, 69, 79, 95, 239
terrestrial forests 400
terrestrial habitats, aquatic and **118–119**
terrestrial inputs 494
terrestrial runoff 494
terrigenous sediment 25
terrigenous sources 18
tertiary consumers 413

Tetraselmis convolutae 321
Thalassia testudinum 442
Thalassiosira 550
Thalassoma pavo 135, 337, *338*, 351, *351*
Thalassonerita naticoidea 520
thalli 173
theca 404
thecae 42
Thelenota anax 97
theory of coral reefs formation 401–402
Theristov melnikov 461
thermal optimum 56
Thermanemertes valens 545
thermoclines *36*, 505
   water masses 572
thermohalines 33
   circulation of global ocean *458*
thermotolerant 99
thiobios 152, 546
Thioploca spp. 526
thiotrophic bacteria 545
three-dimensionality 4
three-dimensional marine ecoregions 576
Thyasiridae 519
tidal excursion 49
tide 49, *49*
Tigriopus revicornis 327
time-based approaches 608
time-stability hypothesis 511
tolerance model 552
Tomopteris sp. *81*
top-down control 158
   on trophic food webs 331–332
top-down effect 286
total organic matter content (TOM) 298
Totoaba macdonaldi 589
Tragedy of the Commons 603
transitional aquatic environments 372, *373*
transitional environments between land and ocean 372–380
Trichechus manatus latirostris 354
Trichodesmium spp. 190
Tridacna crocea 350
trigger fish *88*
Trigriopus 67
tripton 295
Tritonia tetraquetra 353
trochophores 184, 266
trophic cascades 315, 333–338
trophic efficiency 324
trophic food webs
   bottom-up control of 329–331
   top-down control on 331–332
trophic groups 165
trophic levels
   number of 9
   and renewal of biomass 7, 10
trophic networks 323–326, 397–398
   based on dissolved organic matter 327
   and cascades 323–332

trophic value 295
trophic webs and functioning 478–481
trophosome 62, 322
tropical biodiversity, longitudinal gradients in 128
tropical fisheries 493
turbidity currents 136, 513
turbulent mixing 493–494
turgor pressure 65
Turneroconcha magnifica 544, 548
turnover 424
Tursiops truncatus 354
twilight zone 306
type 1 error 343
type 2 error 343
type II disturbance 392
typical cold seep habitats *519*
Tyrrhenian Deep Water (TDW) 574
Tyrrhenian Sea circulation 574

u
ubiquitous 68
ulcer white spots disease (PUWSD) 357–358
Ulkenia amoeboidea 353
ultra-abyssal trachomedusae 533
ultraviolet light 299
Umbellula 559
UN Convention on Biological Diversity 1992 614
UN Convention on the Law of the Sea (UNCLOS) 615
"undamaged sites" 620
UN Decade on Ecosystem Restoration 631
underwater caves 423–427
underwater coral nursery *626*
unicellular eukaryotes, symbiosis between invertebrates and 321
unidentifiable cohorts' method 289
United Nations Convention on Biological Diversity (CBD) 610
United Nations Framework Convention on Climate Change 619
Upogebia pusilla 444
upper bathyal zone 18
upper slope 401
upwelling 48, *48*, 62
urchin barrens 242
urchin proliferation 398
Urodasys anorektoxys 525
Ursus maritimus 242, *242*
US National Research Council 614

v
vagile (mobile) fauna *371*
Vampyrophrya pelagic 353
V. coralliilyticus 355
Velella velella 83
veliger larvae 184
veligers 266
Vema Seamount 539
vent and seep communities, affinities between 554
vent communities 546

vent emissions 542
vent fields 542
vent macrofauna 545
vent species 549
vermetidae reef *94*
vertebrates
  biodiversity *539*
  herbivores *414*
  and invertebrates, symbiosis between 321
  and prokaryotes, symbiosis between 321
vertical and horizontal currents 493
vertical currents 47
vertically homogenous estuary 434
vertical migrations 568–569
  example 202–208
vesicles 80
Vesicomya 519
vesicomyid bivalves 519
vesicomyids 519
vessel size limits 607
vestimentiferans 548
Vibrio fischeri 323, 323
Vibrio harveyi 357
Vibrio shiloi 355
vicarious 172
viral abundances 145, 188
viral lysis 329
viral short circuit 328
viral shunt 157, 328–329
viriobenthos 144
virioplankton 186–189, 197
viriosphere 144
viruses 144
"virus-like particle" (VLP) 145
viscosity 3, 38
viscous sublayer 22
vision and bioluminescence 69–72
"visually limited" pelagic groups 290
vital (life history) strategies 267
viviparous 223
vocalize 55
volcanic islands 19
volcanogenous sediments 25
vulnerable ecosystems 644

w

walruses 244
warm coastal vents 550
wasp-waist environmental variable control *332*
water column
  density, shape, and buoyancy 79–86
  ocean in motion 44–50
  properties and characteristics of seawater 29–44
  swimming and dispersal 86–90
water content 24
water masses 572
water molecule *32*
water movement 392
water scatters 41
water-sediment interface *427*
waves 48, *48*
  and energy, adaptation to 94–95
well-mixed estuaries 434
wentworth grain-size scale **21**
whale carcasses 551–554, *553*
  biodiversity 552
  communities 551
  ecosystems 552
whale carcass systems, functioning of 552–554
whale fall 552
whale pump theory 303
white band disease (WBD) 356
"white corals" 515
white plague (WP) 356, 357
white pox (WPO) 356
white syndrome 357
winds 45
Winkler method 44
worldwide distribution of lagoons *373*
World Wildlife Fund for Nature 608

x

xeronekton 217, 218

y

yellow band disease (YBD) 356
young ice 471

z

Zalophus californianus 354
"zero pollution" 633
zoeal stages 266
zonation within coral reefs 400–401
zooplankton 79, 179, **202**, *210*, 302, 487–490, 532, 578
  abundance 550
  communities 209, 211
  distributions *204*
  migration 302
  nutritional mode 211–213
zooxanthellae 356, 399, 404, 405, 407
  coral feeding and symbiosis with 404–407
Zostera marina 626
zygote 261